# Solid-Phase Extraction

**Handbooks in Separation Science**

The goal of the series and volume editors is to develop a new vehicle for collating, interpreting, and disseminating the essential fundamental and practical information of separation science for future generations of separation scientists and to do this by creating the seminal work in the field. Each volume is designed to cover a specific topic and contains relatively succinct chapters with a sharp focus and clear presentation contributed by leading scientists in the field. The target audience for these volumes is professional scientists with responsibility for managing or participating in research projects in either academia or industry. Included in this group are graduate students and professionals in disciplines other than separation science seeking insight into a topic at a level associated with current capabilities. The current volume follows on from the success of earlier volumes with additional volumes in production or planned for the future.

2012—C.F. Poole (Editor). Gas Chromatography
2013—S. Fanali, P.R. Haddad, C.F. Poole, P. Schoenmakers, D. Lloyd (Editors).
Liquid Chromatography: Fundamentals and Instrumentation
S. Fanali, P.R. Haddad, C.F. Poole, P. Schoenmakers, D. Lloyd (Editors).
Liquid Chromatography: Applications
2015—C.F. Poole (Editor). Instrumental Thin-Layer Chromatography
A. Gorak, E. Sorensen (Editors). Distillation: Fundamentals and Principles
A. Gorak, H. Schoenmakers (Editors). Distillation: Operation and Applications
A. Gorak, Z. Olujic (Editors). Distillation: Equipment and Processes
2017—C.F. Poole (Editor). Supercritical Fluid Chromatography
S. Fanali, P.R. Haddad, C.F. Poole, M.-L. Riekkola (Editors). Liquid Chromatography:
Fundamentals and Instrumentation, Second Edition
S. Fanali, P.R. Haddad, C.F. Poole, M.-L. Riekkola (Editors). Liquid Chromatography:
Applications, Second Edition
C.F. Poole (Editor). Capillary Electromigration Separation Methods
2018—A.F. Ismail, M.A. Rahman, M.H.D. Othman, T. Matsuura (Editors). Membrane Separation Principles and Applications: From Material Selection to Mechanisms and Industrial Uses
2019—C.F. Poole (Editor). Liquid-Phase Extraction

# Solid-Phase Extraction

*Edited by*

*Colin F. Poole*

ELSEVIER

Elsevier
Radarweg 29, PO Box 211, 1000 AE Amsterdam, Netherlands
The Boulevard, Langford Lane, Kidlington, Oxford OX5 1GB, United Kingdom
50 Hampshire Street, 5th Floor, Cambridge, MA 02139, United States



**Library of Congress Cataloging-in-Publication Data**
A catalog record for this book is available from the Library of Congress

**British Library Cataloguing-in-Publication Data**
A catalogue record for this book is available from the British Library

ISBN: 978-0-12-816906-3

For information on all Elsevier publications visit our website at https://www.elsevier.com/books-and-journals

*Publisher:* Susan Dennis
*Acquisition Editor:* Kathryn Eryilmaz
*Editorial Project Manager:* Redding Morse
*Production Project Manager:* Omer Mukthar
*Cover Designer:* Greg Harris

Typeset by TNQ Technologies

# Contents

# Contributors

**Abbi Abdel-Rehim** Faculty of Science and Engineering, University of Manchester, Manchester, United Kingdom

**Mohamed Abdel-Rehim** Department of Clinical Neuroscience, Center for Psychiatry Research, Karolinska Institutet and Stockholm County Council, Stockholm, Sweden

**Martha B. Adaime** Laboratory of Pesticide Residue Analysis (LARP), Chemistry Department, Federal University of Santa Maria, Santa Maria, Rio Grande do Sul, Brazil

**Hossam Al-Suod** Department of Environmental Chemistry and Bioanalytics, Faculty of Chemistry, Nicolaus Copernicus University, Torun, Poland; Centre for Modern Interdisciplinary Technologies, Nicolaus Copernicus University, Torun, Poland

**Beatriz Albero** Departamento de Medio Ambiente y Agronomía, Instituto Nacional de Investigación y Tecnología Agraria y Alimentaria (INIA), Madrid, Spain

**Tahereh Golzari Aqda** Environmental and Bio-Analytical Laboratories, Department of Chemistry, Sharif University of Technology, Tehran, Iran

**Sergio Armenta** Department of Analytical Chemistry, Research Building, University of Valencia, Valencia, Spain

**María Asensio-Ramos** Instituto Volcanológico de Canarias (INVOLCAN), INtech La Laguna, San Cristóbal de La Laguna, Tenerife, Islas Canarias, España

**Sara Asgari** Environmental and Bio-Analytical Laboratories, Department of Chemistry, Sharif University of Technology, Tehran, Iran

**Sanka N. Atapattu** CanAm Bioresearch Inc., Winnipeg, MB, Canada

**Habib Bagheri** Environmental and Bio-Analytical Laboratories, Department of Chemistry, Sharif University of Technology, Tehran, Iran

**A. Ballester-Caudet** Miniaturization and Total Analysis Methods (MINTOTA) Research Group, Departament de Química Analítica, Facultat de Química. Universitat of València, Valencia, Spain

**Francesc Borrull** Department of Analytical Chemistry and Organic Chemistry, Universitat Rovira i Virgili, Tarragona, Spain

**Bogusław Buszewski** Department of Environmental Chemistry and Bioanalytics, Faculty of Chemistry, Nicolaus Copernicus University, Torun, Poland; Centre for Modern Interdisciplinary Technologies, Nicolaus Copernicus University, Torun, Poland

**P. Campíns-Falcó** Miniaturization and Total Analysis Methods (MINTOTA) Research Group, Departament de Química Analítica, Facultat de Química. Universitat of València, Valencia, Spain

**S. Cárdenas** Departamento de Química Analítica, Instituto Universitario de Investigación en Nanoquímica (IUNAN), Edificio Marie Curie (anexo), Campus de Rabanales, Universidad de Córdoba, Córdoba, Spain

**Beibei Chen** Department of Chemistry, Wuhan University, Wuhan, Hubei, China

**Yanlong Chen** School of Chemistry, Sun Yat-sen University, Guangzhou, Guangdong, PR China

**Miguel de la Guardia** Department of Analytical Chemistry, Research Building, University of Valencia, Valencia, Spain

**Júlia A. de Oliveira** Laboratory of Pesticide Residue Analysis (LARP), Chemistry Department, Federal University of Santa Maria, Santa Maria, Rio Grande do Sul, Brazil

**M.C. Díaz-Liñán** Departamento de Química Analítica, Instituto Universitario de Investigación en Nanoquímica (IUNAN), Edificio Marie Curie (anexo), Campus de Rabanales, Universidad de Córdoba, Córdoba, Spain

**Jianwei Dong** School of Chemistry, Sun Yat-sen University, Guangzhou, Guangdong, PR China

**Antonio Martín Esteban** Departamento de Medio Ambiente y Agronomía, INIA, Madrid, Spain

**Francesc A. Esteve-Turrillas** Department of Analytical Chemistry, Research Building, University of Valencia, Valencia, Spain

**Nestor Etxebarria** Department of Analytical Chemistry, Faculty of Science and Technology, University of the Basque Country (UPV/EHU), Basque Country, Spain; Research Centre for Experimental Marine Biology and Biotechnology (PIE), University of the Basque Country (UPV/EHU), Basque Country, Spain

**Núria Fontanals** Department of Analytical Chemistry and Organic Chemistry, Universitat Rovira i Virgili, Tarragona, Spain

**Leon Fuks** Centre of Radiochemistry and Nuclear Chemistry, Institute of Nuclear Chemistry and Technology, Warszawa, Poland

**Kenneth G. Furton** International Forensic Research Institute, Department of Chemistry and Biochemistry, Florida International University, Miami, FL, United States

**Belén González-Gaya** Department of Analytical Chemistry, Faculty of Science and Technology, University of the Basque Country (UPV/EHU), Basque Country, Spain; Research Centre for Experimental Marine Biology and Biotechnology (PIE), University of the Basque Country (UPV/EHU), Basque Country, Spain

**Man He** Department of Chemistry, Wuhan University, Wuhan, Hubei, China

**Irena Herdzik-Koniecko** Centre of Radiochemistry and Nuclear Chemistry, Institute of Nuclear Chemistry and Technology, Warszawa, Poland

**R. Herráez-Hernández** Miniaturization and Total Analysis Methods (MINTOTA) Research Group, Departament de Química Analítica, Facultat de Química. Universitat of València, Valencia, Spain

**Bin Hu** Department of Chemistry, Wuhan University, Wuhan, Hubei, China

**Abuzar Kabir** International Forensic Research Institute, Department of Chemistry and Biochemistry, Florida International University, Miami, FL, United States

**Hian Kee Lee** National University of Singapore Environmental Research Institute, National University of Singapore, Singapore, Singapore; NUS Graduate School for Integrative Sciences and Engineering, National University of Singapore, Singapore, Singapore; Department of Chemistry, National University of Singapore, Singapore, Singapore; Tropical Marine Science Institute, National University of Singapore, Singapore, Singapore

**Gongke Li** School of Chemistry, Sun Yat-sen University, Guangzhou, Guangdong, PR China

**Yanxia Li** School of Chemistry, Sun Yat-sen University, Guangzhou, Guangdong, PR China

**A.I. López-Lorente** Departamento de Química Analítica, Instituto Universitario de Investigación en Nanoquímica (IUNAN), Edificio Marie Curie (anexo), Campus de Rabanales, Universidad de Córdoba, Córdoba, Spain

**R. Lucena** Departamento de Química Analítica, Instituto Universitario de Investigación en Nanoquímica (IUNAN), Edificio Marie Curie (anexo), Campus de Rabanales, Universidad de Córdoba, Córdoba, Spain

**Faranak Manshaei** Environmental and Bio-Analytical Laboratories, Department of Chemistry, Sharif University of Technology, Tehran, Iran

**Rosa M. Marcé** Department of Analytical Chemistry and Organic Chemistry, Universitat Rovira i Virgili, Tarragona, Spain

**Mohammad Mahdi Moein** Department of Clinical Neuroscience, Center for Psychiatry Research, Karolinska Institutet and Stockholm County Council, Stockholm, Sweden

**Y. Moliner-Martinez** Miniaturization and Total Analysis Methods (MINTOTA) Research Group, Departament de Química Analítica, Facultat de Química. Universitat of València, Valencia, Spain

**C. Molins-Legua** Miniaturization and Total Analysis Methods (MINTOTA) Research Group, Departament de Química Analítica, Facultat de Química. Universitat of València, Valencia, Spain

**Rosa Montes** Department of Analytical Chemistry, Nutrition and Food Science, IIAA – Institute for Food Analysis and Research, Universidade de Santiago de Compostela, Santiago de Compostela, Spain

**Aline L.H. Müller** Laboratory of Pesticide Residue Analysis (LARP), Chemistry Department, Federal University of Santa Maria, Santa Maria, Rio Grande do Sul, Brazil

**Nyi Nyi Naing** National University of Singapore Environmental Research Institute, National University of Singapore, Singapore, Singapore

**Oscar Núñez** Department of Chemical Engineering and Analytical Chemistry, University of Barcelona, Barcelona, Spain; Serra Húnter Fellow, Generalitat de Catalunya, Barcelona, Spain

**Maitane Olivares** Department of Analytical Chemistry, Faculty of Science and Technology, University of the Basque Country (UPV/EHU), Basque Country, Spain; Research Centre for Experimental Marine Biology and Biotechnology (PIE), University of the Basque Country (UPV/EHU), Basque Country, Spain

**Rosa Ana Pérez** Departamento de Medio Ambiente y Agronomía, Instituto Nacional de Investigación y Tecnología Agraria y Alimentaria (INIA), Madrid, Spain

**Valérie Pichon** Department of Analytical, Bioanalytical Sciences and Miniaturization, UMR, CBI, ESPCI Paris, PSL Research University, Paris, France; Sorbonne University, Paris, France

**Colin F. Poole** Department of Chemistry, Wayne State University, Detroit, MI, United States

**Osmar D. Prestes** Laboratory of Pesticide Residue Analysis (LARP), Chemistry Department, Federal University of Santa Maria, Santa Maria, Rio Grande do Sul, Brazil

**Ailette Prieto** Department of Analytical Chemistry, Faculty of Science and Technology, University of the Basque Country (UPV/EHU), Basque Country, Spain; Research Centre for Experimental Marine Biology and Biotechnology (PIE), University of the Basque Country (UPV/EHU), Basque Country, Spain

**Maria Eugênia C. Queiroz** Departamento de Química, Faculdade de Filosofia Ciências e Letras de Ribeirão Preto, Universidade de São Paulo, Ribeirão Preto, SP, Brasil

**José Benito Quintana** Department of Analytical Chemistry, Nutrition and Food Science, IIAA – Institute for Food Analysis and Research, Universidade de Santiago de Compostela, Santiago de Compostela, Spain

**María Ramil** Department of Analytical Chemistry, Nutrition and Food Science, IIAA – Institute for Food Analysis and Research, Universidade de Santiago de Compostela, Santiago de Compostela, Spain

**Omid Rezvani** Environmental and Bio-Analytical Laboratories, Department of Chemistry, Sharif University of Technology, Tehran, Iran

**Rosario Rodil** Department of Analytical Chemistry, Nutrition and Food Science, IIAA – Institute for Food Analysis and Research, Universidade de Santiago de Compostela, Santiago de Compostela, Spain

**Miguel Ángel Rodríguez-Delgado** Departamento de Química, Unidad Departamental de Química Analítica, Facultad de Ciencias, Universidad de La Laguna (ULL). Avenida Astrofísico Francisco Sánchez, San Cristóbal de La Laguna, Tenerife, España

**Ruth Rodríguez-Ramos** Departamento de Química, Unidad Departamental de Química Analítica, Facultad de Ciencias, Universidad de La Laguna (ULL). Avenida Astrofísico Francisco Sánchez, San Cristóbal de La Laguna, Tenerife, España

**Jack M. Rosenfeld** Department of Pathology and Molecular Medicine, McMaster University, Hamilton, ON, Canada

**Yoshihiro Saito** Department of Applied Chemistry and Life Science, Toyohashi University of Technology, Toyohashi, Aichi, Japan

**Álvaro Santana-Mayor** Departamento de Química, Unidad Departamental de Química Analítica, Facultad de Ciencias, Universidad de La Laguna (ULL). Avenida Astrofísico Francisco Sánchez, San Cristóbal de La Laguna, Tenerife, España

**Javier Saurina** Department of Chemical Engineering and Analytical Chemistry, University of Barcelona, Barcelona, Spain

**Sònia Sentellas** Department of Chemical Engineering and Analytical Chemistry, University of Barcelona, Barcelona, Spain

**Bárbara Socas-Rodríguez** Departamento de Química, Unidad Departamental de Química Analítica, Facultad de Ciencias, Universidad de La Laguna (ULL). Avenida Astrofísico Francisco Sánchez, San Cristóbal de La Laguna, Tenerife, España

**Israel D. Souza** Departamento de Química, Faculdade de Filosofia Ciências e Letras de Ribeirão Preto, Universidade de São Paulo, Ribeirão Preto, SP, Brasil

**Małgorzata Szultka-Młyńska** Department of Environmental Chemistry and Bioanalytics, Faculty of Chemistry, Nicolaus Copernicus University, Torun, Poland

**José L. Tadeo** Departamento de Medio Ambiente y Agronomía, Instituto Nacional de Investigación y Tecnología Agraria y Alimentaria (INIA), Madrid, Spain

**Sze Chieh Tan** NUS Graduate School for Integrative Sciences and Engineering, National University of Singapore, Singapore, Singapore; Department of Chemistry, National University of Singapore, Singapore, Singapore

**Esther Turiel** Departamento de Medio Ambiente y Agronomía, INIA, Madrid, Spain

**Ikuo Ueta** Department of Applied Chemistry, University of Yamanashi, Kofu, Yamanashi, Japan

**Aresatz Usobiaga** Department of Analytical Chemistry, Faculty of Science and Technology, University of the Basque Country (UPV/EHU), Basque Country, Spain; Research Centre for Experimental Marine Biology and Biotechnology (PIE), University of the Basque Country (UPV/EHU), Basque Country, Spain

**J. Verdú-Andrés** Miniaturization and Total Analysis Methods (MINTOTA) Research Group, Departament de Química Analítica, Facultat de Química. Universitat of València, Valencia, Spain

**Ling Xia** School of Chemistry, Sun Yat-sen University, Guangzhou, Guangdong, PR China

**Renato Zanella** Laboratory of Pesticide Residue Analysis (LARP), Chemistry Department, Federal University of Santa Maria, Santa Maria, Rio Grande do Sul, Brazil

**Shakiba Zeinali** Environmental and Bio-Analytical Laboratories, Department of Chemistry, Sharif University of Technology, Tehran, Iran

**Naiyu Zheng** Department of Bioanalytical Sciences, Bristol-Myers Squibb Company, Princeton, NJ, United States

**Olatz Zuloaga** Department of Analytical Chemistry, Faculty of Science and Technology, University of the Basque Country (UPV/EHU), Basque Country, Spain; Research Centre for Experimental Marine Biology and Biotechnology (PIE), University of the Basque Country (UPV/EHU), Basque Country, Spain

# Core concepts and milestones in the development of solid-phase extraction

# 1

*Colin F. Poole*
Department of Chemistry, Wayne State University, Detroit, MI, United States

## 1.1 Introduction

The origins of solid-phase extraction are as old as chromatography, which in its early days was exploited for the isolation of compounds from mixtures by their selective interaction with a solid stationary phase and subsequent recovery by elution in a mobile phase. Chromatography and extraction have since diverged in their general function in chemical analysis and are regarded as complementary techniques today. Extraction is typically employed for isolation, preconcentration, matrix simplification, or solvent exchange ahead of the separation and identification of compounds by chromatographic-based (and other) techniques. The key to understanding the relationship between these common laboratory techniques is to consider extraction as an enabling technique that modifies sample properties to facilitate a successful separation and detection of target compounds by the most appropriate technique. In the absence of an extraction step the sample would appear to be too complex, too dilute or incompatible with sustaining instrument performance rendering the analysis unsuccessful. Overtime the scale, speed, material costs, and level of automation for the extraction step have adapted to changing laboratory needs. Thus, solid-phase extraction, one variant of extraction methods, is a dynamic field, and while old, it is still heavily researched with the flux of advances far from concluded. At the time of writing, it is reasonable to identify miniaturization, advances in material science, ease of automation, and compatibility with the goals of green analytical chemistry as the primary driving forces maintaining the general interest in advancing the techniques of solid-phase extraction [1–3].

## 1.2 First generation formats

Solid-phase extraction is based on the transfer of target compounds in a gas, liquid, or supercritical fluid matrix to a solid sorbent [4]. Typically, the sample containing the target compounds flows over the solid sorbent which retains the compounds by their favorable interactions with the sorbent. The sorbent is subsequently separated from the sample and the target compounds recovered by solvent displacement or thermal desorption into the gas phase. An early application of solid-phase extraction in the

**Solid-Phase Extraction. https://doi.org/10.1016/B978-0-12-816906-3.00001-7**

1950s was the use of activated carbon-filled columns to isolate organic contaminants from surface waters for toxicity evaluation [5]. The low concentration of contaminants and the poor capability of instrumental methods to identify compounds and assess their toxicity at that time resulted in large-scale operations in which thousands of liters of water were sampled over several days. The introduction of macroreticular porous polymers in the early 1970s was responsible for redirecting interest in solid-phase extraction for both field and laboratory applications as well as extending its scope to air sampling and the isolation of drugs from biological fluids. These sorbents had reasonable mechanical strength, a large surface area, a large sample capacity, low water retention, and provided high recovery of target compounds by solvent or thermal desorption. Compared with activated carbon the overall recovery of target compounds was generally better and irreversible adsorption and catalytic activity greatly diminished. These properties together with further improvements in instrumental methods facilitated a general downsizing of sorbent beds, a reduction in sample size, and increasing use of solid-phase extraction as a general laboratory technique for a wider range of applications than was previously the case [6,7]. Porous polymers of high thermal stability and low water retention were responsible for revolutionizing the analysis of volatile organic compounds in air and purge gas samples from dynamic stripping of volatile organic compounds from aqueous solution. Compounds trapped on the sorbent bed were thermally desorbed directly into a gas chromatograph for analysis eventually leading to fully automated sampling and analysis systems for routine use [8,9]. The general acceptance of solid-phase extraction for sampling liquids, however, occurred later in the early 1980s with the introduction of disposable cartridge devices containing silica-based chemically bonded sorbents of a suitable particle size for sample processing by gently suction [10−14]. Within a few years cartridge-based solid-phase extraction was considered a suitable alternative to liquid-liquid extraction for many applications and entered a period of evolutionary change. Typical cartridge devices consist of short columns (generally an open syringe barrel) containing sorbent with a nominal particle size between 20 and 60 μm, preferably with a narrow particle size range, packed between porous plastic or metal frits, Fig. 1.1. A wide range of sorbent chemistries (silica-based chemically bonded, mixed mode, porous polymer, restricted access media, molecularly imprinted polymers, immunosorbent, bonded cryptands, etc.) are available today providing for the diverse application base of modern cartridge-based solid-phase extraction [12−14]. Low-volume cartridges or precolumn devices soon appeared as the basis of online integrated systems for automation of the sampling and separation processes, in for example, solid-phase extraction (SPE)-liquid chromatography (LC), SPE-gas chromatography (GC), SPE-capillary electrophoresis (CE), SPE inductively coupled plasma spectroscopy (ICP) and LC-SPE-nuclear magnetic resonance spectroscopy (NMR). By the mid-1990s these systems had matured into robust practical systems in use in many laboratories with a high sample workload and little variation in sample matrix, for example, drugs in biological fluids, contaminants in surface waters, target compounds in food extracts, etc. [15−18]. Standard solid-phase extraction procedures lend themselves to automation using robotic platforms or special purpose processing units that simultaneously extract and prepare samples for separation [12,19]. Multiwell plates with a sorbent

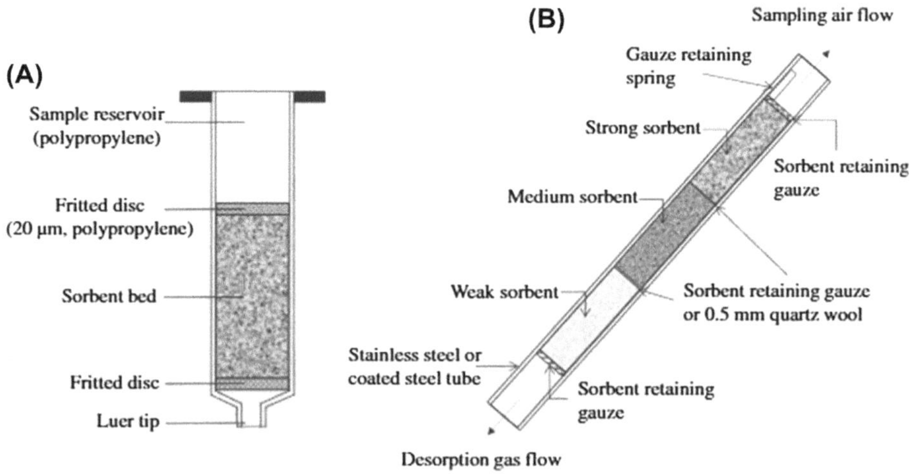


**Figure 1.1** Solid-phase extraction using a cartridge device for liquid samples (A) and gas-phase samples (B).

bed at the bottom of the well combined with liquid handling robots are suitable for high-throughput parallel sample processing [20].

Cartridge-based solid-phase extraction was initially marketed as a replacement for traditional liquid-liquid extraction. Traditional liquid-liquid extraction methods were viewed as labor intensive, difficult to automate, and frequently plagued by practical problems, such as emulsion formation. In addition, they tend to consume large volumes of high purity solvents, some of which are considered significant health hazards with high disposal costs [5,12,21]. Cartridge-based solid-phase extraction procedures, on the other hand, were more economical, afforded shorter sample processing times, consumed less solvent, and allowed for simpler sample handling procedures. They are also more convenient for field sampling because they minimize the transport and storage problems associated with bulk samples, which have to be returned to the laboratory for processing. These arguments, although persuasive at the time, are not as true today. Modern liquid-liquid extraction procedures in their several forms, known collectively as liquid phase microextraction, address many of the problems associated with traditional liquid-liquid extraction and compete effectively and complement solid-phase extraction procedures [22−24].

It should not be overlooked that cartridge-based solid-phase extraction has its own, albeit different, problems to those of liquid-liquid extraction. The extraction properties of solid-phase sorbents are not as reproducible as those for solvents. Typical sorbent surfaces contain more than one functional group in amounts not easily replicated by synthesis. In addition, their sorption properties are typically more adversely affected by contaminants. Sorbents also tend to have a higher level of extractable contaminants originating from the manufacturing process and from packaging materials. This chemical background may interfere in the subsequent sample analysis. Rinsing of the sorbent bed prior to use and using blanks to establish background contamination levels diminishes sample throughput and adds significantly to solvent consumption and sample processing costs. Sorbents are also affected by sample processing conditions, such as overloading, displacement of target compounds by excess matrix, and blocking of sorbent pores. These problems easily go unnoticed resulting in unforeseen changes in the recovery of target compounds.

Solid-phase microextraction (SPME) was introduced in 1990 as an alternative approach to cartridge-based solid-phase extraction, which was well established by this time [3,25]. It is sometimes thought of as a miniaturized version of solid-phase extraction as implied by its name, but this is not the case. The downscaling of solid-phase extraction to accommodate small sample volumes, but otherwise incorporating the same extraction principles, is correctly known as porous membrane protected micro-solid-phase extraction, or simply micro-solid-phase extraction (μSPE) and first appeared in 2006 [26,27]. It employs a small sorbent bed enclosed in a porous membrane allowing solvent and low-mass compounds access to the sorbent bed while excluding macromolecules. The ratio of sample volume to sorbent surface area (or volume) is low in the relative sense but large compared with the fiber format used in solid-phase microextraction, and exhaustive extraction of target compounds remains the general goal. The fiber format employs a thin layer of immobilized extraction phase coated on the outside of a fused-silica or metal-wire support. The extraction fiber is attached to the plunger of a modified microsyringe which both protects and facilitates manipulation of the fiber during sampling, Fig. 1.2 [28]. The main difference between solid-phase microextraction and solid-phase extraction is the extreme ratio of the sample to sorbent volume or surface area. An extremely small amount of sorbent is utilized in comparison to the sample volume in solid-phase microextraction and exhaustive extraction of target compounds is not favored. Typically only a small portion of the target compounds is extracted from the sample (negligible depletion unless the distribution constant is unusually large). This amount increases with the extraction time until equilibrium conditions are reached, as illustrated by Fig. 1.3 [29]. For calibration either the linear portion of the preequilibrium or at or near equilibrium conditions are selected [3,30,31]. The time to reach equilibrium can be long, in which case the linear portion of the preequilibrium curve becomes the preferred choice. Extraction in the preequilibrium region is kinetically controlled and mass transfer is dominated by diffusion (stagnant solution) or by the method of agitation (agitated solution). The selectivity of the extraction process is usually poor in the preequilibrium region compared with equilibrium sampling. The latter is controlled by the differences in the extraction phase-sample solution distribution constants. The small volume of extraction phase favors applications in field sampling (spot and time weighted average), as it is unnecessary to determine the sample volume processed. The amount of extracted target compounds is independent of the sample volume so long as the product of the distribution constant and extraction phase volume (or surface area) is small compared with the sample volume [29,32].

Solid-phase microextraction can be used for sampling by direct immersion, headspace extraction, or extraction with membrane protection [28]. It integrates sampling, extraction, concentration, and sample introduction into a single solvent-free step for gas chromatography and mass spectrometry and with minimal solvent use for liquid chromatography and capillary electrophoresis. Sample processing requires

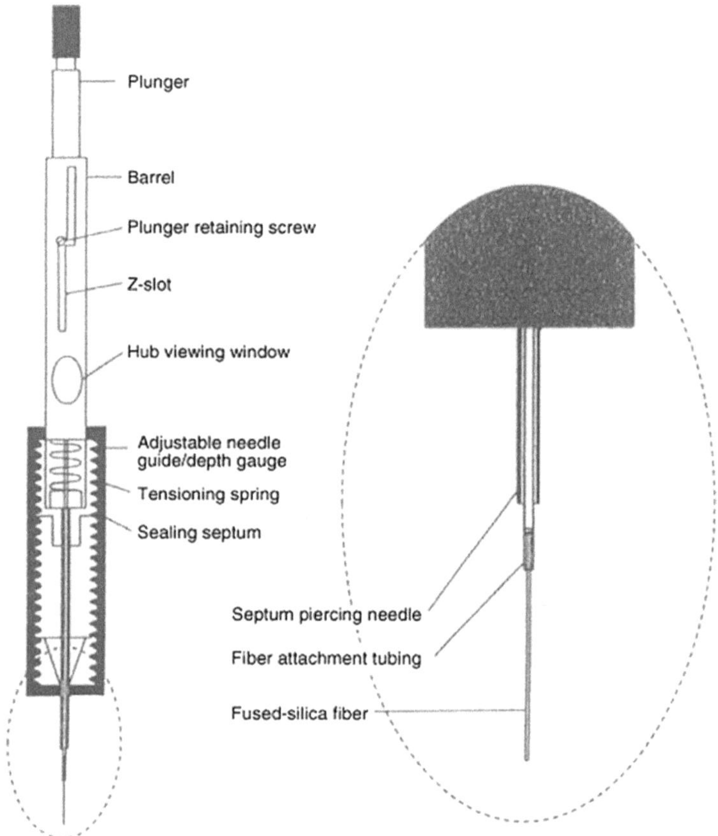


**Figure 1.2** Device for solid-phase microextraction utilizing a fiber format.

two steps: the distribution of target compounds between the extraction phase and the sample matrix and desorption of the extracted compounds by thermal desorption (solventless extraction) or solvent desorption. Only a small volume of solvent is required for solvent desorption due to the small volume of extraction phase. Thus, the concentration of target compounds in the desorption solvent is relatively high even though the amount extracted is considerably less than for exhaustive extraction. The minimal dilution, or complete transfer in the case of gas phase desorption, compensates for the low absolute amount of extracted compounds. The often cited advantages of solid-phase microextraction are its ease of miniaturization, ease of automation, and straightforward coupling with different measurement systems [2,3,33,34]. The main factors affecting sampling efficiency include: the extraction phase chemistry, extraction mode, agitation method, sample modification (pH, ionic strength, presence of organic solvent, etc.), extraction time, and desorption conditions. Limitations include the small number of commercially available extraction phases, the fragile nature of fused-silica fibers, limited fiber reusability due to carryover or matrix contamination, and the short lifetime of physically coated fibers. Problems are more

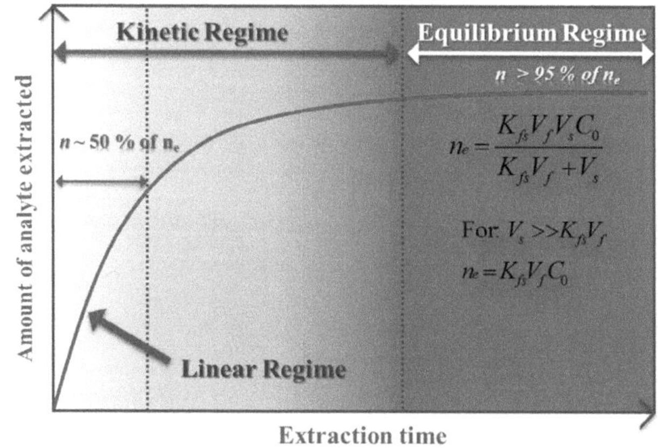


**Figure 1.3** Extraction time profile for sampling with a coated fiber (solid-phase microextraction). Amount of analyte extracted = n ($n_e$ at equilibrium), $K_{fs}$ = analyte distribution constant between the extraction phase and sample, $V_f$ = volume of extraction phase, $V_s$ = sample volume, and $C_s$ = analyte concentration in the sample matrix.
Reproduced from Souza-silva EA, Jiang R, Rodriguez-Lafuente A, Gionfriddo E, Pawliszyn J. A critical review of the state of the art of solid-phase microextraction of complex matrices I. Environmental analysis. Trends Anal Chem 2015;71:224–235 with permission.

common with direct immersion in samples of high complexity, such as biological and food samples, for which the development of biocompatible fiber coatings is an active research area [35–37].

## 1.3 Second generation formats

The low packing density typical of cartridge devices results in the use of longer sorbent beds than necessary to compensate for reduced retention due to channeling [38]. The larger bed mass, in turn, increases nonspecific matrix adsorption resulting in an increase in contamination of extracts. The disk format utilizes smaller particles in a more stable packing configuration to minimize channeling, affording a cleaner chemical background through reduced matrix adsorption. Solid-phase extraction disks are available in at least three different formats. Particle-loaded membranes, introduced in 1990, consist of a web of polytetrafluoroethylene (PTFE) microfibrils, suspended in which are sorbent particles of about 8–12 μm in diameter [39,40]. These membranes are flexible with a homogeneous structure containing upwards of 90% (w/w) sorbent particles in the form of circular disks about 0.5 mm thick with diameters from 4 to 96 mm. For general use they are supported on a sintered glass disk (or other support) in a standard filtration apparatus using suction to generate the desired flow through the membrane. Particle-loaded membranes are also available in a syringe barrel (cartridge) format. In this case, the sorbent bed contains particles of a larger diameter, about 50 μm, in thicker disks, about 1.0 mm, sealed into the base of an open syringe barrel. Particle-embedded glass fiber disks contain 10–30 μm diameter sorbent particles woven into a glass fiber matrix [41]. The small-diameter disks are rigid and self-supporting, while the larger diameter disks require a supporting structure similar to particle-loaded membranes. Laminar disks consist of a sandwich of 10 μm sorbent particles in a consolidated 0.5–1.0 mm bed located between two glass-fiber filters, with a screen to hold the filters in place [10]. The 50 mm diameter laminar disks are typically mounted in an open syringe barrel superficially similar to conventional cartridge devices [10].

The slow sample processing rates for large sample volumes typical of cartridges and their low tolerance to blockage by particles and sorbed matrix components provided the initial interest in disk technology for environmental applications, such as the analysis of surface waters for trace contaminants [40]. On account of their larger cross-sectional area and decreased pressure drop disks allow the use of higher sample flow rates resulting in shorter sample processing times [38]. An integral prefilter attached to the top surface of the disk reduces plugging by suspended particles. Small-diameter disks are suitable for handling samples of restricted volume. Small-diameter disks facilitate integrated sample processing techniques, such as in-vial desorption and on-disk derivatization [42]. The large surface area per unit bed mass of disks facilitates their use for passive sampling in which the disk is suspended in the sample as opposed to the conventional approach of passing the sample through the disk in a manner similar to filtration. The slow equilibrium of the extraction process, even for agitated solutions, however, is a limitation for laboratory applications but is less of a concern for field studies [43]. Particle-loaded membranes have been utilized for biomimetic extraction as surrogate models for bioconcentration and for toxicity risk assessment [44]. Disk technology contributed directly to the automation of solid-phase extraction methods through the development of multiwell extraction plates, used for the cleanup of samples in high-throughput screening techniques in drug development [20]. The characteristic physical properties of disks have supported their application for integrated sampling/detection techniques such as *in situ* radiochemical, phosphorescence, and X-ray fluorescence detection and as a substrate for MALDI mass spectrometry [45–47].

Microextraction by packed sorbent bed (MEPS) was introduced in 2004 as a miniaturized version of cartridge-based solid-phase extraction designed for handling small-sample volumes with a view to easy automation using a laboratory liquid handler [48,49]. The extraction device is typically a 100–250 μL syringe housing a short bed containing 1–2 mg sorbent located either between the plunger and the needle or built into the needle, Fig. 1.4. The MEPS syringe facilitates low-dead volume sample processing by vertical movement of the plunger with sample and solvent flow possible in both directions through the sorbent bed in a cyclic fashion. This is in contrast to conventional solid-phase extraction techniques which utilize a unidirectional flow and a single contact between the sample and solvents and the sorbent bed. Needle-based extraction formats now include internally coated needles [50] and needle trap devices with sorbent-packed needles [50,51]. Internally coated needles typically employ a stainless steel needle internally coated with a 50 μm thick film of immobilized stationary phase resulting in an open structure similar to open tubular columns in gas chromatography. Typical applications include automated headspace sampling of

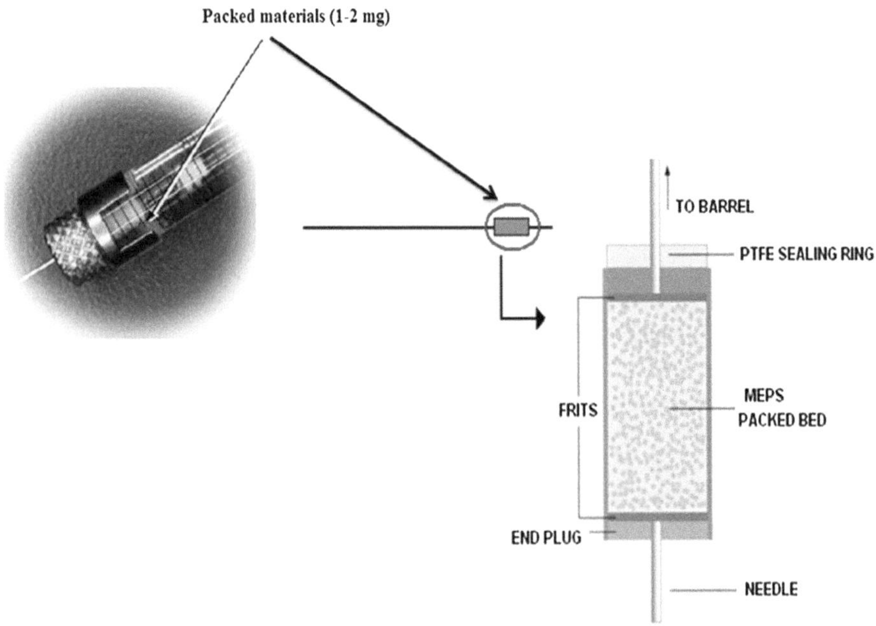


**Figure 1.4** Modified syringe for microextraction by packed sorbent with an expanded view of the sorbent bed (canister). The canister has a dead volume of about 7 μL.
Reproduced from Moein MM, Abdel-Rehim A, Abdel-Rehim M. Microextraction by packed sorbent (MEPS). Trends Anal Chem 2015;67:34–44 with permission.

volatile organic compounds referred to as solid-phase dynamic extraction (SPDE). Needle traps consist of a specially designed stainless steel needle packed with adsorbent. For extraction a fixed volume of a gas or liquid sample is passed through the extraction needle with recovery of target compounds by thermal desorption, or less commonly, by solvent desorption. Needle traps have also been used as passive sampling devices for gas phase samples. Needle traps are more durable than solid-phase microextraction fibers and have a higher extraction capacity with the possibility of exhaustive extraction.

The extraction rate and sensitivity of fiber-based solid-phase microextraction methods can be enhanced by increasing the volume and/or surface area of the extraction phase. Simply increasing the thickness of the extraction phase in the fiber format results in long extraction times and alternative formats have been explored to tackle this issue resulting in the development of in-tube, stir bar, rotating disk, thin-film, and dispersed particle approaches [2,3,52,53]. An important performance parameter for solid-phase microextraction is the equilibrium time (the time required to reach 95% equilibrium where statistically no difference in the amount extracted is observed by extending the sampling time). Shorter equilibration times can be obtained by utilizing a geometry for the extraction phase with a higher surface area-to-volume ratio than the cylindrical geometry of fiber-based solid-phase microextraction devices. The solid-phase microextraction arrow, Fig. 1.5, consists of a steel

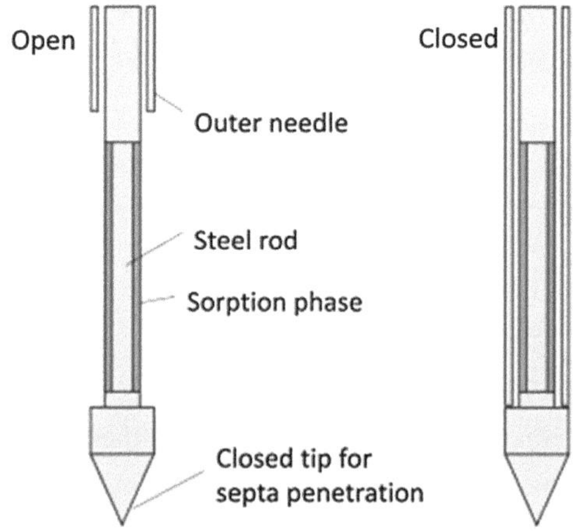


**Figure 1.5** Arrow solid-phase microextraction device with exposed sorbent (sampling mode) left and injection (covered) mode right.
Reproduced from Helin A, Ronnko T, Parshintser K, Hartonen K, Schillin B, Laubli T, Riekkola ML. Solid-phase microextraction arrow for the sampling of volatile amines in wastewater and atmosphere. J Chromatogr A 2015;1426:56−63 with permission.

rod of larger diameter than a conventional fiber coated with a larger amount of extraction phase, while still being compatible with thermal desorption in a standard injection liner of a gas chromatograph on account of its dimensions and sharp, closed tip [50,54]. Alternatively the surface area-to-volume ratio of the extraction phase can be increased using a thin-film format in which the extraction phase is immobilized on the outer surface of a support of suitable geometry, Fig. 1.6 [29,55,56]. The thin-film geometry is the basis for fully automated parallel sample processing in a

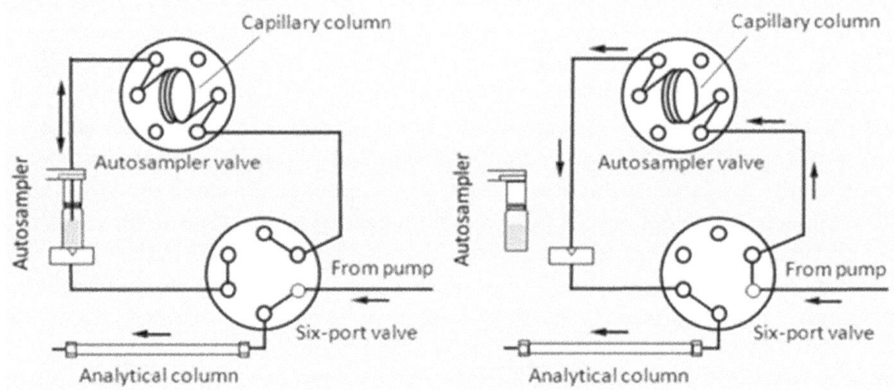


**Figure 1.6** In-tube solid-phase microextraction with the extraction column utilized as the sample loop of the injection valve of a liquid chromatograph. The valve is shown in the extraction position (load) on the left and extract injection position (inject) on the right.
Reproduced from Queiroz MEC, Melo LP. Selective capillary coating material for in-tube solid-phase microextraction coupled to liquid chromatography to determine drugs and biomarkers in biological samples. A review. Anal Chim Acta 2014;826:1−11 with permission.

96-well plate format utilizing an extraction unit fashioned into a 96-blade device in which the extraction phase is coated over the flat end of each blade [33,52,55].

An early approach for automated solid-phase microextraction from 1997 was the coupling of an internally coated capillary column, the extraction device, to a liquid chromatograph for separation and detection known as in-tube solid-phase microextraction [57–61]. Two approaches are typically used today [59]. In the first approach a short capillary column is placed between the injection loop and injection needle of an autosampler. The injection syringe repeatedly draws and ejects samples from a series of vials under computer control cycling the sample through the capillary column in the forward and reverse direction for each sample until equilibrium is reached or extraction is sufficient. The extracted compounds are subsequently desorbed by solvent from the extraction column and transported to the analytical column for separation. Alternatively, the capillary column is used as a sample loop for an injection valve and target compounds are extracted from the sample with the valve in the load position and then transferred to the column by mobile phase after switching the valve to the inject position, Fig. 1.6 [59]. This arrangement is favored for handling larger sample volumes to increase sensitivity. Wire-in-tube or fiber-in-tube configurations can be used to enhance the extraction rate and efficiency by reducing the phase ratio of the extraction column. Capillary columns of the type used for gas chromatography are often used as the extraction device, as well as a wider range of laboratory-made extraction phases to enhance selectivity, for example, poly(pyrrole), restricted access media, immunosorbents, molecularly imprinted polymers, monolithic sorbents, etc. [59]. Samples are processed sequentially through in-tube solid-phase microextraction, which cannot be considered high-throughput when compared with parallel sample processing using the 96-well plate format. Other general limitations include a low extraction efficiency for some compounds, a low sorbent loading capacity, instability of some extraction phases, and long extraction times resulting from poor mass transfer kinetics.

Stir bar sorptive extraction was introduced in 1999 and quickly gained popularity [62–65]. The extraction device consists of a magnetic stir bar in a glass sleeve externally coated with a 0.3–1.0 mm layer of extraction phase. Liquid samples are analyzed by direct immersion of the stir bar with vigorous stirring and gas phase samples by suspending the stir bar in the headspace above the sample. Extracted compounds are recovered by either thermal desorption or solvent desorption. The limited number of commercially available extraction phases [poly(dimethylsiloxane), poly(acrylate), and an ethylene glycol/silicone copolymer] limit general applications to the extraction of compounds of low- and intermediate-polarity from water [63,65]. The larger volume of extraction phase, typically two orders of magnitude or more compared with fiber-based devices, is responsible for the enhanced extraction efficiency as well as the longer extraction times required to reach either equilibrium or sufficient extraction. The relatively low-selectivity of commercially available extraction phases result in significant matrix sorption with possible interference in the separation process [65]. The sampling process is not easy to automate and a special purpose-designed thermal desorption unit is required for the sequential desorption of extracted compounds from stir bars for gas chromatography [62]. Direct contact between the stir bar and the extraction vessel and the high stirring rates employed for direct immersion extraction results in a loss of coating due to friction, thus reducing the lifetime of the stir bar. Rotating disk sorbent extraction attempts to address this problem with a novel stir bar design [66–68]. A PTFE disk with an embedded magnetic bar at its base and extraction phase coated only on the top surface is used. An alternative design has a cavity on the top surface loaded with sorbent and covered by a glass fiber filter or semipermeable membrane. There is no contact between the extraction phase and the extraction vessel during stirring prolonging the lifetime of the disk and the protected cavity on the top surface allows a wider choice of sorbent chemistries to be exploited. Further developments based on stir rod and stir membrane devices have been proposed [67].

Nanoparticles have a large surface area-to-volume ratio and magnetic nanoparticles are promising extraction materials for dispersive solid-phase extraction [69–72]. The sorbent material does not need to be packed in a particular type of sampling device, unlike for traditional solid-phase extraction methods. The small particle size allows their dispersion throughout the sample volume to minimize mass transfer problems and their magnetic core facilitates easy collection with a permanent magnet. The efficient dispersion of the sorbent into the matrix yields a high recovery of target compounds limited only by their distribution constants and clean extracts that depends on the ability of the sorbent to reject matrix components. The sequence of steps in the extraction process is illustrated in Fig. 1.7 [70]. Particles are typically 1–100 nm in diameter and made up of several layers, typically with a magnetic core of iron, nickel, cobalt, or any of their oxides. Magnetite ($Fe_3O_4$) is the most common magnetic core.

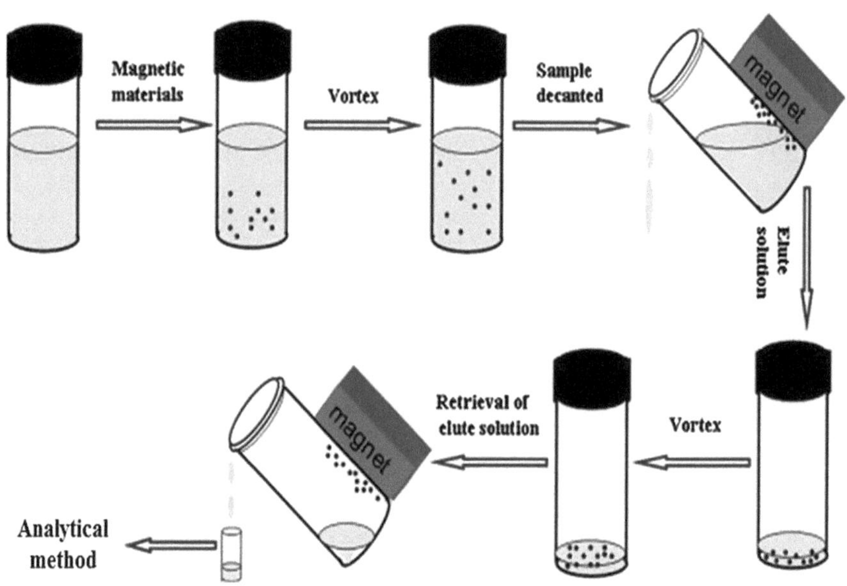


**Figure 1.7** Procedural steps for dispersive solid-phase extraction using magnetic nanoparticles. Reproduced from Wlerucka M, Biziuk M. Application of magnetic nanoparticles for magnetic solid-phase extraction in preparing biological, environmental and food samples. Trends Anal Chem 2014;59:50–58 with permission.

The magnetic core is typically encased in a shell of inorganic oxide (silica, alumina, manganese dioxide, etc.) or carbon. This can be coated with polymer or the surface modified by chemical reaction to modify selectivity. A large number of surface modified nanoparticles have been synthesized for dispersive solid-phase extraction but only a few are commercially available [71,72].

Nanofibers represent an alternative to nanoparticles for some applications in solid-phase extraction [73]. Nanofibers are continuously spun wires with tunable diameters in the nanometer to micrometer range, variable porosity, and large surface area; they are synthesized in the laboratory by applying a high voltage between a viscous solution of a polymer and a collector electrode (in the form of a wire for coated rod devices or sheet of conductive material for membrane devices). A mat of electrospun fibers sealed in a special holder affords a suitable device for conventional solid-phase extraction characterized by a high retention capacity and a low backpressure. Depositing a mat of fibers on the surface of a suitable substrate provides materials for thin-film microextraction. Positioning a nanofiber sheet in the loop of an injection valve affords a suitable interface for in-tube solid-phase microextraction coupled to liquid chromatography. Fiber-coated wires are suitable for solid-phase microextraction. Nanofibers can also be packed into the lower portion of pipette tips, to fashion extraction devices for sampling small volumes. A wide range of single polymers, copolymers, composites, and hybrid nanoparticle loaded fibers have been described so far, but only a few have been used in solid-phase extraction. Limited commercial availability of nanofibers and devices containing nanofibers has restricted their applications to experimental studies.

Monoliths provide an alternative bed structure to particle-packed beds for solid-phase extraction [74–77]. A monolith is a single rod of biporous organic or inorganic polymer with a uniform structure consisting of flow-through pores (1–2 μm in diameter) providing high permeability (low back pressure) and mesopores providing a high surface area for retention and sample capacity. Organic monoliths are easily prepared in (usually) a one-step process in a suitable mold containing monomers (styrene, acrylate, methacrylate, etc.), cross-linking agent (dimethacrylate, divinylbenzene, etc.), porogen, and a free radical initiator. Inorganic monoliths are mostly based on silica but their preparation is more difficult. They are typically prepared from mixtures of a tetraalkyloxysilane, poly(ethylene glycol), and an acid catalyst forming an initial silica sol that is transformed to a gel by aging. Silica monoliths are more mechanically stable, less prone to changing dimensions (swelling) in different solvents, have a higher surface area, and are easily modified by reaction of surface silanol groups. Silica monoliths are commonly used for the extraction of small molecules in cartridge and disk formats. Organic monoliths are a better choice for the extraction of biopolymers due to their relatively low surface areas, greater biocompatibility, and higher tolerance of extreme pH. Hybrid organic silica monoliths combine the positive features of organic and silica polymer monoliths, but are not commercially available [77]. Monoliths incorporating nanoparticles, molecular-organic frameworks, immunosorbents, molecularly imprinted polymers, aptamers, etc., have been developed to enhance selectivity or sample capacity, but only as experimental materials [76–79]. Wide diameter ($\geq$4 mm) rod structures became available only recently. The main applications of mainly narrow diameter silica monoliths or thin-film coatings have been described for in-tube solid-phase extraction, online precolumn liquid chromatography, fiber and stir bar coatings, and microfluidic devices [75,77]. In the disk format monoliths are typically used in pipette tips, spin column (disk solid-phase extraction utilizing centrifugal forces for sample processing), and in multiwell extraction plates [80].

## 1.4 Sorbent chemistries and properties

Common sorbents for solid-phase extraction can be classified as inorganic oxides, low-specificity sorbents (silica-based chemically bonded phases, porous polymers, and carbon) and compound and group-selective or high-specificity sorbents (ion exchange, mixed mode, macrocyclic, restricted access, affinity-based, and molecular imprinted polymers) [13]. Alternative classification schemes are possible as well as a wider range of subcategories [1,3,52,81−84]. There is also a considerable disconnect between the enormous number of experimental sorbents described in the contemporary literature and the much smaller number of commercially available sorbents. Outside of research centers and university laboratories the skills, interest, and experience necessary to synthesize experimental sorbents are not usually available. Further details of experimental sorbents can be found in the chapters that follow this general overview. Industrial and regulatory applications are dominated by commercially available sorbents, which are the focus of this section.

### *1.4.1 Inorganic oxides*

The most important inorganic oxide adsorbent for solid-phase extraction is silica gel and to a lesser extent alumina, titania, zirconia, florisil, and diatomaceous earth. Inorganic oxides have a high concentration of active functional groups on their surface responsible for the adsorption of analytes by polar, ion-exchange, and Lewis acid/base interactions [85]. For aqueous samples ion-exchange and Lewis acid/base interactions dominate while dipole-type and hydrogen-bonding interactions are usually as or more important for nonaqueous samples. Silica gel is in many ways a near ideal adsorbent for the extraction of small polar organic compounds and is available in a wide range of particle sizes for use in different extraction formats, average pore sizes, in the range 4−30 nm, and specific surface areas from 300 to 800 $m^2/g$ [5,12]. Silica gel is also a common substrate for monolithic sorbents [77]. The sol-gel process allows silica gel to be utilized as a thin porous coating with chemical attachment to the supporting substrate to increase durability [25,52]. Silica is soluble in aqueous solutions at $pH > 7.4$ and its use is limited to acidic and neutral solutions. Organic-inorganic hybrid particles can be used to extend the pH operating range of silica-based sorbents [86].

Alumina is available as neutral, acidic, or basic forms determined as an apparent surface pH depending on processing conditions. Typical adsorbents for solid-phase extraction have an average pore size of 6 nm and a specific surface area of 150 $m^2/g$. The surface chemistry of alumina is more complex than silica and is dominated by ion-exchange and Lewis acid/base interactions. Stability to extreme pH favors its use for extraction based on ion-exchange. It also has a high adsorption capacity for metal ions. Titania and zirconia have strong Lewis acid/base sites and are used for the selective isolation of polyoxy anions (phosphates, phosphonates, borates, carboxylates, sulfates, and in particular, phospholipids) [85]. Florisil is a synthetic magnesium silicate with a surface area of about 250–300 $m^2/g$ and an apparent surface pH of about 8.5. It is commonly used for sample cleanup in the isolation of pesticide residues from fats and oils. Diatomaceous earth is a flux-calcined form of silica gel with a low surface area. It is mainly used as a filter aid in sample preparation and as a dispersant for solvent extraction (matrix solid-phase dispersion) [87].

### 1.4.2 Low-specificity sorbents

Low-specificity sorbents for aqueous samples include silica-based chemically bonded sorbents and coatings, porous polymers and polymer coatings, and various forms of carbon [1,10–13]. For gas phase samples porous polymers and various forms of carbon, and occasionally, inorganic oxides or inorganic oxides physically coated with a liquid phase or reactive material are typically used [8,9,88–90]. Low-specificity sorbents are characterized by retention governed by dispersive interactions accompanied by weak or modest polar intermolecular interactions.

Silica-based porous particle, monoliths, and thin films with chemically bonded surfaces are prepared by reacting surface silanol groups with organosilanes forming stable siloxane bonds. By varying the silica substrate and organosilane reagent chemically bonded surfaces with a wide range of properties (pore size, bound ligand type, bonding density, etc.) can be prepared by chemistry developed for organosiloxane-bonded silica stationary phases for liquid chromatography [91,92]. Some examples are indicated in Table 1.1. Chemically bonded sorbents are generally synthesized by reaction of monofunctional or trifunctional silanes with silica, followed by endcapping (reaction of sterically crowded silanol groups with a small-size organosilane) in some cases. Monofunctional reagents result in monomeric surface coverage while trifunctional reagents can react both with the silica surface and hydrolyzed reagent forming extended polymeric layers of higher carbon loading and greater pH stability. Chemically bonded sorbents with high surface areas (500–600 $m^2/g$), long alkyl chains, and high phase loading are generally used for the isolation of low-mass compounds from aqueous solution, while wide pore materials with short alkyl chains are used for macromolecules. For silica-based, alkylsiloxane-bonded sorbents retention increases with solute size, and for compounds of a similar size, is reduced by polar interactions with water (particularly hydrogen bonding) and by ionization for aqueous samples. Siloxane-bonded sorbents containing polar functional groups (e.g., 3-cyanopropyl, 3-aminopropyl, and spacer-bonded propanediol) are used mainly to isolate target compounds from samples dissolved in organic solvents based on their selective interactions with polar functional groups of the target compounds. They are generally less retentive than typical alkylsiloxane-bonded sorbents for aqueous samples.

Silica-based organosiloxane-bonded sorbents have two general limitations in addition to small, and possibly inadequate, breakthrough volumes for low-mass, highly polar organic compounds. The silica-based sorbents contain a low concentration of

**Table 1.1** Structures of porous silica-based, monomeric, organosiloxane-bonded sorbents (≡Si-R).

| Type | Symbol | Functional group | Structure (R) |
|---|---|---|---|
| Alkyl | C30 | Tricontane | $-C_{30}H_{61}$ |
| | C18 | Octadecyl | $-C_{18}H_{37}$ |
| | C8 | Octyl | $-C_8H_{17}$ |
| | C4 | Butyl | $-C_4H_9$ |
| Cyclohexyl | CH | Cyclohexyl | $-C_6H_{11}$ |
| Aromatic | PH | Phenyl | $-C_6H_5$ |
| | BH | Biphenyl | $-C_{12}H_9$ |
| Cyano | CN | 3-Cyanopropyl | $-CH_2CH_2CH_2CN$ |
| Amino | NH2 | 3-Aminopropyl | $-CH_2CH_2CH_2NH_2$ |
| Hydroxyl | DIOL | Spaced-bonded propanediol | $-CH_2CH_2CH_2OCH_2CH(OH)CH_2OH$ |
| Quaternary amine | SAX | Trimethylaminopropyl | $-CH_2CH_2CH_2N(CH_3)^+ Cl^-$ |
| Carboxylic acid | CBA | Carboxylpropyl | $-CH_2CH_2CH_2CO_2H$ |
| Sulfonic acid | PRS | Propanesulfonic acid | $-CH_2CH_2CH_2SO_3^- H^+$ |
| | SCX | Benzenesulfonic acid | $-C_6H_4SO_3^- H^+$ |

ionized silanol groups capable of ion-exchange interactions with basic compounds as well as poor hydrolytic stability at extreme pH ($2 > pH > 8$) [93,94]. Low recovery of basic compounds can result from the inability of the elution solvents to displace these compounds from the ion-exchange sites (addition of a competing base to the elution solvent might solve this problem).

Macroreticular porous polymers are biporous copolymers of styrene-divinylbenzene or acrylic esters, typically, consisting of agglomerates of randomly packed microspheres permeated by a network of holes and channels [91,95]. The highly strained porous structure is swollen to different extents by organic solvents and aqueous-organic solvent mixtures, and is partially responsible for the favorable retention properties of these materials. The addition of small volumes of water-miscible organic solvents, used as a processing aid to minimize dewetting of the polymer during sampling, can have a remarkable effect on the breakthrough volume for these sorbents [96]. Macroreticular porous polymers with surface areas typically less than 600 $m^2/g$, exhibited inadequate retention for low-mass polar compounds, which led to the development of hyper-cross-linked polymers with specific surface areas of 800–2000 $m^2/g$ and commensurate increased retention [97,98]. Hydrophilic copolymer sorbents were developed to reduce interfacial tension between the polymer surface and water, increasing the retention of polar compounds by enhancing their contact with the sorbent. They have the added advantage of being fully water wettable simplifying sample processing. These are typically copolymers of *N*-vinylpyrrolidone and divinylbenzene, methacrylate and divinylbenzene, or surface-modified conventional styrene-divinylbenzene polymers [97].

The main forms of carbon used for solid-phase extraction are activated carbon, graphitized carbon blacks, carbon nanotubes, and porous graphitic carbon [99–104]. Granular activated carbon is prepared by the low temperature oxidation of vegetable charcoals and has a large surface area (300–2000 $m^2/g$), a wide pore distribution and a heterogeneous surface containing active functional groups. Graphitized carbon blacks are (largely) nonporous with moderate surface areas of 5–100 $m^2/g$. Their surfaces are contaminated by oxygen species and other functional groups depending on the origin of the material used for preparation of the carbon sorbent by pyrolysis. The retention mechanism for activated and graphitic carbon is quite complex in which surface functional groups contribute to retention, for example, by ion exchange. Porous graphitic carbon prepared by the silica-template process has a low level of surface contamination but is used less frequently than other forms of carbon because of its relatively high cost. For aqueous samples retention is influenced by hydrophobic forces (increases with solute size), by adsorption on the electronically polarizable surface (dipolar interactions and electron lone pair interactions increase retention), and by contributions related to the microscopically flat surface of graphitic carbon in which planar compounds are retained preferentially over bent and angular compounds [105].

Carbon nanotubes (also cones, disks, horns, fibers) are relatively new materials characterized by a large surface area to volume ratio, high thermal stability, and with a surface that is easily modified to adjust their properties [83,101–104]. Carbon nanofibers have significantly larger dimensions than carbon nanotubes and are preferred for cartridge-based solid-phase extraction [106]. Single-walled carbon nanotubes consist of a single sheet of graphene rolled up in the shape of a cylinder with typical diameters from 0.4 to 3.0 nm and with ends normally capped by fullerenelike structures. Multiwalled carbon nanotubes consist of several rolled up graphene sheets concentrically arranged around a common axis with typical diameters from 1.4 to greater than 100 nm with (usually) capped ends. Carbon nanotubes can be prepared in various lengths up to a few micrometers. Graphene has a high adsorption capacity for compounds of all types including inorganic ions [107]. Surface modification not only enlarges their potential range of applications but equally important, enhances their dispersion in solvents, which otherwise barely occurs due to strong intertube interactions. Experimental materials with a wide range of physically coated and chemically bonded surface functional groups have been described but are not commercially available [103,104]. Graphene is quite reactive and surface modification by halogenation, hydrogenation, oxidation, and radical and nucleophilic addition make it suitable for the preparation of more selective sorbents.

Cross-linked poly(siloxanes) analogous to those used as stationary phases for gas chromatography are commonly used as coatings for film-based formats such as fiber, in-tube and thin-film solid-phase microextraction and stir bar sorptive extraction [108,109]. More polar experimental coatings based on copolymers with polar and hydrogen-bonding monomers, for example, poly(acrylonitrile), poly(pyrrole), poly(-urethane), and poly(dopamine) have also been described [59,109–111]. Currently only a limited number of different fiber coatings for solid-phase microextraction is commercially available, notably single polymeric phases such as poly(dimethylsiloxane) and poly(acrylate) and dispersions of solid adsorbents, such as carboxen (carbon molecular sieve) and poly(divinylbenzene) in a supporting polymeric matrix [28,82]. For stir bar sorptive extraction commercially available coatings are limited to poly(dimethylsiloxane), poly(acrylate), and poly(ethylene glycol)-modified silicone [109]. Poly(dimethylsiloxane) coatings meet most demands for the extraction of low and moderately polar compounds from water but are of limited utility for the extraction of polar compounds and ions. Thermal desorption is typically used for recovery of extracted compounds requiring thermally stable coatings, although liquid desorption is possible. Some polymers absorb organic solvents resulting in a change in dimensions as well as stability, limiting applications employing direct immersion. Thin films of poly(dimethylsiloxane) are used as a secondary coating for other sorbents to improve their biocompatibility for sampling by direct immersion [3,35].

Various porous polymers, carbon-based, and inorganic oxide sorbents as well as liquid-coated sorbents are commonly used for extracting volatile organic compounds from air and for work place monitoring [88,89,92,112,113]. Tenax, a polymer based on 2,6-diphenyl-4-phenylene oxide, revolutionized sorbent trapping of semivolatile organic compounds from air and purge gas from aqueous samples (purge-and-trap devices). Tenax differs from most other polymeric sorbents in that it is a granular powder of low surface area, about 35 $m^2/g$, but has exceptional thermal stability, up to about 375°C, facilitating the recovery of low-volatility compounds by thermal desorption. It exhibits strong retention of semivolatile organic compounds ($> C_7$) at room temperature with minimal adsorption of water vapor [114]. Macroreticular porous copolymers, such as poly(styrene-divinylbenzene), have higher surface areas of 300–750 $m^2/g$ but significantly lower thermal stability ($< 250$°C) limiting the mass range of compounds that can be recovered by thermal desorption compared with Tenax. Besides Tenax, carbon in the form of activated carbon, graphitized carbon blacks and carbon molecular sieves are widely used [99,115,116]. Graphitized carbon blacks are primarily used for trapping volatile organic compounds, either separately or in combination with Tenax. Carbon molecular sieves are prepared by the controlled pyrolysis of porous polymers and consist of cross-linked graphite crystallites with a disordered cavity-aperture structure. They are microporous with large surface areas (500 to $>$1200 $m^2/g$) and are used primarily for trapping volatile organic compounds ($C_1$ to $C_3$) at room temperature often as a component of a multibed adsorbent trap. Since no single adsorbent is ideal for trapping all compounds typically found in contaminated atmospheres, it is common practice to use extraction tubes packed in series with two or more adsorbents [99,113]. Foamed polyurethanes composed of agglomerated spherical micrometer-sized particles, bonded to one another in a rigid and highly permeable structure, are suitable for sampling low-volatility organic compounds at high flow rates [112]. They are frequently used in conjunction with high-volume air samplers on account of their low pressure drop compared with standard sorbent cartridges.

### 1.4.3 High-specificity sorbents

Various sorbents based on ion exchange, bioaffinity, molecular recognition, and restricted access materials are used to achieve higher selectivity in solid-phase extraction to supplement the low-specificity sorbents discussed above [1,5,12,97]. Ion exchange is used to extract ionizable compounds and permanent ions from aqueous samples, usually, with sorbents containing immobilized ionic sites of opposite charge to the target compounds. Ions of a similar charge to the sorbent and most neutral compounds are either unretained or weakly retained and removed during the rinse step. Ion-exchange sorbents are typically classified as weak or strong depending on the identity of the immobilized ionic group and whether its charge is independent of the sample pH (strong ion exchanger) or can be manipulated by changing the sample pH (weak ion exchanger). The most common functional groups are sulfonic and carboxylic acids as strong and weak cation exchangers, respectively, and quaternary amines and secondary and tertiary amines as strong and weak anion exchangers, respectively. For many applications polymer-based and silica-based ion exchangers are interchangeable, although due to differences in nonspecific adsorption of matrix components the chemical background of the extracts may be different. Polymer-based ion-exchangers generally have a higher exchange capacity and a wider pH operating range. Disk-based solid-phase extraction is a popular method of sample cleanup for the removal of nontarget ions that interfere in the separation of the ions of interest by ion chromatography and capillary electrophoresis [117]. Mixed-mode sorbents containing cobonded ion exchange and alkyl groups in cartridge or disk format are popular for the isolation of ionizable drugs and their metabolites from biological fluids [1,12,81,118]. Target compounds are retained by a synergistic combination of ion exchange and reversed-phase interactions. The strong retention and use of efficient rinse solvents provides cleaner extracts for chromatographic analysis compared with single-mode sorbents. Surface-bonded macrocyclic ligands are used for the isolation of metal ions and some anions from aqueous samples, particularly samples of high ionic strength (e.g., sea water) or extreme pH, where ion-exchange sorbents are generally ineffective [72,119]. An enormous number of immobilized ligands with different donor atoms, ring sizes, and ligand geometry have been described. The most important are crown ethers, calixarenes, and cyclic polyamines and polysulfides. The number of donor atoms and spacer arms determine the stability of the macrocycle-metal complex while the type of donor atoms (O, S, N typically) are important for determining the metal ion selectivity. Extracted metal ions are retained in the cavity of the macrocyclic ligand until released by elution with a solution of a complexing agent with a higher binding constant for the metal. Both processes provide a high degree of specificity, allowing sorbents to be developed for single metal ions or group of metal ions. Current research is largely focused on the selective extraction of radioisotopes from nuclear decommissioning sites.

Immunosorbents were used for a long time for extracting target compounds from biological systems and as immobilized reagents in assay systems based on bioaffinity. More general applications in food and environmental analysis, however, are more recent [120,121]. In part, this is due to the difficulty of raising antibodies with a high selectivity to small molecules, as well as a lack of familiarity among analytical chemists of the procedures utilized in the production and standardization of antibodies. Fortunately, some immunosorbents are commercially available for specific compound types, such as mycotoxins, paralytic shellfish toxins, triazines, and some pesticides [121,122]. Immunosorbents are prepared by covalently bonding a suitable antibody to an appropriate sorbent. A high degree of molecular recognition is obtained based on the specificity of the antibody-antigen (target compound) interaction. High specificity allows the extraction of target compounds from complex matrices in a single step with minimal coextraction of matrix interferences. By taking advantage of cross-reactivity, class-specific immunosorbents have been developed. The development of new immunosorbents is time consuming, expensive, and requires expertise in the production, purification, and standardization of antibodies. Oligosorbents and molecularly imprinted polymers have gained traction as suitable alternatives [122].

Aptamers are short, single-stranded, synthetic oligonucleotides capable of shape recognition with a high specificity for target compounds [83,123]. After immobilization on a suitable porous particle support, they can be used as oligosorbents in solid-phase extraction. They are easier to prepare than immunosorbents and afford similar molecular recognition selectivity when optimized for specific target compounds. Aptamers for a selected target compound can be screened and produced *in vitro* by the process known as systematic evolution of ligands by exponential enrichment followed by amplification using the polymerase chain reaction. Matrix components may affect analyte-aptamer interactions reducing the extraction efficiency. Other limitations are limited reusability and the need for specific storage conditions. Dendrimer-based sorbents are another example of synthetic extraction phases based on molecular recognition [124]. Dendrimers are macromolecules with highly repetitive branched structures. Their size, shape, inner core, and functional groups at the outer surface can be tailored to achieve molecular recognition. They are largely exploratory materials at this time.

Molecularly imprinted polymers are synthetic analogs of immunosorbents that are easier and less expensive to prepare [125−128]. They are typically organic copolymers with artificially generated recognition sites able to specifically rebind target compounds in preference to other closely related compounds. The imprint is obtained by the polymerization of functional and cross-linking monomers in the presence of a template molecule and a small amount of solvent, which acts as a porogen. Originally polymers were obtained as monoliths and crushed, ground, and sieved to obtain particles of a suitable size for cartridge-based solid-phase extraction. Increasingly precipitation, multistep swelling, and emulsion polymerization are used to synthesize spherical particles of desired size today. The general approaches for synthesizing molecularly imprinted polymers have been adapted to all contemporary solid-phase extraction formats [3,23,36,53,82,83,129]. A few molecularly imprinted polymers are available commercially for high-volume applications, but otherwise, due to the specific nature of the extraction mechanism, most publications utilize experimental sorbents [129−133].

Choice of the template-monomer system is the key to success. In solution the template molecule complexes with one or more functional monomers, which then become spatially fixed in the solid polymer after polymerization. Methacrylic acid and 4-vinylpyridine are commonly used functional monomers for the extraction of basic and acidic compounds, respectively. Ethylene glycol dimethacrylate is widely used as the cross-linking agent, the purpose of which is to impart mechanical strength to the polymer, stabilize the molecular recognition site, and control the porosity of the polymer. The resultant imprint possesses a steric (size and shape) and chemical (spatial arrangement of complementary functional groups) memory for the template molecule. The template-functional monomer interaction needs to be reversible and if the template molecule binds too strongly to the functional monomer then sample contamination from leakage of the template molecule into the sample due to incomplete removal from the polymer can be a problem. If the functional monomer binds the template molecule too weakly, then lower specificity may result for the extraction step. Low-polarity organic solvents are typically used in the synthesis of the imprint polymers that are usually different to the sample solvent. Shrinking or swelling of the polymers in different solvents during sample processing and recovery can have a profound effect on the extraction efficiency of the target compounds. As water is not generally a suitable solvent for synthesizing molecularly imprinted polymers, the extraction of aqueous samples can be a problem and may require a different protocol for sample processing compared with conventional approaches.

Metal-organic frameworks [134,135] and covalent organic frameworks [136] are a new type of microporous crystalline materials with a uniform pore structure, high surface area, and molecular recognition capability. The molecular recognition capability can be tailored to some extent for specific applications by modifying the cavity size and functional groups available for interacting with target compounds. So far few proven applications have emerged but these experimental sorbents are considered promising for solid-phase extraction.

Restricted access media were developed to facilitate the extraction of low-mass target compounds from complex biological, food, and environmental samples with minimal sample pretreatment [137–140]. They work by excluding macromolecules from the regions of the sorbent where retention occurs by adsorption, partition, or ion exchange. Exclusion from the active sampling region of the sorbent is provided by either a physical diffusion barrier, such as a pore diameter, or by a chemical diffusion barrier, such as a polymer network at the outer surface of the particle. In either case, the outer surface of the particle must be compatible and nonadsorptive of macromolecules. Restricted access media are used in all solid-phase extraction formats but the most common application is in automated online sample processing in liquid chromatography. In this case, a short precolumn packed with the restricted access sorbent is connected to the analytical column through a six-port switching valve. The sample is injected directly onto the precolumn, which retains the compounds of interest. Potentially interfering macromolecules are flushed to waste. The target compounds are then eluted from the precolumn directly onto the analytical column for separation. Simultaneously the precolumn is reconditioned (or exchanged) before processing the next sample.

### 1.4.4 Sorption mechanisms

A fundamental understanding of the extraction mechanism to support sorbent design and selectivity optimization requires a suitable model capable of describing the retention of target compounds under typical sample processing conditions. This quest has not proven simple as indicated by the more detailed studies of retention mechanisms in liquid chromatography [91,141]. Inorganic oxides are considered classic adsorbents with a heterogeneous surface and major contributions to retention by site-specific surface interactions with immobilized active sites [142]. Carbon is a further example of a classic adsorbent with a more homogeneous surface and significant contributions from dispersion and dipole-type interactions [105]. Silica-based chemically bonded sorbents are more complex with properties strongly affected by sample processing solvents. For predominantly aqueous samples, the major obstacles to formulating a simple extraction model are identified as the heterogeneous composition of the stationary phase, the difficulty of providing a working definition for the phase ratio, and the uncertainty as to whether the distribution mechanism for varied compounds is partition, adsorption, or some combination thereof [141,143]. For polymer films, such as poly(dimethylsiloxane) a partition mechanism from the gas phase or solution is generally assumed while contributions from interfacial adsorption is a further possibility. Solid porous polymers are assumed to act as adsorbents although solvent uptake and swelling indicate that they may also show some partition behavior [91,96]. Extraction behavior is explained by interpreting features of the extraction process in terms of an implied model. Simplicity dictates that a single model is utilized in most cases, but in practice this may be misleading, because for individual compounds different mechanisms or different contribution to a mixed mechanism is a possibility. Progress in sorbent extraction design has been made on empirical grounds with acceptable results but with only a limited understanding at a molecular level.

## 1.5 Theoretical contributions to modeling sample processing conditions for cartridge and disk devices

The parameters that describe the processing steps in solid-phase extraction using cartridge and disk devices are amenable to measurement by chromatographic methods or by estimation from general theoretical principles employed in chromatography with suitable modifications [13,38,144–147]. Target compound concentrations are generally low for typical sample processing conditions and a safe sampling volume, corresponding to the maximum amount of a compound that can be extracted, is determined by the breakthrough volume for the sampling device. For matrix simplification the volume and type of rinse solvent is determined by the type and amount of sorbent employed for the isolation step. To select a rinse solvent a common approach is to apply a volume of strong solvent that preserves a certain minimal retention for the least retained of the target compounds in systems using solvent desorption for recovery of extracted compounds. Maximum preconcentration of target compounds is achieved by elution with the minimal solvent volume, typically, two to three holdup volumes for the sampling device. This requires a solvent yielding retention factors $< 2$. These requirements apply equally to thermal desorption except that larger desorption volumes can be employed if cryogenic focusing is used for band reconcentration. If conditions for recovery of the target compounds simultaneously results in some matrix components remaining immobilized on the sorbent, then additional matrix simplification is achieved at the expense of limited reuse of the sampling device. When considering a model for extraction by a cartridge-based of disk-based extraction device, it is necessary to consider that the sorption mechanism for sample application is best described by frontal analysis while rinse and desorption conditions are elution processes. Because typical solid-phase extraction devices contain short sorbent beds with a low plate number, the likelihood that sample retention is affected by kinetic properties of the sampling device cannot be ignored in practice.

### 1.5.1 Breakthrough volumes

The breakthrough volume is defined as the sample application volume at which a defined amount of the target compound(s) drawn through the extraction device in a continuous fashion is eluted at the exit of the extraction device [4,145−150]. To facilitate measurement of the breakthrough volume, the defined amount exiting the sampling device is typically some fraction (e.g., 1%, 5%, 10%, etc.) of the input amount and should be sufficiently large for convenient detection. The breakthrough volume, $V_B$, is determined by its breakthrough curve, Fig. 1.8 [145]. At the start of sample application the target compounds are quantitatively retained by the sorbent up to the time the sample volume exceeds the retention capacity of the sorbent. Further sample passing through the sorbent bed is not quantitatively retained, and eventually the concentration of target compounds entering and exiting the sampling device are the same. As shown in Fig. 1.8 the retention capacity and sorption capacity of the sorbent are not the same. After breakthrough, the sampling device continues to take up the compounds of interest but inefficiently, such that an increasing amount of these compounds escapes from the sampling device in excess of that used to define the breakthrough volume. The point of inflection for the breakthrough curve corresponds to the chromatographic retention volume provided that the plate number for the sampling device is not too small.

In general, there are two common causes of premature breakthrough for frontal chromatography. Sorbent overload caused by a high concentration of adsorbed target compounds or matrix components, or the sorbent bed fails to adequately retain the target compounds due to low kinetic efficiency such that the retention capacity depends on the plate number for the sampling device as well as the sample flow rate.

#### 1.5.1.1 Experimental determination of breakthrough volumes

The simplest methods for determining breakthrough volumes are direct methods employing either online or offline detection. Online detection is especially convenient for precolumn devices used as a component of automated SPE-LC (solid-phase extraction-liquid chromatography) systems [144,145,147,149]. A solution containing

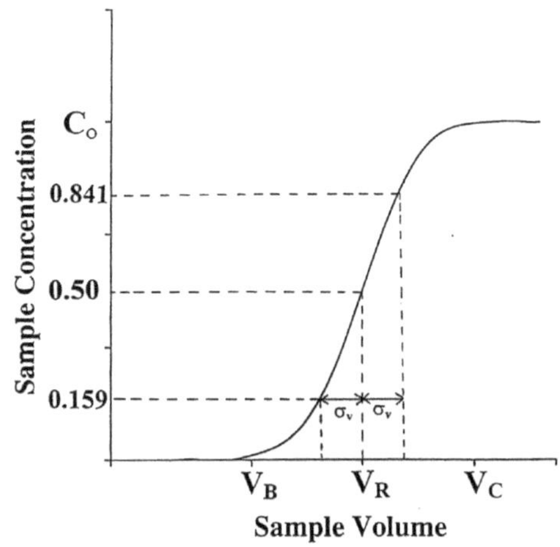


**Figure 1.8** Breakthrough curve indicating the breakthrough volume, $V_B$; the sampling volume corresponding to the saturation capacity of the sorbent, $V_C$; and the elution volume $V_R$ for the sampling device. The standard deviation corresponding to the derivative of the curve is $\sigma_V$. Reproduce from Poole CF, Gunatilleka AD, Sethuraman R. Contributions of theory to method development in solid-phase extraction. J Chromatogr A 2000;885:17−39 with permission.

a low but detectable concentration of target compounds is pumped at a constant flow rate through the precolumn, which is connected directly to the liquid chromatography detector. The detector output is a breakthrough curve similar to Fig. 1.8 from which the breakthrough volume can be determined. For cartridge and disk devices, offline sample processing is typically used [96,149,151−153]. Standard solutions are processed in aliquots by the same protocol for regular samples. An offline detection method is used to determine the concentration of target compounds in the eluent from each sample aliquot. To simplify calculation each aliquot contains the same amount of target compounds but in a different sample volume. Plotting the observed recovery for the complete sampling process against the corresponding aliquot volume results in a breakthrough curve from which the breakthrough volume is estimated. For gas phase sampling, either series coupled cartridges or a serial packed bed with the same sorbent is used. With sampling employing frontal analysis conditions the breakthrough volume is determined by the presence of compounds inadequately retained on the first sorbent bed detected on the second cartridge/sorbent bed according to the direction of flow [8,9,88]. This method has the advantage that it can also be deployed in field studies to indicate breakthrough for the sampling cartridge.

### *1.5.1.2 Estimation methods for breakthrough volumes*

The determination of breakthrough volumes is time consuming and somewhat subjective. Consequently, methods that enable breakthrough volumes to be estimated directly from compound properties or calculated from a small number of experiments are particularly attractive [38,144,147,150,154]. Cartridge-based sorbent beds only provide from 5 to 40 plates per centimeter of bed length depending on flow rate and packing density [38,154]. The main contributions to the plate height are flow anisotropy, a consequence of an inadequate packing density and a relatively large particle size to achieve a low pressure drop, and resistance to mass transfer [145]. Particle-loaded membranes of about 0.5 mm thickness typically provide from 5 to 9 plates with strong flow-rate dependence [38,41,155]. In both cases the conventional models for frontal analysis are not applicable and several empirical models have been developed to account for the low plate number on sample processing conditions. The Lokvist and Jonsson model is generally used for this purpose and allows the breakthrough volume to be estimated as a function of the plate number for the sampling device and a selected breakthrough level [148]. A general model for the breakthrough volume for sorbent beds with a low plate number can be written as

$$\log V_B = \log Q V_M + \log(1 + k_S) \tag{1.1}$$

Calculation of the breakthrough volume requires measurement or knowledge of the plate number for the sampling device, N, its holdup volume, $V_M$, and either measurement or estimation of the retention factor for the target compounds for the sample solvent. The main criteria for optimizing sample processing conditions are the dimensions of the sorbent bed ($V_M$, N), kinetic properties (N, particle size, flow rate), and retention (N, $k_S$, and $V_M$) [145]. The variable Q has values $\leq 1$ and is roughly constant for a defined sampling device and sampling conditions (it is estimated by the method of Lokvist and Jonsson for a specific plate number and breakthrough level). $Q < 1$ indicates reduced sample retention resulting from the low plate number for the sampling device at the sampling flow rate. Larger plate numbers result in a sharper front boundary and a higher breakthrough volume, and while desirable, smaller values are capable of providing useful breakthrough volumes and facilitate miniaturization of sampling devices and cost savings.

For the recovery of target compounds by elution, it is necessary to consider the shape of the front as well as the retention capacity of the sorbent [144]. The required elution volume for 99% recovery, $V_E$, on a cartridge or disk device with a low plate number can be estimated from

$$V_E = V_M(1 + k_S)\left(1 + 2.3/\sqrt{N}\right) \tag{1.2}$$

The only practical way to minimize the elution volume is to use a sorbent device with a small bed volume (minimize $V_M$) and to use a strong solvent ($k_S < 3$). Hendriks et al. [147] described a function based on the exponentially modified Gaussian model to accommodate peak tailing on the calculation of the elution volume.

#### *1.5.1.3 Experimental determination of retention factors*

Sorbent retention factors for sample application, rinse solvents, and elution volumes can be determined by one of three methods. The steady state equilibrium method uses a fixed volume of sample solution recycled by a pump through the sampling device and back to the sample reservoir until a steady state is obtained [150]. The target compounds retained by the sorbent are then recovered by elution and quantified. The retention factors are calculated from the volume and concentration of the sample solution, the holdup volume for the sorbent bed, and the amount of each target compound taken up by the sorbent under steady state conditions assuming a linear sorption isotherm. The shake-vial equilibrium method has lower accuracy but is easier to perform [146,156]. A solution with a known concentration of the target compounds is shaken with a known amount of sorbent until equilibrium is reached. Sampling the solution phase at equilibrium allows the adsorbed amount of each compound to be calculated. The retention factors are then calculated from the solid-liquid distribution constants and the phase ratio for the sampling device (estimated in a separate experiment). The direct measurement of sorbent retention factors by liquid chromatography in columns packed with sorbent is straightforward and uses the same protocols employed in column characterization [147,153,154,157]. Forced-flow planar chromatography can be used to determine retention factors for particle-loaded membranes and particle-embedded membranes [41,151,155]. Compounds with retention factors up to about 10,000 can be determined by the direct method, which is a more than adequate range for sampling purposes, as compounds with larger values will be more than adequately retained and only need be approximated.

### 1.5.1.4 Estimation of retention factors

The measurement of retention factors takes additional time and is not possible at all if standards are unavailable. Methods that allow estimation of retention factors from structure and sorbent retention properties for different conditions are useful for the initial phase of sorbent selection and to identify initial trial conditions for sample processing. For aqueous samples or samples in the gas phase the solvation parameter model is easily adapted to this problem [145,153,154]. For aqueous solutions

$$\log \mathrm{k_S} = c + e\mathrm{E} + s\mathrm{S} + a\mathrm{A} + b\mathrm{B} + v\mathrm{V} \tag{1.3}$$

and for sampling from the gas phase

$$\log \mathrm{k_S} = c + e\mathrm{E} + s\mathrm{S} + a\mathrm{A} + b\mathrm{B} + l\mathrm{L} \tag{1.4}$$

The model equations contain product terms representing target compound properties (descriptors), indicated by capital letters, and the complementary system properties (sorbent and sampling device), indicated by lowercase letters [13,143,159–161]. Each product term defines the relative contribution of a specific intermolecular interaction to the correlated property, in this case $\log \mathrm{k_S}$. The contribution from electron lone pair interactions (or the additional dispersion forces that result from the larger polarizability of compounds with weakly held electrons) is defined by $e\mathrm{E}$, interactions of a dipole-type (orientation and induction) $s\mathrm{S}$, hydrogen-bonding interactions by $a\mathrm{A}$ (sorbent hydrogen-bond basicity and compound hydrogen-bond acidity) and $b\mathrm{B}$ (sorbent hydrogen-bond acidity and compound hydrogen-bond basicity), and the difference in cavity formation and dispersion interactions for transfer of a compound between condensed phases (*v*V) or from the gas phase to a condensed phase (*l*L). The compound descriptors are formally defined as excess molar refraction E, dipolarity/polarizability S, effective hydrogen-bond acidity A, effective hydrogen-bond basicity B, McGowan's characteristic volume V, and the gas-liquid partition constant on hexadecane at 25°C, L [162−164]. The compound descriptors are not discussed further here except to note that values for thousands of compounds are available from the literature; methods for determining their values from solubility, chromatographic retention factors, and liquid-liquid distribution constants are available; and software products (fragmentation methods) for their estimation from structure are also available [162−164]. The constant in Eqs. (1.3) and (1.4), *c* term, is not a compound property but is required to estimate retention factors. It is a complex function of factors of which the most important is the phase ratio of the sampling system. Consequently, model equations are specified for a specific sampling device while the system constants (*e*, *s*, *a*, *b*, *v*, *l*) are characteristic of sorbent properties. The solvation parameter model was devised from a cavity model of solvation based on partition. It is generally applicable to adsorption at organic surfaces with flexible bound groups and polymer surfaces but is unsuitable for describing adsorption on inorganic oxides characterized by site-specific and size-dependent interactions. The model also applies to neutral compounds, and while ionizable compounds and ions can be handled in different ways, this is not a typical application of the model in solid-phase extraction [143,159]. Ionization tends to reduce retention compared with neutral compounds except for ion-exchange sorbents. Ionization suppression using buffers is often a practical solution to increasing retention for ionizable compounds on neutral sorbents.

System maps are typically used for selection or evaluation of sampling system for a chosen application [13,143,145]. A system map is a plot of the system constants of the solvation parameter model as a function of solvent composition (binary or ternary solvents) or temperature (for gas chromatography). A system map for the sorbent Oasis HLB is shown in Fig. 1.9 for methanol-water compositions from 0% to 50% (v/v) methanol [158]. Each system constant changes smoothly with the independent variable and can be fitted to simple linear or second order polynomial functions for computer-aided simulation of sampling conditions. The left-hand side of the map, corresponding to low organic solvent, is the region of interest for establishing a safe sampling volume using Eq. (1.3) to estimate $k_S$ and Eq. (1.1) to compute the breakthrough volume. The intermediate region of the system map is of interest for the selection of rinse solvents using Eqs. (1.2) and (1.3) to estimate a suitable solvent strength to elute matrix components while providing sufficient retention to immobilize the weakest retained of the target compounds. The right-hand side of the map is used to identify a strong solvent to elute the most retained of the target compounds with $k_S < 3$ and preferably $k_S \approx 0$.

The selection of low-specificity sorbents for a particular application is based on identifying an appropriate sorbent chemistry for the extraction that provides the desired breakthrough volume for the target compounds [145,153,158]. Only system constants with a positive sign contribute to sorbent retention. For Oasis HLB, this corresponds to the compound size (*v* system constant) and electron lone pair interactions (*e* system constant). Interactions of a dipole type (*s* system constant) and sorbent hydrogen-bond basicity (*a* system constant) are close to zero and of minor importance for explaining

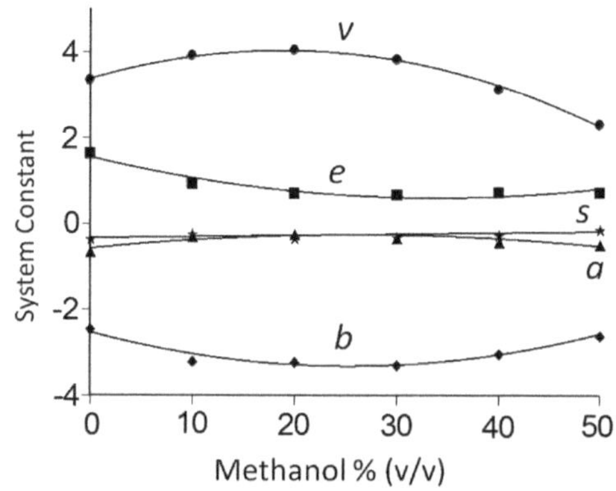


**Figure 1.9** System map for OASIS HLB with methanol-water solvent compositions.

retention on Oasis HLB. The hydrogen-bond acidity (*b* system constant) is large and negative signifying that this parameter is influential in the retention of compounds of roughly the same size but with different hydrogen-bond basicity. Fig. 1.9 is fairly typical of results for aqueous samples in which the retention mechanism is dominated by the characteristic properties of water: its strong cohesion favoring transfer to the sorbent (*v* system constant) and strong hydrogen-bond acidity reducing the retention of hydrogen-bond basic compounds (*b* system constant). Individual sorbents typically vary in the relative contribution of the system constants to extraction but no sorbent is able to decouple the dominant influence of the characteristic properties of water on extraction for either low-specificity sorbents or more selective polar sorbents. For samples soluble in organic solvents specific sorbent interactions are usually competitive with solute-solvent interactions and a wider range of retention properties are observed. However, overall retention is often weaker than for aqueous samples because the cohesion of the sample solution and solvated sorbent are of a similar value and the major driving force observed for aqueous extraction is now missing and replaced by individual polar interactions, which are usually weaker in comparison.

## 1.6 Conclusions

Solid-phase extraction is continuing to develop in response to changing laboratory needs driven by the desire for increased automation, miniaturization, and adoption of green chemical principles. This is reflected in new devices and sorbent chemistry which are active research fields today. While chromatographic sorbents have evolved into specialized and complex formats the synthesis of solid-phase extraction sorbents is less demanding and presents a lower barrier to the development and application of new chemistries for sample preparation. Thus, a large portion of materials that attract the most interest in the contemporary literature are experimental materials in contrast to the large number of purely application papers that employ commercially available materials. This has produced a gap in capabilities between a stable commercial sector supporting established formats of solid-phase extraction and a research dominated sector continuing to invent and develop new possibilities, some of which show potential to cross-over into the commercial sector. These are interesting times as sample preparation is recognized as the bottleneck to higher throughput in instrumental analysis and is gaining increasing attention as the problem to solve in the 21st century. It is also clear that the likely solution(s) will have to be innovative and involve novel technologies.

## References

[1] Wen Y, Chen L, Li J, Liu D, Chen L. Recent advances in solid-phase sorbents for sample preparation prior to chromatographic analysis. Trends Anal Chem 2014;59:26–41.
[2] Andrade-Eiroa A, Canle M, Leroy-Cancellieri V, Cerda V. Solid-phase extraction of organic compounds. A critical review. Trends Anal Chem 2016;80:641–54.
[3] Reyes-Garces N, Gionfriddo E, Gomez-Rios M, Alam N, Boyaci E, Bojko B, Singh V, Gandy J, Pawliszyn J. Advances in solid-phase microextraction and perspective on future directions. Anal Chem 2018;90:302–60.
[4] Poole CF, Mester Z, Miro M, Pedersen-Bjergaard S, Pawliszyn J. Glossary of terms used in extraction (IUPAC recommendations 2016). Pure Appl Chem 2016;88:517–58.
[5] Poole SK, Dean TA, Oudsema JW, Poole CF. Sample preparation for chromatographic separations: an overview. Anal Chim Acta 1990;236:3–42.
[6] Dressler M. Extraction of trace amounts of organic compounds from water with porous organic polymers. J Chromatogr 1979;165:167–206.
[7] Junk GA, Richard JJ, Grieser MD, Witiak D, Witiak JL, Arguello MD, Vick R, Svec HJ, Fritz JS, Caldor GV. Use of macroreticular resins in the analysis of water for trace organic contaminants. J Chromatogr 1974;99:745–62.
[8] Woolfenden E. Sorbent-based sampling methods for volatile and semi-volatile organic compounds in air. Part 1. Sorbent based air monitoring options. J Chromatogr A 2010; 1217:2674–84.
[9] Demeeslere K, Dewulf J, De Witte B, van Longenhove H. Sample preparation for the analysis of volatile organic compounds in air and water matrices. J Chromatogr A 2007; 1153:130–44.
[10] Thurman EM, Mills MS. Solid-phase extraction. Principles and practice. New York: Wiley; 1998.
[11] Fritz JS. Analytical solid-phase extraction. New York: Wiley; 1999.
[12] Simpson NJK, editor. Solid-phase extraction: principles, strategies and applications. New York: Marcel Dekker; 2000.
[13] Poole CF. New trends in solid-phase extraction. Trends Anal Chem 2003;22:362–73.
[14] Buszewski B, Szultka M. Past, present, and future of solid phase extraction: a review. Crit Revs Anal Chem 2012;42:198–213.
[15] Pan J, Zhang C, Zhang Z, Li G. Review of online coupling of sample preparation techniques with liquid chromatography. Anal Chim Acta 2014;815:1–15.
[16] Hyotylainen T. Principles, developments and applications of on-line coupling of extraction with chromatography. J Chromatogr A 2007;1153:14–28.
[17] Das D, Dutta M, Cervera ML, dela Guardia M. Recent advances in on-line solid-phase extraction pre-concentration for mineral elements. Trends Anal Chem 2012;33:35–45.
[18] Wilson SR, Malerod H, Petersen D, Simic N, Bobu M-M, Rise F, Lundanes E, greibrokk T. Controlling LC-SPE-NMR systems. J Sep Sci 2006;29:582–9.

[19] Alexovic M, Dotsikas Y, Bober P, Sabo J. Achievements in robotic automation of solvent extraction and related approaches for bioanalysis of pharmaceuticals. J Chromatogr B 2018;1092:402–21.
[20] Venn RF, Merson J, Cole S, Macrae P. 96-Well solid-phase extraction: a brief history of its development. J Chromatogr B 2005;817:77–80.
[21] Ridgway K, Lalljie SPD, Smith RM. Sample preparation techniques for the determination of trace residues and contaminants in foods. J Chromatogr A 2007;1153:36–53.
[22] Poole CF, editor. Handbooks of separation science: liquid-phase extraction. Amsterdam: Elsevier; 2019.
[23] Sajid M, Plotka-Wasylka J. Combined extraction and microextraction techniques: recent trends and future perspectives. Trends Anal Chem 2018;103:74–86.
[24] Dimpe KM, Nomngongo PN. Current sample preparation methodologies for analysis of emerging pollutants in different environmental matrices. Trends Anal Chem 2016;82: 199–207.
[25] Arthur CL, Pawliszyn J. Solid phase microextraction with thermal desorption using fused silica optical fibers. Anal Chem 1990;62:2145–8.
[26] Basheer C, Ainedhary AA, Rao BSM, Valliyaveettil S, Lee HK. Development and application of porous membrane protected carbon nanotubes micro-solid-phase extraction combined with gas chromatography/mass spectrometry. Anal Chem 2006;78:2853–8.
[27] Sajid M. Porous membrane protected micro-solid-phase extraction: a review of features, advancements and applications. Anal Chim Acta 2017;965:36–53.
[28] Pawliszyn J, editor. Handbook of solid phase microextraction. Amsterdam: Elsevier; 2012.
[29] Souza-silva EA, Jiang R, Rodriguez-Lafuente A, Gionfriddo E, Pawliszyn J. A critical review of the state of the art of solid-phase microextraction of complex matrices I. Environmental analysis. Trends Anal Chem 2015;71:224–35.
[30] Ouyang G, Pawliszyn J. A critical review in calibration methods for solid-phase microextraction. Anal Chim Acta 2008;627:184–97.
[31] Poole C, Mester Z, Miro M, Pedersen-Bjergaard S, Pawliszyn J. Extraction for analytical scale sample preparation (IUPAC technical report). Pure Appl Chem 2016;88:649–87.
[32] Piri-Moghadam H, Ahmadi F, Pawliszyn J. A critical review of solid phase microextraction for the analysis of water samples. Trends Anal Chem 2016;85:133–43.
[33] O'Reilly J, Wang G, Setkova L, Hutchinson JP, Lord HL, Linton CM, Pawliszyn J. Automation of solid-phase microextraction. J Sep Sci 2005;28:2010–22.
[34] Vuckovic D. High-throughput solid-phase microextraction in multi-well-plate format. Trends Anal Chem 2013;45:136–53.
[35] Souza-Silva EA, Gionfriddo E, Pawliszyn J. A critical review of the state of the art of solid-phase microextraction of complex matrices II. Food analysis. Trends Anal Chem 2015;71:236–48.
[36] Souza-Silva EA, Reyes-Garces N, Gomez-Rios GA, Boyaci E, Bojko B, Pawliszyn J. A critical review of the state of the art of solid-phase microextraction of complex matrices III. Bioanalytical and clinical applications. Trends Anal Chem 2015;71:249–64.
[37] Kataoka H, Ishizaki A, Saito K. Recent progress in solid-phase microextraction and its pharmaceutical and biomedical applications. Anal Methods 2016;8:5773–86.
[38] Poole CF, Poole SK, Seibert DS, Chapman CM. Determination of kinetic and retention properties of cartridge and disk devices for solid-phase extraction. J Chromatogr B 1997; 689:245–59.
[39] Hagan DR, Markell CG, Schmitt G, Blevins DD. Membrane approach to solid-phase extraction. Anal Chim Acta 1990;236:157–64.

[40] Erger C, Schmidt TC. Disk-based solid-phase extraction analysis of organic substances in water. Trends Anal Chem 2014;61:74–82.
[41] Mayer ML, Poole CF, Henry MP. Sampling characteristics of octadecylsiloxane-bonded silica particle-embedded glass fiber discs for solid-phase extraction. J Chromatogr A 1995;695:267–77.
[42] Subramanian R, Aslot C, Juhliu L, Nilsson C, Ostin A. Determination of S-2-(N,N-diisopropylaminoethyl)- and S-2-(N,N-diethylaminomethyl)phosphothiolate nerve agent markers in water samples using strong anion-exchange disk extraction, in-vial trimethylsilylation and gas chromatography-mass spectrometry analysis. J Chromatogr A 2012;1229:86–94.
[43] Aran ATK, Hyne RV, Doble P. Calibration of a passive sampling device for time-integrated sampling of hydrophilic herbicides in aqueous environments. Environ Toxicol Chem 2007;26:435–43.
[44] Verbruggen EMJ, Van Loon WMGM, Tonkes M, Van Dujin P, Deinen W, Hermens JLM. Biomimetic extraction as a tool to identify chemicals with high bio-concentration potential: an illustration by two fragrances in sewage treatment plant effluents and surface waters. Environ Sci Technol 1999;33:801–6.
[45] Beals DM, Britt WG, Bibler JP, Brooks DA. Radionucleide analysis using solid-phase extraction disks. J Radioanal Nucl Chem 1998;236:187–91.
[46] Fons-Castells J, Vasile M, Loots H, Bruggeman M, Llaurado M. On the direct measurement of Ra-226 and Ra-228 using 3M Empore™ RAD disk by liquid scintillation spectrometry. J Radioanal Nucl Chem 2016;309:1123–31.
[47] Hagiwara K, Inui T, Koike Y, Aizawa M, Nakamura T. Speciation of inorganic arsenic in drinking water by wavelength-dispersive X-ray fluorescence spectrometry after in-situ preconcentration with a miniature solid-phase extraction disk. Talanta 2015;134:739–44.
[48] Moein MM, Abdel-Rehim A, Abdel-Rehim M. Microextraction by packed sorbent (MEPS). Trends Anal Chem 2015;67:34–44.
[49] Yang L, Said R, Abdel-Rehim M. Sorbent, device and application in microextraction by packed sorbent (MEPS): a review. J Chromatogr B 2017;1043:33–43.
[50] Kedziora-Koch K, Wasiak W. Needle-based extraction techniques with protrected sorbent as powerful sample preparation tools for gas chromatographic analysis: trends in applications. J Chromatogr A 2018;1565:1–18.
[51] Lord HL, Zhan W, Pawliszyn J. Fundamentals and applications of needle trap devices: a critical review. Analy Chim Acta 2010;677:3–18.
[52] Piri-Moghadam H, Alam MN, Pawliszyn J. Review of geometries and coating materials in solid-phase microextraction: oportunities, limitations, and future perspectives. Anal Chim Acta 2017;984:42–65.
[53] Nerin C, Salafranca J, Aznar M, Bartle R. Critical review on recent developments in solventless techniques for extraction of analytes. Anal Bioanal Chem 2009;393:809–33.
[54] Helin A, Ronnko T, Parshintser K, Hartonen K, Schillin B, Laubli T, Riekkola ML. Solid-phase microextraction arrow for the sampling of volatile amines in wastewater and atmosphere. J Chromatogr A 2015;1426:56–63.
[55] Miraghi FS, Hein D, Pawliszyn J. Thin-film microextraction coupled with mass spectrometry and liquid chromatography-mass spectrometry. Chromatogr 2013;76:1215–23.
[56] Cudjoe E, Vuckovic D, Hein D, Pawliszyn J. Investigation of the effect of the extraction phase geometry on the performance of automated solid-phase microextraction. Anal Chem 2009;81:4226–32.
[57] Eisert R, Pawliszyn J. Automated in-tube solid-phase microextraction coupled to high-performance liquid chromatography. Anal Chem 1997;69:3140–7.

[58] Kataoka H, Ishizaki A, Nonaka Y, Saito K. Developments and applications of capillary microextraction techniques: a review. Anal Chim Acta 2009;655:8−29.
[59] Queiroz MEC, Melo LP. Selective capillary coating material for in-tube solid-phase microextraction coupled to liquid chromatography to determine drugs and biomarkers in biological samples. A review. Anal Chim Acta 2014;826:1−11.
[60] Moliner-Martinez Y, Herraz-Hernandez R, Verdu-Andres J, Molins-Legua C, Campins-Falco P. Recent advances of in-tube solid-phase microextraction. Trends Anal Chem 2015;71:205−13.
[61] Fernandez-Amado M, Prieto-Blanco MC, Lopez-Mahia P, Muniategui-Lorenzo S, Prada-Rodriguez D. Strengths and weaknesses of in-tube solid-phase microextraction: a scoping review. Anal Chim Acta 2016;906:41−57.
[62] Prieto A, Basauri O, Rodil R, Usobiaga A, Fernandez LA, Etxebarria N, Zuloaga O. Stir-bar sorptive extraction: a review on method optimization, novel applications, limitations and potential solutions. J Chromatogr A 2010;1217:2642−66.
[63] Nogueira JMF. Stir bar sorptive extraction: 15 years making sample preparation more environment friendly. Trends Anal Chem 2015;71:214−23.
[64] He M, Chen B, Hu B. Recent developments in stir bar sorptive extraction. Anal Bioanal Chem 2014;406:2001−26.
[65] Camino-Sanchez FJ, Rodriguez-Gomez R, Zafra-Gomez A, Santos-Fandila A, Vilchez JL. Stir bar sorptive extraction: recent applications, limitations and future trends. Talanta 2014;130:388−99.
[66] Richter P, Leiva C, Choque C, Giordano A, Sepulveda B. Rotating disk sorptive extraction of nonylphenol from water samples. J Chromatogr A 2009;1216:8598−602.
[67] Lucena R. Extraction and stirring integrated techniques: examples and recent advances. Anal Bioanal Chem 2012;403:2213−23.
[68] Donato FP, Bandeira NMG, dos Santos GC, Prestes OD, Adaime MB, Zanella R. Evaluation of the rotating disk sorptive extraction technique with polymeric sorbent for multiresidue determination of pesticides in water by ultra-high-performance liquid chromatography-tandem mass spectrometry. J Chromatogr A 2017;1516:54−63.
[69] Vasconcelos I, Fernandes C. Magnetic solid phase extraction for determination of drugs in biological matrices. Trends Anal Chem 2017;89:41−2.
[70] Wlerucka M, Biziuk M. Application of magnetic nanoparticles for magnetic solid-phase extraction in preparing biological, environmental and food samples. Trends Anal Chem 2014;59:50−8.
[71] Ahmadi M, Elmongy H, Madrakian T, Abdel-Rehim M. Nanomaterials as sorbents for sample preparation in bioanalysis: a review. Anal Chim Acta 2017;958:1−21.
[72] Hemmati M, Rajabi M, Asghari A. Magnetic nanoparticle based solid-phase extraction of heavy metal ions: a review on recent advances. Microchem Acta 2018;185(160). https://doi.org/10.1007/s00604-018-2670-4 [in press].
[73] Reyes-Gallardo EM, Lucena R, Cardenas S. Electrospun nanofibers as sorptive phases in microextraction. Trends Anal Chem 2016;84:3−11.
[74] Namera A, Nakamoto A, Saito T, Miyazaki S. Monoliths as new sample preparation material: recent devices and applications. J Sep Sci 2011;34:901−24.
[75] Huang X, Yuan D. Recent developments of extraction and micro-extraction technologies with porous monoliths. Crit Revs Anal Chem 2012;42:38−49.
[76] Nema T, Chan ECY, Ho PC. Applications of monolithic materials for sample preparation. J Pharm Biomed Anal 2014;87:130−41.
[77] Masini JC, Svec F. Porous monoliths for on-line sample preparation: a review. Anal Chim Acta 2017;964:24−44.

[78] Fresco-Cala B, Cardenas S. Potential of nanoparticle-based hybrid monoliths as sorbents in microextraction techniques. Anal Chim Acta 2018;1031:15–27.
[79] Lv Y, Svec F. Preparation and application of monolith structures containing metal-organic frameworks. J Sep Sci 2017;40:272–87.
[80] Bordin DCM, Alves MNR, De Campos EG, De Martins BS. Disposable pipette tips extraction: fundamentals, applications and state of the art. J Sep Sci 2016;39: 1168–72.
[81] Augusto F, Hantano LW, Mogollon NGS, Braga SCGN. New materials and trends in sorbents for solid-phase extraction. Trends Anal Chem 2013;43:14–23.
[82] Augusto F, Carasek E, Gomes CSR, Rivellin SR, Batista AD, Marlendal E. New sorbents for extraction and microextraction techniques. J Chromatogr A 2010;1217:2533–42.
[83] Plotka-Wasylka J, Szczepanska N, de la Guardia M, Namiesnik J. Modern trends in solid-phase extraction: new sorbent media. Trends Anal Chem 2016;77:23–43.
[84] Maya F, Cabello CP, Ghani M, Palomino GT, Cerda V. Emerging materials for sample preparation. J Sep Sci 2018;41:262–87.
[85] Xu J, Wu P, Ye EC, Yuan BF, Feng YQ. Metal oxides in sample pretreatment. Trends Anal Chem 2016;80:41–56.
[86] Ayazi Z. Application of nanocomposite-based sorbents in microextraction techniques: a review. Analyst 2017;142:721–39.
[87] Capriotti AL, Cavaliere C, Foglia P, Samperi R, Stampachiacchiere S, Ventura S, Lagana A. Recent advances and developments in matrix solid-phase dispersion. Trends Anal Chem 2015;71:186–93.
[88] Harper M. Sorbent trapping of volatile organic compounds from air. J Chromatogr A 2000;885:129–51.
[89] Dettmer K, Engewald W. Adsorbent materials commonly used in air analysis for adsorptive enrichment and thermal desorption of volatile organic compounds. Anal Bioanal Chem 2002;373:490–500.
[90] Szulejko JE, Kim KH. A review of sampling and pretreatment techniques for the collection of airborne amines. Trends Anal Chem 2014;57:118–34.
[91] Poole CF. The essence of chromatography. Amsterdam: Elsevier; 2003.
[92] Fanali S, Haddad PR, Poole CF, Riekkola ML, editors. Liquid chromatography. 2nd ed., vol. 1. Amsterdam: Elsevier; 2017.
[93] Bocian S, Buszewski B. Residual silanols at reversed-phase silica in high-performane liquid chromatography – a contribution to a better understanding. J Sep Sci 2012;35: 1191–200.
[94] McCalley DV. The challenges of the analysis of basic compounds by high-performance liquid chromatography: some possible approaches for improved separations. J Chromatogr A 2010;1217:858–80.
[95] Masque N, Marce RM, Borrull F. New polymeric and other types of sorbents for solid-phase extraction of polar organic micropollutants from environmental water. Trends Anal Chem 1998;17:384–94.
[96] Poole SK, Poole CF. Influence of solvent effects on the breakthrough volume in solid-phase extraction using porous polymer particle-loaded membranes. Analyst 1995;120: 1733–8.
[97] Fontanals N, Marce RM, Borrull F. New hydrophilic materials for solid-phase extraction. Trends Anal Chem 2005;24:394–406.
[98] Castaldo R, Gentile G, Avella M, Carfagna C, Ambrogi V. Microporous hyper-crosslinked polystyrenes and nanocomposites with high adsorption properties: a review. Polymers 2017;9:651. https://doi.org/10.3390/polym9120651.

[99] Matisova E, Skrabkova S. Carbon sorbents and their utilization for the preconcentration of organic pollutants in environmental samples. J Chromatogr A 1995;707:145–79.
[100] Hennion M-C. Graphitized carbons for solid-phase extraction. J Chromatogr A 2000;885: 73–95.
[101] de Toffoli AL, Maciel EVS, Fumes BH, Lancas FM. The role of graphene-based sorbents in modern sample preparation techniques. J Sep Sci 2018;41:288–302.
[102] Ravelo-Perez LM, Herrera-Herrera AV, Hernandez-Borges J, Rodriguez-Delgado M-A. Carbon nanotubes: solid-phase extraction. J Chromatogr A 2010;1217:2618–41.
[103] Liang X, Liu S, Wang S, Guo Y, Jiang S. Carbon-based sorbents: carbon nanotubes. J Chromatogr A 2014;1357:53–67.
[104] Socas-Rodriguez B, Herrera-Herrera AV, Asensio-Ramos M, Hernandez-Borges J. Recent applications of carbon nanotube sorbents in analytical chemistry. J Chromatogr A 2014;1357:110–46.
[105] Lepont C, Gunatilleka AD, Poole CF. Retention characteristics of porous graphitic carbon in reversed-phase liquid chromatography with methanol-water mobile phases. Analyst 2001;126:1318–25.
[106] Boonjob W, Miro M, Segundo MA, Cerda V. Flow-through dispersed carbon nanofiber-based microsolid-phase extraction coupled to liquid chromatography for automatic determination of trace levels of priority environmental pollutants. Anal Chem 2011;83: 5237–44.
[107] Zhang B-T, Zheng X, Li H-F, Lin J-M. Application of carbon-based nanomaterials in sample preparation. Anal Chim Acta 2013;784:1–17.
[108] Seethapathy S, Gorecki T. Applications of polydimethylsiloxane in analytical chemistry: a review. Anal Chim Acta 2012;750:48–62.
[109] Gilart N, Marcee RM, Borrull F, Fontanals N. New coatings for stir-bar sorptive extraction of polar emerging organic contaminants. Trends Anal Chem 2014;54:11–23.
[110] Che D, Chang J, Ji Z, Zhang S, Li G, Sun Z, You J. Recent advances and applications of polydopamine-derived adsorbents for sample pretreatment. Trends Anal Chem 2017;97: 1–14.
[111] Szultka-Mlynska M, Olszowy P, Buszewski B. Nanoporous conducting polymer-based coatings in microextraction techniques for environmental and biomedical applications. Crit Revs Anal Chem 2016;46:236–7.
[112] Camel V, Caude M. Trace enrichment methods for the determination of organic pollutants in ambient air. J Chromatogr A 1995;710:3–19.
[113] Ramirez N, Cuadras A, Rovira E, Borrull F, Marce RM. Comparative study of solvent extraction and thermal desorption methods for determining volatile organic compounds in ambient air. Talanta 2010;82:719–27.
[114] Helmig D, Vierling L. Water-adsorption capacity of the solid adsorbents Tenax-TA, Tenax –GR, Carbotrap, Carbotrap-C, Carbosieve-SIII, and Carboxen-569 and water management-techniques for the atmospheric sampling of volatile organic trace gases. Anal Chem 1995;67:4380–6.
[115] Bradley RH. Recent developments in the physical adsorption of toxic organic vapors by activated carbons. Adsorp Sci Technol 2011;29:1–28.
[116] Foley C. Carbogenic molecular-sieves – synthesis, properties and applications. Microporous Mater 1995;4:407–33.
[117] Fritz JS, Gjerde DT. Ion chromatography. Weinheim: Wiley-VCH; 2009.
[118] Fontanals N, Cormack PAG, Marce RM, Borrull F. Mixed-mode ion-exchange polymeric sorbents: dual-phase materials that improve selectivity and capacity. Trends Anal Chem 2010;29:765–79.

[119] Rahman IMM, Begum ZA, Hasegawa H. Selective separation of elements from complex solution matrix with molecular recognition plus macrocycles attached to a solid-phase: a review. Microchem J 2013;110:485–93.

[120] Hennion M-C, Pichon V. Immuno-based sample preparation for trace analysis. J Chromatogr A 2003;1000:29–52.

[121] Senyuva HZ, Gilbert J. Immunoaffinity column clean-up techniques in food analysis: a review. J Chromatogr B 2010;878:115–32.

[122] Pichon V, Combes A. Selective tools for the solid-phase extraction of Ochratoxin A from various complex samples: immunosorbents, oligosorbents, and molecularly imprinted polymers. Anal Bioanal Chem 2016;408:6983–99.

[123] Pichon V, Brothier F, Combes A. Aptamer-based-sorbents for sample treatment – a review. Anal Bioanal Chem 2015;407:681–98.

[124] Sajid M. Dendimers based sorbents: promising materials for analytical extractions. Trends Anal Chem 2018;98:114–27.

[125] Haginaka J. Molecularly imprinted polymers as affinity-based separation media for sample preparation. J Sep Sci 2009;32:1548–65.

[126] Beltran A, Borrull F, Cormack PAG, Marce RM. Molecularly imprinted polymers: useful sorbents for selective extraction. Trends Anal Chem 2010;29:1363–75.

[127] Turiel E, Martin-Esteban A. Molecularly imprinted polymers for sample preparation: a review. Anal Chim Acta 2010;688:87–99.

[128] Alvarez-Lorenzo C, Concheiro A, editors. Handbook of molecularly imprinted polymers. Shrewsbury: Smithers Rapara Technology; 2013.

[129] Huang SY, Xu JQ, Zheng JT, Zhu F, Xie LJ, Ouyang GF. Synthesis and applications of magnetic molecular imprinted polymers in sample preparation. Anal Bioanal Chem 2018; 410:3991–4014.

[130] Li GZ, Row KH. Recent applications of molecularly imprinted polymers (MIPs) on micr-extraction techniques. Sep Purif Revs 2018;47:1–18.

[131] Speltini A, Scalabrini A, Marash F, Stierini M, Profumo A. Newest applications of MIPs for extraction of contaminants from environmental and food matrices: a review. Anal Chim Acta 2017;974:1–26.

[132] Ansari S, Karimi M. Novel developments and trends of analytical methods for drug analysis in biological and environmental samples by molecularly imprinted polymers. Trends Anal Chem 2017;89:146–62.

[133] Martin-Esterban A. Recent MIP-based sample preparation techniques in environmental analysis. Trends Env Anal Chem 2016;9:8–14.

[134] Wang YH, Rui M, Lu GH. Recent applications of metal-organic frameworks in sample pretreatment. J Sep Sci 2018;41:180–94.

[135] Hashemi B, Zohrabi P, Raza N, Kim KH. Metal-organic frameworks as advanced sorbents for the extraction and determination of pollutants from environmental, biological and food media. Trends Anal Chem 2017;97:65–82.

[136] Qian H-L, Yang C-X, Wang W-L, Yang C, Yan X-P. Advances in covalent organic frameworks in separation science. J Chromatogr A 2018;1542:1–18.

[137] de Faria HD, de Carvalho Abrao LC, Santos MG, Barbosa AF, Figueiredo EC. New advances in restricted access matrials for sample preparation: a review. Anal Chim Acta 2017;959:43–65.

[138] Souverain S, Rudaz S, Veuthey JL. Restricted access materials and large particle supports for on-line sample preparation: an attractive approach for biological fluid analysis. J Chromatogr B 2004;801:141–56.

[139] Yang SH, Fan H, Classon RJ, Schug KA. Restricted access media as a streamlined approach towards on-line sample preparation: recent advancements and applications. J Sep Sci 2013;36:2922–38.

[140] Popov A, Blinnikova ZK, Tsyurupa P, Davankov VA. Hypercrosslinked polymeric restricted access materials for the analysis of biological fluids. J Sep Sci 2018;41:3302–9.

[141] Poole CF. Influence of solvent effects on the retention of small molecules in reversed-phase liquid chromatography. Chromatographia 2019;82:49–64.

[142] Kiridena W, Poole CF. Influence of solute size and site-specific surface interactions on the prediction of retention in liquid chromatography using the solvation parameter model. Analyst 1998;123:1265–70.

[143] Poole CF. Chromatographic test methods for characterizing alkylsiloxane-bonded silica columns for reversed-phase liquid chromatography. J Chromatogr B 2018;1092:207–19.

[144] Hennion M-C. Solid-phase extraction: method development, sorbents, and coupling with liquid chromatography. J Chromatogr A 1999;856:3–54.

[145] Poole CF, Gunatilleka AD, Sethuraman R. Contributions of theory to method development in solid-phase extraction. J Chromatogr A 2000;885:17–39.

[146] Ortega L, Lopez R, Cacho J, Ferreira V. Use of solid-liquid distribution coefficients to determine retention properties of Porapak-Q resins. Determination of optimal conditions and beta-damascenone from wine. J Chromatogr A 2001;931:31–9.

[147] Hendriks G, Uges DRA, Franke JP. New practical algorithm for modeling analyte recovery in bioanalytical reversed phase and mixed-mode solid-phase extraction. J Pharm Biomed Anal 2008;48:158–70.

[148] Lokvist P, Jonsson J-A. Capacity of sampling and preconcentration columns with a low number of theoretical plates. Anal Chem 1987;59:818–21.

[149] Bielicka-Daszkiewicz K, Voelkel A. Theoretical and experimental methods of determination of the breakthrough volume of solid-phase extraction sorbents. Talanta 2009;80: 614–21.

[150] Gelencser A, Kiss G, Krivacsy Z, Varga-Puchony Z, Hlavay J. A simple method for the determination of capacity factor on solid-phase extraction cartridges. I. J Chromatogr A 1995;693:217–25.

[151] Larrivee ML, Poole CF. A solvation parameter model for the prediction of breakthrough volumes in solid-phase extraction with particle-loaded membranes. Anal Chem 1994;66: 139–46.

[152] Mayer ML, Poole CF. Identification of the procedural steps that affect recovery of semi-volatile compounds by solid-phase extraction using cartridge and particle-loaded membrane (disk) devices. Anal Chim Acta 1994;294:113–26.

[153] Seibert DS, Poole CF. A general model for the optimization of sample processing conditions by solid-phase extraction applied to the isolation of estrogens from urine. J High Resolut Chromatogr 1998;21:481–90.

[154] Miller KG, Poole CF. Methodological approach for evaluating operational parameters and the characterization of a popular sorbent for solid-phase extraction by high pressure liquid chromatography. J High Resolut Chromatogr 1994;17:125–34.

[155] Fernando WPN, Larrivee ML, Poole CF. Investigation of the kinetic properties of particle-loaded membranes for solid-phase extraction by forced flow planar chromatography. Anal Chem 1993;65:588–95.

[156] Ferreira V, Jarauta I, Ortega L, Cacho J. Simple strategy for the optimization of solid-phase extraction procedures through the use of solid-liquid distribution coefficients. Application to the determination of aliphatic lactones in wine. J Chromatogr A 2004; 1025:147–56.

[157] Poole CF, Poole SK. Foundations of retention in partition chromatography. J Chromatogr A 2009;1216:1530–50.
[158] Dias NC, Poole CF. Mechanistic study of the sorption properties of OASIS HLB and its use in solid-phase extraction. Chromatographia 2002;56:269–75.
[159] Poole CF, Poole SK. Column selectivity from the perspective of the solvation parameter model. J Chromatogr A 2002;965:263–99.
[160] Vitha MF, Carr PW. The chemical interpretation and practice of linear solvation energy relationships in chromatography. J Chromatogr A 2006;1126:143–94.
[161] Abraham MH, Poole CF, Poole SK. Classification of stationary phases and other materials by gas chromatography. J Chromatogr A 1999;842:79–114.
[162] Abraham MH, Ibrahim A, Zissimos AM. Determination of sets of solute descriptors from chromatographic measurements. J Chromatogr A 2004;1037:29–47.
[163] Poole CF, Atapattu SN, Poole SK, Bell AN. Determination of solute descriptors by chromatographic methods. Anal Chim Acta 2009;652:32–53.
[164] Poole CF, Ariyasena TC, Lenca N. Estimation of the environmental properties of compounds from chromatographic measurements and the solvation parameter model. J Chromatogr A 2013;1317:85–104.

# Inorganic oxide and chemically bonded sorbents

2

*Małgorzata Szultka-Młyńska* [1], *Hossam Al-Suod* [1,2], *Bogusław Buszewski* [1,2]
[1]Department of Environmental Chemistry and Bioanalytics, Faculty of Chemistry, Nicolaus Copernicus University, Torun, Poland; [2]Centre for Modern Interdisciplinary Technologies, Nicolaus Copernicus University, Torun, Poland

## 2.1 Introduction

Sample preparation is a crucial component of modern environmental, biomedical, and pharmaceutical analysis and often involves elaborate, time-consuming procedures (app. 80% of the total analysis time). Choosing an appropriate sample preparation method is most important in the qualitative and quantitative determination of target compounds. A large number of target compounds are of interest, ranging from highly lipophilic to moderately polar and exhibiting basic, acidic, or neutral properties. In addition, this step is required for several reasons: to eliminate a possible compound interference, to concentrate and stabilize the target compounds, and to achieve the conditions suitable for the final chromatographic or other analysis [1−3].

The most common sample preparation method is solid-phase extraction (SPE). Numerous SPE applications have been developed and are summarized in several reviews [1,4−11]. Generally, during the extraction process, an aqueous sample passes over a solid phase with the transfer of the target compounds to the sorbent from which they are subsequently recovered by elution with a suitable solvent. A successful SPE is based on a suitable and selective distribution of analytes to the solid phase from the sample solvent described by a distribution constant for low analyte concentrations for which a linear sorption isotherm might be assumed. The extraction requires that the target compounds have a greater affinity for the solid phase than for the sample matrix [12]. Insufficient retention by the solid phase results in low breakthrough volumes. An important strategy to overcome inadequate breakthrough volumes is to optimize the sorbent type [13−16].

Due to good reproducibility, precision, relatively low cost, operational simplicity, and the possibility of combining the SPE technique directly (*online* or *off-line*) with other methods, has made it one of the most widely used techniques for the routine analysis of pharmaceutical, cosmetic, biomedical, and environmental samples, in controlling technological processes, and for foods and fodder. Moreover, it is used for selective isolation and enrichment of analytes from gaseous, liquid, and solid matrices.

**Solid-Phase Extraction. https://doi.org/10.1016/B978-0-12-816906-3.00002-9**

## 2.2 Inorganic oxide sorbents

The selection of an appropriate sorbent material is a critical point in SPE since it ensures suitable retention of the analytes and their easy release for high recovery upon elution. Nowadays, a variety of sorbents are available for SPE, which can be divided into three general categories: inorganic oxides, low-specific sorbents, and compound-specific/class-specific sorbents. Silica, alumina ($Al_2O_3$), and Florisil (a synthetic hydrated magnesium silicate) are examples of typical inorganic oxide sorbents. Their large surface area and high activity increase the retention based on the type, number, and location of the complementary analyte functional groups. Nonpolar compounds, such as aromatic and alkenes with polarizable functional groups, are poorly retained by these sorbents, while polar compounds with hydrogen bonding functional groups are highly retained (e.g., sulfonic acid, carboxylic acid, phenol, etc). Typical silica gels used for SPE have surface areas of 300−800 $m^2/g$, a pore size of 4−10 nm, and an apparent pH of 5.5−7.5. Alumina typically used for SPE has a surface area of about 150 $m^2/g$, a pore size of 6 nm, and is available in neutral, weakly acidic, acidic, and basic forms depending on the processing conditions. Florisil has a surface area of about 250−300 $m^2/g$ and an apparent pH of about 8.5. Limited pH stability, especially for samples with a $pH > 8$, is the most important weakness of silica gel. Other metal oxides, such as zirconia ($ZrO_2$) and titania ($TiO_2$), have significantly better pH stability, but their surface chemistry is more complex, and they are not as popular as silica for SPE applications. The solution to the limited pH stability of silica is the synthesis of organic-inorganic hybrid sorbents stable up to pH 12. Common applications of inorganic oxide sorbents include the isolation of compounds such as pesticides or polycyclic aromatic hydrocarbons (PAHs) from aqueous samples [17,18], polychlorinated biphenyls (PCBs) and organochlorine pesticides [19], as well as mycotoxins in herbal matrices [20]. Inorganic oxide sorbents were used for the isolation of N-acyl homoserine lactones from soil [21].

### 2.2.1 *Physicochemical properties*

The term *silica* denotes a substance with a stoichiometric formula $SiO_2$, as well as its hydrated form $SiO_2 \times H_2O$. Proper classification of silica requires consideration of four basic parameters: (1) crystal structure; (2) dispersion (solid silica dispersed in gaseous or liquid dispersion medium), silica hydrogels (commonly called *silica gels*); (3) porosity; and (4) structural and surface heterogeneity [22−25]. Furthermore, there also exist soluble silica, silica sols, and solutions of polymerized silica built up of branched siloxane chains [13,14,22−24].

According to Tanaka [26], the pores in wide-pore materials differ significantly within the same sample. Pore width is a median value calculated on the basis of the pore size distribution. Three basic pore size distribution curves can be distinguished: homogenous, with a size distribution close to a bell-shaped curve; bimodal, with two maxima; and heterogenous, with several peaks corresponding to the populations of pores of different sizes. Pore size is reflected in the specific surface area of the adsorbent. The external surface area $S_e$ corresponds to the surface for porous spherical particles per gram of silica. It is inversely proportional to the bead size $d_p$ and density ρ. In turn, the internal surface area $S_i$ consists of the surfaces of the internal pore walls, with the exception of closed pores. A large specific surface area (the sum of its external and internal surface) is indicative of the presence of pores of a small diameter. In turn, a low specific surface area indicates the presence of a population of macropores [25].

In addition to shape and pore size, another factor characterizing sorbent porosity is the specific volume ($V_p$) described as the amount of liquid that will fill the pores completely, calculated per gram of silica [27]. By analogy to the specific surface area, the total pore volume depends on pore size.

According to the postulate of Brunauer, Emmet, and Teller (the BET method) [28–30], sorption isotherms of different shapes result from different mechanisms for the adsorption and desorption of vapors from pores of a different size. Sorption properties of porous silica are the result of the existence of *globular* or *sponge-like* matrix structure [31–33]. Optimal adsorbent for SPE is characterized by good mechanical properties and a limited number of micropores that adversely affect the mass exchange and diffusion [33–35], but also by a uniform pore shape [22–25]. A silica sample of suitable morphology is illustrated in Fig. 2.1.

Silica gel is the base material commonly used for the synthesis of chemically bonded phases with complementary properties to native silica gel for use in SPE. It is relatively easy to synthesize in commercial quantities with uniform characteristics at a low cost [23,24].

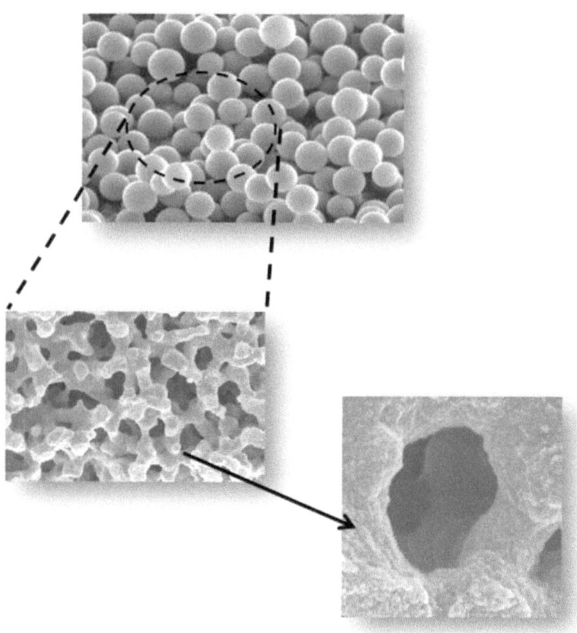

**Figure 2.1** Scanning electron microsope images of a silica adsorbent with a bead diameter of 35 μm (scale 1:1500).

The surface heterogeneity of silica gel is a result of various functional groups – different types of silanols and siloxanes. Structural heterogeneity results from the differences in pore shapes (cylindrical, funnel-shaped, and ink-bottle shaped) associated with the three-dimensional skeleton of $SiO_2$. Silica used for SPE extraction is essentially porous and noncrystalline with an abundance of surface silanol groups. Silica gel is a useful adsorbent, and on account of its rigid skeleton does not swell or shrink in common solvents and has good mechanical strength [13,14]. The silanol groups are responsible for the polar adsorption centers. Three types of silanol groups can be distinguished: isolated (free, single); vicinal or bridged (total content 60%–65%); and geminal (10%–12%) [36–38]. The relative concentration and type of silanol groups depend on the hydrothermal or heat treatment of the silica particles. Some typical physicochemical properties of silica sorbents are summarized in Table 2.1.

**Table 2.1** Physicochemical characterization of silica sorbents [38].

| **Characteristic value** | **Symbol [units]** | **"Typical"** | **Criteria "Theoretically optimal"** | **Measurement technique** |
|---|---|---|---|---|
| Specific surface area | $S_{BET}$ [$m^2g^{-1}$] | 150–500 | 320 | Low-temperature adsorption-desorption of nitrogen or helium |
| Mean pore diameter | $D$ [nm] | 10–30 | 10 | |
| Mean particle size | $d_p$ [μm] | 20–60 | 40 | Laser counter microscopy |
| Interparticle column porosity | ε [-] | 0.4 | ≤0.84 | Chromatography |
| Trace amounts of metals | -[ppm] | – | ≤500 | Inductively coupled plasma (spectroscopy), atomic absorption spectroscopy |
| Powder density | $\sigma_s$ [g $cm^{-1}$] | 0.4–0.6 | 0.45 | Pycnometry |
| Particle shape | – | Spherical | Strictly spheroidal | Microscopy |
| Surface pH | – | – | 2≤pH≤9 | pH-metry of dispersions |
| Concentration of OH groups | $\alpha_{OH}$ [μmol $m^{-2}$] | 8 | $5\leq \alpha_{OH}\leq 8$ | Spectrometry, chemical methods |

### 2.2.2 Silica-based chemically bonded sorbents

Silica can be used as an SPE sorbent without further modification of silanols. However, in order to increase its applicability and selectivity, the silica surface is usually modified by bonding different functional groups to the surface. Silica gel may be modified to obtain chemically bonded sorbents with hydrophobic or hydrophilic functional groups. These modifications may be carried out using different methodologies.

Chemically bonded sorbents can be grouped into several categories based on the chemical nature of the embedded functional groups, the nature of the support, as well as their structure. Based on structure, Unger suggests a division into several categories [36]. The first group includes sorbents obtained by the action of a single modifier (monolayer structure). The multilayer may be bonded but may also be coated onto the silica surface. A second type is chemically bonded sorbents with a diffusion barrier, which protect bonded ligands. Mixed-mode sorbents have various types of functional groups affording multiple interaction mechanisms that facilitate their use in a wider range of applications. Lastly, another category of chemically bonded sorbents has a "sandwich" structure resulting from polymerization.

Generally, the preparation of amorphous silica gel is carried out with the benign and convenient sol−gel methodology. This type of polymerization involves two chemical steps, namely hydrolysis and condensation, which results in the formation of Si−O−Si chemical bonds from a silicic acid monomer, such as sodium silicate, or silicic acid esters (e.g., tetraethylorthosilicate, TEOS). The monomers are polymerized to a specific molecular weight or size. The initial form of the prepolymerized silica is a sol, which is further agglomerated to the shape of the final products.

The silanol groups present on the silica gel surface result in the formation of siloxane bonds by reaction with silane reagents. Depending on the type of silane, the chemically bonded sorbent surface can be monomeric (Fig. 2.2A) or polymeric (Fig. 2.2B). The synthesis procedure may be divided into two stages (Fig. 2.2C): (a) silanization and (b) hydrosilylation. For the reaction of surface silanol groups to form siloxane bonds, chloro-, amino- or methoxy/ethoxy- silanes are used [9,23,31,37−39]. The ratio of the amount of silane reagent to silica and the number and type of silanol groups provides the possibility to obtain sorbents with controlled coverage density [24]. Alternatively, a stable bond between silicon and carbon (Si−C) can be formed while the attachment of the organic moiety takes place. For this purpose, alkynes, cyano, and other ligands are used. The reaction between silicon hydrides and alkenes is the preferred laboratory method to form Si−C bonds [23,24,37,38].

Luo et al. [40] successfully prepared humic acid-bonded silica via an amide linkage between humyl chloride and the amino group of APTS-bonded silica as a novel sorbent for the solid-phase extraction of benzo[a]pyrene in edible oils, Fig. 2.3. A high recovery of benzo[a]pyrene ranging between 93.3% and 100.1% was obtained. In addition, Petrova [41] prepared two new sorbents, silica gel modified with Cystine (Sig-Cys-S-S-Cys) and silica gel modified with N-Benzyloxycarbonyl-L-Methionine (Sig-Z-Met-OH) and evaluated them for the quantitative extraction of Au(III) from hydrochloric acid solutions. Under optimized conditions, less than 43% of Au(III)

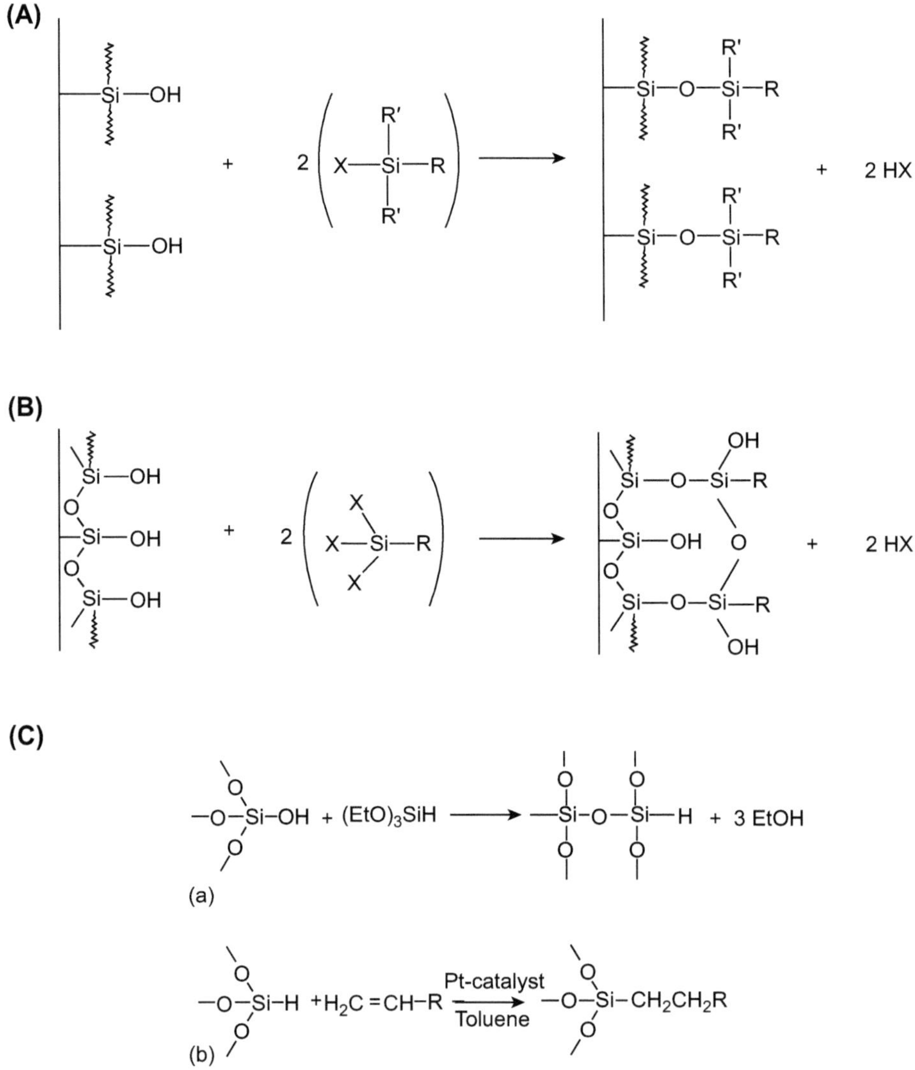


**Figure 2.2** Synthesis of (A) monomeric and (B) polymeric chemically bonded sorbents and (C) silica hydride.

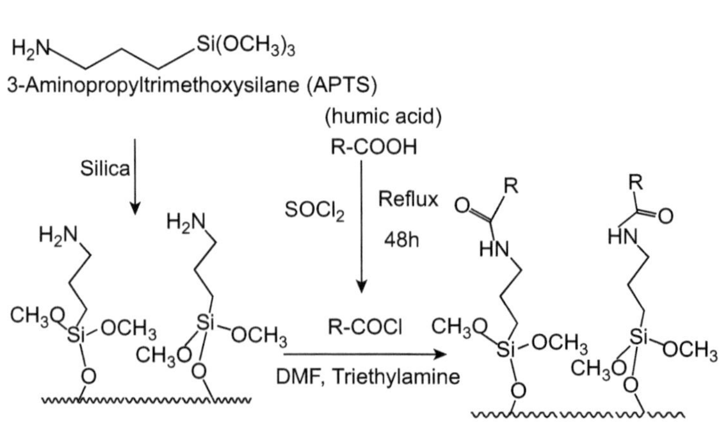


**Figure 2.3** The scheme to prepare humic acid-bonded silica as an SPE adsorbent.

was recovered, whereas Sig-Z-Met-OH enabled fast and quantitative extraction of Au(III) from 0.1 to 0.01 $mol^{-1}$ HCl.

## 2.3 Evaluation of formats, sorbent types, and modes of interaction

SPE sorbents are available in several formats: contained within cartridges, in columns fashioned like syringe barrels, or in disks, Fig. 2.4. Typical columns are manufactured of polypropylene or glass, and the sorbent is held within the column by porous frits made of polyethylene or polytetrafluoroethylene (PTFE) [12]. A typical format is a syringe barrel consisting of a 20 μm frit at the bottom of the barrel, with the relevant sorbent above and an additional frit at the top. Alternatively, disks of various sizes with sorbent particles homogeneously distributed in a support matrix with a diameter generally significantly larger than its height, pipette tips, and 96-well SPE microtiter plates (which use the disk format). There are two common types of disk formats, disc, particle-loaded membranes (PLMs), which are composed of polytetrafluoroethylene microfibrils and enclosed sorbent particles with a diameter of about 8 μm, and particle-embedded glass fiber discs (PEGFDs) [37]. The disk format is advantageous for handling small sample sizes and for sampling at relatively high flow rates with disks of high cross-sectional area. Last, a small column connected *online* to a chromatographic system facilitating automated sample handling and separation. The 96-well plate format also facilitates automation and high-throughput sample preparation using liquid handling devices and parallel sample processing.

Native silica is typically used for the extraction of nonpolar and low-polarity analytes dissolved in organic solvents. Chemically bonded sorbents ($C_1$, $C_2$, $C_8$, $C_{18}$, Ph, CN, $NH_2$) can be used for similar samples but are more commonly used for the extraction of neutral compounds of wide polarity from aqueous samples. The effectiveness of SPE methods depends on several factors, such as the sorbent chemistry, pH (ionizable compounds), sample pretreatment, organic solvents used for washing and eluting steps, as well as the flow rate during the different steps. The elution of the analytes is effected by a suitable solvent, leaving the interfering substances on the column. Strong and weak elution solvents for adsorbed compounds are described in Table 2.2.

To achieve selectivity SPE provides countless diverse choices of sorbents, ranging from the traditional reversed-phase sorbents ($C_{18}$, $C_8$), normal phase (silica, alumina), ion exchange, mixed-mode (ion exchange and reversed phase) and functionalized resins based on styrene-divinylbenzene (SDVB) polymers [1,5,38]. The interactions between the target compounds and the sorption centers on the solid phase include hydrophobic interactions and hydrophilic interactions such as dipole-dipole, induced dipole-dipole, hydrogen bonding, and $\pi-\pi$ interactions. In addition, for analytes with charged functional groups, electrostatic interactions contribute to selectivity on sorbents with complementary charged functional groups. Affinity-type interactions

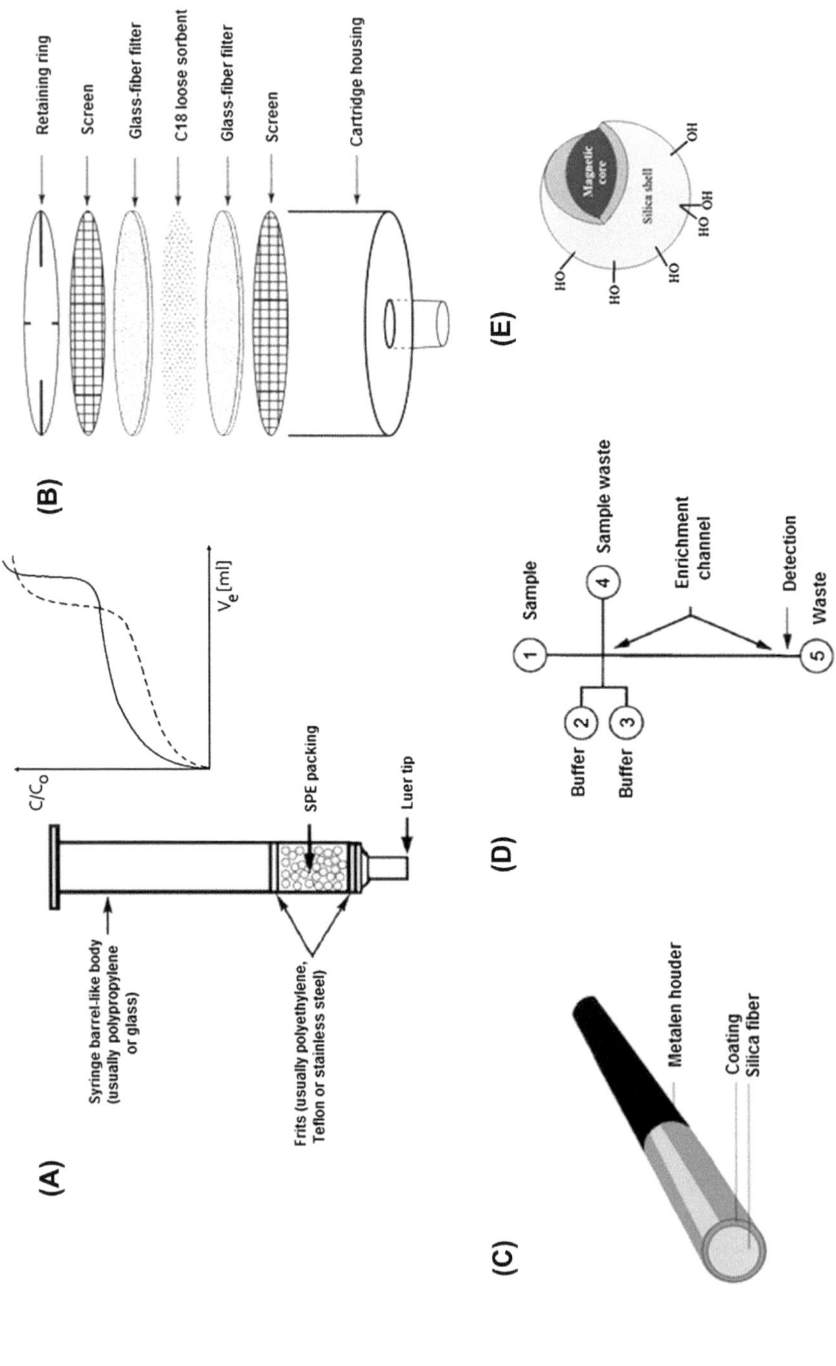


**Figure 2.4** Different formats for SPE: (A) tube, (B) Speedisk, (C) fiber, (D) chip and (E) magnetic silica-based core/shell nanoparticles.

**Table 2.2** Sorbents applied in SPE.

| Type of bed sorbent | Structure of bound ligand | Analyte type | Dissolving solvents | Elution solvents |
|---|---|---|---|---|
| **Reversed phases** | | | | |
| Octadecyl ($C_{18}$) | $-(CH_2)_{17}CH_3$ | Slightly-moderately nonpolar-nonpolar | Methanol/water; Acetonitrile/water | Hexane, chloroform (nonpolar analytes); Methanol (polar analytes) |
| Octadecyl ($C_{18}$) LightLoad | $-(CH_2)_{17}CH_3$ | | | |
| Octadecyl ($C_{18}$) PolarPlus | $-(CH_2)_{17}CH_3$ | | | |
| Octyl ($C_8$) | $-(CH_2)_7CH_3$ | | | |
| Ethyl ($C_2$) | $-CH_2CH_3$ | | | |
| Cyclohexyl | $-CH_2CH_2-$ | | | |
| Phenyl | $-CH_2CH_2CH_2-$ | | | |
| **Normal phases (bonded)** | | | | |
| Cyano (CN) | $-(CH_2)_3CH$ | Slightly-moderately polar-strongly polar | Hexane, chloroform, acetone | Methanol |
| Amino ($NH_2$) | $-(CH_2)NH_2$ | | | |
| Diol (COHCOH) | $-(CH_2)_3OCH_2CH-CH_2OHOH$ | | | |
| **Normal phases (adsorption)** | | | | |
| Silica gel | -SiOH | Slightly-moderately polar-strongly polar | Hexane, chloroform | Methanol (dependent on the type of analyte) |
| Florisil | $Mg_2SiO_3$ | | | |
| Alumina | $Al_2O_3$ | | | |

*Continued*

**Table 2.2** Sorbents applied in SPE.—cont'd

| Type of bed sorbent | Structure of bound ligand | Analyte type | Dissolving solvents | Elution solvents |
|---|---|---|---|---|
| **Ion exchangers (anion and cation exchange)** | | | | |
| Amino (NH2) | $-(CH_2)_3NH_2$ | Anion exchange-ionic acid<br>Cation exchange-ionic base | Water or buffer (pH=pKa+2)<br>Water or buffer (pH=pKa-2) | a) buffer (pH=pKa+2)<br>b) pH where sorbent or analyte is neutral<br>c) solvent with high ionic strength<br>a) buffer (pH=pKa-2)<br>b) pH where sorbent or analyte is neutral<br>c) solvent with ionic strength |
| 1°, 2°-Amino ($NH/NH_2$) | $-(CH_2)_3NHCH_2CH_2NH_2$ | | | |
| Quaternary amine ($N^+$) | $-(CH_2)_2COOH$ | | | |
| Carboxylic acid (COOH) | $-(CH_2)_3N^+(CH_3)_2$ | | | |
| Acid ($SO_2OH$) | $-(CH_2)_3SO_2OH$ | | | |
| Acid ($ArSO_2OH$) | $-(CH_2)_3$- $-SO_2OH$ | | | |
| **SE-Sephadex** | | | | |
| Sephadex G-25 | Dextran | - | - | - |
| **Wide-porous** | | | | |
| RP butyl ($C_4$) | $-(CH_2)_3CH_3$ | - | - | - |
| HI HI-propyl ($C_3$) | $-(CH_2)_2CH_3$ | | | |
| IE CBX (carboxylic acid) | -COOH | | | |
| PEI (polyethyleneimine) | -(CH2CH2NH)n- | | | |

based on molecular recognition sites are possible for specifically designed sorbent surfaces made selective for a particular analyte structural motif.

The strong adsorption sites on unmodified inorganic oxides are generally associated with hydrogen-bonding interactions. These sites are generally deactivated by water which results in weak retention of analytes. Surface bonding of polar functional groups like cyano-, amine- or diol- provide an alternative to the harsh interactions with inorganic oxides for the extraction of polar compounds from organic solvents by hydrophilic interactions. Chemically bonded sorbents for reversed phase extraction are typically modified by octadecyl, octyl, cyclohexyl, or phenyl groups [10,12].

### 2.3.1 Sorbent pretreatment

Generally, SPE consists of four steps: column preparation, sample loading, column postwash, and sample desorption. The above mentioned prewash step is used for conditioning the stationary phase. Additionally, the postwash is utilized to remove undesirable substances. The target analytes are retained by the sorbent after washing out the interfering compounds.

Particle-loaded membranes containing $C_{18}$ is the main disk format in use. One of the drawbacks of disks compared with cartridges is the decrease in the breakthrough volume, mainly for more polar compounds. For this reason, disks are used when there is a strong interaction between the analyte and the sorbent [42]. Poole et al. [43] reviewed models and experimental procedures for predicting sorption isotherms (so-called *breakthrough curves*) for cartridge and disk devices. The breakthrough curves provide sufficient information for identifying the physicochemical processes involved in the solute transport through the packing materials.

### 2.3.2 Miniaturization and automation

Solid-phase extraction has benefited from and contributed to the general progress made in miniaturization and automation of extraction technique. Sampling in SPE occurs in two modes, *off-line* in which sample processing occurs before the chromatographic measurements and *online* in which sample processing is integrated directly with the chromatographic system. The latter is a more complex arrangement but has the advantages of lower contamination, automation, and higher sensitivity due to the analysis of the whole sample rather than an aliquot of the sample. Nowadays, *online* SPE coupled with either gas or liquid chromatography can be considered routine since only minor equipment requirements are necessary to fabricate a suitable system and commercial devices are available. The main advantages are higher throughput, good precision, limited manual processing, as well as lower cost, and greater sensitivity. As an alternative to conventional particle-packed extraction columns, Kang et al. described a packed-fiber SPE device for the analysis of trace pollutants in the environmental water and for drugs and their metabolites in human plasma [44]. They showed that nanofiber sorbents possess numerous advantages in terms of sensitivity, reproducibility, and limited detection, and sample processing time. Recent years have witnessed the development of the microfluidic and on-chip analytical system with integrated solid-phase extraction units [45]. The microfluidic technology integrates injection, reaction, separation, and detection in a single device. Among different sample pretreatment methods carried out on a microchip device, SPE is one of the most important. A typical example is described by Kutter et al. [46] who prepared in an open-tube format a channel coated with C18 on a simple glass chip.

## 2.4 Sorbent characterization techniques

The properties and architecture of sorbents have a significant effect on their chemical modification. Those processes are investigated by the use of various physical and chemical techniques, such as elemental analysis, porosimetry, infrared spectroscopy, nuclear magnetic resonance imaging, microscopy, and tomography. Elemental analysis is one of the basic methods used to estimate the extent of surface coverage for chemically bonded sorbents from the percentage of carbon, hydrogen, and nitrogen, the surface area of the sorbent, and the structure of the bonded phase, usually expressed as $\mu mols/m^2$ or the number of bonded ligands per $nm^2$ [47,48]. However, the elemental analysis does not provide information about the homogeneity of the bonded phase and should be used in conjunction with other methods to infer sorbent properties. Elemental analysis has a precision of approximately 0.2% (w/w) for typical elements.

Low-temperature nitrogen adsorption-desorption facilitates the determination of the quantity of gas adsorbed on the sorbent surface based on the mass and pressure of the analyzed gas. The BET adsorption isotherm provides information on the specific surface area ($m^2/g$), volume (mL/g), and the diameter and size of pores. Owing to the size of nitrogen molecules, nitrogen adsorption is typically used for micropore and mesopore sorbents. Mercury porosimetry is more appropriate for macroporous sorbents.

One of the most frequently used techniques for the determination of the adsorbent surface structure is infrared spectroscopy (IR). This method makes it possible to obtain information for bonded or associated silanol groups on the carrier surface and the surface structure of the chemically bonded phase. The main difficulty in applying this technique is the strong absorption of infrared radiation by silica gel. Advanced spectroscopic techniques, such as diffuse reflectance Fourier transform infrared spectroscopy (DRIFT) are preferable to transmission measurements [48–50]. The DRIFT method has been used to determine the hydrolytic stability of cyano, amino, and carboxyl groups added to the surface of silica gel.

Thermogravimetry (TG), derivative thermogravimetry (DTG), differential thermal analysis (DTA) and differential scanning calorimetry (DSC) are used in the study of phase transformations. DSC and thermal analysis are valuable tools to determine the supramolecular organization of the solid phase. DSC experiments make it possible to precisely determine the interval in which the so-called *transition phases* occur.

They can be interpreted as the formation of more ordered structures, which is influenced by temperature or addition of an organic modifier, or ligands moving from unevenly thick clusters to a more homogenous monolayer. The reliability of the results is limited by the low density of the bonded phase coating. Thermogravimetric analysis, more often used for pure silica gel, provides valuable information for the temperature range in which the thermodesorption of physically adsorbed water or loss of mass connected with the condensation of silanols occurs. However, the characterization of SPE sorbents, involving the use of TGA allows conclusions to be drawn regarding the different location of ligands bound on the matrix surface.

Microscopic methods support observations of the sorbent surface and any impurities adsorbed during synthesis. Pore diameter can be measured with sufficient magnification. The most popular microscopic methods include optical microscopy, scanning electron microscopy (SEM), atomic force microscopy (AFM), and scanning tunneling microscopy (STM) [51].

On the basis of fluorescence spectroscopy of surfaces modified with ligands containing fluorescent groups, Lochmüller [52] suggested three-component model for silica-based chemically bonded phases: (1) fluorophores which are not reached by solvent particles because they are localized in micropores or surrounded by an agglomeration of alkyl chains; (2) fluorophores open for interaction with solvent molecules; and (3) fluorophores relatively able to interact with solvent but bonded by hydrogen bonds to silanols present on the surface.

A comparison of the results of XPS (a surface analysis technique with limited depth of several nanometers) and XRS enables conclusions to be drawn regarding the differences between the external layer of the bead and its internal pore network [53]. $^{29}Si$ and $^{13}C$ CP/MAS NMR is one of the most frequently used techniques providing detailed information on the presence of various ligands on the surface of a silica matrix and their conformation [47]. In contrast to infrared spectroscopy, NMR makes it possible to identify signals from different types of Si-OH groups. Initial problems connected with the limited mobility of ligands resulting in band broadening caused by chemical shift anisotropy and dipolar hydrogen-carbon interactions were solved by using magic angle spinning with dipolar decoupling.

Mainly $C_{18}$ sorbents and their modifications are used in biomedicine, pharmacology, or toxicology for extraction from different biological matrices. Martin et al. [54] studied the influence of different $C_{18}$ silica-based sorbents on the extraction of acidic and basic compounds. The preferred sorbents were those of intermediate carbon loading. In addition, endcapping was important for compounds susceptible to silanophilic interactions. Buszewski et al. [55] prepared new types of silica gel sorbents used for five selected β-receptor antagonist sample preparation methods. They obtained successful preparation of silica gels of various porosities chemically modified by cholesterol ligands. Using optimized conditions, satisfactory recoveries of the analytes from buffers, urine, and blood samples were obtained.

Several applications of these new modified silica sorbents are presented in Table 2.3 [56–65].

**Table 2.3** Applications of new modified silica gel sorbents for SPE [56–65].

| Sorbent | Chelating agent | Sample | Target compounds | Adsorption capacity (mg/g) | Recovery | Ref. |
|---|---|---|---|---|---|---|
| Silica gel | 5-[(E)-(5-sulfonyl-1,3,4-thiadiazol-2-yl) diazenil] pyrimidine-2,4,6 (1H, 3H, 5H)-trione | aqueous solutions | Th(IV) | 24.85 | – | [56] |
| Amino bonded silica gel | Polyelectrolyte | environmental water (drinking water, well water, snow water, waste water) | silver(I) | 8.6 | 94%–105% | [57] |
| Silica gel | Phenothiazine | environmental water | nitrobenzene compounds | 1.15 – 1.24 | 71.4%–124.3% | [58] |
| Silica gel | Pyrenebutyric acid | water | polychlorinated biphenyls | – | 70.42–110.51 | [59] |
| Silica gel | Glycerol | water | cobalt(II) | 0.25 | 96.6–97.4 | [60] |
| Aminopropyl-modified silica gels | Aminopolycarboxylic | water samples | cobalt, nickel, cadmium, lead, zinc | – | 99.4% | [61] |

| | | | | | | |
|---|---|---|---|---|---|---|
| Kiesegel Si-60 and SG-100 | Octadecydimethylchlorosilane, dodecydimethylchlorosilane, Octydimethylchlorosilane, hexyldimethylchlorosilane, trimethylchlorosilane, octadecytrimethylchlorosilane | urine | 5-hydroxyindole-3-acetic acid (5-HIAA) and serotonin (5-HT) | – | 15.2%–92.8% and 1.7%–75.3% | [62] |
| SG-100 | Monofunctional (MC18) and trifunctional (TC18) octadecylchlorosilanes | urine | glucuronide, paracetamol, cycteine and mercapturic acid | – | 98%–110%, 1%–42%, 15%–120% and 5%–110% | [63] |
| Silica gel | Murexide | water | uranium (VI) | – | 98% | [64] |
| Silica gel | 2-((3-silylpropylimino) methyl) phenol | aqueous solutions | Fe(III), Pb(II), Cu(II), Ni(II), Co(II) and Zn(II) ion | 69.8–90.0 | 95.6%–98.0% | [65] |

## 2.5 Conclusions

The above overview demonstrates that sorbent selection makes it possible to isolate target compounds from various matrices by specific and nonspecific interactions. The effectiveness of the isolation process, i.e., selective sorption, and quantitative desorption, and as a consequence recovery, depends on the silica porosity and surface density of the bonded phase, and appropriate solvent selection. Interactions that participate in the process of isolating analytes from the matrix are characterized by varying energy. The selectivity of the process depends not only on the eluting power of solvents or their mixture but also on the selection of sorbent sorption properties. Trends in further development are currently focused on miniaturization and automation with a view to reducing costs, simplifying sample manipulation steps, and providing higher sample throughput. The application of micro- and nanoscale-based extraction and separation techniques will be developed in the future resulting in quick and sensitive analytical methods.

## Acknowledgements

The work was financially supported by the National Science Center, Cracow, Poland in the frame of the project Sonata 12 No. 2016/23/D/ST4/00305 (2017–2020) and Preludium 14 No. 2017/27/N/ST4/00354 (2018–19).

## References

[1] Poole CF. Trends Anal Chem 2003;22:362–73.
[2] Castro R, Natera R, Duran E, Garcia-Barroso C. Eur Food Res Technol 2008;228:1–18.
[3] Zhou J, Xue X, Li Y, Zhang J, Chen F, Wu L, Chen L, Zhao J. Food Chem 2009;115: 1074–80.
[4] Kataoka H. Trends Anal Chem 2003;22:232–44.
[5] Hennion M-C. J Chromatogr A 1999;856:3–54.
[6] Tamayo FG, Turiel E, Martin-Esteban A. J Chromatogr A 2007;1152:32–40.
[7] Huck CW, Bonn GK. J Chromatogr A 2000;885:51–72.
[8] Novakova L, Vlckova H. Anal Chim Acta 2009;656:8–35.
[9] Buszewski B, Szultka M. Crit Rev Anal Chem 2012;42:198–213.
[10] Fontanals N, Galia M, Cormack PAG, Marce RM, Sherrington DC, Borrull F. J Chromatogr A 2005;1075:51–6.
[11] Żwir-Ferenc A, Biziuk MP. J Environ Stud 2006;15:677–90.
[12] Ligor M, Wójcik J, Buszewski B. Pol J Food Nutr Sci 2004;13/54:355–62.
[13] Pesek J,J, Matyska MT. J Sep Sci 2005;28:1845–54.
[14] Buszewski B, Jezierska-Switala M, Kaliszan R, Wojtczak A, Albert K, Bachmann S, Matyska MT, Pesek J. J Chromatogr 2001;53:S204–12.
[15] Lambert S. Chromatogr Today 2009;2(2):12–4.
[16] Peterson DS, Rohr T, Svec F, Fŕechet JM. J Anal Chem 2003;75:5328–35.
[17] Moral A, Sicilia MD, Rubio S, Pérez-Bendito D. Anal Chim Acta 2008;608:61–72.

[18] Ncube S, Madikizela L, Cukrowska E, Chimuka L. Trends Anal Chem 2018;99:101–16.
[19] Muir D, Sverko E. Anal Bioanal Chem 2006;386:769–89.
[20] Zhang L, Dou XW, Zhang Ch, Logrieco AF, Yang MH. Toxins 2018;10. https://doi.org/10.3390/toxins10020065.
[21] Sheng H, Song Y, Bian Y, Wu W, Xiang L, Liu G, Jiang X, Wang F. Anal Methods 2017;9:688–96.
[22] Scott RPW. Silica gel and bonded phases. Their production, properties and use in LC. Chichester: J. Wiley & Sons; 1993.
[23] Skoczylas M, Krzemińska K, Bocian Sz, Buszewski B. Silica gel and its derivatization for liquid chromatography. Ramtech Ltd. In: Meyers RA, editor. Encyclopedia of analytical chemistry. Applications, theory and instrumentation. Tarzana, CA: USA John Wiley & Sons, Ltd; 2017. p. 11450–71. Chapter 12.
[24] Buszewski B, Jurášek A, Garaj J, Nondek L, Novák I, Berek D. J Liq Chromatogr 1987;10:2325–36.
[25] Sandoval J, Pesek J. J Anal Chem 1989;61:2067–75.
[26] Tanaka N, Hashidzuma K, Araki M, Tsuchiya H, Okumo A, Iwaguchi K, Ochnishi S, Takai N. J Chromatogr 1998;448:95.
[27] Gregg SJ, Sing KSW. Adsorption surface area and porosity. London: Academic Press; 1967.
[28] Brunauer S, Emmet PH. J Am Chem Soc 1935;57:1754.
[29] Brunauer S, Emmet PH, Teller E. J Am Chem Soc 1935;60:1938.
[30] Berek D, Novak I. Chromatogr 1990;30:582.
[31] Buszewski B, Leboda R. Chem Stos 1990;34:196.
[32] Buszewski B, Jaroniec M, Staszczuk P, Gilpin RK. Wiad Chem 1995;49:223.
[33] Nawrocki J, Buszewski B. J Chromatogr 1998;449:1.
[34] Novák I, Buszewski B, Garaj J, Berek D. Chem Papers 1990;44:31.
[35] Schwarz H, Shaik S, Li J. J Am Chem Soc 2017;139:17201–12.
[36] Unger KK. Packings and stationary phases in chromatographic techniques. 1990. Basel, New York.
[37] Žuvela P, Skoczylas M, Liu JJ, Bączek T, Kaliszan R, Wong MW, Buszewski B. Chem Rev 2018. https://doi.org/10.1021/acs.chemrev.8b00246.
[38] Buszewski B, Jezierska M, Welniak M, Berek D. J High Resolut Chromatogr 1998;21:267–81.
[39] Galceran MT, Jauregui O. Anal Chem Acta 1995;304:75–84.
[40] Petrova P, Karadjova I, Chochkova M, Dakova I, Karadjov M. Bulg Chem Comm 2017;SI E:95–100.
[41] Kang X-J, Chen L-Q, Wang Y, Zhang Y-Y, Gu Z-Z. Biomed Microdevices 2009;11:723–9.
[42] Poole CF, Gunatilleka AD, Sethuraman R. J Chromatogr A 2000;885:17–39.
[43] Luo D, Yu QW, Yin HR, Feng YQ. Anal Chim Acta 2007;588:261–7.
[44] Oleschuk RD, Shultz-Lockyear LL, Ning Y, Harrison DJ. J Anal Chem 2000;72:585–90.
[45] Kutter JP, Jacobson SC, Ramey JM. J Microcolumn Sep 2000;12:93–7.
[46] Berendsen GE, Pikaart A, de Galan L. J Liq Chromatogr 1980;3:1437–64.
[47] Albert K, Bayer E. J Chromatogr 1991;544:345–70.
[48] Buszewski B, Kasturi P, Gilpin RK, Gandoga ME, Jaroniec M. Chromatogr 1994;39:155–61.
[49] Hetem MJJ. Chemically modyfied silica surfaces in chromatography – a fundamental study. Heidenberg: Hüthing Buch Verlag; 1993.
[50] Sander LC, Callis JB, Field LR. Anal Chem 1983;55:1068–75.

[51] Murthy RSS, Crane LJ, Bronnimann CE. J Chromatogr 1991;542:205–20.

[52] Lochműller CH, Colburn AS, Hunnicutt ML, Harris JM. Anal Chem 1983;55:1344.

[53] Hunnicutt ML, Harris JM. Anal Chem 1986;58:748.

[54] Martin P, Morgan ED, Wilson ID. Anal Chem 1997;69:2972–5.

[55] Buszewski B, Welerowicz T, Tegowska E, Krzemiński TF. Anal Bioanal Chem 2009;393: 263–72.

[56] Kirkan B, Aycik GA. J Radioanal Nucl Chem 2016;308:81–91.

[57] Xiang G, Li L, Jiang X, He L, Fan L. J Chil Chem Soc 2013;4:2182–5.

[58] Peng XT, Zhao X, Feng YQ. J Chromatogr A 2011;1218:9314–20.

[59] Jin S, Zhang X, Gao R, He B, Yi Q. In: International conference on electric technology and civil engineering (ICETCE), Lushan; 2011. p. 5847–50.

[60] Safavi A, Iranpoor N, Saghir N, Momeni S. Anal Chim Acta 2006;569:139–44.

[61] Repo E, Warchoł JK, Sillanpä M. J Sust Develop Energy Wat Environ Sys 2017;5: 89–100.

[62] Buszewski B. J Pharm Biomed Anal 1990;8:645–9.

[63] Mouelhi ME, Buszewski B. J Pharm Biomed Anal 1990;8:651–3.

[64] Sadeghi S, Sheikhzadeh E. J Hazard Mater 2009;163:861–8.

[65] Mortazavi K, Ghaedi M, Roosta M, Montazerozohori M. Ind J Sci Tech 2012;5: 1893–900.

# Porous polymer sorbents

3

*Núria Fontanals, Rosa M. Marcé, Francesc Borrull*
Department of Analytical Chemistry and Organic Chemistry, Universitat Rovira i Virgili, Tarragona, Spain

## 3.1 Introduction

Solid-phase extraction (SPE) is the most widely used technique for liquid samples or liquid extracts from other types of extractions. This is because SPE is versatile and can be used with different materials that cover various types of interactions with compounds having a wide range of properties.

The first sorbents developed were silica-based sorbents and those modified mainly with long alkyl chains, such as $C_{18}$ and $C_8$, but also with phenyl, $NH_2$ or others. However, these sorbents have some drawbacks, such as instability at extreme pHs, the activity of residual silanols, and low retention of polar compounds. Later, carbon-based sorbents appeared, including graphitized carbon blacks (GCBs) and porous graphitic carbon (PGC). However, they have a high retention capacity for some compounds, and thus eluting them is difficult and even irreversible.

Polymer-based materials are one of the main developments with continuous progress appearing over the later years. This is mainly due to their morphological features (high surface area and well-defined porosity) and the diversity of synthetic routes, which facilitates incorporation of various chemical functionalities into the porous framework. This translates into a high retention capacity for different types of compounds and enhanced stability in different SPE conditions. These properties overcome the main disadvantages of silica- and carbon-based materials. Other recent developments in sorbent technology, including carbon nanomaterials, metallic nanoparticles, and metal-organic frameworks have also emerged [1], and will be covered in the other chapters of this handbook. However, over the years, porous polymer sorbents have remained an important class of sorbents for SPE.

Porous polymer sorbents have been developed to accommodate high capacity extraction with the development of hypercrosslinked networks and/or the introduction of hydrophilic moieties, high selective extraction (the emergence of molecularly imprinted polymers (MIPs)) or a combination of these (mixed-mode ion-exchange materials). In this chapter, we cover all types of porous polymers, both commercially available and prepared in-house, except for MIPs, which are covered in a separate individual chapter. We give particular attention to the morphological and chemical properties of porous polymers, which are closely related to the type of polymerization procedure applied. The main technique used for polymer particle manufacturing is suspension polymerization, although precipitation polymerization (PP) and emulsion polymerization are also used. These polymerization techniques result in different

**Solid-Phase Extraction. https://doi.org/10.1016/B978-0-12-816906-3.00003-0**

morphological properties and different particle sizes. Therefore, the technique used should be selected carefully depending on the desired features of the polymer developed. Monolithic polymers are generated by bulk polymerization into the mold. Although porous polymers in particle format are more common in SPE, both particle and monolithic formats are covered in the different sections of this chapter. We also discuss the analytical capabilities of porous polymers and provide some illustrative examples of their applications.

## 3.2 Hydrophobic porous polymers

Porous polymeric materials were initially introduced to solve the drawbacks of silica- and carbon-based materials. From the early developments in the 1980s, research into hydrophobic porous polymers has aimed to make polymers more retentive toward target compounds.

### 3.2.1 Macroporous polymers

Conventional polymeric sorbents are crosslinked polymers obtained by suspension polymerization based on poly(styrene-divinylbenzene) (PS-DVB), which has a hydrophobic structure and specific surface area (SSA) of up to 500 $m^2/g$. The amount of crosslinker (i.e., DVB) and the type of solvent (porogen) used during the suspension polymerization determine the porosity (IUPAC definition: micropores $<2$ nm; mesopores 2–50 nm; macroporous $>50$ nm) of the resin, which in turn determines the SSA (measured by $N_2$ sorption and application of the BET theory). Thus, adding a thermodynamically good solvent for the monomer mixture leads to the formation of micropores and mesopores, resulting in beads with high SSAs. Adding a poor solvent leads to the formation of macropores, and thus beads with significantly lower SSAs [2,3].

> When preparing macroporous polymers, adding thermodynamically good solvent (porogen) for the monomer mixture leads to the formation of micropores and mesopores, which results in beads with high SSAs.

For example, using percentages of up to 70% DVB and toluene as the solvent (a good solvent for the PS-DVB polymer) produces macroporous PS-DVB sorbents with SSAs up to 600 $m^2/g$; however, when cyclohexanone (a poor solvent) is used, a lower SSA (25 $m^2/g$) is obtained [4]. In summary, macroporous SPE resins are formed when a porogen (toluene in this example) is present in the comonomer mixture (PS-DVB), which leads to phase separation of the polymer matrix at the stage where micropores are formed [2]. Moreover, generating beads by suspension polymerization usually results in uniformly shaped polymer particles with polydisperse sizes ranging from $\sim$50 to $\sim$500 μm, and these can be sieved later to obtain the suitable size for SPE.

The hydrophobic structure of the PS-DVB polymer interacts with analytes through van der Waals forces and π-π interactions of the aromatic rings. Macroporous

PS-DVB sorbents result in higher analyte retention compared to silica-based sorbents due to the numerous $\pi$-$\pi$ sites they possess; however, the hydrophobic structure has a poor capacity and low retention for polar analytes [5].

Different macroporous hydrophobic sorbents are currently available: Amberlite XAD series supplied by Rohm & Haas (SSA from 100 to 800 $m^2/g$), PLRP-S-10 (500 $m^2/g$) and Bond Elut ENV (700 $m^2/g$) supplied by Agilent; Strata SDB-L (500 $m^2/g$) supplied by Phenomenex; AttractSPE DVB (600 $m^2/g$) supplied by Affinisep; Purophase PCG900M (600 $m^2/g$); and Isolute 101 (500 $m^2/g$) supplied by Biotage. These sorbents have been used to extract various compounds, mainly those of moderate to low polarity. For instance, when online-SPE-liquid chromatography (LC) with tandem mass spectrometry (MS/MS) was used to determine 22 pesticides in groundwater, PLRP-S cartridges were used to retain 16 of the compounds, whereas a Hysphere GP resin (enhanced retention features) was used for the remaining six pesticides with recoveries higher than 75% in all cases with a 5 mL sample volume [6].

The first study of a polymer monolith for SPE was based on a macroporous ethylstyrene (ES)-DVB. This material was prepared directly inside a polyether ether ketone (PEEK) tube connected online to LC. In this case, the SSA of 400 $m^2/g$ was achieved by increasing the percentage of DVB (80%) while keeping the porogen in 52% dodecanol (poor solvent) and 8% toluene (good solvent) [7]. Svec [8] pointed out that there are considerable differences in pore sizes for identical polymerization mixture depending on whether bulk polymerization (monolith) or suspension polymerization (particles) is used. In addition, the type and proportion of porogens in the monolith preparation should be chosen carefully to ensure that the final monolith has suitable mesopores to enable the flow of different solvents employed in the extraction. Furthermore, it is always of interest to create micropores providing higher SSAs. More research into the effect of the porogen in monolith technology is required to determine the monolith morphology in more detail [9]. Nevertheless, toluene is often mixed with higher alcohols (i.e., dodecanol) to prepare PS-DVB monoliths, although this does not exclude the use of other porogens [9].

### 3.2.2 Hypercrosslinked polymers

The retention capacity of the polymer is closely related to the SSA. Although the percentage of the crosslinking agent is maximized in macroporous resins, a significant number of vinyl groups remain unreacted due to steric impediments of the aromatic groups. In the 1970s, Davankov presented a novel polymerization procedure to obtain polymers with hypercrosslinked (HXL) networks that consist of a postcrosslinking linear or lightly crosslinked (0.3%–2% DVB) PS by means of a Friedel–Crafts reaction using a crosslinking agent and a Friedel–Crafts catalyst (mainly $FeCl_3$) [10]. With this procedure, almost all the aromatic rings can be crosslinked, thus yielding resins with an HXL structure, high micropore content, and ultrahigh SSA (up to 2000 $m^2/g$). Unlike other porous polymers, HXL polymers have permanent porosity, which leads to extremely high SSA. These morphological properties give HXL resins a higher retention capacity than macroporous resins.

Later, Jerábek [11] proposed including vinylbenzyl chloride (VBC) into the polymer chains, so that the chloromethyl ($CH_2Cl$) moiety in VBC acts as an internal electrophile to crosslink the aromatic rings. Sherrington [12] presented a development that consisted of hypercrosslinking both gel-type and macroporous VBC-DVB precursors. In this approach, the crosslinking process is intramolecular, and thus, reactions are extremely efficient because a VBC-DVB precursor can become almost completely HXL in only 15 min [12]. In addition, if the precursor resin is already porous, the resulting HXL resins have a bimodal pore size distribution because they have both the original pore size (macroporous) and that generated during the hypercrosslinking process (microporous). Furthermore, Sherrington's research group demonstrated that this approach is feasible when the precursors are obtained via suspension polymerization [13], which produces polydisperse particles with diameters ranging from 50 to 200 μm, nonaqueous dispersion (NAD) polymerization, which produces particles in the range 4–10 μm, and precipitation polymerization (PP), which produces monodisperse particles of ~4 μm [14]. Later, Svec's group also proposed this synthetic route in bulk polymerization to develop HXL in the form of monoliths [15]. Except in the case of monoliths (which should contain mesopores in order to ensure suitable kinetic and thermodynamic features), the different polymerization procedures produce similar morphological properties, and the main difference between the procedures is the particle size. Different reviews [16,17] detail the variation in the synthetic approaches as well as the morphological properties of the HXL

HXL polymers have permanent porosity that leads to an extremely high SSA.

The morphology of HXL monoliths is best illustrated in Ref. [18], describing the use of an HXL monolith to SPE. In this study, an HLX monolith was prepared from a precursor based on poly(VBC-S-DVB) and its morphology compared to a polyDVB monolith and PS-DVB in particulate form. It was shown that even the HXL monolith displayed larger SSAs (817 $m^2/g$) than the polyDVB monolith (531 $m^2/g$) in the dry state. However, in the solvated state, they found that the area of the HXL monolith decreased considerably (341 $m^2/g$) whereas the area of the polyDVB monolith did not. Moreover, a comparison of the SPE performance of the polyDVB monolith and particulate poly(S-DVB) showed that anisole (as a model compound) was well-retained on the polyDVB monolith regardless of the flow rate (0.1–1 mL/min), but the particulate material behaved better at lower flow-rates. This was also confirmed when the polyDVB monolith was compared to different commercially available sorbents, including Bond Elut-LMS, Strata SDB, and Oasis HLB, which confirmed that the monolith's performance is independent of the flow rate.

Several manufacturers have commercialized HXL resins since they were first introduced as SPE sorbents. Some examples are: Amberchrom GC-161m (900 $m^2/g$, Sigma–Aldrich), Hysphere series (>1000 $m^2/g$, Spark Holland), Hypersol-Macronet sorbents (1200 $m^2/g$, Purolite International), Lichrolut EN (1200 $m^2/g$, Merck), Envi-Chrom P (900 $m^2/g$, Supelco), and Chromabond HR-X (1200 $m^2/g$, Macherey–Nagel).

Other advances in HXL technology for SPE have been made over the years [13,14,19]. In one of these studies [19], three different HXL polymers were synthesized from precursors obtained by precipitation polymerization with areas ranging from 880 to 1320 $m^2/g$ depending on the percentage of VBC feed in the precursor (from 25% to 75%). The performance of these three HXL sorbents was compared to a commercially available HXL sorbent, Lichrolut EN (1200 $m^2/g$), for the online extraction of phenols and pesticides from environmental water samples. The results showed that the in-house prepared HXL sorbents had a higher extraction efficiency and capacity (they attained recoveries between 74% and 105% for a 500 mL sample volume) compared to Lichrolut EN (recoveries of 41%−78%). As all the sorbents had similar morphological (SSAs ~ 1000 $m^2/g$) and chemical properties (hydrophobic), the in-house prepared sorbents' higher extraction efficiency was attributed to the particle size. The particle size was ~4 μm for the in-house prepared sorbents, compared to 40−120 μm for Lichrolut EN. Smaller particle sizes promote better packing in the sorbent bed (especially in online SPE) and lead to a more efficient and reproducible extraction procedure.

Five different sorbents (Chromabond HR-X, Oasis HLB, Bond Elut Plexa, SampliQ $C_{18}$, and Chromabond HR-HAW) were evaluated for the extraction of 24 compounds (including pharmaceuticals, personal care products, endocrine disrupting compounds and artificial sweeteners) by offline SPE/LC-MS/MS in water samples. The hydrophobic Chromabond HR-X was the most effective for the compounds studied (recoveries >60%) for all samples analyzed (groundwater, surface water, and raw water). This sorbent was particularly successful for the sweeteners, which are the most challenging compounds in the group [20]. Further examples of commercial HXL sorbents used to extract different compounds, such as UV filters [21] or sweeteners [22] from environmental water samples have been reported. Nevertheless, there are some applications for other fields. For instance, Lichrolut EN (80 mg) was used for the online extraction of a group of endocrine disrupting compounds from 5 mL of urine, 1 mL of blood, and 1 mL of breast milk with good results achieved [23]. It should be noted that although the above analytes have polar functional groups, they can only be retained on the sorbent through hydrophobic interactions. In view of this, introducing polar features into these sorbents might improve the retention of polar compounds.

> When hydrophobic HXL sorbents are used in SPE, the analytes can only be retained on the sorbent through hydrophobic interactions.

## 3.3 Hydrophylic porous polymers

Introducing polar moieties into a sorbent promotes polar interactions with the analytes, and thus enhances the retention capacity of the sorbent as well as reducing its high hydrophobicity. Hydrophilic porous polymers can be prepared by chemical modification of the hydrophobic network or by copolymerization with hydrophilic monomers.

The following sections cover these two strategies for commercially available and in-house sorbents.

> Hydrophilic porous polymers can be prepared by chemical modification of the macroporous or hypercrosslinked hydrophobic network with polar moieties or by copolymerization with hydrophilic monomers.

## 3.3.1 Functionalized polymers with polar moieties

Generally, the PS-DVB polymeric network is chemically modified by means of a Friedel–Crafts reaction. With this approach, polymeric materials with optimal morphological features can be obtained from the precursor. Fig. 3.1 shows different polar functional groups, such as acetyl [24,25], hydroxymethyl [24], benzoyl [26], o-carboxybenzoyl [27], 2-carboxy-3/4-nitrobenzoil, and 2,4-dicarboxybenzoyl [28], that have been introduced into an HXL PS-DVB network. The main limitation of this approach is the low degree of modification due to the restricted accessibility of the functional group in the crosslinked matrix. In this sense, the degree of modification for the benzoyl moiety was 60%, whereas it was only 6% for 2,4-dicarboxybenzoyl (larger moiety) due to restricted accessibility of the reactive sites [28]. However, the polymers chemically modified with polar moieties showed higher retention of polar compounds during extraction than their unmodified analogs. More recently, Amberlite XAD-4 was functionalized with chelating agents, so it could be used for the preconcentration of metal ions from aqueous solutions [29]. It is noteworthy that, in all instances, the precursor polymer has the same morphological and particle size characteristics after the functionalization.

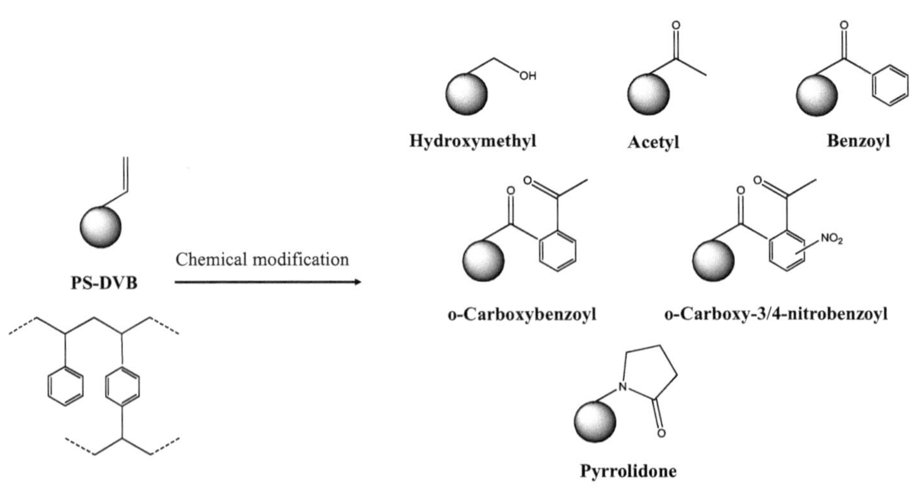


**Figure 3.1** Scheme of some chemical modifications of PS-DVB polymer.

Several commercial chemically modified sorbents, both macroporous and HXL, are available: Strata-X (800 $m^2/g$, Phenomenex), Bond Elut Plexa and Bond Elut Focus (550 $m^2/g$, Agilent Technologies), HyperSep Retain PEP (550–750 $m^2/g$, Thermo Scientific), ExtraBond PolyU and EB2 (850 and 700 $m^2/g$, respectively, Scharlau). A more exhaustive list of these sorbents and their features is provided in previous reviews [5,30].

Bond Elut PPL, Bond Elut ENV, and Strata X were compared for the extraction of trihalomethane precursors from water samples. All three sorbents were found to be suitable for onsite monitoring of trihalomethane precursors since they all successfully extracted trihalomethanes [31]. In another example, Strata X was used for the clean-up step during the extraction of antibiotics from porcine edible tissues [32]. Another study evaluated different strategies and different SPE sorbents, such as Oasis HLB and Bond Elute Nexus; Bond Elut Plexa as the clean-up sorbent for analysis of cephalosporins and quinolones in milk [33].

Some approaches have been conducted in monoliths to functionalize the hydrophobic network with polar moieties, although most of these materials act as ion-exchangers (summarized in Section 3.4). In one study [34], poly(DVB) was grafted with two polyethylenglycol dimethacrylate (pEDMA) monomers with different molar mass distributions (Mn360 and Mn950) to isolate analytes from protein-rich samples. One of the critical points during the preparation was optimizing the percentage (from 5% to 20%) of grafting monomer so that its density was balanced to prevent the proteins from interacting with the hydrophobic surface (DVB) but sufficiently thin and permeable to ensure that the analyte would interact with the inner surface. Finally, poly90%DVB-g-10%pEDMA950 showed the best performance [34].

### 3.3.2 Copolymerization with a hydrophilic monomer

This approach consists of copolymerizing a hydrophilic monomer with a crosslinking monomer. The hydrophilic monomer favors hydrophilic interactions, whereas the crosslinking monomer, usually DVB, increases the SSA and favors hydrophobic interactions. This strategy has been widely applied to both commercial and in-house SPE sorbents. In most cases, these hydrophilic copolymers are obtained by suspension polymerization, which produces macroporous materials with 500–800 $m^2/g$ SSAs and particle sizes ranging from ~50 to 200 μm.

> In hydrophilic sorbents prepared by copolymerizing a hydrophilic monomer with a crosslinker, the hydrophilic monomer favors the hydrophilic interactions, whereas the crosslinking monomer increases the SSA and favors the hydrophobic interactions.

Different hydrophilic sorbents with macroporous structures have been commercialized: Amberlite XAD-7 and XAD-8 (methacrylic acid (MAA)-DVB, 450 and 310 $m^2/g$, respectively) supplied by Rohm and Haas, and Bond Elut Nexus (no data) supplied by Agilent Technologies; Sep-pak Porapak RDX (N-vinylpyrrolidone

(NVP)-DVB, 550 $m^2/g$) and Oasis HLB (NVP-DVB, 830 $m^2/g$) supplied by Waters, or Chromabond HLB (NVP-DVB, 750 $m^2/g$) supplied by Macherey−Nagel, or Extrabond EHB (no data) from Sharlau. Discovery DPA 6S (Polyamide (PA)-DVB, no data) and SampliQ OPT (PA-DVB) are both from Agilent Technologies. Table 3.1 shows selected examples of the commercially available hydrophilic sorbents. More information is provided in previous reviews [5,30].

Oasis HLB is the most widely used commercially available sorbent, used to extract different types of compounds (i.e., pharmaceuticals, proteins, pesticides, sweeteners, high volume production chemicals, illicit drugs, etc.) from matrices, such as food, environmental waters, and biological fluids and tissues [5,30]. Moreover, it has been used not only to enrich analytes from the matrix but also eliminate matrix interferences from complex samples. In order to accommodate all types of applications, the commercially available sorbents are presented in different formats (cartridge, precolumns, 96-well plate, disks) with different bed-volumes (usually, from 30 to 500 mg). For instance, Oasis HLB in a cartridge format (150 mg) was used in the multiresidue screening of around 1500 organic pollutants from 250 mL of environmental water [40] with a preconcentration factor of 500. In contrast, a 96-well plate format (10 mg) was used for the cleanup of different growth promoters from 0.5 g of bovine meat, which was washed six times to remove the matrix interferents [41].

In addition, different comparative studies were conducted to select the best SPE sorbent. For example, a study compared the performance of 11 SPE sorbents, including polymeric (Oasis HLB, SampliQ OPT, SampliQ Polymer SCX, Strata SDB-L, and Strata-X) and silica-based (Sep-Pak C18, SampliQ $C_8$, SampliQ $C_8$/Si-SCX, SampliQ $C_{18}$, Strata $C_8$, and Strata $C_{18}$) sorbents, for the enrichment of a group of pharmaceuticals from 100 mL of surface water [42]. The recovery (%R) values for the different SPE sorbents showed that polymeric sorbents perform better and have a higher retention capacity than silica-based sorbents. Of the polymeric sorbents, Strata-X (recovery values > 95%), Strata SDB-L (%R > 86%) and Oasis HLB (%R > 99%) provided the best SPE performances. This can be attributed to the presence of DVB, which promotes $\pi$-$\pi$ interactions and polar moieties, thus enhancing the retention of the studied analytes. Therefore, when the porous material contained polar moieties, both functionalization and copolymerization had similar outcomes.

Although several hydrophilic sorbents are commercially available, several research groups have synthesized alternative copolymeric hydrophilic sorbents with different hydrophilic monomers and degrees of crosslinking. Our research group has synthesized most of the in-house hydrophilic copolymeric sorbents, but the Bagheri and Trochimczuk research groups have also been working in this area. Table 3.1 provides some examples of the in-house synthesized hydrophilic copolymeric sorbents.

A series of conductive linear polymers (without crosslinking agent) based on polyaniline (PANI) [35], poly-N-methylaniline (PNMA) [43], polydiphenylaniline (PDPA) [43], and polypyrrole (PPy) [44] have been synthesized. As they are linear polymers, their SSAs are limited (32−48 $m^2/g$), which is possibly the cause of the low recoveries achieved for the extraction of polar compounds. In another study [45], copolymers were prepared from di(methacryloyloxymethyl) naphthalene (DMN), p,p'-dihydroxydiphenylmethane diglycidil methacrylic ester (MEMDE),

**Table 3.1** Structure and properties of some sorbents obtained by copolymerization with hydrophilic monomer.

| Sorbent | Copolymeric structure | | SSA ($m^2/g$) | Supplier/reference |
|---|---|---|---|---|
| Amberlite XAD-7 | Methacrylate-divinylbenzene (MA-DVB) | 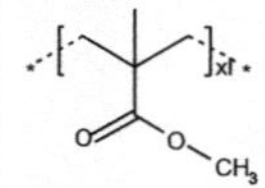 | 450 | Applied separations |
| Amberlite XAD-8 | | | 310 | |
| Bond Elut Nexus | | | 575 | Agilent Technologies |
| Oasis HLB | N-vinylpyrrolidone-divinylbenzene (NVP-DVB) | 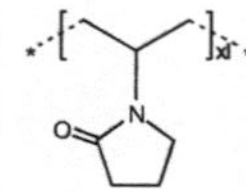 | 830 | Waters |
| Porapak RDX | | | 550 | |
| Chromabond HLB | | | 750 | Macherey–Nagel |
| SampliQ OPT | Polyamide-DVB | 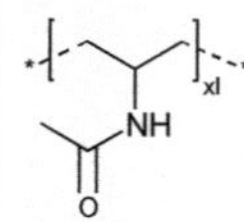 | n.d. | Agilent Technologies |
| PANI | R=H; polyaniline (PANI) | 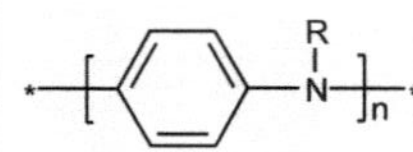 | 48 | [35,36] |
| AN-DVB | R=$H_3$; acrylonitrile (AN)-DVB | 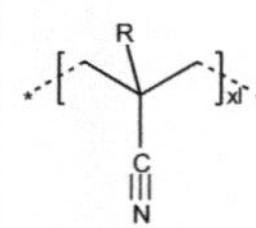 | 460 | [37] |
| MAN-DVB | R=$CH_3$; methacrylonitrile (MAN)-DVB | | 560 | [37] |
| NVIm-DVB | N-vinylimidazole-DVB | 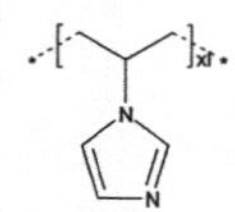 | 626 | [38] |
| MAA-EDMA | Methacrylic acid-ethylene dimethacrylate | 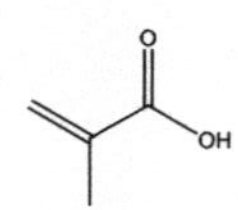 | 50 | [39] |

*n*, linear polymer; *n.d.*, no data; *xl*, crosslinked.

p,p′-dihydroxydiphenylpropane diglycidil methacrylic ester (MEDDE) and bis(maleimido)diphenylmethane (BM), all polymerized with DVB; however, although they are polar, they have low SSAs (up to 100 $m^2/g$) due to the low DVB content. More recently, 2-hydroxyethyl methacrylate (HEMA) copolymerized with the hydrophilic crosslinking agent EDMA, was synthesized and applied as a sorbent in a tip connected to a syringe to extract antibiotics from milk. However, its morphological and polar features were not described, and therefore, it cannot be compared [46].

The Trochimczuk group also prepared copolymers based on acrylonitrile (AN) [37], methacrylonitrile (MAN) [37] and cyanomethylstyrene (CMSt) [47], which are all hydrophilic monomers crosslinked with DVB. These sorbents had larger SSAs (300–700 $m^2/g$) due to a higher amount of DVB. Moreover, the authors concluded that both the hydrophilic content and the SSA are equally important for retaining polar compounds. Therefore, copolymers with a 1:1 ratio provided the best performance. A proper balance between hydrophilicity (from the polar monomer) and SSAs (from the crosslinker) was also demonstrated in studies aimed at the development of new hydrophilic copolymeric sorbents for SPE. These sorbents were based on 4-vinylpyridine-divinylbenzene (4VP-DVB) [48,49], N-vinylimidazole-divinylbenzene (NVIm-DVB) [38,50] and 4-vinylimidazole-divinylbenzene (4VIm-DVB) [51]. They were used to extract different polar compounds from environmental water samples. Table 3.2 summarizes the recovery for different sample volumes of ultrapure water spiked with phenolic compounds for different in-house prepared macroporous hydrophilic copolymer sorbents.

Apart from the ratio of the crosslinking agent, the type and proportion of the porogen also plays an important role in the pore formation, and eventually, in the SSA. Nonetheless, there are no exact rules for selecting the polymerization conditions for the monomer system, and each system should be evaluated individually. In most cases, toluene was used as the main component of the preferred porogens because DVB or other related styrenic monomers are used in the copolymerization [54]. For instance, in the preparation of 4VP-DVB resins [48], as 4VP is a styrenic monomer, the porogen system was the same as that used in the preparation of PS-DVB consisting of toluene (88%) and dibutyl phatalate (12%). However, when dibutyl phatalate was replaced by cyclohexanol (a poor solvent for PS-DVB) the SSA dropped from 115 to $<2$ $m^2/g$, and when the porogen only consisted of toluene, the SSA increased to 500 $m^2/g$. Nevertheless, for the comonomer system based on HEMA-EDMA, cyclohexanol was selected as it provided the best SSA (250 $m^2/g$) [54]. In a recent study [39], different polymers (MAA-DVB (1:2), MAA-EDMA-DVB (1:1:1), MAA-HEMA-EDMA (0.5:0:5:2), and MAA-EDMA (1:2)) were prepared by bulk polymerization in chloroform, and the polymers were evaluated for the extraction of nerve agents. The results showed that MAA-EDMA (1:2) provides the best recovery results. In a further step, the MAA-EDMA polymer was prepared in a more polar solvent (acetonitrile-ACN) to prove that the presence of a polar solvent during the synthesis promotes the exposure of the polar parts to the outer periphery of the polymer. The polymer's morphology was different when it was prepared in chloroform (SSA 100 $m^2/g$) than when it was prepared in ACN (SSA 50 $m^2/g$), and it retained polar compounds better when it was prepared in a polar solvent (ACN) [39].

**Table 3.2** Recovery values (%) obtained using different hydrophilic copolymeric resins as SPE sorbents to enrich phenol, 4-chlorophenol (4-CP) and 4-nitrophenol (4-NP) from different volumes of ultrapure water.

| Sorbents (mg) | SSA ($m^2/g$) | Recovery values (%) | | | Sample volume (mL) | References |
|---|---|---|---|---|---|---|
| | | Phenol | 4-CP | 4-NP | | |
| PANI (120) | 48 | 0 | 62 | – | 200[a] | [43] |
| PNMA (120) | 32 | 32 | 106 | – | 200[a] | [43] |
| PDPA (120) | 38 | 0 | 72 | – | 200[a] | [43] |
| PPy (35) | 40 | 84 | 94 | – | 25[b] | [52] |
| BM-DVB (200) | 35 | 85 | – | – | 50[a] | [45] |
| DMN-DVB (200) | 100 | 90 | – | – | 50[a] | [45] |
| MEMDE-DVB (200) | 70 | 22 | – | – | 50[a] | [45] |
| MEDDE-DVB (200) | 20 | 20 | – | – | 50[a] | [45] |
| 4-VP-DVB (40) | 710 | 70 | – | 85 | 100[b] | [49] |
| NVIm-DVB (40) | 626 | 88 | – | 84 | 100[b] | [38] |
| 4VIm-DVB (40) | 504 | 70 | – | 87 | 100[b] | [51] |
| HEMA-VBC-DVB (200) | 850 | 86 | – | 91 | 1000[a] | [53] |
| Oasis HLB (200) | 830 | 78 | – | 86 | 1000[a] | [53] |
| Isolute ENV+ (200) | 1100 | 92 | – | 95 | 1000[a] | [53] |
| Lichrolut EN (200) | 1200 | 79 | – | 85 | 1000[a] | [53] |

–, not included in the study.
[a]Offline SPE.
[b]Online SPE.

In addition to macroporous networks, there have been other attempts to obtain hydrophilic copolymer sorbents using HXL networks. This was first achieved by hydrolyzing the VBC isomer used as a precursor during the synthesis of the precursor (VBC-DVB) by suspension polymerization [13]. Thus, when a mixture of VBC isomers (~70% meta and ~30% para) were used as a precursor, the HXL sorbent (HXLGmix) had a higher SSA (~1900 $m^2/g$) and lower hydroxyl group content (1.5 mmol/g). However, when paraVBC was used as the precursor, the HXL sorbent had a lower SSA (~900 $m^2/g$) and a larger hydroxyl moiety content (2.4 mmol/g). These differences in morphology and chemistry of the two HXL sorbents were also reflected in the SPE performance, and higher recoveries were obtained for HXLGp. Once again, the best SPE performance is obtained when the sorbent combines a suitable SSA and a certain amount of polar moieties [19]. In fact, the commercially available Isolute ENV+ (1100 $m^2/g$, International Sorbent Technology) is claimed to be a hydroxylated HXL polymer, presumably obtained by a similar synthetic route.

> The best SPE performance is obtained when the sorbent combines a suitable SSA and a certain amount of polar moieties.

A hydrophilic HXL terpolymer based on HEMA-VBC-DVB (25/25/50, mole ratio) with an SSA of 850 $m^2/g$ was prepared [53]. This material was evaluated for the offline extraction of several polar compounds from an environmental water sample. The results showed that the presence of polar moieties together with a high SSA provided higher recoveries (>88%) than for Lichrolut EN (>79%) Oasis HLB (>66%) or Isolute ENV+ (>80%). A triphenylamine-based HXL polymer was prepared using self-condensation of triphenylamine in a typical Friedel–Crafts reaction. The polymer attained an SSA of 720 $m^2/g$ with the presence of nitrogen moieties; thus, this sorbent presented both hydrophobic and hydrophilic interactions and was evaluated for the extraction of pesticides from food samples [55].

It should be noted that most of the monoliths applied in sample preparation are based on the copolymerization of at least one, but often two, hydrophilic monomers. Table 3.3 lists some examples of the main features for monolith preparation based on hydrophilic monomers. EDMA (polar monomer) is the most commonly used crosslinker [56–59,64–67], although DVB is also used [34,68,69]. These crosslinkers are copolymerized with different polar monomers, such as glycidyl methacrylate (GMA) [59,67–69], MAA [57,65], or butylmethacrylate (BMA) [56,66], or polymers in the form of termonomers, by combining two functional monomers, such as GMA-MAA-triethylene glycol dimethacrylate (TEGDMA) [70] or GMA-S-DVB [68,69]. Moreover, the ratio of the different monomers varies (details in Table 3.3). This ratio affects not only the composition of the materials but also the pore formation.

The mixture of cyclohexanol and dodecanol is popular as a porogen for the-preparation of hydrophilic monoliths, especially those based on GMA-EDMA [9,62,67,71]. However, other binary mixtures, such as 1-propanolol and 1,4-butanediol [56,60,62,64,72] or single porogens 1-propanolol [61] or n-hexanolol [73] have been used. Monoliths have also been prepared using polyethylene glycol (PEG) with a different molecular weight in combination with alcohols [58,59,65,66]. In addition, the higher the PEG molecular weight, the larger the pore size [9]. As stated before, the choice and the proportion of these porogens affect the morphology of the monolith. For example, different porogens were selected to prepare alkyl methacrylate-based monoliths by copolymerization of EDMA (crosslinker) and GMA, BMA, or LMA to improve the final porosity. Cyclohexanol and 1-dodecanol were selected to prepare a GMA-EDMA monolith; however, this monolith did not have large SSAs (due to the lack of mesopores and micropores), and thus, had a low loading capacity. The BMA-EDMA and LMA-EDMA monoliths were prepared using 1-propanolol and 1,4-butanediol as porogens, providing higher porosity [62].

Monolithic polymers were prepared in different molds and used accordingly in different SPE formats, including columns [58–60,63,65–67,71,73] that were later connected online to LC, syringes or cartridges [18,34,57,61,62,64], or pipette tips [56,68], among others. Table 3.3 summarizes the information on the monolith format.

**Table 3.3** Structure and properties of some "in-house" hydrophilic and mixed-mode ion-exchange monolithic materials used in SPE.

| Material | Copolymer structure | Porogen | SSA ($m^2/g$) | Format | References |
|---|---|---|---|---|---|
| **Hydrophilic monoliths** | | | | | |
| VBC-S-DVB | Different % | Toluene 18%<br>1-Dodecanol 42% | 340–817 | Bulk<br>PE tubing<br>PP cartridges | [18] |
| BMA-EDMA | BMA 30%<br>EDMA 70% | 1-Propanolol 65%<br>1,4-Butanediaol 25%<br>Water 10% | <100 | Pipette tip | [56] |
| MAA-EDMA | MAA 12%<br>EDMA 88% | Toluene 12%<br>Dodecanol 88% | – | Syringe | [57] |
| PEDMA | PEDMA | 1-Propanolol 54%<br>PEG-400 46% | 135 | 10 mm × 2.1 mm precolum | [58] |
| Poly(IL-GMA-EGDMA) | poly(1-vinyl-3-Butylimidazolium 14%<br>GMA32%<br>EDMA 54% | 1,2-Propanediol 60%<br>PEG200 40% | 19 | 50 mm × 4.6 mm precolumn | [59] |
| **Mixed-mode ion-exchange monoliths** | | | | | |
| GMA-EDMA-IDA | GMA 75%<br>EDMA 25%<br>Iminodiacetic acid | 1-Propanolol 65%<br>1,4-Butanediaol 35% | – | 50 mm × 2.1 mm precolumn | [60] |
| GMA-EDMA-EDA | GMA 75%<br>EDMA 25% dimethacrylate<br>Ethylenediamine | 1-Propanolol 65%<br>1,4-Butanediaol 35% | – | 50 mm × 2.1 mm precolumn | [60] |

*Continued*

**Table 3.3** Structure and properties of some "in-house" hydrophilic and mixed-mode ion-exchange monolithic materials used in SPE.—cont'd

| Material | Copolymer structure | Porogen | SSA ($m^2/g$) | Format | References |
|---|---|---|---|---|---|
| GMA-EDMA-COOH | GMA 75%<br>EDMA 25% dimethacrylate<br>Chloroacetate | 1-Propanolol 65%<br>1,4-Butanediaol 35% | – | 50 mm × 2.1 mm precolumn | [60] |
| GMA-EDMA-$SO_3$ | GMA 75%<br>EDMA 25% sulfonic | 1-Propanolol 65%<br>1,4-Butanediaol 35% | – | 50 mm × 2.1 mm precolumn | [60] |
| Poly(MAA-VPBA-EGMDA) | MAA 8%,<br>VPBA 8%<br>EDMA 84% | 1-Propanolol | – | Plastic syringe | [61] |
| Poly(GMA-EDMA-EDA)<br>Poly(BMA-EDMA-EDA)<br>Poly(LMA-EDMA-EDA) | GMA or BMA or LMA 60%<br>EDMA 60%<br>Ethylenediamine (EDA) | Cyclohexanol 33%<br>1-Dodecanol 66%<br>1-Propanolol 56%<br>1,4-Butanediol 44%<br>1-Propanolol 25%<br>1,4-Butanediol 75% | – | PP syringe | [62] |
| AETAC-EDMA<br>Anion-exchange | 1.7 wt of the total polyme mix 2-(acryloyloxy) ethyl] trimethylammonium chloride-*co*-EDMA | 1-Propanolol 65%<br>PEG-400 33%<br>Water 2% | – | 10 mm × 2.1 mm precolumn | [63] |

It should be noted that, to the best of our knowledge, no porous monolith is commercially available in any of these formats.

Other approaches have emerged to enhance the retention of porous polymeric sorbents. Hybrid magnetic nanoparticles (MNPs) were developed as a novel material for SPE [74,75]. Due to size restrictions, however, these polymers are mainly used in other SPE formats, such as μSPE or dispersive SPE. Monolithic polymers with various types of nanoparticles, such as gold, silver, graphene, fullerene or metal-organic frameworks, were recently reviewed in Ref. [76] and further information can be found in other chapters of this handbook. As an illustrative example, a suspension of multi-walled carbon nanotubes (MWCNTs) was admixed to a prepolymerization mixture of MAA-EDMA, and the mixture used to fill a 100 mm × 2.1 mm i.d. stainless steel tube, which was used for the online enrichment of proteins. This admixed approach achieved a homogeneous distribution of the MWCNTs throughout the column and provided an increased SSA (86 $m^2/g$) compared to the SSA obtained when conventional in situ polymerization was used (14 $m^2/g$) [65]. Other examples in monolithic form include functionalization with ionic liquids (ILs) into a GMA-EDMA polymer [59,67], incorporation of a porous organic cage (CC3) into a poly(EDMA) network [64], and grafting of a poly(MAA) onto a GMA-TEGDMA polymer that is later functionalized with $Fe^{3+}$ p-phthalic acid [70]. Furthermore, the Herrero-Martínez group recently prepared a variety of monolithic polymethacrylate-based polymers modified with metal nanoparticles [77,78], magnetic nanoparticles [79] or oxidized single-walled carbon nanohorns [80], which were used as SPE sorbents. However, in some approaches, the monolithic rod was ground and packed in SPE cartridges, and in others, the hybrid materials were used in a spin column [80] or inside a pipette to perform μSPE [81].

## 3.4 Mixed-mode ion-exchange polymers

Mixed-mode ion-exchange polymeric sorbents are prepared by introducing ionic functional groups into the polymeric backbone. They can be obtained by functionalizing the polymeric backbone with ion-exchange moieties or through a monomer that already contains this ion-exchange group. These sorbents are classified into four groups according to the ionic functional group attached and are cation or anion exchangers with strong and weak features. Strong cation-exchangers (SCX) are usually functionalized with sulfonic acid, while weak cation-exchangers (WCX) are functionalized with carboxylic acid. Strong anion-exchangers (SAX) usually have quaternary amine groups, whereas ternary, secondary and primary amines are used for weak anion-exchangers (WAX). Thus, the main difference between strong and weak exchangers is the chargeability of the materials. Strong ion-exchange materials remain charged regardless of the pH of the media, whereas weak ion-exchange materials may be fully or partially charged or neutral depending on the pH. Thus, these mixed-mode sorbents have ion-exchange interactions (specific) for ionizable compounds and reversed-phase interactions (nonspecific) with neutral compounds [82].

The benefit of these specific interactions is that in the course of the SPE protocol, these interactions are not disrupted unless there is a change in pH. This allows the use of solvents of high elution strength for neutral compounds to be used as a sample clean-up step.

> Mixed-mode ion-exchange polymeric sorbents have ion-exchange interactions (specific) with ionizable compounds and reversed-phase interactions (nonspecific) with neutral compounds.

There are several commercially available mixed-mode ion-exchange sorbents, and most of them have macroporous networks. To the best of our knowledge, all of them are based on previously commercialized polymeric sorbents modified with an ion-exchange group. In particular, Oasis MCX is prepared by sulfonation of the aromatic rings of Oasis HLB, whereas Oasis WCX is obtained from an intermediate chlorinated resin subsequently oxidized to carboxylic groups. Oasis MAX and Oasis WAX were also obtained from the chlorinated intermediate reacted with dimethylbutylamine (DMBA) and piperazine, respectively [83]. Likewise, other mixed-mode polymeric sorbents have been produced using an undisclosed synthetic approach from Strata-X (Phenomenex), Bond elut Plexa (Agilent), AttractSPE (Polyintell) and Chromabond HR (Macherey-Naagel). Fig. 3.2 lists the mixed-mode ion-exchange polymer sorbents classified into four groups. Fig. 3.3 gives examples of the network structure as well as the different ionic moieties used to attain SCX, SAX, WCX, and WAX. More details on their features and ionic moieties are compiled in reviews [82,84].

There are numerous examples in which these commercial mixed-mode sorbents have been used to exploit the selectively toward ionizable compounds and to minimize interferences from complex matrices such as biological fluid, foodstuff, or sewage water. A selection of these studies can be found in recent reviews [82,84,85]. As an example, Oasis MAX was used to selectively and simultaneously extract a group of organophosphorus compounds (acidic behavior) from environmental water samples. It is an alternative to Oasis HLB but is unable to extract all of the compounds in a single step [86]. In a further step, two mixed-mode cartridges (Strata-X-AW followed by Strata-X-C) were combined to obtain a suitable extraction method that covers the urinary metabolome [87].

In-house mixed-mode ion-exchange sorbents have also been prepared based on both macroporous and HXL polymers. Fig. 3.2 provides some examples of these in-house mixed-mode ion exchange sorbents. The macroporous NVIm-DVB synthesized by suspension polymerization [88] was applied as SAX. A copolymer based on 2-acrylamido-2-methylpropanesulphonic acid (AMPSA) (contains sulfonic acid moieties), HEMA and pentaerythritol triacrylate (PETRA), and a HEMA-DVB copolymer modified with $H_2SO_4$ were prepared and used as SCX mixed-mode sorbents [89]. They were all benchmarked for the offline SPE of drugs from environmental samples. VBC-EDMA was functionalized with trimethylamine (TEA), imidazole, piperidine and pyrrolidine to obtain four mixed-mode sorbents with anion exchange capabilities [90]. 2-(diethylamino)ethyl methacrylate (DEAEMA) (a ternary amine)

**SCX**

- Oasis MCX (Waters)
- Strata X-C (Phenomenex)
- BondElut Plexa PCX (Agilent Technologies)
- Evolute CX (Biotage)
- Cleanert PCX (Bonna-Agela Technologies)
- AttractSPE SCX (Polyintell)
- Extrabond ECX (Scharlau)
- CHROMABOND HR-XC (Macherey-Nagel)
- HyperSep SCX (Thermo Scientific)
- HXLPP-SCX (In-house)
- AMPSA/HEMA/PETRA (In-house)
- GMA-DVB-$SO_3$

**SAX**

- Oasis MAX (Waters)
- Strata X-A (Phenomenex)
- BondElut Plexa PAX (Agilent Technologies)
- Evolute AX (Biotage)
- Cleanert PAX (Bonna-Agela Technologies)
- AttractSPE SAX (Polyintell)
- Extrabond EAX (Scharlau)
- CHROMABOND HR-XA (Macherey-Nagel)
- HyperSep SAX (Thermo Scientific)
- HXLPP-SAX (In-house)
- Supported Ionic Liquids (In-house)
- DEAEMA-DVB-TEA
- VBC-EDMA-TEA

**WCX**

- Oasis WCX (Waters)
- Strata X-WC (Phenomenex)
- Abselut NEXUS WCX (Agilent Technologies)
- Not available (J.T. Baker)
- Evolute WCX (Biotage)
- Cleanert PWCX (Bonna-Agela Technologies)
- AttractSPE WCX (Polyintell)
- CHROMABOND HR-XCW (Macherey-Nagel)
- Styrene screen CCX (UTC)
- HXLPP-WCX (In-house)

**WAX**

- Oasis WAX (Waters)
- Strata X-AW (Phenomenex)
- Not available (Agilent Technologies)
- Speedisk H2O Phobic WA-DVB (J.T. Baker)
- Evolute WAX (Biotage)
- Cleanert PWAX (Bonna-Agela Technologies)
- AttractSPE SAX (Polyintell)
- CHROMABOND HR-WAX (Macherey-Nagel)
- Styrene screen THC (UTC)
- HXLPP-WAX (In-house)
- VBC-EDMA-pyrrolidone/piperidine

**Figure 3.2** Classification of the mixed-mode ion-exchange polymeric sorbents commercially available and in-house prepared.

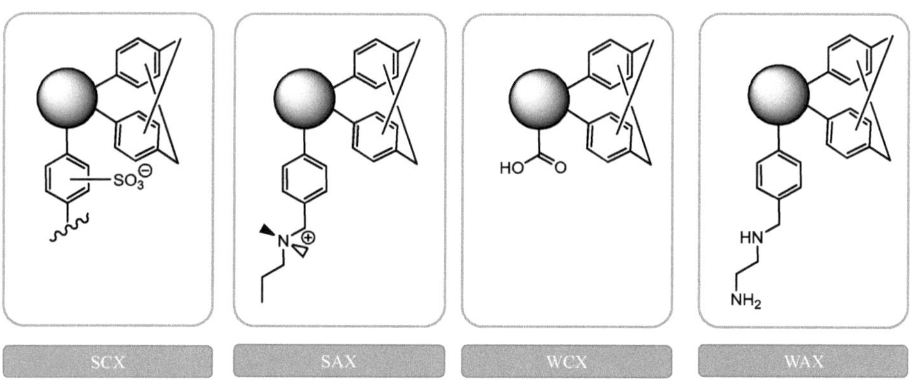


**Figure 3.3** Examples of the structure of each type of mixed-mode ion-exchange polymer.

was first copolymerized with DVB (DEAEMA-DVB) and then functionalized with diglycidyl ether derivatives followed by quaternization with TEA [91] or N,N-dimethylethanolamine (DMEA) [92] to produce two mixed-mode SAX sorbents. Both SAX sorbents were used to extract pharmaceuticals from wastewater [91] or urine samples [92], with results comparable to those for Oasis MAX [91,92]. GMA (which has an epoxy ring whose reactivity is exploited to produce functionalized materials) previously copolymerized with DVB was also sulfonated to obtain an SCX sorbent used to extract alkylate-purine adducts in human urine [93]. PS-DVB grafted with GMA was also functionalized with ethylenediamine (EDA) moieties and then used as a chelating sorbent for the adsorption of $Cu^{2+}$ [94].

Our research group is a pioneer in functionalizing HXL resins prepared from VBC-DVB precursors via precipitation polymerization (HXLPP). This introduces moieties through the chloromethyl moiety with a different ion-exchange capacity to provide SCX, SAX, WCX, and WAX mixed-mode polymers. HXLPP was modified with EDA and piperazine groups to produce HXLPP-WAX sorbents. Fig. 3.4A shows the synthetic approach for preparing HXLPP-WAX-EDA. These HXLPP-WAX sorbents outperformed the commercial sorbents Oasis WAX and Strata-X-AW for the online extraction of pharmaceuticals from environmental water samples [95]. These results were clearly due to the enhanced morphological features of the HXL resin (SSA $\sim$ 1000 $m^2/g$) compared to the commercial resins, which have a macroporous structure (SSA $\sim$ 800 $m^2/g$). For the synthesis of HXLPP-SCX, the sulfonic acid moieties were introduced with acetyl sulfonate and lauroyl sulfate. The results showed that the HXLPP-SCX prepared with 50% lauroyl sulfate had the highest ion-exchange capacity (IEC) (2.5 meq/g) and the largest SSA (1370 $m^2/g$), and had the best features as an SPE sorbent [96]. In contrast, prefunctionalization was used for HXLPP-SAX since the quaternary amine (DBMA) was introduced before the hypercrosslinking process. This was because amination was less efficient after hypercrosslinking due to restricted access of the DBMA group to react with the free chloromethyl groups. Thus, the VBC-DVB (PP) precursor was functionalized (PP-SAX) and then hypercrosslinked (HXLPP-SAX) [97]. Nonetheless, recently in the preparation of HXLPP-SAX functionalized with trimethylamine (TMA), the quaternization and the hypercrosslinking reaction were completed in a single step [98]. In a different synthetic route, for the HXLPP-WCX sorbent, the carboxylic moiety was introduced via the MAA in the terpolymer MAA-VBC-DVB, which was used as a precursor for the following hypercrosslinking reaction [99].

Other approaches for obtaining ionically modified materials have also been presented. For instance, VBC-DVB copolymers were functionalized with ILs based on N-methylimidazole [100,101] or N-butylimidazole [102] and used as SAX materials. MWCNT was functionalized with poly(diallyaldimethylammonium chloride) to obtain another SAX material [103]. It should be noted that these materials do not have morphological features similar to the above-mentioned porous polymers and so their retention mechanisms are limited.

As most of the monoliths prepared for sample preparation are based on GMA. There are several examples where the GMA is functionalized with EDA, which has WAX properties. GMA can be part of the comonomeric mixture [60,62,71] or grafted on the polymer so that it also contains the epoxy group [62]. Fig. 3.4B shows a scheme for the postfunctionalization of a GMA-based monolith with EDA. Ribeiro et al. [60] functionalized GMA-EDMA with ionic moieties that included EDA (WAX properties), iminodiacetic acid (WCX properties), and sulfonic acid (SCX properties). These

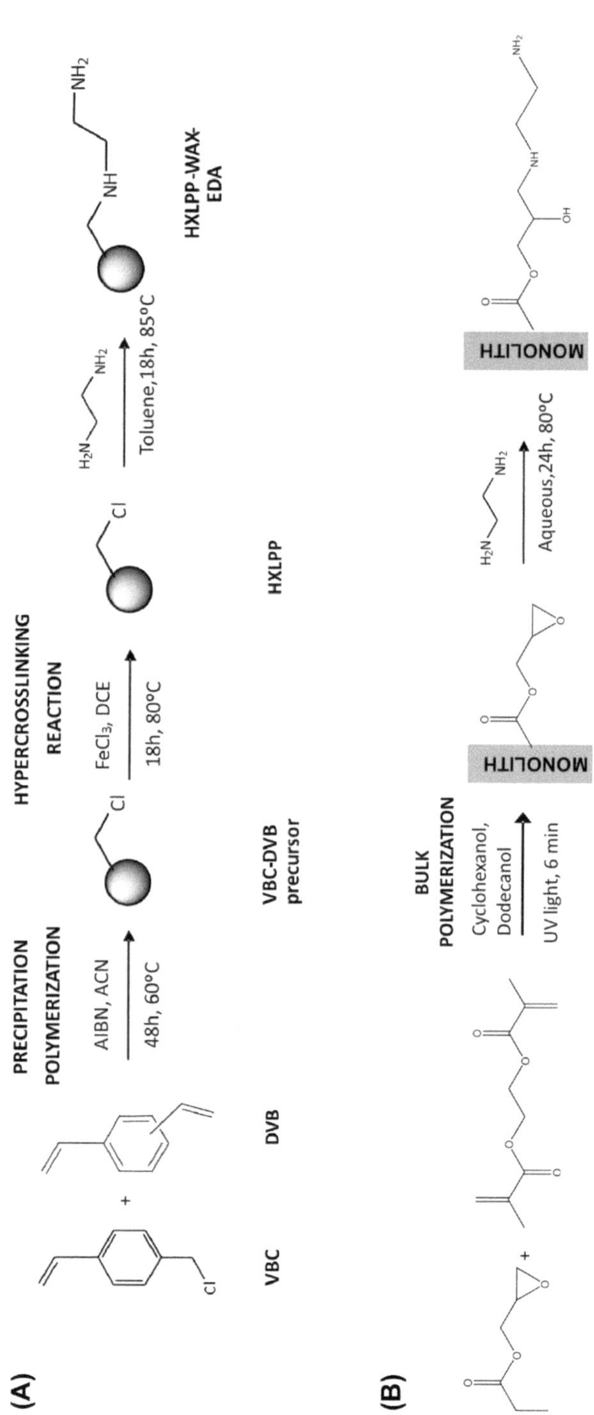


**Figure 3.4** Synthetic approach to prepare in-house mixed-mode WAX polymeric material in form of particles for SPE (A) and monolithic form (B).

monoliths were used for online SPE, followed by spectrophotometric detection to determine $Cu^{2+}$. Other strategies for introducing ion-exchange moieties into monoliths have also been presented. For example, by copolymerization with monomers that contain the ion-exchange moiety, such as MAA, which provides carboxyl acid moieties for WCX [61,71], or acryloyloxy ethyltrimethylammonium, which contains quaternary amine moieties for SAX [63]. In these cases, the ratio between ion-exchange moieties and morphological properties should be properly balanced. Table 3.3 provides examples of functionalized monoliths that have been used as mixed-mode ion-exchange sorbents.

## 3.5 Conclusions

Porous polymers and their application in SPE are great achievements in sample preparation. This is demonstrated by the long list of commercially available and in-house prepared sorbents based on porous polymers. Significant developments include hypercrosslinked networks, hydrophilicity, and ion-exchange moieties to enhance the morphology and chemistry of the porous polymer sorbents, and thus, improve their retention capacity and performance in SPE. Porous polymers will continue to be used in SPE, and other improvements will emerge in the future to fine-tune their properties.

## Acknowledgements

The authors would like to thank the *Ministerio de Ciencia, Innovacion y Universidades* and the European Regional Development Fund (ERDF) (Project: CTQ2017-88548-P) for the financial support given.

## References

[1] Maya F, Palomino Cabello C, Ghani M, Turnes Palomino G, Cerdà V. Emerging materials for sample preparation. J Sep Sci 2018;41:262–87. https://doi.org/10.1002/jssc.201700836.

[2] Sherrington DC. Preparation, structure and morphology of polymer supports. Chem Commun 1998:2275–86.

[3] Gokmen MT, Du Prez FE. Porous polymer particles—a comprehensive guide to synthesis, characterization, functionalization and applications. Prog Polym Sci 2012;37:365–405. https://doi.org/10.1016/j.progpolymsci.2011.07.006.

[4] Deleuzel H, Schultze X, Sherrington DC. Porosity analysis of some poly(styrene/divinylbenzene)beads by nitrogen sorption and mercury intrusion porosimetry. Polym Bull 2000;44:179–89. https://doi.org/10.1007/s002890050590.

[5] Fontanals N, Marcé RM, Borrull F. Trends in polymeric sorbents for solid-phase extraction. In: Taylor JC, editor. Adv. Chem. Res., vol. 4. New York: Nova Publisher; 2010. p. 218−57.

[6] Postigo C, López de Alda MJ, Barceló D, Ginebreda A, Garrido T, Fraile J. Analysis and occurrence of selected medium to highly polar pesticides in groundwater of Catalonia (NE Spain): An approach based on on-line solid phase extraction−liquid chromatography−electrospray-tandem mass spectrometry detection. J. Hydrol. 2010; 383:83−92. https://doi.org/10.1016/j.jhydrol.2009.07.036.

[7] Xie S, Svec F, Fréchet JM. Porous polymer monoliths: preparation of sorbent materials with high-surface areas & controlled surface chemistry for high-throughput, online SPE of polar org compounds. Chem Mater 1998;10:4072−8.

[8] Svec F, Fréchet JMJ. Kinetic control of pore formation in macroporous polymers. Formation of "molded" porous materials highflow characteristics for separation or catalysis. Chem Mater 1995;7:707−15.

[9] Svec F. Porous polymer monoliths: amazingly wide variety of techniques enabling their preparation. J Chromatogr A 2010;1217:902−24. https://doi.org/10.1016/j.chroma.2009.09.073.

[10] Davankov V, Rogozhin V, Tsjurupa M. Macronet polystyrene structures for ionites and method of producing same. US Pat, 3729457. 1969.

[11] Veverka P, Jerábek K. Mechanism of hypercrosslinking of choromethylated St-DVB copolymers. React Funct Polym 1999;41:21−5.

[12] Ahn J-H, Jang J-E, Oh C-G, Ihm S-K, Cortés J, Sherrington DC. Rapid generation and control of microporosity, bimodal pore size distribution, and surface area in Davankov-type hyper-cross-linked resins. Macromolecules 2006;39:627−32.

[13] Fontanals N, Cortés J, Galià M, Cormack PAG, Marcé RM, Borrull F, et al. Synthesis of Davankov-type hypercrosslinked resins using different isomer compositions of vinylbenzyl chloride monomer, and application in the solid-phase extraction of polar compounds. J Polym Sci Part A Polym Chem 2005;43:1718−28.

[14] Fontanals N, Manesiotis P, Sherrington DC, Cormack PAG. Synthesis of spherical ultra-high surface area monodisperse amphipathic polymer sponges in the low micron range. Adv Mater 2008;20:1298−302.

[15] Urban J, Svec F, Fréchet JMJ. Hypercrosslinking: new approach to porous polymer monolithic capillary columns with large surface area for the highly efficient separation of small molecules. J Chromatogr A 2010;1217:8212−21.

[16] Fontanals N, Marcé RM, Borrull F, Cormack PAG. Hypercrosslinked materials: preparation, characterisation and applications. Polym Chem 2015;6:7231−44. https://doi.org/10.1039/C5PY00771B.

[17] Xu S, Luo Y, Tan B. Recent development of hypercrosslinked microporous organic polymers. Macromol Rapid Commun 2013;34:471−84. https://doi.org/10.1002/marc.201200788.

[18] Candish E, Wirth H-J, Gooley AA, Shellie RA, Hilder EF. Characterization of large surface area polymer monoliths and their utility for rapid, selective solid phase extraction for improved sample clean up. J Chromatogr A 2015;1410:9−18. https://doi.org/10.1016/j.chroma.2015.07.065.

[19] Fontanals N, Galià M, Cormack PAG, Marcé RM, Sherrington DC, Borrull F. Evaluation of a new hypercrosslinked polymer as a sorbent for solid-phase extraction of polar compounds. J Chromatogr A 2005;1075:51−6.

[20] Tran NH, Hu J, Ong SL. Simultaneous determination of PPCPs, EDCs, and artificial sweeteners in environmental water samples using a single-step SPE coupled with HPLC−MS/MS and isotope dilution. Talanta 2013;113:82−92. https://doi.org/10.1016/j.talanta.2013.03.072.
[21] Salemi A, Vosough M, Mazlumi E. Determination of uv-filter compounds in highly contaminated river water by hplc-dad: a comparative study on extraction efficiency and applicability of different solid phase extraction sorbents. J Liq Chromatogr Relat Technol 2012;35:2113−24. https://doi.org/10.1080/10826076.2011.627752.
[22] Scheurer M, Brauch H-J, Lange FT. Analysis and occurrence of seven artificial sweeteners in German waste water and surface water and in soil aquifer treatment (SAT). Anal Bioanal Chem 2009;394:1585−94. https://doi.org/10.1007/s00216-009-2881-y.
[23] Azzouz A, Rascón AJ, Ballesteros E. Simultaneous determination of parabens, alkylphenols, phenylphenols, bisphenol A and triclosan in human urine, blood and breast milk by continuous solid-phase extraction and gas chromatography−mass spectrometry. J Pharm Biomed Anal 2016;119:16−26. https://doi.org/10.1016/j.jpba.2015.11.024.
[24] Sun JJ, Fritz JS. Chemically modified resins for solid-phase extraction. J Chromatogr A 1992;590:197−202. https://doi.org/10.1016/0021-9673(92)85382-4.
[25] Masqué N, Galià M, Marcé RM, Borrull F. Chemically modified polymeric resin used as sorbent in a solid-phase extraction process to determine phenolic compounds in water. J Chromatogr A 1997;771:55−61. https://doi.org/10.1016/s0021-9673(97)00125-8.
[26] Masqué N, Galià M, Marcé RM, Borrull F. Solid-phase extraction of phenols and pesticides in water with a modified polymeric resin. Analyst 1997;122:425−8.
[27] Masqué N, Galià M, Marcé RM, Borrull F. New chemically modified polymeric resin for solid-phase extraction of pesticides and phenolic compounds from water. J Chromatogr A 1998;803:147−55.
[28] Masqué N, Galià M, Marcé RM, Borrull F. Influence of chemical modification of polymeric resinon retention of polar compounds in solid-phase extraction. Chromatographia 1999;50:21−6.
[29] Karadaş C, Turhan O, Kara D. Synthesis and application of a new functionalized resin for use in an on-line, solid phase extraction system for the determination of trace elements in waters and reference cereal materials by flame atomic absorption spectrometry. Food Chem 2013;141:655−61. https://doi.org/10.1016/j.foodchem.2013.03.042.
[30] Gilart N, Borrull F, Fontanals N, Marcé RM. Selective materials for solid-phase extraction in environmental analysis. Trends Environ Anal Chem 2014;1:e8−18.
[31] Goss CD, Wiens R, Gorczyca B, Gough KM. Comparison of three solid phase extraction sorbents for the isolation of THM precursors from manitoban surface waters. Chemosphere 2017;168:917−24. https://doi.org/10.1016/j.chemosphere.2016.10.118.
[32] Xie S, Pan Y, Chen D, Qu W, Sattar A, Cheng G, et al. Quantitative analysis of bacitracin in porcine edible tissues by high-performance liquid chromatography–electrospray ionization tandem mass spectrometry and its application to residue depletion study. Food Anal Methods 2017;10:539−48. https://doi.org/10.1007/s12161-016-0613-4.
[33] Karageorgou E, Myridakis A, Stephanou EG, Samanidou V. Multiresidue LC−MS/MS analysis of cephalosporins and quinolones in milk following ultrasound-assisted matrix solid-phase dispersive extraction combined with the quick, easy, cheap, effective, rugged, and safe methodology. J Sep Sci 2013;36:2020−7. https://doi.org/10.1002/jssc.201300194.

[34] Candish E, Khodabandeh A, Gaborieau M, Rodemann T, Shellie RA, Gooley AA, et al. Poly(ethylene glycol) functionalization of monolithic poly(divinyl benzene) for improved miniaturized solid phase extraction of protein-rich samples. Anal Bioanal Chem 2017; 409:2189–99. https://doi.org/10.1007/s00216-016-0164-y.
[35] Bagheri H, Saraji M. New polymeric sorbent for the solid-phase extraction of chorophenols from water samples followed by gas chromatography-electron capture detection. J Chromatogr A 2001;910:87–93.
[36] Bagheri H, Saraji M, Barceló D. Evaluation of polyaniline asa sorbent for SPE of a variety of polar pesticides from water followed by CD-MEKC-DAD. Chromatographia 2004;59: 283–9.
[37] Trochimczuk AW, Streat M, Korlarz B. Highly polar polymeric sorbents. Characterization & sorptive properties towards phenol and its derivates. React Funct Polym 2001;46: 259–71.
[38] Fontanals N, Marcé RM, Galià M, Borrull F. Synthesis of hydrophilic sorbents from N-vinylimidazole-divinylbenzene and the evaluation of their sorption properties in the solid-phase extraction of polar compounds. J Polym Sci Part A Polym Chem 2004;42: 2019–25.
[39] Roy KS, Purohit AK, Chandra B, Goud DR, Pardasani D, Dubey DK. Polymeric sorbent with controlled surface polarity: an alternate for solid-phase extraction of nerve agents and their markers from organic matrix. Anal Chem 2018;90:7025–32. https://doi.org/10.1021/acs.analchem.8b01428.
[40] Pitarch E, Cervera MI, Portolés T, Ibáñez M, Barreda M, Renau-Pruñonosa A, et al. Comprehensive monitoring of organic micro-pollutants in surface and groundwater in the surrounding of a solid-waste treatment plant of Castellón, Spain. Sci Total Environ 2016; 548–549:211–20. https://doi.org/10.1016/j.scitotenv.2015.12.166.
[41] van Tricht F, Essers M, Groot M, Sterk S, Blokland M, van Ginkel L. A Fast quantitative multi-analyte method for growth promoters in bovine meat using bead-disruption, 96-well spe clean-up and narrow-bore UHPLC-MS/MS analysis. Food Anal Methods 2018;11:2206–17. https://doi.org/10.1007/s12161-018-1164-7.
[42] Pavlović DM, Ašperger D, Tolić D, Babić S. Development and optimization of the determination of pharmaceuticals in water samples by SPE and HPLC with diode-array detection. J Sep Sci 2013;36:3042–9. https://doi.org/10.1002/jssc.201300486.
[43] Bagheri H, Saraji M. Conductive polymers as new media for solid-phase extraction: isolation of chrolophenols from water sample. J Chromatogr A 2003;986:111–9.
[44] Bagheri H, Mohammadi A. Pyrrole-based conductive polymer as solid-phase extraction medium for the preconcentration of environmental pollutants in water samples followed by gas chromatography with flame ionization and mass spectrometry detection. J Chromatogr A 2003;1015:23–30.
[45] Bielicka-Daszkiewicz K, Voelkel A, Szejner M, Osypiuk J. Extraction properties of new polymeric sorbents in SPE/GC analysis of phenol and hydroquinone from water samples. Chemosphere 2006;62:890–8. https://doi.org/10.1016/j.chemosphere.2005.05.032.
[46] Gao M, Yan H, Sun N. Water-compatible poly (hydroxyethyl methacrylate) polymer sorbent for miniaturized syringe assisted extraction of sulfonamides in milk. Anal Chim Acta 2013;800:43–9. https://doi.org/10.1016/j.aca.2013.09.026.
[47] Drechny D, Trochimczuk AW. Synthesis and some sorptive properties of highly cross-linked cyanomethyl styrene/divinylbenzene copolymers. React Funct Polym 2006;66: 323–33.

[48] Fontanals N, Marcé RM, Galià M, Borrull F. Preparation and characterization of highly polar polymeric sorbents from styrene-divinylbenzene and vinylpyridine-divinylbenzene for the solid-phase extraction of polar organic pollutants. J Polym Sci Part A Polym Chem 2003;41:1927–33.

[49] Fontanals N, Puig P, Galià M, Marcé RM, Borrull F. New hydrophilic polymeric resin based on 4-vinylpyridine-divinylbenzene for solid-phase extraction of polar compounds from water. J Chromatogr A 2004;1035:281–4.

[50] Fontanals N, Galià M, Marcé RM, Borrull F. Solid-phase extraction of polar compounds with hydrophilic copolymer sorbent. J Chromatogr A 2004;1030:63–8.

[51] Fontanals N, Galià M, Marcé RM, Borrull F. Comparison of hydrophilic polymeric sorbents for on-line solid-phase extraction of polar compounds from aqueous samples. Chromatographia 2004;60:511–5. https://doi.org/10.1365/s10337-004-0419-1.

[52] Bagheri H, Mohammadi A, Salemi A. On-line trace enrichment of phenolic compounds from water using a pyrrole-based polymer as solid-phase extraction sorbent coupled with high-performance liquid chromatography. Anal Chim Acta 2004;513:445–9.

[53] Bratkowska D, Fontanals N, Cormack PAG, Sherrington DC, Borrull F, Marcé RM. Hydrophilic hypercrosslinked sorbents for the solid-phase extraction of polar contaminants from water. J Chromatogr A 2010;1217:3238–43.

[54] Mohamed MH, Wilson LD. Porous copolymer resins: tuning pore structure and surface area with non reactive porogens. Nanomaterials (Basel Switz) 2012;2:163–86. https://doi.org/10.3390/nano2020163.

[55] Wu J, Ma R, Hao L, Wang C, Wu Q, Wang Z. Triphenylamine-based hypercrosslinked organic polymer as adsorbent for the extraction of phenylurea herbicides. J Chromatogr A 2017;1520:48–57. https://doi.org/10.1016/j.chroma.2017.09.012.

[56] Altun Z, Blomberg LG, Jagerdeo E, Abdel-Rehim M. Drug screening using microextraction in a packed syringe (MEPS)/mass spectrometry utilizing monolithic-, polymer-, and silica-based sorbents. J Liq Chromatogr Relat Technol 2006;29:829–40.

[57] Zhang M, Wei F, Zhang Y-F, Nie J, Feng Y-Q. Novel polymer monolith microextraction using a poly(methacrylicacid-ethylene glycol dimethacrylate) monolith and its application tosimultaneous analysis of several angiotensin II receptor antagonists inhuman urine by capillary zone electrophoresis. J Chromatogr A 2006;1102:294–301.

[58] Li H, Ai L, Fan S, Wang Y, Sun D. Rapid determination of 18 glucocorticoids in serum using reusable on-line SPE polymeric monolithic column coupled with LC-quadrupole/orbitrap high-resolution mass spectrometer. J Chromatogr B 2017;1065–1066:79–86. https://doi.org/10.1016/j.jchromb.2017.09.025.

[59] Liu S, Wang C, He S, Bai L, Liu H. On-line SPE using ionic liquid-based monolithic column for the determination of antihypertensives in human plasma. Chromatographia 2016;79:441–9. https://doi.org/10.1007/s10337-016-3045-9.

[60] Ribeiro LF, Lopes Martins R, de Souza Costa DM, Masini JC. Poly glycidyl methacrylate-co-ethylene dimethacrylate porous monolith as a versatile platform for the development of separations and solid-phase extractions in sequential injection analyzers. J Sep Sci 2018;41:4449–57. https://doi.org/10.1002/jssc.201800995.

[61] Chen Y, Meng J, Zou J, An J. Selective extraction based on poly(MAA-VB-EGMDA) monolith followed by HPLC for determination of hordenine in plasma and urine samples. Biomed Chromatogr 2015;29:869–75. https://doi.org/10.1002/bmc.3367.

[62] Wang H, Zhang H, Lv Y, Svec F, Tan T. Polymer monoliths with chelating functionalities for solid phase extraction of metal ions from water. J Chromatogr A 2014;1343:128–34. https://doi.org/10.1016/j.chroma.2014.03.072.

[63] Wang M-M, Li N, Ai L-F, Li L, Chen X-L, Jia Y-Q, et al. Poly[(2-(acryloyloxy) ethyl] trimethylammonium chloride-co-ethylene dimethacrylate monolith on-line solid-phase extraction coupled with liquid chromatography and tandem mass spectrometry for the fast determination of salicylic acid in foodstuffs. J Sep Sci 2018;41:3432−40. https://doi.org/10.1002/jssc.201800464.

[64] Wang H, Tan J, Shang X, Zheng X, Liu X, Wang J, et al. Porous organic cage incorporated monoliths for solid-phase extraction coupled with liquid chromatography-mass spectrometry for identification of ecdysteroids from Chenopodium quinoa Willd. J Chromatogr A 2019;1583:55−62. https://doi.org/10.1016/j.chroma.2018.11.019.

[65] Zhou C, Du Z, Li G, Zhang Y, Cai Z. Oligomers matrix-assisted dispersion of high content of carbon nanotubes into monolithic column for online separation and enrichment of proteins from complex biological samples. Analyst 2013;138:5783−90.

[66] Li X, Wang M-M, Zheng G-Y, Ai L-F, Wang X-S. Fast and online determination of five avermectin residues in foodstuffs of plant and animal origin using reusable polymeric monolithic extractor coupled with LC-MS/MS. J Agric Food Chem 2015;63:4096−103. https://doi.org/10.1021/acs.jafc.5b00739.

[67] Liu S, Zong J, Wei Z, Zhang H, Bai L, Liu H, et al. Determination of trace macrolide antibiotics in milk with online solid-phase extraction with an ionic-liquid-based monolithic column. J Appl Polym Sci 2016;133. https://doi.org/10.1002/app.43943.

[68] Du F, Zheng X, Sun L, Qin Q, Guo L, Ruan G. Development and validation of polymerized high internal phase emulsion monoliths coupled with HPLC and fluorescence detection for the determination of trace tetracycline antibiotics in environmental water samples. J Sep Sci 2015;38:3774−80. https://doi.org/10.1002/jssc.201500497.

[69] Su R, Ruan G, Nie H, Xie T, Zheng Y, Du F, et al. Development of high internal phase emulsion polymeric monoliths for highly efficient enrichment of trace polycyclic aromatic hydrocarbons from large-volume water samples. J Chromatogr A 2015;1405:23−31. https://doi.org/10.1016/j.chroma.2015.05.067.

[70] Li N, Song Y, Qiu J, Zhao Y-C, Qian Y-Z. Polymer brushes-containing coordination polymer networks on monolith for rapid solid phase extraction of multi-class drug residues in meat samples. Talanta 2018;185:573−80. https://doi.org/10.1016/j.talanta.2018.03.077.

[71] Yang G, Feng S, Liu H, Yin J, Zhang L, Cai L. On-line clean-up and screening of oxacillin and cloxacillin in human urine and plasma with a weak ion exchange monolithic column. J Chromatogr B 2007;854:85−90. https://doi.org/10.1016/j.jchromb.2007.04.001.

[72] Potter O, Hilder E. Porous polymer monoliths for extraction: diverse applications and platforms. J Sep Sci 2008;31:1881−906.

[73] Wang H, Liu H, Guo B, Lan D, Pang X, Yan H, et al. A novel poly (NMA-co-DEA-co-EDMA) monolithic column as a sorbent for online solid-phase extraction and its application in the determination of β-sitosterol in plant oil samples. Food Chem 2019;278:594−600. https://doi.org/10.1016/j.foodchem.2018.11.104.

[74] Azzouz A, Kailasa SK, Lee SS, Rascón AJ, Ballesteros E, Zhang M, et al. Review of nanomaterials as sorbents in solid-phase extraction for environmental samples. Trends Anal Chem 2018;108:347−69. https://doi.org/10.1016/j.trac.2018.08.009.

[75] Ahmadi M, Elmongy H, Madrakian T, Abdel-Rehim M. Nanomaterials as sorbents for sample preparation in bioanalysis: a review. Anal Chim Acta 2017;958:1−21. https://doi.org/10.1016/j.aca.2016.11.062.

[76] Svec F, Lv Y. Advances and recent trends in the field of monolithic columns for chromatography. Anal Chem 2015;87:250–73. https://doi.org/10.1021/ac504059c.

[77] Vergara-Barberán M, Lerma-García MJ, Simó-Alfonso EF, Herrero-Martínez JM. Polymeric sorbents modified with gold and silver nanoparticles for solid-phase extraction of proteins followed by MALDI-TOF analysis. Microchim Acta 2017;184:1683–90. https://doi.org/10.1007/s00604-017-2168-5.

[78] Catalá-Icardo M, Gómez-Benito C, Simó-Alfonso EF, Herrero-Martínez JM. Determination of azoxystrobin and chlorothalonil using a methacrylate-based polymer modified with gold nanoparticles as solid-phase extraction sorbent. Anal Bioanal Chem 2017;409: 243–50. https://doi.org/10.1007/s00216-016-9993-y.

[79] Meseguer-Lloret S, Torres-Cartas S, Catalá-Icardo M, Simó-Alfonso EF, Herrero-Martínez JM. Extraction and preconcentration of organophosphorus pesticides in water by using a polymethacrylate-based sorbent modified with magnetic nanoparticles. Anal Bioanal Chem 2017;409:3561–71. https://doi.org/10.1007/s00216-017-0294-x.

[80] Fresco-Cala B, Cárdenas S, Herrero-Martínez JM. Preparation of porous methacrylate monoliths with oxidized single-walled carbon nanohorns for the extraction of nonsteroidal anti-inflammatory drugs from urine samples. Microchim Acta 2017;184:1863–71. https://doi.org/10.1007/s00604-017-2203-6.

[81] Fresco-Cala B, Mompó-Roselló Ó, Simó-Alfonso EF, Cárdenas S, Herrero-Martínez JM. Carbon nanotube-modified monolithic polymethacrylate pipette tips for (micro)solid-phase extraction of antidepressants from urine samples. Microchim Acta 2018;185:127. https://doi.org/10.1007/s00604-017-2659-4.

[82] Fontanals N, Cormack PAG, Marcé RM, Borrull F. Mixed-mode ion-exchange polymeric sorbents: dual-phase materials that improve selectivity and capacity. Trends Anal Chem 2010;29:765–79.

[83] Brousmiche DW, O'Gara JE, Walsh DP, Lee PJ, Iraneta PC, Trammell BC, et al. Functionalization of divinylbenzene/N-vinylpyrrolidone copolymer particles: ion exchangers for solid phase extraction. J Chromatogr A 2008;1191:108–17.

[84] Fontanals N, Borrull F, Marcé RM. Mixed-mode sorbents in solid-phase extraction. In: Dragana E, editor. Adv. sep. by Spec. Sorbents, vol. 10. New York: CRC Press; 2015. p. 245–83.

[85] Fontanals N. Application of novel materials in sample treatment and separation. Cleanup and chromatographic improvements. In: Nollet L, Lambropoulou D, editors. Chromatogr. Anal. Environ. Mass Spectrom. based approaches. 4th ed. New York: CRC Press; 2017. p. 197–218.

[86] García-López M, Rodríguez I, Cela R. Mixed-mode solid-phase extraction followed by liquid chromatography-tandem mass spectrometry for the determination of tri- and di-substituted organophosphorus species in water samples. J Chromatogr A 2010;1217: 1476–84.

[87] Chetwynd AJ, Abdul-Sada A, Hill EM. Solid-phase extraction and nanoflow liquid chromatography-nanoelectrospray ionization mass spectrometry for improved global urine metabolomics. Anal Chem 2015;87:1158–65. https://doi.org/10.1021/ac503769q.

[88] Fontanals N, Trammell BC, Galià M, Marcé RM, Iraneta PC, Borrull F, et al. Comparison of mixed-mode anion-exchange performance of N-vinylimidazole-divinylbenzene sorbent. J Sep Sci 2006;29:1622–9.

[89] Gilart N, Cormack PAG, Marcé RM, Fontanals N, Borrull F. Selective determination of pharmaceuticals and illicit drugs in wastewaters using a novel strong cation-exchange solid-phase extraction combined with liquid chromatography–tandem mass spectrometry. J Chromatogr A 2014;1325:137–46. https://doi.org/10.1016/j.chroma.2013.12.012.

[90] Meischl F, Kirchler CG, Stuppner SE, Rainer M. Comparative study of substituted poly(4-vinylbenzyl chloride/ethylene glycol dimethacrylate) sorbents for enrichment of selected pharmaceuticals and estrogens from aqueous samples. J Hazard Mater 2018;355: 180–6. https://doi.org/10.1016/j.jhazmat.2018.05.016.

[91] Huang C, Li Y, Yang J, Peng J, Jin J, Dhanjai, et al. Preparation of a reversed-phase/ anion-exchange mixed-mode spherical sorbent by Pickering emulsion polymerization for highly selective solid-phase extraction of acidic pharmaceuticals from wastewater. J Chromatogr A 2017;1521:1–9. https://doi.org/10.1016/j.chroma.2017.09.021.

[92] Huang C, Li Y, Yang J, Peng J, Tan J, Fan Y, et al. Hyperbranched mixed-mode anion-exchange polymeric sorbent for highly selective extraction of nine acidic non-steroidal anti-inflammatory drugs from human urine. Talanta 2018;190:15–22. https://doi.org/ 10.1016/j.talanta.2018.07.033.

[93] Hu K, Zhao G, Liu J, Jia L, Xie F, Zhang S, et al. Simultaneous quantification of three alkylated-purine adducts in human urine using sulfonic acid poly(glycidyl methacrylate-divinylbenzene)-based microspheres as sorbent combined with LC-MS/MS. J Chromatogr B 2018;1081–1082:15–24. https://doi.org/10.1016/j.jchromb.2018.02.028.

[94] Wang L, Li F, Yao M, Qiu T, Jiang W, Fan L-J. Atom transfer radical polymerization of glycidyl methacrylate followed by amination on the surface of monodispersed highly crosslinked polymer microspheres and the study of cation adsorption. React Funct Polym 2014;82:66–71. https://doi.org/10.1016/j.reactfunctpolym.2014.06.002.

[95] Fontanals N, Cormack PAG, Sherrington DC. Hypercrosslinked polymer microspheres with weak anion-exchange character. Preparation of the microspheres and their applications pH-tunable, selective extractions of analytes from complex environmental samples. J Chromatogr A 2008;1215:21–9.

[96] Cormack PAG, Davies A, Fontanals N. Synthesis and characterisation of microporous polymers microspheres with strong-cation exchange character. React. Func. Polym. 2012; 72:939–46.

[97] Bratkowska D, Davies A, Fontanals N, Cormack PAG, Borrull F, Sherrington DC, et al. Hypercrosslinked strong anion-exchange resin for extraction of acidic pharmaceuticals from environmental water. J Sep Sci 2012;35:2621–8.

[98] Liang X, Hu P, Zhang H, Tan W. Hypercrosslinked strong anion-exchange polymers for selective extraction of fluoroquinolones in milk samples. J Pharm Biomed Anal 2019; 166:379–86. https://doi.org/10.1016/j.jpba.2018.12.047.

[99] Bratkowska D, Marcé RM, Cormack PAG, Sherrington DC, Borrull F, Fontanals N. Synthesis and application of hypercrosslinked polymers with weak cation-exchange character for the selective extraction of basic pharmaceuticals from complex environmental samples. J Chromatogr A 2010;1217:1575–82.

[100] Fontanals N, Ronka S, Borrull F, Trochimczuk AW, Marcé RM. Supported imidazolium ionic liquid phases: a new material for solid-phase extraction. Talanta 2009;80:250–6.

[101] Bratkowska D, Fontanals N, Ronka S, Borrull F, Trochimczuk AW, Marcé RM. Comparison of different imidazolium supported ionic liquid polymeric phases with strong anion-exchange character for the extraction of acidic pharmaceuticals from complex environmental samples. J Sep Sci 2012;35:1953–8.

[102] Zhu L, Deng Y, Zhang J, Chen J. Adsorption of phenol from water by N-butylimidazolium functionalized strongly basic anion exchange resin. J Colloid Interface Sci 2011;364:462−8. https://doi.org/10.1016/j.jcis.2011.08.068.

[103] Kanaujia PK, Pardasani D, Purohit AK, Tak V, Dubey DK. Polyelectrolyte functionalized multi-walled carbon nanotubes as strong anion-exchange material for the extraction of acidic degradation products of nerve agents. J Chromatogr A 2011;1218:9307−13. https://doi.org/10.1016/j.chroma.2011.10.036.

# Carbon-based adsorbents

**4**

*Álvaro Santana-Mayor[1], Ruth Rodríguez-Ramos[1], Bárbara Socas-Rodríguez[1], María Asensio-Ramos[2], Miguel Ángel Rodríguez-Delgado[1]*
[1]Departamento de Química, Unidad Departamental de Química Analítica, Facultad de Ciencias, Universidad de La Laguna (ULL). Avenida Astrofísico Francisco Sánchez, San Cristóbal de La Laguna, Tenerife, España; [2]Instituto Volcanológico de Canarias (INVOLCAN), INtech La Laguna, San Cristóbal de La Laguna, Tenerife, Islas Canarias, España

## 4.1 Introduction

In spite of the new advances in instrumentation, sample preparation is still the most important step of the whole analytical procedure when the target compounds are in complex matrices. Solid-phase extraction (SPE), as well as other sorbent-based techniques, have found a greater use in this field since its introduction in the 1970s, given its ease of use, selectivity, extraction capacity, the commercial availability of cartridges and disk devices and the number of variables that can be modified to improve the quality of the extraction [1]. For years, the main field of research in SPE has been the development of new stationary phases [2]. In this respect, nanomaterials have aroused a particular interest due to their exceptional properties (structure and specific surface area) over other sorbents [3]. In particular, they show exceptional mechanical, electronic and thermal properties, ease of derivatization, and high surface-to-volume ratio [4].

In the 1970s, activated carbon became popular, and subsequently, carbon-based nanomaterials have expanded their applications in SPE, to become the second most important materials after silica-based stationary sorbents [5]. Fullerenes, carbon nanotubes (CNTs), graphene, nanohorns (NHs), nanodiamonds (NDs) and quantum dots (QDs) are all part of this remarkable sorbent family and considering the possibility of functionalization, the options are almost endless [4]. These materials can interact with organic molecules by noncovalent forces (van der Waals forces, hydrogen bonds, π-π interactions, etc.), which together with their particular hollow or layered structures, makes them favorable sorbent materials for SPE [4].

This chapter offers an overall vision of the state of the art of the use of carbonaceous materials, including their functionalized forms, as well as their combination with other materials in the different modes of SPE, i.e., SPE, micro-SPE (μ-SPE), dispersive SPE (dSPE), magnetic micro-dispersive SPE (m-μ-dSPE) and micro-dispersive SPE (μ-dSPE). The advantages or disadvantages of each carbon sorbent will be described as well as representative applications with an emphasis on modern carbon-based sorbents.

**Solid-Phase Extraction. https://doi.org/10.1016/B978-0-12-816906-3.00004-2**

## 4.2 Carbon-based adsorbent for solid- phase extraction

### 4.2.1 Fullerenes

Fullerenes were the first carbon-based nanomaterials discovered in 1985 by Kroto et al. [6]. Their characteristic polyhedral structure, in which the carbon atoms are organized in five or six member rings as hollow spheres or cylinders called *buckyballs* or *buckytubes*, respectively [7], provides the particular features of these materials that promote their use as adsorbents in SPE and other sorbent-based extraction procedures [8,9]. Among their favorable properties that should be highlighted are a low tendency to aggregate, high electron affinity, low solubility in both aqueous and organic solvents, and a high surface-to-volume ratio. However, compared to other carbon allotropic forms, such as CNTs and graphene, their high cost and low accessibility limited their application [10].

Despite the small number of reports in which fullerenes were used as adsorbents in SPE, Table 4.1 summarizes some of the most notable applications including m-μ-dSPE [11−13], μ-dSPE [14] and on-line SPE [15] both as a raw material [15] and combined with other nanomaterials [11−14]. In general, fullerenes have been used for the extraction of organic compounds [11,12,14,15] with fewer applications for inorganic analytes [13]. As an example of the latter, Kilinç [13] developed a methodology based on an m-μ-dSPE using $C_{60}$-fullerene@γ-$Fe_2O_3$ sorbent for the extraction of bismuth (III) from milk, water, and energy drinks. Recovery values in the range 97%−99% and a limit of detection (LOD) of 0.06 μg/L were obtained, with high selectivity against typical interfering ions without further sorbent functionalization or use of a chelating reagent.

However, the main applications are for the extraction of organic compounds, which can be physically adsorbed through dispersive interactions, especially those with aromatic structures that allow π-π electron interactions [7,16,17]. This feature led to their use for the extraction of important groups of contaminants, such as polycyclic aromatic hydrocarbons (PAHs) [11], BTEX compounds (benzene, toluene, ethylbenzene, and xylene) [15] and azo dyes, [12] and for the separation of aliphatic and aromatic compounds [17]. As an example, Zhou et al. [11] used a $C_{60}$-fullerene sorbent in combination with magnetic nanoparticles ($Fe_3O_4$, m-NPs) to extract 16 PAHs from tea samples by m-μ-dSPE with only 40 mg of the sorbent. The fullerene composite was synthesized as shown in Fig. 4.1 starting from $Fe_3O_4$ nanoparticles with a $SiO_2$ coating and surface immobilization of a functionalized fullerene material. The fullerene composite demonstrated a high extraction efficiency for PAHs (recovery range 92%−107%), LODs in the range 0.8−14.3 ng/L with high selectivity over n-alkanes.

Fullerenes were also shown to be suitable for the extraction of phosphorylated peptides due to the strength of electron donor-acceptor complexes, which results in enhanced binding of the phosphorylated peptides. Fischnaller et al. [14] extracted nine phosphorylated peptides from a bovine serum albumin digest with a sorbent prepared from 3-aminopropylsilica modified with $C_{60}$-fullerene for analysis by matrix-assisted laser desorption/ionization (MALDI)-time-of-flight (ToF)-mass spectrometry (MS). The combination of the binding with fullerene and interactions

**Table 4.1** Applications of fullerene-based materials in solid-phase extraction.

| Analyte | Sample (amount) | SPE mode | Sorbent (amount) | Determination technique | Recovery (%) | Detection limits | Comments | Reference |
|---|---|---|---|---|---|---|---|---|
| 16 PAHs | Tea (100 mL) | m-μ-dSPE | $Fe_3O_4$@$SiO_2$@$C_{60}$-fullerene (40 mg) | GC-Q-MS | 92–107 | 0.8–14.3 ng/L | Selectivity of $Fe_3O_4$@$SiO_2$@fullerene was evaluated in presence of alkanes.<br>Reusability of the sorbent was tested. | [11] |
| 3 azo dyes | Wastewater (500 μL) | m-μ-dSPE | $Fe_3O_4$@fullerene@activated carbon (5 mg) | CE-PDA | 95–106 | 1.0–2.0 mg/L | Method optimization was evaluated through a simplex centroid design | [12] |
| Bismuth | Milk (5 mL)<br>Tap, mineral, and river water (100 mL)<br>Energy drink (-) | m-μ-dSPE | $C_{60}$-fullerene@γ-$Fe_2O_3$ (50 mg) | ICP-OES | 97–99 | 0.06 μg/L | Selectivity of the procedure was tested.<br>Reusability of the sorbent was tested.<br>A standard reference material was used for method validation.<br>Real samples were evaluated. | [13] |

*Continued*

**Table 4.1** Applications of fullerene-based materials in solid-phase extraction.—cont'd

| Analyte | Sample (amount) | SPE mode | Sorbent (amount) | Determination technique | Recovery (%) | Detection limits | Comments | Reference |
|---|---|---|---|---|---|---|---|---|
| 9 multi-, mono-, nonphosphorylated peptides | Milk (200 μL) | μ-dSPE | $C_{60}$-fullerene@aminopropyl@$SiO_2$ (5 mg) | MALDI-ToF-MS | 74 | - | Reusability and selectivity of the sorbent were tested. | [14] |
| Bencene, toluene, ethylbencene, xylene | Sea and wastewater (15 mL) | Online-SPE | $C_{60}$-fullerene (60 mg) | GC-Q-MS | - | 0.04 μg/L | Sorbent was compared with Tenax TA and RP-$C_{18}$. Reusability and selectivity of the sorbent were tested. Fluorobencene was used as IS. | [15] |

μ-dSPE: micro-dispersive solid-phase extraction; $C_{18}$: octadecylsilane; CE: capillary electrophoresis; ICP: inductively coupled plasma; GC: gas chromatography; IS: internal standard; MALDI: matrix-assisted laser desorption/ionization; m-μ-dSPE: magnetic-micro-dispersive solid phase extraction; MS: mass spectrometry; OES: optical emission spectrometry; PAH: polycyclic aromatic hydrocarbon; PDA: photodiode array detector; Q: single quadrupole; RP: reverse phase; SPE: solid-phase extraction; ToF: time-of-flight.

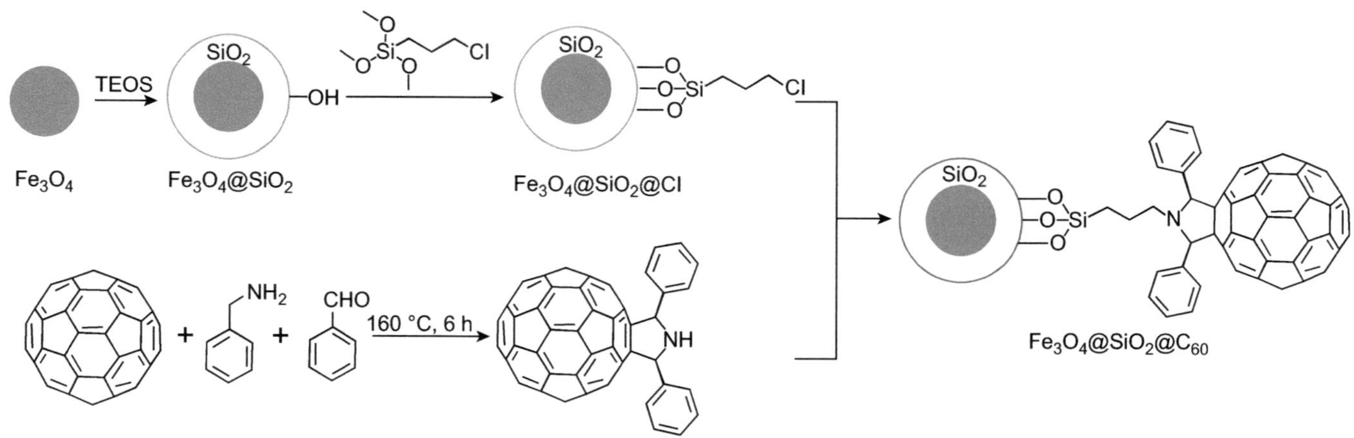


**Figure 4.1** Scheme of the $Fe_3O_4@SiO_2@C_{60}$-fullerene sorbent synthesis.
Reprinted from Zhou DB, Sheng X, Han F, Hu YY, Ding L, Lv YL, Song W, Zheng P. Magnetic solid-phase extraction based on [60] fullerene functionalization of magnetic nanoparticles for the determination of sixteen polycyclic aromatic hydrocarbons in tea samples. J Chromatogr A 2018; 1578:53–60 with permission from Elsevier.

between the peptides and protonated amino groups of the 3-aminopropylsilica provided enhanced recovery values around 70% using a μ-dSPE with only 5 mg of the sorbent.

Although, as shown in Table 4.1, the dispersive mode has generally been preferred for extraction with fullerene-based sorbents, applications in on-line SPE have also been reported. Serrano et al. [15] developed an automated system in which a laboratory-made sorption column (length 1.1 cm and internal diameter 3 mm) packed with 60 mg of $C_{60}$-fullerene was used for the extraction of benzene, toluene, ethylbenzene, and xylene from sea and wastewater samples. The analytes were eluted with only 150 μL of ethyl acetate prior to analysis by gas chromatography (GC)-MS (LOD 0.04 μg/L). The results demonstrated clear advantages over Tenax TA and octadecylsiloxane-bonded silica sorbents for this particular application. This fact was attributed to the moderately strong π—π interactions, the larger specific surface area, extended reusability and the selectivity of the fullerene-based sorbent for nonpolar aromatic compounds.

### 4.2.2 Carbon nanotubes

Since their discovery in 1991 [18], CNTs have become one of the most widely used carbon allotropes in sample preparation. CNTs are made of graphene sheets, rolled up into a cylinder shape and capped by fullerene-like structures with a typical diameter from a few tenths to tens of nanometers and a length of several micrometers. They exist mainly as three types: single-walled CNTs (SWCNTs), double-walled CNTs (DWCNTs) and multi-walled CNTs (MWCNTs). In general, all forms have large surface areas between 150 and 1500 $m^2/g$ [4,5], suitable for be used as adsorbents in SPE. In addition, CNTs exhibit high stability, favorable mechanical strength, and high thermal and electrical conductivity. Their structures and electronic properties allow them to interact with both organic compounds and inorganic species (in this case, usually, after some modifications), through electrostatic-type interactions, hydrogen bonds, hydrophobic interactions, or π-π stacking, among others. However, for SPE, SWCNTs, and DWCNTs are not as widely used as MWCNTs because of their larger size and lower surface area/volume ratio that makes them less efficient in extraction procedures [2]. Like fullerenes, CNTs exhibit limited solubility in most solvents, but the lack of functional groups on their surface limits their application for dSPE and for the extraction of metal ions [19,20]. Besides, CNTs have a large tendency to aggregate, reducing their adsorption capacity. When packed into SPE cartridges or columns, agglomeration tends to increase the back pressure for the systems generating poor reproducibility during sample loading and the elution of analytes. To minimize this problem, Tomai et al. [21] described a μ-SPE method as an alternative to conventional packed cartridge SPE based on a rotating disc of buckypaper (BP) enwrapped in a polypropylene mesh pouch. The BP is a macroscopic aggregate of CNTs randomly intertwined by strong van der Waals forces that forms a network structure, which allows it to act as an adsorbent membrane with high porosity, high specific adsorption capacity, good permeability, and high mechanical stability [22]. The μ-SPE device was used for the extraction of five F2-isoprostanes from maternal and cord plasma. The extraction procedure (illustrated in Fig. 4.2) consists of three steps: the activation and conditioning of BP; deproteinization of plasma samples; and the μ-SPE step. In this method, 1 mL of plasma sample was pretreated to precipitate the proteins, and the extraction was diluted to 50 mL with Milli-Q water. The aqueous extract was then submitted to the μ-SPE procedure under magnetic stirring overnight. Afterward, the device was placed in a bottle, and two 2-mL fractions of methanol (MeOH) were employed to re-extract the analytes. The extract was dried with anhydrous sodium sulfate and reduced in volume to 100 μL for analysis by liquid chromatography (LC). Recovery values between 30% and 120% and LODs of 1.47−4.06 μg/L were obtained.

Other strategies to solve the problem of agglomeration include functionalizing the CNTs or their combination with other materials. Functionalization, which can be carried out by both covalent and noncovalent methods with different functional groups, biomolecules, polymers, bacteria, among others, provides a more selective interaction with the analytes of interest and increases the extractive capacity of the sorbent. As demonstrated by Table 4.2, CNTs have been used alone [23], after simple

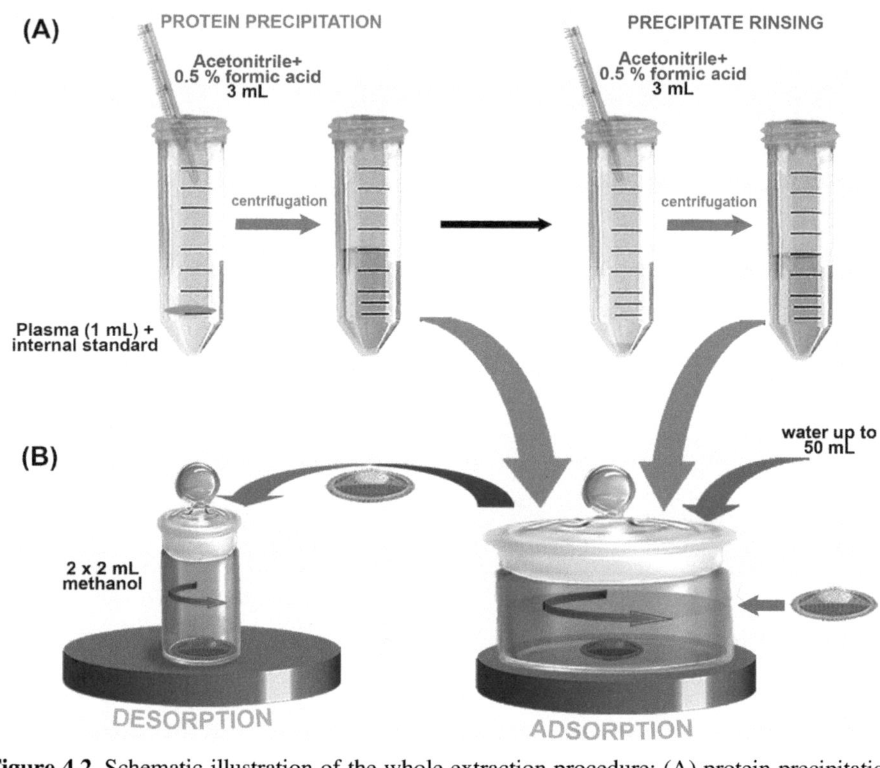


**Figure 4.2** Schematic illustration of the whole extraction procedure: (A) protein precipitation (B) and operational steps of rotating-disc micro-SPE clean-up.
Reprinted from Tomai P, Martinelli A, Gasperi T, Bianchi M, Purcaro V, Teofili L, Papacci P, Cori MS, Vento G, Curini R, Fanali S, Gentili A. Rotating-disc micro-solid phase extraction of F2-isoprostanes from maternal and cord plasma by using oxidized buckypaper as sorbent membrane. J Chromatogr A 2019;1586:30-39 with permission from Elsevier.

**Table 4.2** Applications of carbon nanotubes in solid-phase extraction.

| Analyte | Sample (amount) | SPE mode | Sorbent (amount) | Determination technique | Recovery (%) | Detection limits | Comments | Reference |
|---|---|---|---|---|---|---|---|---|
| 5 $F_2$-isoprostanes | Cord and maternal plasma (1 mL) | μ-SPE | *o*-MWCNTs (-) | HPLC-QTrap-MS/MS | 30–120 | 1.47–4.06 μg/L | 8-iso-$PGF_{2\alpha}$-$d_4$ as surrogate.<br>Protein precipitation of plasma samples was previously carried out.<br>Extracts were diluted with Milli-Q water to 50 mL.<br>μ-SPE device consisted in a rotating disc of *o*-MWCNTs enwrapped in a polypropylene mesh pouch. A commercial dispersant was employed to prepare the device.<br>Plasma samples from eight mother/child pairs were analyzed. | [21] |

| | | | | | | | | |
|---|---|---|---|---|---|---|---|---|
| Hg(II) and methyl-Hg | Fish oil dietary supplements (0.5 g)<br>Antartic water (20 mL) | Online-μ-SPE | Poly-L-methionine@MWCNTs (3 mg) | HPLC-ICP-MS | 93−114 | 15−17 ng/L | Dietary supplements were pretreated with HCl, and the final extract (10 mL) was submitted to the online-μ-SPE procedure.<br>Tuna fish and lobster hepatopancreas CRMs were used to check the accuracy. | [23] |
| 10 PAHs | Water (100 mL) | SPE | Poly-(benzyl methacrylate)@MWCNTs (-) | Nano-LC-VWD | 81−95 | 0.02−0.22 μg/L | 3.0 mg/mL of MWCNTs (0.3 wt % with respect to monomers) was added to the polymerization mixture.<br>Nano-LC columns were prepared with the same mixtures in different weight percentages (0.1 wt % of MWCNTs respect to monomers). | [24] |

*Continued*

**Table 4.2** Applications of carbon nanotubes in solid-phase extraction.—cont'd

| Analyte | Sample (amount) | SPE mode | Sorbent (amount) | Determination technique | Recovery (%) | Detection limits | Comments | Reference |
|---|---|---|---|---|---|---|---|---|
| 3 NSAIDs | Plasma, urine and wastewater (10 mL) | μ-dSPE | Zn-Al-LDH@ *o*-MWCNTs (8 mg) | HPLC-UV | 93–103 | 0.1–0.2 μg/L | The developed method was composed by a μ-dSPE coupled to a SFOD-DLLME method.<br>μ-dSPE was assisted by sonication and syringe nanofilter was employed to avoid centrifugation steps. | [25] |
| Pb(II), Cd(II), Zn(II), Mn(II), Fe(III) | Rice (500 mg) | μ-dSPE | *o*-MWCNTs (1 mg) | FAAS | 93–103 | 0.13–0.35 μg/L | White and wild rice were previously milled, sieved, and microwave-assisted digested and the extract (50 mL) was submitted to the μ-dSPE procedure.<br>Batophenanthroline was used as cationic chelating agent.<br>Method accuracy was evaluated using bush branches, leaves, and rice CRMs. | [26] |

| | | | | | | | | |
|---|---|---|---|---|---|---|---|---|
| 4 Sudan dyes and 2 metabolites | Salted duck egg yolk and tomato sauce (1 g) | μ-dSPE | Crown ether@MWCNTs (60 μL) | UHPLC-(Q-ToF)-MS | 80−104 | 0.084−13.13 μg/kg | Samples were previously extracted with 10 mL of ACN and 200 μL of the extract was submitted to the μ-dSPE procedure.<br>Extraction suspension was prepared with 5 mg of crown ether and 1 mg of MWCNTs in 10 mL of water. | [27] |
| 4 antidepressants | Urine (50 mL) | m-dSPE | $Fe_3O_4$@PS-DVB@MWCNTs (500 mg) | CE-DAD | > 90 | 0.014−0.041 mg/L | Adsorbent was conditioned with (2 x 3) mL of MeOH and deionized water.<br>Extraction capacity of m-MWCNTs, three nonmagnetic polymeric sorbents and three magnetic polymeric sorbents was compared. | [28] |
| 3 steroid hormones | Milk (40 mL) | m-μ-dSPE | $Fe_3O_4$@$SrTiO_3$@MWCNTs (4 mg) | HPLC-VWD | 60−113 | 0.03−0.34 μg/L | Milk samples were previously deproteinized with 1:1 milk:ACN volume ratio.<br>Sorbent was synthesized with 1:1 w/w ratio of m-MWCNTs:$SrTiO_3$. | [29] |

*Continued*

**Table 4.2** Applications of carbon nanotubes in solid-phase extraction.—cont'd

| Analyte | Sample (amount) | SPE mode | Sorbent (amount) | Determination technique | Recovery (%) | Detection limits | Comments | Reference |
|---|---|---|---|---|---|---|---|---|
| 6 Sudan dyes and Para Red | Chili powder, chili tomato sauces and ketchup (10 g) | m-μ-dSPE | γ-$Fe_2O_3$@ MWCNTs-COOH (40 mg) | HPLC-DAD | - | 0.13–0.84 μg/L | Samples were pretreated with 30 mL of acetone:DCM:MeOH (3:2:1, v/v/v), diluted with ACN:MeOH (80:20, v/v) to 50 mL, and the extract was submitted to the m-μ-dSPE procedure. | [30] |
| 18 chiral pesticides | Water (200 mL)<br>Soil and river sediments (2 g) | m-μ-dSPE | $Fe_3O_4$@ MWCNTs-$NH_2$ (75 mg) | UHPLC-QqQ-MS/MS | 80–106 | 0.02–0.62 ng/L | Solid samples were pretreated with 10 mL of ACN and 10 mL of ultrapure water. 10 mL of supernatant was diluted with 50 mL of ultrapure water and the extract was submitted to the m-μ-dSPE procedure.<br>Pristine MWCNTs and MWCNTs-$NH_2$ were compared. | [31] |

| | | | | | | | | |
|---|---|---|---|---|---|---|---|---|
| 5 mycotoxins | *Salviae miltiorrhizae* Radix et Rhizoma (Danshen) (2 g) | μ-SPE | $Fe_3O_4$@ MWCNTs (20 mg) | UHPLC-QqQ-MS/MS | 74−92 | 0.45−1.80 μg/kg | Samples were pretreated with 10 mL of ACN: water (84:16, v/v). 5 mL of supernatant was dried, reconstituted in 0.5 mL of ACN:water (20:80, v/v) and diluted with 4.5 mL of water. The extract (5 mL) was submitted to the μ-SPE procedure. $Fe_3O_4$ enhanced the adsorption ability of MWCNTs. | [32] |
| 7 PCBs | Environmental water (1 mL) | μ-dSPE | MWCNTs-COOH (2 mg) | GC-ECD | 99−106 | 1.4−3.5 ng/L | SWCNTs, MWCNTs, and MWCNTs-COOH were compared. | [33] |

*Continued*

**Table 4.2** Applications of carbon nanotubes in solid-phase extraction.—cont'd

| Analyte | Sample (amount) | SPE mode | Sorbent (amount) | Determination technique | Recovery (%) | Detection limits | Comments | Reference |
|---|---|---|---|---|---|---|---|---|
| 9 PAHs, Cr(II), Cd(III) and Pb(II) | Water (250 mL) | μ-dSPE | MWCNTs-COOH (50 mg) | (i) GC-Q-MS<br>(ii) AAS | (i) 68–102<br>(ii) 75–88 | (i) 0.003 μg/L<br>(ii) 0.030 μg/L | After extraction procedure, the mixture was transferred to a PTFE column containing two PTFE frits, and MWCNTs were dried under vacuum.<br>(i) PAHs were reextracted with 8 mL of DCM and analyzed by GC-MS.<br>(ii) Metal ions were reextracted with 5 mL of $HNO_3$ and analyzed by AAS.<br>Helical MWCNTs, MWCNTs (> 50 nm), MWCNTs (< 8 nm), MWCNTs-OH, and MWCNTs-COOH were tested as sorbents. | [34] |

| | | | | | | | | |
|---|---|---|---|---|---|---|---|---|
| Cr(III) | Water (200 mL) | m-μ-dSPE | $Fe_3O_4$@MWCNTs-COOH (50 mg) | Laser-induced fluorescence spectroscopy detection | 91−102 | 0.094 nM | Cr(III) analysis were performed in a flow injection microfluidic chip.<br>Rhodamine derivative as fluorescent derivatizing agent.<br>Total chromium was determined as Cr(III) by reducing Cr(VI) to Cr(III). | [35] |
| 18 OCPs | Environmental water (50 mL) | m-μ-dSPE | ChCl/urea (DES) @$Fe_3O_4$@MWCNTs (-) | LVI-GC-μ-ECD | 91−102 | 0.04−0.27 ng/L | Bucky gel sorbent was formed by 5 mg of m-MWCNTs dispersed in 50 μL of ChCl/urea DES.<br>Analytes were reextracted from m-MWCNTs with 150 μL of ACN.<br>Equilibrium time for analytes extraction had no significant effect on the extraction efficiency. | [36] |

μ-dSPE: micro-dispersive solid-phase extraction; μ-ECD: micro-electron capture detector; μ-SPE: micro−solid-phase extraction; 8-iso-$PGF_{2\alpha}$-$d_4$: 8-iso-prostaglandin-$F_{2\alpha}$-$d_4$; AAS: atomic absorption spectrometry; ACN: acetonitrile; CE: capillary electrophoresis; ChCl: choline chloride; CRM: certified reference material; DAD: diode-array detector; DCM: dichloromethane; DLLME: dispersive liquid−liquid microextraction; ECD: electron capture detector; FAAS: flame atomic absorption spectrometry; GC: gas chromatography; HPLC: high-performance liquid chromatography; ICP: inductively coupled plasma; LC: liquid chromatography; LDH: layered double hydroxide; LVI: large-volume injection; m-μ-dSPE: magnetic-microdispersive solid-phase extraction; m-dSPE: magnetic-dispersive solid-phase extraction; MeOH: methanol; m-MWCNT: magnetic-multiwalled carbon nanotube; MS/MS: tandem mass spectrometry; MS: mass spectrometry; MWCNT: multiwalled carbon nanotube; MWCNT-COOH: carboxylated-multiwalled carbon nanotube; MWCNT-$NH_2$: amine functionalized-multiwalled carbon nanotube; MWCNT-OH: hydroxylated-multiwalled carbon nanotube; NSAID: nonsteroidal antiinflammatory drugs; OCP: organochlorine pesticide; *o*-MWCNT: *oxidized*-multiwalled carbon nanotube; PAH: polycyclic aromatic hydrocarbon; PCB: polychlorinated biphenyl; PS-DVB: poly-(styrene-co-divinylbenzene); PTFE: polytetrafluoroethylene; Q: single quadrupole; QqQ: triple quadrupole; QTrap: triple quadrupole linear ion trap; SFOD: solidified floating organic droplets; SPE: solid-phase extraction; SWCNT: single-walled carbon nanotube; ToF: time-of-flight; UHPLC: ultrahigh-performance liquid chromatography; UV: ultraviolet detector; VWD: variable-wavelength detector; wt: weight.

oxidation [21] or functionalized with diphenylcarbazide [24], pyridine [25], bovine serum albumin [26], poly-L-methionine [27], poly-(benzyl methacrylate) [28]; and in combination with m-NPs [29] or graphene oxide (GO) [30] for the extraction of metal ions [24–27], pesticides [23], $F_2$-isoprostanes [21], PAHs [28], mycotoxins [29]; and for drugs [30] from plant sources [24,29], environmental waters [23,24,27,28], fruits [25], blood serum [26], cord and maternal plasma [21], fish oil dietary supplements [27], and pork [30] using a conventional packed cartridge SPE format.

For use in dSPE, CNTs have been used alone [31,32], after functionalizing with hydroxyl [33] or carboxyl groups [34,35], with organic molecules, including crown ethers [36], or in combination with other nanomaterials, such as layered double hydroxide (LDH) [37]. They have been used for the extraction of a wide variety of organic and inorganic compounds including antibiotics [32], fungicides [33], mycotoxins [31], nonsteroidal antiinflammatory drugs (NSAIDs) [37], metal ions [38], dyes [36], polychlorinated biphenyls (PCBs) [34], and PAHs [35] from water [31–35,37], fruit juices [33], infant milk [31], plasma [37], urine [37], rice [38], tomato sauce [36], and salted duck egg yolk [36], among others. As an example, Arghavani-Beydokhti et al. [37] developed a nanohybrid material formed by *oxidized* MWCNTs (*o*-MWCNTs) and a Zn-Al-LDH for μ-dSPE for the extraction of three NSAIDs from plasma, urine and wastewater samples. In order to avoid the characteristic centrifugation steps of dSPE procedures, μ-dSPE coupled to solidified floating organic droplets (SFOD)-dispersive liquid-liquid microextraction (DLLME) was carried out. The first step consisted of the ultrasound-assisted extraction of 10 mL of aqueous sample solution with 8 mg of Zn-Al-LDH@*o*-MWCNTs. To re-extract the NSAIDs, 100 μL of trifluoroacetic acid (TFAA) (8% v/v) was sucked into the syringe holding the sample, and the plunger slowly moved to fully dissolve the Zn-Al-LDH. Afterward, the residue was transferred to a glass tube, diluted to 5 mL with deionized water and submitted to the SFO-DLLME method using 60 μL of 1-undecanol as the extraction phase. The mixture obtained was transferred to another tube containing 5 g of NaCl to induce the salting-out effect. The 1-undecanol floating drops with extracted target analytes were solidified in an ice bath, collected with a spatula, melted at room temperature, and analyzed by LC. The sample preparation procedure was completed in 8 min without using centrifugation with a recovery range of 90%–103% and LODs of 0.1–0.2 μg/L.

One of the most relevant modifications employed in recent years is the combination of CNTs with m-NPs, not only because it improves the dispersion of the carbon-based nanomaterials in solvents, but also because it greatly simplifies the extraction procedure compared to conventional SPE and dSPE. In addition to its use with m-NPs (both magnetite [39–42] and maghemite [43]) to carry out magnetic dSPE (m-dSPE), its combination with other materials, such as graphene [44] or reduced-GO (r-GO) [45], $SrTiO_3$ [40] or even deep eutectic solvents (DESs) [46], and also functionalization, including carboxyl [42,43] and amino [41] groups or poly(styrene-*co*-divinylbenzene) (PS-DVB) [39] have been used. These sorbents were applied for the extraction of drugs [39,45], steroid hormones [40], dyes [43], pesticides [41,46], PAHs [44], and metal ions [42] from urine [39], milk [40,45], chili powder [43] and tomato [43], ketchup [43], water [41,42,46], soil [41] and sediments [41], among others. Yousefi et al. [46] combined the favorable characteristic properties of MWCNTs and m-NPs, with a new type of environmentally friendly solvents, namely DESs. In this way, they created a bucky gel with increased dispersibility of magnetic MWCNTs (m-MWCNTs) by combining them with a DES prepared from choline chloride (ChCl) and urea. In addition, ChCl can act as a salting-out agent, enhancing the extraction efficiency of the procedure. The above sorbent was used for the extraction of 18 organochlorine pesticides (OCPs) from environmental water samples for determination by GC For the m-μ-dSPE step, 5 mg of m-MWCNTS were dispersed in 50 μL of DES for the extraction of up to 50 mL of the sample solution. The m-MWCNTs were isolated by an external magnet, and the analytes desorbed with 150 μL of acetonitrile (ACN) for analysis. Recovery values between 91% and 102% and LODs of 0.04−0.27 ng/L were suitable for the determination of OCPs in environmental water samples.

Other applications of CNTs in SPE involve their use as a clean-up sorbent in the QuEChERS (*quick, easy, cheap, effective, rugged, and safe*) method [47] to reduce matrix effects caused by co-extractive materials from food samples.

### *4.2.3 Graphene*

Graphene, a nanomaterial consisting of a carbon monolayer packed in a bidimensional honeycomb network, is considered the basic building block of carbon-based materials including graphite, fullerenes, and CNTs, etc [48]. First isolated in 2004 [49], it has become of great interest in many scientific fields, mainly due to its ultrahigh theoretical surface area ($2630\ m^2\ g^{-1}$ for a single layer); high mechanical strength, elasticity, and thermal conductivity; and high electron conductivity [50]. In particular, graphene has played an important role in analytical methods due to its high hydrophobicity and ease of functionalization, as well as its electron delocalization that allows strong π-π interactions with benzene-ring structures [10,51]. In addition, its planar structure provides graphene a high adsorption capacity with analyte adsorption able to occur on both sides of the sheet [52]. These features make it possible to obtain high extraction efficiencies using small amounts of graphene as a sorbent [51]. Apart from that, graphene can be easily obtained by oxidation of graphite without the formation of residual heterogeneous materials derived from catalytic processes [8].

As shown in Table 4.3, graphene-based sorbents have been used in several SPE formats for the analysis of both organic compounds [45,53−63] and metal ions [64−66] in environmental [53,54,58,60,64,66], food [45,55,57,59,61−63,65] and biological matrices [56].

On account of its tendency to agglomerate (that reduces the theoretical surface area) and its poor dispersibility in water, raw graphene has not been widely used as a sorbent for SPE [67,68]. However, GO, which contains large quantities of oxygen atoms as epoxy, hydroxyl, and carboxyl groups on its surface has attracted more attention as a sorbent for extraction since it shows a better dispersibility in aqueous solutions and additional capability to establish hydrogen bonds and electrostatic interactions with organic compounds and metal ions [51]. Similarly, r-GO, which is an

**Table 4.3** Applications of graphene-based materials in solid-phase extraction.

| Analyte | Sample (amount) | SPE mode | Sorbent (amount) | Determination technique | Recovery (%) | Detection limits | Comments | Reference |
|---|---|---|---|---|---|---|---|---|
| Ni(II) | Tap, river, lagoon water, and wastewater (200 mL) | SPE | GO@P-*p*-AP (150 mg) | FAAS | 97–98 | 0.7 μg/L | A comparison with GO was carried out. The sorbent could be reused for seven cycles. Accuracy and precision of the method was tested by analyzing a CRM. | [53] |
| Pb(II) | Trout, salmon (200 mg) River and sea water (100 mL) | μ-dSPE | GO@Phen (1 mg) | ICP-OES | 96–104 | 0.046 μg/L | Fish samples were acid-digested with nitric acid and the extract was submitted to the μ-dSPE extraction procedure. Accuracy and precision of the method was tested by analyzing CRMs. | [54] |
| 7 PAHs | Tap, river, well, and wastewater (50 mL) | μ-dSPE | Graphene@MOF-199 (20 mg) | GC-FID | 92–100 | 0.003-0.01 μg/L | A comparison with MOF-199 and Fullerene@MOF-199 was carried out. | [55] |
| 3 NSAIDs | Tap, river water, and groundwater (10 mL) | m-μ-dSPE | $Fe_3O_4$@PEI-r-GO (5 mg) | HPLC-DAD | 91–101 | 0.2 μg/L | A comparison with $Fe_3O_4$@PEI, $Fe_3O_4$@r-GO and $Fe_3O_4$@PEI-GO was carried out. The sorbent could be reused for 10 cycles. | [56] |

| | | | | | | | | |
|---|---|---|---|---|---|---|---|---|
| Imidacloprid and 2, 4-dichloro phenoxyacetic acid | Tap water (50 mL)<br>Cucumber and tomato (1 g) | m-μ-dSPE | TPN@ $Fe_3O_4$@ GO (11 mg) | HPLC-UV | 91–102 | 0.17–1.7 μg/L | Vegetable samples were previously extracted using ACN/water and the extract was submitted to the m-μ-dSPE extraction procedure.<br>The m-μ-dSPE procedure was optimized using an experimental design. | [57] |
| SAs | Milk (10 mL) | m-μ-dSPE | $Fe_3O_4$@r-GO-CNTs (20 mg) | HPLC-UV | 88–106 | 0.35–1.32 μg/L | Milk samples were deproteinized before m-μ-dSPE the extraction procedure.<br>The m-μ-dSPE procedure was assisted by vortex. | [46] |
| 4 PFAAs | Blood serum (0.5 mL) | SPE | GO-$NH_2$@ S-DVB (500 mg) | IC-Q-MS | 86–101 | - | A comparison with commercial HLB sorbent was carried out.<br>TFAA was used as surrogated.<br>Blood serum samples were acid-digested and deproteinized before the SPE extraction procedure.<br>The sorbent could be reused for five cycles.<br>LOQs were found in the range 0.50–2.0 μg/L.<br>The method was applied to real samples. | [58] |

*Continued*

**Table 4.3** Applications of graphene-based materials in solid-phase extraction.—cont'd

| Analyte | Sample (amount) | SPE mode | Sorbent (amount) | Determination technique | Recovery (%) | Detection limits | Comments | Reference |
|---|---|---|---|---|---|---|---|---|
| 9 Mycotoxins | Milk (1 mL) | μ-SPE | r-GO@Au (10 mg) | UHPLC-QqQ-MS/MS | 70–111 | 0.01–0.07 μg/L | Milk samples were deproteinized before the μ-SPE extraction procedure.<br>A comparison with GO and r-GO was carried out.<br>The method was applied to real samples. | [59] |
| Cr(III), Cu(II) Zn(II), and Pb(II) | Mineral, spring, river, and lake water (50 mL) | μ-dSPE | GO@IDA (1 mg) | EDXRF | 95–102 | 0.06–0.11 μg/L | Accuracy and precision of the method was tested by analyzing CRMs. | [60] |
| 10 allergenic disperse dyes | Industrial, environmental, and wastewater (10 mL) | μ-dSPE | r-GO@PS-DVB (20 mg) | SFC-UV | 89–100 | 1.1–15.6 μg/L | The μ-dSPE procedure was assisted by sonication.<br>A comparison with PS-DVB and two commercial sorbents (silica $C_{18}$ and Oasis HLB) was carried out.<br>The sorbent could be reused for 20 cycles. | [61] |
| Melatonin, L-Tryptophan and derivates | Black sesame seeds (2 g) | μ-dSPE | GO@$SiO_2$ (15 mg) | HPLC-DAD | 89–115 | 50–100 μg/L | The μ-dSPE procedure was assisted by vortex.<br>The results obtained were confirmed by UHPLC-MS/MS. | [62] |
| 14 PAEs | Mineral, pond and wastewater (25 mL) | m-μ-dSPE | $Fe_3O_4$@r-GO (60 mg) | UHPLC-QqQ-MS/MS | 70–120 | 0.006–0.178 μg/L | DBP-$d_4$ was used as surrogated. | [63] |

|  |  |  |  |  |  |  |  |  |
|---|---|---|---|---|---|---|---|---|
| Sildenafil citrate | Herbal products (1 g) | m-μ-dSPE | $Fe_3O_4$@ND@GO (30 mg) | HPLC-DAD | 94–104 | 1.46 μg/L | An extraction with MeOH was previously developed.<br>The extraction procedure was assisted by vortex.<br>The sorbent could be reused for 10 cycles.<br>The method was applied to real pharmaceutical herbal products samples. | [64] |
| 8 pesticides | Cabbage (10 g) | m-μ-dSPE | $Fe_3O_4$@$SiO_2$@r-GO@$VOIm^+$ $NapSO_3^-$ (60 mg) | UHPLC-QqQ-MS/MS | 79–96 | 0.01–0.15 μg/L | 34 preservatives were initially selected as target analytes.<br>An extraction with ACN/hexane was previously developed.<br>The extraction procedure was assisted by vortex.<br>A comparison between four types of ILs (benzene-, naphthalene-, naphthoquinone-, and anthraquinone-ring-bound imidazole structures) combined with $Fe_3O_4$@$SiO_2$@r-GO was carried out.<br>The method was applied to real vegetable samples (cabbage, cucumber, tomato, potato, and carrot). | [65] |

*Continued*

**Table 4.3** Applications of graphene-based materials in solid-phase extraction.—cont'd

| Analyte | Sample (amount) | SPE mode | Sorbent (amount) | Determination technique | Recovery (%) | Detection limits | Comments | Reference |
|---|---|---|---|---|---|---|---|---|
| 5 carbamate pesticides | Fruit juices (3 mL) | PT-SPE | r-GO (4 mg) | UHPLC-QqQ-MS/MS | 81−125 | 0.0022−0.033 μg/L | A comparison with commercial sorbents ($C_{18}$, PRS, and GCB) was carried out. The method was applied to real juice samples (grape, pear, and lemon juices). | [66] |

μ-dSPE: micro-dispersive solid-phase extraction; μ-SPE: microsolid-phase extraction; ACN: acetonitrile; $C_{18}$: octadecylsilane; CNT: carbon nanotube; CRM: certified reference material; DAD: diode-array detector; DBP-$d_4$: dibutyl phthalate-3,4,5,6-$d_4$; EDXRF: energy dispersive X-ray fluorescence spectrometry; FAAS: flame atomic absorption spectrometry; FID: flame ionization detector; GC: gas chromatography; GCB: graphitized carbon black; GO: graphene oxide; GO-NH2: amino-functionalized−graphene oxide; HLB: hydrophilic-lipophilic balanced; HPLC: high-performance liquid chromatography; IC: ion chromatography; ICP: inductively coupled plasma; IDA: 2,2′-iminodiacetic acid; IL: ionic liquid; m-μ-dSPE: magnetic-microdispersive solid-phase extraction; MeOH: methanol; MOF: metal-organic framework; MS/MS: tandem mass spectrometry; MS: mass spectrometry; ND: nanodiamond; NSAID: nonsteroidal antiinflammatory drug; OES: optical emission spectrometry; PAE: phthalic acid ester; PAH: polycyclic aromatic hydrocarbon; PEI: polyethylenimine; PFAA: perfluorinated alkyl acid; Phen: 5-amino-1,10-phenanthroline; P-*p*-AP: poly-(*para*-aminophenol); PRS: propylsulfonic acid silica; PS-DVB: poly-(styrene-co-divinylbenzene); PT: pipette tip; Q: single quadrupole; QqQ: triple quadrupole; r-GO: reduced-graphene oxide; SA: sulfonamide; S-DVB: styrene-divinylbenzene; SFC: supercritical fluid chromatography; SPE: solid-phase extraction; TFAA: trifluoroacetic acid; TPN: triazine-based polymeric network; UHPLC: ultrahigh-performance liquid chromatography; UV: ultraviolet detector; $VOIm^+NapSO_3^-$: 1-vinyl-3-octylimidazole-2-naphthalene sulfonate.

intermediate material between graphene and GO obtained by chemical reduction using typical reducing agents such as hydrazine ($N_2H_4$) or sodium borohydride ($NaBH_4$), among others [10,69], constitutes another alternative for graphene. Shi et al. [63] developed a method for the simultaneous determination of five carbamate pesticides residues in juice samples using pipette tip (PT)-SPE with r-GO as a sorbent for analysis by LC. Recovery values in the range 81%−125% and LODs between 0.0022 and 0.033 μg/L demonstrated the suitability of the method in which only 4 mg of adsorbent was consumed for the extraction of the target compounds. In addition, r-GO proved more effective than octadecylsiloxane-bonded silica, a silica-based propylsulfonic acid material and graphitized carbon black (GCB) for the extraction of the carbamate contaminants.

In addition to the above advantages, GO and r-GO are easier to functionalize than graphene, and these modifications result in more specific interactions with target compounds resulting in more selective extraction procedures [52]. Jiang et al. [57] used r-GO in combination with gold nanoparticles (NPs) for the simultaneous μ-SPE extraction of nine mycotoxins in milk. The r-GO@Au composite showed to be a more effective adsorbent than either GO or r-GO due to the gold NPs acting as nanospacers that minimize the agglomeration of the carbon-based material.

The fact that small sheets of graphene, GO and r-GO can cause high back pressure and escape from the SPE cartridge limits their application in conventional SPE [51]. However, these carbon-based materials can be added directly to the sample solution providing strong interactions with target compounds in the dispersive SPE mode. Niu et al. [59] prepared GO immobilized on a silica surface by a sol-gel technique for the simultaneous clean-up and preconcentration of melatonin and structurally related compounds in black sesame seeds by μ-dSPE. Only 15 mg of GO@$SiO_2$ were required for extraction. Graphene-based materials functionalized with biomolecules [70], polymers [54−56,58,64], ionic liquids (ILs) [62], molecularly imprinted polymers (MIPs) [71−73], and other nanomaterials (e.g., metal-organic frameworks (MOFs) [53], CNTs [45] or NDs [61]) have been employed to improve extraction selectivity. Amiri et al. [53] used a hybrid nanocomposite prepared from a copper-based MOF (MOF-199) and graphene or fullerene for the μ-dSPE of PAHs from water samples, Fig. 4.3. The graphene@MOF-199 nanocomposite (20 mg) offered the highest extraction efficiencies for the target compounds. The higher adsorption capacity of graphene@MOF-199 is due to its greater porosity and higher surface area than MOF-199 and fullerene@MOF-199.

Graphene-based materials can also be modified with m-NPs, which allow their easy separation from the sample solution using a magnet after the extraction process, avoiding filtration or centrifugation steps [20,74]. The magnetic core is usually composed of $Fe_3O_4$ because of their easy synthesis and surface modification, low cost, and suitable stability to typical extraction conditions [10]. Li et al. [54], used a magnetic polyethyleneimine (PEI) modified r-GO composite ($Fe_3O_4$@PEI@r-GO) for the extraction of polar NSAIDs from different water samples. The introduction of PEI as r-GO modifier not only improved its affinity for the polar compounds of interest but also increased the number of available adsorption sites. The adsorption mechanism of $Fe_3O_4$@PEI@r-GO was compared with those of

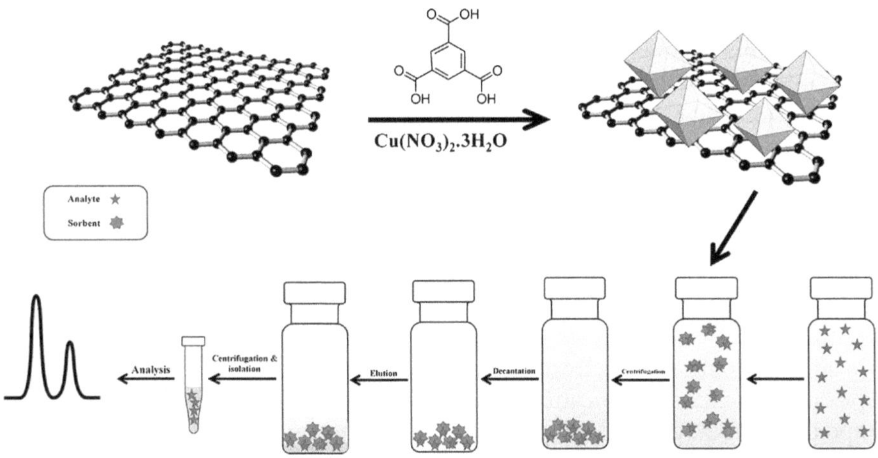


**Figure 4.3** Scheme of the graphene@MOF-199 nanocomposite synthesis and subsequent application on the μ-dSPE of PAHs from water samples.
Reprinted from Amiri A, Ghaemi F, Maleki B. Hybrid nanocomposites prepared from a metal-organic framework of type MOF-199(Cu) and graphene or fullerene as sorbents for dispersive solid phase extraction of polycyclic aromatic hydrocarbons. Microchim Acta 2019;186:1−8 with permission from Springer-Verlag Wien.

$Fe_3O_4$@PEI, $Fe_3O_4$@r-GO, and $Fe_3O_4$@PEI@GO. The results demonstrated that hydrogen-bond interactions with amino groups and hydrophobic interactions with the polymer chains for the compounds of interest were responsible for its improved extraction performance compared with GO and r-GO. Graphene-based materials have also been used for on-line SPE [75,76] and in combination with other sample preparation techniques, such as DLLME [72,77].

### 4.2.4 Nanohorns, nanodiamonds, nanofibres, and quantum dots

In recent years, a large number of novel carbon-based materials have been evaluated as extraction sorbents in SPE, including carbon NHs (CNHs), NDs, carbon nanofibers (CNFs) and QDs as well as their modifications. CNHs are conical carbon nanostructures formed from a $sp^2$ carbon sheet. Conceptually, single-walled CNHs (SWCNHs) are related to SWCNTs but are shaped like a horn with a narrower cone angle composed of five pentagonal rings at its apex. These structures form stable spherical aggregates of ~100 nm in diameter bound by van der Waals forces [78]. CNHs are a good alternative to CNTs in many applications due to their high porosity and large surface area, high affinity for organic compounds [4], low-toxicity (absence of a toxic metal catalyst used in their synthesis compared to CNTs) and the possibility of mass production of pure material at ambient temperature. However, SWCNHs have not been as widely used as other carbon-based nanomaterials possibly owing to the difficulty of surface functionalization and limited commercial availability [78]. Table 4.4 summarizes the applications

**Table 4.4** Applications of carbon-based nanohorns, nanodiamons, nanofibers, and quantum dots in solid-phase extraction.

| Analyte | Sample (amount) | SPE mode | Sorbent (amount) | Determination technique | Recovery (%) | Detection limits | Comments | Reference |
|---|---|---|---|---|---|---|---|---|
| **Nanohorns** | | | | | | | | |
| Benzophenone-3 | Swimming pool water (200 mL) | µ-SPE | *o*-SWCNHs (< 10 mg) | UHPLC-DAD | 78–110 | 0.16 µg/L | *o*-SWCNHs were immobilized on the pores of an activated borosilicate disk.<br>The disk was coupled to a rotating metallic axle that was assembled in a drill to form a stirring extraction device. | [79] |
| 11 triazines | Water (10 mL) | µ-dSPE | *o*-SWCNHs (0.2 mg) | GC-Q-MS | 63–110 | 15–100 ng/L | A dispersion of 0.2 g/L of *o*-SWCNHs in Milli-Q water was used.<br>Extraction was assisted by vortex.<br>*o*-SWCNHs were recovered by filtration on a 0.45 µm Nylon filter and analytes were eluted with 200 µL of MeOH.<br>River, tap, and bottled mineral water samples were analyzed. | [80] |

*Continued*

**Table 4.4** Applications of carbon-based nanohorns, nanodiamons, nanofibers, and quantum dots in solid-phase extraction.—cont'd

| Analyte | Sample (amount) | SPE mode | Sorbent (amount) | Determination technique | Recovery (%) | Detection limits | Comments | Reference |
|---|---|---|---|---|---|---|---|---|
| **Nanodiamonds** | | | | | | | | |
| β-Sitosterol | Edible oils (-) | Online-μ-SPE | Poly-(ND@co-$C_{12}$-co-TEGDMA-co-TAIC) (2 mg) | HPLC-UV | 90–106 | 3.0 mg/L | The column was connected with $C_{18}$ column.<br>The separation performance of the monolith column was evaluated for the separation of small acidic, basic, and neutral aromatic compounds.<br>Extraction procedure was applied to commercial phytosterol mixtures, corn, corn germ, and peanut oil, and the extractive of peanut cake.<br>Monolithic column with and without NDs were compared. | [81] |

| | | | | | | | | |
|---|---|---|---|---|---|---|---|---|
| Doxorubicin | Urine (5 mL) | μ-dSPE | β-CD@NDs (2 mg) | FLD | 93–94 | 18 μg/L | ACN was previously added to urine samples to precipitate proteins.<br>Doxorubicin was eluted from the sorbent using a mixture of γ-CDs in phosphate buffer at 0.5 g/L with 10 % of MeOH to form a more stable host–guest inclusion complex.<br>Fluorometric determination was compared with voltammetric and chromatographic methods. | [82] |
| Ziram | Water (tap and lake) (10 mL)<br>Foodstuff (rice, cracked wheat) (1 g) | m-μ-dSPE | $Fe_3O_4$@NDs-COOH (25 mg) | FAAS | 93–101 | 5.3 μg/L | Crushed foodstuffs (1 g) were previously treated with 15 mL of ACN and the final extract was submitted to the m-μ-dSPE procedure.<br>Extraction was assisted by vortex.<br>DTC synthetic mixtures containing ziram were used to evaluate the influence of other DTCs in the determination of ziram. | [83] |
| **Nanofibers** | | | | | | | | |
| 3 PAHs | Water (50 mL) | m-μ-dSPE | $Fe_3O_4$@ CNFs (10 mg) | GC-FID | 90–101 | 0.008–0.030 μg/L | Desorption of analytes from sorbent was assisted by sonication.<br>Tap, river, well, and wastewater were analyzed. | [84] |

*Continued*

**Table 4.4** Applications of carbon-based nanohorns, nanodiamons, nanofibers, and quantum dots in solid-phase extraction.—cont'd

| Analyte | Sample (amount) | SPE mode | Sorbent (amount) | Determination technique | Recovery (%) | Detection limits | Comments | Reference |
|---|---|---|---|---|---|---|---|---|
| 4 aromatic amines | Wastewater (30 mL) | μ-dSPE | CNFs@PVA (10 mg) | HPLC-UV | 70–108 | 0.009–0.081 μg/L | CNFs@PVA solution was electrospun into fiber mats.<br>Desorption of analytes from sorbent was assisted by sonication.<br>Tap, river, well, and wastewater were analyzed. | [85] |
| La(III), Ce(III), Sm(III), Eu(III), Dy(III), Y(III) | Human hair (0.5 g) | μ-SPE | *o*-CNFs (20 mg) | ICP-MS | 95–115 | 0.2–1.2 ng/L | Samples were previously decomposed with concentrated nitric acid and hydrogen peroxide, and then digested in a microwave oven.<br>200 mL of sample extract were submitted to μ-SPE procedure.<br>Method validation was carried out using a human hair CRM. | [86] |

| | | | | | | | | |
|---|---|---|---|---|---|---|---|---|
| **Quantum dots** | | | | | | | | |
| Aflatoxin $B_1$ | Peanut (5 g) | SPE | C-QDs@MIP (-) | HPLC-FLD | 80–91 | 0.118 μg/L | C-QDs were doped in MIP monolithic column.<br>5,7-dimethoxycoumarin as alternative template molecule instead aflatoxin $B_1$.<br>Peanut sample was extracted with MeOH:water (8:2, v/v) mixture and the extract was submitted to the SPE procedure.<br>C-QDs@MIP and C-QDs@NIP absorption capacity were compared showing that the first one exhibited a higher affinity.<br>Barley, maize, and barley and nutmeg traditional Chinese medicine samples were analyzed. | [87] |
| 3 BFRs | Water (10 mL) | μ-dSPE | $MoS_2$@C-QDs (40 mg) | HPLC-DAD | 80–91 | 0.01–0.06 μg/L | Extraction and desorption of analytes were assisted by sonication.<br>Desorption step was repeated twice.<br>Single $MoS_2$ and C-QDs, and $MoS_2$@C-QDs were firstly compared. | [88] |

*Continued*

**Table 4.4** Applications of carbon-based nanohorns, nanodiamons, nanofibers, and quantum dots in solid-phase extraction.—cont'd

| Analyte | Sample (amount) | SPE mode | Sorbent (amount) | Determination technique | Recovery (%) | Detection limits | Comments | Reference |
|---|---|---|---|---|---|---|---|---|
| BPA | Deionized water (600 mL) | m-μ-dSPE | $Fe_3O_4$@G-QDs (50 mg) | HPLC-UV | 96–105 | 12.3 ng/L | Adsorption efficiencies of $Fe_3O_4$, $Fe_3O_4$@graphene, and $Fe_3O_4$@G-QDs were compared.<br>A study of the migration of BPA in mineral water samples stored in polyethylene bottles subjected to sunlight for a week was carried out.<br>Mineral water was analyzed. | [89] |

β-CD: beta-cyclodextrin; γ-CD: gamma-cyclodextrin; μ-dSPE: micro-dispersive solid-phase extraction; μ-SPE: micro-solid-phase extraction; ACN: acetonitrile; BFR: brominated flame retardant; BPA: bisphenol A; $C_{12}$: 1-dodecene; $C_{18}$: octadecylsilane; CNF: carbon nanofiber; C-QD: carbon-quantum dot; CRM: certified reference material; DAD: diode-array detector; DTC: dithiocarbamate; FAAS: flame atomic absorption spectrometry; FID: flame ionization detector; FLD: fluorescence detector; GC: gas chromatography; G-QD: graphene-quantum dot; HPLC: high-performance liquid chromatography; ICP: inductively coupled plasma; m-μ-dSPE: magnetic-microdispersive solid-phase extraction; MeOH: methanol; MIP: molecularly imprinted polymer; MS: mass spectrometry; ND: nanodiamond; ND-COOH: carboxylated nanodiamond; NIP: nonimprinted polymer; *o*-CNF: *oxidized*-carbon nanofiber; *o*-SWCNH: *oxidized*-single-walled carbon nanohorn; PAH: polycyclic aromatic hydrocarbon; PVA: poly-(vinyl alcohol); Q: single quadrupole; SPE: solid-phase extraction; TAIC: triallyl isocyanurate; TEGDMA: triethylene glycol dimethacrylate; UHPLC: ultrahigh-performance liquid chromatography; UV: ultraviolet detector; ziram: zinc bis-(dimethyldithiocarbamate).

of SWCNHs and *oxidized* SWCNHs (*o*-SWCNHs) in dSPE [4]. Roldán-Pijuán et al. [79] developed a method based on *o*-SWCNHs (less than 10 mg) immobilized on a stir borosilicate disk extraction device for the pre-concentration of benzophenone-3 from swimming pool water. Recovery values of 78%−110% and a LOD of 0.16 μg/L were obtained, although the high time-consuming preparation of the extraction device and lack of automation are likely to limit its use for routine analysis.

Apart from conventional SPE, other modes, such as μ-dSPE, have been used with *o*-SWCNHs for the extraction of PAHs [80], triazines [81] and bilirubin [82] from water [80,81] and plasma samples [82], respectively. Jiménez-Soto et al. [81] employed μ-dSPE for the extraction of 11 triazines in river, tap, and bottled mineral water. 10 mL of water samples were treated with 1 mL of *o*-SWCNHs in Milli-Q water at a concentration of 0.2 g/L and stirred for 2 min using a vortex mixer. Afterward, the sorbent was isolated by filtration on a 0.45 μm nylon filter, and the analytes desorbed with 200 μL of MeOH for GC analysis (recovery range 63%−110% and LODs of 15−100 ng/L).

NDs are characterized by high hardness and chemical stability. In addition, NDs can have a high specific surface area and, although the surface of common synthetic diamonds does not have functional groups, it can be modified by coating them with polymers or by introducing different functional groups to improve the selectivity [4,83]. Currently, applications of NDs for SPE are scarce, perhaps due to their high cost in comparison with other sorbents [83]. As shown in Table 4.4, NDs have been combined with other materials, such as GO [61] or $MoS_2$ [84], functionalized by oxidation [85] or coated with polyarginine [86], single/double-arm amide-thiourea ligands [87], β-cyclodextrin (β-CD) [88], bacteria [89] and polymeric mixtures [90] to improve their extraction performance for a wider range of compounds. Cui et al. [90] prepared a monolithic column with NDs modified in multiple steps by a triethylene glycol dimethacrylate (TEGDMA) and triallyl isocyanurate (TAIC) polymer, poly(ND@*co*-$C_{12}$-*co*-TEGDMA-*co*-TAIC) for the online (μ-SPE) of β-sitosterol in edible oil (recovery range 90%−106% and LOD of 3 mg/L). The extraction procedure was also applied to other vegetable samples, such as corn, corn germ, peanut oil, and peanut cake. The polymer monolith containing NDs provided better extraction performance than the naked polymer monolith. *Bacillus altitudinis* immobilized on NDs was used for the pre-concentration of metals ions in water and food samples (apple juice, strawberry juice, energy drink, meat, chicken, flour, honey, milk, olive, white cheese, corn, tomato, potato, and black tea samples) by SPE [89]. The modification of the ND surface with the specific bacteria provided low LODs in the range 0.016−0.071 μg/L. Most applications employing NDs used the dSPE mode [84−88]. The NDs were generally functionalized [85−88] or used in combination with other sorbents [84] improving the selectivity and versatility of the materials. They were used for the extraction of drugs [88], oligosaccharides [86] membrane proteins [85], and metal ions [84,87] in diluted solutions of artificial cerebrospinal fluid [86], *methylococcus capsulatus* and *E. coli* cells [84,85], water [87] and urine samples [88].

Yilmaz et al. [91] used magnetic carboxylated-NDs (NDs-COOH) for the extraction of ziram fungicide (zinc bis-(dimethyldithiocarbamate) complex) from water, foodstuffs, and synthetic mixtures using a vortex-assisted m-μ-dSPE procedure with 10 mL of the sample solution and 25 mg of $Fe_3O_4$@NDs-COOH. The solution was vortexed for 2 min, the sorbent isolated by a magnet, and the ziram eluted with 1 mL of 0.25 M $HNO_3$ in acetone. The total extraction time was less than 10 min per sample, recovery values between 93% and 101% and a LOD of 5.3 μg/L. In another application, Yilmaz et al. [61] used a magnetic sorbent based on NDs and GO for the extraction of sildenafil from alleged herbal aphrodisiacs. In this case, the aim of NDs was to avoid the problems of aggregation and restacking of GO nanosheets by their insertion between the GO sheets.

CNFs lack a hollow cavity, and their diameters are generally larger than those of CNTs. They are solid carbon fibers a few microns long with diameters less than 100 nm. An attraction for SPE is their uniform mesoporous structure with a high specific surface area, up to 1877 $m^2/g$, among the highest values recorded for a nanostructured material [4]. CNFs are expected to have a high adsorption capacity. However, untreated CNFs are nonpolar and have a low affinity for the extraction of polar compounds. Their surface can be treated to form certain functional groups that favor the establishment of specific interactions [92]. Generally, CNFs have been used alone as sorbents in SPE [93–97] and in a few cases after surface functionalization [92,98,99] or in combination with other sorbents, such as m-NPs [100,101] or graphene [101]. Chen et al. [92] extracted trace rare earth elements (including La(III), Ce(III), Sm(III), Eu(III), Dy(III), and Y(III)) in certified reference material (CRM) of human hair using *oxidized* CNFs (*o*-CNFs) as a sorbent in μ-SPE (recovery range 95%–115% and LODs around 0.2–1.2 ng/L). The adsorption capacity of *o*-CNFs was compared with MWCNTs and activated carbon under the same conditions, demonstrating a higher adsorption capacity for the *o*-CNFs due to the formation of polar groups on its surface. In addition to the SPE of metal ions [92,94,95,97], CNFs were also used for the extraction of triazine herbicides [93], fungicides [96], and explosive compounds [99] in various samples, such as soil [93], water [94,99], CRMs for tea leaves [95], human hair [92] and mussels [95], apple juice [96], lemon [96] and cucumber [96]. Although CNFs have been used mainly as sorbents for conventional SPE, they have also been used in the μ-dSPE [98] and m-μ-dSPE [100] formats for the extraction of aromatic amines and PAHs, respectively, from water samples.

QDs first identified in 2004 [102] during the purification of SWCNTs by electrophoresis, are zero-dimensional, chemically inert nanomaterial with high fluorescence activity, highly water-soluble, small particle sized (less than 10 nm), have a high surface-to-volume ratio, negligible toxicity, and excellent photostability. In addition, QDs are highly biocompatible and environmentally friendly. Also, these NPs represent a good alternative to conventional inorganic semiconductor QDs [103,104]. In general terms, QDs can be classified into two groups that differ in some structural features: carbon-QDs (C-QDs) and graphene-QDs (G-QDs). While C-QDs are quasi-spherical NPs with diameters lower than 10 nm, G-QDs consist of graphene nanosheets of less than 10 nm thickness with single, double, or multiple layers [103]. Since most of the synthesized QDs are NPs that contain carboxyl, hydroxyl, carbonyl, epoxide, and amino functional groups, among others, on its surface, they have a high adsorption capacity due to the presence and number of active sites [103,105]. Despite their theoretical advantages as extraction sorbents, they have scarcely been exploited in this role [106] compared with potential applications as sensors or biosensors [104]. Table 4.4 summarizes some recent applications for SPE. Liang et al. [107] prepared a MIP-based monolithic column doped with C-QDs for the selective extraction of aflatoxin $B_1$ from food samples (peanut, barley, maize, nutmeg) and traditional herbal medicine. The C-QDs in the C-QDs@MIP monolith increased the adsorption capacity of the column.

QDs have been applied as sorbents either as raw materials [108] or functionalized [106,109,110] and in combination with other types of sorbents [105–107,109–115]. Razmi et al. [108] provided the only example of the use of single G-QDs, in the form of eggshell, for the extraction of PAHs from water. The high hydrophilicity resulting from the presence of surface functional groups increased the dispersion capacity of QDs in aqueous solution but hindered their separation from the medium, which limits their application as an adsorbent in dSPE [105,115]. Therefore, in general, the QDs have been supported on other materials for applications in μ-dSPE [105,106,115] or combined with m-NPs for m-dSPE [113] and m-μ-dSPE [109,111,112,114] formats. Dong et al. [105] combined the advantages of C-QDs with those of $MoS_2$ for the μ-dSPE of three brominated flame retardants (BFRs) from water samples. The intrinsic properties of C-QDs added to the layered structure of $MoS_2$ provide favorable conditions for the extraction with the recovery values of 80%–91% and LODs of 0.01–0.06 μg/L for the brominated flame retardants. To demonstrate the extraction capacity of the prepared composite, single $MoS_2$, and C-QDs, and $MoS_2$@ C-QDs were compared, demonstrating the synergy of interactions for the $MoS_2$@ C-QDs sorbent.

Although for conventional SPE, QDs have only been used for the extraction of PAHs [108], pesticides [110] and aflatoxins [107], their potential for the extraction of other organic compounds, such as bisphenol A (BPA) [111], BFRs [105], and metal ions [106,109,112,114,115] by μ-dSPE or m-μ-dSPE has received attention. Mohammad-Rezaei et al. [111] synthesized a magnetic nanocomposite composed of m-NPs and G-QDs by the hydrothermal method. The $Fe_3O_4$@G-QDs adsorbent was used for the preconcentration of BPA in drinking water samples (recovery range 96%–105% and LOD of 12.3 ng/L). Fig. 4.4 [116] illustrates the synthesis of G-QDs and graphene (or GO) by pyrolysis of citric acid. The G-QDs consist of graphene sheets less than 10 nm thick with a lateral size of 100 nm. The transformation of the graphene sheets into the G-QDs considerably increases the extraction capacity of the sorbent due to the high surface-to-volume ratio and high adsorption capacity for aromatic compounds due to π-π interactions.

QDs have high water solubility and can be used to improve the suitability of other sorbents for extraction from aqueous solution. Yang et al. [117] used a C-QDs-based composite to enhance the dispersion and hydrophilicity of an IL in water and milk samples. The nanomaterial (a ferrofluid) was formed between m-NPs coated with oleic acid, C-QDs, and 1-octyl-3-methylimidazolium hexafluorophosphate ($[C_8MIM]PF_6$) IL. The C-QDs not only enhanced the dispersibility of the sorbent material but also improved the reproducibility and precision of the method for the extraction of four phenolic compounds.

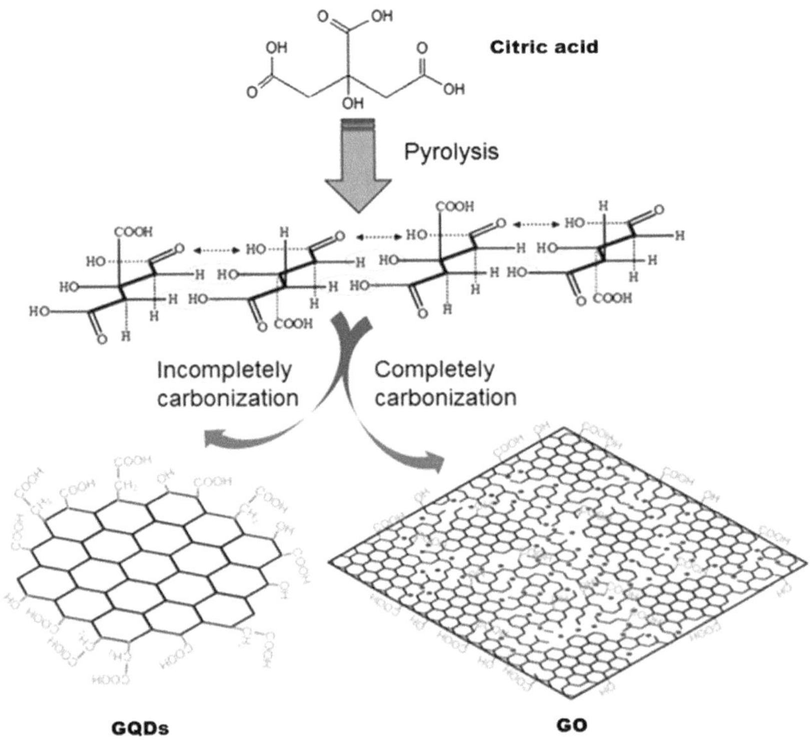


**Figure 4.4** Diagram for the synthesis of GQDs and GO. The black dots in the GO represent oxygen atoms.
Reprinted from Dong Y, Shao J, Chen C, Li H, Wang R, Chi Y, Lin X, Chen G. Blue luminescent graphene quantum dots and graphene oxide prepared by tuning the carbonization degree of citric acid. Carbon 2012;50:4738–4743 with permission from Elsevier.

## 4.3 Conclusions

The search for suitable materials as adsorbents to be used in SPE is an important research area in material science and sample preparation. Among the numerous materials evaluated for SPE carbon-based nanomaterials is an important group with several advantages: favorable mechanical stability, high surface to volume ratio, and the capability to modify selectivity by surface modification processes for the extraction of organic and inorganic compounds from a variety of sample types. Undoubtedly, CNTs and graphene have been the most widely used carbon-based nanomaterials for SPE due to their ease of modification and a large number of commercial products available. However, other materials such as fullerenes, NHs, NDs, CNFs, and C-QDs are less well established but have demonstrated a potential for future developments.

# List of abbreviations

| | |
|---|---|
| **β-CD** | Beta-cyclodextrin |
| **γ-CD** | Gamma-cyclodextrin |
| **$[C_8MIM]PF_6$** | 1-octyl-3-methylimidazolium hexafluorophosphate |
| **μ-dSPE** | Micro-dispersive solid-phase extraction |
| **μ-ECD** | Micro-electron capture detector |
| **μ-SPE** | Micro-solid-phase extraction |
| **8-iso-$PGF_{2\alpha}$-$d_4$** | 8-iso-prostaglandin-$F_{2\alpha}$-$d_4$ |
| **AAS** | Atomic absorption spectrometry |
| **ACN** | Acetonitrile |
| **BFR** | Brominated flame retardant |
| **BP** | Buckypaper |
| **BPA** | Bisphenol A |
| **$C_{12}$** | 1-dodecene |
| **$C_{18}$** | Octadecylsilane |
| **CE** | Capillary electrophoresis |
| **ChCl** | Choline chloride |
| **CNF** | Carbon nanofiber |
| **CNH** | Carbon nanohorn |
| **CNT** | Carbon nanotube |
| **C-QD** | Carbon-quantum dot |
| **CRM** | Certified reference material |
| **DAD** | Diode-array detector |
| **DBP-$d_4$** | Dibutyl phthalate-3,4,5,6-$d_4$ |
| **DCM** | Dichloromethane |
| **DLLME** | Dispersive liquid-liquid microextraction |
| **dSPE** | Dispersive solid-phase extraction |
| **DTC** | Dithiocarbamate |
| **DWCNT** | Double-walled carbon nanotube |
| **ECD** | Electron capture detector |
| **EDXRF** | Energy dispersive X-ray fluorescence spectrometry |
| **FAAS** | Flame atomic absorption spectrometry |
| **FID** | Flame ionization detector |
| **FLD** | Fluorescence detector |
| **GC** | Gas chromatography |
| **GCB** | Graphitized carbon black |
| **GO** | Graphene oxide |
| **GO-$NH_2$** | Amino-functionalized-graphene oxide |
| **G-QD** | Graphene-quantum dot |
| **HLB** | Hydrophilic-lipophilic balanced |
| **HPLC** | High performance liquid chromatography |
| **IC** | Ion chromatography |
| **ICP** | Inductively coupled plasma |
| **IDA** | 2,2′-iminodiacetic acid |
| **IL** | Ionic liquid |
| **IS** | Internal standard |
| **LC** | Liquid chromatography |

| | |
|---|---|
| **LDH** | Layered double hydroxide |
| **LOD** | Limit of detection |
| **LVI** | Large-volume injection |
| **m-μ-dSPE** | Magnetic micro-dispersive solid-phase extraction |
| **MALDI** | Matrix-assisted laser desorption/ionization |
| **m-dSPE** | Magnetic dispersive solid-phase extraction |
| **MeOH** | Methanol |
| **MIP** | Molecularly imprinted polymer |
| **m-MWCNT** | Magnetic multi-walled carbon nanotube |
| **m-NPs** | Magnetic nanoparticles |
| **MOF** | Metal-organic framework |
| **MS** | Mass spectrometry |
| **MS/MS** | Tandem mass spectrometry |
| **MWCNT** | Multi-walled carbon nanotube |
| **MWCNT-COOH** | Carboxylated multi-walled carbon nanotube |
| **MWCNT-$NH_2$** | Amine functionalized-multi-walled carbon nanotube |
| **MWCNT-OH** | Hydroxylated multi-walled carbon nanotube |
| **ND** | Nanodiamond |
| **ND-COOH** | Carboxylated-nanodiamond |
| **NH** | Nanohorn |
| **NIP** | Nonimprinted polymer |
| **NP** | Nanoparticle |
| **NSAID** | Nonsteroidal antiinflammatory drug |
| ***o*-CNF** | *Oxidized* carbon nanofiber |
| **OCP** | Organochlorine pesticide |
| **OES** | Optical emission spectrometry |
| ***o*-MWCNT** | *Oxidized* multi-walled carbon nanotube |
| ***o*-SWCNH** | *Oxidized* single-walled carbon nanohorn |
| **PAE** | Phthalic acid ester |
| **PAH** | Polycyclic aromatic hydrocarbon |
| **PCB** | Polychlorinated biphenyl |
| **PDA** | Photodiode array detector |
| **PEI** | Polyethylenimine |
| **PFAA** | Perfluorinated alkyl acid |
| **Phen** | 5-amino-1,10-phenanthroline |
| **P-*p*-AP** | Poly-(*para*-aminophenol) |
| **PRS** | Propylsulfonic acid silica |
| **PS-DVB** | Poly-(styrene-co-divinylbenzene) |
| **PT** | Pipette tip |
| **PTFE** | Polytetrafluoroethylene |
| **PVA** | Poly-(vinyl alcohol) |
| **Q** | Single quadrupole |
| **QD** | Quantum dot |
| **QqQ** | Triple quadrupole |
| **QTrap** | Triple quadrupole linear ion trap |
| **r-GO** | Reduced-graphene oxide |
| **RP** | Reverse phase |
| **SA** | Sulfonamide |
| **S-DVB** | Styrene-divinylbenzene |

| | |
|---|---|
| **SFC** | Supercritical fluid chromatography |
| **SFOD** | Solidified floating organic droplets |
| **SPE** | Solid-phase extraction |
| **SWCNH** | Single-walled carbon nanohorn |
| **SWCNT** | Single-walled carbon nanotube |
| **TAIC** | Triallyl isocyanurate |
| **TEGDMA** | Triethylene glycol dimethacrylate |
| **TFAA** | Trifluoroacetic acid |
| **ToF** | Time-of-flight |
| **TPN** | Triazine-based polymeric network |
| **UHPLC** | Ultra-high performance liquid chromatography |
| **UV** | Ultraviolet detector |
| **$VOIm^+NapSO_3^-$** | 1-vinyl-3-octylimidazole-2-naphthalene sulfonate |
| **VWD** | Variable-wavelength detector |
| **Wt** | Weight |
| **Ziram** | Zinc bis-(dimethyldithiocarbamate) |

# References


[1] Andrade-Eiroa A, Canle M, Leroy-Cancellieri V, Cerdà V. Solid-phase extraction of organic compounds: a critical review (part I). TrAC-Trends Anal Chem 2016;80:641–54.

[2] Płotka-Wasylka J, Szczepańska N, de la Guardia M, Namieśnik J. Modern trends in solid phase extraction: new sorbent media. TrAC-Trends Anal Chem 2016;77:23–43.

[3] Xu L, Qi X, Li X, Bai Y, Liu H. Recent advances in applications of nanomaterials for sample preparation. Talanta 2016;146:714–26.

[4] Zhang BT, Zheng X, Li HF, Lin JM. Application of carbon-based nanomaterials in sample preparation: a review. Anal Chim Acta 2013;784:1–17.

[5] Socas-Rodríguez B, Herrera-Herrera AV, Asensio-Ramos M, Hernández-Borges J. Recent applications of carbon nanotube sorbents in analytical chemistry. J Chromatogr A 2014;1357:110–46.

[6] Kroto HW, Heath JR, Brien SCO, Curl RF, Smalley RE. $C_{60}$: buckminsterfullerene. Nature 1985;318:162–3.

[7] Gupta VK, Saleh TA. Sorption of pollutants by porous carbon, carbon nanotubes and fullerene-an overview. Environ Sci Pollut Res 2013;20:2828–43.

[8] Wen Y, Chen L, Li J, Liu D, Chen L. Recent advances in solid-phase sorbents for sample preparation prior to chromatographic analysis. TrAC-Trends Anal Chem 2014;9:26–41.

[9] He M, Huang L, Zhao B, Chen B, Hu B. Advanced functional materials in solid phase extraction for ICP-MS determination of trace elements and their species – a review. Anal Chim Acta 2017;973:1–24.

[10] Socas-Rodríguez B, González-Sálamo J, Hernández-Borges J. Carbon nanomaterials in sample preparation. In: Garcia CD, Crevillén AG, Escarpa A, editors. Carbon-based nanomaterials in analytical chemistry. Cambridge: The Royal Society of Chemistry; 2015. p. 37–68.

[11] Zhou DB, Sheng X, Han F, Hu YY, Ding L, Lv YL, Song W, Zheng P. Magnetic solid-phase extraction based on [60] fullerene functionalization of magnetic nanoparticles for the determination of sixteen polycyclic aromatic hydrocarbons in tea samples. J Chromatogr A 2018;1578:53–60.

[12] Rodriguez JA, Ibarra IS, Miranda JM, Barrado E, Santos EM. Magnetic solid phase extraction based on fullerene and activated carbon adsorbents for determination of azo dyes in water samples by capillary electrophoresis. Anal Methods 2016;8:8466−73.

[13] Kilinç E. Determination of trace Bi by ICP-OES after magnetic solid phase extraction with fullerene $C_{60}$ modified $\gamma$-$Fe_2O_3$ superparamagnetic iron oxide nanoparticles. Anal Methods 2015;7:10306−11.

[14] Fischnaller M, Bakry R, Vallant RM, Huber LA, Bonn GK. $C_{60}$-fullerene bound silica for the preconcentration and the fractionation of multiphosphorylated peptides. Anal Chim Acta 2013;761:92−101.

[15] Serrano A, Gallego M. Fullerenes as sorbent materials for benzene, toluene, ethylbenzene, and xylene isomers preconcentration. J Sep Sci 2006;29:33−40.

[16] Wen Y, Chen L, Li J, Liu D, Chen L. Recent advances in solid-phase sorbents for sample preparation prior to chromatographic analysis. TrAC-Trends Anal Chem 2018;108: 347−69.

[17] Jurado-Sánchez B, Ballesteros E, Gallego M. Fullerenes for aromatic and non-aromatic N-nitrosamines discrimination. J Chromatogr A 2009;1216:1200−5.

[18] Iijima S. Helical microtubules of graphitic carbon. Nature 1991;354:56−8.

[19] Valcárcel M, Cárdenas S, Simonet BM, Moliner-Martínez Y, Lucena R. Carbon nanostructures as sorbent materials in analytical processes. TrAC-Trends Anal Chem 2008;27: 34−43.

[20] Herrero-Latorre C, Barciea-García J, García-Martín S, Peña-Crecente RM. Graphene and carbon nanotubes as solid phase extraction sorbents for the speciation of chromium: a review. Anal Chim Acta 2018;1002:1−17.

[21] Tomai P, Martinelli A, Gasperi T, Bianchi M, Purcaro V, Teofili L, Papacci P, Cori MS, Vento G, Curini R, Fanali S, Gentili A. Rotating-disc micro-solid phase extraction of F2-isoprostanes from maternal and cord plasma by using oxidized buckypaper as sorbent membrane. J Chromatogr A 2019;1586:30−9.

[22] López-Lorente AI, Simonet BM, Valcárcel M. The potential of carbon nanotube membranes for analytical separations. Anal Chem 2010;82:5399−407.

[23] Ma J, Lu X, Xia Y, Yan F. Determination of pyrazole and pyrrole pesticides in environmental water samples by solid-phase extraction using multi-walled carbon nanotubes as adsorbent coupled with high-performance liquid chromatography. J Chromatogr Sci 2015;53:380−4.

[24] Behbahani M, Bagheri A, Amini MM, Sadeghi O, Salarian M, Najafi F, Taghizadeh M. Application of multiwalled carbon nanotubes modified by diphenylcarbazide for selective solid phase extraction of ultra traces Cd(II) in water samples and food products. Food Chem 2013;141:48−53.

[25] Rezvani M, Ebrahimzadeh H, Aliakbari A, Khalilzadeh A, Kasaeian M, Amini MM. Novel modified carbon nanotubes as a selective sorbent for preconcentration and determination of trace copper ions in fruit samples. J Sep Sci 2014;37:2559−65.

[26] Barbosa VMP, Barbosa AF, Bettini J, Luccas PO, Figueiredo EC. Direct extraction of lead (II) from untreated human blood serum using restricted access carbon nanotubes and its determination by atomic absorption spectrometry. Talanta 2016;147:478−84.

[27] Londonio A, Hasuoka PE, Pacheco P, Gil RA, Smichowski P. Online solid phase extraction-HPLC-ICP-MS system for mercury and methylmercury preconcentration using functionalised carbon nanotubes for their determination in dietary supplements. J Anal At Spectrom 2018;33:1737−44.

[28] Al-Rifaia A, Aqel A, Wahibia LA, ALOthman ZA, Badjah-Hadj-Ahmed A-Y. Carbon nanotube-based benzyl polymethacrylate composite monolith as a solid phase extraction adsorbent and a stationary phase material for simultaneous extraction and analysis of polycyclic aromatic hydrocarbon in water. J Chromatogr A 2018;1535:17–26.

[29] Jiang K, Huang P, Luan L, Fan K, Guo W, Zhao Z, Wu Y, Han Z. Iron (II, III) oxide/multi-walled carbon nanotube composite as solid-phase extraction sorbent followed by ultra-high performance liquid chromatography tandem mass spectrometry for simultaneous determination of zearalenone and type A trichothecenes in *Salviae miltiorrhizae* Radix et Rhizoma (Danshen). J Chromatogr A 2017;1482:1–10.

[30] Yuan Y, Jiao X, Han Y, Bai L, Liu H, Qiao F, Yan H. One-pot synthesis of ethylenediamine-connected graphene/carbon nanotube composite material for isolation of clenbuterol from pork. Food Chem 2017;230:154–63.

[31] Socas-Rodríguez B, González-Sálamo J, Hernández-Borges J, Rodríguez-Delgado MÁ. Application of multiwalled carbon nanotubes as sorbents for the extraction of mycotoxins in water samples and infant milk formula prior to high performance liquid chromatography mass spectrometry analysis. Electrophoresis 2017;37:1359–66.

[32] Herrera-Herrera AV, Hernández-Borges J, Afonso MM, Palenzuela JA, Rodríguez-Delgado MÁ. Comparison between magnetic and non magnetic multi-walled carbon nanotubes-dispersive solid-phase extraction combined with ultra-high performance liquid chromatography for the determination of sulfonamide antibiotics in water samples. Talanta 2013;116:695–703.

[33] Xu X, Long N, Lv J, Wang L, Zhang M, Qi X, Zhang L. Functionalized multiwalled carbon nanotube as dispersive solid-phase extraction materials combined with high-performance liquid chromatography for thiabendazole analysis in environmental and food samples. Food Anal Methods 2016;9:30–7.

[34] Piramoon S, Azar PA, Tehrani MS, Mohammadiazar S, Tavassoli A. Solid-phase nanoextraction of polychlorinated biphenyls in water and their determination by gas chromatography with electron capture detector. J Sep Sci 2017;40:449–57.

[35] Paszkiewicz M, Caban M, Bielicka-Giełdoń A, Stepnowski P. Optimization of a procedure for the simultaneous extraction of polycyclic aromatic hydrocarbons and metal ions by functionalized and non-functionalized carbon nanotubes as effective sorbents. Talanta 2017;165:405–11.

[36] Du L-J, Hu Y-H, Wang Q-Y, Zhang Q-D, Chen Y-B, Peng L-Q, Pan S-L, Li Q, Cao J. Crown ether microfunctionalized carbon nanotubes for dispersive microsolid-phase extraction of Sudan dyes and their metabolites. Food Chem 2018;262:118–28.

[37] Arghavani-Beydokhti S, Rajabi M, Asghari A. Coupling of two centrifuge less ultrasound-assisted dispersive solid/liquid phase microextractions as a highly selective, clean, and efficient method for determination of ultra-trace amounts of non-steroidal anti-inflammatory drugs in complicated matrices. Anal Chim Acta 2018;997:67–79.

[38] Feist B, Sitko R. Method for the determination of Pb, Cd, Zn, Mn and Fe in rice samples using carbon nanotubes and cationic complexes of batophenanthroline. Food Chem 2018; 249:38–44.

[39] Murtada K, de Andrés F, Ríos A, Zougagh M. Determination of antidepressants in human urine extracted by magnetic multiwalled carbon nanotube poly(styrene-co-divinylbenzene) composites and separation by capillary electrophoresis. Electrophoresis 2018;39: 1808–15.

[40] Razmkhah K, Sereshti H, Soltani S, Nodeh HR. Extraction and determination of three steroid molecules in milk using functionalized magnetic carbon nanotube-based solid phase extraction coupled with HPLC. Food Anal Methods 2018;1:3179–89.
[41] Zhao P, Wang Z, Li K, Guo X, Zhao L. Multi-residue enantiomeric analysis of 18 chiral pesticides in water, soil and river sediment using magnetic solid-phase extraction based on amino modified multiwalled carbon nanotubes and chiral liquid chromatography coupled with tandem mass spectrometry. J Chromatogr A 2018;1568:8–21.
[42] Peng G, He Q, Lu Y, Huang J, Lin J-M. Flow injection microfluidic device with on-line fluorescent derivatization for the determination of Cr(III) and Cr(VI) in water samples after solid phase extraction. Anal Chim Acta 2017;955:58–66.
[43] Kılınç E, Çelik KS, Bilgetekin H. $\gamma$-$Fe_2O_3$ magnetic nanoparticle functionalized with carboxylated multi walled carbon nanotube for magnetic solid phase extractions and determinations of Sudan dyes and Para Red in food samples. Food Chem 2018;242: 533–7.
[44] Chen K, Jin R, Luo C, Song G, Hu Y, Chen H. Synthesis of polydopamine-functionalized magnetic graphene and carbon nanotubes hybrid nanocomposites as an adsorbent for the fast determination of 16 priority polycyclic aromatic hydrocarbons in aqueous samples. J Sep Sci 2018;41:1847–55.
[45] Feng Y, Hu X, Zhao F, Zeng B. $Fe_3O_4$/reduced graphene oxide-carbon nanotubes composite for the magnetic solid-phase extraction and HPLC determination of sulfonamides in milk. J Sep Sci 2019;42:1058–66.
[46] Yousefi SM, Zhemirani F, Ghorbanian SA. Deep eutectic solvent magnetic bucky gels in developing dispersive solid phase extraction: application for ultra trace analysis of organochlorine pesticides by GC-micro ECD using a large-volume injection technique. Talanta 2017;168:73–81.
[47] González-Curbelo MÁ, Socas-Rodríguez B, Herrera-Herrera AV, González-Sálamo J, Hernández-Borges J, Rodríguez-Delgado MÁ. Evolution and applications of the QuEChERS method. TrAC-Trends Anal Chem 2015;71:169–85.
[48] Geim AK, Novoselov KS. The rise of graphene. Nat Mater 2007;6:183–91.
[49] Novoselov KS, Geim AK, Morozov SV, Jiang D, Zhang Y, Dubonos SV, Grigorieva LV, Firsov AA. Electric field effect in atomically thin carbon films. Science 2004;306:666–9.
[50] Crevillén AG, Escarpa A, García CD. Carbon-based nanomaterials in analytical chemistry. In: Garcia CD, Crevillén AG, Escarpa A, editors. Cambridge: The Royal Society of Chemistry; 2015. p. 1–36.
[51] Sitko R, Zawisza B, Malicka E. Graphene as a new sorbent in analytical chemistry. TrAC-Trends Anal Chem 2013;51:33–43.
[52] Tian J, Xu J, Zhu F, Lu T, Su C, Ouyang G. Application of nanomaterials in sample preparation. J Chromatogr A 2013;1300:2–16.
[53] Amiri A, Ghaemi F, Maleki B. Hybrid nanocomposites prepared from a metal-organic framework of type MOF-199(Cu) and graphene or fullerene as sorbents for dispersive solid phase extraction of polycyclic aromatic hydrocarbons. Microchim Acta 2019;186: 1–8.
[54] Li N, Chen J, Shi YP. Magnetic polyethyleneimine functionalized reduced graphene oxide as a novel magnetic sorbent for the separation of polar non-steroidal anti-inflammatory drugs in waters. Talanta 2019;191:526–34.
[55] Shahrebabak SM, Saber-Tehrani M, Faraji M, Shabanian M, Aberoomand-Azar P. Simultaneous magnetic solid phase extraction of acidic and basic pesticides using triazine-based polymeric network modified magnetic nanoparticles/graphene oxide nanocomposite in water and food samples. Microchem J 2019;146:630–9.

[56] Zhong Z, Li G, Guo L, Yao J, Liu Z, Deng J. Solid-phase extraction based on amino-functionalized graphene oxide nanocomposites for analysis of short-chain perfluorinated alkyl acids in human serum by ion chromatography mass spectrometry. Biomed Chromatogr 2018;32:4223–32.

[57] Jiang K, Huang Q, Fan K, Wu L, Nie D, Guo W, Wu Y, Han Z. Reduced graphene oxide and gold nanoparticle composite-based solid-phase extraction coupled with ultra-high-performance liquid chromatography tandem mass spectrometry for the determination of 9 mycotoxins in milk. Food Chem 2018;264:218–25.

[58] Lou C, Wu C, Zhang K, Guo D, Jiang L, Lu Y, Zhu Y. Graphene-coated polystyrene-divinylbenzene dispersive solid-phase extraction coupled with supercritical fluid chromatography for the rapid determination of 10 allergenic disperse dyes in industrial wastewater samples. J Chromatogr A 2018;1550:45–56.

[59] Niu J, Zhang X, Qin P, Yang Y, Tian S, Yang H, Lu M. Simultaneous determination of melatonin, L-tryptophan, and two L-tryptophan-derived esters in food by HPLC with graphene oxide/$SiO_2$ nanocomposite as the adsorbent. Food Anal. Methods 2018;11: 2438–46.

[60] Santana-Mayor A, Socas-Rodríguez B, Afonso MM, Palenzuela-López JA, Rodríguez-Delgado MÁ. Reduced graphene oxide-coated magnetic-nanoparticles as sorbent for the determination of phthalates in environmental samples by micro-dispersive solid-phase extraction followed by ultra-high-performance liquid chromatography tandem mass spectrometry. J Chromatogr A 2018;1565:36–47.

[61] Yilmaz E, Ulusoy Hİ, Demir Ö, Soylak M. New magnetic nanodiamond/graphene oxide hybrid ($Fe_3O_4$@ND@GO) material for pre-concentration and sensitive determination of sildenafil in alleged herbal aphrodisiacs by HPLC-DAD system. J Chromatogr B 2018; 1084:113–21.

[62] Cao S, Chen Y, Zhang L, Xi C, Li X, Tang B, Wang G, Chen Z. Designed multifunctional ionic liquids-magnetic graphene nanocomposites as the adsorbent of MSPE for the determination of preservatives. Anal Methods 2018;10:1420–30.

[63] Shi Z, Li Q, Xu D, Huai Q, Zhang H. Graphene-based pipette tip solid-phase extraction with ultra-high performance liquid chromatography and tandem mass spectrometry for the analysis of carbamate pesticide residues in fruit juice. J Sep Sci 2016;39:4391–7.

[64] Kojidi MH, Aliakbar A. Synthesis of graphene oxide-based poly(p-aminophenol) composite and its application in solid phase extraction of trace amount of Ni(II) from aquatic samples. Environ Monit Assess 2019;191:1–12.

[65] Feist B, Pilch M, Nycz J. Graphene oxide chemically modified with 5-amino-1,10-phenanthroline as sorbent for separation and preconcentration of trace amount of lead(II). Microchim Acta 2019;186:1–8.

[66] Pytlakowska K, Matussek M, Hachula B, Pilch M, Kornaus K, Zubko M, Pisarskia WA. Graphene oxide covalently modified with 2,2′-iminodiacetic acid for preconcentration of Cr(III), Cu(II), Zn(II) and Pb(II) from water samples prior to their determination by energy dispersive X-ray fluorescence spectrometry. Spectrochim Acta Part B 2018;147: 79–86.

[67] Mazaher A, Elmongy H, Madrakian T. Nanomaterials as sorbents for sample preparation in bioanalysis: a review. Anal Chim Acta 2017;958:1–21.

[68] Mehdinia A, Rouhani S, Mozaffari S. Microwave-assisted synthesis of reduced graphene oxide decorated with magnetite and gold nanoparticles, and its application to solid-phase extraction of organochlorine pesticides. Microchim Acta 2016;183:1177–85.

[69] Li N, Jiang H-L, Wang X, Wang X, Xu G, Zhang B, Wang L, Zhao RS, Lin JM. Recent advances in graphene-based magnetic composites for magnetic solid-phase extraction. TrAC-Trends Anal Chem 2018;102:60−74.
[70] Seidi S, Moosavi NS, Shanehsaz M, Abdolhosseini M, Sadeghi SJ. Rapid ultrasound-assisted dispersive solid-phase extraction of nonsteroidal anti-inflammatory drugs in urine using oleic acid functionalized magnetic graphene oxide. J Sep Sci 2018;41: 4370−8.
[71] Guo L, Ma X, Xie X, Huang R, Zhang M, Li J, Zeng G, Fan Y. Preparation of dual-dummy-template molecularly imprinted polymers coated magnetic graphene oxide for separation and enrichment of phthalate esters in water. Chem Eng J 2019;361:245−55.
[72] Xie X, Ma X, Guo L, Fan Y, Zeng G, Zhang M. Novel magnetic multi-templates molecularly imprinted polymer for selective and rapid removal and detection of alkylphenols in water. Chem Eng J 2019;357:55−65.
[73] Han S, Li X, Wang Y, Chen S. Graphene oxide-based fluorescence molecularly imprinted composite for recognition and separation of 2,4,6-trichlorophenol. RSC Adv 2015;5: 2129−36.
[74] Mahpishanian S, Sereshti H. Three-dimensional graphene aerogel-supported iron oxide nanoparticles as an efficient adsorbent for magnetic solid phase extraction of organophosphorus pesticide residues in fruit juices followed by gas chromatographic determination. J Chromatogr A 2016;1443:43−53.
[75] Chen X-L, Ai L-F, Nian Q-X, Jia Y-Q, Hao Y-L, Wang M-M. Rapid determination of sulfonamides in chicken muscle and milk using efficient graphene oxide-based monolith on-line solid-phase extraction coupled with liquid chromatography-tandem mass spectrometry. Food Anal Methods 2019;12:271−81.
[76] Cui B, Guo B, Wang H, Zhang D, Liu H, Bai L, Yan H, Han D. Graphene oxide-based composite monolith as new sorbent for the on-line solid phase extraction and high performance liquid chromatography determination of β-sitosterol in food samples. Talanta 2018;186:200−5.
[77] Yao L, Zhu Y, Xu W, Wang H, Wang X, Zhang J, Liu H, Lin C. Combination of dispersive solid phase extraction with dispersive liquid-liquid microextraction for the sequential speciation and preconcentration of Cr(III) and Cr(VI) in water samples prior to graphite furnace atomic absorption spectrometry determination. J Ind Eng Chem 2019;72: 189−95.
[78] Karousis N, Suarez-Martinez I, Ewels CP, Tagmatarchis N. Structure, properties, functionalization, and applications of carbon nanohorns. Chem Rev 2016;116:4850−83.
[79] Roldán-Pijuán M, Lucena R, Cárdenas S, Valcárcel M. Micro-solid phase extraction based on oxidized single-walled carbon nanohorns immobilized on a stir borosilicate disk: application to the preconcentration of the endocrine disruptor benzophenone-3. Microchem J 2014;115:87−94.
[80] Jiménez-Soto JM, Cárdenas S, Valcárcel M. Evaluation of single-walled carbon nanohorns as sorbent in dispersive micro solid-phase extraction. Anal Chim Acta 2012;714: 76−81.
[81] Jiménez-Soto JM, Cárdenas S, Valcárcel M. Dispersive micro solid-phase extraction of triazines from waters using oxidized single-walled carbon nanohorns as sorbent. J Chromatogr A 2012;1245:17−23.
[82] Yamazaki K, Shinke K, Ogino T. Selective adsorption of bilirubin against albumin to oxidized single-wall carbon nanohorns. Colloids Surf B Biointerfaces 2013;112:103−7.
[83] Nesterenko PN, Haddad PR. Diamond-related materials as potential new media in separation science. Anal Bioanal Chem 2010;396:205−11.

[84] Baghban N, Yilmaz E, Soylak M. Nanodiamond/$MoS_2$ nanorod composite as a novel sorbent for fast and effective vortex-assisted micro solid phase extraction of lead(II) and copper(II) for their flame atomic absorption spectrometric detection. J Mol Liq 2017;234: 260–7.
[85] Pham MD, Yu SS-F, Han C-C, Chan SI. Improved mass spectrometric analysis of membrane proteins based on rapid and versatile sample preparation on nanodiamond particles. Anal Chem 2013;85:6748–55.
[86] Hsieh C-C, Guo JY, Hung S-U, Chen R, Nie Z, Chang H-C, Wu C-C. Quantitative analysis of oligosaccharides derived from sulfated glycosaminoglycans by nanodiamond-based affinity purification and matrix-assisted laser desorption/ionization mass spectrometry. Anal Chem 2013;85:4342–9.
[87] Zhao X, Zhang S, Bai C, Li B, Li Y, Wang L, Wen R, Zhang M, Ma L, Li S. Nanodiamond particles functionalized with single/double-arm amide-thiourea ligands for adsorption of metal ions. J Colloid Interface Sci 2016;469:109–19.
[88] Soriano ML, Carrillo-Carrion C, Ruiz-Palomero C, Valcárcel M. Cyclodextrin-modified nanodiamond for the sensitive fluorometric determination of doxorubicin in urine based on its differential affinity towards β/γ-cyclodextrins. Microchim Acta 2018;185:115.
[89] Ozdemir S, Kilinc E, Celik KS, Okumus V, Soylak M. Simultaneous preconcentrations of $Co^{2+}$, $Cr^{6+}$, $Hg^{2+}$ and $Pb^{2+}$ ions by *Bacillus altitudinis* immobilized nanodiamond prior to their determinations in food samples by ICP-OES. Food Chem 2017;215:447–53.
[90] Cui B, Liu H, Yu H, Pang X, Yan H, Bai L. Monolithic material prepared with nanodiamond as monomer for the enrichment of β-sitosterol in edible oil. Food Anal Methods 2019;12:697–704.
[91] Yılmaz E, Soylak M. Preparation and characterization of magnetic carboxylated nanodiamonds for vortex-assisted magnetic solid-phase extraction of ziram in food and water samples. Talanta 2016;158:152–8.
[92] Chen S, Xiao M, Lu D, Zhan X. Use of a microcolumn packed with modified carbon nanofibers coupled with inductively coupled plasma mass spectrometry for simultaneous on-line preconcentration and determination of trace rare earth elements in biological samples. Rapid Commun Mass Spectrom 2007;21:2524–8.
[93] Boonjob W, Miró M, Segundo MA, Cerdà. Flow-through dispersed carbon nanofiber-based microsolid-phase extraction coupled to liquid chromatography for automatic determination of trace levels of priority environmental pollutants. Anal Chem 2011;83: 5237–44.
[94] Chen S, Zhan X, Lu D, Liu C, Zhu L. Speciation analysis of inorganic arsenic in natural water by carbon nanofibers separation and inductively coupled plasma mass spectrometry determination. Anal Chim Acta 2009;634:192–6.
[95] Chen S, Xiao M, Lu D, Zhan X. Carbon nanofibers as solid-phase extraction adsorbent for the preconcentration of trace rare earth elements and their determination by inductively coupled plasma mass spectrometry. Anal Lett 2007;40:2105–15.
[96] Wang L, Zhang M, Zhang D, Zhan L. New approach for the simultaneous determination fungicide residues in food samples by using carbon nanofiber packed microcolumn coupled with HPLC. Food Control 2016;60:1–6.
[97] Chen S, Xiao M, Lu D, Wang Z. The use of carbon nanofibers microcolumn preconcentration for inductively coupled plasma mass spectrometry determination of Mn, Co and Ni. Spectrochim Acta Part B 2007;62:1216–21.
[98] Vadukumpully S, Basheer C, Jeng CS, Valiyaveettil S. Carbon nanofibers extracted from soot as a sorbent for the determination of aromatic amines from wastewater effluent samples. J Chormatogr A 2011;1218:3581–7.

[99] Tavengwa NT, Hintsho N, Durbach S, Weiersbye I, Cukrowska E, Chimuka L. Extraction of explosive compounds from aqueous solutions by solid phase extraction using β-cyclodextrin functionalized carbon nanofibers as sorbents. J Environ Chem Eng 2016; 4:2450–7.
[100] Sarafraz-Yazdi A, Rokhian T, Amiri A, Ghaemi F. Carbon nanofibers decorated with magnetic nanoparticles as a new sorbent for the magnetic solid phase extraction of selected polycyclic aromatic hydrocarbons from water samples. New J Chem 2015;39: 5621–7.
[101] Rezvani-Eivari M, Amiri A, Baghayeri M, Ghaemi F. Magnetized graphene layers synthesized on the carbon nanofibers as novel adsorbent for the extraction of polycyclic aromatic hydrocarbons from environmental water samples. J Chromatogr A 2016;1465: 1–8.
[102] Xu X, Ray R, Gu Y, Ploehn HJ, Gearheart L, Raker K, Scrivens WA. Electrophoretic analysis and purification of fluorescent single-walled carbon nanotube fragments. J Am Chem Soc 2004;126:12736–7.
[103] Benítez-Martínez S, Valcárcel M. Graphene quantum dots in analytical science. TrAC-Trends AnalChem 2015;72:93–113.
[104] Molaei MJ. A review on nanostructured carbon quantum dots and their applications in biotechnology, sensors, and chemiluminescence. Talanta 2019;196:456–78.
[105] Dong S, Lou Q, Huang G, Guo J, Wang X, Huang T. Dispersive solid-phase extraction based on $MoS_2$/carbon dot composite combined with HPLC to determine brominated flame retardants in water. Anal Bioanal Chem 2018;410:7337–46.
[106] Yang T, Li Y-K, Chen M-L, Wang J-H. Supported carbon dots decorated with metallothionein for selective cadmium adsorption and removal. Chin Chem Lett 2015;26: 1496–501.
[107] Liang G, Zhai H, Huang L, Tan X, Zhou Q, Yu X, Lin H. Synthesis of carbon quantum dots-doped dummy molecularly imprinted polymer monolithic column for selective enrichment and analysis of aflatoxin B1 in peanut. J Pharm Biomed Anal 2018;149: 258–64.
[108] Razmi H, Abdollahi V, Mohammad-Rezaei R. Graphene quantum dots-eggshell nanocomposite to extract polycyclic aromatic hydrocarbons in water. Environ Chem Lett 2016;14:521–6.
[109] Mashkani M, Mehdinia A, Jabbari A, Bide Y, Nabid MR. Preconcentration and extraction of lead ions in vegetable and water samples by N-doped carbon quantum dot conjugated with $Fe_3O_4$ as a green and facial adsorbent. Food Chem 2018;239:1019–26.
[110] Li S, Xu M, Wu X, Luo J. Synergetic recognition and separation of kelthane and pyridaben base on magnetic molecularly imprinted polymer nanospheres. J Sep Sci 2016;39: 4019–26.
[111] Mohammad-Rezaei R, Razmi H, Abdollahi V, Matin AA. Preparation and characterization of graphene quantum dots-$Fe_3O_4$ nanocomposite as an efficient adsorbent in magnetic solid phase extraction: application to determination of Bisphenol A in water samples. Anal Methods 2014;6:8413–9.
[112] Limchoowong N, Sricharoen P, Areerob Y, Nuengmatcha P, Sripakdee T, Techawongstien S, Chanthai S. Preconcentration and trace determination of copper (II) in Thai food recipes using $Fe_3O_4$@Chi-GQDs nanocomposites as a new magnetic adsorbent. Food Chem 2017;230:388–97.
[113] Li Y, Zhang X, Wang Y, Lin Y, Zhou J. Loading controlled magnetic carbon dots for microwave-assisted solid-phase extraction: preparation, extraction evaluation and applications in environmental aqueous samples. J Sep Sci 2018;41:3622–30.

[114] Sricharoen P, Limchoowong N, Areerob Y, Nuengmatcha P, Techawongstien S, Chanthai S. $Fe_3O_4$/hydroxyapatite/graphene quantum dots as a novel nano-sorbent for preconcentration of copper residue in Thai food ingredients: optimization of ultrasound-assisted magnetic solid phase extraction. Ultrason Sonochem 2017;37:83–93.

[115] Li Y-K, Yang T, Chen M-L, Wang J-H. Supported carbon dots serve as high-performance adsorbent for the retention of trace cadmium. Talanta 2018;180:18–24.

[116] Dong Y, Shao J, Chen C, Li H, Wang R, Chi Y, Lin X, Chen G. Blue luminescent graphene quantum dots and graphene oxide prepared by tuning the carbonization degree of citric acid. Carbon 2012;50:4738–43.

[117] Yang D, Li X, Meng D, Yang Y. Carbon quantum dots-modified ferrofluid for dispersive solid-phase extraction of phenolic compounds in water and milk samples. J Mol Liq 2018; 261:155–61.

# Restricted access media

# 5

*Maria Eugênia C. Queiroz, Israel D. Souza*
Departamento de Química, Faculdade de Filosofia Ciências e Letras de Ribeirão Preto, Universidade de São Paulo, Ribeirão Preto, SP, Brasil

## 5.1 Introduction

Trace analysis of drugs and their metabolites or biomarkers in biological matrixes is a challenge that always has to be faced during pharmaceutical, clinical, omics, and toxicological studies. Considering the physicochemical properties of these analytes, liquid chromatography coupled with tandem mass spectrometry (LC-MS/MS) has become the most sensitive and selective analytical technique for their analysis in these complex samples. Conversely, direct injection of biological samples into conventional LC-MS/MS systems is inappropriate. Endogenous macromolecules, primarily proteins, can adsorb onto the column, leading to back-pressure build-up, modifying retention times, decreasing column efficiency, and suppressing ions during electrospray ionization mass spectrometry. In this context, a sample preparation step preceding the chromatographic separation is essential to isolate the analytes and to remove the endogenous interferents.

Recent trends in sample preparation techniques include miniaturization, automation, and development of selective extraction sorbents. In this scenario, restricted access media (RAM) sorbents integrated with an LC system allows the direct high-throughput online analysis of biological samples.

## 5.2 Restricted access media

Desilets et al. introduced the term restricted access media (RAM) to describe a group of phases that enable direct and repetitive injection of untreated biological samples into a reversed-phase LC system by limiting the accessibility of macromolecules (sample matrix) to the adsorption sites of the porous supports [1]. Access is restricted on the basis of size; size exclusion combined with conventional hydrophobic or ion exchange interactions allows small molecules to pass while restricting the access of macromolecules. Therefore, macromolecules such as proteins from biological matrixes are excluded (by physical or chemical means, or a combination of both) in the void volume, whereas small hydrophobic analytes are retained [1–4].

The excluded macromolecules only interact with the outer surface of the support coated with hydrophilic groups, minimizing matrix protein adsorption. Considering the protein exclusion mechanism (or the barrier nature), RAM can be classified into (a) ***physical diffusion barrier:*** macromolecules are excluded due to pore diameter,

**Solid-Phase Extraction. https://doi.org/10.1016/B978-0-12-816906-3.00005-4**

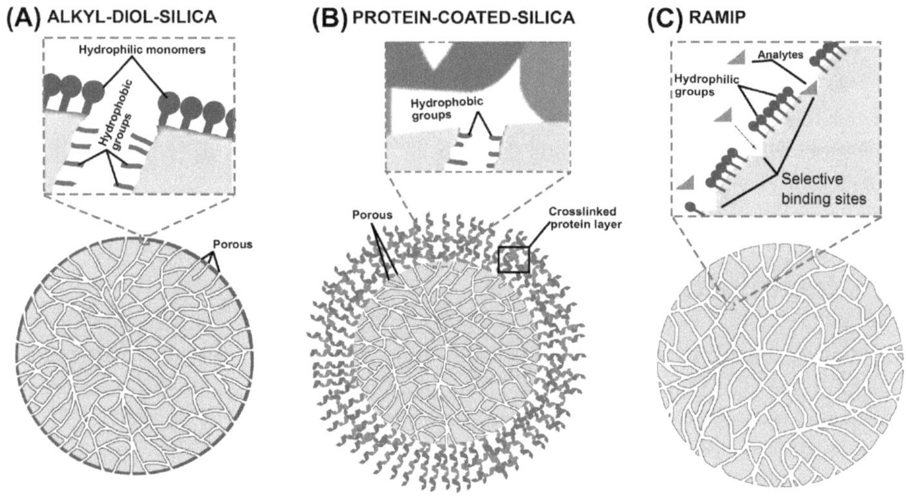


**Figure 5.1** Schematic representation of (A) alkyl-diol-silica, (B) protein-coated silica and (C) RAMIP material.

and (b) ***chemical diffusion barrier:*** macromolecules are excluded by a network covering the outer surface of the particle, which consists of synthetic/natural polymers or protein covalently bound or adsorbed at the outer surface of silica [1−4], Fig. 5.1.

Apart from this classification, these phases have been characterized according to the sorbent surface structure. Bimodal phases consist of a hydrophilic exterior and hydrophobic internal surfaces, while unimodal phases comprise inner and outer surfaces displaying the same properties [1,4]. A further class of RAM is (c) ***RAMIPs***. These combine an efficient ability at removing macromolecules with a higher selectivity based on molecular recognition of the analyte, Fig. 5.1.

## 5.2.1 RAM with a physical barrier

### 5.2.1.1 Alkyl-diol-silica material

The alkyl-diol-silica (ADS) sorbent is the most common RAM material. Hydrophilic electroneutral groups (glycerylpropyl; i.e., diol moieties) bound at the outer surface of the particles exclude macromolecules through a physical barrier (size-exclusion process, pore size = 6 nm), avoiding macromolecule adsorption onto the support, Fig. 5.1A. The inner surface of the porous particles (reversed phase or strong cation-exchanger, SCX) adsorbs the target compounds [1−4]. Table 5.1 summarizes different approaches for online extraction systems with LiChropher RP8-ADS [5−9] and RP18-ADS [10−12] cartridges, which are suitable for direct analysis of biological fluids with minimal matrix effects. Cation-exchange RAMs have been synthesized to extract basic compounds for pharmaceutical analysis [13]. A simple system that combines online diol silica SCX-RAM extraction with two-dimensional RP × RP LC-MS/MS was used for the determination of peptides in human serum [14].

**Table 5.1** LC methods to analyze compounds by direct injection of biological or environmental matrixes into RAM columns.

| Analytes | Matrix | RAM sorbent | Ext. tech. | Sample volume (μL) | Ext. conditions | | Analytical Column | LC system | Linear range (ng/mL) | Ref. |
|---|---|---|---|---|---|---|---|---|---|---|
| | | | | | Sorption | Elution | | | | |
| **RAM with a chemical barrier** | | | | | | | | | | |
| α1-adrenergic receptor antagonists | Rat feces | Capcell Pak MF (C8-protein) | CS | – | Phosphate buffer pH 7.4/ACN (92:8, v/v) | Acetate buffer pH 5.3/ MeOH/ACN (22:45:33, v/v/v) | Chiralpak IA column (250 × 4.6 mm, 5 μm) | LC-FD | 22.5–15,000 | [17] |
| Olanzapine, fluoxetine and norfluoxetine | Plasma | Capcell MF Ph-1 (MM) | CS | 200 | 10 mM ammonium formate/ACN (99:1, v/v) | 10 mM ammonium formate with 0.01% formic acid/ACN (65:35 v/v) | Hypersil Gold C18 column (150 × 4.6 mm, 5 μm) | LC-MS/ MS | 0.5–100 | [18] |
| Sterols | Serum | BioTrap 500C18 (MM) | CS | 300 | Water | Water/2-propanol (60: 40, v/v) | XTerras MS C18 column (4.6 × 100 mm, 3.5 μm) | LC-MS/ MS | 17–2700 | [19] |
| Albendazole metabolites | In-vitro microsomal culture | Phenyl-BSA | 2D | 350 | Water | Water/MeOH (50:50, v/v) | Amylose tris (3,5-dimethylphenyl carbamate) (150 × 4.6 mm, 7 μm) | LC-UV-vis | – | [21] |
| Statins | Plasma | C18-BSA | CS | 500 | 100 mM potassium phosphate pH 2.5/ACN (60: 40, v/v) | 100 mM phosphate buffer pH 2.5/ACN (40: 60, v/v) | Lichrocart 125–4 Purospher StarC8 (125 × 4.0 mm, 5 μm) | LC-PDA | 125–876 | [22] |
| Fluoroquinolones | Water | BSA-octyl-silica | SC | 2000 | (Water with 0.1% formic acid)/ ACN (90:10, v/v) | 0.1% formic acid/ACN (10:90, v/v) | – | LC-MS/ MS | 20–800 | [23] |

Continued

**Table 5.1** LC methods to analyze compounds by direct injection of biological or environmental matrixes into RAM columns.—cont'd

| Analytes | Matrix | RAM sorbent | Ext. tech. | Sample volume (μL) | Ext. conditions | | Analytical Column | LC system | Linear range (ng/mL) | Ref. |
|---|---|---|---|---|---|---|---|---|---|---|
| | | | | | Sorption | Elution | | | | |
| Antihypertensives | Serum | CNTs-BSA | CS | 100 | 10 mM ammonium formate | 10 mM ammonium formate pH 5.0)/MeOH (90:10, v/v) | Shim-pack XR-ODS C18 (150 × 4.6 mm, 5 μm) | LC-MS/MS | 200–500 | [26] |
| Anticonvulsants | Plasma | CNTs-BSA | CS | 100 | Water | 0.01 mol/L phosphate buffer pH 6.0/ACN/MeOH (55:25:20; v/v/v) | C18 (250 × 4.6 mm, 5 μm) | LC-UV | 2000–40,000 | [27] |
| Tetracyclines | Milk | CNTs-BSA | CS | 500 | 0.10 mM oxalic acid/ACN (80:20, v/v) | 10 mM oxalic acid/ACN (70:30, v/v) | Shim-pack XR-ODS C18 (150 × 4.6 mm, 5 μm) | LC-DAD | 50,000–200,000 | [28] |
| Enrofloxacin andgatifloxacin | Milk | Sil-g-p(St/DVB)-Diol/BSA | CS | 1000 | 100 mM phosphate buffer pH 7.0/ACN/trifluoro acetic acid/isopropanol (90:10:1:1, v/v/v) | Water/ACN with 0.1% formic acid (95/5, v/v) | Luna (150 × 4.6 mm, 5 μm) | LC-DAD | 10–300 | [29] |
| **RAM with a physical barrier** | | | | | | | | | | |
| Melamine and cyromazine | Milk | Sil-g-p(SPM/EDMA)-Diol | CS | 1000 | 5 mM ammonium acetate pH 3.0/ACN (70/30, v/v) | 10 mM ammonium acetate pH 8.0/ACN (15/85, v/v) | HILIC Luna (150 × 4.60 mm, 5 μm) | LC-PDA | 5–200,000 | [15] |

| | | | | | | | | | | |
|---|---|---|---|---|---|---|---|---|---|---|
| Sulfonamides | Honey | Hybrid silica monolith (C3-ADS) | 2D | 4.2 g | Water/ACN (75:25, v/v) | 100 mM phosphate buffer/ACN (15:85, v/v) | Venusil XBP C18 (250 × 4.6 mm, 5 μm) | LC-UV | 0.05–0.2 | [16] |
| Mercapturic acids derived | Urine | LiChrospher (RP8-ADS) | CS | 200 | 0.05% formic acid | 0.05% formic acid/ACN (50:50, v/v) | Zorbax Eclipse XDB C18 (150 × 2.1 mm, 3.5 μm) | LC-MS/MS | 0.10–20.0 | [5] |
| Mercapturic acids | Urine | LiChrospher (RP8-ADS) | CS | 100 | Water with 0.1% formic acid | 0.1% formic acid/ACN 0.1% formic acid (88:12, v/v) | Luna C8 (150 × 4.6 mm, 3 μm) | LC-MS/MS | 0.03–150 | [6] |
| 1-vinyl-2-pyrrolidone-mercapturic acid | Urine | LiChrospher (RP8-ADS) | CS | 100 | Water pH 2.5/ACN with 0.1% formic acid (88:12, v/v) | Water pH 2.5/ACN with 0.1% formic acid (75:25, v/v) | Luna 3u C8(2) 100A (150 × 4.6 mm, 3 μm) | LC-MS/MS | 500–500,000 | [7] |
| Cortisol and cortisone | Hair | LiChrospher (RP8-ADS) | CS | 50 mg | 0.02% ammonium hydroxide/MeOH (90:10, v/v) | Water/MeOH (20:80, v/v) | Poroshell 120 EC-C18 (50 × 4.6 mm, 2.7 μm) | LC-MS/MS/MS | 2–200 pg/mg | [8] |
| Psychotropics | Plasma | LiChrospher (RP8-ADS) | CS | 200 | Water | 4 mM ammonium acetate with 0.1% formic acid/ACN (63:37, v/v) | Kinetex C18 (100 × 2.1 mm, 1.7 μm) | LC-MS/MS | 0.025–1.25 | [9] |
| Atenolol and propranolol | Plasma | LiChrospher (RP18-ADS) | multisyringe flow injection analysis | 200 | Water/ACN (98:2, v/v) | ACN | AV-5-MS (30 m × 0.25 mm, 0.25 mm film thickness) | GC-MS | 6–1000 | [10] |

*Continued*

**Table 5.1** LC methods to analyze compounds by direct injection of biological or environmental matrixes into RAM columns.—cont'd

| Analytes | Matrix | RAM sorbent | Ext. tech. | Sample volume (μL) | Ext. conditions | | Analytical Column | LC system | Linear range (ng/mL) | Ref. |
|---|---|---|---|---|---|---|---|---|---|---|
| | | | | | Sorption | Elution | | | | |
| Alprenolol and propranolol | Blood | LiChrospher (RP18-ADS) | 2D | 100 | Water/ACN (98:2, v/v) | TFA solution with 0.1% triethylamine/ACN (74:26, v/v) | Luna PFP2 (150 × 4.6 mm, 3 μm) | LC-FD | 5–200 | [11] |
| Mercapturic acids of naphthalene | Urine | LiChrospher (RP18-ADS) | 2D | 1000 | 1% acetic acid | 100 mM ammonium acetate buffer pH 4.5/MeOH (60:40, v/v) | KinetexBiphenyl Column (100 × 2.1 mm, 2.6 μm) | LC-MS/MS | 800–20,000 | [12] |
| Peptides | Serum | Lichroprep ($SO_3$-Diol) | 2D | 400 | 20 mM ammonium acetate pH3.0 | 1000 mM ammonium acetate pH 3.0 | Kromasil-ODS 300 Å (150 × 4.6 mm, 5 μm) | LC-MS/MS | – | [14] |
| Xenobiotics and endogenous metabolites | Urine | MSpak 4A (styrene-PVA) | 2D | 10 | Water | 2.5 mM formic acid/ACN (10:90, v/v) | ZIC-HILIC and Luna PFP (150 × 4.60 mm, 3.5 μm) | LC-MS/MS | 2–10 | [51] |
| Vanillin and metabolites | Plasma | Strata-X (MM) | CS | 100 | Water/ACN (90:10, v/v) | 0.2% formic acid/ACN (10:90, v/v) | Kinetex core–shell C18 column (50 × 4.6 mm, 2.6 μm) | LC-MS/MS | 5–1000 | [52] |
| Lipid mediators | Skeletal muscles | SUPELCOSIL LC-HISEP (MM) | 2D | 100 mg | 10 mM ammonium acetate | 0.1% formic acid | Ultra C8 (150 × 2.1 mm, 3 μm) | LC-MS/MS | 0.031–32 | [53] |

| | | | | | | | | | | |
|---|---|---|---|---|---|---|---|---|---|---|
| **RAMIPs** | | | | | | | | | | |
| 2-Arylpropionic acid derivatives | Serum | RAMIP (hydroxyl) | CS | 20 | 20 mM phosphoric acid/ acetonitrile pH 2.2 (78:22, v/v) | 20 mM sodium phosphate buffer pH 7.3/ACN (75:25, v/v) | Cosmosil 5C18-MS-II (150 × 4.6 mm, 5 μm) | LC-UV | 200–50,000 | [31] |
| β-blockers | Plasma | RAMIP (hydroxyl) | CS | 0.05 | 20 mM sodium phosphate buffer pH 5.8/ ACN (80:20, v/v) | 20 mM sodium phosphate buffer pH 3.7/ACN (78:22, v/v) | ULTRON ES-PhCD (150 × 6 mm, 5 μm) | LC-UV | 12.5–250 | [32] |
| Phenobarbital, amobarbital, and phenytoin | River water | RAMIP (hydroxyl) | CS | 50,000 | 2 mM ammonium acetate | 2 mM ammonium acetate/ACN (60:40, v/v) | Cosmosil 5C18-MS-II (150 × 4.6 mm, 5 μm) | LC-MS | 0.5–50 | [36] |
| Methylthiotriazine herbicides | River water | RAMIP (hydroxyl) | CS | 100,000 | Water | 50 mM potassium phosphate buffer pH 7.0/ACN (62:38, v/v) | Cosmosil 5C18-MS-II (150 × 4.6 mm, 5 μm) | LC-UV | 0.05–0.5 | [37] |
| Sulfonyl urea residue | Soil | RAMIP (hydroxyl) | CS | 10 g | 50 mM phosphate buffer pH 5.0/ ACN (90:10, v/v) | 50 mM phosphate buffer pH 3.3/ACN (60:40, v/v) | Gemini C8 (150 × 4.6 mm, 5 μm) | LC-UV-vis | 10–5000 | [38] |
| Clenbuterol | Serum | RAMIP (hydroxyl) | CS | 1000 | 10 mM phosphate buffer pH 7.0/ ACN (80:20, v/v) | 10 mM phosphate buffer pH 3.0/ACN (75:25, v/v) | Kromasil C18 column (250 × 4.6 mm, 5 μm) | LC- UV-vis | 2–1000 | [39] |
| Sulfonamides | Milk | RAMIP (hydroxyl) | CS | 100 | 100 mM phosphate buffer pH 6.0/ MeOH (95:5, v/v) | 100 mM phosphate buffer pH 7.0/ACN (83:17, v/v) | Phenomenex C18 (250 × 4.6 mm, 5 μm) | LC-UV | 2–400 | [45] |

*Continued*

**Table 5.1** LC methods to analyze compounds by direct injection of biological or environmental matrixes into RAM columns.—cont'd

| Analytes | Matrix | RAM sorbent | Ext. tech. | Sample volume (μL) | Ext. conditions | | Analytical Column | LC system | Linear range (ng/mL) | Ref. |
|---|---|---|---|---|---|---|---|---|---|---|
| | | | | | Sorption | Elution | | | | |
| Oxacillin and cloxacillin | Urine and plasma | RAMIP (hydroxyl) | CS | 100 | Water | 25 mM phosphate buffer pH 4.5/ACN (74:26, v/v) | C18 diomonsil column (250 × 4.6 mm, 5 μm) | LC-UV | 1000–60,000 | [46] |
| Chlorpromazine | Plasma | RAMIP (hydroxyl/BSA) | CS | 100 | Water | 20 mM acetate pH 4.1/ACN 55:45 (v/v) | Phenomenex C18 (250 × 4.6 mm, 5 μm) | LC-UV-vis | 30,000–350,000 | [47] |
| Serotonin reuptake inhibitors | Plasma | RAMIP (hydroxyl/BSA) | CS | 200 | – | 20 mM acetate buffer pH 4.5/MeOH/ACN (60:3:37 v/v/v) | Phenomenex C18 (250 × 4.6 mm, 5 μm) | LC-UV-vis | 20–500 | [48] |
| Ivermectin | Meat | RAMIP (hydroxyl/BSA) | CS | 500 mg | – | ACN/MeOH/water, (60:30:10 v/v/v) | Phenomenex C18 (250 × 4.6 mm, 5 μm) | LC-UV-vis | 50–500 μg kg $^{-1}$ | [49] |
| Oxprenolol | – | RAMIP (hydroxyl/BSA) | CS | 100 | | 10 mM ammonium formate pH 3.7/MeOH (95:5, v/v) | Shim-PackXR-ODS C18 (100 × 3 mm, 2.2 μm) | LC-MS | – | [50] |
| Tricyclic antidepressants | Plasma | BSA-MIP | SC | 1000 | Water | 0.01% acetic acid/ACN (30:70, v/v) | – | -MS | 15,000–500,000 | [54] |
| Phenobarbital | Serum | RAMIP (glycopolymer) | CS | 20 | Water | MeOH | Comatex-AB C18 column (250 × 4.6 mm, 5 μm) | LC-UV | – | [55] |

*ACN*, acetonitrile; *ADS*, alkyl-diol silica; *CS*, column-switching; *MM*, mixed mode material; *MeOH*, methanol; *SC*, single mode column; *TFA*, trifluoro acetic acid.

#### 5.2.1.2 SCX-RAM synthesized by surface-initiated atom transfer radical polymerization

A new SCX-RAM has been developed by sequentially grafting different polymers on silica via the surface-initiated atom transfer radical polymerization (ATRP) technique. During the synthesis, poly(3-sulfopropyl methacrylate-co-ethylene dimethacrylate) [p(SPM/EDMA)] was grafted onto the silica surface initially through the ATRP "grafting form". Then, a hydrophilic chemical barrier, poly(glycerol mono-methacrylate) [p(GMMA)], was established by using ATRP to graft glycidyl methacrylate (GMA) on the p(SPM/EDMA)-grafted silica, which was followed by hydrolysis of the epoxy groups. The resulting Sil-g-p(SPM/EDMA)-g-pGMMA can exclude proteins during HPLC analysis [15].

#### 5.2.1.3 Monolithic phases

Monolithic phases, which are porous rod structures characterized by mesopores and macropores, are also suitable for direct injection of biological fluids. Due to the high permeability of monolithic phases, extractions can be performed at a high flow rate without generating high backpressure. A novel RAM hybrid monolithic column (RAM-HMC) prepared by in situ synthesis (sol-gel procedure) in a stainless-steel column was developed to clean-up honey samples online and to preconcentrate the sulfonamide residues. This RAM-HMC can effectively exclude macromolecules and enrich the samples containing the target analytes. A hydrophobic monolithic column (HMC) was prepared by the acid catalyzed reaction of methyltrimethoxysilane with tetraethoxysilane. Hydrophilic structures originated on the through-pore surface of the HMC by grafting of 3-(2,3-epoxypropoxy)propyltrimethoxysilane. The optimized synthetic conditions provided a uniform microchannel and a stable skeletal structure [16].

Monolithic columns developed by the in-situ sol-gel polymerization technology do not require frits at column extremities, which are often the main source of endogenous material adsorption.

### 5.2.2 RAM with a chemical barrier

#### 5.2.2.1 Mixed-functional material

A mixed-functional extraction RAM sorbent is commercialized under the trade name Capcell Pak MF. Both the internal and external surfaces comprise a mixture of hydrophilic poly(oxyethylene) and hydrophobic styrene groups grafted on a porous silicone polymer-coated silica substrate. The long poly(oxyethylene) chains limit access of macromolecules to retentive surfaces. Therefore, matrix components only interact with the hydrophilic, nonadsorptive polymer network, and elute in the void volume [2,3].

By using a Capcell Pak MF C8 column in the first dimension, an online column-switching chiral high-performance liquid chromatography (HPLC) method was developed to determine naftopidil and the enantiomers of its O-desmethyl metabolites simultaneously in rat feces. Direct and multiple injections of the supernatant from rat feces homogenate through the column-switching system was possible [17]. An online solid-phase extraction (SPE) (MF Ph-1)-LC-MS/MS method, was developed to quantify olanzapine, fluoxetine, and norfluoxetine in human plasma [18].

#### 5.2.2.2 Protein-coated silica

Protein-coated silica RAM phases consist of a protein-like human plasma protein, a1-acid glycoprotein (AGP), or bovine serum albumin (BSA), covalently bound to a reversed phase (RP) sorbent. This makes the external hydrophilic surface (protein network) of the particles compatible with a protein sample, which cannot penetrate the small pores, Fig. 5.1B. Hydrophobic groups at the inner surface are responsible for interacting with the analytes [2,3].

In 1994, Hermansson and Graham introduced extraction supports bearing AGP for direct injection of biological fluids. These supports are commercialized under the trade name BioTrap [2]. An automated method involving BioTrap 500C18 cartridges was developed to determine free noncholesterol sterols in human serum (treated serum) by online SPE-LC-MS [19]. In this method, a lipid extraction step (Bligh−Dyer method) is essential to determine sterols even when a RAM cartridge is used in the online system. The sample preparation time is reduced by 75%, and 25 samples can be handled in only 2 h. Matrix effects have been thoroughly studied and sufficiently suppressed that external standard calibration can be used [19].

BSA has been used to provide the external hydrophilic layer, resulting in a RAM-RP-BSA extraction support. On the basis of the protocol described by Menezes and Felix [20], BSA can be immobilized in situ by reaction with glutaraldehyde as the crosslinking agent, which is followed by a washing step with sodium borohydride solution. The BSA amine groups are interconnected by glutaraldehyde, which affords a crosslinked BSA layer that is unbound to the silica core. Two multidimensional LC methods (fluorescence detector) based on chiral and achiral separations have been described for direct analysis of microsomal fractions obtained from rat livers. Phenyl (RAM-phenyl-BSA) and octyl (RAM-C8-BSA) restricted access media columns were evaluated in the first dimension for sample clean up [21]. Table 5.1 describes different methods using RAM-RP-BSA columns (first dimension) for column-switching liquid chromatography. These methods share the advantages of lower organic solvent consumption (environmental friendly methods) and short analysis time. RAM-RP-BSA columns present high protein exclusion capability and appropriate sorption retention for acidic, neutral, and basic compounds [22,23]. A BSA shield eliminates macromolecules more efficiently because the negative charge density on the polymer surface increases when the medium pH is higher than the BSA isoelectric point (4.7) [4].

#### 5.2.2.3 BSA-coated carbon nanotubes

Carbon nanotubes (CNTs) that are useful extraction sorbents due to their high surface area and mechanical strength allow $\pi$-$\pi$ interactions to be established, and easy chemical modification [24]. However, protein can be retained on CNT surfaces, reducing their adsorption capacity and prevent their use in direct analysis of biological fluids. In 2015, Figueiredo et al. synthesized the first CNT-based RAM sorbent, which was called *restricted access carbon nanotubes* (RA-CNTs) [25]. This material was obtained by modifying the CNTs with a chemically crosslinked BSA layer. When RA-CNTs are used, proteins in the biological sample are excluded due to electrostatic repulsion between sample proteins and the BSA coating. RA-CNT columns (first dimension) were evaluated for direct column-switching LC analysis of antihypertensive drugs and some of their metabolites in human serum [26], anticonvulsants in human plasma [27], and tetracyclines in bovine milk [28]. The RA-CNT sorbent can exclude about 100% of the proteins (bovine milk and plasma samples) in less than 2 min [26,28] and has a long lifetime (about 250–300 analytical cycles) [26,27].

#### 5.2.2.4 BSA-coated polymer

A novel RAM sorbent for direct analysis of enrofloxacin and gatifloxacin in milk samples was prepared by combining the hydrophilic polymer poly(glycerol monomethacrylate) (pGMMA) and BSA as the crosslinking agent. During preparation, poly(styrene-*co*-divinylbenzene) and pGMMA were successively grafted on silica by atom transfer radical polymerization. Then, BSA was adsorbed onto the material and crosslinked through an in-column process. The protein exclusion efficiency was higher than 99.3%. The RAM column good stability and was used for at least 5 months without significant changes in its efficiency [29].

### 5.2.3 Restricted access material combined with molecularly imprinted polymers (RAMIP)

Molecularly imprinted polymers (MIPs) are biomimetic synthetic materials (microporous matrix) with microcavities (binding sites) having a three-dimensional structure that can recognize target compounds (used as a template) in terms of size, shape, and chemical functionality [4]. However, when these sorbents are applied to biological samples, the MIP hydrophobic surface can adsorb nonspecific proteins from biological samples. Different strategies have been evaluated to add a protective hydrophilic coating onto MIP surfaces. In this context, the RAMIP sorbent combines the high selectivity of MIPs and the ability of RAMs to exclude macromolecules, thereby avoiding macromolecule adsorption, Fig. 5.1C. Basically, to obtain a RAMIP sorbent, the MIP surface can be covered with (1) hydrophilic monomers, (2) comonomers, followed by additional chemical treatment to convert these comonomers into their hydrophilic form, or (3) hydrophilic monomers associated with a crosslinked BSA layer [4]. Table 5.1 summarizes the main studies reporting RAMIPs for online sample preparation.

#### 5.2.3.1 RAMIP synthesis using hydrophilic monomers

In Haginaka's pioneering work [30], a uniform-sized MIP with a hydrophilic external layer was prepared by the multistep swelling and polymerization method.

(S)-naproxen (template), 4-vinylpyridine (functional monomer), and EDMA (cross-linker) were added after the third swelling step to create the MIP material. The RAM layer was grafted on the MIP surface by polymerization of the hydrophilic monomers GMMA and glycerol dimethacrylate (GDMA), which were added to the reaction medium. The resulting RAMIP particles can selectively extract (S)-naproxen and exclude macromolecules from serum samples [30].

The general concepts of RAMIP sorbent multistep swelling and polymerization were applied for the determination of 2-arylpropionic acid derivatives [31], β-blockers [32], endocrine disruptors [33,34], nonsteroidal antiinflammatory drugs [35], antiepileptic drugs [36], and methylthiotriazine herbicides [37]. Biological fluids [31,33] and environmental [34−37] samples were directly injected into the chromatographic system, using RAMIP sorbents.

More recently, the reversible addition-fragmentation chain transfer radical polymerization technique (RAFT) was reported to incorporate hydrophilic monomers into the MIP surface [38,39]. This methodology uses a chain transfer agent (RAFT agent; e.g., thiocarbonylthiol compounds) to ensure a controlled and milder radical polymerization [40−42]. RAFT polymerization was employed to prepare RAMIP to determine clenbuterol in biological samples [39]. This RAMIP column presented higher mechanical strength and could be used for 15 months without losing extraction/exclusion efficiency. Yang et al. [38] evaluated two types of RAFT agent, dibenzyltrithiocarbonate and cumyl dithiobenzoate, to prepare a bifunctional pyrazosulfuron-ethyl imprinted RAMIP sorbent to determine sulfonylurea residue in soil samples. The authors concluded that cumyl dithiobenzoate yields sorbents with the best extraction/exclusion efficiency.

#### *5.2.3.2 RAMIP synthesis using comonomers*

Unfortunately, in some cases, the addition of hydrophilic monomers to the reaction medium can decrease the molecular imprinting process efficiency. Hydrophilic monomers possess active hydroxyl groups that can establish hydrogen bonds with some types of template molecules or functional monomers. In this case, the prepolymerization complex can be affected during the polymerization process. To overcome this drawback, prohydrophilic comonomers followed by their conversion to their hydrophilic form is more convenient. GMA is the most commonly used prohydrophilic comonomer [43].

To illustrate this approach, Pouci et al. [44] developed a RAMIP sorbent to determine p-acetaminophenol in gastrointestinal simulating fluids, and Xu et al. [45] synthesized a MIP polymer on the surface of modified silica gel particles to determine sulfonamides in bovine milk. In both cases, after the poly(GMA) was grafted on the surface of MIP-silica particles, it was converted to GMMA via hydrolysis. During the grafting process, the pGMA chain length was controlled by the reaction time. A higher grafting density was obtained by this approach. Yang et al. [46] used an ethylenediamine/tetrahydrofuran mixture to open the epoxide groups grafted on pGMA instead of the conventional hydrolysis procedure. This reaction generates a monolithic RAMIP sorbent with weak ion exchange properties. This sorbent was used to determine oxacillin and cloxacillin in human urine and plasma.

### 5.2.3.3 RAMIP synthesis with hydrophilic monomer combined with a BSA layer

Besides hydrophilic monomers, other protective coatings, such as proteins anchored around the polymer, can also lead to good macromolecule exclusion. RAMIP sorbents using a dual RAM layer consisting of hydrophilic monomers combined with a crosslinked BSA layer have been reported [47–50]. This methodology was used to develop a RAMIP sorbent to determine chlorpromazine in human plasma [47]. After synthesizing the MIP particles, GDMA and HEMA monomers were used to create a hydrophilic surface on the MIP particles. Then, the hydrophilic surface was covered with a BSA layer, crosslinked by means of glutaraldehyde monomers. This study revealed that the macromolecule exclusion efficiency increased when the hydrophilic monomers were combined with the crosslinked BSA layer. Additional studies used the same methodology to ascertain the RAMIP (hydrophilic monomers/BSA) efficiency for antidepressants [48] and ivermectin [49] in biological samples.

Santos et al. [50] published a comparative study evaluating the extraction/exclusion efficiency of different RAMIP sorbents. It was concluded that both hydrophilic monomers and BSA layer contribute to improve molecular recognition in an aqueous medium. Moreover, the authors attested that hydrophilic monomers combined with a BSA layer on the MIP surface lead to the best macromolecule exclusion. When the sample pH is higher than the protein isoelectric point, the BSA layer and the sample proteins are negatively charged, which results in electrostatic repulsion and avoids protein adsorption. At the same time, low-molecular-weight analytes diffuse through the BSA chains and bind to selective sites in the polymer. However, hydrophobic low-molecular-weight interferents that diffuse through the BSA layer cannot adsorb onto the hydrophilic MIP surface [48,49].

### 5.2.3.4 Nonconventional RAMIP material

Besides the main synthetic strategies mentioned above, alternative approaches have been employed to design innovative RAMIP sorbents. Hua et al. [49] reported a novel RAMIP obtained by incorporating sugar groups into MIP particles to create a hydrophilic surface. The authors used a modified multistep swelling and polymerization method to prepare this sorbent. The modification consisted of using lactose octa-acetate functionalized with polymerizable groups. Through copolymerization of this modified molecule, after the third swelling step, the glycopolymer was incorporated into the MIP polymer. The results revealed that the recognition sites for phenobarbital remained unchanged after modification with the glycol layer. The biocompatible RAMIP sorbent presented more than 90% efficiency for protein exclusion.

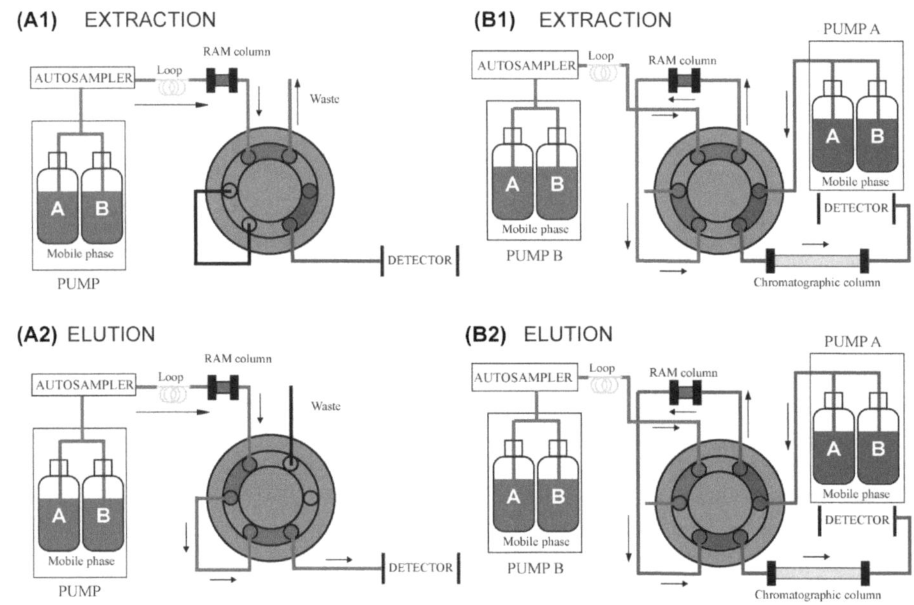


**Figure 5.2** (A) single column mode and (B) coupled-column mode system configuration.

## 5.3 RAM sorbents interfaced with LC systems

After the synthesis procedure, RAM sorbents need to be packed in an appropriate column before they are coupled to the chromatography system. Normally, particulate sorbents are packed in stainless-steel columns ($\sim 10 \times 4.0$ mm I.D.; $50 \times 4.6$ mm I.D.) by means of the slurry technique, during which the particles are homogeneously suspended in an appropriate liquid to avoid particle-to-particle bridging or sedimentation. Then, this suspension is forced into the stainless-steel column under pressure and at a high flow-rate [51,52]. These RAM columns allow complex matrixes to be directly injected into the chromatography system. Typical instrument set-ups are (1) the single column mode, in which the RAM column is directly connected to a detector, and (2) the column-switching mode, in which a switching valve connects the RAM column to the separation column (Fig. 5.2).

### *5.3.1 RAM sorbents in the single column mode*

In the single column configuration, the RAM column is connected directly to a detector. The sample is injected into the column for the extraction of target compounds and the exclusion of macromolecules. Subsequently, the analytes are eluted from the column to the detector. The single column mode has more promising applications when it is combined with mass spectrometric detectors—the high sensitivity and selectivity of these detectors has significant advantageous over UV, DAD, or fluorescence detectors. Moreover, the mass resolving power of the mass spectrometer compensates to some extent for incomplete analyte separation. Therefore, shorter RAM columns and higher flow rates can be used, reducing analysis times [53,54]. RAM columns direct coupling to MS/MS systems, for example RAMIP-MS [55] and a RAM-MS/MS [23] methods were developed for tricyclic antidepressants and fluoroquinolones, respectively. The RAMIP-MS and RAM-MS/MS methods provided a lower limit of quantification (LLOQ) of 15,000 and 20 ng $mL^{-1}$, respectively.

### 5.3.2 RAM sorbents hyphenated with column-switching and multidimensional systems

Column-switching and multidimensional systems have been used with two or more columns. The target fractions are selectively transferred online from the first column to a second column for further separation [56]. Two approaches have been described: the column-switching and the multidimensional methods.

In the column-switching approach, the RAM column is installed in the six-port valve or sometimes placed in the loop position. Target analytes in aqueous samples are extracted (partition or adsorption) and concentrated on the RAM column. Subsequently, the extracted analytes are directly desorbed from the RAM column by the initial flow of the mobile phase (dynamic desorption) after the six-port valve (inject mode) is switched. Finally, the desorbed analytes are transported to the LC column for separation and detection [57]. In this context, many LC (in the column-switching mode) methods using packed RAM columns have been reported for different analytes by direct injection of environmental and biological samples (Table 5.1). Typically, the use of RAM columns containing fully porous silica particles with alkyl [5–12] or phenyl [21] groups, porous silica with alkyl groups combined with a glycoprotein [29] or BSA [19,22] layer, reversed-phase sorbents using ADS particles [5–12], CNTs combined with crosslinked BSA [26,27], and RAMIPs [31,32,36–39,45–50,58] have been reported. These methods were validated for LC-UV [16,21,27,31,32,37–39,45–49,58], LC-DAD [15,22,28,29], LC-FD [11,17], LC-MS [36,50], LC-MS/MS [5–7,9,18,19,23,26,59], LC-MS/MS/MS [8], and GC-MS [10] systems.

In the multidimensional approach, two or more independent separation modes are combined to separate complex mixtures. The main focus of this approach is to obtain chromatographic separation on both columns—on the RAM column, in the first dimension, and on the separation column, in the second dimension. Regarding the separation method, a single desired segment (heart-cutting mode) or all the eluent (comprehensive mode) from the first column is transferred to the second column for further separation, which typically involves a different mechanism. A switching valve is used to couple the two dimensions [57]. Multidimensional methods with different column combination such as RAM (Phenyl-BSA) × chiral column [21], RAM (SO3-Diol) × reversed phase (C18) [14], RAM (Alkyl-Diol) × reversed phase (C18) [16], RAM (polyvinyl alcohol) × HILIC [60], RAM (C18-ADS) × normal phase (pentafluorophenylpropyl linkage) [11], and RAM (hydrophobic-hydrophilic monomers) × reversed-phase (C8) [12] have been reported to determine target analytes in complex samples.

## 5.4 Conclusion and future challenges

Online column-switching or multidimensional (heart cutting or bidimensional) LC methods with RAM sorbents not only enhance analytical accuracy and precision but also decrease the matrix effect. This chapter aimed to emphasize innovative RAM sorbents; that is, hybrid materials with BSA coating, monolithic phases, carbon nanotubes, and molecularly imprinted polymer. Tuning of the sorbent structure leads to significantly enhanced selectivity and sensitivity for online LC methods. Future challenges are related to interfacing RAM columns with miniaturized chromatographic systems (CapLC and NanoLC) or their direct coupling to MS/MS systems.

## Acknowledgements

The authors would like to acknowledge FAPESP (Fundação de Amparo à Pesquisa do Estado de São Paulo, process 2017/02147−0 and 2016/01082−9) and INCT-TM (465458/2014−9) (Instituto Nacional de Ciência e Tecnologia Translacional em Medicina).

## References

[1] Cassiano NM, Lima VV, Oliveira RV, de Pietro AC, Cass QB. Development of restricted-access media supports and their application to the direct analysis of biological fluid samples via high-performance liquid chromatography. Anal Bioanal Chem 2006;384:1462−9. https://doi.org/10.1007/s00216-005-0253-9.

[2] Souverain S, Rudaz S, Veuthey J-L. Restricted access materials and large particle supports for on-line sample preparation: an attractive approach for biological fluids analysis. J Chromatogr B 2004;801:141−56. https://doi.org/10.1016/j.jchromb.2003.11.043.

[3] Sadílek P, Šatínský D, Solich P. Using restricted-access materials and column switching in high-performance liquid chromatography for direct analysis of biologically-active compounds in complex matrices. TrAC Trends Anal Chem 2007;26:375−84. https://doi.org/10.1016/j.trac.2007.02.002.

[4] de Faria HD, Abrão LC de C, Santos MG, Barbosa AF, Figueiredo EC. New advances in restricted access materials for sample preparation: a review. Anal Chim Acta 2017;959: 43−65. https://doi.org/10.1016/j.aca.2016.12.047.

[5] Zhang X, Xiong W, Shi L, Hou H, Hu Q. Simultaneous determination of five mercapturic acid derived from volatile organic compounds in human urine by LC−MS/MS and its application to relationship study. J Chromatogr B 2014;967:102−9. https://doi.org/10.1016/j.jchromb.2014.07.013.

[6] Pluym N, Gilch G, Scherer G, Scherer M. Analysis of 18 urinary mercapturic acids by two high-throughput multiplex-LC-MS/MS methods. Anal Bioanal Chem 2015;407:5463−76. https://doi.org/10.1007/s00216-015-8719-x.

[7] Bertram J, Schettgen T, Kraus T. Isotope-dilution method for the determination of 1-vinyl-2-pyrrolidone-mercapturic acid as a potential human biomarker for 1-vinyl-2-pyrrolidone via online SPE ESI-LC/MS/MS in negative ionization mode. J Chromatogr B 2016; 1033−1034:321−7. https://doi.org/10.1016/j.jchromb.2016.08.026.

[8] Quinete N, Bertram J, Reska M, Lang J, Kraus T. Highly selective and automated online SPE LC−MS 3 method for determination of cortisol and cortisone in human hair as biomarker for stress related diseases. Talanta 2015;134:310−6. https://doi.org/10.1016/j.talanta.2014.11.034.

[9] Acquaro VR, Domingues DS, Costa Queiroz ME. Column switching UHPLC−MS/MS with restricted access material for the determination of CNS drugs in plasma samples. Bioanalysis 2017;9:555−68. https://doi.org/10.4155/bio-2016-0301.

[10] Brunetto M del R, Clavijo S, Delgado Y, Orozco W, Gallignani M, Ayala C, et al. Development of a MSFIA sample treatment system as front end of GC−MS for atenolol and propranolol determination in human plasma. Talanta 2015;132:15−22. https://doi.org/10.1016/j.talanta.2014.08.056.

[11] Gonçalves VMF, Rodrigues P, Ribeiro C, Tiritan ME. Quantification of alprenolol and propranolol in human plasma using a two-dimensional liquid chromatography (2D-LC). J Pharm Biomed Anal 2017;141:1−8. https://doi.org/10.1016/j.jpba.2017.03.064.

[12] Zobel M, Klotz K, Göen T. LC−MS/MS procedure for the simultaneous determination of N -acetyl- S -(1-naphthyl)cysteine and N -acetyl- S -(2-napthyl)cysteine in human urine. J Chromatogr B 2018;1074−1075:139−45. https://doi.org/10.1016/j.jchromb.2017.12.023.

[13] Chiap P, Rbeida O, Christiaens B, Hubert P, Lubda D, Boos K-S, et al. Use of a novel cation-exchange restricted-access material for automated sample clean-up prior to the determination of basic drugs in plasma by liquid chromatography. J Chromatogr A 2002;975:145−55. https://doi.org/10.1016/S0021-9673(02)01334-1.

[14] Hu L, Boos K-S, Ye M, Zou H. Analysis of the endogenous human serum peptides by on-line extraction with restricted-access material and HPLC-MS/MS identification. Talanta 2014;127:191−5. https://doi.org/10.1016/j.talanta.2014.04.011.

[15] Zhang Y, Lin S, Jiang P, Zhu X, Ling J, Zhang W, et al. Determination of melamine and cyromazine in milk by high performance liquid chromatography coupled with online solid-phase extraction using a novel cation-exchange restricted access material synthesized by surface initiated atom transfer radical polymeriz. J Chromatogr A 2014;1337:17−21. https://doi.org/10.1016/j.chroma.2014.02.038.

[16] Lv Y-K, Guo Z-Y, Wang J-Z, Guo M-M, Yu L-K, Fang H. Preparation of a restricted access material−macroporous hybrid monolithic column − for on-line solid-phase extraction of the sulfonamides residues from honey. Anal Methods 2015;7:1563−71. https://doi.org/10.1039/C4AY02787F.

[17] Zhang Y, Liu X, Zhu L, Yuan M. Simultaneous stereoselective analysis of naftopidil and O -desmethyl naftopidil enantiomers in rat feces using an online column-switching high-performance liquid chromatography method. Biomed Chromatogr 2014;28:1030−5. https://doi.org/10.1002/bmc.3112.

[18] Fan Y, Shen G, Li P, Xi X, Wu H, Tian H, et al. A simple and automated online SPE-LC-MS/MS method for simultaneous determination of olanzapine, fluoxetine and nor-fluoxetine in human plasma and its application in therapeutic drug monitoring. RSC Adv 2015;5:34342−52. https://doi.org/10.1039/C5RA02489G.

[19] Mendiara I, Bentayeb K, Nerín C, Domeño C. Talanta online solid-phase extraction − liquid chromatography − mass spectrometry to determine free sterols in human serum. Talanta 2015;132:690−7. https://doi.org/10.1016/j.talanta.2014.10.029.

[20] Menezes ML, Felix G. Analysis of organochlorine pesticides in plain milk using direct injection on an ISRP column, with column switching. J Liq Chromatogr Relat Technol 1996;19:3221−8. https://doi.org/10.1080/10826079608015820.

[21] Belaz KRA, Pereira-Filho ER, Oliveira RV. Development of achiral and chiral 2D HPLC methods for analysis of albendazole metabolites in microsomal fractions using multivariate analysis for the in vitro metabolism. J Chromatogr B 2013;932:26–33. https://doi.org/10.1016/j.jchromb.2013.06.005.

[22] Fagundes VF, Leite CP, Pianetti GA, Fernandes C. Rapid and direct analysis of statins in human plasma by column-switching liquid chromatography with restricted-access material. J Chromatogr B 2014;947–948:8–16. https://doi.org/10.1016/j.jchromb.2013.12.002.

[23] Denadai M, Cass QB. Simultaneous determination of fluoroquinolones in environmental water by liquid chromatography – tandem mass spectrometry with direct injection: a green approach. J Chromatogr A 2015;1418:177–84. https://doi.org/10.1016/j.chroma.2015.09.066.

[24] Zhang J, Terrones M, Park CR, Mukherjee R, Monthioux M, Koratkar N, et al. Carbon science in 2016: status, challenges and perspectives. Carbon N Y 2016;98:708–32. https://doi.org/10.1016/j.carbon.2015.11.060.

[25] Barbosa AF, Barbosa VMP, Bettini J, Luccas PO, Figueiredo EC. Restricted access carbon nanotubes for direct extraction of cadmium from human serum samples followed by atomic absorption spectrometry analysis. Talanta 2015;131:213–20. https://doi.org/10.1016/j.talanta.2014.07.051.

[26] de Faria HD, Bueno CT, Krieger JE, Krieger EM, Pereira AC, Santos PCJL, et al. Online extraction of antihypertensive drugs and their metabolites from untreated human serum samples using restricted access carbon nanotubes in a column switching liquid chromatography system. J Chromatogr A 2017;1528:41–52. https://doi.org/10.1016/j.chroma.2017.10.072.

[27] dos Santos RC, Kakazu AK, Santos MG, Belinelli Silva FA, Figueiredo EC. Characterization and application of restricted access carbon nanotubes in online extraction of anticonvulsant drugs from plasma samples followed by liquid chromatography analysis. J Chromatogr B 2017;1054:50–6. https://doi.org/10.1016/j.jchromb.2017.02.025.

[28] de Faria HD, Rosa MA, Silveira AT, Figueiredo EC. Direct extraction of tetracyclines from bovine milk using restricted access carbon nanotubes in a column switching liquid chromatography system. Food Chem 2017;225:98–106. https://doi.org/10.1016/j.foodchem.2017.01.004.

[29] Huang W, Wang P, Jiang P, Dong X, Lin S. Preparation and application of a restricted access material with hybrid poly(glycerol mono-methacrylate) and cross-linked bovine serum albumin as hydrophilic out layers for directly on-line high performance liquid chromatography analysis of enrofloxacin a. J Chromatogr A 2018;1573:59–65. https://doi.org/10.1016/j.chroma.2018.08.067.

[30] Haginaka J, Takehira H, Hosoya K, Tanaka N. Uniform-sized molecularly imprinted polymer for (S)-naproxen selectively modified with hydrophilic external layer. J Chromatogr A 1999;849:331–9. https://doi.org/10.1016/S0021-9673(99)00570-1.

[31] Haginaka J, Sanbe H. Uniform-sized molecularly imprinted polymers for 2-arylpropionic acid derivatives selectively modified with hydrophilic external layer and their applications to direct serum injection analysis. Anal Chem 2000;72:5206–10. https://doi.org/10.1021/ac0005215.

[32] Sanbe H, Haginaka J. Restricted access media-molecularly imprinted polymer for propranolol and its application to direct injection analysis of β-blockers in biological fluids. Analyst 2003;128:593–7. https://doi.org/10.1039/B301257N.

[33] Sambe H, Hoshina K, Hosoya K, Haginaka J. Direct injection analysis of bisphenol A in serum by combination of isotope imprinting with liquid chromatography-mass spectrometry. Analyst 2005:38−40.

[34] Sambe H, Hoshina K, Hosoya K, Haginaka J. Simultaneous determination of bisphenol A and its halogenated derivatives in river water by combination of isotope imprinting and liquid chromatography-mass spectrometry. J Chromatogr A 2006;1134:16−23. https://doi.org/10.1016/j.chroma.2006.08.072.

[35] Hoshina K, Horiyama S, Matsunaga H, Haginaka J. Simultaneous determination of non-steroidal anti-inflammatory drugs in river water samples by liquid chromatography−tandem mass spectrometry using molecularly imprinted polymers as a pretreatment column. J Pharm Biomed Anal 2011;55:916−22. https://doi.org/10.1016/j.jpba.2011.03.014.

[36] Hoshina K, Horiyama S, Matsunaga H, Haginaka J. Molecularly imprinted polymers for simultaneous determination of antiepileptics in river water samples by liquid chromatography-tandem mass spectrometry. J Chromatogr A 2009;1216:4957−62. https://doi.org/10.1016/j.chroma.2009.04.071.

[37] Sambe H, Hoshina K, Haginaka J. Molecularly imprinted polymers for triazine herbicides prepared by multi-step swelling and polymerization method. Their application to the determination of methylthiotriazine herbicides in river water. J Chromatogr A 2007;1152: 130−7. https://doi.org/10.1016/j.chroma.2006.09.003.

[38] Yang M, Zhang Y, Lin S, Yang X, Fan Z, Yang L, et al. Preparation of a bifunctional pyrazosulfuron-ethyl imprinted polymer with hydrophilic external layers by reversible addition−fragmentation chain transfer polymerization and its application in the sulfonylurea residue analysis. Talanta 2013;114:143−51. https://doi.org/10.1016/j.talanta.2013.03.078.

[39] Li X, Zhou M, Turson M, Lin S, Jiang P, Dong X. Preparation of clenbuterol imprinted monolithic polymer with hydrophilic outer layers by reversible addition−fragmentation chain transfer radical polymerization and its application in the clenbuterol determination from human serum by on-line solid-phase extraction/HPLC analysis. Analyst 2013;138: 3066. https://doi.org/10.1039/c3an36801g.

[40] Sixto A, Fiedoruk-Pogrebniak M, Rosende M, Cocovi-Solberg D, Knochen M, Miró M. A mesofluidic platform integrating restricted access-like sorptive microextraction as a front end to ICP-AES for the determination of trace level concentrations of lead and cadmium as contaminants in honey. J Anal At Spectrom 2016;31:473−81. https://doi.org/10.1039/C5JA00387C.

[41] Moad G. RAFT polymerization − then and now. ACS Symp Ser 2015:211−46. https://doi.org/10.1021/bk-2015-1187.ch012.

[42] Chiefari J, Chong (Bill) YK, Ercole F, Krstina J, Jeffery J, Le TPT, et al. Living free-radical polymerization by reversible addition−fragmentation chain transfer: the RAFT process. Macromolecules 1998;31:5559−62. https://doi.org/10.1021/ma9804951.

[43] Chen L, Xu S, Li J. Recent advances in molecular imprinting technology: current status, challenges and highlighted applications. Chem Soc Rev 2011;40:2922. https://doi.org/10.1039/c0cs00084a.

[44] Puoci F, Iemma F, Cirillo G, Curcio M, Parisi OI, Spizzirri UG, et al. New restricted access materials combined to molecularly imprinted polymers for selective recognition/release in water media. Eur Polym J 2009;45:1634−40. https://doi.org/10.1016/j.eurpolymj.2009.01.021.

[45] Xu W, Su S, Jiang P, Wang H, Dong X, Zhang M. Determination of sulfonamides in bovine milk with column-switching high performance liquid chromatography using surface imprinted silica with hydrophilic external layer as restricted access and selective extraction material. J Chromatogr A 2010;1217:7198−207. https://doi.org/10.1016/j.chroma.2010.09.035.

[46] Yang G, Feng S, Liu H, Yin J, Zhang L, Cai L. On-line clean-up and screening of oxacillin and cloxacillin in human urine and plasma with a weak ion exchange monolithic column. J Chromatogr B 2007;854:85−90. https://doi.org/10.1016/j.jchromb.2007.04.001.

[47] De Oliveira Isac Moraes G, Da Silva LMR, Dos Santos-Neto ÁJ, Florenzano FH, Figueiredo EC. A new restricted access molecularly imprinted polymer capped with albumin for direct extraction of drugs from biological matrices: the case of chlorpromazine in human plasma. Anal Bioanal Chem 2013;405:7687−96. https://doi.org/10.1007/s00216-013-7275-5.

[48] da Silva KKMS, Boralli VB, Wisniewski C, Figueiredo EC. On-line restricted access molecularly imprinted solid-phase extraction of selective serotonin reuptake inhibitors directly from untreated human plasma samples followed by HPLC-UV analysis. J Anal Toxicol 2016;40:108−16. https://doi.org/10.1093/jat/bkv121.

[49] de Lima MM, Vieira AC, Martins I, Boralli VB, Borges KB, Figueiredo EC. On-line restricted access molecularly imprinted solid phase extraction of ivermectin in meat samples followed by HPLC-UV analysis. Food Chem 2016;197:7−13. https://doi.org/10.1016/j.foodchem.2015.10.082.

[50] Santos MG, Moraes G de OI, Nakamura MG, dos Santos-Neto ÁJ, Figueiredo EC. Restricted access molecularly imprinted polymers obtained by bovine serum albumin and/or hydrophilic monomers' external layers: a comparison related to physical and chemical properties. Analyst 2015;140:7768−75. https://doi.org/10.1039/C5AN01482D.

[51] Chapter 5 silica columns−packing procedures and performance characteristics. 1979. p. 169−86. https://doi.org/10.1016/S0301-4770(08)60809-X.

[52] Claessens HA, Aben G, Vonk N. Packing procedure of silica columns for HPLC with aqueous slurries. J High Resolut Chromatogr 1982;5:250−4. https://doi.org/10.1002/jhrc.1240050505.

[53] Mullett WM. Determination of drugs in biological fluids by direct injection of samples for liquid-chromatographic analysis. J Biochem Biophys Methods 2007;70:263−73. https://doi.org/10.1016/j.jbbm.2006.10.001.

[54] Huclová J, Šatínský D, Maia T, Karlíček R, Solich P, Araújo AN. Sequential injection extraction based on restricted access material for determination of furosemide in serum. J Chromatogr A 2005;1087:245−51. https://doi.org/10.1016/j.chroma.2004.11.055.

[55] Santos MG, Tavares IMC, Barbosa AF, Bettini J, Figueiredo EC. Analysis of tricyclic antidepressants in human plasma using online-restricted access molecularly imprinted solid phase extraction followed by direct mass spectrometry identification/quantification. Talanta 2017;163:8−16. https://doi.org/10.1016/j.talanta.2016.10.047.

[56] Kataoka H, Saito K. Recent advances in column switching sample preparation in bioanalysis. Bioanalysis 2012;4:809−32. https://doi.org/10.4155/bio.12.28.

[57] Costa Queiroz ME, Donizeti de Souza I, Marchioni C. Current advances and applications of in-tube solid-phase microextraction. TrAC Trends Anal Chem 2019;111:261−78. https://doi.org/10.1016/j.trac.2018.12.018.

[58] Hua K, Zhang L, Zhang Z, Guo Y, Guo T. Surface hydrophilic modification with a sugar moiety for a uniform-sized polymer molecularly imprinted for phenobarbital in serum. Acta Biomater 2011;7:3086−93. https://doi.org/10.1016/j.actbio.2011.05.006.

[59] Li D, Zhang Z, Yang X, Zhou C, Qi J. Online restricted-access material combined with high-performance liquid chromatography and tandem mass spectrometry for the simultaneous determination of vanillin and its vanillic acid metabolite in human plasma. J Sep Sci 2016;39:3318–26. https://doi.org/10.1002/jssc.201600466.

[60] García-Gómez D, Rodríguez-Gonzalo E, Carabias-Martínez R. Design and development of a two-dimensional system based on hydrophilic and reversed-phase liquid chromatography with on-line sample treatment for the simultaneous separation of excreted xenobiotics and endogenous metabolites in urine. Biomed Chromatogr 2015;29:1190–6. https://doi.org/10.1002/bmc.3407.

# Aptamer-based and immunosorbents

6

*Valérie Pichon* [1,2]
[1]Department of Analytical, Bioanalytical Sciences and Miniaturization, UMR CBI, ESPCI Paris, PSL Research University, Paris, France; [2]Sorbonne University, Paris, France

## 6.1 Introduction

Regardless of technological advances leading to a gain in sensitivity of analytical instruments, a sample processing step is generally required to extract and isolate compounds from environmental samples, biological fluids, or food matrices. Solid-phase extraction (SPE) is nowadays one of the most widely used methods for sample preparation. Conventional SPE sorbents are based on hydrophobic, hydrophilic, or ionic interactions that allow the retention and extraction of the target compounds from the liquid samples or extracts before their analysis by the separation methods. However, while high extraction recoveries can be expected owing to the large variety of sorbents available in various formats, the coextraction of the compounds also present in the sample can lead to coelution with the analytes of interest during the separation step, which prevents their quantification. In order to overcome the lack of selectivity of these sorbents, other sorbents involving antibodies or aptamers, thus promoting a selective retention mechanism based on molecular recognition, have been developed.

Antibodies are proteins produced in the mammalian immune system in response to the presence of a foreign substance (called *antigen*). Among the five major types of immunoglobulins (IgA, IgD, IgD, IgE, IgG, and IgM), IgGs are the simplest ones with a molecular weight of about 150 kDa. They contain two identical heavy chains and light chains (containing 450−650 and 214 amino acids respectively) that are covalently linked by Y-shaped disulfide bonds in and forming two identical antigen-binding sites located at the upper ends of the Y structure (Fig. 6.1A). A large number of interactions (hydrogen bonds, ionic, hydrophobic, and Van der Waals forces) and an appropriate geometric adjustment between the binding site of the antibody and the antigenic determinant provide an antigen-antibody complex that is formed with dissociation equilibrium constant from nM to pM.

Aptamers are single-stranded oligonucleotides usually between 20 and 100 bases long and may have, like antibodies, a specificity toward a ligand (small organic molecules, peptides, nucleic acids, proteins, intact cells). They have a specific and complex 3-D shape, characterized by various elements, such as stems, loops, bulges, hairpins, pseudoknots, triplexes, or quadruplexes [1], which allows them to be bound to a wide variety of targets (Fig. 6.1B). Indeed, they can be generated against various targets, such as divalent metal ions, small organic molecules, proteins, and cells.

**Solid-Phase Extraction. https://doi.org/10.1016/B978-0-12-816906-3.00006-6**

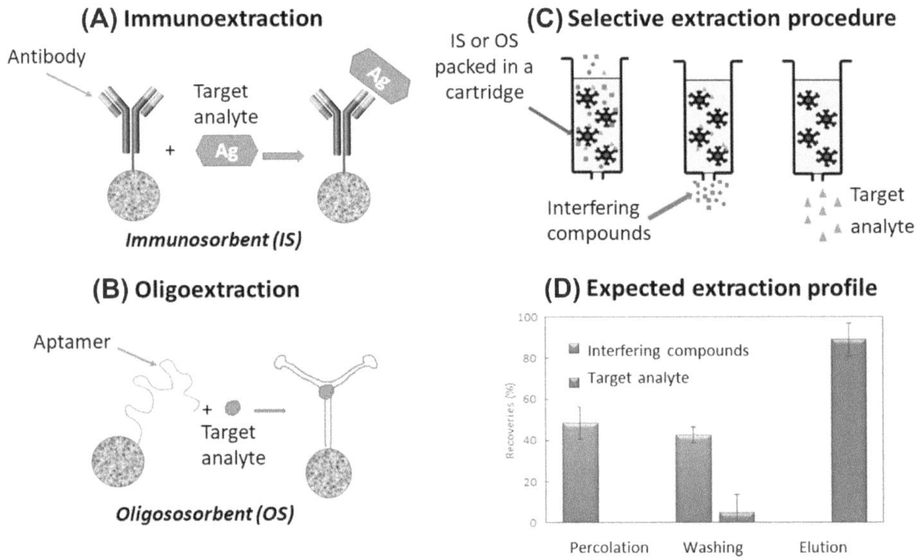


**Figure 6.1** Representation of the specific entrapment of a compound by (A) an immunosorbent (IS) or (B) an oligosorbents (OS), (C) selective SPE procedure on IS/OS and (D) expected extraction profiles by comparing the retention of the targeted analyte specifically retained on IS or OS this retention of compounds from the matrix sample.

They are selected from a random bank containing up to $10^{15}$ different sequences by an *in vitro* combinatorial selection method called *Systematic Evolution of Ligands by EXponential enrichment* (SELEX) according to their ability to recognize a target [2,3]. Once the sequence is selected, aptamers are produced by a chemical process without requiring mammals. Some aptamers are characterized by dissociation equilibrium constants similar to those of antibodies as recently reported for different types of target molecules [4] and in particular for marine biotoxins [5].

While antibodies have been widely used for the development of bioassays, such as the well-known ELISA methods or sensors, the immunosorbents have already largely proven their effectiveness in selectively retaining or extracting different types of molecules in terms of size and of physicochemical properties. Indeed, many studies have described their potential in separation (chromatography, electrophoresis) [6,7] and solid-phase extraction [8−11]. Numerous immunosorbents are now available on the market, and their use has been validated for numerous applications, particularly for food analysis [12].

Various analytical aptamer-based methods have been developed, including enzyme-linked oligonucleotide assay (ELONA or ALISA, a variant of ELISA with aptamers instead of antibodies) [3] or biosensors ("aptasensors") [5,13,14]. The use of aptamers as SPE sorbents (i.e., of oligosorbents) is more recent than the use of antibodies but seems to be a very promising approach [4,15,16].

This chapter describes the development, characterization, and application of both sorbents to the selective extraction of target analytes from real samples. Thus, the principle, advantages, limitations, and complementaries of these sorbents will be presented. The introduction of these selective tools in miniaturized devices will also be discussed.

## 6.2 Grafting of antibodies and aptamers on a solid support

When used as selective tools for the extraction of target molecules, antibodies and aptamers must be bound to solid supports (particles of different sizes −from large beads to nanoparticles, rods, fibers, internal capillary walls, chip channel surfaces, etc.) according to the method to be developed.

These supports must possess a chemical and biochemical inertness, good mechanical stability, and homogeneity in terms of particle size or surface. They must be easily activated to allow effective binding of biomolecules and have large pore sizes, particularly to promote the accessibility of antibodies, which are large molecules compared to aptamers (10−15 nm vs. 1−2 nm). They must be hydrophilic in order to avoid any nonspecific interactions. Finally, the immobilization procedure must preserve the affinity of antibodies/aptamers toward their target analyte.

If the trapping of the antibodies by their immobilization in the pores of a hydrophilic glass matrix has been proposed, the covalent grafting of the antibodies onto a sorbent by coupling accessible chemical functions of antibodies amino acids with reactive groups on the surface of the support is the most frequently described approach. This type of covalent immobilization on traditional supports, including silica, agarose, cellulose, and synthetic polymers, such as polymethacrylate derivatives, is often based on the reaction between the amino groups of lysine residues of antibodies. This leads to random immobilization of antibodies that can affect the accessibility of the antigen to their binding sites [9]. Concerning aptamers, their chemical synthesis makes it possible to introduce modifications at the 5′ or 3′ end of the oligonucleotide sequence to facilitate their immobilization. This modification is chosen according to the nature of the bonding. Modification by an amino group has often been reported resulting in immobilization procedures similar to those applied for the random immobilization of antibodies with the advantage that the aptamer sequence is oriented directly toward the analyte. In addition, a spacer arm can be introduced to maintain the binding properties of the aptamer when it is bounded to a surface. It may be an n-alkyl chain in C6 or C12, an ethylene glycol derivative, as illustrated by the data reported in Table 6.1. The aptamers have also been functionalized by thiol-groups, thus favoring their binding to a gold surface or, as often described for antibodies, to biotin for achieving a noncovalent grafting to streptavidin activated particles that are commercially available. Nevertheless, noncovalent binding procedure presents some drawbacks, in terms of reusability and sorbent life-time, particularly when using organic modifiers in the SPE procedures that affect the biotin-streptavidin affinity [24,41].

Several studies have attempted to optimize the antibody immobilization through oriented binding procedures by exploiting the carbohydrate moieties located on the

**Table 6.1** Applications of aptamers in extraction devices for the analysis of targets in real samples.

| Target analyte | Sample | Amount or dimension of the extraction device/$V_{sample}$ | Sorbent and reagent used for the grafting | Aptamer modification/ Spacer | Ref. |
|---|---|---|---|---|---|
| **Off-line SPE** | | | | | |
| Aflatoxin B2 | Peanut extract (diluted) | 60 mg/5 mL | CNBr-sepharose | $NH_2/C_7$ | [17] |
| CEA | Serum | 4 μL (PDMS chip)/200 μL | Carboxylated-polystyrene beads (40–50 μm), NHS | $NH_2$/- | [18] |
| Cocaine | Plasma (diluted ×2) | 35 mg/200 μL | CNBr-sepharose | $NH_2/C_6$ | [19] |
| | Post mortem blood (after PP) | 35 mg/100 μL | | $NH_2/C_{12}$ | [20] |
| Cocaine + diclofenac | Drinking water | 50 × 8 mm i.d./0.2–1 L | CNBr-sepharose | $NH_2/C_6$ | [21] |
| Ochratoxin A | Wheat extract | 300 μL in cartridge/ 10–12 mL | DADPA (noncovalent) | –/– | [22,23] |
| | Red wine | 35 mg/1 mL | CNBr-sepharose | $NH_2/C_6$ | [24] |
| | Wheat extract (diluted ×10) | | | $NH_2/C_{12}$ | [25] |
| | Ginger powder (diluted ×10) | 200 μL/3 mL | NHS-sepharose | $NH_2/C_6$ | [26] |
| Tetracycline | Plasma (after PP) and urine (diluted ×3) | 35 mg/200 μL | CNBr-sepharose | $NH_2/C_6$ | [27] |

| | | | | | |
|---|---|---|---|---|---|
| Thrombin | Human serum | 0.53 mm i. d./100 μL | Poly(TMOS-co-γ-MAPS) monolith | $SH/C_6$ | [28] |
| | Human serum (diluted ×50) | 0.53 mm i. d. ×2 cm/1 mL | Poly(AEAPTES-co-TEOS) monolith modified with AuNP | $SH/C_6$ | [29] |
| **On-line SPE/LC** | | | | | |
| OH-deoxyguanosine | Urine (diluted ×5) | 200 × 0.5 mm i. d./2 mL | Amino-mNP, GTD | $NH_2$/- | [30] |
| Ochratoxin A | Beer (diluted ×2) | 100 × 0.1 mm i.d./100 μL | POSS-acrylate-based monolith (one-pot synthesis) | SH/- | [31] |
| | Beer, wine (diluted ×9) | 10 cm × 75 μm i. d./20 μL | Poly(TMOS-co-MPTMS) monolith modified with AuNPs (25 nm) | $SH/C_6$ | [32] |
| **On-line SPE/nanoLC** | | | | | |
| Ochratoxin A | Beer (diluted ×2) | 70 × 100 μm i. d./250 nL | Poly(APTES-co-TEOS) monolith, GTD | $NH_2/C_{12}$ | [33] |
| **In-line SPE/CE** | | | | | |
| Ochratoxin A | Beer and white wine (diluted ×2) | 15 mm × 75 μm i. d./ 2.65 μL | Poly(TMOS-co-MTMS) monolith, VTMS | SH/- | [34] |
| **On-column concentration– direct injection of real samples** | | | | | |
| Adenosine | Dialysated from rat cortex | 5 cm × 150 μm i. d./4 μL | Streptavidin activated - porous glass beads | Biotin/TEG | [35] |
| Cytochrome C | Serum (diluted ×10) | 8.5 cm × 100 μm i. d./- | Streptavidin- modified poly(GMA-co-TRIM) monolith | Biotin/- | [36] |

*Continued*

**Table 6.1** Applications of aptamers in extraction devices for the analysis of targets in real samples.—cont'd

| Target analyte | Sample | Amount or dimension of the extraction device/$V_{sample}$ | Sorbent and reagent used for the grafting | Aptamer modification/ Spacer | Ref. |
|---|---|---|---|---|---|
| Doxorubicin, epirubicin | Serum, urine | 10 cm × 75 μm i. d./20 μL | Poly(TMOS-co-γ-MAPS) monolith (one-pot synthesis) | SH/- | [37] |
| Lysozyme | Chicken egg white | 10 × 4.6 mm i. d./5 μL | Poly(GMA-co- EDMA, ethylenediamine, GTD | $NH_2/C_6$ | [38] |
| L-selectin | CHO cell-conditioned medium | 5 × 0.5 mm i. d./- | Streptavidin-polyacrylamide resin | Biotin/- | [39] |
| Thrombin | Serum (diluted ×20) | 250 μm i. d./- | Poly(APTES-co-TEOS) monolith, GTD | $NH_2/C_6$ | [40] |

heavy chains, at the lower ends of the Y structure (i.e., Fc region) after the mild oxidation of these moieties in aldehyde residues, which may react with a sorbent containing a hydrazide or amine. The use of antibody fragments was also recently proposed to limit their size while facilitating their oriented grafting. This approach seems particularly interesting for the development of miniaturized extraction devices. It must provide greater antigen-binding capacity because immobilization occurs far from the antigen-binding sites and results in a more oriented reaction. However, despite the theoretical advantages of the oriented immobilization procedure, the greater binding capacity generally observed using random immobilization that can result from a higher number of active groups greatly reduces its interest. This oriented immobilization can also be ensured by the use of a sorbent activated by proteins A or G, which also bind the antibody in the Fc region. This noncovalent immobilization strategy is based on the high affinity between the IgGs and these proteins but can be easily disrupted by lowering the pH or using an organic solvent, thus affecting the reusability of the resulting ISs as for streptavidin activated supports. A recent review on miniaturized immunoextraction-based sorbents summarized, in detail, the different approaches proposed to improve antibodies immobilization based on this reduced-size extraction format [42].

## 6.3 Extraction procedure using IS and OS

During the extraction step on the immunosorbent (IS), the antibody-antigen complex is formed primarily by electrostatic forces, which attract and orient both entities. Then, this causes the formation of secondary hydrogen bonds, bringing the molecules closer together, and excluding water. Finally, Van der Waals forces are initiated to form a stable noncovalent bond. The establishment of these interactions is promoted in the aqueous media and then well-adapted to aqueous samples (real waters, biological fluids). For other types of samples, such as organic extracts of solid samples, dilution of the extracts with an aqueous buffer is recommended [8]. Analytes are further recovered by the disruption of these complexes that occurs by applying conditions that may differ from the nature of the target. Indeed, the desorption of proteins is often carried out by chaotropic agents (chloride, iodide, perchlorate, and thiocyanate ions at concentration between 1.5 and 8 mol/L), temperature increase, pH variations, and the use of denaturing agents, such as water-organic solvent mixtures being applied to the desorption of small size targets [8,9].

For oligosorbents (OSs), effective trapping of the target analyte is promoted by a sample whose composition is close to the composition of the buffer used during the selection of aptamers (often called the *binding buffer* (BB)) to favor the interactions between both entities. The strength of the binding between an aptamer and its target, which is highly dependent on the conformation of the aptamer molecule, may be affected by a variety of cations. As an example, it is well known that some guanosine-rich aptamers can adopt inter or intramolecular quadruplex structure that is stabilized by the presence of G-quartets (square arrangement of guanines) that are favored by the presence of $K^+$ as often reported for thrombin aptamers. This example shows the need to study the effect of different cations to determine the optimal conditions for strong binding of the target on the OS during the extraction step. Some parameters, such as temperature [43–47], ionic strength, and pH [43] also strongly affect the conformation of aptamers. These parameters must be controlled and adapted to ensure high extraction recoveries or favor the elution [48]. Different approaches are possible to induce elution: water-organic modifier mixtures and chaotropic agents such as $NaClO_4$, denaturing agents such as urea or guanidine hydrochlorides for antibodies, and scavenging agents, such as EDTA for aptamers sensitive to cations, temperature increase, and pH variations. The choice of those conditions strongly depends on the nature of the analyte(s) and of the aptamers. The use of organic solvents never reported for antibodies (such as hexane [49], a hexane-ethyl acetate mixture [50] or dichloromethane [51,52]) was reported without preventing the reusability of the OS. The addition of trypsin for the direct digestion of a target protein [53] or of recombinant DNase that damages the aptamers [54] have also been described.

When not in use, both ISs and OSs are stored at 4°C in a buffer solution, a PBS buffer for ISs and the BB for OSs. Sodium azide, a bacteriostatic agent, is generally added in the storage buffer. This solution can be removed by pure water or by renewing BB solution during the conditioning step.

## 6.4 Various extraction format and methods

Immunoaffinity cartridge of 1–3 mL is the most commonly used format for immunoextraction [8,9,12]. Many oligosorbents were also developed in this format too. However, during these last years, IS and OS available in other formats and allowing the development of other extraction methods than SPE have been developed and are described below.

### 6.4.1 SPE in cartridge or column

Many ISs have been developed since the early 1990s in cartridges to carry out off-line SPE [8], such as the commercialized ISs dedicated to mycotoxins analysis, which constitute the most important ISs market [12]. For application to real samples, 30–60 mg of ISs are generally packed between two frits into disposable cartridges or columns as a conventional sorbent for SPE as shown in Fig. 6.1C. The advantage of the cartridge format is that it is adapted to all the consumable and automated methods developed for conventional SPE sorbents. For OSs, their application in SPE to real samples is more recent. Those works are summarized in Table 6.1. As commercially available ISs, most of the OSs were prepared by immobilizing aptamers on activated Sepharose beads and by applying a similar extraction procedure.

After conditioning them with a few ml of aqueous media that favors the interaction of the target analyte with the antibodies/aptamers, the sample is percolated. After a washing step that can be applied to remove some matrix components slightly retained during the sample percolation, the target analyte is then desorbed by a solution affecting the interactions with the antibodies/aptamers. As schematized in Fig. 6.1D, matrix components, not recognized by antibodies/aptamers, and therefore, not retained on the IS/OS must be removed after the washing step, thus giving rise to a final elution fraction that contains only the targeted analyte.

An important parameter in this sequence, as in conventional SPE procedure, is the volume that can be percolated without loss of the analyte during this percolation step to ensure high recoveries. This volume is highly dependent on the affinity of the antibodies/aptamers for the target analyte(s). It can be drastically reduced by choosing a percolation medium that does not promote this affinity, as previously mentioned. The dilution of the real samples with a buffer is often performed to limit the risk of loss of affinity caused by matrix components.

If dissociation constants for antibodies and their antigens are in the nanomolar range, the Kd values for aptamers may fluctuate [2], and high Kd values limit the possibility to percolate high sample volumes and then to reach high enrichment factors. This was well illustrated by Hu et al. who reported the possibility to percolate up to 2 L of drinking water containing diclofenac without observing a decrease of recovery, while the sample volume was limited to 0.5 L of water for the extraction of cocaine. Indeed, the breakthrough volume was lower for cocaine than for diclofenac because of the higher Kd value of cocaine aptamer (Kd = 5 μM) compared to diclofenac aptamer (Kd = 47.2 nM) [21]. In return, for antibodies, the Kd value can be low for its antigen but higher for a structural analog, thus giving rise to a decrease of recovery for this molecule for too high percolated volume.

An example of evaluation of this breakthrough volume is given in Fig. 6.2 A, which illustrates the effect of the volume of sample percolated on an anticiprofloxacin (CIP) IS (2 mL gel) on the recovery of extraction of CIP. The affinity of the antiCIP antibodies allows the percolation of up to 5.5 mL of the sample while the recovery of quinine (QUIN), not recognized by these antibodies, decreases from 2 mL of the sample [55]. The optimization of an extraction procedure also consists in applying strong elution conditions to limit the volume of eluent, as illustrated by the elution of CIP with different ratios of methanol in water (Fig. 6.2B), thus favoring high enrichment factors without requiring in some cases additional evaporation steps of the final extract.

The selectivity brought by ISs and OSs is illustrated by comparing the chromatograms of elution fractions from an OS (Fig. 6.3A), an IS (Fig. 6.3C), and hydrophobic support (C18 grafted silica, Fig. 6.3B) during the purification of a cereal extract. The IS and OS allow the elimination of a large part of the interfering compounds and eliminate the visible coelution at the OTA peak.

To miniaturize these selective SPE devices, the immobilization of antibodies and aptamers onto monoliths in-situ synthesized into capillaries of 75−100 μm internal diameter or chip channel has also been proposed, as illustrated in Table 6.1 for aptamers and Table 6.2 for antibodies. These miniaturized extraction devices can be directly coupled on(in)-line with CE, nanoLC or on-chip separations, thus facilitating the automation of the whole analytical procedure while reducing the volume of sample and reagent consumption. This last point is particularly important when relatively expensive reagents such as antibodies are used.

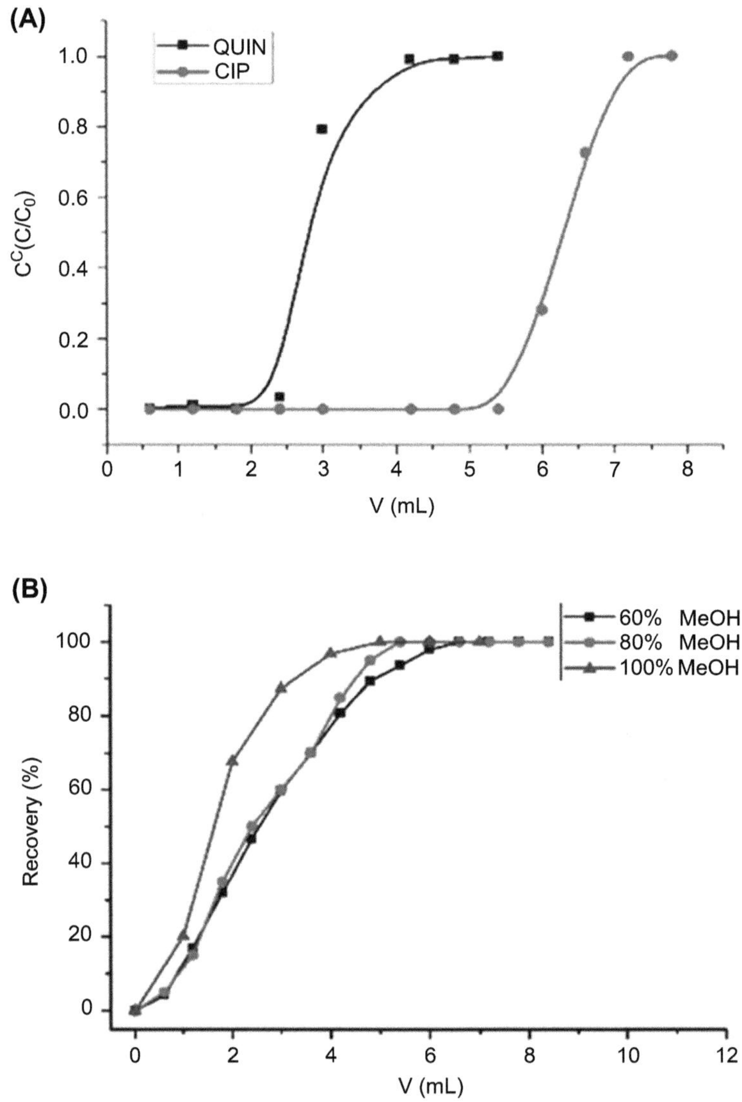


**Figure 6.2** (A) Breakthrough curves of CIP and QUIN obtained in antiCIP pAb immobilized column. (A) Elution profiles of CIP with a different percent of methanol.
Reproduced from Zhang XH, Deng Y, Zhao MZ, Zhou YL, Zhang XX. Highly-sensitive detection of eight typical fluoroquinolone antibiotics by capillary electrophoresis-mass spectroscopy coupled with immunoaffinity extraction. RSC Adv.2018;8(8):4063–71 with permission from the Royal Society of Chemistry.

At last, some applications of a column containing grafted aptamers and directly connected to UV detection are listed in Table 6.1. In these works, if the concentration effect cannot be expected, limiting thus the sensitivity of the method, the aptamers are used to remove the matrix components at the beginning of the chromatogram before

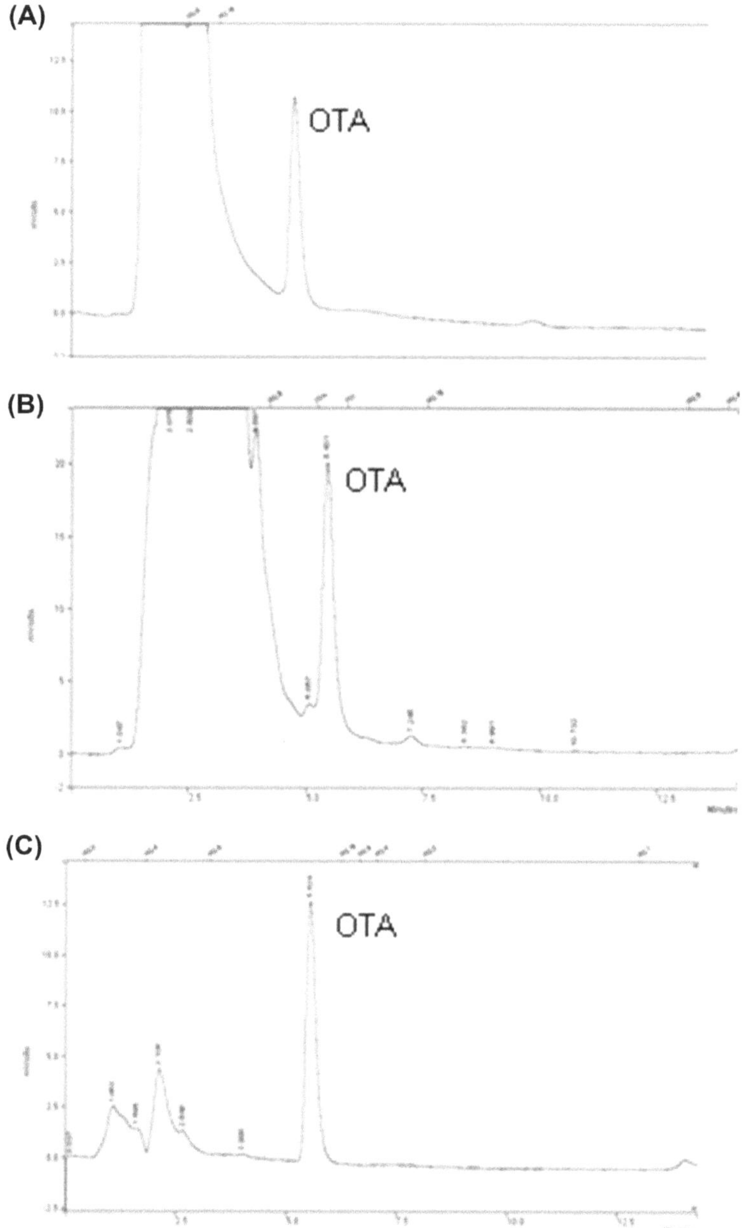


**Figure 6.3** HPLC chromatograms resulting from the analysis of elution fractions after the extraction of a red wine sample spiked at 2 μg/L with Ochratoxin A (OTA) on the oligosorbent based on a covalent immobilization (A) on the conventional C18 silica cartridge (B) and on the immunosorbent (C).
Reproduced from Chapuis-Hugon F, du Boisbaudry A, Madru B, Pichon V. New extraction sorbent based on aptamers for the determination of ochratoxin A in red wine. Anal Bioanal Chem 1 mai 2011;400(5):1199–1207. With permission from Springer Nature.

**Table 6.2** Immunoaffinity monolith coupled to CE, nanoLC or integrated on-chip and applied to real samples.

| Target | Sample | Amount or dimension of the extraction device/$V_{sample}$ | Sorbent and reagent used for the grafting | Antibody grafting | Ref |
|---|---|---|---|---|---|
| Haptoglobin | Serum | 100 μm i.d. capillary/ | Poly (GMM-co-PETA) (diol), poly (GMA-co EDMA) (epoxy), poly GMM-co-EDMA, poly (aminopropylacrylamide hydrochloride-co- EDMA) monoliths | Reduced pAbs (aldehyde groups) grafted on diol or epoxy-based sorbent or pAbs and Fabs (amino group) grafted using GTD | [56] |
| Microcystin-LR | Algae extract | 4.5 cm × 100 μm i. d. Capillary/ 150 nL | poly(APTES-co-TEOS) hybrid monolith | mAbs grafted using GTD | [57] |
| Labeled -proteins (ferritin, lactoferrin) | Labeled human serum | 0.6 mm length/ | poly(GMA-co-EGDMA) monolith polymerized in the COC chip activated with PEGdiacrylate | Covalent grafting (epoxy) of pAbs | [58] |
| Labeled protein (ferritin) | Human serum (Diluted ×5) | 0.6 mm length/ | poly(GMA-co-EGDMA) monolith prepared in a 3D printed chip (45 μm x 50 μm) | Covalent grafting (epoxy) of pAbs | [59] |
| Labeled ProGRP digest | Human serum digest | 15 cm × 180 μm i. d. Capillary/ 20 μL | Poly(EDMA-co-VDM) monolith | Covalent grafting (via VDM) of mAbs | [60] |

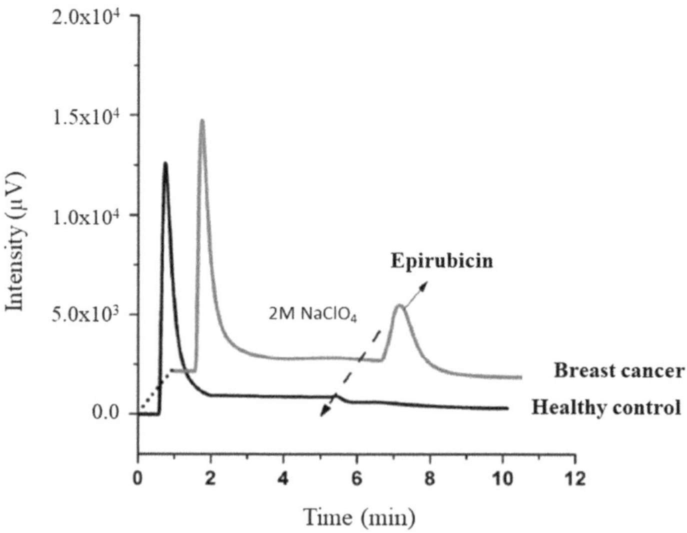


**Figure 6.4** Chromatograms for the determination of epirubicin in the serum of breast cancer patients treated with epirubicin by intravenous injection. Facile one-pot synthesis of an aptamer-based organic–silica hybrid monolithic capillary column by "thiol–ene" click chemistry for detection of enantiomers of chemotherapeutic anthracyclines.
Reproduced from Jiang H-P, Zhu J-X, Peng C, Gao J, Zheng F, Xiao Y-X, et al. Facile one-pot synthesis of a aptamer-based organic–silica hybrid monolithic capillary column by "thiol–ene" click chemistry for detection of enantiomers of chemotherapeutic anthracyclines. Analyst 2014; 139(19):4940–4946 with permission from the Royal Society of Chemistry.

the increase in the eluent strength of the mobile phase allowing the elution of the target molecule. An example is provided in Fig. 6.4. It shows the potential of this simple approach to differentiate the serum of breast cancer patients treated with epirubicin and serum from a healthy control.

## 6.4.2 Dispersive SPE

In the dispersive mode, dSPE, the extraction is carried out by introducing a given amount of IS/OS into the sample. After a sufficient extraction time under stirring, the particles are recovered mainly by centrifugation or magnetic field (if magnetic core particles are used) and then introduced into a suitable desorption solvent after a possible washing step. Application of dSPE methods using antibodies and aptamers are summarized in Table 6.3. In these works, antibodies/aptamers were grafted on conventional beads, magnetic beads and also nanoparticles (in the order of tens of nanometers instead of tens of micrometers). As with conventional dSPE sorbents, the key-parameters to be optimized are the sample volume/sorbent ratio, the extraction time, the stirring rate, the nature of the solvent, and the time required for desorption [77].

While some studies have reported the use of commercially available streptavidin-magnetic agarose beads that make it really easy to immobilize biotinylated antibodies/aptamers and apply them, numerous works reported the use of NPs.

**Table 6.3** Some applications of ISs and OSs in dSPE for the analysis of targets in real samples.

| Target analyte | Samples | Amount or dimension of the extraction device/$V_{sample}$ | Sorbent and reagent used for the grafting | Grafting method | Ref |
|---|---|---|---|---|---|
| **Immunosorbent** | | | | | |
| Anatoxin A | Water | 10 μL/20 mL | NHS-sepharose magnetic beads (27 μm) | Grafting of mAbs via amino groups | [61] |
| Epitestosterone | Urine | 10 mg/20 mL | $Fe_3O_4$-Au NPs (50 nm) | Grafting through Au-S bonds of half mAb (reduction of S-S bond) | [62] |
| Human chorionic gonadotropin | Dried blood spot, serum, urine | 20 μL/1 mL | Tosylactivated magnetic beads (2.8 μm) | Grafting of mAbs (previously stored at pH 2.5) | [63,64] |
| Low-density lipoprotein | Plasma (diluted 1/500) | 10 μL/10 μL | Au-NP (28 nm) | Noncovalent grafting; covalent grafting of mAbs to carboxy terminated-pegylated NPs (amide), oriented immobilization of mAb via oxidized carbohydrate moiety on hydrazide derivatized NP, noncovalent via cysteine-tagged protein A-NPs | [65] |
| Microcystins | Urine | 5 μg/100 μL | Streptavidin -magnetic beads | Biotinylated antibodies | [66] |
| Soy proteins | Soy milk (500 μL) | 23 mg/500 μL | $Fe_3O_4$-Au NPs (12 nm) | Oxydized pAbs on amino-functionalized Au-NPs | [67] |
| **Oligosorbent** | | | | | |
| ATP | Deproteinated cell lysate | 25 μL/250 μL | AuNP (13 nm) | SH- Apt | [68] |
| Adenosine | Urine | 8 mg/200 μL | Silanized mNPs, APTES, GTD | $NH_2$-$C_6$- Apt | [69] |

| | | | | | |
|---|---|---|---|---|---|
| Aflatoxin M1 | Milk extract (purified by PP and LLE) | 16 mg/15 mL | Silanized mNPs, APTES, GTD | – | [51] |
| Aflatoxin B1, B2 | Maize extract | 1 mL/1 mL | Streptavidin-magnetic agarose beads (28 μm) | Biotin-Apt | [41] |
| Amphenicol antibiotics | Milk (diluted ×100) | 50 mg/100 mL | Silanized mNPs, APTES, EDC, Sulfo-NHS | $NH_2$- Apt | [43] |
| Arsenic | Groundwater | 2 mg/1 mL | Streptavidin-agarose beads | Biotin-Apt | [70] |
| Berberine | Herb extract | 20 mg/1 mL | Avidin -activated amino-mNPs, GTD | Biotin-Apt | [50] |
| Bisphenol A | Human serum, urine (diluted ×2) treated with enzymes | 5 mg/1 mL | Avidin- mNPs | Biotin-Apt | [71] |
| 17β-estradiol | River water | 50 beads/ 100 μL | NCS-modified glass beads (250 μm), APTMS, PDITC | $NH_2$- Apt | [48] |
| $His_6$-Tag protein | E Coli lysates | 5 mg/100 μL | $NH_2$-magnetic beads, GTD | $NH_2$-$C_6$ or $C_7$-Apt | [72] |
| Histones | Cell lysate | –/– | Streptavidin agarose beads | Biotin-Apt | [73] |

*Continued*

**Table 6.3** Some applications of ISs and OSs in dSPE for the analysis of targets in real samples.—cont'd

| Target analyte | Samples | Amount or dimension of the extraction device/$V_{sample}$ | Sorbent and reagent used for the grafting | Grafting method | Ref |
|---|---|---|---|---|---|
| Ochratoxin A | Coffee extract (diluted ×2) | 100 μL/100 μL | Carboxylated-mNPs (100–250 nm)+ EDC, NHS | $NH_2$-$C_6$- Apt | [44] |
| | Corn, peanut extract diluted in BB | 10 mg/- | Streptavidin - magnetic MOF | Biotin-Apt | [74] |
| PCBs | Soil extract | 30 mg/100 mL | mNP coated by amino-mMOF, GTD | $NH_2$-$C_6$- Apt | [49] |
| 0H-PCBs | Human serum | 30 mg/40 mL | mNP (18 nm), APTES | $NH_2$- Apt | [75] |
| Sulfanilamide | Milk | 10 μL/990 μL | Magnetic carboxylated-silica particles | $NH_2$- Apt | [45] |
| Thrombin | Serum, blood | –/– | Magnetic beads coated with AuNP | SH-$C_6$- Apt | [53] |
| | Plasma | 100 μL/100 μL | Au-nanorods (77 × 17 nm) | SH-$(EG)_6$- Apt | [76] |
| Thyroid transcription factor 1 (TTF1) | Bacterial lysate | 2 mg/100 μL | Streptavidin-magnetic beads | Biotin-Apt | [54] |

The works reported in Table 6.3 show that the quantities of supports used are variable, but the use of nanoparticles leads to a decrease in these quantities, and therefore, to a decrease in the quantities of aptamers and antibodies used, thus decreasing the cost of the methods while reducing sample volumes. Immuno-dSPE, also known as *immunocapture*, was applied to sample volumes from 10 μL to 20 mL, depending on the expected enrichment factor, with extraction time between 5 min and overnight. For OS-based dSPE, ratios of 10 μL of beads for the extraction of 100 μL of sample to 50 mg of beads dispersed in 100 mL sample with extraction times in the range 3 min to 2 h were reported. With both sorbents, recoveries higher than 80% were reported. The possibility of performing this extraction method in 96-well plates for easy automation has recently been reported [66].

### 6.4.3 SPME, SBSE, and associated methods

Other methods based on the equilibrium of the target between the sample and the immobilized antibodies/aptamers were developed by replacing particles by other formats, such as fibers or rods of different sizes up to thin films resulting in selective extraction methods named solid-phase microextraction (SPME), stir-bar sorptive extraction (SBSE), or thin-film microextraction (THME), respectively. A quite exhaustive view of the applications related to the use of OS and IS in these formats is provided by Table 6.4.

Mostly developed for the extraction of the volatile compounds before their analysis by gas chromatography (GC), the applications of SPME have been extended to biological fluids with an off-line coupling with liquid chromatography (LC) [87]. Therefore, its combination with the high antibodies or aptamers selectivity which limits the coextraction of interfering compounds present in complex samples, such as biological fluids, can be an interesting approach to improve the sensitivity of the method.

SPME generally consists in the extraction of compounds by immersing a fused silica fiber (100 μm diameter) usually coated by an organic polymer (7−100 μm thickness) into a sample, and different parameters can be optimized to extract high amounts of analytes, such as the extraction time, stirring speed, nature, and the volume of the desorption solution or the desorption time, as for conventional SPME. The same parameters have to be optimized in SBSE and THME that are only supposed to differ by the device sizes. Nevertheless, the dimensions of these different devices are not so different as demonstrated by those reported in Table 6.4, the classification in this Table being in agreement with the name of the method provided by the authors of the reported works.

The preparation of the extraction devices mostly consists in the direct activation of silica-based material, cellulose paper, and graphene oxide sheet allowing the covalent grafting with $NH_2$-functionalized aptamers or amino-groups of antibodies. The immobilization of activated particles or the polymerization of a monolith onto rods were also proposed, those approaches being used to improve the sorbent capacity, as it will be discussed later on.

In most of the cases, the extraction times are about 30 and 60 min and desorption times between 20 and 40 min as for conventional SPME sorbent. Reported recoveries

**Table 6.4** Applications of ISs and OSs in SPME, SBSE, and associated.

| Method | Target | Samples | Amount or dimension of the extraction device/ $V_{sample}$ | Sorbent used for the grafting | Grafting method | Ref |
|---|---|---|---|---|---|---|
| **Immunosorbent** | | | | | | |
| SPME | Theophylline | Serum (diluted with PBS, 1/100, v/v) | Fiber (1.8 mm, 2.3 cm)/- | Silanized fiber | pAbs grafted with GTD | [78] |
| | Benzodiazepines [3] | Urine | Rods (4 mm, 2.5 cm)/- | Silanized borosilicate glass rods | pAbs purified pAbs, mAbs grafted with GTD | [79,80] |
| | Non-steroidal estrogens [3] | Environmental waters (dilution 1/2) | Stainless steel rods (2 × 18 mm)/1 mL | 5 μm-porous silica particles coated on rods | mAbs grafted with GTD | [81] |
| SBSE | Quinolones [11] | Bovine milk (centrifuged to remove fat) | bars (5 mm, 3 cm)/- | Silanized borosilicate glass bars | mAbs grafted with GTD | [82] |
| **Oligosorbent** | | | | | | |
| SPME | Adenosine | Human plasma | 1 cm length fiber/- | Wire (125 μm diameter) coated with poly (TMOS-co-APTES) | COOH-Apt + EDC, NHS | [83] |
| | Thrombin | Human plasma (diluted ×20) | 55 × 1.5 mm i.d./ 2 mL | Stainless steel rod coated with PANCMA (electrospun microfiber) | $NH_2$-$C_6$-Apt + EDC, NHS | [84] |

| SBSE | PCBs | Purified (silica gel) fish extract | 50 × 2 mm/- | Wire coated with MOF-5, PDDA | Noncovalent | [52] |
|---|---|---|---|---|---|---|
| TFME | Codeine | Urine (diluted ×100) | -/10–15 mL | Oxidized cellulose paper | $NH_2$-$C_6$-Apt. + $NaCNBH_3$, | [46] |
| | Codeine, Acetamiprid | Pharmaceutic tablets, urine | Triangle (14 mm × 4 mm)/ 15 mL | | $NH_2$-$C_6$-Apt. + $NaIO_4$ | [47] |
| | Methamphetamine | Saliva (diluted ×10), precipitated plasma (diluted ×20) | Triangle (14 mm × 4 mm)/ 700 μL | Cellulose treated by carbon dots | $NH_2$-$C_6$-Apt. + EDC, NHS | [85] |
| | Thrombin | Plasma (diluted) | 20 μg/1 mL | Graphene oxide sheet | $NH_2$-$C_6$-Apt. +EDC | [86] |

are between 13% and 67% with associated RSD values lower than 14% for IS based methods and higher than 80% with RSD values lower than 10% for OS-based methods.

In addition to these methods, immunoaffinity-based capillary microextraction method, also called *in-tube SPME* (IT-SPME) was proposed. With this approach, the analytes are extracted by antibodies fixed at the inner surface of a capillary before being desorbed and transferred directly to the separation device. It presents several advantages compared to SPME, such as an online coupling with separation methods (capillary electrophoresis (CE) and more frequently, LC). Applications of immunoaffinity IT-SPME to real samples are reported in Table 6.5. To my knowledge, no similar developments were achieved using aptamers. In 1998, Phillips and Dickens proposed an "immunoaffinity capillary electrophoresis" method that consists in the covalent grafting of antibodies fragments (Fabs) at the inner surface of approximatively 6 cm of a 100 μm silica capillary, with the remaining part of the 30 cm-capillary being used for the CE separation [88]. Since this pioneering work, other methods were proposed. They can perform based on the injection at a fixed flow-rate of a fixed volume of sample or by repeated aspirations (draw) and ejections of a fixed volume samples through the capillary. The sample draw/ejection volume and the number of draw/ejection cycles have to be adapted to the capillary capacity [89] and to the volume comprised between the injection needle and the capillary. All the extractions performed by this approach were achieved in less than 10 min, the shortest time than using the previous methods.

In addition to the developments using capillaries, the coating of the surface of a chip channel was also proposed by polymerizing a thin film of a polymer on a 0.6 mm-length channel to couple IT-SPME online with a chip-based CE separation [91,92]. This device was first developed for the trapping of a single molecule, α-fetoprotein (AFP), from human serum thanks to the immobilization of antiAFP antibodies on the thin film of the polymer [91]. Four antibodies specific of four different biomarkers were further simultaneously immobilized in the same way, thus enabling the simultaneous extraction of these biomarkers from the same human serum sample, their individual identifications being achieved after their transfer to the separation channel [92].

## 6.5 Capacity

The capacity is defined as the maximal amount of a molecule that can be retained by the sorbent during the percolation step. For these selective sorbents, it depends on the number of active immobilized antibodies/aptamers, which is directly related to their bonding density, usually expressed in mg/mL of sorbent bed or mg/mg of the sorbent.

This parameter strongly depends on the concentration of biomolecules introduced in the grafting solutions and of the specific surface area of the solid support accessible for the immobilization of the biomolecules [4,8]. Supports with small pore sizes have a high surface area, but low accessibility for the large biomolecules. On the other hand, supports with large pore sizes have good accessibility but a small surface

Table 6.5 Applications of immunoaffinity in-tube SPME.

| Target analyte | Sample | Dimension of the extraction device/$V_{sample}$ | Extraction sorbent | Grafting method | Ref |
|---|---|---|---|---|---|
| Cytokines | Urine, plasma, CSF, saliva | 6 cm of a 30 cm capillary, 100 μm i. d./30 nL | Activated (APTES) silica capillary | Grafting of Fabs from mAbs, SSMCC | [88] |
| Fluoxetine | Serum | 70 cm, 250 μm i. d./ 20 × 50 μL | Activated (APTES) silica capillary | Covalent grafting of pAbs, GTD | [89] |
| Interferon α | Plasma | 60 cm, 250 μm i. d./ 20 × 150 μL | Activated (APTES) silica capillary | Covalent grafting of mAbs, GTD | [90] |
| FITC-labeled AFP, four labeled-biomarkers (proteins) | Labeled-human serum | 3 μm thickness, 6 mm/10 μL | PMMA-chip channel coated with poly(GMA-co-PEGDA) monolith film | Covalent grafting of Abs | [91,92] |

area. The bonding density can be calculated experimentally by measuring the concentration of the antibodies/aptamers that remains in the binding solution by UV spectrophotometry. Nevertheless, this approach only allows the determination of the amount of grafted biomolecules, including also those that are not accessible for the target analyte.

The real capacity of ISs/OSs can be determined by measuring the amount of retained target analyte as a function of its concentration in the percolated sample after applying the optimized extraction procedure. By plotting the amount of target analyte retained by the sorbent (i.e., recovered in the elution fraction) as a function of its concentration in the percolated sample, a curve is obtained that is characterized by two different parts [17,19,20,24,25]. First, for the lowest percolated amounts/concentration levels, a linear part is obtained and corresponds to a range of concentrations for which a constant recovery yield of extraction is obtained, this recovery yield being given by the slope of this linear part. Working in this range ensures the possibility to carry out quantitative analyses. In the second part of the curve, for higher percolated amount, the recovery yields decrease, and the amounts of target analyte retained by the sorbent tend to reach a plateau. This decrease in recovery yield is caused by the overloading of the sorbent capacity, i.e., the saturation of the antibodies/aptamers-binding sites. The capacity of the sorbent is given by the amount/concentration corresponding to the upper limit of the linear range. Knowing this amount, it is then possible to evaluate the part of active aptamers or antibodies among the total amount of immobilized ones, thus allowing a better evaluation of the efficiency of the grafting procedure(17,40).

Concerning immunosorbents, both polyclonal (pAbs) and monoclonal (mAbs) antibodies can be used for their preparation. The pAbs are cheaper to obtain, but their production suffers from a lack of reproducibility in terms of the time of response of an animal, quantity, and even specificity. In contrast, the production of mAbs is costly but guarantees a long-term production of reproducible antibodies that does not require animals for further large-scale production. Some authors also proposed the use of purified pAbs allowing to increase the amount of specific immobilized antibodies [79,80], of Abs fragment (Fabs) [88] or of half-antibodies obtained by splitting them into two (disruption of disulfide bonds between the two heavy chains) [62], the objective being to increase the density of the recognition sites.

Concerning the OSs, the binding density of the aptamers must be facilitated by the possibility to introduce a chemical function during their synthesis according to the nature of the solid-phase selected for their immobilization. In addition to the introduction of this functionality, it was also shown that the length of the spacer affects the binding density. Indeed, when studying the immobilization of aptamers, specific to OTA, on CNBr-activated sepharose, a higher capacity was obtained using an aptamer linked with a C12 spacer arm to the amino-group than using a C6 spacer arm [25]. In return, the immobilization of an aptamer through the 3′-end side and from the 5′-end side gave rise to similar capacities [25].

Values of capacity found in the literature for target molecule on OSs available as SPE cartridges are quite homogeneous and are in the range 0.26–69 nmoL/g of OS [4]. Such capacity values are associated with binding density values in the range

19%−37% on Sepharose beads, a binding density of active aptamers of 68% being recently reported [17]. Those capacity values for OSs are very close to those obtained with immunosorbents (from 4 to 93.6 nmoL/g of sorbent, e.g., immunosorbents specific to pesticides) [93].

While it is sufficient to adapt the quantity of IS/OS introduced into an extraction cartridge to ensure sufficient capacity, the miniaturization of extraction devices has led to the evaluation and comparison of new immobilization approaches. As an example, as the grafting of antibodies directly immobilized on the inner wall of a capillary was low, the use of NP-coated capillary was proposed and has allowed to reach a bonding density to increase the sorbent capacity by a factor 5 [94]. The same group recently proposed to modify first the capillary surface by NPs further functionalized with antibodies that were immobilized in an orientated way (immobilization of the pAbs through the oxidized carbohydrate chain located on their Fc part). The orientation of the antibodies allowed to improve the capacity of the immunosorbent by a factor three compared to a random immobilization and the use of NPs by a factor 1.5 [95]. To improve the capacity of such miniaturized OSs, the introduction of NPs grafted by aptamers and further immobilized on monolith was also proposed [29,32].

## 6.6 Contribution in the selectivity of the OS/IS and control of nonspecific interactions

Nonspecific interactions (NSIs) can occur between the analyte and the sorbent used for the immobilization of aptamers/antibodies and those macromolecules too. These NSIs can contribute to the unexpected retention of interfering compounds present in the sample, thus affecting the selectivity of the procedure.

Agarose gel, such as Sepharose and hydrophilic monoliths have been often selected as solid-phase to reduce NSIs as much as possible, i.e., mainly hydrophobic and/or electrostatic interactions favored in aqueous media. The comparison of the retention of the target analyte on the IS/OS with the nongrafted solid-support, named control sorbent [17,19,20,24−27,29−32,44,69,70,72−75,83], or grafted with another antibody/oligonucleotide sequence, having no affinity for the target [17,19,20,27,32,33,39−41,44,48,69,83], is now often reported to evaluate the contribution of these NSIs. The use of these sorbents in parallel to IS/OS is particularly helpful for optimizing the washing step and then obtaining an optimal selectivity [33,57]. At least the evaluation of the selectivity can be achieved by controlling the lack of retention of compounds often present in the studied matrices and not recognized by antibodies/aptamers [25,30−33,41,45,46,48−51,53,54,71,73,84].

## 6.7 Specificity toward structural analogs

Oligosorbents and immunosorbents largely differ by their ability to retain target structural analogs. Indeed, antibodies may present a broad cross-reactivity, particularly when they have been produced for small molecules, thus allowing the trapping of a structural group of compounds. This is illustrated by the commercialization of ISs during the last decade for the clean-up of samples for the analysis of classes of mycotoxins (aflatoxins, ochratoxins, fumonisins, etc.) in food, of the class of veterinary drugs (clenbuterol and analogs) and of drugs of abuse (LSD and its metabolites, etc). The commercialization of ISs containing several antibodies specific for several toxins for their simultaneous trapping is now also proposed to expand this specific retention to compounds from different classes present in the same samples on a single cartridge. In return, the ability of aptamers to retain structural analogs can be poor. As an example, OTA aptamers present 100-fold less affinity for Ochratoxin B than for OTA, a difference of only 3-fold being observed for antiOTA monoclonal antibodies [22]. Moreover, a slight change in a sequence leads to loss of molecular recognition as it was shown by the loss of cocaine retention on an OS prepared with a DNA sequence similar to the DNA sequence of the cocaine aptamer but presenting only one mutation [20]. Concerning theophylline aptamers, they have no affinity for caffeine despite the structural analogy of these molecules that only differ by a methyl group [96]. However, in this last case, the SELEX procedure was carried out in order to remove oligonucleotides that could present an affinity for caffeine, using a counter selection process, thus showing that the affinity of an aptamer to recognize some structural analogs depends strongly on the SELEX process.

Indeed, the SELEX procedure can also be adapted in order to select aptamers showing an affinity for different targets by using them alternatively during the selection. This shows the possibility offered by the SELEX technology when preparing aptamers to control their specificity that cannot be achieved with antibodies.

## 6.8 Reusability, regeneration

The reuse of ISs is not recommended by the suppliers of commercial immunoaffinity cartridges. However, many studies show that their regeneration is possible, which is interesting considering their cost. Indeed, despite their susceptibility to irreversible denaturation in some conditions, numerous studies have shown the possibility to reuse ISs after their storage at 4°C in a PBS solution, which is close to the physiological conditions and in the presence of an antimicrobial agent, such as sodium azide. This possible regeneration was confirmed recently by a detailed study related to the reusability of commercial ISs dedicated to mycotoxin analysis [97]. In the same way, despite their sensitivity to nuclease, OSs can be stored at 4°C in a buffer solution, whose composition is close to the binding buffer, and also containing sodium azide. Their reusability can be explained by their high stability to low and high temperature, broad pH range, and salt, these different parameters affecting the binding of the target without causing irreversible denaturation. This can be explained by the stability of the phosphodiester bond and by the possibility to improve their stability by chemical modification [98]. In addition to their less expensive preparation, aptamers have short renaturation times, and can, therefore, be reused after a few minutes in a buffer a few minutes against 24 or 48 h for antibodies. Thus, the reusability of OS-SPE cartridges has been demonstrated, as for IS, by their successive application, 5 times [23] and even 15 times [25], to the purification of cereal extracts.

The reusability also depends on the nature of the bonding, noncovalent bonding being less stable than covalent bonding as previously mentioned. It was also reported that the use of orientated immobilization of antibodies also gave rise to higher stability of the IS compared to random immobilization [95]. Of course, the stability of these sorbents depends on the number of applied real samples.

The reusability of OS and IS was also demonstrated for miniaturized devices. As an example, an OS-SPME fiber was applied up to 20 times to plasma samples [83]. For IS-SPME, Yao et al. mentioned that the reusability was evaluated once every 3 days for 45 days by column capacity determination showing a loss of the capacity by a factor two during this period [82]. This is in good agreement with another study mentioning that only 58% of the binding capacity remained after 10 uses [81]. Similar results were reported for IS monoliths that were reused up to 50 times without observing any loss of trapping efficiency [60]. In return, some authors observed losses after 5 [99,100] or even 3 [101] uses on pure samples. This loss of performance after 3 uses was observed on 6 monoliths (whose performance was similar for the first uses) confirming this loss of performance. The results of the first use nevertheless illustrate the repeatability of the monolith preparation also reported by other groups [57,60]. Concerning this last point, it can be considered that the commercialization of Sepharose based-ISs illustrates the reproducibility of their preparation. Concerning other formats, such as miniaturized ones, the reproducibility of the synthesis of the monolith and of the grafting of antibodies or of aptamers was already demonstrated [33,57].

## 6.9 Conclusion

The potential of using immunosorbents for the selective extraction of molecules from complex samples is no longer demonstrated. In order to reduce the development cost of these selective supports, it seems interesting to replace antibodies by aptamers, which allows the developing of equally varied extraction tools with similar potential in terms of extraction yields and selectivity. However, it is important to note some properties that make them different, such as their ability to recognize different analogs, as a result of an immunization choice for some or as a result of selection criteria for others. The possibility of easily introducing certain chemical functions to assist in their grafting can increase the interest of aptamers. However, it is still important to continue work on identifying aptamers sequences of sufficient affinity, of the nanomolar order as is generally obtained with antibodies, to allow the development of supports with high retention potential.

## List of abbreviations

| | |
|---|---|
| **AFP** | α-fetoprotein |
| **APTES** | (3-aminopropyl)-triethoxysilane |
| **BB** | binding buffer |
| **APTMS** | aminopropyltrimethoxysilane |
| **CEA** | carcinoembryonic antigen |
| **CNBr** | cyanogen bromide |
| **DADPA** | diaminodipropylamine |
| **dSPE** | dispersive solid-phase extraction |
| **EDC** | ethyl-dimethylaminopropyl-carbodiimide |
| **EDMA** | ethylene dimethacrylate |
| **FITC** | fluorescein isothiocyanate GMA: glycidylmethacrylate |
| **GMM** | glycerylmethacrylate; GTD: glutaraldehyde |
| **IS** | immunosorbent |
| **γ-MAPS** | γ-methylmethacrylate trimethoxysilane |
| **MOF** | metal organic framework |
| **MPTMS** | mercaptopropyltrimethoxysilane |
| **MTMS** | methyltrimethoxysilane |
| **NHS** | N-hydroxysuccinimide |
| **NP** | nanoparticle |
| **OS** | oligosorbents |
| **PANCMA** | poly(acrylonitrile-co-maleic acid) |
| **PBS** | phosphate buffer saline |
| **PDDA** | poly(diallyldimethylammonium) |
| **PDITC** | phenylene diisocyanate |
| **PDMS** | polydimethylsiloxane |
| **PEGDA** | polyethyleneglycoldiacrylate |
| **PETA** | pentaerythritoltriacrylate PMMA: polymethylmethacrylate |
| **POSS** | polyhedral oligomeric silsesquioxane |
| **PP** | protein precipitation |
| **SBSE** | stir bar sorptive extraction |
| **SPE** | solid-phase extraction |
| **SPME** | solid-phase microextraction |
| **SSMCC** | sulfosuccinimidyl-4-(N-maleimido-methyl) cyclohexane-1 −carboxylate |
| **TEG** | triethyleneglycol TEOS: tetraethyl orthosilicate |
| **TFME** | thin-film micro-extraction |
| **TMOS** | tetratrimethoxysilane |
| **VTMS** | vinyltrimethoxysilane |

## References

[1] Mascini M, Palchetti I, Tombelli S. Nucleic acid and peptide aptamers: fundamentals and bioanalytical aspects. Angew Chem Int Ed February 6,2012;51(6):1316−32.

[2] Stoltenburg R, Reinemann C, Strehlitz B. SELEX—a (r)evolutionary method to generate high-affinity nucleic acid ligands. Biomol Eng 2007;24(4):381−403.

[3] Sharma TK, Bruno JG, Dhiman A. ABCs of DNA aptamer and related assay development. Biotechnol Adv March 1, 2017;35(2):275−301.

[4] Pichon V, Brothier F, Combès A. Aptamer-based-sorbents for sample treatment—a review. Anal Bioanal Chem 2015;407(3):681−98.
[5] Zhao L, Huang Y, Dong Y, Han X, Wang sai, Liang X. Aptamers and aptasensors for highly specific recognition and sensitive detection of marine biotoxins: recent advances and perspectives. Toxins 2018;10(11):427.
[6] Li Z, Rodriguez E, Azaria S, Pekarek A, Hage DS. Affinity monolith chromatography: a review of general principles and applications. Electrophoresis May 5, 2017;38(22−23): 2837−50.
[7] Gunasena DN, El Rassi Z. Organic monoliths for hydrophilic interaction electro-chromatography/chromatography and immunoaffinity chromatography. Electrophoresis 2012;33(1):251−61.
[8] Hennion M-C, Pichon V. Immuno-based sample preparation for trace analysis. J Chromatogr A 2003;1000(1−2):29−52.
[9] Pichon V, Chapuis-Hugon F, Hennion MC. 2.19 - bioaffinity sorbents. In: Pawliszyn J, éditeur, editors. Comprehensive sampling and sample preparation [internet]. Oxford: Academic Press; 2012. p. 359−88. Disponible sur: http://www.sciencedirect.com/science/article/pii/B9780123813732000454.
[10] Tsikas D. Quantitative analysis of biomarkers, drugs and toxins in biological samples by immunoaffinity chromatography coupled to mass spectrometry or tandem mass spectrometry: a focused review of recent applications. Immunoaffinity Tech Anal Janaury 15, 2010;878(2):133−48.
[11] Schänzer W, Thevis M. Immunoaffinity techniques coupled to mass spectrometry for the analysis of human peptide hormones: advances and applications AU - Thomas, Andreas. Expert Rev Proteomics September 2, 2017;14(9):799−807.
[12] Şenyuva HZ, Gilbert J. Immunoaffinity column clean-up techniques in food analysis: a review. Immunoaffinity Tech Anal Janaury 15, 2010;878(2):115−32.
[13] Feng C, Dai S, Wang L. Optical aptasensors for quantitative detection of small biomolecules: a review. Biosens Bioelectron September 15, 2014;59:64−74.
[14] Hasanzadeh M, Shadjou N, de la Guardia M. Aptamer-based assay of biomolecules: recent advances in electro-analytical approach. TrAC Trends Anal Chem. April 1, 2017; 89:119−32.
[15] Zhao Q, Wu M, Chris Le X, Li X-F. Applications of aptamer affinity chromatography. TrAC Trends Anal Chem December 2012;41:46−57.
[16] Du F, Guo L, Qin Q, Zheng X, Ruan G, Li J, et al. Recent advances in aptamer-functionalized materials in sample preparation. TrAC Trends Anal Chem April 1, 2017;67:134−46.
[17] Liu H, Luan Y, Lu A, Li B, Yang M, Wang JH. An oligosorbent-based aptamer affinity column for selective extraction of aflatoxin B2 prior to HPLC with fluorometric detection. Microchimica Acta 2018;185:71.
[18] Koh Y, Lee B-R, Yoon H-J, Jang Y-H, Lee Y-S, Kim Y-K, et al. Bead affinity chromatography in a temperature-controllable microsystem for biomarker detection. Anal Bioanal Chem November 2012;404(8):2267−75.
[19] Madru B, Chapuis-Hugon F, Peyrin E, Pichon V. Determination of cocaine in human plasma by selective solid-phase extraction using an aptamer-based sorbent. Anal Chem 2009;81(16):7081−6.
[20] Madru B, Chapuis-Hugon F, Pichon V. Novel extraction supports based on immobilised aptamers: evaluation for the selective extraction of cocaine. Talanta Jul 15, 2011;85(1): 616−24.

[21] Hu X, Mu L, Zhou Q, Wen J, Pawliszyn J. ssDNA aptamer-based column for simultaneous removal of nanogram per liter level of illicit and analgesic pharmaceuticals in drinking water. Environ Sci Technol June, 2011;45(11):4890−5.

[22] Cruz-Aguado JA, Penner G. Determination of ochratoxin a with a DNA aptamer. J Agric Food Chem 2008;56:10456−61.

[23] De Girolamo A, McKeague M, Miller JD, DeRosa MC, Visconti A. Determination of ochratoxin A in wheat after clean-up through a DNA aptamer-based solid phase extraction column. Food Chem August 1, 2011;127(3):1378−84.

[24] Chapuis-Hugon F, du Boisbaudry A, Madru B, Pichon V. New extraction sorbent based on aptamers for the determination of ochratoxin A in red wine. Anal Bioanal Chem May 1, 2011;400(5):1199−207.

[25] Hadj Ali W, Pichon V. Characterization of oligosorbents and application to the purification of ochratoxin A from wheat extracts. Anal Bioanal Chem February 1, 2014;406(4): 1233−40.

[26] Yang X, Kong W, Hu Y, Yang M, Huang L, Zhao M, et al. Aptamer-affinity column clean-up coupled with ultra high performance liquid chromatography and fluorescence detection for the rapid determination of ochratoxin A in ginger powder. J Sep Sci April 1, 2014;37(7):853−60.

[27] Aslipashaki SN, Khayamian T, Hashemian Z. Aptamer based extraction followed by electrospray ionization-ion mobility spectrometry for analysis of tetracycline in biological fluids. J Chromatogr B April 15, 2013;925:26−32.

[28] Wang Z, Zhao J, Lian H, Chen H. Aptamer-based organic-silica hybrid affinity monolith prepared via "thiol-ene" click reaction for extraction of thrombin. Talanta June 1, 2015; 138:52−8.

[29] Zhao J, Zhu Q, Zhao L, Lian H, Chen H. Preparation of an aptamer based organic−inorganic hybrid monolithic column with gold nanoparticles as an intermediary for the enrichment of proteins. Analyst 2016;141(16):4961−7.

[30] Gan H, Xu H. A novel aptamer-based online magnetic solid phase extraction method for the selective determination of 8-hydroxy-2′-deoxyguanosine in human urine. Anal Chim Acta May 30, 2018;1008:48−56.

[31] Chen Y, Chen M, Chi J, Yu X, Chen Y, Lin X, et al. Aptamer-based polyhedral oligomeric silsesquioxane (POSS)-containing hybrid affinity monolith prepared via a "one-pot" process for selective extraction of ochratoxin A. J Chromatogr A August 17, 2018;1563:37−46.

[32] Chi J, Chen M, Deng L, Lin X, Xie Z. A facile AuNPs@aptamer-modified mercaptosiloxane-based hybrid affinity monolith with an unusually high coverage density of aptamer for on-column selective extraction of ochratoxin A. Analyst 2018; 143(21):5210−7.

[33] Brothier F, Pichon V. Miniaturized DNA aptamer-based monolithic sorbent for selective extraction of a target analyte coupled on-line to nanoLC. Anal Bioanal Chem 2014; 406(30):7875−86.

[34] Marechal A, Jarrosson F, Randon J, Dugas V, Demesmay C. In-line coupling of an aptamer based miniaturized monolithic affinity preconcentration unit with capillary electrophoresis and laser induced fluorescence detection. J Chromatogr A 2015;1406: 109−17.

[35] Deng Q, Watson CJ, Kennedy RT. Aptamer affinity chromatography for rapid assay of adenosine in microdialysis samples collected in vivo. J Chromatogr A 2003;1005(1−2): 123−30.

[36] Zhao Q, Li X-F, Le XC. Aptamer-Modified monolithic capillary chromatography for protein separation and detection. Anal Chem May 1, 2008;80(10):3915–20.
[37] Jiang H-P, Zhu J-X, Peng C, Gao J, Zheng F, Xiao Y-X, et al. Facile one-pot synthesis of a aptamer-based organic–silica hybrid monolithic capillary column by "thiol–ene" click chemistry for detection of enantiomers of chemotherapeutic anthracyclines. Analyst 2014;139(19):4940–6.
[38] Han B, Zhao C, Yin J, Wang H. High performance aptamer affinity chromatography for single-step selective extraction and screening of basic protein lysozyme. J Chromatogr B August 15, 2012;903:112–7.
[39] Romig TS, Bell C, Drolet DW. Aptamer affinity chromatography: combinatorial chemistry applied to protein purification. J Chromatogr B 1999;731:275–84.
[40] Deng N, Liang Z, Liang Y, Sui Z, Zhang L, Wu Q, et al. Aptamer modified organic–inorganic hybrid silica monolithic capillary columns for highly selective recognition of thrombin. Anal Chem December 4, 2012;84(23):10186–90.
[41] Liu H, Lu A, Fu H, Li B, Yang M, Wang J, et al. Affinity capture of aflatoxin B1 and B2 by aptamer-functionalized magnetic agarose microspheres prior to their determination by HPLC. Microchim Acta June 12, 2018;185(7):326.
[42] Pichon V, Combès A, Delaunay N. Immunosorbents in microextraction. Trends Anal Chem 2019;113:246–55.
[43] Huang S, Gan N, Liu H, Zhou Y, Chen Y, Cao Y. Simultaneous and specific enrichment of several amphenicol antibiotics residues in food based on novel aptamer functionalized magnetic adsorbents using HPLC-DAD. J Chromatogr B August 15, 2017;1060:247–54.
[44] Wu X, Hu J, Zhu B, Lu L, Huang X, Pang D. Aptamer-targeted magnetic nanospheres as a solid-phase extraction sorbent for determination of ochratoxin A in food samples. J Chromatogr A October 14, 2011;1218(41):7341–6.
[45] Fischer C, Kallinich C, Klockmann S, Schrader J, Fischer M. Automated enrichment of sulfanilamide in milk matrices by utilization of aptamer-linked magnetic particles. J Agric Food Chem Dec 7, 2016;64(48):9246–52.
[46] Hashemian Z, Khayamian T, Saraji M. Anticodeine aptamer immobilized on a Whatman cellulose paper for thin-film microextraction of codeine from urine followed by electrospray ionization ion mobility spectrometry. Anal Bioanal Chem December 19, 2014; 407(6):1615–23.
[47] Zargar T, Khayamian T, Jafari MT. Immobilized aptamer paper spray ionization source for ion mobility spectrometry. J Pharm Biomed Anal Janaury 5, 2017;132:232–7.
[48] Huy GD, Jin N, Yin B-C, Ye B-C. A novel separation and enrichment method of 17β-estradiol using aptamer-anchored microbeads. Bioprocess Biosyst Eng Feb 2011; 34(2):189–95.
[49] Lin S, Gan N, Cao Y, Chen Y, Jiang Q. Selective dispersive solid phase extraction-chromatography tandem mass spectrometry based on aptamer-functionalized UiO-66-$NH_2$ for determination of polychlorinated biphenyls. J Chromatogr A May 13, 2016; 1446:34–40.
[50] Jiang L-F, Chen B-C, Chen B, Li X-J, Liao H-L, Zhang W-Y, et al. Aptamer-functionalized $Fe_3O_4$ magnetic nanoparticles as a solid-phase extraction adsorbent for the selective extraction of berberine from cortex phellodendri. J Sep Sci 2017;40(14): 2933–40.
[51] Khodadadi M, Malekpour A, Mehrgardi MA. Aptamer functionalized magnetic nanoparticles for effective extraction of ultratrace amounts of aflatoxin M1 prior its determination by HPLC. J Chromatogr A August 24, 2018;1564:85–93.

[52] Lin S, Gan N, Zhang J, Qiao L, Chen Y, Cao Y. Aptamer-functionalized stir bar sorptive extraction coupled with gas chromatography−mass spectrometry for selective enrichment and determination of polychlorinated biphenyls in fish samples. Talanta Mars 1, 2016; 149:266−74.

[53] Zhang X, Zhu S, Deng C, Zhang X. Highly sensitive thrombin detection by matrix assisted laser desorption ionization-time of flight mass spectrometry with aptamer functionalized core-shell $Fe_3O_4$@C@Au magnetic microspheres. Talanta Janauary 15, 2012;88:295−302.

[54] Murphy MB, Fuller ST, Richardson PM, Doyle SA. An improved method for the in vitro evolution of aptamers and applications in protein detection and purification. Nucleic Acids Res September 15, 2003;31(18):e110.

[55] Zhang X-H, Deng Y, Zhao M-Z, Zhou Y-L, Zhang X-X. Highly-sensitive detection of eight typical fluoroquinolone antibiotics by capillary electrophoresis-mass spectroscopy coupled with immunoaffinity extraction. RSC Adv 2018;8(8):4063−71.

[56] Gunasena DN, El Rassi Z. Hydrophilic diol monolith for the preparation of immunosorbents at reduced nonspecific interactions. J Sep Sci 2011;34(16−17):2097−105.

[57] Brothier F, Pichon V. Immobilized antibody on a hybrid organic−inorganic monolith: capillary immunoextraction coupled on-line to nanoLC-UV for the analysis of microcystin-LR. Anal Chim Acta August 20, 2013;792(0):52−8.

[58] Sonker M, Parker EK, Nielsen AV, Sahore V, Woolley AT. Electrokinetically operated microfluidic devices for integrated immunoaffinity monolith extraction and electrophoretic separation of preterm birth biomarkers. Analyst 2018;143(1):224−31.

[59] Parker EK, Nielsen AV, Beauchamp MJ, Almughamsi HM, Nielsen JB, Sonker M, et al. 3D printed microfluidic devices with immunoaffinity monoliths for extraction of preterm birth biomarkers. Anal Bioanal Chem 2019;411(21):5405−13.

[60] Levernæs MCS, Brandtzaeg OK, Amundsen SF, Reubsaet L, Lundanes E, Halvorsen TG, et al. Selective fishing for peptides with antibody-immobilized acrylate monoliths, coupled online with NanoLC-MS. Anal Chem December 4, 2018;90(23):13860−6.

[61] Le T, Esteve-Turrillas FA, Armenta S, de la Guardia M, Quiñones-Reyes G, Abad-Fuentes A, et al. Dispersive magnetic immunoaffinity extraction. Anatoxin-a determination. J Chromatogr A December 22, 2017;1529:57−62.

[62] Qiu S, Xu L, Cui Y-R, Deng Q-P, Wang W, Chen H-X, et al. Pseudo-homogeneous immunoextraction of epitestosterone from human urine samples based on gold-coated magnetic nanoparticles. Talanta May 15, 2010;81(3):819−23.

[63] Lund H, Løvsletten K, Paus E, Halvorsen TG, Reubsaet L. Immuno−ms based targeted proteomics: highly specific, sensitive, and reproducible human chorionic gonadotropin determination for clinical diagnostics and doping analysis. Anal Chem September 18, 2012;84(18):7926−32.

[64] Rosting C, Tran EV, Gjelstad A, Halvorsen TG. Determination of the low-abundant protein biomarker hCG from dried matrix spots using immunocapture and nano liquid chromatography mass spectrometry. J Chromatogr B Mars 1, 2018:1077−8. 44-51.

[65] Haller E, Lindner W, Lämmerhofer M. Gold nanoparticle−antibody conjugates for specific extraction and subsequent analysis by liquid chromatography−tandem mass spectrometry of malondialdehyde-modified low density lipoprotein as biomarker for cardiovascular risk. Anal Chim Acta Feb 1, 2015;857:53−63.

[66] Wharton RE, Ojeda-Torres G, Cunningham B, Feyereisen MC, Hill KL, Abbott NL, et al. Quantification of microcystin-LR in human urine by immunocapture liquid chromatography tandem mass spectrometry. Chem Res Toxicol September 17, 2018;31(9): 898−903.

[67] Molina-Delgado MÁ, Aguilar-Caballos MP, Gomez-Hens A. Determination of soy proteins in food samples by dispersive solid-phase immunoextraction and dynamic long-wavelength fluorometry. Microchim Acta October 1, 2013;180(13):1279–86.
[68] Huang Y-F, Chang H-T. Analysis of adenosine triphosphate and glutathione through gold nanoparticles assisted laser desorption/ionization mass spectrometry. Anal Chem Jul 1, 2007;79(13):4852–9.
[69] Najafabadi ME, Khayamian T, Hashemian Z. Aptamer-conjugated magnetic nanoparticles for extraction of adenosine from urine followed by electrospray ion mobility spectrometry. J Pharm Biomed Anal Mars 25, 2015;107:244–50.
[70] Kim M, Um H-J, Bang S, Lee S-H, Oh S-J, Han J-H, et al. Arsenic removal from Vietnamese groundwater using the arsenic-binding DNA aptamer. Environ Sci Technol December 15, 2009;43(24):9335–40.
[71] Su Y, Shao C, Huang X, Qi J, Ge R, Guan H, et al. Extraction and detection of bisphenol A in human serum and urine by aptamer-functionalized magnetic nanoparticles. Anal Bioanal Chem Mars 1, 2018;410(7):1885–91.
[72] Kökpinar Ö, Walter J-G, Shoham Y, Stahl F, Scheper T. Aptamer-based downstream processing of his-tagged proteins utilizing magnetic beads. Biotechnol Bioeng October 1, 2011;108(10):2371–9.
[73] Shao N, Zhang K, Chen Y, He X, Zhang Y. Preparation and characterization of DNA aptamer based spin column for enrichment and separation of histones. Chem Commun June 6, 2012;48(53):6684–6.
[74] Zhang Q, Yang Y, Zhi Y, Wang X, Wu Y, Zheng Y. Aptamer-modified magnetic metal-organic framework MIL-101 for highly efficient and selective enrichment of ochratoxin A. J Sep Sci 2019;42(3):716–24.
[75] Jiang D, Hu T, Zheng H, Xu G, Jia Q. Aptamer-functionalized magnetic conjugated organic framework for selective extraction of Traces of hydroxylated polychlorinated biphenyls in human serum. Chem Eur J July 20, 2018;24(41):10390–6.
[76] Yasun E, Gulbakan B, Ocsoy I, Yuan Q, Shukoor MI, Li C, et al. Enrichment and detection of rare proteins with aptamer-conjugated gold nanorods. Anal Chem Jul 17, 2012;84(14):6008–15.
[77] Khezeli T, Daneshfar A. Development of dispersive micro-solid phase extraction based on micro and nano sorbents. TrAC Trends Anal Chem April 1, 2017;89:99–118.
[78] Yuan H, Mullett WM, Pawliszyn J. Biological sample analysis with immunoaffinity solid-phase microextraction. Analyst 2001;126(8):1456–61.
[79] Lord HL, Rajabi M, Safari S, Pawliszyn J. Development of immunoaffinity solid phase microextraction probes for analysis of sub ng/mL concentrations of 7-aminoflunitrazepam in urine. Hyphenated Tech Pharm Biomed Anal February 24, 2006;40(3):769–80.
[80] Lord HL, Rajabi M, Safari S, Pawliszyn J. A study of the performance characteristics of immunoaffinity solid phase microextraction probes for extraction of a range of benzodiazepines. Hyphenated Tech Pharm Biomed Anal June 28, 2007;44(2):506–19.
[81] Wang C, Yang L, Li N, Zhang X, Guo Y, Li C. Development of immunoaffinity solid phase microextraction rods for analysis of three estrogens in environmental water samples. J Chromatogr B September 1, 2017;1061–2: 41-8.
[82] Yao K, Zhang W, Yang L, Gong J, Li L, Jin T, et al. Determination of 11 quinolones in bovine milk using immunoaffinity stir bar sorptive microextraction and liquid chromatography with fluorescence detection. J Chromatogr B October 15, 2015;1003:67–73.
[83] Mu L, Hu X, Wen J, Zhou Q. Robust aptamer sol–gel solid phase microextraction of very polar adenosine from human plasma. J Chromatogr A Mars 1, 2013;1279:7–12.

[84] Du F, Alam MN, Pawliszyn J. Aptamer-functionalized solid phase microextraction—liquid chromatography/tandem mass spectrometry for selective enrichment and determination of thrombin. Anal Chim Acta October 3, 2014;845:45—52.

[85] Zargar T, Khayamian T, Jafari MT. Aptamer-modified carbon nanomaterial based sorption coupled to paper spray ion mobility spectrometry for highly sensitive and selective determination of methamphetamine. Microchim Acta Janauary 10, 2018; 185(2):103.

[86] Xu Y, Tan S, Liang Q, Ding M. One-step facile synthesis of aptamer-modified graphene oxide for highly specific enrichment of human A-thrombin in plasma. Sensors 2017; 17(9).

[87] Kumazawa T, Lee X-P, Sato K, Suzuki O. Solid-phase microextraction and liquid chromatography/mass spectrometry in drug analysis. Appl Liq Chromatogr Coupled Mass Spectrom Pharmacol September 15, 2003;492(1):49—67.

[88] Phillips TM, Dickens BF. Analysis of recombinant cytokines in human body fluids by immunoaffinity capillary electrophoresis. Electrophoresis 1998;19(16-17):2991—6.

[89] Queiroz MEC, Oliveira EB, Breton F, Pawliszyn J. Immunoaffinity in-tube solid phase microextraction coupled with liquid chromatography—mass spectrometry for analysis of fluoxetine in serum samples. J Chromatogr A 2007;1174:72—7.

[90] Chaves AR, Queiroz MEC. Immunoaffinity in-tube solid phase microextraction coupled with liquid chromatography with fluorescence detection for determination of interferon $\alpha$ in plasma samples. J Chromatogr B June 1, 2013;928:37—43.

[91] Yang W, Sun X, Wang H-Y, Woolley AT. Integrated microfluidic device for serum biomarker quantitation using either standard addition or a calibration curve. Anal Chem 2009;81(19):8230—5.

[92] Yang W, Yu M, Sun X, Woolley AT. Microdevices integrating affinity columns and capillary electrophoresis for multibiomarker analysis in human serum. Lab Chip 2010; 10(19):2527—33.

[93] Pichon V, Bouzige M, Miège C, Hennion M-C. Immunosorbents: natural molecular recognition materials for sample preparation of complex environmental matrices. TrAC Trends Anal Chem 1999;18(3):219—35.

[94] Xu B, Cheng S, Wang X, Wang D, Xu L. Novel polystyrene/antibody nanoparticle-coated capillary for immunoaffinity in-tube solid-phase microextraction. Anal Bioanal Chem April 1, 2015;407(10):2771—5.

[95] Ying L-L, Ma Y-C, Xu B, Wang X-H, Dong L-Y, Wang D-M, et al. Poly(glycidyl methacrylate) nanoparticle-coated capillary with oriented antibody immobilization for immunoaffinity in-tube solid phase microextraction: preparation and characterization. J Chromatogr A August 4, 2017;1509:1—8.

[96] Zimmermann GR, Jenison RD, Wick CL, Simorre J-P, Pardi A. Interlocking structural motifs mediate molecular discrimination by a theophylline-binding RNA. Nat Struct Mol Biol 1997;4:644—9.

[97] Liu X, Liu X, Huang P, Wei F, Ying G, Lu J, et al. Regeneration and reuse of immunoaffinity column for highly efficient clean-up and economic detection of ochratoxin a in malt and ginger. Toxins 2018;10:462.

[98] Keefe AD, Pai S, Ellington A. Aptamers as therapeutics. Nat Rev Drug Discov July, 2010;9(7):537—50.

[99] Hodgson RJ, Brook MA, Brennan JD. Capillary-scale monolithic immunoaffinity columns for immunoextraction with in-line laser-induced fluorescence detection. Anal Chem July, 2005;77(14):4404—12.

[100] Chamieh J, Faye C, Dugas V, Moreau T, Vandenabeele-Trambouze O, Demesmay C. Preparation and full characterization of a micro-immunoaffinity monolithic column and its in-line coupling with capillary zone electrophoresis with ochratoxin A as model solute. J Chromatogr A 2012;1232:93–100.
[101] Sun X, Yang W, Pan T, Woolley AT. Affinity monolith-integrated poly(methyl methacrylate) microchips for on-line protein extraction and capillary electrophoresis. Anal Chem 2008;80:5126–30.

# Metal-selective sorbents

7

*Leon Fuks, Irena Herdzik-Koniecko*
Centre of Radiochemistry and Nuclear Chemistry, Institute of Nuclear Chemistry and Technology, Warszawa, Poland

## 7.1 Introduction

### 7.1.1 *Metals*

A stair-stepped line, starting at aluminum (Al; Z = 13) and going down to polonium (Po; Z = 84), separates into two groups of elements in the periodic table: metals and nonmetals. The only exceptions within the group of metals are germanium (Ge; Z = 32) and antimony (Sb; Z = 51). In the following scheme, border-line elements between metals and nonmetals are presented in green (Scheme 7.1)

To characterize metals in short, one must state, that:

- metals are solids with the exception of mercury (Hg), which is a liquid
- they are shiny
- they are good conductors of electricity and heat
- they are *ductile* (i.e., they can be drawn into thin wires)
- they are *malleable* (which means that they can be easily flattened into thin sheets)
- all metals have a tendency to lose a number of electrons depending on their position in the periodic chart, and this feature determines their chemical properties.

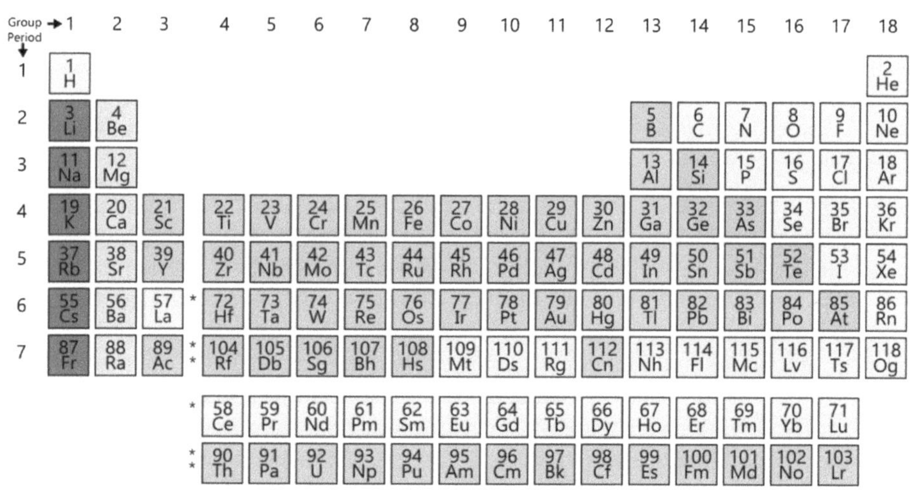

**Scheme 7.1** Metals and nonmetals in the periodic table of elements.

Solid-Phase Extraction. https://doi.org/10.1016/B978-0-12-816906-3.00007-8

The whole group of metals may be divided into several subgroups, i.e., the heavy metals, transition metals (including the rare earth and the transuranic), alkaline earth, and the alkali metals. Below, we present the characteristics of each group in a condensed form:

- The term *heavy metal* is applied for the metals which are comparatively of high density (it is assumed as more than 5 g/cm$^3$) and less reactive than lighter metals. Physical and chemical characterization of heavy metals needs to be made with great care, as the group is not always strictly defined. The most popular definition was proposed by S.J. Hawkes, which says that *heavy metals comprise a block of all the metals in Groups 3 to 16 that are in periods four and greater* [1]. Typically, this group contains: aluminum (Al), antimony (Sb), arsenic (As), barium (Ba), cadmium (Cd), cobalt (Co), chromium (Cr), copper (Cu), gold (Au), iron (Fe), lithium (Li), nickel (Ni), lead (Pb), manganese (Mn), mercury (Hg), molybdenum (Mo), platinum (Pt), rubidium (Rb), silver (Ag), scandium (Sc), selenium (Se), strontium (Sr), tin (Sn), titanium (Ti), tungsten (W), uranium (U), vanadium (V) and zinc (Zn). Heavy metals are natural components of the Earth's crust, but anthropogenic activities, such as the mining of heavy metals, their industrial production and usage, agricultural use, and use in household applications are the main sources of heavy metal pollution.
- According to the IUPAC nomenclature, any element with a partially filled *d*-electron subshell is called a *transition metal.* The metals are placed in groups 3 through 12 in the periodic table. In turn, the block of electron elements (lanthanides and actinides, situated below the main body of the periodic table) are called the *inner transition metals*. Transition metals exhibit a wide range of oxidation states. Complex formation by transition metals is crucial to the separation of these metals. According to the broader definition, coinage metals (Cu(II), Ag(II), and Au(III)) are generally considered transition elements. The elements in group 12, in turn, are not typically considered as transition metals, although chemically they frequently resemble the transition elements.
- Beryllium (Be), magnesium (Mg), calcium (Ca), strontium (Sr), barium (Ba), and radium (Ra) are all called as the *alkaline earth metals*—chemical elements in the *s*-block of the periodic table with very similar properties. All of the alkaline earth metals are commercially valuable and do not occur as a free metal in nature as they are very reactive.
- The alkali metals are the elements placed in the first group of the periodic table. Namely, lithium (Li), sodium(Na), potassium (K), rubidium (Rb), cesium (Cs), and francium (Fr). Alkali metals are even more reactive than alkaline earth metals, and none of them occur as a free metal. Many of the alkali metal compounds (NaCl, KCl, $Na_2CO_3$, NaOH) are commercially important, so the process of their separation is the subject of many studies.

A large number of different technologies have been developed for sequestering metals from wastewaters and for separation of individual metals. These methods are based on the processes of coagulation, filtration, ion exchange, solvent extraction, foam flotation, activated sludge, aerobic and anaerobic treatment, microbial reduction, electrolysis and, ultimately, the process of adsorption [2]. Adsorption is as effective as the other techniques but does not possess certain of their limitations. The most important of these is the formation of a large amount of sludge, low efficiency, sensitivity of the operating conditions to various physical or chemical factors, high cost of operation, and difficult waste disposal. In addition, the adsorption technology is easy to adapt to specific conditions, offers high-purity of both the resulting solution and adsorbate and the adsorbent can be regenerated [3]. According to the Scopus database, during the last 10 years, more than 4500 reports were published in the field of the heavy metal sorption.

## 7.1.2 Basic terms used in sorption science and technology

Generally speaking, sorption is a physical and/or chemical process, leading to binding of certain components of the liquid or gaseous mixtures (*sorbates*) with the immiscible solid matter (*sorbent*). The common term *sorption* includes three similar, but different, phenomena:

- *Absorption*, which occurs when the sorbate enters a sorbent material. In the course of this process, molecules of the sorbate are dissolved or diffused in the absorbent and form a solution. Once dissolved, the molecules cannot be easily separated from the absorbent.
- Term *adsorption* describes the process when the molecules are attached to the surface of the adsorbent and can be easily removed from it.
- *Ion exchange* process consists of an exchange of ions between two electrolytes or between a solution containing the sorbate and the chemical groups present on the surface of the sorbent. The so-called *sorbing groups* should show the ability to release ions of the same sign (electric charge) as the ions of the sorbate.

As the list of basic terms used in sorption science and technology can be found in many publications [e.g., Refs. [4,5]], here we present them only in the visual form of Fig. 7.1.

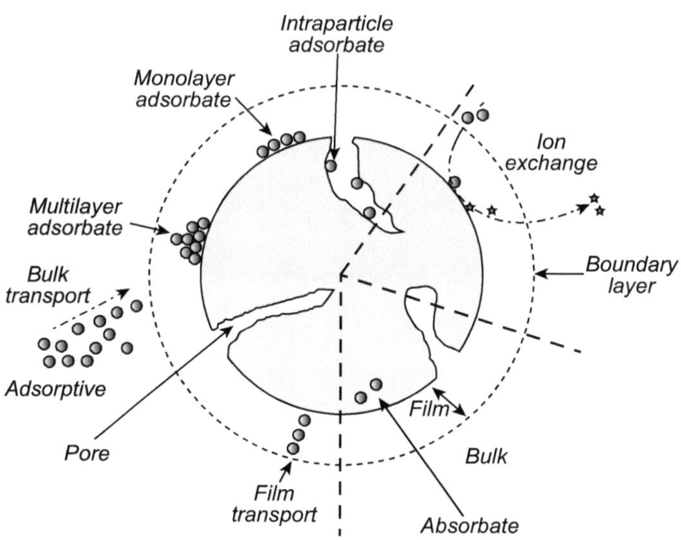


**Figure 7.1** Some basic terms used in sorption science and technology.

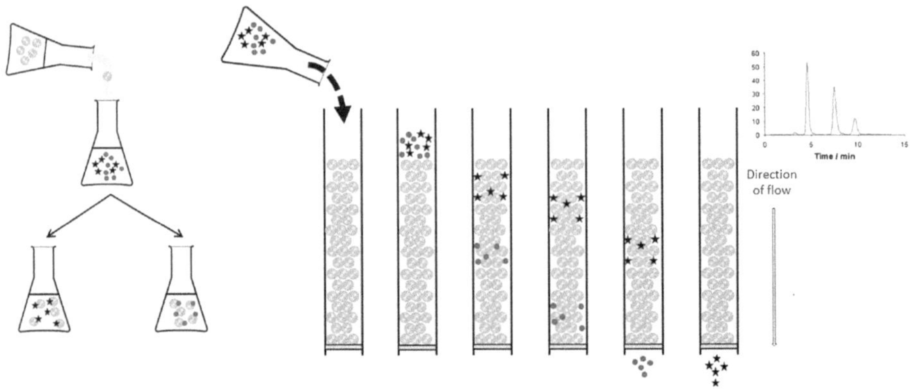


**Figure 7.2** Two types of the metal-selective sorption processes: batch mode (left), and dynamic mode (right).

The sorption of metals with the aim of decontamination of aqueous solutions and possibly further re-use of the metals may be done in several ways (see, Fig. 7.2):

- **Batch** (static) method: due to the extensive contact between the immiscible phases, the sorbate migrates from the solution to the sorbent. Afterward, both immiscible phases may be separated either by gravity (*sedimentation*) or by centrifugation. Certain sorbents, with magnetic properties, can be separated by an external magnetic field. For sorbent recovery and to obtain the sorbate in a pure form, some supplementary operations are necessary (e.g., *desorption*).
- **Column** (flow) method, sometimes called *column chromatography*, is preferred for the separation of mixtures of several substances. In this technique, a liquid or gaseous solution is forced to flow through a tube filled with a thoroughly ground solid (*sorbent*). If necessary, the solid column filling may be coated with a liquid phase. Dissolved components of the solution may be separated because they move along the column at different rates, which depend on the strength of their interaction with the stationary phase.

Results of the sorption process are usually expressed in terms of the equilibrium partition constant $K_p$, which is defined as the ratio of the concentrations of sorbate in both heterogeneous phases in equilibrium with each other. It is related to the specific adsorption capacity of the dry sorbent ($q_M$, $mg \cdot g^{-1}$), which depends mainly on the solution pH, initial metal concentration, presence or absence of other solution components (often with competitive affinity for the sorbent) and temperature of the process, as well as on dimension of the sorbent particles and the pretreatment method of the sorbent [6–8]. The last two factors may significantly increase the porosity of the sorbent surface, a crucial parameter for determining sorption efficiency. For the sorption of metals the most decisive factor, however, is the speciation of the metal ions compared with the protonation of the binding sites of the sorbent. Both these features depend predominantly on the acidity and composition of the purified solution.

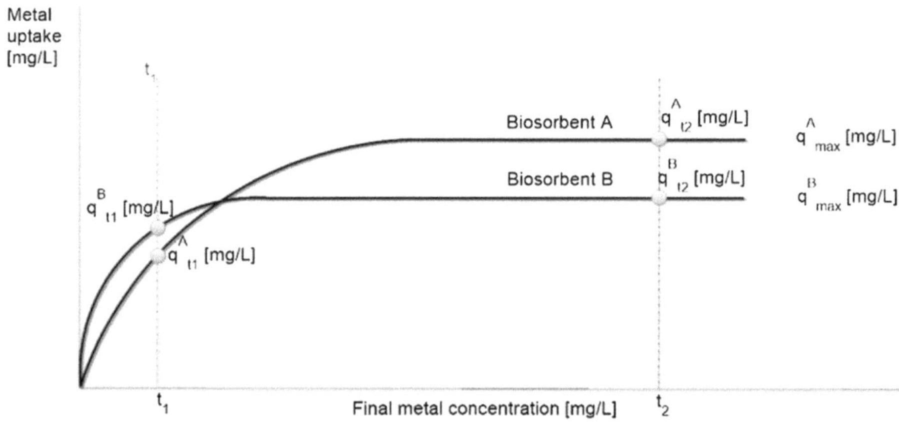


**Figure 7.3** Examples of sorption isotherm curves. Sorbent B has better sorption properties (higher q) than sorbent A at lower equilibrium concentrations but worse (lower q) at higher concentrations.

The functional relationship between the sorbate amount and the sorbent in the equilibrium state is called the *sorption isotherm* [9] as shown in Fig. 7.3.

In turn, the kinetics of the overall sorption process is typically represented as a sum of several consecutive stages. In the first, metal ions are transported to the sorbent surface controlled by the diffusion process. In the next, their diffusion on the outer surface of the sorbent (called as the *film diffusion*) is followed by penetration inside the pores (*pore diffusion*). At last, on the inner surface of the pores, metal ions interact with the unoccupied active sites of the sorbent. So, the cumulative rate of the sorption is called a *pseudo model* of the rate [10].

From the chemical point of view, the affinity of the metal ion for the binding sites of the sorbent, as well as the selectivity of the separation in the case of the multimetal solutions, is typically expressed in terms of the Pearson's theory of hard and soft acids and bases [11]. It means that the *hard acids* (i.e., metals of low polarizability, such as Mg(II), La(III), Fe(III), Cr(III), U(VI), V(IV) and so on) favor bonding with *hard bases* (i.e., groups containing the electron donor atoms) ranked in the following order: N $\gg$ P, O $\gg$ S, F $\gg$ Cl. On the contrary, *soft acids* (e.g., Ag(I), Cd(II), Hg(II), Au(III), or Sn(II)) have a preference to bind with *soft bases* (i.e., P $\gg$ N, S $\gg$ O, I $\gg$ F). A significant number of hard and soft borderline metals have intermediate binding properties, and are generally categorized with the Irving–Williams series for their stability constants: Mn(II) < Fe(II) < Co(II) < Ni(II) < Cu(II) < Zn(II). Therefore, the affinity of alginates for metals forms the following series: Pb(II) > Cu(II) > Zn(II) > Co(II) > Ni(II) > Mn(II) > Fe(II) [12,13]. On the other hand, the selectivity of metal separations on chitosan, due to several types of binding groups with different electron donor atoms (see, Fig. 7.4), do not form a logical series with respect to either ionic size or hardness of the metal cations [14]. Therefore, even today,

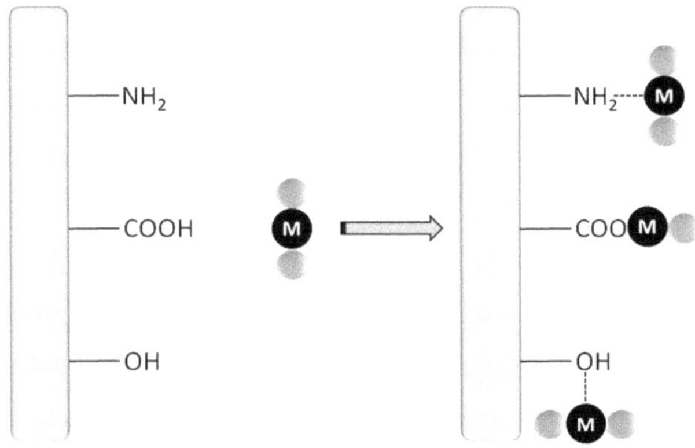


**Figure 7.4** Selected binding types occurring in the adsorption of divalent metal compounds by sorbents with different binding groups.

there is no uniform theory that allows the prediction of the sorption results in different systems. Consequently, our present-day knowledge of sorption characteristics is based mainly on tabularized experimental data.

## 7.2 Sorbents for sequestering metal ions from aqueous solutions

Sorbents are generally fine-grained solids, the particles of which show a large surface area and, as a result, noticeable sorption properties. Depending on their origin, sorbents may be divided into two main groups: natural and synthetic materials. Furthermore, within these groups, sorbents can be classified as mineral sorbents (also called *inorganic materials*), carbons (activated carbons, carbon nanostructures, fullerenes, and graphenes) and organic sorbents (usually obtained from so-called *biomass*). For heavy metals, the classification of Fig. 7.5 is proposed.

**Natural mineral sorbents** are produced from the crushed rocks, thermally treated if necessary. The thermal treatment of the raw, granular materials is to increase the surface area and porosity of the sorbent. Particles of 0.1−1 mm are considered small granules, while these of 1−3 mm are treated as rough. Depending on the particle size, mineral sorbents are used for internal use (small grains) or external practice (large grains). Rough-grained materials are more massive and more chemically resistant than finely ground materials. Natural inorganic sorbents include clay, sand, and even volcanic ashes.

**Natural organic sorbents** are produced from either locally available raw biomass, e.g., from the tree cortex or sawdust, peat, lichens, etc., or any natural product that contains carbon, mainly, industrial, agricultural, or pharmaceutical wastes. Typically, they indicate better sorption properties than mineral wastes, but also lower stability in contact with aggressive chemicals or from mechanical impact. Thus, they are not

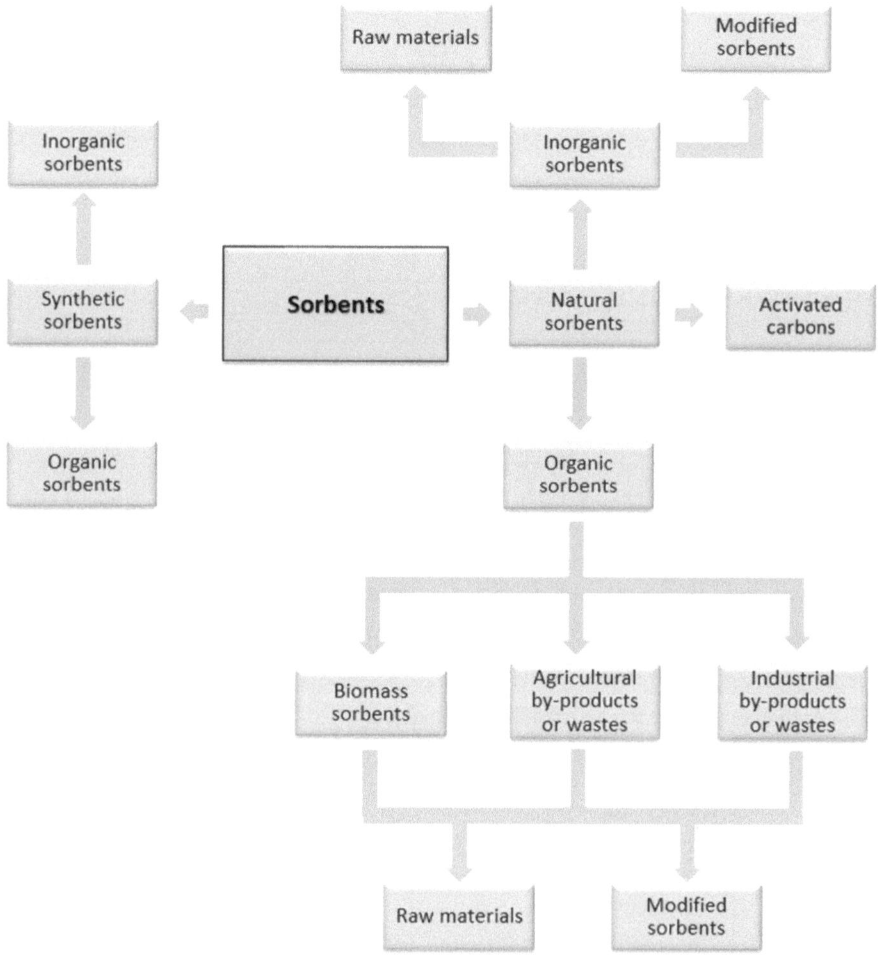


**Figure 7.5** Classification of sorbents used for metal recovery or concentration.

commonly used in the removal of metals from aggressive fluids such as acids, alkali, and oxidizing substances.

**Synthetic sorbents** represent the man-made materials that generally resemble plastics and are designed to adsorb harmful or reusable constituents from solution. Some other synthetic sorbents, crosslinked polymers or rubber materials, absorb liquids into their solid matrix, and as a result of this process, the sorbent swells. Synthetic sorbents, usually, are more efficient than natural sorbents [15], which should be activated in order to acquire the appropriate physical, chemical, catalytic, and adsorptive properties.

Most of the commercial sorbents produced on a large scale, e.g., silica gels, zeolites, aluminosilicate minerals, activated carbons, cyanoferrate complexes with different metals, and others [e.g., Refs. [16–18]] with numerous advantages, demonstrate certain limitations. Among these, one should mention an unfavorable surface area

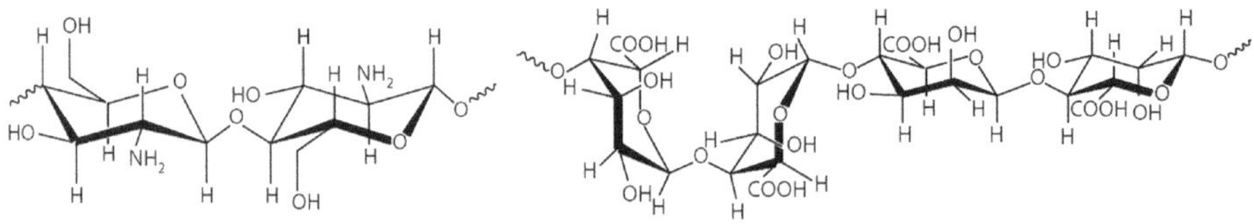

**Figure 7.6** Basic units of the biopolymers: chitosan (left) and alginic acid (right).

and morphology, insufficient pore size distribution, as well as the type of the functional groups attached to the adsorbent surface [19]. In addition, the cost of these materials is generally high [20]. Therefore, there is a need to find novel sorbents, in order to decrease material costs for sequestering metals or modifying the existing sorbents with intent to improve their sorption properties. These novel sorbents often do not occur naturally, but as a rule, are relatively inexpensive [21].

Contemporary database for metal sorption using different sorbents contain a large number of records, and new information is added continuously. In addition, some sorbents are identified only by their name and lack important chemical information. For example, chitosan of different origin differs significantly in the degree of deacetylation or the alginates differ by the ratio of the guluronic and the mannuronic acids (see Fig. 7.6). In spite of the large amount of experimentally derived information, their structural diversity does not allow even the most general conclusions to be drawn.

**Table 7.1** Methods of modification of the sorbents [22].

| **Type of the modification** | | **Examples** |
|---|---|---|
| Physical modification | By addition of the components which improve the physical properties | Application of such additives as:<br>✓ plasticizers,<br>✓ fillers |
| | Action of the physical factors | Application of such impulsions as:<br>✓ thermal energy (thermal treatment),<br>✓ ultrasound,<br>✓ electric field<br>✓ magnetic field<br>✓ ionizing radiation |
| Chemical modification | Modification of the existing sorbent | Synthesis of e.g.,:<br>✓ phenyltrimethylammonium (PTMA)- or benzyltrimethylammonium (BTMA)-bentonites<br>✓ Crosslinking the polymers |
| | Modification in the course of the synthesis | ✓ Sulfation of sodium alginate prior to its gelation<br>✓ Acidic treatment or impregnation of the activated carbon |

Modification of sorbents allows a number of their physicochemical and/or process properties to be corrected. Such methods are generally more efficient than the synthesis of novel materials. Typically, it can be performed in two ways: chemical and/or physical modification (see Table 7.1). Physical modification is usually easier and less expensive compared with chemical modification. However, the mechanism of the process is sometimes complex. It is also known as a structural modification, because the action of certain physical factors results in, mainly, modification of the molecular structure of the material.

Chemical modification leads to improved properties by controlled alteration of their molecular composition. Common reactions are used to introduce new functional groups, to convert specific functional groups to a different form (e.g., esterification, deacetylation), oxidation or reduction reactions, grafting, crosslinking of polymer chains or their degradation. These modifications may be introduced during synthesis or are made to an existing sorbent.

### 7.2.1 Biomass, industrial and agricultural by-products and wastes

Removal of toxic metals or recovery of valuable metals from aqueous solutions using biosorbents, products obtained from plants, living organisms, or from the industrial and agricultural by-products, is the driving force in the search for novel and environment-friendly sorbents [23]. The initial challenge is the selection of the most promising types of biomass from the extremely large group of locally available and inexpensive types of biomass [23]. To facilitate this process, the biosorption capacity of various biomasses is generally determined [24–26]. However, the application of biosorbents for metal recovery and separation rarely extends beyond the laboratory scale at present. In addition, the isolation of the metal-loaded biosorbent from the solution is often difficult. Recently, the synthesis of magnetic composite biosorbents has been initiated [27].

Biosorbents can be assigned to one of the following groups: bacteria, fungi, algae, industrial wastes, agricultural wastes, natural residues, and other biomaterials [23]. There are many review papers on this topic, so in Table 7.2, we present only an extract of the data concerning the sorption of lead(II) by popular plant materials [28].

**Table 7.2** Pb(II) sorption capacity of some plant-derived low-cost biosorbents; pH 5 [28].

| Biomass type | $q_M$ (mg.g$^{-1}$) | Biomass type | $q_M$ (mg.g$^{-1}$) |
|---|---|---|---|
| Grass clippings | 28.2 | Peepul leaves | 107.5 |
| Tea leaves | 33.1 | Mango leaves | 30.0 |
| Banana peels | 63.6 | Rice husk | 29.5 |
| Teak sawdust | 38.8 | Rice stem | 45.1 |
| Peanut hulls | 56.4 | Coir fibers | 48.2 |

**Table 7.3** Values of $q_{max}$ ($mg \cdot g^{-1}$) for the main monometallic systems for different types of algae [30].

| Algae | Ni(II) | Cu(II) | Zn(II) | Cd(II) | Pb(II) |
|---|---|---|---|---|---|
| | ($mg \cdot g^{-1}$) | | | | |
| *Codium vermilara* | 13.2 | 16.9 | 23.8 | 21.8 | 63.3 |
| *Spirogyra insignis* | 17.5 | 19.3 | 21.1 | 22.9 | 51.5 |
| *Asparagopsis armata* | 17.1 | 21.3 | 21.6 | 32.3 | 63.7 |
| *Chondrus crispus* | 37.2 | 40.5 | 45.7 | 75.2 | 204.1 |
| *Ascophyllum nodosum* | 43.3 | 58.8 | 42.0 | 87.7 | 178.6 |
| *Fucus spiralis* | 50.0 | 70.9 | 53.2 | 114.9 | 204.1 |

**Table 7.4** Sorption capacity of some metals on unmodified biosorbent (a mixture of tea waste, maple leaves and mandarin peel) *versus* chemically modified [31].

| | Cd ($mg.g^{-1}$) | Cu ($mg.g^{-1}$) | Pb ($mg.g^{-1}$) | Zn ($mg.g^{-1}$) |
|---|---|---|---|---|
| A mixture of tea waste, maple leaves and Mandarin peel | 31.73 | 41.06 | 76.25 | 26.63 |
| Raw material acid treated in the elevated temperature | 69.56 | 127.70 | 245.20 | 70.55 |

At the same time it is worth mentioning that the sorption properties studied for the same sorbent, but of different origin, may differ significantly from each other. One such example is presented in Table 7.3 [29,30].

Sorption properties of most biosorbents may be improved by simple methods, such as boiling with concentrated mineral acids or bases. One such example is illustrated by Table 7.4 [31].

To improve their activity for different metals, biosorbents may be functionalized by binding different chemical compounds. Two examples employing chitosan are illustrated in Tables 7.5 and 7.6 [32].

## 7.2.2 Activated carbons

Activated carbon (also referred to as activated charcoal or active carbon) is one of the oldest and widely used forms of carbon for the removal of contaminants for wastewater treatment. Due to the high porosity, activated carbons have surface area surpassing 3000 $m^2/g$. This high porosity facilitates the sorption of metals and dies. The material

**Table 7.5** Sorption capacity for Cu(II) by chitosan and chitosan-based.

| Materials | Sorption capacity $mg \cdot g^{-1}$ |
|---|---|
| Chitosan | 80.71 |
| Chitosan-crosslinked GLA | 59.67 |
| Chitosan-crosslinked ECH | 62.47 |
| Chitosan-crosslinked EGDE | 45.94 |
| Copper(II)-imprinted chitosan | 201.66 |
| Nonimprinted chitosan without template | 89.51 |
| Microspheres crosslinked carboxyl-grafted chitosan | 5.01 6 |
| N,O-carboxymethyl-chitosan | 162.5 |
| Xanthate-modified magnetic chitosan | 34.5 |
| Carboxymethylated chitosan-bound $Fe_3O_4$ nanoparticles | 21.5 |
| Chitosan-tripolyphosphate beads | 26.06 |
| Chitosan-tripolyphosphate chelating resin | 200 3.15 |
| Gel beads crosslinked magnetic chitosan–isatin Schiffs base resin (CSIS) | 103.16 |
| Chitosan electrospun nanofiber mats | 485.44 |
| Porous chitosan functionalized with histidine | 127–190 |
| Different chitosan–zeolite composites | 14.75–51.32 |
| Bacillus subtilis immobilized on chitosan | 100.7 1 |

**Table 7.6** Chitosan and derivatives of metal affinity order.

| Sorbent types | Affinity or selectivity order |
|---|---|
| Magnetic chitosan microspheres modified with thiourea | $Mg \cdot g^{-1}$: 625.2 ($Hg^{2+}$), 66.7 ($Cu^{2+}$), 15.3 ($Ni^{2+}$) |
| Xanthate-modified magnetic chitosan beads | Pb(II) > Cu(II) > Zn(II) |
| Chitosan film | Cu(II) > Hg(II) > Zn(II) > Cd(II) > Ni(II) >Co(II) ≈ Ca(II) and Eu(III) PNd(III) > Cr(III) ≈ Pr(III). |
| Crown ether grafted chitosan | Pd (II) or Ag(I) >Pb(II) and Cr(III) |
| EDTA–chitosan | $Cu^{2+} \approx MoO_2^{2+} > Ni^{2+} > VO_2{}^{+} > Zn^{2+} \geq Co^{2+} > Al^{3+}$ |

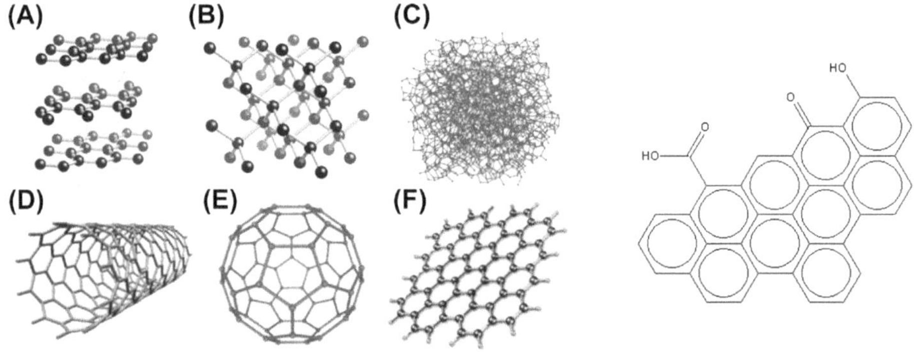


**Figure 7.7** Left: Type of typical carbon sorbents (A) graphite (B) diamond (C) amorphous carbon (D) nanotube (E) fullerene and (F) graphene. Right: Main oxygen-containing groups on the surface of activated carbon suitable for modification: carboxyl, carbonyl, and phenol.

obtained from coal is generally referred to as activated coal, while that obtained from coke as activated coke. Different functional groups exist on the surface of the sorbent (right part of Fig. 7.7), but further chemical treatment enhances its adsorption properties. As an example, the maximum sorption capacity for lead(II) ions was found to be about 19 $mg\ g^{-1}$ for activated carbons. Acid-base treatment of these sorbents was found to double the sorption capacity. Immobilization of baker's yeast on the surface of activated carbon sorbents generally improves the sorption capacity of lead, up to 120 $mg\ g^{-1}$ [33].

Metal adsorption on activated carbons takes place mainly in micropores and to a small extent in the mesopores. Macropores, act only as flow-channels for the adsorbate transported toward the smaller pores. The method of activation and nature of precursors for the production of active carbons strongly influences the type of surface functional groups and the pore structure of the activated carbon. Various methods, such as acid or base treatment, surface impregnation, ozone or surfactant treatment, plasma or microwave treatment, have been used. These methods have been reviewed, e.g., in Ref. [34].

Several methods have been proposed for the synthesis of graphene (G) and its modification to graphene oxide (GO) [35,36]. GO contains numerous functional groups containing oxygen atoms, especially epoxy, hydroxyl, and carboxylic acid groups [37]. Presence of these groups determines the hydrophilicity and the extent of the negative charge density of GO. Several structural models of GO have been already proposed, all of which show the existence of regular lattices formed by the repeating units, for example, Fig. 7.8 [38]. GO is currently the most actively researched carbon sorbent for the sorption of metal ions [39—41]. A few representative examples are summarized in Table 7.7 [42]. Carbon nanotubes are not as widely used, and a few representative examples are summarized in Table 7.8 [3].

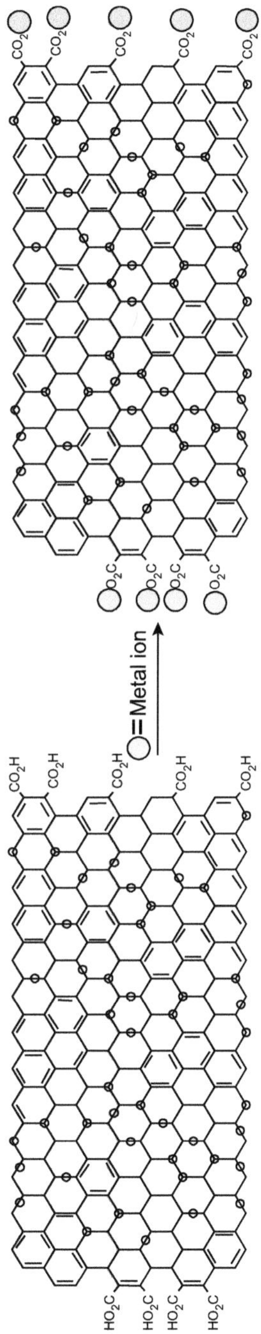


**Figure 7.8** Structure of the graphene oxide GO (left) and its interaction with heavy metal cations (right).

**Table 7.7** Metal sorption capacities on the GO and some modified GO materials. Selected data cited from Ref. [42].

| Adsorbent | Cr(VI) | Co(II) | Cu(II) | Zn(II) | Cd(II) | Pb(II) |
|---|---|---|---|---|---|---|
| | (mg·g$^{-1}$) | | | | | |
| Graphene oxide (GO) | | 68.2 | 46.6 | 30.1 | 14.9<br>106.3 | 367 |
| GO/$Fe_3O_4$ | | 13.0–22.7 | | | | |
| GO-EDTA | | | | | | 479 |
| GO-Chitosan | | | | | | 99 |
| rGO-ED | 5 | | | | | |

**Table 7.8** Maximum sorption capacities of metal ions on the carbon nanotubes [3].

| Adsorbent[a] | Cr(VI) | Co(II) | Ni(II) | Cu(II) | Cd(II) | Pb(II) |
|---|---|---|---|---|---|---|
| | (mg·g$^{-1}$) | | | | | |
| CNTs | | | | | | 102.04 |
| MWCNTs | | | 18.08 | | | |
| Acidified MWCNTs | | | | | | 85 |
| Oxidized CNTs | | | 49.26 | | | |
| Oxidized MWCNTs | 4.262 | | | | 25.7 | |
| CNTs immobilized by calcium alginate | | | | 67.9 | | |
| MWCNTs/iron oxide composite | | | 9.18 | | | |

[a]CNTs - carbon nanotubes; MWCNTs - multiwalled carbon nanotubes.

## 7.2.3 Mineral adsorbents

A number of mineral sorbents are commonly used for metal separation because of their advantages compared with biomaterials. As they have a high specific gravity, they are also known as sinking sorbents. The most important qualities are their incombustibility, chemical inertness, are relatively inexpensive, and widely available. Generally, prior to their use as sorbents minerals, they are ground to a particle size of several nm to several mm (usually, 3 mm) and a bulk density of about 0.45–0.90 kg dm$^{-3}$. These sorbents are not often used for the removal of metals from water because of their low buoyancy and absorbability comparable with polymers or natural organic sorbents. However, after modification with organic compounds, they are suitable for water media.

### 7.2.3.1 Clay minerals, zeolites, silica minerals

Natural layered aluminosilicates, e.g., the bentonite or vermiculite, are important raw materials for the production of mineral sorbents. They are widely available with many methods for their modification already developed. These procedures allow the design sorbents with specific properties for the selective separation of metals. The most commonly known representatives of the clay minerals are kaolinite, montmorillonite, and sepiolite. The presence of hydroxyl groups and weak electrostatic interactions between layers/sheets and the exchangeable cations determine their sorption properties.

Zeolites are natural minerals with others obtained synthetically. Natural zeolites were formed as the result of weathering of volcanic rocks. At present, about 40 groups of zeolites are known, and most of them have been evaluated as sorbents for metals.

Silica minerals are a partially dehydrated polymeric form of colloidal silicic acid with the formula $SiO_2 \cdot nH_2O$. This amorphous material comprises spherical particles 2–20 nm in diameter, which may aggregate to form adsorbents with pore diameters of 6–25 nm. The silica sorbents group includes rocks (siliceous earth, diatomaceous earth, diatomites) and perlite minerals. Most are mineraloids (e.g., opal and chalcedony) with some having a noticeable crystal structure (e.g., cristobalite, quartz, clay minerals, and carbonates). Silica sand is widely used in sorbent production for water purification and metal separations because of its favorable adsorption properties [43].

Aluminum silicates form a group of minerals in which silicon is present in the form of tetrahedral $[SiO_4]^{4-}$ connected via oxygen forming structurally complex two- and three-dimensional systems. They can be distinguished depending on the number of layers present, i.e., the 1:1 or 1:2 minerals. Sometimes, we also find amorphous forms of aluminosilicates, the so-called *allophane* $Al_2 \cdot SiO_2 \cdot nH_2O$. Zeolites are crystalline aluminosilicates of the alkali metals and different divalent metals. Their basic structural units are tetrahedral silicon dioxide $\left(SiO_4{}^{4-}\right)$. Tetrahedrons, linked by common oxygen atoms, form a three-dimensional structure containing channels with diameters between 0.3 and 3 nm, and are known as *molecular sieves*. Some of the silicon atoms can be replaced by aluminum atoms of lower valence with the formation of a negative charge that determines their sorption properties. Representative structures of both mineral types are shown in Fig. 7.9. The hydroxyl groups offer surface sites ready for chemisorption of metal ions [44,45].

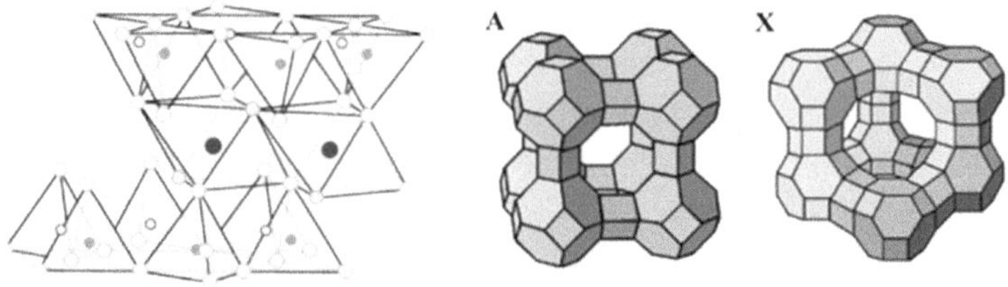


**Figure 7.9** Structure of aluminosilicate (left) and zeolites of the A and X type (right).

**Table 7.9** The removal capacity of various natural clays and clay minerals toward various heavy metal ions [46].

| Adsorbent | Ag(I) | Cr(III) | Cr(VI) | Co(II) | Cu(II) | Ni(II) | Zn(II) | Cd(II) | Hg(II) | Pb(II) |
|---|---|---|---|---|---|---|---|---|---|---|
| | $(mg \cdot g^{-1})$ | | | | | | | | | |
| *Palygorskite* | | | | | | | | | | 104.3 |
| *Montmorillonite* | | | | | | 21.10 | | 30.70 | | 31.10 |
| *Kaolinite* | | | | | | | | | | 11.52 |
| *Cankiri bentonite* | | | | | | | 80.64 | | | |
| *Illitic clay* | | 39.9 | | | | | | 52.5 | | |
| *Expanded perlite* | 8.46 | | | | 1.95 | | | | 0.35 | |
| *Brazilian smectite* | | | 60.32 | | | | | | | |
| *Brazilian sepiolite* | | | | | | | | | | |
| *Brazilian kaolinite* | | | | | | | | | | |
| *Clinoptilolite* | | | | | | | | | | 80.93 |
| *Clinoptilolite* [47] | | 2.4 | | 1.5 | 3.8 | 0.9 | 2.7 | 3.7 | | 6.0 |
| *Chabazite* [47] | | 3.6 | | 5.8 | 5.1 | 4.5 | 5.5 | 6.7 | | 6.0 |
| *Kaolinite* | | | | | | 7.10 | | 6.80 | | 11.5 |
| *Palygorskite* | | | | 8.88 | | | | | | |

| | | | | | | | | | | |
|---|---|---|---|---|---|---|---|---|---|---|
| *Diatomite* | | | | | | | | | | 25.01 |
| *Dolomite* | | | | | | | | | | 19.69 |
| *Perlite* | | | | | | | | | | 8.91 |
| *Apatite* | | | | | | | | | | 82.88 |
| *Sericite* | | | | | | 44,00 | | | | |
| *Kaolinite* | | 3.01 | | | | | | | | |
| *Sericite* | | | | | 1.67 | | | | | 4.69 |
| *Cankiri bentonite* | | | | | 44.84 | | | | | 4.69 |
| *Natural zeolite* [29]* | | | | | 1.12 | | 1.32 | | | |

The sorption behavior of natural zeolites for metal ions has been intensively studied, and there are a number of excellent reviews, e.g., Ref. [48]. As an example, cadmium(II), copper(II), nickel(II), lead(II) and zinc(II) are sorbed on clinoptilolite [49]. The zeolite sorbs about 32%, 75%, 28%, 99%, and 59% of each metal, respectively, from solutions containing metals in the concentration range 50–300 mg $g^{-1}$. The selectivity sequence of the metals is Pb(II) > Cu(II) > Zn(II) > Cd(II) > Ni(II). Certain examples of the natural mineral sorbents are shown in Table 7.9.

#### *7.2.3.2 Modified mineral adsorbents*

Many attempts have been made to improve the selectivity of mineral sorbents through chemical modification. Generally, these modified sorbents have been prepared by thermal treatment or functionalization of the mineral's surface with organic compounds. Physical modification of the sorbents is possible by calcination (e.g., diatomite, perlite, or vermiculite treated at a temperature above 1000°C). Although production costs are relatively high an advantage of these materials is their facile desorption of the adsorbed metals [50]. In recent years, several methods of surface modification of mineral sorbents by different surfactants have been described [51–53]. This method is associated with either changing the character of the surface from hydrophilic to hydrophobic or the creation of metal chelating centers. The process of chemical modification of mineral sorbents increases the cost of the final product and organo-mineral sorbents are commonly used at the industrial scale. Due to the potential toxicity risk related to the use of frequently harmful modifiers, the disposal of metal-loaded sorbents can be a problem [54]. Certain examples of modified sorbents are shown in Table 7.10.

### *7.2.4 Synthetic sorbents*

Synthetic sorbents are more widely utilized materials for the recovery, removal, or concentration of heavy metals by solid-phase extraction (SPE) method than natural materials. The development of the production of synthetic sorbents for better sorptive properties of metals is still on-going. Among synthetic sorbents, three basic categories can be distinguished, namely, inorganic sorbents (e.g., metal oxides), organic sorbents (e.g., polymers), and hybrid sorbents (mixtures of inorganic-organic sorbents).

To the synthetic inorganic adsorbents belong silica gels, metal oxides (such as aluminum oxides, magnesium oxides, ferric oxides, manganese oxides, titanium oxides, cerium oxides), zeolites, aluminosilicate minerals, activated carbons, cyanoferrate complexes with different metals and others. Synthetic organic sorbents are a group of polymers with ion exchange or chelating functional groups. They have large surface areas with a high uptake capacity, good pore structures, and efficiently and selectively adsorb metals from aqueous solutions. Solid-phase extraction processes with synthetic adsorbents allow reducing the amount of solvent and safer procedures compared with liquid-liquid extraction methods. A further advantage of synthetic adsorbents is their stability in both acidic and alkaline solutions and in organic solvents. Synthetic sorbents can be easily regenerated for further use. The attrition resistance of synthetic sorbents is also better than that for natural materials allowing their use in a larger

**Table 7.10** Sorption of metals on the modified minerals (selected data).

| | Mn(II) | Ni(II) | Cu(II) | Zn(II) | Cd(II) | Pb(II) |
|---|---|---|---|---|---|---|
| **Adsorbent** | $q_L$ (mg·g$^{-1}$) | | | | | |
| **Smectites** | | | | | | |
| Montmorillonite [46] | | 21.1 | | | 30.7 | 31.1 |
| Acid-modified montmorillonite (AMM) [54] | 2.2 | 4.0 | 2.8 | 76.9 | 0.6 | 1.6 |
| **Zeolites** | | | | | | |
| Natural Turkish clinoptilolite [55] | | | 40.9 | 24 | 0.6 | 23 |
| Natural Sardinian clinoptilolite [56] | | | | 24.1 | 63.7 | 181 |
| Natural Ukrainian clinoptilolite [57] | | 13.03 | 25.8 | | 4.2 | 27.7 |
| Natural Turkish clinoptilolite [58] | | | | | 13; 20.2 | |
| NaCl pretreated Turkish clinoptilolite [58] | | | | | 80.6; 45.9 | |
| NaCl pretreated Mexican zeolite-rich tuff [59] | | | | | 7.86; 9.18 | |
| Thiourea pretreated Mexican zeolite-rich tuff [59] | | | | | 1.20; 11.60 | |
| **Clays** | | | | | | |
| Natural Slovak bentonite clay [60] | | | | | 27.7 | |
| Iron coated magnetic Slovak bentonite [60] | | | | | 29.7 | |

number of cycles resulting in a reduction of the amount of waste sorbent. Finally, the economic-environmental impact of synthetic sorbents is also favorable.

### 7.2.4.1 *Inorganic synthetic sorbents*

#### 7.2.4.1.1 Silica gel

Silica gel is composed of amorphous $SiO_2$ in the form of hard irregular granules. The selectivity of commercially available sorbents depends on several factors, such as the type of modifier, its size, and specific hard-soft acid-base character. The common method of modifying silica gel is by reaction with an organosilane containing a chelating functional group (e.g., 2,5-Dimercapto-1,3,4-thiadiazole; 2-Pyridinecarboxaldehyde phenylhydrazone; or others) [61]. For example, chemically modified silica gel containing aminothioamidoanthraquinone groups was used for sorption of Pb(II), Cu(II), Ni(II), Co(II) and Cd(II) at pH $\geq 3$ with a selectivity order Pb(II) > Cu(II) > Cd(II); while for pH $\geq 4$ Ni(II) > Co(II) with sorption capacities of 0.56, 0.30, 0.15, 0.12 and 0.067 mmol/g for Pb(II), Cu(II), Ni(II), Co(II) and Cd(II), respectively [61]. Silica gel impregnated with a mixture of Aliquat 336 and Eriochrome Blue SE was used for the separation of Cu(II), Cd(II), Pb(II), and Zn(II) with the relative capacity of the sorbent toward these metal increasing at higher pH [62]. A new type of modified silica sorbent was used for the selective extraction of Fe(III) from water using a pyridinium ionic liquid [63].

#### 7.2.4.1.2 Metal oxides (e.g., aluminum oxides, magnesium oxides, ferric oxides, manganese oxides, titanium oxides, cerium oxides)

Metal oxide sorbents at the nanometer scale, such as aluminum oxide, iron oxide, titanium oxide, manganese oxide, magnesium oxide, zirconium oxide, or cerium oxide have received increasing interest due to their sorption capability and high surface area [64]. Their sorption properties can be adjusted by varying the experimental conditions, such as sample properties (ionic strength, ion types, pH), experimental procedure (batch extraction or column flow), and temperature. An abbreviated comparison of the metal adsorption capacity for different nanosized metal oxides is given in Table 7.11 [64].

**Table 7.11** Adsorption capacities of nanosized metal oxides (NMOs) for the extraction of heavy metals from aqueous samples [64].

| | Pb | Cu | Cr | Cd | Zn | Ni |
|---|---|---|---|---|---|---|
| NMOs | Adsorption capacities of heavy metals [$mg \cdot g^{-1}$] | | | | | |
| Goethite ($\alpha$-FeOOH) | | 149.25 | | | | |
| Hematite ($\alpha$-$Fe_2O_3$) | | 84.46 | | | | |
| Hydrous ferric oxide (HFO) | 20.27 | | | | | |
| Hydrous aluminum oxide (HAO) | 20.27 | | | | | |
| Hydrous manganese oxide (HMO) | 324.32 | | | 143.31 | 57.21 | |
| Maghemite ($\gamma$-$Fe_2O_3$) | | 26.8 | 19.2 | | | 23.6 |
| $\alpha$-$MnO_2$ | | 82.6 | | | | |
| $TiO_2$ | 81.3 | | | 7.9−15.2 | 15.3 | 67.4 |
| $Al_2O_3$ | | | | | | 176.1 |
| Modified $Al_2O_3$ | 100 | 16.3 | 100 | 83.33 | | 18.18 |
| ZnO | 6.7 | 1600 | | | | |
| $CeO_2$ | 9.2 | 15.4−121.95 | | | | |

#### 7.2.4.1.3 Aluminosilicates

Mesoporous silica-alumina oxide (97% $SiO_2$, 3% $Al_2O$) is a suitable sorbent for the extraction of heavy metals [65]. Toxic heavy metal ions, such as cadmium(II) and lead(II) can be extracted with a high adsorption capacity suitable for use at an industrial scale [66].

#### 7.2.4.1.4 Cyanoferrates and hexacyanoferrates

For the selective recovery of metal ions (also radionuclides), metal hexacyanoferrates are effective sorbents. Metal hexacyanoferrates are produced by the reaction of metal salts with potassium hexacyanoferrate [67,68]. A novel magnetic composite material based on magnetite and cobalt cyanoferrate nanoparticles ($K_2[CoFe(CN)]$) was evaluated for the solid-phase extraction of radionuclide ions from aqueous solutions for the management of radioactive waste. The nanocomposite removes Cs(I), Eu(III) Co(II), Am(III) and Tc(VII) radionuclides from aqueous solutions with acceptable efficiency for waste management purposes [69].

#### 7.2.4.1.5 Zeolites

Synthetic zeolites are microporous crystalline aluminosilicates containing certain extra-framework cations. They are cation exchangers similar to natural zeolites but generally have a higher capacity [70]. They are more expensive than natural zeolites, but it is easier to tailor their physicochemical properties for specific applications [71]. The extraction of divalent metal ions (e.g., $Cu^{2+}$, $Ni^{2+}$, $Cd^{2+}$, and $Pb^{2+}$) from aqueous samples by two types of zeolites (synthetic Faujasite zeolite NaX) and zeolite A, LTA, was evaluated. The selectivity order for the NaX zeolite was $Pb^{2+} > Cu^{2+} > Cd^{2+} > Ni^{2+}$ while for LTA $Pb^{2+} > Cd^{2+} > Cu^{2+} > Ni^{2+}$ [70]. High sorption capacities were obtained, 715, 430, and 250 mg $g^{-1}$ for Pb(II), Cd(II), and Cu(II), respectively [71].

### *7.2.4.2 Organic synthetic sorbents*

#### 7.2.4.2.1 Ion-exchange resins

Ion-exchange resins usually contain strong and weak cation and anion functional groups, which are bound to silica gel. They are classified based on the functionality of the ion-exchange sorbent. Classic ion-exchange sorbents are typically crosslinked styrene-divinylbenzene polymers with attached functional groups. Four types of ion-exchangers are distinguished: strongly acidic cation-exchange resins, weakly acidic cation-exchange resins, strongly basic anion-exchange resins, and weakly basic anion-exchange resins. Sorbents of the first group contain ion-exchange sites consisting of sulfonic acid groups. Weakly acidic cation-exchange resins usually contain carboxylic acid groups. Strongly basic anion-exchange resins contain quaternary ammonium functional groups. Weakly basic anion-exchange resins contain one or more primary, secondary and/or tertiary amine functional groups.

The use of ion-exchange resins has been described numerous times, and here we present only a few representative results for the main groups of ion-exchangers. Dowex HCR S/H, a strong acid cation-exchange resin, was proposed for the separation of Cd(II) and Zn(II) [72], Dowex 50W synthetic resin - for $Cu^{2+}$,$Zn^{2+}$,$Ni^{2+}$,$Cd^{2+}$ and $Pb^{2+}$ [73],

Dowex 50WX4 nanocomposite - for Cr(VI), Ni(II), Cu(II), Cd(II), and Pb(II) [74], Dowex M-4195 - for Cu(II), Ni(II), Pb(II), Fe(III), Co(II), Mn(II) [75]) from water samples. The weak acid ion-exchanger Lewatit CNP 80 and a chelating derivative Lewatit TP 207, were evaluated for the extraction of Pb(II), Cu(II), Zn(II), Cd(II), and Ni(II) from aqueous solution [76]. The ion-exchange resin provided better performance than the chelating resin for the removal of the metal ions. It also showed higher selectivity for Ni and Cu than for Cd, Zn, and Pb [76]. The strong acid cation-exchange resin Lewatit S100 was suitable for the extraction of Cu(II), Zn(II), and Pb(II) ions [77].

Amberlite IR-120 cation-exchange resin, which has the sulfonate functional groups, appeared to be enough reactive and extremely pH sensitive. It was also proposed to be applied for adsorption and separation of Cu(II), Zn(II), Ni(II), Pb(II), and Cd(II) with an aim to their analytical determination in a wastewater sample. The sorbent shows better adsorption of Ni(II) and Zn(II) than Cu(II), Pb(II), and Cd(II) [78]. The Diphonix resin (containing diphosphonic, sulfonic, and carboxylic groups on a polystyrene-divinylbenzene matrix) was used to for the adsorption of Nd(III), Ce(III), La(III), Fe(III), Co(II), Ni(II), Cu(II), and Zn(II) metal ions in mixture from the regeneration of the Ni-MH batteries [79]. This resin had good adsorptive properties for rare earth elements and a lower capacity for other metals ions (i.e., almost 100% for the REEs and below 50% for the others). The selectivity series appeared to be La(III) > Fe(III) > Nd(III) > Ce(III) > Cu(II) > Zn(II) > Co(II) > Ni(II) [79]. The adsorption capacity was 59.41 (La), 26.69 (Ni), 0.65 (Nd), 1.00 (Ce), 6.65 (La), 3.48 (Fe), 1.74 (Co), 14.86 (Ni), 0.15 (Cu), 0.41 $mg \cdot g^{-1}$ (Zn).

#### 7.2.4.2.2 Chelating resins

The most popular chelating sorbents contain iminodiacetic acid, polyamine, and glucamine groups bound to a highly porous styrene-divinylbenzene matrix or thiourea-based resins. Different kinds of chelating sorbents have been designed for the selective extraction of heavy metals, alkali, alkaline earth, and transition metal ions. They bind selectively all metal ions which form complexes by coordination with specific chelation groups [80]. A number of methods have been used for their synthesis [80,81]. A selective chelating sorbent with iminodiacetate groups was evaluated for the extraction of heavy metal ions from water by batch experiments [82]. The sorption capacity followed the order: Pb(II) > Cd(II) > Zn(II) with a maximum adsorption capacity of 106 mg $g^{-1}$ Pb(II), 90 mg $g^{-1}$ for Cd(II), and only 15 mg $g^{-1}$ Zn(II) [82]. Examples of the use of chelating sorbents for the preconcentration and separation of trace metal ions, especially heavy metals, are summarized in Table 7.12 [83].

### 7.2.4.3 Hybrid sorbents

Synthetic hybrid inorganic-organic sorbents are a kind of artificial resins containing a mixture of inorganic and organic components. Hybrid sorbents typically possess favorable mechanical properties, enhanced chemical stability, and specific properties for the selective extraction of heavy metal ions [84]. A cellulose-nanoscale-manganese oxide composite (C-NMOC) was described for the extraction of Pb(II) ions from aqueous solution [85]. The adsorption capacity of C−NMOC (per gram of Mn present)

**Table 7.12** Chelating resins based on silica gel for the separation/or preconcentration of metal ions [83].

| Functional group | Metal ions | Functional group | Metal ions |
|---|---|---|---|
| 1,5-Bis(di-2-pyridyl) methylene thiocarbohydrazide | Hg, Zn, Cd, Co, Pb, Ni, Pt, Cr | Dendrimer-like polyamidoamine | Pd, Pt |
| 1,8-Dihydroxyanthraquinone | Fe, Co, Ni, Cu, Pb, Zn, Cd | Didecylaminoethyl-beta-tridecylammonium | Zn, Cu, Co, Pb |
| 1-Aminoanthraquinone | Cu, Cr | Diethylenetriamine, mono-naphthaldehyde, Mono-salicyaldehyde | Fe, Ni, Cu, Zn, Cd, Pb |
| 1-Nitroso-2-naphthol | Co | Dithioacetal derivatives | Hg |
| 2-Amino-1-cyclopentene-1-dithiocarboxylic acid | Ag, Hg, Pd, Cu, Ni, Cd Zn, Pt | Eriochrome black-T | Zn, Mg, Ca |
| 2-Hydroxy-1-naphthaldehyde | Cu, Zn, Cd, Hg, Pb | Formylsalicylic acid | Fe |
| 2-Mercaptobenzothiazole | Bi, Pb, Cd, Hg, Au, Pt, Pd, Ag, Cu | Iminosalicyl group | Fe, Co, NiCo, Zn, Cd |
| 3-Hydroxy-2-methyl-1,4-naphthoquinone | Co, Cu, Zn, Fe | Methylthiosalicylate | Cd, Pb |
| 3-Mercaptopropyl group | Hg, Au, Pd, Se,As | Morin | Zr, Be, Al, Sn |
| 4-(2-Pyridylazo)resorcinol or 2-(2-Pyridylazo)-5-dimethylaminophenol | Cd | N-propyl-Nc-[1-(2-thiobenzothiazole)-2,2c,2cc-trichloroethyl]urea | Ag, Au, Pd |
| 4-Amino-3-hydroxy-2-(2-chlorobenzene)-azo-1-naphthalene | Cr, Ni, Cu, Zn, Cd, Pb | N-tripropionate tetraazamacrocycles | U |
| 4-Aminoantipyrene | Hg, Cd, Pb, Cu, Zn | o-Dihydroxybenzene | Cu, Pb, Fe, Zn, Co, Ni, Cd |

*Continued*

**Table 7.12** Chelating resins based on silica gel for the separation/or preconcentration of metal ions [83].—cont'd

| Functional group | Metal ions | Functional group | Metal ions |
|---|---|---|---|
| 8-Hydroxyquinoline [quinolin-8-ol] | Cd, Pb, Zn, Cu, Fe, Mn, Ni, Co | o-Vanillin | Cu, Co, Fe, Zn |
| Acid red 88 | Mg, Ni, Cu, Fe | Purpurogallin | Fe |
| Amidinothiourea | Ag, Au, Pd | Pyridinium ion | Co, Cu, Zn, Cd, Hg |
| Aminophenol, aminobenzoic acid | Fe, Co, Ni, Cu, Cd, Pb | Pyrocatechol violet | Cu, Ni, Mn, Co, Cu, Hg, Fe, Zn, Pd, Au, Pt, Ag |
| Aminopropyltriethoxysilane | Bi, Pb, Ni, Cu, Cd, Cr, V, Mn, Ag | Quaternary ammonium salts | Bi, Fe |
| Aurin tricarboxylic acid | Mo, Sn, Cu | Quercetin | Mo, Sn |
| Benzimidazole | Mn, Ni, Cd, Zn, Cu | Resacetophenone | Cu, Ni, Co, Zn, Fe |
| Calix-[4] arene tetrahydroxamate | Ni, Zn, Co, Pb, Cu, Mn | Thiosemicarbazide | Pd |
| Crown ether carboxylic acid | Se, Th, Rb, Na, K | Thiourea | Au, Ag, Pd |
| Cyanex 301 | Bi | | |

was several-fold greater than manganese oxide (β-$MnO_2$) and even nanoscale-manganese oxide (NMO). Physisorption plays a dominant role in Pb(II) adsorption by both NMO and C-NMOC [85]. An organic-inorganic hybrid sorbent was prepared by self-hydrolysis, self-condensation, and cocondensation of the cross-linking agent (tetraethoxysilicate) and 3-(2- aminoethylamino)propyltrimethoxysilane monomer followed by gelation. This sorbent was used for the selective separation of Cd(II) from Zn(II) in aqueous solution. The selectivity coefficient for Cd(II)/Zn(II) was over 100 [86]. A mercapto-functionalized hybrid sorbent was used for sequestering Sb(III) species from aqueous samples [87]. A nanosized hybrid, biodegradable polymeric colloid was synthesized from nanosized MnO particles and polysaccharide [88]. The eco-friendly sorbent had a sorption capacity of 114 mg $g^{-1}$ for Ni(II).

## 7.3 Conclusions

- ✔ The literature on sorption, in general, and on metal sorption, in particular, is rich and constantly growing.
- ✔ So far, no theory has been formulated, which allows the prediction of the sorption properties of sorbents.
- ✔ For the choice of a specific sorbent (both for metals and for other compounds), a review of existing sorption databases should be searched prior to starting experimental studies.

## References

[1] Hawkes SJ. What is a "heavy metal?". J Chem Educ 1997;74:1374.

[2] Nguyen TAH, Ngo HH, Guo WS, Zhang J, Liang S, Yue QY, et al. Applicability of agricultural waste and by-products for adsorptive removal of heavy metals from wastewater. Bioresour Technol 2013;148:574−85.

[3] Fu F, Wang Q. Removal of heavy metal ions from wastewaters: a review. J Environ Manag 2011;92:407−18.

[4] Wang S, Xu Y, Norbu N, Wang Z. Remediation of biochar on heavy metal polluted soils. IOP Conf Ser Earth Environ Sci 2018;108:042113.

[5] Limousin G, Gaudet J-P, Charlet L, Szenknect S, Barthes V, Krimissa M. Sorption isotherms: a review on physical bases, modeling and measurement. Appl Geochem 2007;22: 249−75.

[6] Milonjić SK. Comments on "factors influencing the removal of divalent cations by hydroxyapatite". J Hazard Mater 2009;162:1588−9.

[7] Milonjic S. Comments on "Removal of uranium (VI) from aqueous solution by adsorption of hematite", by X. Shuibo, Z. Chun, Z. Xinghuo, Y. Jing, Z. Xiaojian, W. Jingsong. J Environ Radioact 2009;00:921−2.

[8] Milonjic SK. Comments on the authors' response to the comments on "factors influencing the removal of divalent cations by hydroxyapatite", by Smiciklas et al. J Hazard Mater 2010;176:1126−7.

[9] Ho Y-S. Pseudo-isotherms using a second order kinetic expression constant. Adsorption 2004;10:151−8.

[10] Crini G, Badot P-M. Application of chitosan, a natural aminopolysaccharide, for dye removal from aqueous solutions by adsorption processes using batch studies: a review of recent literature. Prog Polym Sci 2008;33:399−447.

[11] (a) Pearson RG. Hard and soft acids and bases, HSAB, part 1: fundamental principles. J Chem Educ 1968;45:581−6.(b) Pearson RG. Hard and soft acids and bases, HSAB, part II: underlying theories. J Chem Educ 1968;45:643−8.

[12] Chen JP, Hong LA, Wu SN, Wang L. Elucidation of interactions between metal ions and Ca alginate-based ion-exchange resin by spectroscopic analysis and modeling simulation. Langmuir 2002;18:9413−21.

[13] Karagunduz A, Unal D. New method for evaluation of heavy metal binding to alginate beads using pH and conductivity data. Adsorption 2006;12:175−84.

[14] Rhazi M, Desbrieres J, Tolaimate A, Rinaudo M, Vottero P, Alagui A, et al. Influence of the nature of the metal ions on the complexation with chitosan. Application to the treatment of liquid waste. Eur Polym J 2002;38:1523−30.

[15] Vengris T, Binkiene R, Sveikauskaite A. Nickel, copper and zinc removal from waste water by a modified clay sorbent. Appl Clay Sci 2001;18:183–90.

[16] Karnib M, Kabbani A, Holail H, Olama Z. Heavy metals removal using activated carbon: silica and silica activated carbon composite. Energy Procedia 2014;50:113–20.

[17] Pires C, Marques A, Guerreiro A, Magan N, Castro P. Removal of heavy metals using different polymer matrixes as support for bacterial immobilisation. J Hazard Mater 2011; 191:277–86.

[18] Wang P, Du M, Zhu H, Bao S, Yang T, Zou M. Structure regulation of silica nanotubes and their adsorption behaviors for heavy metal ions: pH effect, kinetics, isotherms and mechanism. J Hazard Mater 2015;286:533–44.

[19] Ewecharoena A, Thiravetyan P, Wendel E, Berlagnolli H. Nickel adsorption by sodium polyacrylate-grafted activated carbon. J Hazard Mater 2009;171:335–9.

[20] https://www.alibaba.com/showroom/activated-carbon-price-in-kg.html.

[21] https://www.alibaba.com/product-detail/Competitive-factory-Price-Zeolite-for-water_60802216570.html?spm=a2700.7724857.normalList.7.653e3ca5ALV5hm&s=p.

[22] Ostrowska-Czubenko J, Pierog M, Gierszewska M. Modification of chitosan - a concise overview. Wiadomości chemiczne 2016;70:657–79 [Polish].

[23] Park D, Yun Y-S, Park JM. The past, present, and future trends of biosorption. Biotechnol Bioproc Eng 2010;15:86–102.

[24] Park D, Yun Y-S, Jo JH, Park JM. Effects of ionic strength, background electrolytes, heavy metals, and redox-active species on the reduction of hexavalent chromium by *Ecklonia* biomass. J Microbiol Biotechnol 2005;15:780–6.

[25] Salman M, Athar M, Farooq U, Nazir H, Noor A, Nazir S. Microwave-assisted urea-modified sorghum biomass for Cr(III) elimination from aqueous solution. Korean J Chem Eng 2013;30:1257–64.

[26] Seo H, Lee M, Wang S. Equilibrium and kinetic studies of the biosorption of dissolved metals on *Bacillus drentensis* immobilized in biocarrier beads. Environ Eng Res 2013;18: 45–53.

[27] Fuks L, Herdzik-Koniecko I, Polkowska-Motrenko H, Oszczak A. Novel procedure for removal of the radioactive metals from aqueous wastes by the magnetic calcium alginate. Int J Environ Sci Technol 2018;15:2657–68.

[28] Gupta S, Kumar D, Gaur JP. Kinetic and isotherm modeling of lead(II) sorption onto some waste plant materials. Chem Eng J 2009;148:226–33.

[29] Abdel Salam OE, Reiad NA, ElShafei MM. A study of the removal characteristics of heavy metals from wastewater by low-cost adsorbents. J Adv Res 2011;2:297–303.

[30] Romera E, Gonzalez F, Ballester A, Blazquez ML, Munoz JA. Comparative study of biosorption of heavy metals using different types of algae. Bioresour Technol 2007;98: 3344–53.

[31] Abdolali A, Ngo HH, Guo W, Zhou JL, Du B, Wei Q, et al. Characterization of a multi-metal binding biosorbent: chemical modification and desorption studies. Bioresour Technol 2015;193:477–87.

[32] Wang J, Chen C. Chitosan-based biosorbents: modification and application for biosorption of heavy metals and radionuclides. Bioresour Technol 2014;160:129–41.

[33] Mahmoud ME, Abdel-Fattah TM, Osman MM, Ahmed SBJ. Chemically and biologically modified activated carbon sorbents for the removal of lead ions from aqueous media. J Environ Sci Health - Part A Toxic/Hazard Subst Environ Eng 2012;47:130–41.

[34] Bhatnagar A, Hogland W, Marques M, Sillanpää M. An overview of the modification methods of activated carbon for its water treatment applications. Chem Eng J 2013;219: 499–511.

[35] Novoselov KS, Geim AK, Morozov SV, Jiang D, Zhang, et al. Electric field effect in atomically thin carbon films. Science 2004;306:666–9.
[36] McAllister MJ, Li J-L, Adamson DH, Schniepp HC, Abdala AA, Liu J, et al. Single sheet functionalized graphene by oxidation and thermal expansion of graphite. Chem Mater 2007;19:4396–404.
[37] Dreyer DR, Park S, Bielawski CW, Ruoff RS. The chemistry of graphene oxide. Chem Soc Rev 2010;39:228–40.
[38] Lerf A, He H, Forster M, Klinowski J. Structure of graphite oxide revisited. J Phys Chem B 1998;102:4477–82.
[39] Chowdhury S, Balasubramanian R. Recent advances in the use of graphene-family nanoadsorbents for removal of toxic pollutants from wastewater. Adv Colloid Interface Sci 2014;204:35–56.
[40] Lim JY, Mubarak NM, Abdullah EC, Nizamuddin S, Inamuddin MK. Recent trends in the synthesis of graphene and graphene oxide based nanomaterials for removal of heavy metals - a review. J Ind Eng Chem 2018;66:29–44.
[41] Xu J, Cao Z, Zhang Y, Yuan Z, Lou Z, Xu X, et al. A review of functionalized carbon nanotubes and graphene for heavy metal adsorption from water: preparation, application, and mechanism. Chemosphere 2018;195:351–64.
[42] Nupearachchi CN, Mahatantila K, Vithanage M. Application of graphene for decontamination of water; implications for sorptive removal. Groundwater Sustain Dev 2017;5:206–15.
[43] El-Bayaa AA, Badawy NA, Gamal AM, Zidan IH, Mowafy AR. Purification of wet process phosphoric acid by decreasing iron and uranium using white silica sand. J Hazard Mater 2011;190:324–9.
[44] Awan MA, Qazi IA, Khalid I. Removal of heavy metals through adsorption using sand. J Environ Sci 2003;15:413–6.
[45] Unob F, Wongsiri B, Phaeon N, Puanngam M, Shiowatana J. Reuse of waste silica as adsorbent for metal removal by iron oxide modification. J Hazard Mater 2007;142:455–62.
[46] Kurniawan A, Ismadji S, Soetaredjo FE, Ayucitra A. Natural clays/clay minerals and modified forms for heavy metals removal, Ch. 11. In: Sharma SK, editor. Heavy metals in water: presence, removal and safety. Cambridge, UK: Royal Society of Chemistry; 2015. p. 357.
[47] Ouki SK, Kavannagh M. Treatment of metals-contaminated wastewaters by use of natural zeolites. Water Sci Technol 1999;39:115–22.
[48] Natural microporous materials in environmental technology. In: Misaelides P, Macasek F, Pinnavaia TJ, Colella C, editors. Proceedings of the NATO advanced research workshop on the application of natural microporous materials for environmental technology (Smolenice Castle, Slovakia, 26–30 October, 1998). Springer Science+Business Media, B.V; 1999. https://doi.org/10.1007/978-94-011-4499-5.
[49] Shaheen SM, Derbalah AS, Moghanm FS. Removal of heavy metals from aqueous solution by zeolite in competitive sorption system. Intern J Environ Sci Dev 2012;3:362–7.
[50] Adebajo MO, Frost RL, Kloprogge JT, Carmody O, Kokot S. Porous materials for oil spill cleanup: a review of synthesis and absorbing properties. J Porous Mater 2003;10:159–70.
[51] Gładysz-Płaska A, Majdan M, Tarasiuk B, Sternik D, Grabias E. The use of halloysite functionalized with isothiouronium salts as an organic/inorganic hybrid adsorbent for uranium(VI) ions removal. J Hazard Mater 2018;354:133–44.
[52] Ayari F, Manai G, Khelifi S, Trabelsi-Ayadi M. Treatment of anionic dye aqueous solution using Ti, HDTMA and Al/Fe pillared bentonite. Essay to regenerate the adsorbent. J Saudi Chem Soc 2019;23:294–306.

[53] Kostenko LS, Tomashchuk II, Kovalchuk TV, Zaporozhets OA. Bentonites with grafted aminogroups: synthesis, protolytic properties and assessing Cu(II), Cd(II) and Pb(II) adsorption capacity. Appl Clay Sci 2019;172:49–56.

[54] Sarkar B, Megharaj M, Shanmuganathan D, Naidu R. Toxicity of organoclays to microbial processes and earthworm survival in soils. J Hazard Mater 2013;261:793–800.

[55] Turkman AE, Aslan S, Ege I. Treatment of metal containing wastewaters by natural zeolites. Fresenius Environ Bull 2004;13:574–80.

[56] Cincotti A, Mameli A, Locci AM, Orru R, Cao G. Heavy metals uptake by Sardinian natural zeolites: experiment and modeling. Ind Eng Chem Res 2006;45:1074–84.

[57] Sprynskyy M, Buszewski B, Terzyk AP, Namieśnik J. Study of the selection mechanism of heavy metal (Pb2+, Cu2+, Ni2+, and Cd2+) adsorption on clinoptilolite. J Colloid Interface Sci 2006;304:21–8.

[58] Gedik K, Imamoglu I. Affinity of clinoptilolite-based zeolites towards removal of Cd from aqueous solutions. Separ Sci Technol 2008;43:1191–207.

[59] Barragán PP, Macedo MG, Olguín MT. Cadmium sorption by sodium and thiourea-modified zeolite-rich tuffs. J Environ Sci 2017;52:39–48.

[60] Mockovčiaková A, Orolínová Z. Adsorption properties of modified bentonite clay. Chem Technol 2009;50:47–50 [Czech].

[61] Ngeontae W, Aeungmaitrepirom W, Tuntulani T. Chemically modified silica gel with aminothioamidoanthraquinone for solid phase extraction and preconcentration of Pb(II), Cu(II), Ni(II), Co(II) and Cd(II). Talanta 2007;71:1075–82.

[62] Kocjan R, Swieboda R. Silica gel modified with eriochrome blue SE as a sorbent in trace analysis for metal ions. Mikrochim Acta 1996;122:175–82.

[63] Abdolmohammad-Zadeh H, Galeh-Assadi M, Shabkhizan S, Mousazadeh H. Sol-gel processed pyridinium ionic liquid-modified silica as a new sorbent for separation and quantification of iron in water samples. Arabian J Chem 2016;9:S587–94.

[64] Hua M, Zhang S, Pan B, Zhang W, Lv L, Zhang Q. Heavy metal removal from water/wastewater by nanosized metal oxides: a review. J Hazard Mater 2012;211–212:317–31.

[65] Wawrzkiewicz M, Wiśniewska M, Wołowicz A, Gun'ko VM, Zarko VI. Mixed silica-alumina oxide as sorbent for dyes and metal ions removal from aqueous solutions and wastewaters. Microporous Mesoporous Mater 2017;250:128–47.

[66] Xu M, Hadi P, Ning C, Barford J, Jin AK, McKay G. Aluminosilicate-based adsorbent in equimolar and non-equimolar binary-component heavy metal removal systems. Water Sci Technol 2015;72:2166–78.

[67] Vincent T, Vincent C, Guibal E. Immobilization of metal hexacyanoferrate ion-exchangers for the synthesis of metal ion sorbents-a mini-review. Molecules 2015;20:20582–613.

[68] Taj S, Muhammad D, Chaudhry MA, Mazhar M. Lithium, rubidium and cesium ion removal using potassium iron(III) hexacyanoferrate(II) supported on poly-methylmethacrylate. J Radioanal Nucl Chem 2011;288:79–88.

[69] Fuks L, Wawszczak D, Smolinski T, Herdzik-Koniecko I, Chmielewski AG. Novel magnetic cobalt cyanoferrate nanoparticles as potential sorbent for solid-phase extraction of radionuclides from aqueous samples. Radiochemistry 2018;60(4):400–8.

[70] Ezzeddine Z, Batonneau-Gener I, Pouilloux Y. Cation exchange mechanism of divalent metals ions in synthetic NaX and LTA zeolites: efficiency and selectivity. Eur Chem Bull 2018;7(3):93–8.

[71] Ezzeddine Z, Batonneau-Gener I, Pouilloux Y, Hamad H, Saad Z. Synthetic nax zeolite as a very efficient heavy metals sorbent in batch and dynamic conditions. Colloids Interfaces 2018;2:22.

[72] Fil BA, Yilmaz AE, Boncukcuoglu R, Bayar S. Removal of divalent heavy metal ions from aqueous solutions by Dowex HCR-S synthetic resin. Bulg Chem Commun 2012;44(3): 201−7.
[73] Pehlivan E, Altun T. The study of various parameters affecting the ion exchange of $Cu^{2+}$,$Zn^{2+}$,$Ni^{2+}$,$Cd^{2+}$, and Pb2+from aqueous solution on Dowex 50W synthetic resin. J Hazard Mater 2006;B134:149−56.
[74] Lasheen MR, Iman Y, El-Sherif, Shaimaa T, El-Wakeel, Sabry DY, El-Shahat MF. Heavy metals removal from aqueous solution using magnetite Dowex 50WX4 resin nanocomposite. J Mater Environ Sci 2017;8(2):503−11.
[75] Diniz CV, Ciminelli VST, Doyle FM. The use of the chelating resin Dowex M-4195 in the adsorption of selected heavy metal ions from manganese solutions. Hydrometallurgy 2005; 78:147−55.
[76] Pehlivan E, Altun T. Ion-exchange of $Pb^{2+}$, $Cu^{2+}$, $Zn^{2+}$, $Cd^{2+}$, and $Ni^{2+}$ ions from aqueous solution by Lewatit CNP 80. J Hazard Mater 2007;140:299−307.
[77] Pavolová H, Bakalár T, Sabo S, Puškárová P. Adsorption efficiency of selected natural and synthetic sorbents. Geol Geophys Environ 2016;42(1):114−5.
[78] Demirbas A, Pehlivan E, Gode F, Altun T, Arslan G. Adsorption of Cu(II), Zn(II), Ni(II), Pb(II), and Cd(II) from aqueous solution on Amberlite IR-120 synthetic resin. J Colloid Interface Sci 2005;282:20−5.
[79] Fila D, Hubicki Z, Kołodyńska D. Recovery of metals from waste nickel-metal hydride batteries using multifunctional Diphonix resin. Adsorption 2019:1−16.
[80] Roy PK, Rawat AS, Choudhary V, Rai PK. Synthesis and analytical application of a chelating resin based on a crosslinked styrene/maleic acid copolymer for the extraction of trace-metal ions. J Appl Polym Sci 2004;94:1771−9.
[81] Ravikumar Reddy A, Hussain Reddy K. Removal of heavy metal ions using novel polystyrene supported chelating polymer resin. Indian J Chem Technol 2002;9:97−102.
[82] Dinu MV, Dragan ES, Trochimczuk A. Removal of heavy metal ions from aqueous solutions using chelating resins with iminodiacetate groups. XXIII ARS Separatoria 2008: 61−4. Toruń, Poland.
[83] Zougagh M, Cano Pavon JM, Garcia de Torres A. Chelating sorbents based on silica gel and their application in atomic spectrometry. Anal Bioanal Chem 2005;381:1103−13.
[84] Samiey B, Cheng C-H, Wu J. Organic-inorganic hybrid polymers as adsorbents for removal of heavy metal ions from solutions: a review. Materials 2014;7:673−726.
[85] Maliyekkal SM, Lisha KP, Pradeep T. A novel cellulose-manganese oxide hybrid material by in situ soft chemical synthesis and its application for the removal of Pb(II) from water. J Hazard Mater 2010;181:986−95.
[86] Lu Y-K, Yan X-P. An imprinted organic-inorganic hybrid sorbent for selective separation of cadmium from aqueous solution. Anal Chem 2004;76:453−7.
[87] You N, Liu T-H, Fan H-T, Shen H. An efficient mercapto-functionalized organic-inorganic hybrid sorbent for selective separation and preconcentration of antimony(III) in water samples. RSC Adv 2018;8:5106−13.
[88] Krivoshapkin PV, Ivanets AI, Torlopov MA, Mikhaylov VI, Srivastava V, Sillanpää M, Prozorovich VG, Kouznetsova TF, Koshevaya ED, Krivoshapkina EF. Nanochitin/manganese oxide-biodegradable hybrid sorbent for heavy metal ions. Carbohydr Polym 2019; 210:135−43.

# Molecularly imprinted polymers

8

*Esther Turiel, Antonio Martín Esteban*
Departamento de Medio Ambiente y Agronomía, INIA, Madrid, Spain

## 8.1 Introduction

Nowadays, sophisticated and powerful analytical instrumentation is available, eventually allowing the determination of an organic compound in any kind of sample. Typically, chromatographic techniques coupled to common detectors (UV, fluorescence) or, more recently, mass spectrometry (MS) are routinely used in analytical laboratories for the determination of target compounds. However, direct injections of crude extracts are not possible even when using selective detection provided by MS, since matrix components can affect the ionization of analytes, hampering accurate quantification. Thus, a clean sample is generally recommended to improve separation and detection. Typically, several sample treatments are necessary for the unequivocal identification, confirmation, and quantification of target compounds.

The main objectives of sample preparation are the removal of potential interferents, the preconcentration of target compounds (especially in environmental water samples), the derivatization (if needed) of analytes into a form more suitable for detection or separation, and finally, providing a robust and reproducible method independent of variations in the sample matrix. More recently, new objectives have emerged, including smaller sample sizes, higher selectivity, automation compatibility, and minimization of the amount of glassware, and organic solvents used [1]. Traditional liquid−liquid extraction does not fulfill current requirements and has been displaced from laboratory practice by new extraction techniques, such as solid-phase extraction (SPE). Several sorbents with different properties are commercially available, and it is usually possible to find a description of an adequate SPE procedure for nearly all common target compounds. However, these sorbents are not highly selective and retain other components from the matrix along with the target compounds, which might hinder their quantification in the determinant step.

Molecularly imprinted polymers (MIPs) are tailor-made synthetic materials with artificially generated recognition sites able to specifically rebind a target compound in preference to other closely related compounds. As shown in Fig. 8.1, these materials are obtained by polymerizing functional and crosslinking monomers around a template molecule, leading to a highly crosslinked three-dimensional network polymer. The monomers are chosen based on their capability to interact with the functional groups of the template molecule. Once polymerization is complete, the template molecule is extracted, leaving binding sites with shape, size, and functionalities within the polymer network complementary to the target compound. The resulting imprinted polymers are stable, robust, and resistant to a wide range of pH, solvents, and temperature.

**Solid-Phase Extraction. https://doi.org/10.1016/B978-0-12-816906-3.00008-X**

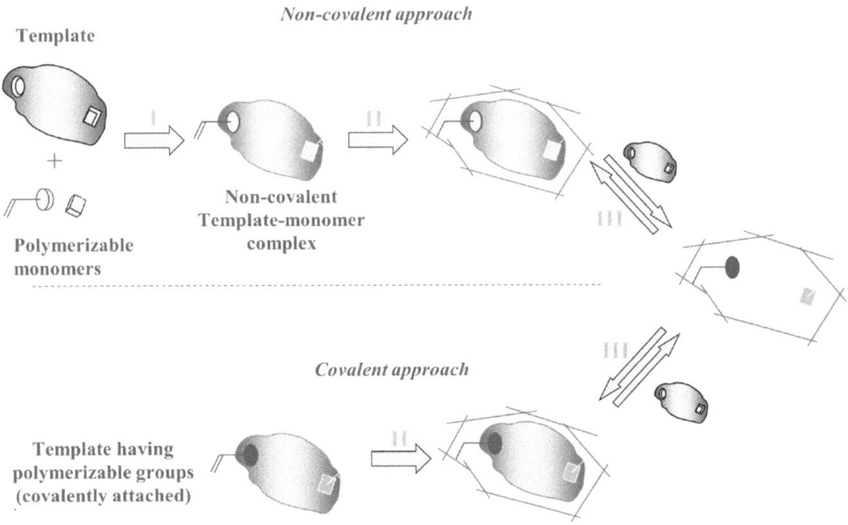


**Figure 8.1** Preparation of MIPs.

Therefore, the behavior of MIPs emulates the interactions established by natural receptors to selectively retain a target molecule (i.e., antibody–antigen) but without the associated stability limitations. In addition, the synthesis of MIPs is relatively simple and inexpensive, providing a clear alternative to the use of natural receptors.

Three general approaches have been described for the synthesis of MIPs, namely, covalent, noncovalent, and semi-covalent approaches. The covalent approach [2] involves the formation of reversible covalent bonds between the template molecule and monomers before polymerization. Then, the template is removed from the polymer by cleavage of the corresponding covalent bonds, which are reformed upon rebinding of the target compound. This approach leads to a polymer with a rather homogenous population of binding sites, minimizing the existence of nonspecific sites, due to the high stability of template-monomer interactions. However, the covalent approach is rather restrictive due to the difficulty of designing an appropriate template-monomer complex in which covalent bond formation and cleavage are readily reversible under mild conditions. The semi-covalent approach offers an alternative intermediate approach [3,4]. In this case, the template is also covalently bound to a functional monomer, but the template rebinding is based only on noncovalent interactions. Finally, the noncovalent approach was introduced by Arshady and Mosbach [5] and is based on the formation of relatively weak noncovalent interactions (i.e., hydrogen bonds, ionic interactions, etc.) between the template molecule and selected monomers before polymerization. This approach is by far the most common for the preparation of MIPs owing to its simplicity and the availability of a wide variety of monomers.

Regardless of how a MIP is prepared, it is evident that they are ideal materials for selective sample treatment prior to the determination of target compounds. Thus, their use in SPE, so-called *molecularly imprinted solid-phase extraction* (MISPE), is by far the most advanced technical application of MIPs. There is a large and expanding

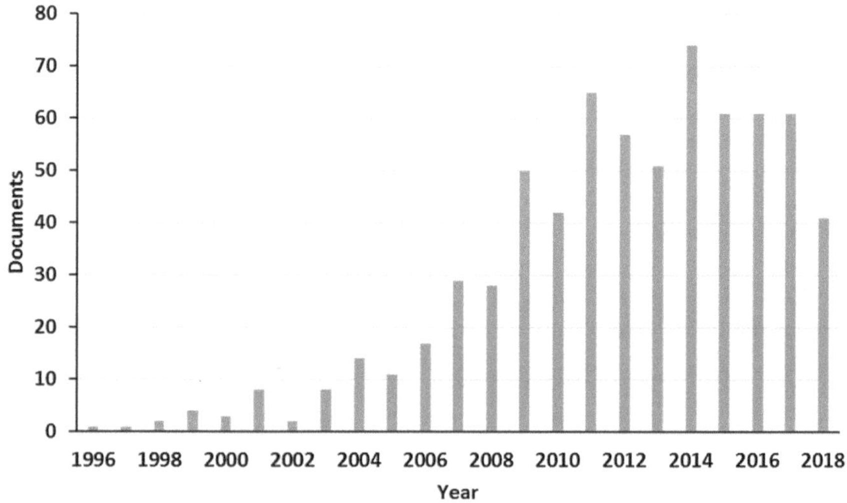


**Figure 8.2** Number of publication on molecularly imprinted solid-phase extraction (MISPE) from 1996 to 2018 according to the Scopus database using the terms "molecularly imprinted solid phase extraction" or "MISPE."

research activity in this area, evidenced by the number of papers published per year during the period 1996–2018, Fig. 8.2.

# 8.2 Preparation of molecularly imprinted polymers

## *8.2.1 General considerations*

The proper selection of the template, monomer(s), crosslinker, and solvent (porogen) is key for the synthesis of a suitably selective MIP. Each variable has a strong influence on the overall performance of a MIPs in terms of affinity, selectivity, and loading capacity, etc. Their proper selection guarantees to a larger extent that MIPs with the appropriate properties are obtained for a particular application.

### *8.2.1.1 Template and monomer(s)*

The first step in the preparation of MIPs consists of prearranging the template and the monomer(s) in a solvent. Monomer selection depends upon the characteristics of the target compound. The template molecule must contain functional groups capable of interacting with the monomer(s) to form a stable complex. For template molecules containing acidic groups, monomers with basic functional groups (e.g., 2- or 4-vinylpyridine [VPy], diethylaminoethylmethacrylate [DEAEMA]) are preferred, whereas acidic monomers (e.g., methacrylic acid [MAA], trifluoromethylacrylic acid [TFM], itaconic acid [ITA]) are used to target bases. For carboxylic acids and amides, high selectivity was observed with primary amide-containing monomers (e.g., methacrylamide [MAAM]). Other neutral solvating monomers that commonly enhance the imprinting effect are N-vinylpyrrolidone (NVP) and hydroxyethylmethacrylate (HEMA). The structures for some common monomers are indicated in Fig. 8.3.

It is important to note that since the template-monomer interactions are governed by an equilibrium process, a relatively large amount of monomer is used in order to displace the equilibrium to form the template-monomer complex. In general, a template-monomer molar ratio of 1:4 provides suitable stability to the template-monomer complex, assuring the desired imprint effect. However, since the excess of free monomers leads to the formation of nonspecific binding sites, the loading, washing, and elution conditions for SPE must be correctly selected, as described later.

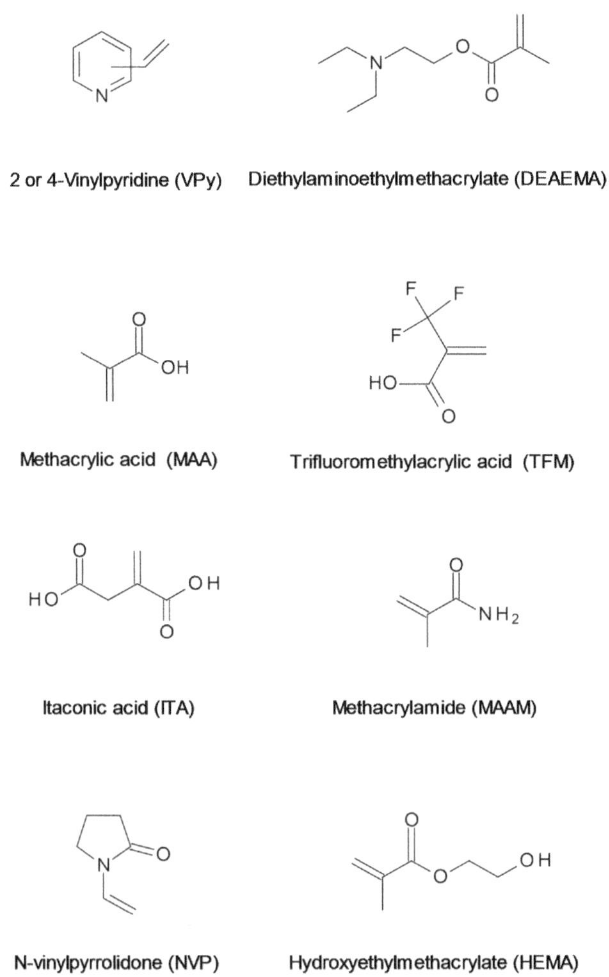


**Figure 8.3** Structures of monomers commonly used in molecular imprinting.

The solvent used during the prepolymerization step is also of prime importance since it has a direct influence on the strength of the template-monomer interaction. In general, solvents with a low dielectric constant, such as chloroform and toluene, are favored to stabilize hydrogen bonding and/or electrostatic interactions between monomer(s) and templates. Solvents with higher dielectric constants (e.g., acetonitrile) have also been used, but usually, the polymers obtained show a lower affinity to rebind the template. On the other hand, protic solvents, such as water and methanol, are not recommended, as they disrupt the template-monomer hydrogen-bonding interactions. Besides, apart from the above considerations, the template has to exhibit moderate to high solubility in the polymerization medium, As a consequence, in some cases, a compromise between the necessary solubility of the template and the strength of template-monomer interactions must be reached.

Finally, the template size and shape has a strong influence on the selectivity of the MIPs, although it is not possible to establish a general rule for all systems. In general, slight structural differences near the functional group responsible for the interaction with the monomer lead to highly selective polymers preventing the binding of structurally related compounds [6,7]. However, in some cases, the absence or presence of groups far from the functional groups binding with the target compound has also allowed the preparation of highly selective imprinted polymers [8].

#### 8.2.1.2 Crosslinker

The presence of a crosslinker is necessary to guarantee the stability of the template-monomer complex during polymerization as well as to increase polymer porosity. It was shown that at least 50% of the total monomer in an MAA-ethylene glycol dimethacrylate (EDMA) system has to be EDMA; otherwise, no recognition can take place [9]. It is important to stress that the presence of a crosslinker not only preserves the binding sites but also has a direct influence on the physical and chemical properties of the polymer matrix. From this perspective, EDMA is the crosslinker most often used in methacrylate-based systems, since it provides mechanical and thermal stability, good wettability, and rapid mass transfer. Other crosslinkers, such as divinylbenzene (DVB) or trimethylolpropane trimethacrylate (TRIM), have also been used showing a similar performance to EDMA for certain applications.

It is important to stress that the porogen (the solvent), together with the crosslinker, influences the polymer morphology in terms of the specific surface area and pore diameter. In general, a low surface area and low macroporosity are not desirable since they might lead to slow diffusion of the analyte to sites located in micro-pores, which reduces template recognition. In other words, a MIP with an inadequate morphology might prevent template recognition even when a solvent capable of stabilizing the template-monomer complex during the pre-polymerization step was employed.

### 8.2.2 Optimization of MIP formulations

As discussed above, several variables (kind and amount of monomer or nature of crosslinker and solvent) affect the final characteristics of a MIP in terms of capacity, affinity, and selectivity for the target compound. Thus, the identification of the optimum formulation for a particular MIP to be used for SPE might take several weeks of trial-and-error experiments. However, in most of the reports, such exhaustive optimization is rarely carried out, and certain standard formulations (e.g., 1:4:20 template:monomer:crosslinker molar ratio) are typically used [10]. For most applications, the MIPs present good affinity and selectivity for target compounds, but it is likely better performance would have been obtained if the MIP formulations were properly optimized. Therefore, some approaches dealing with the optimization of MIP formulations in a simple, fast, and rational way have been proposed to improve molecular recognition capabilities.

#### 8.2.2.1 Computational approach

In this approach, uses molecular modeling software to design and screen a virtual library of monomers against the desired template. In this manner, it is possible to predict template-monomer interaction positions by calculating binding energies, facilitating the selection of the best functional monomer. Following this methodology, polymers with high binding capacity and selectivity have been obtained for different analytes [11−13]. In spite of the evident advantages of computational approaches in MIP design, its practice is far from routine. This situation might change in the future due to MIRATE (MIps RATional dEsign, http://mirate.di.univr.it/), a new open access online resource for the virtual design and refinement of molecularly imprinted polymer binding sites, described by Busato et al. [14], which does not require specific knowledge of molecular modeling techniques.

#### 8.2.2.2 Combinatorial approach

The combinatorial approach [15,16] consists of the preparation of a large number of polymers in HPLC vials as small monoliths (mini-MIPs) as shown in Fig. 8.4. Then, the mini-MIPs are screened by determining the template release after incubation in the presence of a suitable solvent and by rebinding experiments. Due to the variety of parameters influencing the molecular recognition properties of MIPs, this approach can be a useful tool for facilitating the rapid screening and optimization of MIP formulations. This methodology was successfully employed for the selection of the best MIP formulation for different analytes, such as triazines [15,16], sulphonylurea herbicides [17], and bisphenol A [18], among others.

It is important to note that the screening of mini-MIPs is restricted to equilibrium rebinding experiments, which differ from the nonequilibrium process that takes place in SPE. Besides, the fate of a mini-MIP (monolith) might differ from that of a polymer obtained by bulk polymerization after crushing and sieving. Thus, these limitations have to be considered, since it is not clear as to how well the results obtained from the screening of mini-MIPs can be extrapolated to typical SPE conditions.

#### 8.2.2.3 Polymerization strategies

Bulk polymerization was the first strategy used to synthesize imprinted polymers, requiring grinding and sieving of the polymer mass to the desired particle size.

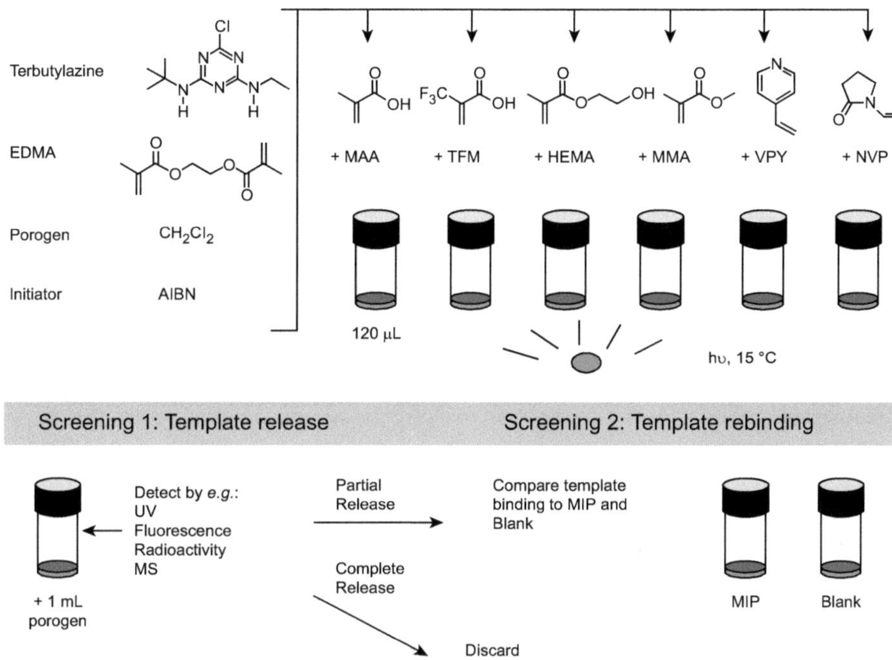


**Figure 8.4** Scheme for the synthesis of mini-MIPs and subsequent screening of their recognition properties.
Reprinted from Lanza F, Sellergren B. Method for synthesis and screening of large groups of molecularly imprinted polymers. Anal Chem 1999;71:2092–6, Copyright (1999) with permission from the American Chemical Society.

This process is tedious and time-consuming, and the particles obtained possessed a heterogeneous particle size distribution with poor binding site accessibility for the target compound. Although these polymers can be useful for most SPE applications, their drawbacks prevent their large-scale production. Accordingly, new polymerization methodologies to obtain MIP beads with proper physical characteristics (size, porosity, pore volume, surface area) have been developed over the last few years. Among these new techniques, multistep swelling and polymerization [19], precipitation polymerization [20] and polymerization into the pores of silica beads, with the latter etched away with $NH_4HF_2$ after polymerization [21], have been used for the preparation of MIPs for SPE. Of these, precipitation polymerization is the most straightforward strategy for the rapid preparation of MIP in high yield. Simply, this method consists of carrying out the polymerization of monomers in a large amount of a suitable solvent (~4%–6% w/v). However, it is important to stress that the compatibility between the crosslinker and the porogen (solvent) is of vital importance for the success of this methodology; otherwise, agglomerates instead of independent beads might be obtained. In any case, the simplicity of the experimental procedure and the elimination of crushing and sieving steps, which eases large-scale production, justify the use of this methodology for the preparation of MIPs for SPE.

#### 8.2.2.4 Template bleeding

Once the MIP has been prepared, regardless of the used polymerization strategy, the template must be removed in order to obtain free binding sites. Template removal is usually carried out by washing the polymer repeatedly with a solvent capable of disrupting the template-monomer interactions or by Soxhlet extraction. However, it is known that even after exhaustive washing by different approaches (thermal annealing, microwave assisted extraction, Soxhlet extraction, and supercritical fluid template desorption) [22], traces of template remain in the polymer network. Consequently, some template leakage might be detected during the elution step in MISPE protocols, although in most applications, no template bleeding was observed at the concentration levels of the study. Thus, although template bleeding exits, it is not necessarily a major problem for typical applications.

Nevertheless, minimizing template bleeding by using an analog of the target compound as a template is recommended. In this manner, the bleeding of the template does not interfere in the quantification of the target compound. The use of a closely related compound to target compound as template was originally proposed by Andersson et al. [23] for the extraction of sameridine from human plasma. Later, this approach was followed by others allowing the quantification of different analytes in a precise and accurate manner.

In spite of such success, template-analog imprinted polymers possess an inferior molecular recognition capability compared to MIPs prepared using the analyte as a template. The best way to improve the molecular recognition capability is to use a template analog whose shape and functionality are as similar to those of the target compound as possible. An interesting alternative is the use of a stable isotope labeled compound as the template molecule, so-called *isotope imprinting*. This approach was originally proposed by Sambe et al. [24] for the preparation of a restricted access material-MIP (RAM-MIP) for bisphenol A-d16 (BPAd16). The molecular recognition ability of this polymer for BPA, BPA-d6, BPA-d16, and related analogs was evaluated for the direct injection of BPA in serum by column-switching LC−MS. In this case, although template bleeding did occur, BPA was detected by selective ion monitoring. Obviously, although the potential of isotope imprinting to overcome the template bleeding problem is evident, its applicability is limited by the availability of the corresponding isotopic labeled compound analog and the use of MS for detection.

## 8.3 Molecularly imprinted solid-phase extraction (MISPE)

### 8.3.1 Off-line protocols

Typically, as in other SPE procedures, a small amount (15−500 mg) of imprinted polymer particles is packed into conventional polyethylene cartridges. Then, the MIP cartridge is conditioned with an appropriated solvent, and the sample is loaded. After a washing step for removal of nonspecifically bound matrix components, the analytes are eluted, ideally free of coextractives. The sample is usually loaded onto the MIP cartridge in a low-polarity solvent (e.g., acetonitrile, dichloromethane, toluene), which in most cases, is the same solvent as the porogen for the MIP synthesis to maximize media specific interactions. The target compounds are eluted with a solvent capable of disrupting the typical noncovalent interactions between the target compounds and the imprinted polymer.

Nevertheless, aqueous samples can also be loaded directly onto MIP cartridges [25]. In this case, MIPs behave like a reversed-phase sorbent controlled by nonspecific interactions between the target compounds and matrix components and the polymer matrix. Then, during the washing step, matrix components are removed using an appropriate solvent capable of redistributing nonspecifically bound analytes to the selective binding sites. Unfortunately, the success of this procedure is not always achieved promoting interest in the synthesis of water-compatible MIPs with recognition abilities in aqueous media. This is achieved by incorporating hydrophilic surface properties to the polymer to reduce nonspecific hydrophobic interactions. Water-compatible MIPs have been prepared using polar porogens [26], hydrophilic comonomers (e.g., 2-hydroxyethyl methacrylate, acrylamide) or cross-linkers (e.g., pentaerythritoltriacrylate, methylenebis(acrylamide)) [27−29] and/or specially designed monomers capable of stoichiometric interactions with the template functional groups [30,31]. The proposed water-compatible MIPs provided recognition of target compounds to a certain extent in aqueous media and further development in this field can be expected in coming years.

As mentioned above, the inclusion of a washing step is necessary to maximize the specific interactions between the target compound and the imprinted polymer with the simultaneous elution of nonspecifically bound interfering compounds. The presence of nonspecific binding sites is attributed to the high amount of monomer used during MIP synthesis by the noncovalent approach. Consequently, the excess of free monomers, randomly incorporated in the polymer matrix leads to the formation of nonselective binding sites. These nonspecific binding sites can lead to the coextraction of some matrix components which can interfere in the chromatographic determination of target compounds after MISPE, even after an exhaustive washing step. Besides, the loss of MIP performance due to matrix compounds strongly bound to the polymeric matrix has been observed [32,33]. These drawbacks can be circumvented by introducing a previous clean-up step using a nonimprinted polymer (NIP), a polymer prepared in the same conditions as the corresponding MIP but in the absence of the template molecule, so-called *two-step* MISPE, applied successfully to the extraction of triazines and phenylurea herbicides from vegetable extracts [33,34]. Briefly, in this approach, the sample is loaded onto the NIP with nonspecific retention of both target compounds and matrix components. Then, during the washing steps, the target compounds (and a small amount of interfering compounds) are eluted, while the majority of matrix compounds remain bound to the NIP. Subsequently, the washing solution is loaded onto the corresponding MIP for further cleanup. In addition, by this approach the MIP is protected from high loading of matrix compounds, thus increasing its durability and preservation of its recognition properties.

In spite of the success of the two-step MISPE approach, the best way of reducing nonspecific interactions is by using imprinted polymers synthesized by the covalent or semi-covalent approaches. This statement was nicely demonstrated by Cacho et al. [35] for the MISPE of triazines from soil and vegetable extracts using a semi-covalent imprinted polymer prepared with propazine methacrylate as a template molecule. Compared to a noncovalent MIP using propazine as a template, a markedly lower nonspecific binding of matrix components was observed leading to lower detection limits, much lower than those obtained using a noncovalent MIP.

### 8.3.2 Online protocols

The first online MISPE procedure coupled to HPLC was proposed by Masqué et al. [36] for the selective extraction of 4-nitrophenol from a mixture of phenolic compounds in river water samples. In general, in this format, a small precolumn packed with the imprinted polymer (~25–50 mg) is placed in the loop of a six-port injection valve. After loading the sample and washing out interfering compounds, the target compounds are eluted by the mobile phase and then separated by liquid chromatography. However, in most cases, there is poor compatibility of the mobile-phase required for the chromatographic separation and the elution solvent required to desorb the target compounds from the MIP pre-column. Consequently, the number of reports utilizing online MISPE is scarce. Alternatively, different online MISPE-HPLC systems, as illustrated in Fig. 8.5, can be employed to direct the elution solvent from the MIP precolumn to the injection loop (Fig. 8.5A) and subsequently injected into the chromatographic system [37] or by mixing of the elution solvent with an aqueous-rich solvent (Fig. 8.5B) before reaching the separation column [38]. Both approaches increase the complexity of the online system requiring the incorporation of additional instrument components (e.g., pumps). Besides, in online systems, MIPs must provide adequate permeability and fast mass transfer of analytes, requiring the synthesis of MIPs with improved performance characteristics. In this regard, molecularly imprinted polymer filaments (MIPFs) as the extraction phase have been proposed, providing improved permeability compared with monolithic or particle-packed columns with finely dispersed sorbents [39]. Besides, the small internal diameter tubing used could produce high linear flow rates at the surface of the coatings, resulting in the fast mass transfer of target compounds between the solid and liquid phases.

### 8.3.3 In-line protocols

The direct coupling of a MIP column in-line with the detection system might allow the extraction, enrichment, separation, and determination of target compounds in a single step owing to the high selectivity of the MIPs. The very first in-line MISPE protocol coupled with UV-detection was described by Sellergren for the determination of pentamidine in urine samples [40]. Dilute urine, 100 mL, was loaded on a column packed with MIP particles, washed with 100 mL of a buffer solution (pH 9), and the target compound pentamidine eluted in ~1.5 mL of buffer at pH 3 with UV

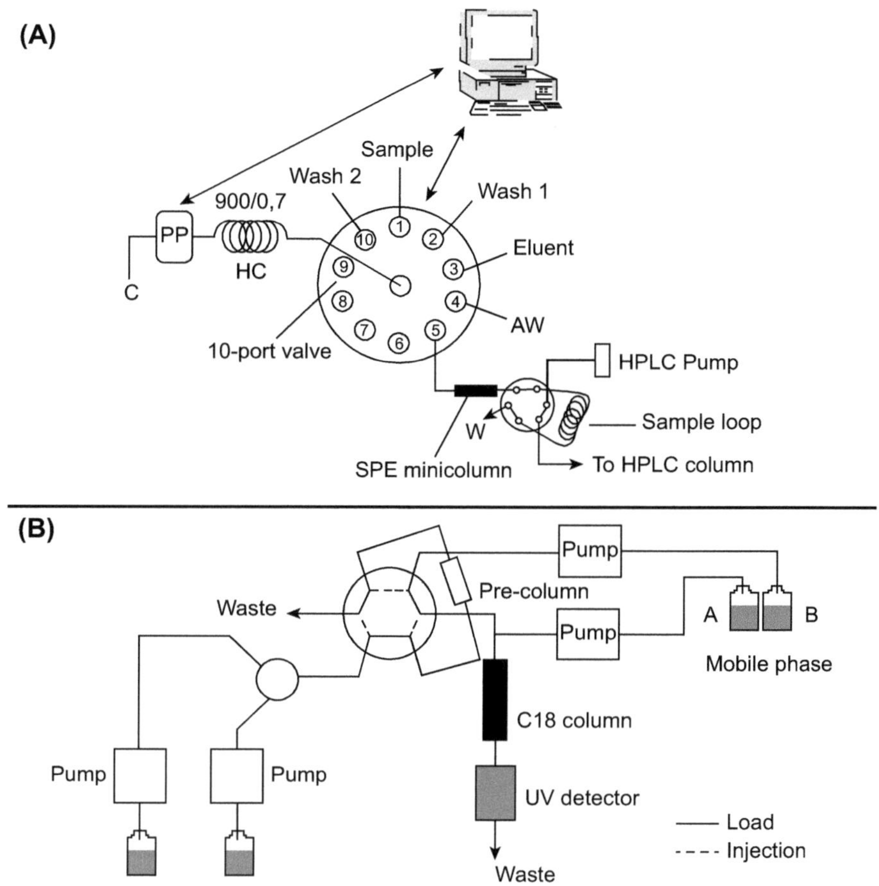


**Figure 8.5** Alternative online MISPE-HPLC set-ups.
Reprinted from Theodoridis G, Zacharis C, Tzanavaras P, Themelis D, Economou A. Automated sample preparation based on the sequential injection principle: solid-phase extraction on a molecularly imprinted polymer coupled on-line to high-performance liquid chromatography. J Chromatogr A 2004;1030:69–76, Copyright (2004), and Ou J, Hu L, Hu L, Li X, Zou H. Determination of phenolic compounds in river water with on-line coupling bisphenol A imprinted monolithic precolumn with high performance liquid chromatography. Talanta 2006;69:1001–6, Copyright (2006) with permission from Elsevier.

detection at 270 nm. However, a large volume of washing solvent was necessary to remove a high amount of matrix compounds retained by the MIP column by hydrophobic interactions, requiring a long analysis time.

Later, since the recognition capabilities of MIPs are enhanced in the organic media, Turiel et al. developed a method for the determination of the fungicide thiabendazole (TBZ) in acetonitrile-based extracts of fruit samples [41]. In this case, the extracts (50 μL) were directly injected onto a column packed with TBZ-imprinted polymer beads using methanol as a mobile-phase. After 2.4 min, the mobile-phase was switched to a methanol-acetic acid mixture (80:20, v/v) in 0.1 min, maintaining these conditions constant for 5 min before returning to the initial conditions. Due to the high selectivity of the MIP column, the target compound was retained on the column, while interferences were rapidly eluted, facilitating the quantification of TBZ in less than 15 min.

Peak broadening and peak tailing phenomena are typically observed in the elution profiles of in-line MISPE procedures due to slow mass transfer and the existence of a heterogeneous binding-site distribution, even for imprinted beads. As an alternative, surface imprinting, the formation of a thin film of imprinted polymer grafted onto the surfaces of preformed beads, may solve some of these drawbacks. MIP microspheres with core-shell morphology and narrow particle-size distributions were proposed for in-line MISPE of TBZ from citrus fruits and orange juice samples [42,43]. By using core-shell particles, a clear improvement in the chromatographic performance was achieved, demonstrating that the slow mass-transfer kinetics associated with conventional MIP stationary phases can be dramatically improved by surface-imprinting techniques. As an example, Fig. 8.6 shows the chromatogram obtained after the direct injection of a nonspiked orange juice sample onto a column packed with core-shell TBZ-imprinted microspheres. The core-shell MIP column permitted retention of the target analyte while the interferences were eluted rapidly.

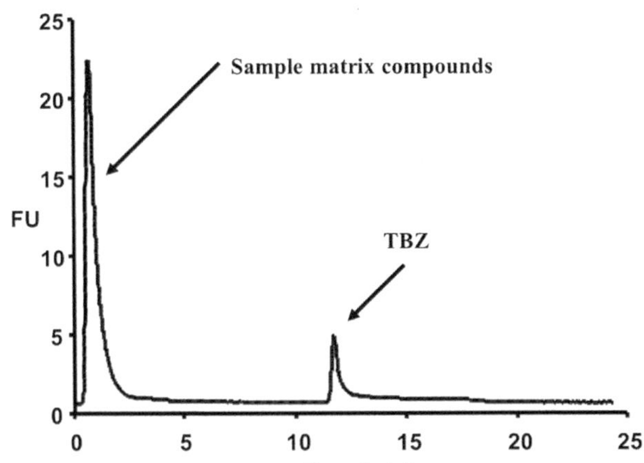


**Figure 8.6** Chromatogram for the direct injection of a nonspiked orange juice sample onto a column packed with core-shell TBZ-imprinted microspheres. Chromatographic conditions were as follows: mobile phase consisting of acetonitrile, 2 min; mobile phase consisting of acetonitrile-methanol (95:5), 8 min; mobile phase consisting of methanol-acetic acid (50:50), 5 min; linear gradient to initial conditions, 5 min; conditioning under initial conditions, 5 min. Reprinted from Barahona F, Turiel E, Cormack PAG, Martin-Esteban A. Synthesis of core-shell molecularly imprinted polymer microspheres by precipitation polymerization for the in-line molecularly imprinted solid-phase extraction of thiabendazole from citrus fruits and orange juice samples. J Sep Sci 2011;34:217–24, Copyright (2011) with permission from John Wiley & Sons.

In the above in-line MISPE procedures, analyte elution is performed by a step solvent-switch. Alternatively, pulsed elution (MISPE-PE), in which a small volume of elution solvent is used, was proposed for the determination of theophylline in chloroform diluted serum samples [44]. In this first work, samples were injected onto a theophylline-imprinted polymer packed into a stainless steel column using chloroform as a mobile phase. After interfering compounds were washed out from the column, theophylline was eluted free of coextracted interferents by injection of 20 μL of methanol and determined spectrophotometrically at 270 nm. Later, MISPE-PE evolved to MISPE with differential pulsed elution (MISPE-DPE), in which successive 20 μL pulses of different solvents are applied to wash out interfering compounds and the analyte fraction [45]. It is important to note that analyte enrichment through injection of large volumes of sample is possible, providing a high sensitivity due to the narrow band obtained by pulsed elution.

### 8.3.4 Improved batch protocols

Batch MISPE was first used for the extraction of sameridine, where the MIP particles were incubated for 1 h with spiked human plasma. After the washing step, sameridine was eluted for quantification by gas chromatography [23]. However, such protocols require additional centrifugation and/or filtration steps and have largely been displaced by the conventional off-line SPE methods. However, batch MISPE is experiencing a rebirth, with the use of MIP-modified magnetic nanoparticles, since after extraction, such particles can be separated from the matrix by a magnet, thereby avoiding tedious filtration and/or centrifugation steps. Several procedures have been described for the synthesis of magnetic core-shell MIP particles. In general, magnetite nanoparticles are initially encapsulated in silica through a typical sol-gel reaction using tetraethyl orthosilicate (TEOS) forming hybrid $Fe_3O_4@SiO_2$ particles. Polymerizable double bonds, necessaries for the subsequent surface imprinting, are introduced on particle surface by the reaction of $Fe_3O_4@SiO_2$ particles with suitable reagents. Finally, the MIPs are formed on the surface of $Fe_3O_4@SiO_2$ by the copolymerization of vinyl end groups with the typical polymerization mixtures used in molecular imprinting. Once the magnetic MIP beads are obtained, they can be used in batch MISPE procedures. This approach was used for the extraction of various analytes in different samples, such as tetracycline antibiotics from egg and tissue [46], microcystins in environmental waters [47], and triazines in soils [48], among others. For further details, ref. [49] is recommended.

Micro solid-phase extraction (μ-SPE) can be considered a variant of batch SPE, which consists of retaining a sorbent material in a poly(propylene) (PP) flat-sheet membrane envelope. Target analytes diffuse freely through the membrane pores and are retained by the sorbent material while other matrix compounds are hindered by the membrane. For improving selectivity for the extraction of phenolic compounds from environmental water samples, MIP particles were employed as the sorbent material in the so-called *molecularly imprinted polymer-based μ-SPE* (MIP-μ-SPE) [50] and for the extraction of ochratoxin A from coffee, grape juice, and urine [51], and cocaine and its metabolites from human urine [52].

Alternatively, the use of a poly(propylene) capillary instead of a flat-sheet membrane envelope has been proposed to reduce the amount of MIP particles and the consumption of organic solvents [53,54]. Although few papers have been published so far, the growing interest in MI-μ-SPE suggests that this should change in the near future driven by the simplicity of operations.

## 8.4 Selected applications

Most of the reports highlight the great selectivity provided by MIPs, but few properly report the advantages of MISPE compared to the traditional methods using nonselective sorbents. A notable exception is a work by Andersson et al. [55] comparing MISPE with mixed-mode SPE for the analysis of benzodiazepines in postmortem hair samples. First, hair samples were washed and then incubated in methanol-25% ammonium hydroxide solution (20:1) overnight. The extract was then reconstituted in a buffer for SPE or toluene for MISPE. After extraction and cleanup, sample extracts were reconstituted in the mobile phase and analyzed by LC-MS/MS. The MISPE method provided higher recovery with lower limits detection limits for diazepam while the SPE method failed to detect it. The authors conclude that not only did MISPE provide superior extraction performance but was also a simpler and more time-efficient procedure.

Another significant contribution to the analysis of environmental wastewater samples is the analysis of β-blockers by liquid chromatography−quadrupole-linear ion trap mass spectrometry [56]. In this work, the performance of a commercially available β-blockers-receptor MIP was compared with a generic mixed-mode SPE phase (Oasis HLB). Recoveries of β-blockers by both sorbents were similar, but the MISPE method required a significantly smaller sample volume (25 mL) compared with conventional SPE (100−200 mL). As a consequence, a reduced matrix effect on the separation and detection was observed. It was found that the extracts from MIP resulted in minimal signal suppression, whereas those from Oasis HLB showed significant suppression of the analyte signal.

In general, the effective clean-up achieved by MISPE are associated with considerable savings in time and costs. This was clearly demonstrated by Mohamed et al. [57] for the determination of chloramphenicol in milk (raw milk, skimmed milk, and milk powder). The advantages of a commercial MIP were evaluated by comparing its performance with that from a standard generic SPE method. The traditional approach involved several steps, including protein precipitation, SPE cleanup, and three liquid-liquid extraction steps. However, the MIP-based method only required a prior centrifugation step. Obviously, it allowed a significant increase in sample throughput (18 samples processed within 3 h compared with 8 h with the classical approach). In addition, the generic SPE method required LC-electrospray ionization (ESI)-MS/MS to achieve the required sensitivity and selectivity, whereas the MISPE protocol allowed the use of LC-UV or fluorescence detectors to achieve the same performance.

A further example of the superior performance of MIPs in food analysis was demonstrated for the extraction of fluoroquinolones from baby food [58]. In this work, the target compounds were isolated by ultrasound-assisted extraction, and the extracts loaded onto a ciprofloxacin-imprinted polymer or a strong anion-exchange resin. After, the corresponding washing steps, the target compounds were eluted and for analysis by HPLC-UV. An example of typical chromatograms for the analysis of spiked and unspiked baby food samples is shown in Fig. 8.7 for conventional SPE using strong anion-exchange cartridges (Fig. 8.7A and B) and by MISPE (Fig. 8.7C and D). A clean baseline was obtained using MISPE allowing an-order-of magnitude reduction of detection limits compared with conventional SPE using strong anion-exchange cartridges.

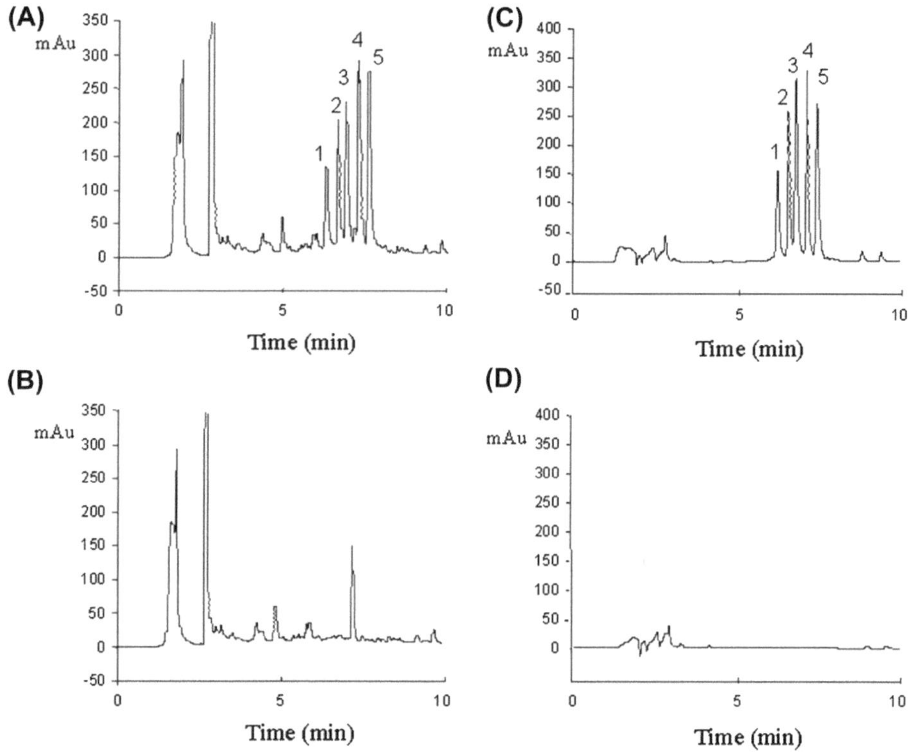


**Figure 8.7** Chromatograms obtained after ultrasound-assisted extraction of baby-food samples and clean-up using strong anion exchange (A and B) and MIP cartridges (C and D). (A and C) Sample spiked at 1 $\mu g\ g^{-1}$; (B and D) unspiked sample. Peak assignment: 1 = Enoxacin; 2 = Norfloxacin; 3 = Ciprofloxacin; 4 = Danofloxacin; 5 = Enrofloxacin.
Reproduced from Díaz-Alvarez M, Turiel E, Martín-Esteban A. Selective sample preparation for the analysis of (fluoro) quinolones in baby food: molecularly imprinted polymers versus anion-exchange resins. Anal Bioanal Chem 2009;393:899−905, Copyright (2009) with permission from Springer.

## 8.5 Conclusions

MIPs are excellent sorbents for customized sample preparation by SPE. MIPs are capable of cleaning up complex samples facilitating the selective isolation of target compounds in a rapid and robust manner. In general, the selectivity provided by MIPs allows lower detection limits to be obtained with conventional detectors (spectrophotometers, spectrofluorimeters, etc.)

Although MISPE can be considered a robust technique, some improvements remain necessary, especially those related to the selection of polymerization mixtures, industrial scale-up through alternative polymerization strategies, and the development of water-compatible sorbents, among others. The development of new SPE formats with improved mass-transfer characteristics and the coupling of MIPs to other extraction techniques are expected in the near future. The development of micro-MISPE devices is likely to be one of the main areas of activity in the coming years.

## References

[1] Smith RM. Before the injection – modern methods of sample preparation for separation techniques. J Chromatogr A 2003;1000(1–2):3–27.

[2] Wulff G, Sarchan A. The use of polymers with enzyme-analogous structures for the resolution of racernates. Angew Chem Int Ed Engl 1972;11. 341–341.

[3] Sellergren B, Andersson LI. Molecular recognition in macroporous polymers prepared by a substrate analog imprinting strategy. J Org Chem 1990;55(10):3381–3.

[4] Whitcombe MJ, Rodriguez ME, Villar P, Vulfson E. A new method for the introduction of recognition site functionality into polymers prepared by molecular imprinting: synthesis and characterization of polymeric receptors for cholesterol. J Am Chem Soc 1995;117(27): 7105–11.

[5] Arshady R, Mosbach K. Synthesis of substrate-selective polymers by host-guest polymerization. Makromol Chem 1981;182:687–92.

[6] Sellergren B, Lepistö M, Mosbach K. Highly enantioselective and substrate-selective polymers obtained by molecular imprinting utilizing noncovalent interactions. NMR and chromatographic studies on the nature of recognition. J Am Chem Soc 1988;110(17): 5853–60.

[7] Ramström O, Yu C, Mosbach K. Chiral recognition in adrenergic receptor binding mimics prepared by molecular imprinting. J Mol Recognit 1996;9:691–6.

[8] Martín-Esteban A, Turiel E, Stevenson D. Effect of template size on the selectivity of molecularly imprinted polymers for phenylurea herbicides. Chromatographia 2001;53: S434–7.

[9] Sellergren B. Molecular imprinting by noncovalent interactions. enantioselectivity and binding capacity of polymers prepared under conditions favoring the formation of template complexes. Macromol Chem 1989;190:2703–11.

[10] Batra D, Shea KJ. Combinatorial methods in molecular imprinting. Curr Opin Chem Biol 2003;7:434–42.

[11] Cowen T, Karim K, Piletsky S. Computational approaches in the design of synthetic receptors – a review. Anal Chim Acta 2016;936:62–74.

[12] Marć M, Kupka T, Wieczorek PP, Namieśnik J. Computational modeling of molecularly imprinted polymers as a green approach to the development of novel analytical sorbents. Trends Anal Chem 2018;98:64–78.

[13] Nicholls IA, Olsson GD, Karlsson BCG, Suriyanarayanan S, Wiklander JG. Theoretical and computational strategies in molecularly imprinted polymer development. In: Kutner W, Sharma PS, editors. Molecularly imprinted polymers for analytical chemistry applications. London: RSC; 2018. p. 197–226.

[14] Busato M, Distefano R, Bates F, Karim K, Bossi AM, López-Vilariño JM, Piletsky S, Bombieri N, Giorgetti A. MIRATE: MIps RATional dEsign science gateway. J Integra Bioinform 2018;15(4). https://doi.org/10.1515/jib-2017-0075.

[15] Lanza F, Sellergren B. Method for synthesis and screening of large groups of molecularly imprinted polymers. Anal Chem 1999;71:2092–6.

[16] Takeuchi T, Fukuma D, Matsui J. Combinatorial molecular imprinting: an approach to synthetic polymer receptors. Anal Chem 1999;71:285–90.

[17] Zhu QZ, Haupt K, Knopp D, Niessner R. Molecularly imprinted polymer for metsulfuron-methyl and its binding characteristics for sulfonylurea herbicides. Anal Chim Acta 2002; 468:217–27.

[18] Martin-Esteban A, Tadeo JL. Selective molecularly imprinted polymer obtained from a combinatorial library for the extraction of bisphenol A. Comb Chem High Throughput Screen 2006;9:747–51.

[19] Hosoya K, Yoshizako K, Shirasu Y, Kimata K, Araki T, Tanaka N, Haginaka J. Molecularly imprinted uniform-size polymer-based stationary phase for high-performance liquid chromatography structural contribution of cross-linked polymer network on specific molecular recognition. J Chromatogr A 1996;728:139–47.

[20] Wang J, Cormack PAG, Sherrington DC, Khoshdel E. Monodisperse, molecularly imprinted polymer microspheres prepared by precipitation polymerization for affinity separation applications. Angew Chem 2003;115:5494–6.

[21] Abou-Hany RAG, Urraca JL, Descalzo AB, Gómez-Arribas LN, Moreno-Bondi MC, Orellana G. Tailoring molecularly imprinted polymer beads for alternariol recognition and analysis by a screening with mycotoxin surrogates. J Chromatogr A 2015;1425:231–9.

[22] Ellwanger A, Berggren C, Bayoudh S, Crecenzi C, Karlsson L, Owens PK, Ensing K, Cormack P, Sherrington D, Sellergren B. Evaluation of methods aimed at complete removal of template from molecularly imprinted polymers. Analyst 2001;126:784–92.

[23] Andersson LI, Paprica A, Arvidsson T. A highly selective solid phase extraction sorbent for pre-concentration of sameridine made by molecular imprinting. Chromatographia 1997;46(1–2):57–62.

[24] Sambe H, Hoshina K, Hosoya K, Haginaka J. Direct injection analysis of bisphenol A in serum by combination of isotope imprinting with liquid chromatography-mass spectrometry. Analyst 2005;130:38–40.

[25] Ferrer I, Lanza F, Tolokan A, Horvath V, Sellergren B, Horvai G, Barcelo D. Selective trace enrichment of chlorotriazine pesticides from natural waters and sediment samples using terbuthylazine molecularly imprinted polymers. Anal Chem 2000;72:3934–41.

[26] Sellergren B, Wieschemeyer J, Boos K, Seidel D. Imprinted polymers for selective adsorption of cholesterol from gastrointestinal fluids. Chem Mater 1998;10:4037–46.

[27] Dirion B, Cobb Z, Schillinger E, Andersson L, Sellergren B. Water-compatible molecularly imprinted polymers obtained via high-throughput synthesis and experimental design. J Am Chem Soc 2003;125:15101–9.

[28] Benito-Pena E, Martins S, Orellana G, Moreno-Bondi M. Water-compatible molecularly imprinted polymer for the selective recognition of fluoroquinolone antibiotics in biological samples. Anal Bioanal Chem 2009;393:235–45.

[29] Lv T, Yan H, Cao J, Liang S. Hydrophilic molecularly imprinted resorcinol–formaldehyde–melamine resin prepared in water with excellent molecular recognition in aqueous matrices. Anal Chem 2015;87:11084–91.

[30] Urraca J, Hall A, Moreno-Bondi M, Sellergren B. A stoichiometric molecularly imprinted polymer for the class-selective recognition of antibiotics in aqueous media. Angew Chem 2006;45:5158–61.

[31] Urraca J, Moreno-Bondi M, Hall A, Sellergren B. Direct extraction of penicillin G and derivatives from aqueous samples using a stoichiometrically imprinted polymer. Anal Chem 2007;79:695–701.

[32] Theodoridis G, Manesiotis P. Selective solid-phase extraction sorbent for caffeine made by molecular imprinting. J Chromatogr A 2002;948:163–9.

[33] Tamayo FG, Casillas JL, Martin-Esteban A. Clean up of phenylurea herbicides in plant sample extracts using molecularly imprinted polymers. Anal Bioanal Chem 2005;381: 1234–40.

[34] Cacho C, Turiel E, Martin-Esteban A, Perez-Conde C, Camara C. Clean-up of triazines in vegetable extracts by molecularly-imprinted solid-phase extraction using a propazine-imprinted polymer. Anal Bioanal Chem 2003;376:491–6.

[35] Cacho C, Turiel E, Martín-Esteban A, Ayala D, Pérez-Conde C. Semi-covalent imprinted polymer using propazine methacrylate as template molecule for the clean-up of triazines in soil and vegetable samples. J Chromatogr A 2006;1114:255–62.

[36] Masqué N, Marcé RM, Borrull F, Cormack PAG, Sherrington DC. Synthesis and evaluation of a molecularly imprinted polymer for selective on-line solid-phase extraction of 4-nitrophenol from environmental water. Anal Chem 2000;72:4122–6.

[37] Theodoridis G, Zacharis C, Tzanavaras P, Themelis D, Economou A. Automated sample preparation based on the sequential injection principle: solid-phase extraction on a molecularly imprinted polymer coupled on-line to high-performance liquid chromatography. J Chromatogr A 2004;1030:69–76.

[38] Ou J, Hu L, Hu L, Li X, Zou H. Determination of phenolic compounds in river water with on-line coupling bisphenol A imprinted monolithic precolumn with high performance liquid chromatography. Talanta 2006;69:1001–6.

[39] Zhong Q, Hu Y, Hu Y, Li G. Dynamic liquid-liquid-solid microextraction based on molecularly imprinted polymer filaments on-line coupling to high performance liquid chromatography for direct analysis of estrogens in complex samples. J Chromatogr A 2012;1241:13–20.

[40] Sellergren B. Direct drug determination by selective sample enrichment on an imprinted polymer. Anal Chem 1994;66:1578–82.

[41] Turiel E, Tadeo JL, Cormack PAG, Martin-Esteban A. HPLC imprinted-stationary phase prepared by precipitation polymerisation for the determination of thiabendazole in fruit. Analyst 2005;130:1601–7.

[42] Barahona F, Turiel E, Cormack PAG, Martin-Esteban A. Chromatographic performance of molecularly imprinted polymers: core-shell microspheres by precipitation polymerization and grafted MIP films via iniferter-modified silica beads. J Polym Sci Part A-Polym Chem 2010;48:1058–66.

[43] Barahona F, Turiel E, Cormack PAG, Martin-Esteban A. Synthesis of core-shell molecularly imprinted polymer microspheres by precipitation polymerization for the in-line molecularly imprinted solid-phase extraction of thiabendazole from citrus fruits and orange juice samples. J Sep Sci 2011;34:217–24.

[44] Mullett W, Lai EPC. Determination of theophylline in serum by molecularly imprinted solid-phase extraction with pulsed elution. Anal Chem 1998;70:3636–41.

[45] Mullett W, Lai EPC. Molecularly imprinted solid phase extraction micro-column with differential pulsed elution for theophylline determination. Microchem J 1999;61:143–55.

[46] Chen L, Liu J, Zeng Q, Wang H, Yu A, Zhang H, Ding L. Preparation of magnetic molecularly imprinted polymer for the separation of tetracycline antibiotics from egg and tissue samples. J Chromatogr A 2009;1216:3710–9.

[47] Pan S, Chen X, Li X, Cai M, Shen H, Zhao Y, Jin M. Double-sided magnetic molecularly imprinted polymer modified graphene oxide for highly efficient enrichment and fast detection of trace-level microcystins from large-volume water samples combined with liquid chromatography-tandem mass spectrometry. J Chromatogr A 2015;1422:1–12.

[48] Patiño-Ropero MJ, Díaz-Álvarez M, Martín-Esteban A. Molecularly imprinted core-shell magnetic nanoparticles for selective extraction of triazines in soils. J Mol Recognit 2017; 30:e2593.

[49] Ansari S. Application of magnetic molecularly imprinted polymer as a versatile and highly selective tool in food and environmental analysis: recent developments and trends. TrAC Trends Anal Chem 2017;90:89–106.

[50] Feng Q, Zhao L, Lin J. Molecularly imprinted polymer as micro-solid phase extraction combined with high performance liquid chromatography to determine phenolic compounds in environmental water samples. Anal Chim Acta 2009;650:70–6.

[51] Lee T, Saad B, Khayoon W, Salleh B. Molecularly imprinted polymer as sorbent in micro-solid phase extraction of ochratoxin A in coffee, grape juice and urine. Talanta 2012;88: 129–35.

[52] Sanchez-Gonzalez J, Tabernero M, Bermejo A, Bermejo-Barrera P, Moreda-Pineiro A. Porous membrane-protected molecularly imprinted polymer micro-solid-phase extraction for analysis of urinary cocaine and its metabolites using liquid chromatography – tandem mass spectrometry. Anal Chim Acta 2015;898:50–9.

[53] Díaz-Álvarez M, Barahona F, Turiel E, Martín-Esteban A. Supported liquid membrane-protected molecularly imprinted beads for micro-solid phase extraction of sulfonamides in environmental waters. J Chromatogr A 2014;1357:158–64.

[54] Díaz-Álvarez M, Martín-Esteban A. Hollow fiber membrane-protected molecularly imprinted microspheres for micro solid-phase extraction and clean-up of thiabendazole in citrus samples. J Chromatogr A 2018;1531:39–45.

[55] Anderson RA, Ariffin MM, Cormack PAG, Miller EI. Comparison of molecularly imprinted solid-phase extraction (MISPE) with classical solid-phase extraction (SPE) for the detection of benzodiazepines in post-mortem hair samples. Forensic Sci Int 2008; 174(1):40–6.

[56] Gros M, Pizzolato TM, Petrovic M, Lopez de Alda MJ, Barcelo D. Trace level determination of beta-blockers in waste waters by highly selective molecularly imprinted polymers extraction followed by liquid chromatography-quadrupole-linear ion trap mass spectrometry. J Chromatogr A 2008;1189:374–84.

[57] Mohamed R, Richoz-Payot J, Gremaud E, Mottier P, Yilmaz E, Tabet JC, Guy Philippe A. Advantages of molecularly imprinted polymers LC-ESI-MS/MS for the selective extraction and quantification of chloramphenicol in milk-based matrixes. Comparison with a classical sample preparation. Anal Chem 2007;79:9557–65.

[58] Díaz-Alvarez M, Turiel E, Martín-Esteban A. Selective sample preparation for the analysis of (fluoro) quinolones in baby food: molecularly imprinted polymers *versus* anion-exchange resins. Anal Bioanal Chem 2009;393:899–905.

# Magnetic nanoparticle sorbents

9

*Bin Hu, Man He, Beibei Chen*
Department of Chemistry, Wuhan University, Wuhan, Hubei, China

## 9.1 Introduction

For magnetic solid-phase extraction (MSPE), the magnetic sorbents employed have a great influence on the separation and enrichment speed, enrichment factor, selectivity, antiinterference ability, and reproducibility of the MSPE-based methodologies. An ideal magnetic sorbent should have the following advantages: (i) strong magnetism to achieve fast magnetic separation; (ii) good dispersion, so as to improve the adsorption/desorption kinetics; (iii) large specific surface area, suitable porosity, and easy to modify, to provide abundant adsorption sites, which help to improve the adsorption capacity and extraction efficiency/recovery of target compounds; (iv) good selectivity/antiinterference capability, which helps to improve the ability of the method to tolerate complex matrixes; (v) good stability, can withstand acid and alkali environments, ultrasound, stirring, and oscillation treatment; (vi) reusable, namely, reversible adsorption; (vii) mild adsorption and desorption conditions; (viii) easy preparation and good reproducibility of preparation; (ix) low cost, and easy to obtain raw materials and low sorbents consumption during extraction; (x) environmentally friendly with low reagent consumption in its preparation and in the extraction process.

Magnetic nanoparticle (MNP) sorbents used for MSPE are nanometer-sized structures with good magnetism and available with various functional groups. MNP sorbents with superparamagnetism can be easily and rapidly separated from aqueous solution with the aid of an external magnetic field, facilitating the isolation of the target compounds adsorbed on the MNPs from the sample matrix. MNP sorbents can be functionalized with specific groups, e.g., hydroxyl, carboxyl, sulfonic acid, amino, mercapto, or chelating functional groups, resulting in favorable selectivity for various target compounds. So far, a variety of materials have been introduced, including silica-based, carbon-based, metal/metal oxides, metal-organic frameworks, porous organic polymers, mesoporous materials, and imprinted and restricted access materials.

## 9.2 Preparation techniques

### 9.2.1 *Synthesis of magnetic nanoparticles*

MNP sorbents with core-shell structures consisting of a magnetic core and a coating or magnetic composites prepared from magnetic particles and other materials are the most common types. There are many types of MNPs that can act as a magnetic core, including nickel (Ni), iron (Fe), cobalt (Co), and their alloys/oxides. Of these, iron

**Solid-Phase Extraction. https://doi.org/10.1016/B978-0-12-816906-3.00009-1**

**Table 9.1** Saturation magnetization of some magnetic metal oxides [1].

| Oxides | Magnetization saturation (emu $g^{-1}$) |
|---|---|
| $\gamma$-$Fe_2O_3$ | 74 |
| $Fe_3O_4$ | 84 |
| $Fe_2O_3$-$Fe_3O_4$ | 80 |
| $CoO \cdot Fe_2O_3$ | 65 |

oxide ($\gamma$-$Fe_2O_3$, $Fe_3O_4$) MNPs are the most widely used [1]. Table 9.1 lists the saturation magnetization of common metal oxides used in MNP sorbents; a saturation magnetization of above 20 emu $g^{-1}$ is beneficial for magnetic separations.

Normal synthesis methods for MNPs, include coprecipitation, hydrothermal/solvent thermal method, microemulsion method, sol-gel method, oxidation method, hydrolysis method, thermal decomposition method, and ultrasonic precipitation method. Coprecipitation is currently the most widely used method [2].

(i) Coprecipitation method: During the precipitation reaction, the metal precursor ($Fe^{2+}$, $Fe^{3+}$) is dissolved in an aqueous solution, followed by the addition of precipitating reagent (e.g., alkali solution) to form insoluble solids. MNPs prepared by this strategy, usually have a wide size distribution and irregular morphology. This method can produce various magnetic materials, including spinel ferrite, perovskites, metals, and alloys.

Based on the simple coprecipitation method, Lin et al. [3] prepared $Fe_3O_4$ nanoparticles (NPs) in a rotating packed bed, and relevant factors affecting the particle size of the NPs were investigated. It was shown that $Fe_3O_4$ NPs with smaller particle size could be obtained when higher rotational speeds and higher reactant and precipitant flow rates were employed. However, the concentration of the reactants and precipitant hardly affect the particle size of $Fe_3O_4$ NPs. The prepared $Fe_3O_4$ NPs had an average particle size of about 5.1 nm and a saturation magnetization of 50 emu $g^{-1}$.

The main advantage of the coprecipitation method is that large quantities of products can be produced by a single synthesis. The simple operation and short reaction time make it suitable for industrial production. However, the particle size is difficult to regulate, and the NPs are prone to agglomeration producing a wide distribution of particle sizes.

(ii) Hydrothermal method: It is a solution reaction carried out in a reaction vessel or high-pressure reactor, usually at temperatures of about 200°C and pressures greater than 2000 psi. In a supercritical state, water works as a reactant and increases the kinetics of the hydrolysis reaction. At high temperatures, the solubility of most metal ions increases, and the viscosity of water decreases, with the metal ions exhibiting increased fluidity. This allows Ostwald ripening to proceed at a faster rate improving the uniformity of the precipitation.

In the hydrothermal reaction, the size and morphology of the product can be adjusted by controlling the reaction time and temperature. Choice of precursor and the reaction condition such as pH significantly affect the phase purity of the NPs.

Ding et al. [4] synthesized $Fe_3O_4$ nanorods by an ethylenediamine-assisted hydrothermal reaction. The $Fe_3O_4$ nanorods have a cubic spinel crystal form, with a particle diameter of about 5 nm and length of about 5 μm. The ethylenediamine acts as a mineralizer and template in the synthesis of the $Fe_3O_4$ nanorods. Its concentration plays a decisive role in controlling the morphology and composition of the nanorods. Li et al. [5] synthesized Mn-Zn ferrite nanoparticles using the hydrothermal method. It was found that a low Fe content and the hydrothermal temperature was important for the synthesis of single-phase Mn-Zn ferrite nanoparticles. The hydrothermal temperature significantly affected the morphology and magnetic properties of the product. Mn-Zn ferrite nanoparticles synthesized at less than 180°C exhibited a saturation magnetic strength of 45.5 emu $g^{-1}$.

The NPs prepared by the hydrothermal method have high purity and controllable crystal forms. Since the reactor needs to be able to withstand high temperatures and pressures, the hydrothermal reaction requires suitable equipment for the reaction, which increases the cost. Furthermore, the reaction is carried out in a closed reactor, which makes it difficult to observe the growth of crystals.

(iii) Microemulsion method: In this method, a solution containing surfactants spontaneously form thermally stable and isotropic globular aggregates at the oil/water interface. Microemulsions can be divided into two types, oil-in-water (O/W) and water-in-oil (W/O). By changing the ratio of the aqueous phase, the oil phase, and the surfactant, the morphology of the product can be effectively regulated. Inouye et al. [6] used the microemulsion method to prepare $Fe_3O_4$ and $\gamma$-$Fe_2O_3$ nanoparticles with a narrow size range from ferrous oxide salts. The reaction was carried out in an emulsion containing sodium dioctyl sulfosuccinate and isooctane. Kelly et al. [7] prepared Mn-Zn ferrite NPs with a spinel structure and narrow particle size distribution in an O/W microemulsion. By changing the composition of the O/W microemulsion and the type/concentration of the precipitant, the particle size, crystallinity, and magnetic properties of the products can be effectively regulated.

Compared with other methods, the microemulsion method produces magnetic nanoparticles with narrow and controllable particle size, resulting in better uniformity. Since the particles are encapsulated by the surfactant, the resultant MNPs are less prone to agglomeration and have better monodispersity. Besides, the use of various surfactants would endow the magnetic NPs with different surface properties. On the other hand, the use of surfactants makes it essential to incorporate a cleaning step in the synthesis to remove surfactants from the surface of the magnetic NPs, which increases the cost.

(iv) Sol-gel method: A metal alkoxide is dissolved in a solvent to form a sol, which is hydrolyzed and condensed to form a gel, and then subjected to heat treatment, annealing, and/or sintering, resulting in NPs. In the sol-gel process, factors such as solvent type, temperature, choice of precursor, catalyst, pH, and additives, need to be considered. They affect the kinetics of the hydrolysis and condensation reactions and the growth of the magnetic NPs. Besides the hydrolysis and condensation reactions, pH affects the isoelectric point and stability of the sol, which in turn affects particle aggregation and particle size. By changing the factors affecting the rate of the hydrolysis and condensation reaction, the structure and properties of the gel can be effectively regulated.

Cui et al. [8] synthesized nearly monodisperse $\alpha$-$Fe_2O_3$, $\gamma$-$Fe_2O_3$, and $Fe_3O_4$ nanoparticles through a low-temperature sol-gel route. Two steps were involved in the preparation process. The first step was the reaction between $FeCl_2$ and propylene oxide in boiling ethanol to form a sol-gel followed by a drying step. The structure of iron oxide can be adjusted by changing the drying condition for the sol. The reaction between propylene oxide and $[Fe(H_2O)_6]^{2+}$ gradually promotes the hydrolysis and condensation of metal ions, resulting in uniform nucleation of iron oxide in the solution. In addition, this method has a high yield with about 60 g of nanoparticles obtained in a single synthesis. Zhang et al. [9] synthesized $Fe_3O_4$ aerogels by the sol-gel method using a supercritical carbon dioxide drying technique. $FeCl_3.6H_2O$, $FeCl_2.4H_2O$, and propylene oxide were used as a precursor and gelling accelerator, respectively. The as-prepared aerogels had a three-dimensional network structure consisting of interconnected NPs. The $Fe_3O_4$ aerogels possessed a low density ($0.25-0.39\ g\ m^{-3}$), large specific surface area ($60-330\ m^2\ g^{-1}$), and superior saturation magnetization ($15-52$ emu $g^{-1}$). The structure and magnetic properties of the aerogels can be controlled by varying the solution concentration, molar ratio of propylene oxide/$Fe^{3+}$, and the calcination temperature.

MNPs prepared by the sol-gel method have high crystallinity and controllable particle size with good uniformity. On the other hand, the preparation process is tedious, and the reaction period is relatively long.

(v) Oxidation method: Adding stabilizer (water-soluble polymer, e.g., polyethylene glycol, polyvinyl alcohol) and oxidant to the aqueous solution containing metal ions can produce metal oxide. For example, a magnetic fluid can be obtained by adding an alkaline solution to a solution containing $Fe^{2+}$ while simultaneously adding $H_2O_2$ or introducing air. MNPs prepared by this method are close to a spherical shape with a large particle size distribution. The average particle diameter of MNPs obtained by this method is often larger than that prepared by the coprecipitation method. By controlling conditions, such as air flow rate and reaction temperature, MNPs with different particle sizes can be obtained.

(vi) Hydrolysis method: This approach can be divided into two main routes, titration hydrolysis, and a Massart hydrolysis [10]. In the former, a mixed solution of ferric and ferrous salts with a molar ratio of n($Fe^{3+}$): n($Fe^{2+}$) of 2:1 is prepared by the dropwise addition of dilute alkali solution gradually increasing the pH of the mixed solution, hydrolyzing the iron ions to form $Fe_3O_4$ NPs. In the latter case, the mixed solution containing ferric and ferrous salts (n($Fe^{3+}$):n($Fe^{2+}$) of 2:1) is directly added into an alkaline aqueous solution, and the iron ions are immediately hydrolyzed to form $Fe_3O_4$ NPs. The reaction conditions are mild and do not require special equipment. on the other hand, the reaction temperature, reactant concentration, precipitant concentration, precipitant addition rate, agitation, and pH affect the particle size distribution of the products.

(vii) Thermal decomposition method: In this method, the organic metal precursor is decomposed at high temperature to prepare NPs. It is one of the simplest methods for NPs preparation. Compared with the other methods, the main advantage is that the decomposition temperature is low and controls the growth of the nanoparticles. Typically, polymers, organic blocking agents, or structural entities are used to limiting the size of the resultant NPs.

Jana et al. [11] prepared nearly monodisperse $Fe_3O_4$ NPs by the pyrolysis of iron fatty acid salts. Regulation of the size and shape of the NPs can be achieved by changing the reactivity and concentration of the precursors. The reactivity can be tuned by changing the chain length and concentration of fatty acid. Asuha et al. [12] synthesized $Fe_3O_4$ powder by the thermal decomposition of iron-urea complex ($[Fe(CON_2H_4)_6](NO_3)_3$) in a closed vessel. The method is simple with good reproducibility. One of the decomposition products of $[Fe(CON_2H_4)_6](NO_3)_3$ is $CON_2H_4$, which is a reducing agent, and plays an important role in the formation of $Fe_3O_4$. The particle size of $Fe_3O_4$ NPs can be adjusted by changing the reaction temperature. Increasing the reaction temperature from 200 to 300°C increases the average particle size of the nanoparticles from 37 to 50 nm. As the saturation magnetic strength of $Fe_3O_4$ NPs is closely related to the particle size, the saturation magnetic strength also increases from 71 to 89 emu $g^{-1}$.

MNPs prepared by thermal decomposition method have the advantages of good dispersibility, adjustable particle size, and narrow size distribution, while, the use of organic reagents during the preparation process may increase environmental pollution.

(viii) Ultrasonic precipitation method: Compared to conventional mixing techniques, ultrasonication makes it easier to obtain a homogeneous dispersion. The "ultrasonic vaporization bubbles" [13] produced by ultrasonic waves can form a local high temperature and pressure environment to accelerate chemical reactions. Ultrasonic precipitation avoids the uneven concentration dispersion of liquids, reduces the occurrence of agglomeration, and greatly increase the reaction speed, favoring the formation of small particles. Vijayakumar et al. [14] prepared $Fe_3O_4$ NPs with a particle size of about 10 nm from an aqueous solution of iron acetate under high-intensity ultrasonic radiation in an argon atmosphere.

## 9.2.2 Surface modification of magnetic nanoparticles

In theory, there are four types of interparticle forces in magnetic nanoparticles: van der Waals forces, magnetic forces, electrostatic repulsion forces, and steric hindrance. The first two forces tend to agglomerate MNPs, and the latter two are beneficial for stabilizing MNPs. Surface modification is an approach that introduces protective molecules on the particle surface that increases the repulsive force between particles. It can also reduce the surface energy, improve hydrophilic/hydrophobic properties, and avoid the oxidation of $Fe_3O_4$ to $\gamma$-$Fe_2O_3$. Meanwhile, by surface modification, various functional groups such as hydroxyl, carboxyl, sulfonic acid, amino, and mercapto can be introduced into the MNPs. Alternatively, various materials such as silicon dioxide, metal oxides, and quantum dots can be coated on the particle surface to adjust sorbent selectivity.

Surface modification methods include physical modification and chemical bonding. The physical modification uses physical means (e.g., adsorption, coating, or wrapping) by such techniques as surface adsorption and surface deposition. The surface modification of MNPs by ultraviolet and plasma radiation is also a physical modification. The chemical modification involves a change in the surface state of MNPs

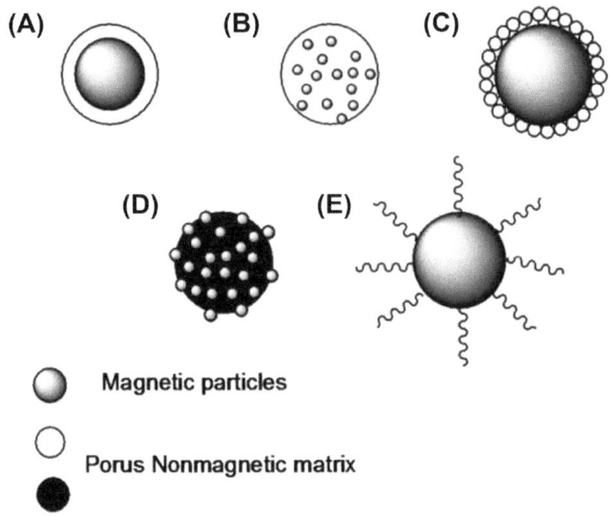


**Figure 9.1** Different morphologies and structures for magnetic beads. (A) core-shell type, (B) multicore, (C and D) bead on bead and (E) brush (hair) morphology [15].

through chemical reactions. Fig. 9.1 illustrates common structures of chemically modified magnetic composites.

### 9.2.2.1 Modification with inorganic materials

Amorphous silica is insoluble in water and many other solvents but is soluble in alkali solutions and hydrofluoric acid. Its favorable chemical/thermal stability, specific composition, and physical structure result in its common use as an adsorbent. Generally, three methods are used for coating silica on the surface of MNPs, namely the Stöber reaction [16,17], embedding [18], and microemulsion [19,20]. In Stöber reaction, the silica coating is prepared by the hydrolysis of silane coupling reagents or silicates by alkali or acid catalysis. Liu et al. [21] dissolved sodium silicate in deionized water, adjusted the pH to 12−13, followed by the mixing of $Fe_3O_4$ NPs with the sodium silicate solution with ultrasonication for 30 min. The mixture was then heated to 80°C with mechanical agitation, followed by dropwise addition of hydrochloride acid to a pH of 6−7, and collection of the silica-coated MNPs. Deng et al. [22] used a sol-gel method to modify the surface of MNPs with silica by the acid-catalyzed hydrolysis of tetraethoxysilane (TEOS). In the embedding strategy, MNPs are embedded in the porous structure of silica. Wu et al. [18] mixed solid $Fe(NO_3)_3$ with ethylene glycol and spiked a porous silica substrate with this mixture, followed by heating the mixture under nitrogen, converting $Fe(NO_3)_3$ into $Fe_3O_4$ NPs in porous silica. In the microemulsion method, the formation, thickness, and morphology of silica on the surface of MNPs are controlled by the action of micelles. Ding et al. [19] prepared silica coated $Fe_3O_4$ NPs with uniform size and regular core-shell structure by the inverse microemulsion method. For obtaining a silica coating of defined thickness, the concentration of $Fe_3O_4$ NPs in the microemulsion, and the amount of added TEOS was precisely controlled.

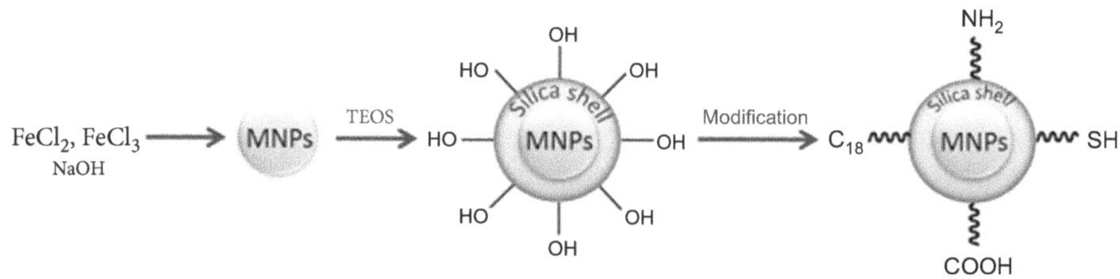


**Figure 9.2** Schematic diagram of the synthesis of silica-coated MNPs modified with different functional groups [31].

MNPs with a silica coating are the most widely used intermediates for preparing sorbents with a wide range of organic surface modifications. It is often used as the protective layers of the magnetic core. $Fe_3O_4$@$SiO_2$ NPs are easily modified by small organic molecule (e.g., bismuthiol-II [23], zincon [24], 3-mercaptopropionic acid [25], 5-sulfosalicylic acid [26]), ionic liquids [27], and polymer [28–30]) by bonding to the silica surface through reactive functional groups, Fig. 9.2.

Suleiman et al. [23] employed $Fe_3O_4$@$SiO_2$ NPs modified by bismuthiol-II for the separation and enrichment of Cr(III), Cu(II), and Pb(II) in environmental water samples. Cao et al. [26] used 5-sulfosalicylic acid modified $Fe_3O_4$@$SiO_2$ NPs for MSPE of Se(VI), Se(IV), and two selenium amino acids in Pueraria lobata extracts. Cui et al. prepared methyl trioctyl ammonium chloride ([$MTOA^+$][$Cl^-$]) ILs and chitosan modified $Fe_3O_4$@$SiO_2$ NPs for the extraction of trace elements from biological samples [27] and inorganic Cr speciation in environmental water samples [29], respectively.

In addition to silica, surface modification of MNPs with other inorganic materials, such as $TiO_2$, Au, and Ag, have been employed. Li et al. [32] spiked $Fe_3O_4$ MNPs and chloroauric acid into sodium citrate solution at 99°C forming a wine red solution. After cooling and magnetic separation, a $Fe_3O_4$/Au nanoparticle composite was obtained with an average particle size of 35 nm. A layer of nano-gold was formed as a shell on the surface of $Fe_3O_4$ MNPs. Ma et al. [33] prepared $Fe_3O_4$@m$TiO_2$ magnetic microspheres. First, magnetic colloidal nanocrystals clusters (MCNCs) of around 280 nm diameter consisting of nanocrystalline (7–11 nm) particles, were synthesized by the solvent-thermal method. Then $TiO_2$ was deposited on the surface of the MCNCs by the sol-gel method and the $TiO_2$ shell transformed into a mesoporous structure by hydrothermal treatment. Magnetic nanoparticle composite sorbents with a shell thickness of about 100 nm, saturation magnetic strength of 17.8 emu $g^{-1}$, $TiO_2$ content of 74wt%, a pore volume of 0.45 $cm^3$ $g^{-1}$ and a mesopore size of 16.4 nm were obtained.

#### 9.2.2.2 Modification with small organic molecules

In addition to inorganic materials, MNPs can be modified with small organic molecules, such as surfactants, ionic liquids, and silane coupling reagents.

#### 9.2.2.2.1 Surfactant modification

Surfactant modification prevents MNPs from being oxidized by air. Also, the addition of appropriate surfactants into the reaction system during synthesis is often used to adjust the size, morphology, and properties of the MNPs and to modify agglomeration. Su et al. [34] prepared $Fe_3O_4$ MNPs by coprecipitation with the addition of oleic acid resulting in oleic acid coated MNPs with improved lipophilicity. When the oleic acid was reacted with potassium permanganate, the double bond was oxidized to form carboxyl groups, improving the hydrophilicity of the MNPs. Takami et al. [35] heated a ferrous sulfate solution to 200°C in the presence of decanoic acid ($C_9H_{19}COOH$) and decylamine ($C_{10}H_{21}NH_2$), respectively. Decanoic acid modified MNPs had an $\alpha$-$Fe_2O_3$ structure with a cubic crystal shape with an average particle size of 25 nm; decylamine modified MNPs had a $Fe_3O_4$ structure with a spherical shape and an average particle size of 14 nm. These results indicate that the addition of surfactants affects the crystal shape, morphology, and size of the resultant MNPs.

#### 9.2.2.2.2 Silane coupling reagent modification

There are an abundant number of hydroxyl groups (Fe-OH) on the surface of ferrite magnetic nanoparticles suitable for reaction with silicon hydroxyl groups (Si-OH) formed by hydrolysis of a silane coupling reagent, such as TEOS, resulting in the formation of Fe-O-Si bonds. At the same time, Si-OH groups can react with each other forming Si-O-Si bonds. This process of hydrolysis and condensation proceed continuously with the formation of a silica coating on the surface of the MNPs, typically, a few nanometers thick. Besides protection from oxidation, these surface modifications can provide active sites for the introduction of other functional groups.

Some silane coupling reagents containing specific functional groups (e.g., (3-mercaptopropyl)trimethoxysilane (MPTMS), 3-aminopropyltriethoxysilane (APTES), 3-(trihydroxysilyl)propyl methylphosphate) react with silica-coated MNPs installing mercapto [36–39], amino [40–44], n-octyl [45], ethylenediamine tetraacetic acid (EDTA) [46], iminodiacetic acid (IDA) [47], thiosemicarbazide [48], and phosphoric acid group [49] on the silica surface of the NPs. Most of these magnetic sorbents have good dispersibility favorable for faster extraction kinetics and facilitate selective enrichment of target compounds based on their affinity for the installed functional groups. These functional groups also allow for further modification of the particle surfaces based on their characteristic reactivity. Click reaction chemistry on mercaptosilane modified MNPs was used to obtain sulfonic acid modified MNPs [50]. The sulfonic modified MNPs exhibited enhanced adsorption capacity for Cd(II) and Pb(II) ions. Schiff base functionalization was performed on an amino-silane modified MNPs for extraction of Pb(II) and Cd(II) [51]. Chloropropyltriethoxysilane (CPTS) modified MNPs were further grafted with mono-2-ethylhexyl 2-ethylphosphonate for selective extraction of rare earth elements [52]. 3-Vinyltriethoxysilane (VTES) [53] or 3-(methacryloxy)propyltrimethoxysilane (MPS) [54] modified MNPs provided suitable substrates for attachment of imprinted polymer coatings by polymerization.

#### 9.2.2.2.3 Ionic liquid modification

Ionic liquids (ILs) consist of asymmetric organic cations and inorganic/organic anion. They are characterized by low volatility, good chemical and thermal stability, and adjustable water solubility. Zhang et al. [55] spiked $Fe_3O_4$ NPs and 1-hexadecyl-3-methylimidazolium bromide ($C_{16}$mimBr) ILs simultaneously into an aqueous solution for the extraction of polycyclic aromatic hydrocarbons (PAHs). The adsorption of $C_{16}$mimBr on the $Fe_3O_4$ NPs was driven by electrostatic interactions. The positively-charged head group of $C_{16}$mimBr, an imidazole ring, strongly interacts with the negatively-charged $Fe_3O_4$ NPs. The extraction of PAHs by the ILs coated MNPs is possibly based on hydrophobic interactions and positively correlated with the alkyl chain length of the IL. Cui et al. [27] prepared methyltrioctylammonium chloride ([$MTOA^+$] [$Cl^-$]) modified MNPs by mixing $Fe_3O_4$ NPs and [$MTOA^+$] [$Cl^-$] under vortex for 30 min. The ILs modified MNPs were used for the extraction of metal ions. Bouri et al. [56] synthesized ILs functionalized MNPs by the reaction silica-coated $Fe_3O_4$ NPs with a silane coupling reagent containing N-methylimidazolium groups. This ILs functionalized MNPs were used for the extraction of sulfonylurea herbicides from water.

### *9.2.2.3 Organic polymer modification*

The hydrophilic/hydrophobic properties of MNPs can be adjusted by coating with organic polymers. Modification by organic polymers increases the Zeta potential and surface charge density of MNPs stabilizing their dispersion in water/oil emulsions. A variety of organic polymers are available for the modification of MNPs, including polysaccharides (e.g., starch, cellulose, glucans, chitosan, alginate, agarose), amino acids (e.g., peptides, enzymes, proteins, gelatin), natural biological macromolecules, and polymer/copolymer composites prepared from a variety of monomers (e.g., styrene (St), acrylic acid (AA), methacrylic acid (MAA), methyl methacrylate (MMA) and acrylamide (PAM)) [12]. Organic polymer modified MNPs generally exhibit three types of structures (Fig. 9.3): (A) MNPs with a uniform coating of polymer on the surface generally referred to as the core-shell-1 structure; (B) Organic polymer wrapped around the surface of MNPs generally referred to as the core-shell-2 structure, and (C) MNPs uniformly distributed in a polymer microsphere generally referred to as the dispersion structure.

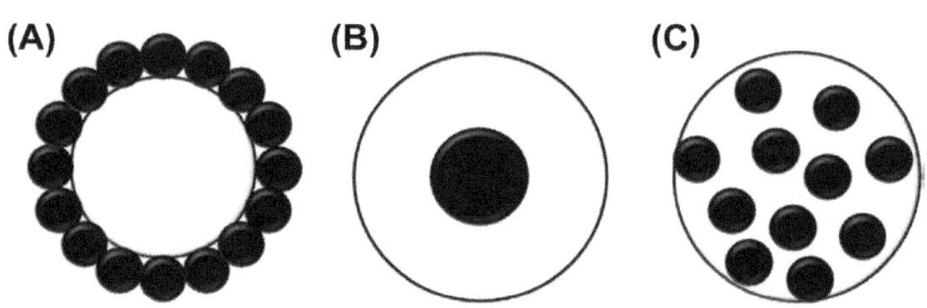


**Figure 9.3** Schematic representation of different structures of magnetic polymer beads: (A) polymer core–magnetic shell, (B) magnetic core-polymer shell, (C) magnetic particles homogeneously distributed in the polymer bead [1].

Natural biological macromolecules can provide good biocompatibility and are typically widely available and inexpensive. Ma et al. [57] dissolved ferric and ferrous chloride in water containing sodium hydroxide under nitrogen and heated the mixture to 60°C followed by the addition of sodium alginate with stirring at 80°C to obtain alginate modified MNPs. For MNPs modified with synthetic polymers, common synthetic routes include embedding (in situ), chemical conversion, dispersion polymerization, suspension polymerization, and emulsion polymerization. All processes involve two procedures: nucleation and nuclear growth. The particle size, size distribution, molecular weight, and conversion rate are determined by the specifics of the two processes [58]. Harris et al. [59] prepared copolymer modified MNPs by the embedding method. Polyethylene oxide (PEO), 3-isocyanate methylene-l, 3, 5, 5-trimethylcyclohexyl isocyanate (IPDI), and 2, 2-dihydroxymethyl propionic acid (DMPA) were initially copolymerized. The copolymer was then dissolved in ethylene dichloride followed by mixing with MNPs for 30 min (pH 8.5−9) with stirring. After removing the excess solvent by nitrogen blowing, the pH of the magnetic suspension was adjusted by hydrochloric acid to 6.5−7. Xiong et al. [60] prepared magnetic sorbents by the in situ method. First, ferric chloride and ferrous chloride were dissolved in $H_2O$ and the mixture heated to 60°C after the addition of polyethylene glycol 4000. The mixture was adjusted to pH 11 with sodium hydroxide solution and stirred for 0.5 h to obtain PEG-4000 modified magnetic sorbents. According to XRD characterization, the particle size of the MNPs decreased from 20 to 11.5 nm after PEG addition. Liu et al. [61] used dispersion polymerization to prepare magnetic polymer microspheres by first synthesizing macromolecular polyoxyethylene monomer and then carrying out a copolymerization reaction between the monomer and styrene for 12 h at 70°C. PEG-4000 was used as the dispersive reagent and a mixture of water and ethanol as the reaction medium. Generally, the products obtained by dispersion polymerization exhibit relatively large sizes, in this case, around 11 μm. Du et al. [62] dispersed PEG-4000 modified $Fe_3O_4$ NPs in a solution of ethanol and styrene; after the addition of styrene and benzoyl peroxide (initiator), the mixture was heated for 6 h at 75°C. Polystyrene-coated MNPs were obtained with a size of about 80 nm and magnetism of 20.7 emu $g^{-1}$. Liu et al. [63] used emulsion polymerization to prepare polymer-coated polymer MNPs, as shown in Fig. 9.4. First, $Fe_3O_4$ NPs of about 10 nm were synthesized by the coprecipitation method and coated with oleic acid. Then, the oleic acid coated MNPs were dispersed in a mixture of styrene, methacrylic acid (MAA), divinylbenzene (DVB), and acrylamide (AM); benzoyl peroxide

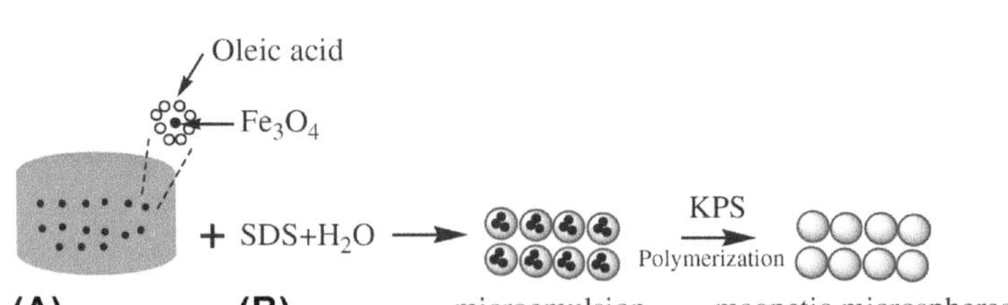


**Figure 9.4** Formulation of polymer coated magnetite nanoparticles. In the first step, hydrophobic magnetite nanoparticles were produced and dispersed in styrene. In the second step, magnetic microspheres were obtained by microemulsion polymerization [63].

(BPO) was added as the initiator to form solution A. Hexadecanols and sodium hexadecylsulfate were dissolved in water to form Solution B. Solution A was added dropwise to solution B and after 0.5 h, ammonium persulfate (initiator) was added and the mixture stirred for a further 3 h forming $Fe_3O_4$@P(St-co-MAA-co-AM) nanoparticles. The polymer coated MNPs were of relatively large size (1−5 μm) and weak magnetic strength [47 emu $g^{-1}$ (oleic acid coated MNPs) versus 1.6 emu $g^{-1}$ (copolymer coated MNPs)].

## 9.3 Characterization methods

### 9.3.1 Crystal structure analysis

X-ray diffraction (XRD) is the most common technique for determination of the crystal structure of MNPs. $Fe_3O_4$ has six characteristic diffraction peaks at $2\theta = 30.1°$, 35.5°, 43.1°, 53.4°, 57.0°, and 62.6° assigned to the (220), (311), (400), (422), (511), and (440) crystal planes of $Fe_3O_4$, respectively. Cui et al. [64] characterized the crystal structure of chitosan (CTS) modified $Fe_3O_4$ NPs by XRD. As can be seen from the XRD pattern, Fig. 9.5, strong peaks at $2\theta = 30.13°$, 35.22°, 43.2°, 53.7°, 57.37°, 62.77° and 74.4° for the diffraction peaks of $Fe_3O_4$ demonstrate the existence of $Fe_3O_4$. The peak at 22.97° is attributed to an amorphous diffraction peak for $SiO_2$, and CTS coated on $Fe_3O_4$.

### 9.3.2 Morphological characterization

#### 9.3.2.1 Electron microscopy

Scanning electron microscopy (SEM) uses a focused electron beam to scan the surface of the sample in a specific order of time and space to evaluate the morphology of

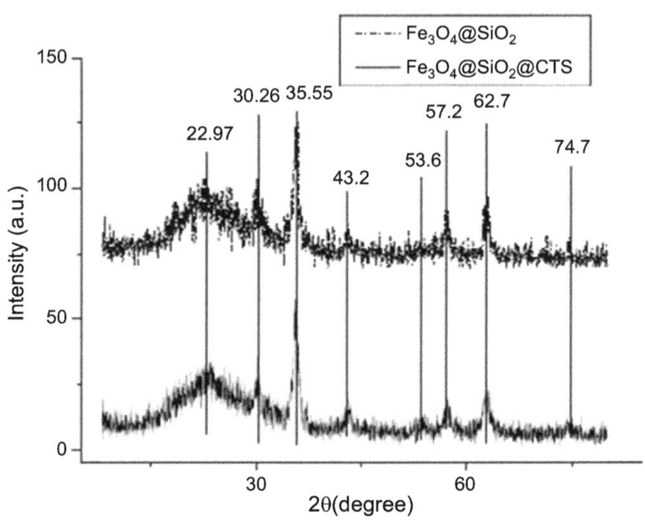


**Figure 9.5** XRD spectra of $Fe_3O_4@SiO_2$ and $Fe_3O_4@SiO_2$@CTS [64].

the MNPs. Its merits include high resolution, wide magnification range (generally 10−150,000 times), large depth of field, 3-D image capability, and simple sample preparation. Nonconductive samples need platinum(Pt)/Gold(Au) sputtering treatment of the surface before characterization.

Transmission electron microscopy (TEM) provides information on the morphology, structure, and size of MNPs. When an electron beam is focused on the MNPs, some electrons pass through the samples while others are scattered. The former beams are amplified and imaged on a fluorescent screen, providing a magnification of 50−800,000 times. TEM offers high resolution and magnification but poor 3-D image capability. It is difficult to observe the real size of agglomerated particles.

Li et al. [65] synthesized molecular imprinting (MIP) MNPs and characterized them by using SEM and TEM. Fig. 9.6 reveals that the uncoated, silica coated, and MIP modified MNPs had a spherical shape. $Fe_3O_4$ NPs has a uniform particle size of about 330 nm. The thickness of the silica coating on the surface of $Fe_3O_4$ is about 18 nm and the MIP coating about 22 nm.

## 9.3.3 Elemental analysis

### 9.3.3.1 Electron spectroscopy

X-ray photoelectron spectroscopy (XPS) is a surface analysis technique widely used to determine the elemental composition and oxidation states of elements at the surface of MNPs by excitation of inner orbital and bonding electrons by a focussed X-ray beam. The XPS spectrum is obtained by measuring the kinetic energy and quantity of electrons. Energy dispersive spectroscopy (EDS) is a surface microanalysis technology in which nanoparticles are irradiated by an electron beam to excite characteristic X-rays. EDS provides information about the elemental analysis, chemical composition, and predicted particle structure.

Fang et al. [66] synthesized $\alpha$- $Fe_2O_3@SiO_2@Au$ and characterized the composition of the material with XPS and EDS, Fig. 9.7. The XPS spectrum contains peaks at 285.03 eV attributed to the binding energy of C1s, 103.29 eV to the binding energy of Si2p, 83.98 and 87.7 eV to the binding energy of $Au4f_{7/2}$ and $Au4f_{5/2}$, respectively, and 532.68 eV to the binding energy of O1s. The EDS spectrum demonstrates the existence of Fe, Si, O, Au, and Cu (Cu signal is an instrumental artifact). XPS data combined with EDS analysis is consistent with the material's composition.

## 9.3.4 Surface area analysis

The Brunauer-Emmett-Teller (BET) method is commonly used for evaluating the surface area and surface morphology of MNPs by nitrogen adsorption. The larger the specific surface area, the more active sites and stronger adsorption capacity, the materials have. Ding et al. [67] synthesized mesoporous MNPs and octadecyl phosphate modified MNPs with a specific surface area and porosity distribution shown in Fig. 9.8. The $N_2$ adsorption/desorption isotherms (Fig. 9.8A) reveals a type IV adsorption isotherm for the materials with an H3 type hysteresis loop and a

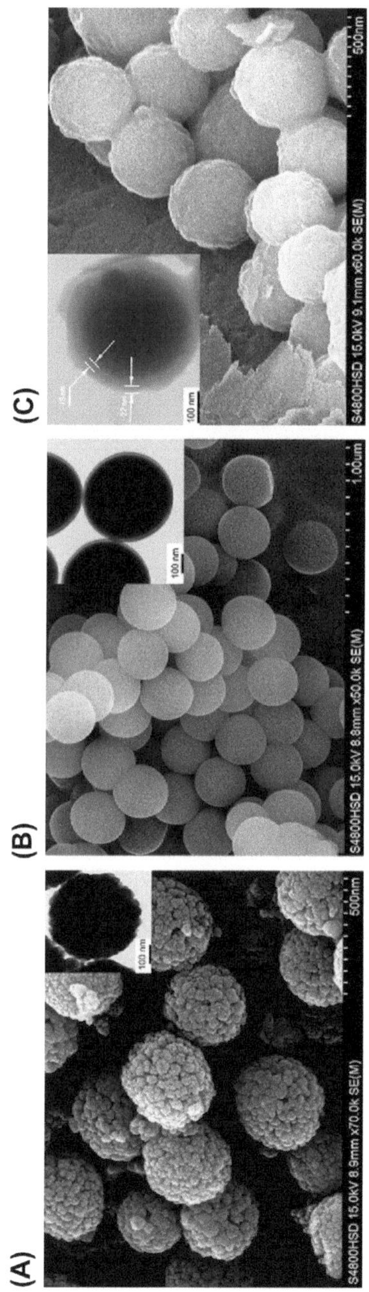


**Figure 9.6** The SEM and TEM images of $Fe_3O_4$ (A), $Fe_3O_4@SiO_2$ (B), and $Fe_3O_4@SiO_2$-MIP (C) [65].

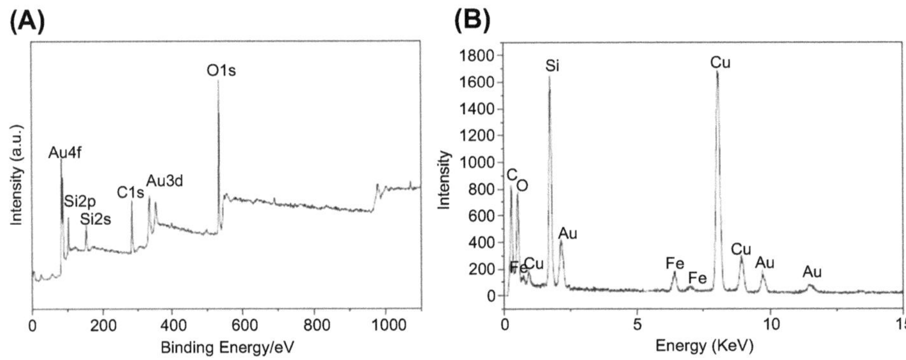


**Figure 9.7** (A) XPS spectrum of the $\alpha$- $Fe_2O_3@SiO_2@Au$ nanocomposites. (B) EDS data from an $\alpha$- $Fe_2O_3@SiO_2@Au$ sphere. The Cu signal is from the Cu grids of the TEM sample [66].

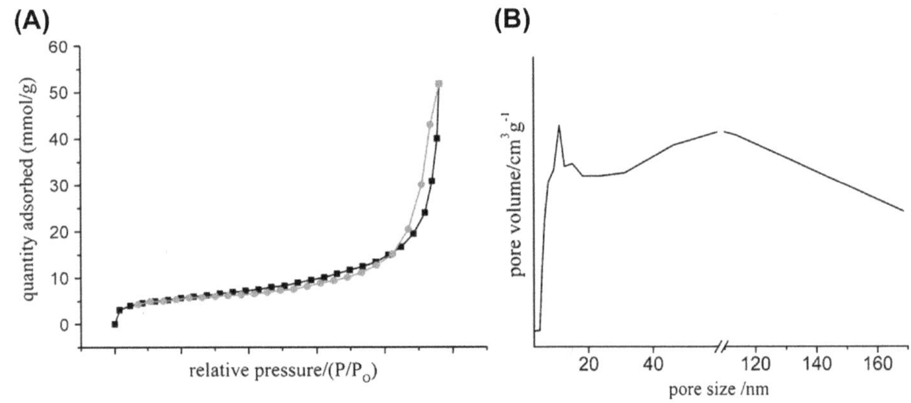


**Figure 9.8** $N_2$ adsorption/desorption isotherms (A) and BJH pore size distribution curve (B) of MMNPs [67].

low-pressure end of the Y-axis. Together, this indicates that the material has a strong interaction with nitrogen, and there are slit holes formed by stacking flaky particles. The pores in the materials are mesoporous.

## 9.3.5 Spectral analysis

### 9.3.5.1 Fourier transform infrared spectroscopy

The Fourier transform infrared (FTIR) spectra is based on the characteristic absorption of infrared radiation due to molecular vibration and is generally nondestructive. Huang et al. [68] synthesized magnetic MOF composites and characterized them by several spectroscopic techniques. The FTIR spectra of Bi-I, $Fe_3O_4@SiO_2$, $Fe_3O_4@SiO_2@Cu(OH)_2$, $Fe_3O_4@SiO_2@$HKUST-1, and Bi-I-functionalized $Fe_3O_4@SiO_2@$HKUST-1 are shown in Fig. 9.9. In the FTIR spectra of $Fe_3O_4@SiO_2$, a broad peak around 1090 $cm^{-1}$ is observed for the stretching vibration of O-Si-O demonstrating

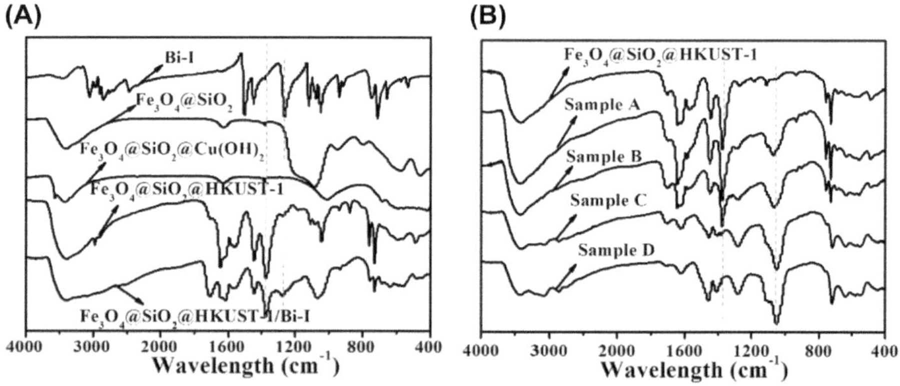


**Figure 9.9** (A) FTIR spectra of Bi-I, $Fe_3O_4$@$SiO_2$, $Fe_3O_4$@$SiO_2$@$Cu(OH)_2$, $Fe_3O_4$@$SiO_2$@HKUST-1, and Bi-I-functionalized$Fe_3O_4$@$SiO_2$@HKUST-1; (B) $Fe_3O_4$@$SiO_2$@HKUST-1 and Bi-I-functionalized $Fe_3O_4$@$SiO_2$@HKUST-1 with a different ratio, sample A(50 mg), B(100 mg), C(150 mg) and D(200 mg) [68].

the successful preparation of $Fe_3O_4$@$SiO_2$. The peaks at 587, 1000 and 3576 $cm^{-1}$ correspond to the Cu-O stretching, Cu-O-H bending, and O-H stretching vibrations, respectively, identified with the $Cu(OH)_2$ coating on the $Fe_3O_4$@$SiO_2$ nanoparticle core. For $Fe_3O_4$@$SiO_2$@HKUST-1, the asymmetric stretching vibration at 3050 $cm^{-1}$ is attributed to the aromatic ring C-H in $H_3BTC$, absorption peaks for the C=O bonds (1709 and 1645 $cm^{-1}$) and the stretching vibration of the C=C bonds of the benzene ring (1447 and 1374 $cm^{-1}$) are also observed. The absence of Cu-O stretching at 1000 $cm^{-1}$ indicates the complete in situ conversion of $Cu(OH)_2$ into HKUST-1. In the spectrum of Bi-I-functionalized $Fe_3O_4$@$SiO_2$@HKUST-1, a distinct peak at 1052 $cm^{-1}$ is assigned to the stretching vibration of Cu-S, indicating a modification of Bi-I on $Fe_3O_4$@$SiO_2$@HKUST-1. The vibration of S-H at 2472 $cm^{-1}$ was not very strong, however. The absorption bands for Bi-I-functionalized $Fe_3O_4$@$SiO_2$@HKUST-1 are different to those of $Fe_3O_4$@$SiO_2$@HKUST-1 in the region of 2750–4000 $cm^{-1}$, clearly revealing that Bi-I molecules are grafted onto the unsaturated metal sites of the framework, rather than just adsorbed on the external surface of HKUST-1.

### *9.3.5.2 Absorption spectroscopy*

Absorption spectroscopy is suitable for determining the properties of nanoparticles surface functionalized by organic compounds (mainly) or inorganic substrates with different oxidation states that absorb light between 200 and 800 nm. Fang et al. [66] synthesized $\alpha$-$Fe_2O_3$@$SiO_2$@Au NPs. The UV-vis spectra of their solutions, for $\alpha$-$Fe_2O_3$@$SiO_2$@Au, $\alpha$- $Fe_2O_3$ and Au are shown in Fig. 9.10. As can be seen, the maximum absorption wavelengths of $\alpha$- $Fe_2O_3$ and Au are 370 and 519 nm, respectively. The maximum absorption wavelengths of $\alpha$-$Fe_2O_3$@$SiO_2$@Au are at 370 and 524 nm, probably a result of a strong interaction between Au and adjacent atoms.

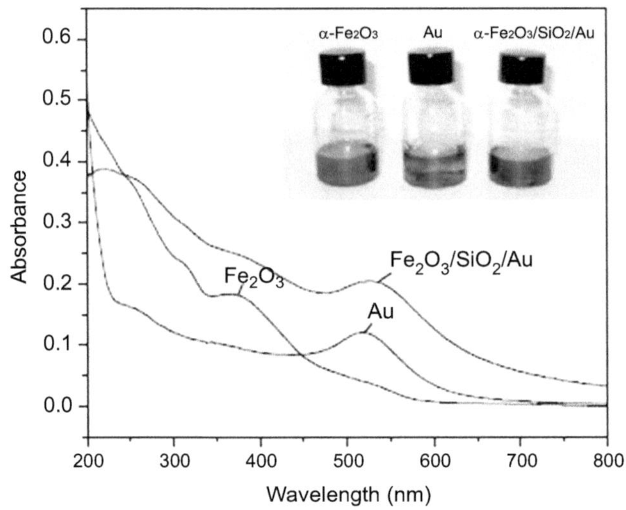


**Figure 9.10** UV-vis spectra of $\alpha$-$Fe_2O_3$@$SiO_2$@Au nanocomposites, $\alpha$- $Fe_2O_3$ nanocubes, and Au nanoparticles dissolved in distilled water. The inset shows a photograph of the samples [66].

## 9.3.6 Thermal gravimetric analysis (TGA)

Thermal gravimetric analysis (TGA) is used to investigate the relationship between the mass variation of MNPs and temperature, providing information about the coating composition. Kan et al. [69] synthesized a magnetic aspirin MIP NPs and characterized it by TGA. Fig. 9.11 shows that the amount of MIP in the magnetic sorbent accounts for 72.5% (w/w).

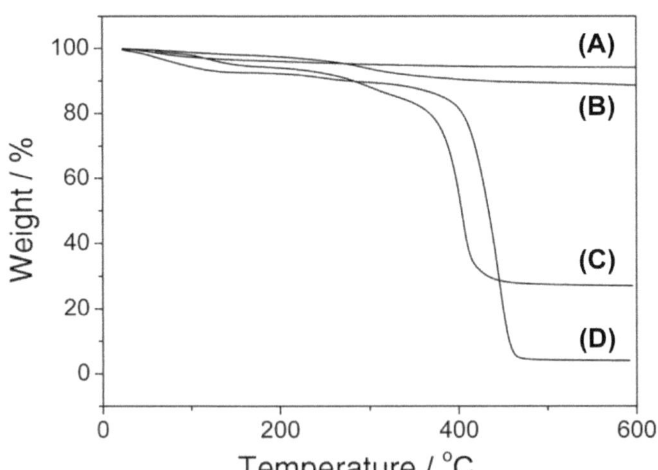


**Figure 9.11** TGA curves of $Fe_3O_4$ (A), double-bond-functionalized $Fe_3O_4$ (B), magnetic MIPs (C) and pure MIPs (D) [69].

### 9.3.7 Zeta potential

A zetasizer can provide information about the surface charge of a material. The isoelectric point (pI) is the pH at which the surface charge of the suspended MNPs is 0. The surface charge density of the materials changes with the pH of the solution, which affects the surface adsorption of magnetic sorbents. When the pH is higher than the pI, the surface of the materials is negatively charged, and when the pH is below the pI the surface is positively charged. Zhang et al. [55] investigated the pI of $Fe_3O_4$ MNPs and reported it to be about 6.5.

### 9.3.8 Magnetism

A vibrating sample magnetometer (VSM) is used to investigate the saturation magnetization of MNPs, construct a magnetization curve, determine the relationship between magnetization strength and external magnetic field intensity, and to identify superparamagnetism. At room temperature, when the magnetization intensity (H) of the material is zero, the hysteresis-free magnetization curve indicates that the material is superparamagnetic, and the stationary point of the magnetization curve is the saturation magnetization strength. When a magnetic sorbent is coated with a polymer, the paramagnetism is hardly affected, but its magnetization strength will decline. Thus the core-shell structure of the material can be verified by VSM. Ma et al. [70] characterized $Fe_3O_4$@PGMA-$NH_2$ with VSM, and the magnetization strength was found to be 16.3 emu $g^{-1}$. It indicates that the nanoparticle coating did not destroy the superparamagnetism of the $Fe_3O_4$ MNPs. Luo et al. [70] synthesized $Fe_3O_4$@$SiO_2$/graphene and characterized its magnetization strength with VSM. Fig. 9.12 shows that the magnetization strength of $Fe_3O_4$, $Fe_3O_4$@$SiO_2$, $Fe_3O_4$@$SiO_2$/graphene are 72, 43, and 21 emu $g^{-1}$, respectively. It indicates that silica and graphene affect the magnetization strength of $Fe_3O_4$ MNPs but do not destroy its superparamagnetism.

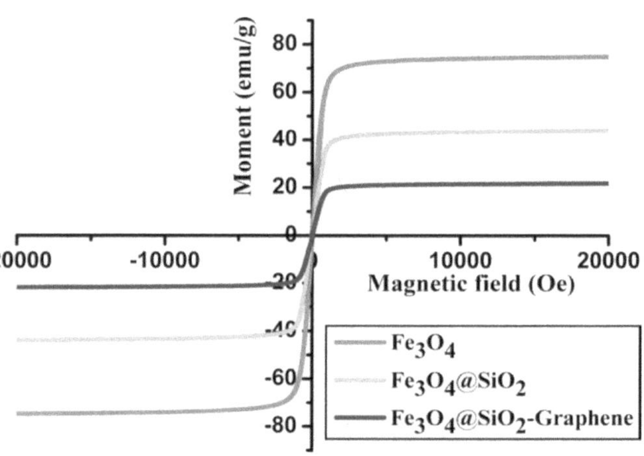


**Figure 9.12** Room-temperature magnetization curves of $Fe_3O_4$, $Fe_3O_4$@$SiO_2$, and $Fe_3O_4$@$SiO_2$/graphene [70].

## 9.4 Magnetic carbon materials

There are many kinds of carbon nano-sorbents, including fullerenes, carbon nanotubes (CNTs), carbon nanofibers (CNFs), graphene (G), and graphene oxide (GO). They are thermally, chemically, and mechanically stable with large specific surface areas. Abundant modifiable sites facilitate easy modification. The integration of MNPs with carbon nanomaterials provides a source of carbon nanomaterials easy to separate from aqueous solution. Carbon nano-sorbents can adsorb aromatic compounds through van der Waals, hydrophobic, and/or $\pi$-$\pi$ stacking interactions. For metal ions surface modification of carbon nano-sorbents is necessary to improve their adsorption capacity and selectivity.

### 9.4.1 Magnetic carbon nanotubes (CNTs)

Carbon nanotubes (CNTs) are usually curled by single or multiple layers of graphene to form hollow tubular structures. According to the number of graphene layers, CNTs can be divided into single-walled carbon nanotubes (SWCNTs) and multi-walled carbon nanotubes (MWCNTs). Since pure CNTs are hydrophobic and poorly dispersed in water, it is difficult to separate and recover from the solution. The use of magnetic CNTs composites is a common method to overcome such difficulties. Magnetic CNTs preparation can be divided into two approaches, namely one-step and two-step reactions.

The two-step reaction was the first approach to be used for the preparation of magnetic CNTs. In the first step, the nanoparticles are synthesized and then combined with the CNTs is a second step forming a magnetic CNTs composite. Wang et al. [71] prepared a magnetic CNTs composite by the electrostatic assembly of $Fe_3O_4$ NPs and MWCNTs. The magnetic MWCNTs were used to extract Se (IV) from water. However, the adsorbents exhibit low adsorption capacity (2.5 mg $g^{-1}$) for Se (IV) with slow kinetics. To improve the adsorption performance of CNTs for metal ions strong acids (e.g., concentrated nitric acid) are often used to oxidize CNTs, resulting in abundant oxygen-containing functional groups (e.g., hydroxyl, carbonyl, and carboxyl groups) on the end and side walls of the CNTs. These functional groups assist in the adsorption of metal ions and improve the dispersion of the CNTs in solvents. These functional groups also provide abundant active sites for further modification. Sadegh et al. [72] mixed oxidized MWCNTs with $Fe_3O_4$ NPs and obtained magnetic MWCNTs with a large specific surface area (92 $m^2$ $g^{-1}$) through a simple self-assembly reaction. These magnetic CNTs exhibited an adsorption capacity of 239 mg $g^{-1}$ for Hg (II).

The synthesis of MNPs in a one-step process employs an in situ reaction on the surface of the CNTs. Demir et al. [73] introduced $Fe_3O_4$ NPs into MWCNTs by the one-step method. Oxidized MWCNTs were spiked into a solution of iron(II) sulfate hexahydrate in $H_2O$: hydrazine (3:1) followed by heating under reflux for 2 h, resulting in black magnetic MWCNTs. Huiqun et al. [74] mixed MWCNTs with an acidic solution of iron nitrate with heating under reflux for 4.5 h. The subsequent addition of a basic solution ($NH_4OH$, 2.5 wt%) led to the precipitation of iron oxide onto the surface of the MWCNTs. Tarigh et al. [75] synthesized $Fe_3O_4$ NPs in situ on oxidized MWCNTs by the coprecipitation method for rapid extraction of Pb (II) and Mn (II) in the presence of the chelating agent PAN. Zhang et al. [76] prepared magnetic CNTs by a one-step solvothermal method and then grafted a sulfhydryl-containing silane reagent onto their surface. The magnetic sorbents had a large specific surface area (97 $m^2\ g^{-1}$), and a relatively high adsorption capacity (65.4 and 65.5 $mg\ g^{-1}$) for Pb (II) and Hg(II) ions.

The magnetic composites can be further functionalized by postfunctionalization. Fayazi et al. [77] synthesized $Fe_3O_4$ NPs containing oxidized MWCNTs by the coprecipitation method subsequently modified with silane coupling reagents containing double bonds them with a double bond. The magnetic composites had good selectivity and high capacity (48.1 $mg\ g^{-1}$) for the adsorption of Pb(II). In addition, 8-hydroxyquinoline [78] and melamine [79] functional groups were grafted on the surface of magnetic MWCNTs after the introduction of chloropropyl groups to improve the selectivity for metal ions. This in situ synthesis procedure with different iron salts is now commonly used to obtain magnetic CNTs due to its simplicity. These materials were also used for the solid-phase extraction of various organic compounds (e.g., polycyclic aromatic hydrocarbons [80] or atrazine [81]) as well as metal ions [78,79].

### 9.4.2 Magnetic graphene and graphene oxide adsorbents

Graphene (G) is a carbon-based material with a single-layer two-dimensional honeycomb lattice structure. Graphene is the basic framework of many graphite-based carbon materials, including CNTs and fullerenes. It has a large surface area and abundant delocalized $\pi$-electrons. It can adsorb aromatic compounds through $\pi$-$\pi$ stacking. However, G contains few functional groups which can interact with metal ions; it needs appropriate surface modification [82] or addition of a chelating agent before use as a sorbent for metal ions [83].

Graphene oxide (GO) is the oxidation product of graphene. Compared with G or reduced GO, GO has more oxygen-containing groups, such as carboxyl, hydroxyl, carbonyl, and epoxy groups on its surface. It is more hydrophilic and easier to chemically modify than G-based materials. Therefore, GO is widely used for the extraction of metal ions. However, GO has good water solubility and is difficult to separate directly from solution. Magnetic GO can avoid the loss of GO during the extraction process.

Magnetic graphene/graphene oxide nanoparticles can be prepared by the in situ growth of MNPs on the G/GO surface through chemical coprecipitation and hydrothermal treatments. Sun et al. [84] blended GO dispersions with a solution containing ferrous and ferric salts. Magnetic GO was obtained by coprecipitation in one step. The magnetic GO was used for the analysis of heavy metal ions in biological samples. Kazemi et al. [85] synthesized magnetic GO by a similar method and used it for the extraction of Au in environmental water samples. The adsorption capacity of this sorbent, 9.8 $mg\ g^{-1}$, is comparatively low. To further improve the adsorption capacity of magnetic GO for metal ions, postfunctionalization is usually employed to introduce additional functional groups on magnetic GO surface. Guo et al. [86] synthesized magnetic GO by a one-step coprecipitation method, then bonded 1,2-cyclohexanediamine tetraacetic acid to the GO surface using ethylenediamine. The functionalized magnetic GO has a larger specific surface area (310 $m^2\ g^{-1}$) and a higher adsorption capacity (77.3 $mg\ g^{-1}$ for Cr (VI) than unmodified magnetic GO and can be used for the removal of Cr (VI) from water. Cui et al. [87] synthesized magnetic GO by the solvothermal method and modified its surface with EDTA. The magnetic GO-EDTA can efficiently extract Pb, Hg, and Cu with an adsorption capacity of 508, 268, and 301 $mg\ g^{-1}$, respectively. Amino-modified magnetic GO was prepared by the solvothermal method with the addition of diethylenetriamine [88]. Compared with the unmodified magnetic GO, the modified magnetic GO has a higher adsorption capacity (123 $mg\ g^{-1}$) for Cr (VI). In addition, various polymers [89–91], ionic liquids [92], small molecules [93], and other functional components [94] have been used for the surface modification of GO.

In the two-step reaction process, MNPs and G are mixed, and magnetic G composites (MGC) obtained by a simple ultrasonication or oscillation treatment. Since MNPs are attached to the surface of G/GO by physical adsorption, the products are not stable for reuse. Su et al. [95] prepared silica coated $Fe_3O_4$ NPs followed by the in situ polymerization of aniline monomers to obtained magnetic polyaniline (MPANI), which was mixed with a GO suspension with ultrasonication. The sorbent composite could be regenerated and reused. It was used for the preconcentration of rare earth elements in water samples based on chelation by surface functional groups on GO (hydroxyl, epoxide, carboxyl and carbonyl groups) with the rare earth elements.

Coupling MNPs and G/GO via chemical bonding is expected to produce more stable products. Owing to the rich oxygen-containing functional groups on its surface GO has plenty of sites for a further chemical reaction. Generally, the MNPs are modified with silane coupling reagents [e.g., tetraethyl orthosilicate (TEOS) and (3-aminopropyl) triethoxysilane (APTES)], producing functionalized MNPs. Then, the amino-modified MNPs are attached to the GO surface by amidation reactions using crosslinking reagents, such as 1-(3-dimethylaminopropyl)-3-ethylcarbodiimide hydrochloride (EDC) and n-hydroxysuccinimide (NHS) [96]. These magnetic sorbents were used for the extraction of phenolic compounds from water. They can be reused at least 15 times without a significant decrease in sorption capacity.

### 9.4.3 Magnetic porous carbon

Typical porous carbons (PC) include activated carbon, CNFs, CNTs, G/GO and their modified materials, as well as new porous carbon materials prepared by high-temperature carbonization and template etching using different carbon-based substrates. This section mainly introduces porous carbon materials prepared by simple high-temperature carbonization process. These materials have large specific surface areas, high porosity, good thermal and chemical stability, and a simple preparation process. The sources of carbon precursors used in calcination are extensive, e.g., biological materials, polymers, CNTs and combinations of these with various templates (e.g., metal oxide and silica NPs, mesoporous silica, polymer, MOFs, etc.) [97–99].

Generally, magnetic porous carbon (MPC) can be prepared by mixing and calcining a magnetic source (Co, $Fe_3O_4$, $Fe_3O_4@SiO_2$, etc.) and carbon source [100–102]. Habila et al. [101] used $Fe_3O_4$ NPs as a magnetic source and a mixture of glucose and polyacrylamide as a carbon source followed by a high-temperature reaction. The resultant amide modified MPC was used for MSPE of several elements from environmental water and food samples. Peng et al. [102] prepared sulfur-doped MPC by carbonization at high temperature (600°C) with $Fe_3O_4@SiO_2$ as a magnetic source, sucrose as the carbon source and basic magnesium sulfate (513MOS) as template and sulfur source. During calcination, 513MOS and sucrose form sulfur ether bonds providing functional groups for selective adsorption. Magnesium oxide produced by the decomposition of 513MOS acts as a porogen and can be removed by acid washing after carbonization. It contributes to high specific surface area and porosity of the sorbents and fast adsorption/desorption kinetics in the extraction process. Due to the high S content, the MSPC possesses a high adsorption capacity for Hg(II) (343 mg $g^{-1}$) with good selectivity. It was used for the analysis of Hg in environmental water samples.

Another strategy is to mix magnetic metal salts and carbon precursors to simultaneously produce magnetic materials and PC after calcination. It provides a simpler way to introduce magnetic sources into PC. Wen et al. [103] used ferric chloride as a magnetic source and tea residues as a carbon source, which was mixed and dried, followed by calcination. Tea residue was carbonized to form PC while iron salt was pyrolyzed forming $\gamma$-$Fe_2O_3$ on the PC, producing MPC materials used to extract As(V), Cr(V), humic acid and organic dyes from water. Chen et al. [104] prepared a magnetic nitrogen-doped mesoporous carbon material using ferric chloride as a magnetic source, mesoporous silica SBA-15 as a template, ionic liquids (1-cyanoethyl-3-methylimidazolium chloride) as carbon and nitrogen sources. After homogeneous mixing, calcination at high temperature and template removal, MPC was obtained with a mesoporous structure and a large specific surface area and pore volume. It had a good adsorption capacity, 23.6 mg $g^{-1}$, for Cu(II). The MPC prepared by using mesoporous silica as a template has good porosity and can be used as a substrate for further nitrogen-doped mesoporous porous carbon materials. Yang et al. [105] used mesoporous silica SBA-15 as a template and ferric nitrate as a magnetic source. Then polyaniline (PANI) was synthesized on its surface and carbonized at a high temperature, producing nitrogen-functionalized magnetic mesoporous carbon N-Fe/OMC, Fig. 9.13. The material has a large specific surface area and good adsorption capacity of 158 and 114 mg $g^{-1}$ for Pb(II) and phenol, respectively. Liu et al. [107] prepared magnetic nitrogen-sulfur codoped mesoporous carbon nanoparticles. Silica NPs modified by L-cysteine and phenolic resin as a carbon, sulfur, and nitrogen source were mixed with iron nitrate solution followed by drying and calcination to produce MPC. The codoped magnetic mesoporous carbon had good adsorption properties for Hg(II).

Due to their three-dimensional network structure, MOFs can be used as a template for pore formation during high-temperature carbonization, and the organic ligands in MOFs can act as a carbon source. In recent years, many researchers have tried to prepare porous carbon using MOF as a precursor [98]. Zeolite imidazole frameworks

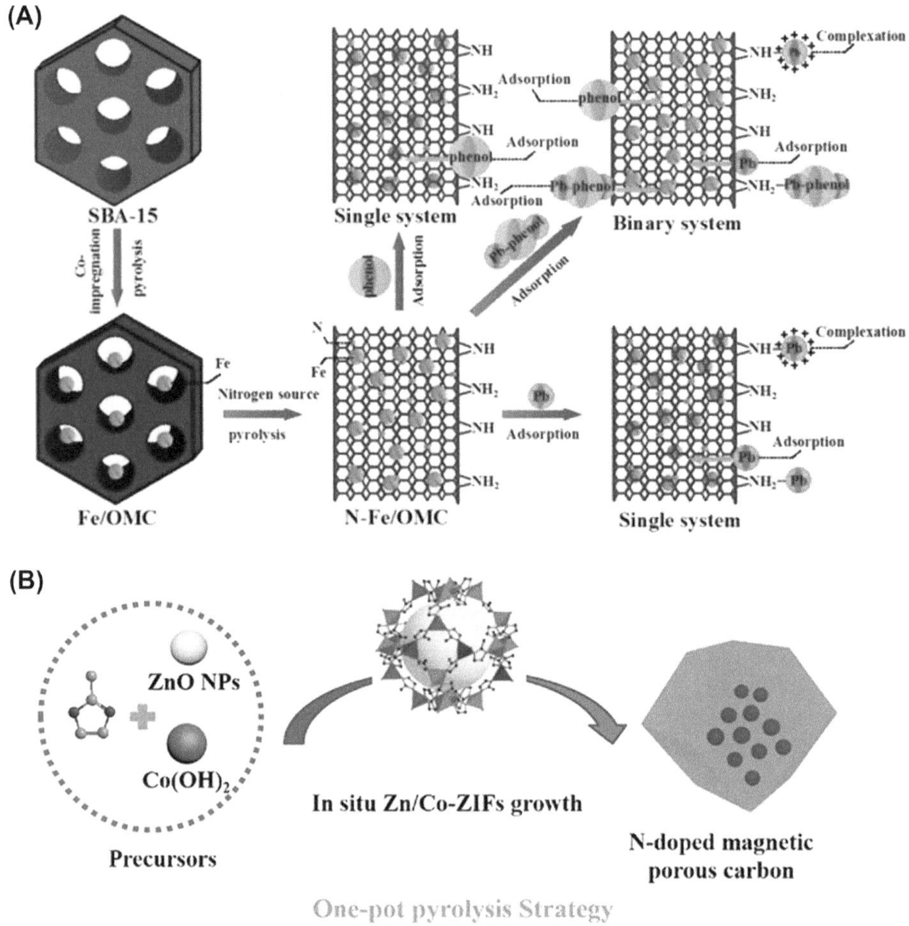


**Figure 9.13** Schematic diagram of the preparation of N-Fe/OMC for the adsorption of Pb(II) and phenol [105] (A) and magnetic N-doped porous carbon by one-pot pyrolysis strategy with in situ synthesized magnetic Co nanoparticles [106] (B).

(ZIFs) have a porous network structure rich in nitrogen, which can act as a nitrogen-doped carbon source, as well as Co metal that acts as a magnetic source [106,108,109]. Huang et al. [106] mixed $Co(OH)_2$ and precursors for ZIF-8 (ZnO NPs and 2-methylimidazole) followed by calcination. During the heating process, ZnO and 2-methylimidazole form ZIF-8 and are carbonized to form nitrogen-doped PC. $Co(OH)_2$ is pyrolyzed to form magnetic Co NPs and dispersed in porous carbon. The one-step method based on ZIF-8 is shown in Fig. 9.13. It is simple, without the need to synthesize ZIF-8 precursors or magnetic source in advance. The synthesized materials contain a large amount of pyridine and pyrrolidone nitrogen contributing to the favorable adsorption capacity, 429 mg $g^{-1}$, for Hg(II). Postfunctionalization is also used to improve the selectivity of MPC for metal ions. Guo et al. [110] prepared magnetic mesoporous carbon with nickel nitrate as a magnetic source, sucrose as a carbon source and mesoporous silica SBA-15 as a template. After calcination at high temperature, magnetic mesoporous carbon was obtained and then mixed with 8-hydroxyquinoline (8-HQ) solution by ultrasound, producing 8-HQ modified magnetic mesoporous carbon. The sorbent was used for the extraction of Cu(II), Pb(II), Sr(II), and Co(II) from water. Samadi et al. [111] prepared magnetic porous carbon Fe@C by one-step carbonization of ferric chloride and urea and then mixed it with 1-(2-thiazolylazo)-2-naphthol (TAN) solution under ultrasound. The TAN-modified Fe@C MPC was used for the extraction of Cu(II) in water and food samples. However, the stability and reusability of the materials obtained by physical blending were poor. At present, the most common methods for functionalizing MPC employ doping by a nitrogen/sulfur source in the calcination process. The obtained sorbents are relatively stable. In the preparation process, it is necessary to mix the source of carbon, nitrogen/sulfur, and magnetism homogeneously prior to calcination to obtain MPC with uniform nitrogen/sulfur doping and uniform distribution of MNPs.

## 9.5 Magnetic metal and metal oxides

Nanometer-sized metal and metal oxide particles have unique surface active sites and physicochemical properties. Coating MNPs with metal/metal oxides can provide additional active sites for adsorption or as intermediates for subsequent grafting with other functional groups or coatings. A variety of nanometer-sized metal/metal oxides have been introduced into magnetic functional sorbents, including silver, gold, zirconia ($ZrO_2$), titanium oxide ($TiO_2$), alumina ($Al_2O_3$), manganese oxide ($MnO_2$), copper oxide (CuO), zinc oxide (ZnO), and magnesium oxide (MgO). Of these, $ZrO_2$, $TiO_2$ are usually prepared by hydrolysis of Zr or Ti containing precursors onto MNPs by the sol-gel method. Wu et al. [112] prepared $Fe_3O_4@ZrO_2$ NPs by the sol-gel method. Because $ZrO_2$ coating is negatively charged under alkaline conditions, $Fe_3O_4@ZrO_2$ NPs can adsorb positively charged Cr (III), but hardly Cr (VI), in the form of $HCr_2O_7^-$, $Cr_2O_7^{2-}$ and $CrO_4^{2-}$ anions. Zhang et al. [113] prepared $Fe_3O_4@SiO_2@TiO_2$ by hydrolysis of tetrabutyl titanate on a $Fe_3O_4@SiO_2$ surface. The $Fe_3O_4@SiO_2@TiO_2$ surface is rich in titanium hydroxyl groups (Ti-OH) and bridging hydroxyl groups. Under alkaline conditions, these oxygen-containing groups are negatively charged, which can realize the extraction of Cu(II), Cd(II), Mn(II), and Cr(III) in tap and lake water. Habila et al. [114] also prepared $Fe_3O_4@SiO_2@TiO_2$ by the sol-gel method. The material has a high adsorption capacity for Cu(II), Zn(II), Cd(II), and Pb(II).

MNPs modified by metal oxides, such as $Al_2O_3$, $MnO_2$, CuO, and MgO can be prepared by a one-step coprecipitation or solvothermal method with the addition of the corresponding metal salts. They were employed for the extraction of metal ions and/or their different species [115–122]. In addition, layered dihydroxides formed by bimetallic oxides with a petal-like pore structure are efficient general adsorbents [115,117,121]. Moreover, the nanometer-sized metal/metal oxide coatings on MNPs can be loaded with chelating agents or functional groups to further improve the selectivity of the sorbents.

Lopez-Garcia et al. [123] prepared nano-silver coatings on $Fe_3O_4$ by in situ reduction of grafted sodium dimercaptoethanesulfonate. The materials are selective to Sb(III) and total Sb at different pH conditions, facilitating the speciation of Sb(III)/(V). By a similar method, Hsu et al. [124] prepared $Fe_3O_4$@AuNPs modified with L-cysteine. The magnetic sorbents were used for the extraction of Hg(II). Yan et al. [125] modified the surface of $Fe_3O_4$@$TiO_2$ with the organophosphorus extractant P204 [di(2-ethylhexyl) phosphate]. $Fe_3O_4$@$TiO_2$@P204 can extract rare earth elements over a wide pH range (4–7). It was used for trace and ultratrace analysis of rare earth elements in environmental samples.

Based on the positively charged surface of nano-$Al_2O_3$ under acidic condition, Karami et al. [126] adsorbed negatively charged sodium dodecyl sulfate (SDS) micelles on the surface of $Fe_3O_4$@$Al_2O_3$. This was then further modified by adsorption dithizone assisted by the preadsorbed micelles. These sorbents were used for the adsorption of Ag(I) by -SH and -NH functional groups on dithizone. Tavallali et al. [127] adsorbed a TX-114 micellar phase on the surface of $Fe_3O_4$@$Al_2O_3$ then coadsorbed 1-(2-pyridylazo)-2-naphthol (PAN). Cr(III) forms a complex with PAN is selectively retained on the sorbents, while Cr(VI) is barely adsorbed. Munonde et al. [128] modified 3-aminopropyltrimethoxysilane (AAPTS) on MNPs modified by $MnO_2$ and $Al_2O_3$ ($Fe_3O_4$@$MnO_2$, $Al_2O_3$). The resultant $Fe_3O_4$@$MnO_2$,$Al_2O_3$@$NH_2$ can selectively adsorb Cr(VI) under acidic conditions.

## 9.6 Magnetic metal-organic frameworks

Metal-organic frameworks (MOFs) are a crystalline hybrid material with a porous structure formed by coordination bonds between metal ions or clusters and organic ligands. The structure of MOFs includes both inorganic and organic building units. Some common structures are illustrated in Fig. 9.14. The organic ligands, mainly include oxygen- and nitrogen-containing heterocyclic compounds. Oxygen-containing ligands are mostly carboxylic, phosphoric and sulfonic acids, while nitrogen-containing compounds are mainly cyanide, imidazole, azole, and pyridine. Compared with traditional carbon-based and zeolite materials, MOFs have several advantages. They are porous materials of relatively low density with a larger specific surface area (10,000 $m^2\,g^{-1}$) and a higher porosity (90%). The pore size of the materials can be adjusted by varying either the metal ion or ligands. MOFs have good thermal and chemical stability. Finally, MOFs are relatively simple to prepare and easy to postfunctionalize.

Magnetic metal-organic framework composites (MFCs) are composed of magnetic components (e.g., $Fe_3O_4$, $\gamma$-$Fe_2O_3$, $NiFe_2O_4$, $CoFe_2O_4$ [129]) and MOFs. MFCs inherit the advantages of easy and rapid magnetic separation, diversity of MOFs components, and large specific surface area. They have been widely used in catalysis, drug controlled release, biomedical analysis, and as adsorbents in complex matrices (Table 9.2). Various synthetic strategies have been used to prepare MFCs and can

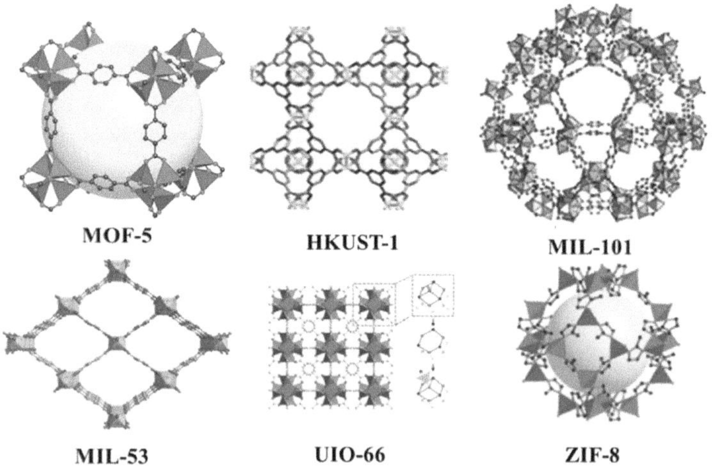


**Figure 9.14** Different types of commonly used MOFs.

be divided into the following categories: embedding method, encapsulation method, layer-by-layer self-assembly, physical mixing, postfunctionalization, and other preparation methods.

## 9.6.1 Embedding method

In this method, MNPs are embedded into the framework of MOFs retaining the original crystal morphology of MOFs. The commonly used preparation method is to add preprepared MNPs into a solution of MOFs precursor containing inorganic metal salts and organic ligands with ultrasound agitation. The MOFs are grown in situ by the hydrothermal or solvothermal methods. The morphology of MFCs synthesized by this method is basically the same as the pure MOFs without MNPs addition. Because of its relatively simple operation, this method can be used to prepare a wide variety of magnetic composites with different MOFs. Ke et al. [133] used $Fe_3O_4$ nanorods as a magnetic source for the preparation of $Fe_3O_4$ nanoparticles intercalated with HKUST-1. Doherty et al. [138] placed the easily prepared magnetic fibers $CoFe_2O_4$ or $NiFe_2O_4$ in the precursor reaction solution for MOF-5. After the hydrothermal reaction, the magnetic fibers with MOF-5-like morphology were obtained. The MFCs was used for the extraction of polycyclic aromatic hydrocarbons (PAHs) from aqueous solution.

## 9.6.2 Encapsulation method

The encapsulation method is used to encapsulate the magnetic components in the MOFs structure. Typically, a layer of organic polymer or carboxylic acid, compatible with MOFs, is modified on the surface of the magnetic components; then the modified material is added to the precursor solution for the MOFs synthesis, resulting in the growth of the MOFs on the surface of the magnetic components controlled by the

**Table 9.2** Some MFCs and their applications.

| Magnetic component | MOFs | Metal | Organic linkers | Application | Reference |
|---|---|---|---|---|---|
| $\gamma$-$Fe_2O_3$ | DUT HKUST-1 | $Cu^{2+}$ or $Al^{3+}$ | $H_2NDC$ $H_2BPDC$ $H_3BTC$ | Catalysis | [130] |
| $Fe_3O_4$ | Cu- carboxylate MOF | $Cu^{2+}$ | $H_2BPDCc$ | Catalysis | [131] |
| $Fe_3O_4$ | ZIF-8 | $Zn^{2+}$ | 2-HmIm | Catalysis | [132] |
| $Fe_3O_4$ | HKUST-1 | $Cu^{2+}$ | $H_3BTC$ | Drug delivery | [133] |
| $Fe_3O_4$ | UiO-66 | $Zr^{4+}$ | $H_2BDC$ | Phosphoproteome research | [134] |
| $Fe_3O_4$-Pyridine)/ | $Cu_3(BTC)_2$ | $Cu^{2+}$ | $H_3BTC$ | $Pd^{2+}$ preconcentration | [135] |
| $Fe_3O_4/Cu_3(BTC)_2{-}H_2Dz$ | $Cu_3(BTC)_2$ | $Cu^{2+}$ | $H_3BTC$ | $Pb^{2+}$ detection | [136] |
| $Fe_3O_4$/MIL-53(Al) | MIL-53(Al) | $Al^{3+}$ | $H_2BDC$ | $Pb^{2+}$ removal | [137] |
| $CoFe_2O_4$ Or $NiFe_2O_4$ | MOF-5 | $Zn^{2+}$ | $H_2BDC$ | PAH sequestration | [138] |
| $Fe_3O_4$ | MIL-101 | $Cr^{3+}$ | $H_2BDC$ | PAH sequestration | [139] |
| $Fe_3O_4$ $Cu_3(BTC)_2$ | $Cu_3(BTC)_2$ | $Cu^{2+}$ | $H_3BTC$ | Dye removal | [140] |

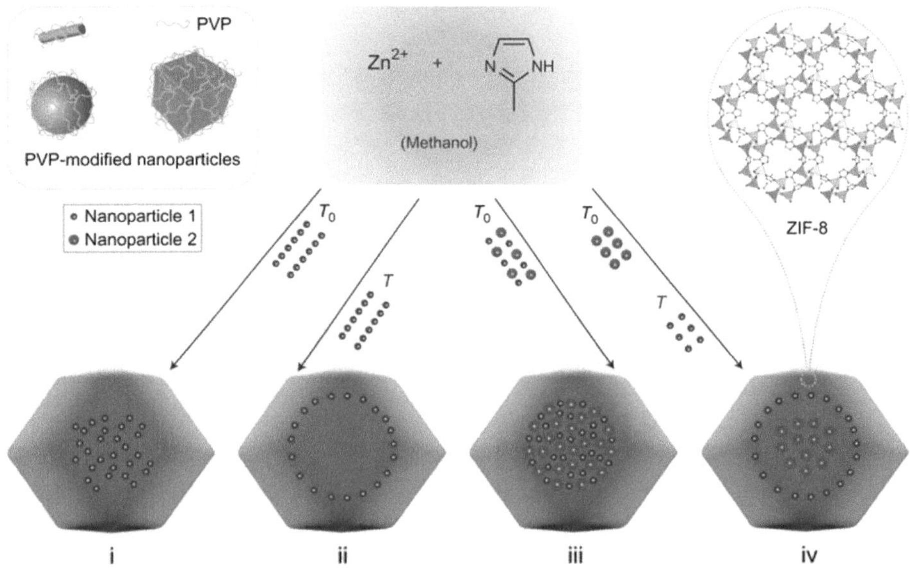


**Figure 9.15** Scheme for the controlled encapsulation of nanoparticles in ZIF-8 crystals [141].

surface modifier groups, and finally the magnetic components are encapsulated in the interior of MOFs structure. By using this method, MFCs with core-shell structures are easily obtained. Lu et al. [141] used PVP-modified NPs as the core, and encapsulated a variety of NPs into ZIF-8 by in situ growth, based on the strong interaction between PVP and $Zn^{2+}$, Fig. 9.15. Zhang et al. [132] formed a layer of sodium polystyrene sulfonate on the surface of $Fe_3O_4$ NPs, which was then added to the precursor solution for ZIF-8; the $Fe_3O_4$@ZIF-8 composite with a core-shell structure was obtained by in situ growth. Zhao et al. [134] coated $Fe_3O_4$ NPs with a layer of polydopamine (PDA) based on self-polymerization of dopamine in alkaline buffer solution; then $Fe_3O_4$@PDA was dispersed in the precursor reaction solution of UiO-66, and the in situ growth of UiO-66 was realized by the strong interaction between PDA and $Zr^{4+}$. A similar method was used to prepare other MFCs with different core-shell structures, such as $Fe_3O_4$@PDA@[$Cu_3(btc)_2$] [142].

## 9.6.3 Layer-by-layer self-assembly

Layer-by-layer self-assembly (also known as liquid phase epitaxy) uses MNPs modified with specific functional groups, which are added to a solution containing metal ions or organic ligands in a repetitive process to realize the layer-by-layer growth of MOFs and finally form MFCs. This method was first proposed by Fisher and Kitagawa's group [143−147] for the preparation of thin films of MOFs. The substrate was modified by molecules containing carboxyl, mercapto or amino functional groups; then sequential deposits of thin films of MOFs were deposited on the substrate from a solution of metal sources and organic ligands through molecular self-assembly. The MFCs prepared by this method maintain the original morphology of MNPs while acquiring a core-shell structure. In 2012, Ke et al. [148] prepared MFCs with a core-shell structure, $Fe_3O_4$@[$Cu_3(BTC)_2$] and $Fe_3O_4$@MIL-100(Fe), by using thioglycolic acid to modify the surface of the MNPs.

### 9.6.4 Physical mixing

In this method, ultrasound is used to mix solutions containing MNPs and MOFs. This method requires the prior preparation of the magnetic and MOFs components. The key point is that there should be a strong interaction between the magnetic and MOFs components, such as electrostatic interactions, to ensure the stability of the MFCs and to meet the requirements for practical application. In 2012, Yan et al. proposed the mixing method strategy for the preparation of MFCs [139]. $Fe_3O_4$@ $SiO_2$ NPs and MIL-101 were mixed in aqueous solution under ultrasound; based on the strong electrostatic interaction between negatively charged $Fe_3O_4$@$SiO_2$ and MIL-101 with positive surface charge, $Fe_3O_4$@$SiO_2$-MIL-101 composites were obtained. These MFCs were used for the extraction of PAHs from aqueous solution. Currently, the physical mixing method, as a simple, fast, and efficient preparation technology, has been widely used [149–152].

### 9.6.5 Postmagnetization

This method involves an in situ growth of magnetic particles on the surface of MOFs for the preparation of MFCs. Usually, the preprepared MOFs materials are added to the precursor solution of $Fe_3O_4$, and then the magnetic components are loaded on to the MOFs by in situ precipitation. Saikia et al. [153] synthesized MIL-101 (Cr) which was then dispersed in an aqueous solution of $FeCl_3$ and $FeCl_2$; the $Fe_3O_4$ NPs were synthesized in situ on the MOFs by coprecipitation, producing MFCs of $Fe_3O_4$@ MIL-101. By using a similar strategy, Jabbari et al. [154] prepared magnetic MOF@GO ternary composites with good adsorption properties for the adsorption of methylene blue in aqueous solution, Fig. 9.16.

### 9.6.6 Other methods

In recent years, some new synthetic strategies have been developed for the preparation of MFCs. Panda et al. [155] used MIL-53 (Fe) as a template and converted part of its skeleton structure into $Fe_2O_3$ by heating, and $Fe_2O_3$/MIL-53(Fe) composites with good magnetic properties were obtained. During the preparation of nano-Fe-MOFs by using different organic reagents as reaction solvents, Sethi et al. [156] found that the magnetic properties of the products obtained in different solvents differed greatly, although the nano-Fe-MOFs have a similar crystal structure to MIL-88B. When dimethyl sulfoxide was used as the reaction solvent, the MOFs had no magnetism, but with *N*,*N*-dimethylformamide good magnetism (up to 50 emu $g^{-1}$) was obtained. Fan et al. [157] prepared a core-shell MFCs, Cu-$CuFe_2O_4$@HKUST-1, by using

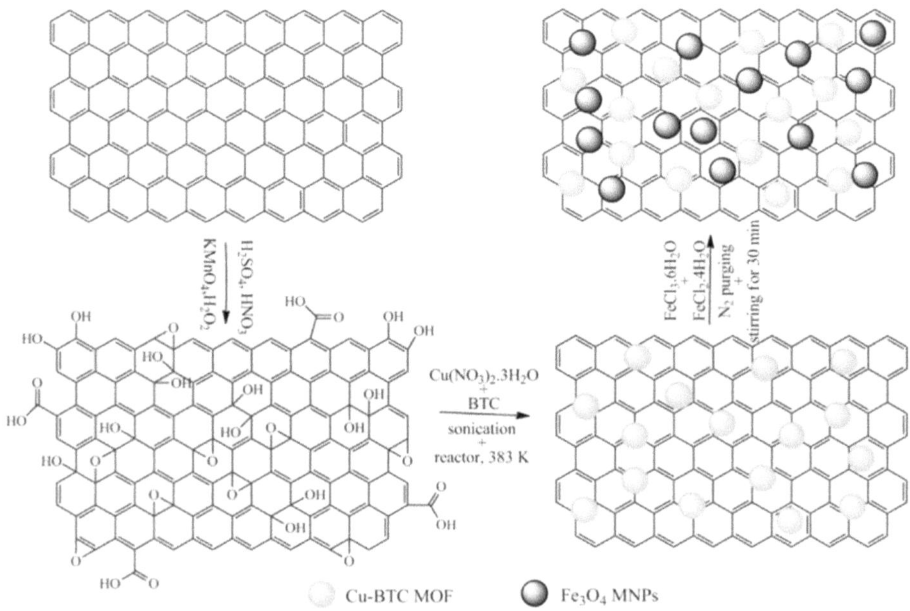


**Figure 9.16** Schematic view of the formation of Cu-BTC MOF and $Fe_3O_4$ MNPs on the GO layer [154].

$Cu\text{-}CuFe_2O_4$ as the magnetic source for MFCs and metal source for the MOFs. When $H_3BTC$ solution was added into the dispersed solution of the $Cu\text{-}CuFe_2O_4$ MNPs, the Cu(0) component in the $Cu\text{-}CuFe_2O_4$ MNPs produced $Cu^{2+}$ inducing MOF nucleation on the $Cu\text{-}CuFe_2O_4$ MNPs surface. With the consumption of the Cu(0) component and the crystallization of HKUST-1, the $Cu\text{-}CuFe_2O_4$ core was gradually covered by the HKUST-1 crystal shell.

Overall, although progress has been made in the preparation of MFCs materials, there are still some problems in need of an urgent solution. (1) When MFCs are prepared by the embedding or mixing methods, the materials are often doped with unmagnetized MOFs or magnetic particles, which are not functionalized by MOFs, which are difficult to purify; (2) For MFCs prepared by encapsulation or layer-by-layer self-assembly methods, it is usually necessary to prefunctionalize the surface of MNPs with certain functional groups, requiring additional preparation steps; (3) MFCs prepared by physical mixing rely on electrostatic, hydrogen bonding, or van der Waals interactions between the magnetic particles and MOFs, which are relatively unstable in applications requiring harsh conditions; (4) In postmagnetization methods, the MNPs grown in situ block the MOFs pore structure, resulting in a lower specific surface area (in addition the MNPs growing in situ are bare without any protection and easily oxidized or corroded in acid solution resulting in poor stability); (5) it is difficult to achieve large-scale production of MFCs by the current synthetic methods.

## 9.7 Magnetic porous organic polymers

Porous organic polymers (POPs) are porous network polymer materials prepared by covalent bonding of organic monomers composed of lightweight elements (e.g., C, H, O, N, B). According to the structure and crystalline characteristics, POPs can be divided into crystalline covalent organic frameworks (COFs) and amorphous hyper-crosslinked polymers (HCPs), intrinsic microporous polymers (PIMs), and conjugated microporous polymers (CMPs). There are also some POPs based on special structures such as Schiff base network polymers (SNWs), covalent triazine frameworks (CTFs), and porous aromatic frameworks (PAFs). Compared with inorganic porous materials, POPs have good stability, large specific surface areas, high porosity, low skeleton density, more adsorption sites per unit mass, and no metal components. However, POPs are difficult to separate and recover because of their lightweight, and many POPs use organic ligands with an aromatic structure to form hydrophobic conjugated networks with poor dispersion properties making their use for extraction somewhat difficult. The combination of POPs and MNPs with good dispersibility and magnetism endows the magnetic materials with abundant pore and adsorption sites and favorable properties for applications in extraction.

Typical preparation method for magnetic POPs (MOPs) can be categorized as (i) physical mixing; (ii) one pot, simultaneous synthesis of POPs and MNPs; (iii) in situ growth of POPs on MNPs; (iv) post magnetization in situ growth of MNPs on POPs; and (v) chemical bridging.

### 9.7.1 Physical mixing

The preparation of MOPs by physical mixing relies on physical interactions between POPs and MNPs. Although this method is simple and easy to operate, the instability of MOPs often affects the extraction reproducibility. Hypercrosslinked polystyrene (HCP) and $Fe_3O_4$ NPs were mixed by vortex in methanol and assembled spontaneously into HCP/$Fe_3O_4$ nanoparticles isolated by a magnet, Fig. 9.17. The HCP/$Fe_3O_4$ NPs were used for the extraction of four sulfonamides from water and milk samples [158].

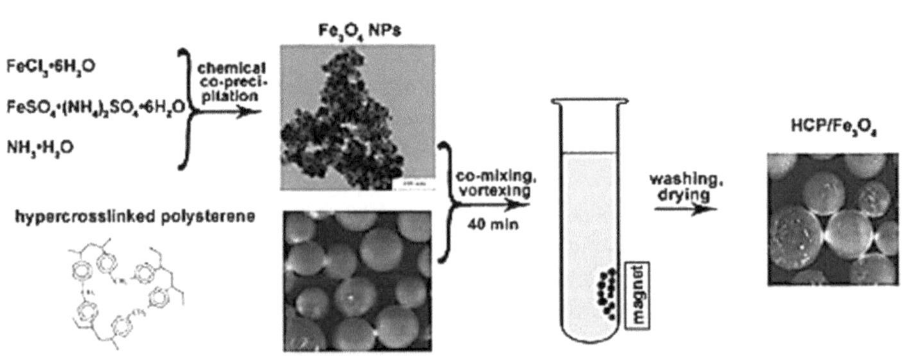


**Figure 9.17** Synthesis of hypercrosslinked polystyrene $Fe_3O_4$ NPs [158].

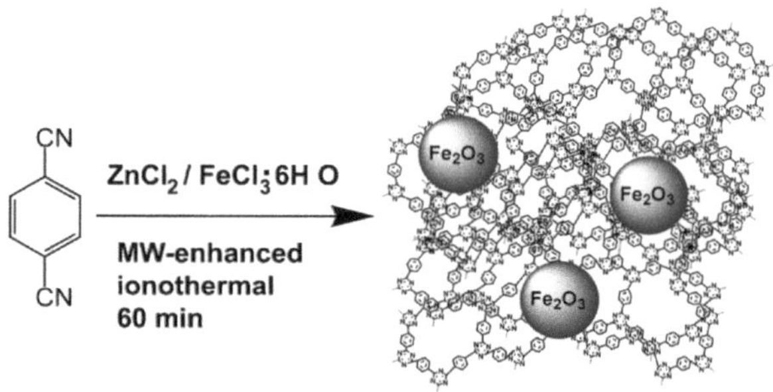


**Figure 9.18** Synthesis of the CTF/$Fe_2O_3$ composite by the microwave-enhanced high-temperature ionothermal method [159].

## 9.7.2 One-pot synthesis

The one-pot synthesis of MOPs requires the simultaneous synthesis of the POPs and MNPs. Although a few steps are involved, the reaction conditions for POPs and MNPs must be compatible, which significantly limits the scope of this approach. Zhang et al. [159] spiked $FeCl_3 \cdot 6H_2O$ into a mixture of zinc chloride and terephthalonitrile to form amorphous CTF/$Fe_2O_3$ nanoparticles with microwave assistance, Fig. 9.18. The CTF/$Fe_2O_3$ composites had high surface areas (930−1149 $m^2\ g^{-1}$) and a saturation magnetization at 300K from 1.1 to 5.9 emu $g^{-1}$ depending on the $Fe_2O_3$ content (6.43−12.4wt%). These composites were used for the extraction of methyl orange from aqueous solution. Ren et al. [160] prepared CTF/$Fe_2O_3$ composites by a similar method for the extraction of perfluorinated compounds from environmental water samples. Under the optimum conditions, this method has a lower detection limit (0.62−1.39 ng/L).

## 9.7.3 In situ growth synthesis

In this strategy, MOPs are synthesized by adding the MNPs to the monomers for the POP synthesis with a coating of the MNPs with the POPs; POPs monomers need to be modified for the subsequent POPs growth on the MNPs. To avoid damaging the MNPs during the synthesis procedure, this strategy is used only for the magnetization of those POPs, which can be obtained under mild condition, e.g., TpPa-1 synthesized at room temperature [161]. 1,3,5-Triformylphoroglucinol (Tp) was first bonded onto the amino-functionalized MNPs and then added into a solution containing Tp and *p*-phenylenediamine (Pa-1). After mechanical agitation at room temperature for 30 min, the brown colored TpPa-1 was magnetically isolated and washed with *N,N*-dimethylformamide, Fig. 9.19.

Huang et al. [162] synthesized POPs directly on the surface of $Fe_3O_4@SiO_2$ under mild reaction conditions forming MNPs encapsulated with POPs, Fig. 9.20. The synthesis involves a 12 h reaction in water at ice-bath temperature, avoiding the use of organic solvents and high-temperature reaction conditions. The MOPs have a moderate specific surface area (270 $m^2\ g^{-1}$) and an adsorption capacity for Hg(II)

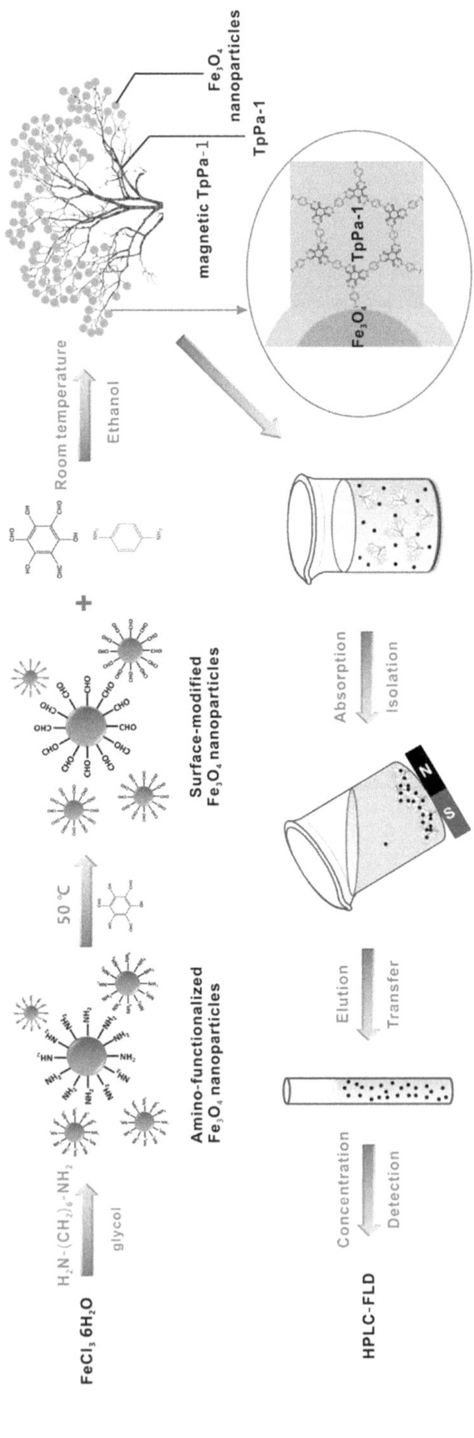


**Figure 9.19** Synthesis and Application of the magnetic TpPa-1 Sorbent [161].

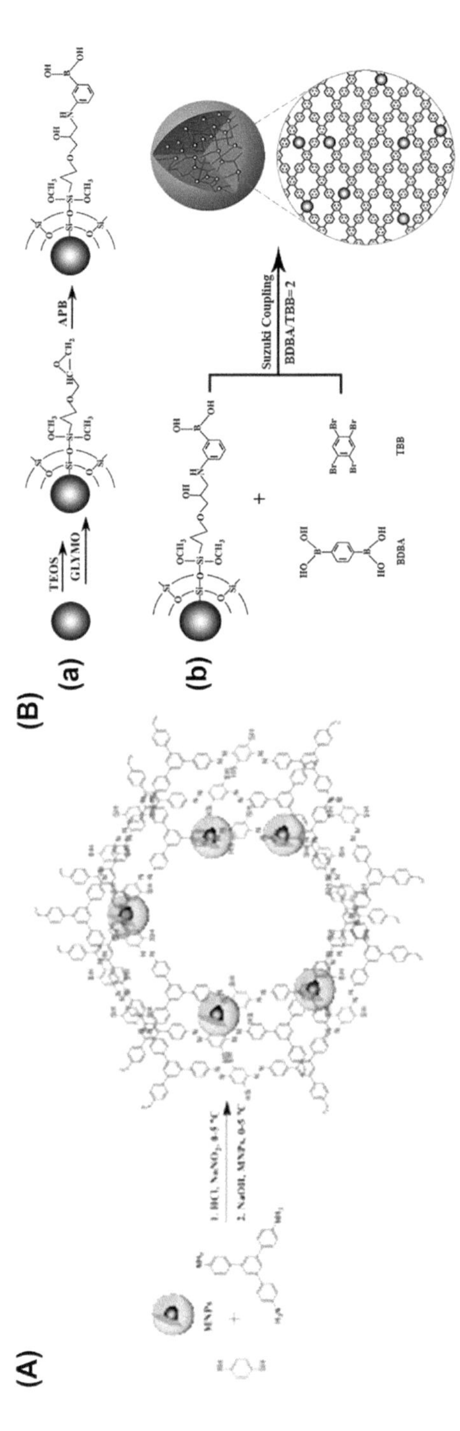


**Figure 9.20** Direct synthesis of POPs on the $Fe_3O_4$@$SiO_2$ MNPs [162] (A) and the linkage strategy for the synthesis of PP-CMP on the APB modified MNPs [163] (B).

up to 703 mg $g^{-1}$. Zhou et al. [163] synthesized polyphenylene conjugated microporous polymer (PP-CMP) on the surface of MNPs through the coupling reaction of phenylboronic acid and bromobenzene monomers. Specifically, aminophenylboric acid (APB)-modified MNPs were added to a solution of terephthalic acid and 1,2,4,5-tetrabromobenzene, Fig. 9.20. The PP-CMP magnetic sorbent was used for the extraction of PAHs in human urine.

### 9.7.4 Postmagnetization and chemical bridging

This method is simple and universal, avoiding the destruction of MNPs under the harsh conditions for POPs synthesis. Rengaraj et al. [164] synthesized $Fe_3O_4$ NPs on a covalent triazine polymer (CTPs) by adding ferrous and ferric salts to the preprepared CTPs solution. The magnetic CTPs were used for the selective extraction of Sr(II) from seawater with an adsorption capacity of 128 mg $g^{-1}$. Wang et al. [165] prepared $Fe_3O_4$ NPs on a porphyrin-based porous organic polymer by the coprecipitation method. The magnetic porphyrin-based porous organic polymer (MP-POP) was used for the extraction of benzoyl urea insecticides from vegetables. The adsorption capacity for the target analytes was 1.90–2.00 mg $g^{-1}$. Yan et al. [166] reduced nickel ions in situ on CTFs by a high-temperature hydrothermal method (Fig. 9.19A). The magnetic CTFs/Ni composites exhibit good extraction performance for phthalate esters in plastic packaging materials.

Chemical bridging offers an alternative strategy to connect preprepared MNPs and POPs. Huang et al. [167] prepared amino-modified $Fe_3O_4$ NPs and N and S-rich POPs, polythiocyanate polymer (PTMT), respectively. The PTMT was connected to the surface of the MNPs by crosslinking of glutaraldehyde, producing magnetic $Fe_3O_4$@PTMT. The sorbent exhibited good adsorption performance for Hg(II), Pb(II), and Cd(II), with an adsorption capacity of 217–603 mg $g^{-1}$ and fast adsorption kinetics. Additional applications are given in [209–213].

## 9.8 Others magnetic nanoparticle sorbents

### 9.8.1 Magnetic mesoporous materials

Mesoporous materials are porous materials with pore diameters ranging from 2 to 50 nm. They have large specific surface areas, adjustable mesoporous channels, large pore volume, and easy surface modification. Magnetic mesoporous materials have demonstrated a good potential for sample pretreatment, such as the extraction of metal ions, adsorption of organic compounds, and the selective enrichment of proteins or peptides [168].

Mesoporous silica gel is a commonly used mesoporous material because of its simple preparation and low cost. It is often used in combination with MNPs to prepare magnetic mesoporous silica. The specific preparation process involves the sol-gel reaction on the surface of MNPs, in the presence of a porogen/template (e.g., surfactant micelle template), and subsequent template removal. Due to the variety of silane coupling reagents, different functional groups (e.g., sulfhydryl, amino, IDA) can be introduced into the mesoporous silica coatings. Li et al. [169] synthesized mercaptosilane modified magnetic mesoporous silica for the extraction of Hg(II) and Pb(II) from water. Yuan et al. [170] prepared an aminosilane-modified magnetic mesoporous silica, $Fe_3O_4$@$SiO_2$@meso-$SiO_2$-$NH_2$, Fig. 9.21, suitable for the extraction of Pb(II), Cu(II), and Cd(II). Zhang et al. [172] prepared IDA modified magnetic mesoporous silica for the extraction of Cd(II), Mn(II), and Pb(II) from environmental water and biological samples. Zhao et al. [171] synthesized cyanosilane modified magnetic mesoporous silica (MMS-AN), and then ammoniated it to obtain ammonia oxime modified composite (MMS-AO), Fig. 9.21. Compared with the unmodified sorbents, MMS-AO exhibited high adsorption capacity for U(VI) (277 mg $g^{-1}$). Zhang et al. [173] immobilized a polyethylene imide polymer on the surface of magnetic mesoporous silica. The sorbents had good adsorption properties for Ag NPs with an adsorption capacity up to 909 mg $g^{-1}$.

Magnetic mesoporous $TiO_2$ is often used for extraction of phosphorylated peptides [33,174] and metal ions [175,176]. Zhao et al. [176] prepared mesoporous $TiO_2$ on the surface of MNPs coated with a phenolic resin coated. The magnetic mesoporous $TiO_2$ was used for the extraction As (V) from the water with an adsorption capacity of 139 mg $g^{-1}$. Because of the good compatibility between titanium and silica sols, $TiO_2$ coating can be modified on the surface of magnetic mesoporous silica. These sorbents are commonly used for the extraction of phosphorylated peptides [177].

### 9.8.2 *Magnetic molecular/ion imprinting polymer*

Molecular/Ion Imprinted Polymers (MIPs/IIPs) are novel adsorbents with specific recognition capability for template molecules or ions prepared by imprinting technology. Under prescribed reaction conditions, monomers, crosslinking reagents, initiators and template molecules or ions form highly crosslinked three-dimensional network polymers. Removal of the template molecules or ions leaves a cavity in the polymer structure, which matches the size, spatial structure, and binding sites of the template. MIPs/IIPs have a high selectivity for the template molecules or ions. They are relatively easy to prepare and stable and with many reported applications. However, repeated filtration and centrifugation steps are often required for phase separation and recovery of products in extraction studies. These operations are not conducive to their rapid preparation and separation. The combination of magnetic materials and MIPs/IIPs is a good way to solve this problem.

Bulk polymerization is the most common and method for their preparation. Typically weak polar solvents are used to dissolve template molecules and functional monomers with the subsequent addition of crosslinking reagents and an initiator. After ultrasonic degassing, the polymerization reaction is initiated by thermal initiation or UV irradiation for the required time. The polymer is crushed, ground, screened and washed to remove template molecules. This method is time-consuming and laborious, cannot support large-scale production, and the product particles are not uniform. Sorribes-Soriano et al. [178] prepared magnetic MIPs (MMIPs) for cocaine by bulk polymerization in the presence of MNPs. Polyethylene glycol (PEG) and

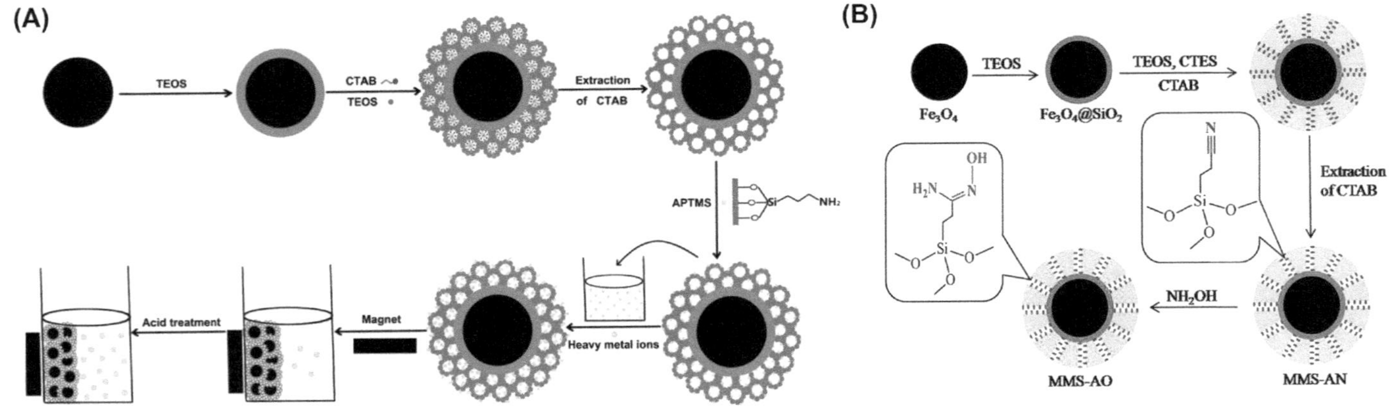


**Figure 9.21** Synthesis of $Fe_3O_4@SiO_2@meso\text{-}SiO_2\text{-}NH_2$ microsphere and its use for the extraction of heavy metal ions [170] (A) and schematic illustration of the preparation procedure of MMS-AO [171] (B).

3-(trimethoxysilyl)propyl methacrylate (V) were used for surface modification of the MNPs. A mixture of cocaine (template), MAA (monomer), EDMA (crosslinker), AIBN (initiator), and modified MNPs were dispersed in acetonitrile (porogen), and the polymerization carried out at 60°C for 24 h with magnetic stirring. The MMIPs were subsequently crushed in an agate mortar, dried at 80°C overnight, and sieved to obtain a homogeneous particle size.

In conventional molecular imprinting, there are still several problems such as incomplete template removal, low binding capacity, poor site accessibility, and slow mass transfer. Surface imprinting techniques have been proposed as a way to overcome these limitations by localizing the cavities close to the polymer's surface. At present, magnetic MIPs/IIPs are mostly prepared by surface imprinting technology, since surface imprinting can localize imprinting sites at the surface of MNPs. These materials have good selectivity and fast adsorption kinetics for template molecules and ions. Ming et al. [179] prepared core-shell structured composite of $Fe_3O_4$ NPs and MIPs by a surface imprinting technique in combination with precipitation polymerization for the selective detection of 17β-estradiol (17β-$E_2$). The detection was based on the competitive desorption of fluorescein from the magnetic MIPs. The competitive rebinding of 17β-$E_2$ to the corresponding recognition sites regulated the release of fluorescein and resulted in an enhanced fluorescence signal. Li et al. [180] proposed a protocol for the preparation of glycoproteins-imprinted MNPs by boronate affinity-based controllable oriented surface imprinting, Fig. 9.22. Self-polymerization of 2-anilinoethanol was used to form an imprinting coating of appropriate thickness. This imprinting protocol allows for nearly complete removal of template molecules and provided good accessibility to target molecules, which is especially suitable for the imprinting of proteins. In addition, excessive binding sites on the boronic acid functionalized MNPs are almost completely covered by the imprinting coating, which effectively suppresses non-specific adsorption.

Luo et al. [181] prepared magnetic Cu(II) IIP by a combination of surface imprinting technology and the sol-gel method. Since the imprinted sites are on the surface of MNPs, the prepared materials have fast adsorption kinetics and higher selectivity for Cu(II) over nonimprinted sorbents. The selectivity coefficients are 29.2 and 38.2 for Cu(II)/Zn(II) and Cu(II)/Ni (Cu), respectively, and the adsorption capacity of Cu(II) is 24.2 mg $g^{-1}$, which is higher than for nonimprinted materials (5.2 mg $g^{-1}$). The magnetic Cu(II)-IIP was used for the selective extraction of trace of Cu(II) in water samples [182].

Hierarchical imprinting (also known as double imprinting) can improve mass transfer kinetics of imprinting materials, facilitate template elution, and improve the leakage of target compounds to some extent. In this technology, sacrificial templates of different sizes are often used in the imprinting reactions. Removing sacrificial templates can leave relatively large holes in imprinted materials, which facilitates the rapid mass transfer of template molecules/ions. Cen et al. [183] used Cd(II) and CTAB surfactant micelles as templates and aminosilane and TEOS as functional monomers and crosslinking reagents, respectively, in a sol-gel polymerization reaction carried out in the presence of $Fe_3O_4$ NPs. Magnetic mesoporous silica Cd(II) IIPs were obtained. There are large holes left by the micelle template around the ion

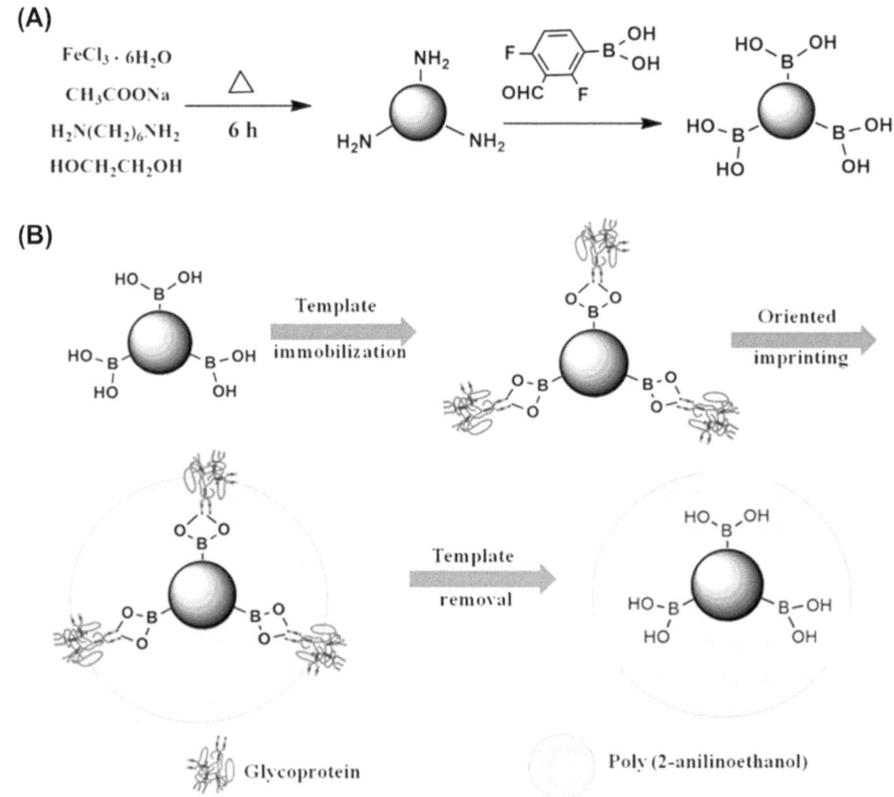


**Figure 9.22** Synthesis of (A) DFFPBA-MNPs and (B) glycoprotein-imprinted MNPs using oriented surface imprinting [180].

imprinting sites, which is beneficial for the rapid mass transfer of the template Cd(II) ion. Wang et al. [184] prepared a double-template MIP on the surface of magnetic mesoporous silica ($Fe_3O_4$@m$SiO_2$@DT-MIP) using dibenzothiophene (DBT) and 4-methyldibenzothiophene (4-MDBT) as template molecules. The adsorption capacity of the magnetic sorbents for DBT and 4-MDBT reached 104.2 and 113.6 mg $g^{-1}$, respectively. $Fe_3O_4$@m$SiO_2$@DT-MIP was used for the selective and simultaneous removal of dibenzothiophene (DBT) and 4-methyldibenzothiophene (4-MDBT) from gasoline. Wei et al. [185] prepared double imprinting magnetic materials with Pb(II) and Cd(II) as the templates. This material demonstrated favorable selectivity for both template ions and a high adsorption capacity. It was used for the selective extraction of Pb(II) and Cd(II) from food and environmental samples.

In the template inspired polymer preparation process, even after repeated washing, not all of the template may be removed. When imprinted materials are used for extraction the residual template may gradually be released, resulting in leakage that affects the accuracy by which the concentration of the template can be determined. Dummy template can be used to prepare imprinted materials, which can effectively avoid the leakage problem. A substance with a similar structure to the target analytes is selected as the template, such as an analog. Due to the small radius of metal ions and the small differences in radius between metal ions, the dummy template method is used quite commonly. Bagheri et al. [186] prepared magnetic dummy (D) MIPs with propanamide as a dummy template molecule for the extraction of acrylamide in biscuit samples. The preparation of the magnetic sorbent involved encapsulation of $Fe_3O_4$ NPs by anchoring hydrophilic PEG chains on the surface of the NPs then polymerization and imprinting of chitosan on the modified $Fe_3O_4$ NPs surface to produce specific molecular binding sites. The selectivity of magnetic DMIP was evaluated by comparing the adsorption capacity, imprinting factors (IF) and selectivity factors (SF) for acrylamide and similar compounds. The IF values for acrylamide, propanamide, L-asparagine, 6-aminocaproic acid, and N-tert-butylacrylamide were 3.49, 3.54, 2.14, 1.73 and 1.5, respectively, and the SF of four analogues 0.90, 1.67, 2.13 and 2.48, respectively.

Luminescent nanoparticles (NPs) incorporated in MIPs have shown great potential as sensors for a wide range of applications. Luo et al. [187] synthesized a magnetic imprinting material ($Fe_3O_4$@CdTe-IIP) by preparing CdTe quantum dots (QDs) on the MNPs surface and then eluting Cd(II) with EDTA. The $Fe_3O_4$@CdTe-IIP sorbent had good selectivity for Cd(II). When Cd(II) was adsorbed on MNPs surface, the fluorescence of CdTe QDs gradually recovers and is enhanced when Cd(II) is combined with Te(II). It can be used as a visual monitor for the real-time adsorption of Cd(II) on materials. Li et al. [188] fabricated multifunctional magnetic luminescent MIP nanocomposites via a one-pot emulsion reaction using polystyrene-*co*-methacrylic acid polymer, $Fe_3O_4$ NPs, and luminescent $LaVO_4:Eu^{3+}$ NPs as building blocks with a phenanthrene template. The resulting nanocomposites were used for the luminescence detection of phenanthrene.

## 9.8.3 Magnetic restricted access materials

Restricted access materials (RAM) are biocompatible sorbents with a physical or chemical diffusion barrier. They limit access of macromolecules such as proteins to the inner surface of porous materials by either size exclusion (physical barrier dependent on the pore structure) and or hydrophilicity (chemical barrier dependent on the formation of a surface hydrophilic layer). Small target compounds can be selectively retained by the sorbent by their transport through the pore structure and favorable interactions with sorption sites on the inner surface of the material. RAM materials avoid the need for protein precipitation when working with biological fluids, which can easily cause the loss of analytes. They can be used directly for sample treatment after simple dilution, saving time, and simplifying the treatment steps [189]. Combining RAM with MNPs, the resultant restricted accessed MNPs can be used to analyze biological samples with complex matrixes (such as urine and serum). The installation of a porous hydrophilic layer on the surface of some adsorbents (e.g., silica, CNTs, magnetic materials, etc.) is a common method to prepare RAM materials. Ye et al. [190] immobilized a nonionic surfactant TW20 on the surface of a $C_{12}$ functionalized $Fe_3O_4$ NPs to extract steroid hormones from environmental water and human urine samples. Yan et al. [191] synthesized a restricted accessed MNPs $Fe_3O_4$@$SiO_2$@PAR@TW20 for the extraction of metal ions from biological fluids. The magnetic RAM material was coated by TW20 and modified by 4-2-pyridylazoresorcinol (PAR) on the inner surface. The magnetic RAM sorbent was used for the extraction of Cd(II), Pb(II), Cr(III), La(III), and Nd(II) from human serum and urine samples. Nanoparticle coatings containing mesoporous silica can provide mesoporous channels for restricted access materials as well as introducing functional groups into those channels. Liu et al. [192] modified the surface of $Fe_3O_4$ NPs with mesoporous silica coatings containing $C_{18}$ groups as an external RAM barrier. The magnetic mesoporous RAM materials were used for the extraction of small molecules from mouse serum. The combination of RAM and imprinted materials can further improve the selectivity toward target compounds. Lv et al. [193] prepared a magnetic microsphere restricted access media-molecularly imprinted (RAM-MIMM) material through layer by layer modification.

# References


[1] Philippovaa O, Molchanov V, Khokhlov A. Magnetic polymer beads: recent trends and developments in synthetic design and applications. Eur Polym J 2011;47(4):542−59.

[2] Liu X, et al. Preparation and characterization of amino−silane modified superparamagnetic silica nanospheres. J Magn Magn Mater 2004;270(1):1−6.

[3] Lin CC, Ho JM. Structural analysis and catalytic activity of $Fe_3O_4$ nanoparticles prepared by a facile co-precipitation method in a rotating packed bed. Ceram Int 2014;40(7):10275−82.

[4] Ding Y, et al. 12-Hydrothermal synthesis and characterization of $Fe_3O_4$ nanorods. J Inorg Organomet Polym Mater 2013;23(2):379−84.

[5] Li X, et al. Synthesis and magnetic properties of manganese−zinc ferrite nanoparticles obtained via a hydrothermal method. J Mater Sci Mater Electron 2017;28(16):12268−72.

[6] Inouye K, et al. Oxygenation of ferrous ions in reversed micelle and reversed microemulsion. J Phys Chem 1982;86(8):1465−9.

[7] Pemartin K, et al. Synthesis of Mn−Zn ferrite nanoparticles by the oil-in-water microemulsion reaction method. Colloid Surf Physicochem Eng Asp 2014;451:161−71.

[8] Cui H, Liu Y, Ren W. Structure switch between $\alpha$-$Fe_2O_3$, $\gamma$-$Fe_2O_3$ and $Fe_3O_4$ during the large scale and low temperature sol−gel synthesis of nearly monodispersed iron oxide nanoparticles. Adv Powder Technol 2013;24(1):93−7.

[9] Zhang Y, et al. Synthesis, structure and electromagnetic properties of mesoporous $Fe_3O_4$ aerogels by sol−gel method. Mater Sci Eng B 2014;188:13−9.

[10] Massart R. Preparation of aqueous magnetic liquids in alkaline and acidic media. IEEE Transact Magn 1981;17(2):1247−8.

[11] Jana NR, Chen Y, Peng X. Size-and shape-controlled magnetic (Cr, Mn, Fe, Co, Ni) oxide nanocrystals via a simple and general approach. Chem Mater 2004;16(20):3931−5.

[12] Asuha S, et al. Direct synthesis of $Fe_3O_4$ nanopowder by thermal decomposition of Fe−urea complex and its properties. J Alloy Comp 2011;509(6):2870−3.

[13] Gedanken A. Using sonochemistry for the fabrication of nanomaterials. Ultrason Sonochem 2004;11(2):47−55.

[14] Vijayakumar R, et al. Sonochemical synthesis and characterization of pure nanometer-sized $Fe_3O_4$ particles. Mater Sci Eng A 2000;286(1):101−5.

[15] Aguilar-Arteaga K, Rodriguez JA, Barrado E. Magnetic solids in analytical chemistry: a review. Anal Chim Acta 2010;674(2):157–65.

[16] Lu Y, et al. Modifying the surface properties of superparamagnetic iron oxide nanoparticles through a sol-gel approach. Nano Lett 2002;2(3):183–6.

[17] Deng YH, et al. Investigation of formation of silica-coated magnetite nanoparticles via sol-gel approach. Colloid Surf Physicochem Eng Asp 2005;262(1–3):87–93.

[18] Wu JH, et al. Titania coated magnetic mesoporous hollow silica microspheres: fabrication and application to selective enrichment of phosphopeptides. Chem Commun 2010; 46(47):9031–3.

[19] Ding HL, et al. $Fe_3O_4$@$SiO_2$ core/shell nanoparticles: the silica coating regulations with a single core for different core sizes and shell thicknesses. Chem Mater 2012;24(23): 4572–80.

[20] Ma Q, et al. Multilayered, core/shell nanoprobes based on magnetic ferric oxide particles and quantum dots for multimodality imaging of breast cancer tumors. Biomater 2012; 33(33):8486–94.

[21] Liu XQ, et al. Preparation and characterization of amino-silane modified superparamagnetic silica nanospheres. J Magn Magn Mater 2004;270(1–2):1–6.

[22] Deng Y, et al. Superparamagnetic high-magnetization microspheres with an $Fe_3O_4$@$SiO_2$ core and perpendicularly aligned mesoporous $SiO_2$ shell for removal of microcystins. J Am Chem Soc 2008;130(1):28. +.

[23] Suleiman JS, et al. Separation/preconcentration of trace amounts of Cr, Cu and Pb in environmental samples by magnetic solid-phase extraction with Bismuthiol-II-immobilized magnetic nanoparticles and their determination by ICP-OES. Talanta 2009;77(5):1579–83.

[24] Jiang HM, et al. Magnetic solid-phase extraction combined with graphite furnace atomic absorption spectrometry for speciation of Cr(III) and Cr(VI) in environmental waters. Talanta 2013;116:361–7.

[25] Mashhadizadeh MH, Amoli-Diva M. Atomic absorption spectrometric determination of $Al^{3+}$ and $Cr^{3+}$ after preconcentration and separation on 3-mercaptopropionic acid modified silica coated-$Fe_3O_4$ nanoparticles. J Anal Atomic Spectrom 2013;28(2):251–8.

[26] Cao YP, et al. Selenium speciation in radix puerariae using ultrasonic assisted extraction combined with reversed phase high performance liquid chromatography-inductively coupled plasma-mass spectrometry after magnetic solid-phase extraction with 5-sulfosalicylic acid functionalized magnetic nanoparticles. Spectrochim Acta B Atom Spectrosc 2016;122:172–7.

[27] Cui C, et al. Ionic liquid-based magnetic solid phase extraction coupled with inductively coupled plasma-optical emission spectrometry for the determination of Cu, Cd, and Zn in biological samples. J Anal Atomic Spectrom 2013;28(7):1110–7.

[28] Zhang F, et al. Removal of heavy metal ions from aqueous solution using $Fe_3O_4$-$SiO_2$-poly(1,2-diaminobenzene) core-shell sub-micron particles. J Colloid Interface Sci 2012; 387:205–12.

[29] Cui C, et al. Chitosan modified magnetic nanoparticles based solid phase extraction combined with ICP-OES for the speciation of Cr(iii) and Cr(vi). Anal Methods 2014; 6(21):8577–83.

[30] Shan C, et al. Removal of Hg(II) by poly(1-vinylimidazole)-grafted $Fe_3O_4$@$SiO_2$ magnetic nanoparticles. Water Res 2015;69:252–60.

[31] Giakisikli G, Anthemidis AN. Magnetic materials as sorbents for metal/metalloid preconcentration and/or separation. A review. Anal Chim Acta 2013;789:1–16.

[32] Huang KP, et al. Novel stationary phase for complexation gas chromatography originating from ionic liquid and metallomesogen. J Chromatogr A 2008;1215(1–2):177–84.
[33] Ma WF, et al. Tailor-made magnetic $Fe_3O_4$@mTiO(2) microspheres with a tunable mesoporous anatase shell for highly selective and effective enrichment of phosphopeptides. ACS Nano 2012;6(4):3179–88.
[34] Su PF, Chen G, Zhao J. Convenient preparation and characterization of surface carboxyl-functioned $Fe_3O_4$ magnetic nanoparticles. Chem J Chinese Univ-Chinese 2011;32(7): 1472–7.
[35] Takami S, et al. Hydrothermal synthesis of surface-modified iron oxide nanoparticles. Mater Lett 2007;61(26):4769–72.
[36] Huang CZ, Hu B. Silica-coated magnetic nanoparticles modified with γ-mercaptopropyltrimethoxysilane for fast and selective solid phase extraction of trace amounts of Cd, Cu, Hg, and Pb in environmental and biological samples prior to their determination by inductively coupled plasma mass spectrometry. Spectrochim Acta B Atom Spectrosc 2008;63(3):437–44.
[37] Huang CZ, Hu B. Speciation of inorganic tellurium from seawater by ICP-MS following magnetic SPE separation and preconcentration. J Sep Sci 2008;31(4):760–7.
[38] Ma SS, et al. Magnetic solid phase extraction coupled with inductively coupled plasma mass spectrometry for the speciation of mercury in environmental water and human hair samples. Talanta 2016;146:93–9.
[39] Li Y, et al. Displacement solid-phase extraction on mercapto-functionalized magnetite microspheres for inductively coupled plasma mass spectrometric determination of trace noble metals. Anal Chim Acta 2011;692(1–2):42–9.
[40] Wang JH, et al. Amino-functionalized $Fe_3O_4$@$SiO_2$ core-shell magnetic nanomaterial as a novel adsorbent for aqueous heavy metals removal. J Colloid Interface Sci 2010;349(1): 293–9.
[41] Zhang JM, et al. Pb(II) removal of $Fe_3O_4$@$SiO_2$-$NH_2$ core-shell nanomaterials prepared via a controllable sol-gel process. Chem Eng J 2013;215:461–71.
[42] Naghizadeh M, et al. Simultaneous preconcentration of bismuth and lead ions on modified magnetic core-shell nanoparticles and their determination by ETAAS. Chem Eng J 2015;281:444–52.
[43] Huang CZ, et al. Speciation of inorganic arsenic in environmental waters using magnetic solid phase extraction and preconcentration followed by ICP-MS. Microchim Acta 2011; 173(1–2):165–72.
[44] Wang H, et al. Chip-based array magnetic solid phase microextraction on-line coupled with inductively coupled plasma mass spectrometry for the determination of trace heavy metals in cells. Analyst 2015;140(16):5619–26.
[45] Li P, et al. Magnetic solid phase extraction for the determination of trace antimony species in water by inductively coupled plasma mass spectrometry. Talanta 2015;134:292–7.
[46] Liu Y, et al. Multifunctional nanocomposites $Fe_3O_4$@$SiO_2$-EDTA for Pb(II) and Cu(II) removal from aqueous solutions. Appl Surf Sci 2016;369:267–76.
[47] Wei WY, et al. Iminodiacetic acid functionalized magnetic nanoparticles for speciation of Cr(III) and Cr(VI) followed by graphite furnace atomic absorption spectrometry detection. RSC Adv 2017;7(14):8504–11.
[48] Cui Y, et al. Magnetic solid-phase extraction of trace-level mercury(II) ions using magnetic core-shell nanoparticles modified with thiourea-derived chelating agents. Microchim Acta 2015;182(7–8):1337–44.
[49] Wang Y, et al. Functionalized magnetic silica nanoparticles for highly efficient adsorption of $Sm^{3+}$ from a dilute aqueous solution. Langmuir 2018;34(8):2674–84.

[50] Chen K, et al. Removal of cadmium and lead ions from water by sulfonated magnetic nanoparticle adsorbents. J Colloid Interface Sci 2017;494:307–16.
[51] Zhao JJ, et al. Synthesis of Schiff base functionalized superparamagnetic $Fe_3O_4$ composites for effective removal of Pb(II) and Cd(II) from aqueous solution. Chem Eng J 2018;347:574–84.
[52] Wu D, Sun Y, Wang Q. Adsorption of lanthanum (III) from aqueous solution using 2-ethylhexyl phosphonic acid mono-2-ethylhexyl ester-grafted magnetic silica nanocomposites. J Hazard Mater 2013;260:409–19.
[53] Asgharinezhad AA, et al. A simple and fast method based on new magnetic ion imprinted polymer nanoparticles for the selective extraction of Ni(II) ions in different food samples. RSC Adv 2015;5(56):45510–9.
[54] Qian J, et al. Synthesis of surface ion-imprinted magnetic microspheres by locating polymerization for rapid and selective separation of uranium(VI). RSC Adv 2015;5(6): 4153–61.
[55] Zhang Q, et al. Ionic liquid-coated $Fe_3O_4$ magnetic nanoparticles as an adsorbent of mixed hemimicelles solid-phase extraction for preconcentration of polycyclic aromatic hydrocarbons in environmental samples. Analyst 2010;135(9):2426–33.
[56] Bouri M, et al. Ionic liquids supported on magnetic nanoparticles as a sorbent preconcentration material for sulfonylurea herbicides prior to their determination by capillary liquid chromatography. Anal Bioanal Chem 2012;404(5):1529–38.
[57] Ma HL, et al. Preparation and characterization of superparamagnetic iron oxide nanoparticles stabilized by alginate. Int J Pharm 2007;333(1–2):177–86.
[58] Ma GH, Su ZG. Polymer microspheres. Chemical Industry Press 2005:1–68.
[59] Harris LA, et al. Magnetite nanoparticle dispersions stabilized with triblock copolymers. Chem Mater 2003;15(6):1367–77.
[60] Xiong LR, et al. Synthesis and stability of PEG coated $Fe_3O_4$ magnetic fluid. Material Report 2007;(S1):195–7.
[61] Liu XY, et al. Preparation and characterization of amphiphilic polymer magnetic microspheres containing amino groups. Acta Polym Sin 2005;(4):519–23.
[62] Du XY, Geng LP. Preparation and characterization of magnetic polystyrene microspheres. Material Report 2010;24(S1):170–171+176.
[63] Liu ZL, et al. Preparation and characterization of magnetic P(St-co-MAA-co-AM) microspheres. J Magn Magn Mater 2006;302(2):529–35.
[64] Cui C, Hu B. Chitosan modified magnetic microsphere solid phase extraction combined with inductively coupled plasma-optical emission spectrometry for the analysis of chromium (III) and chromium (VI) in environmental water samples. Anal Methods 2014; 6(21):8577–83.
[65] Li Y, et al. Synthesis of core-shell magnetic molecular imprinted polymer by the surface RAFT polymerization for the fast and selective removal of endocrine disrupting chemicals from aqueous solutions. Environ Pollut 2010;158(6):2317–23.
[66] Fang CL, et al. Monodisperse alpha-$Fe_2O_3$@$SiO_2$@Au core/shell nanocomposite spheres: synthesis, characterization and properties. Nanotechnol 2008;19(12):125601.
[67] Ding J, et al. n-Octadecylphosphonic acid grafted mesoporous magnetic nanoparticle: preparation, characterization, and application in magnetic solid-phase extraction. J Chromatogr A 2010;1217(47):7351–8.
[68] Huang LJ, et al. A designable magnetic MOF composite and facile coordination-based post-synthetic strategy for the enhanced removal of $Hg^{2+}$ from water. J Mater Chem A 2015;3(21):11587–95.

[69] Kan X, et al. Magnetic molecularly imprinted polymer for aspirin recognition and controlled release. Nanotechnol 2009;20(16).
[70] Ma ZY, Guan YP, Liu HZ. Synthesis and characterization of micron-sized monodisperse superparamagnetic polymer particles with amino groups. J Polym Sci A Polym Chem 2005;43(15):3433–9.
[71] Wang Y, et al. Determination of trace amounts of Se(IV) by hydride generation atomic fluorescence spectrometry after solid-phase extraction using magnetic multi-walled carbon nanotubes. Talanta 2013;112:123–8.
[72] Sadegh H, et al. MWCNTs-$Fe_3O_4$ nanocomposite for Hg(II) high adsorption efficiency. J Mol Liq 2018;258:345–53.
[73] Demir A, et al. Low temperature magnetic investigation of $Fe_3O_4$ nanoparticles filled into multiwalled carbon nanotubes. Synth Met 2014;187:75–80.
[74] Cao HQ, Zhu MF, Li YG. Novel carbon nanotube iron oxide magnetic nanocomposites. J Magn Magn Mater 2006;305(2):321–4.
[75] Tarigh GD, Shemirani F. Magnetic multi-wall carbon nanotube nanocomposite as an adsorbent for preconcentration and determination of lead (II) and manganese (II) in various matrices. Talanta 2013;115:744–50.
[76] Zhang C, et al. Efficient removal of heavy metal ions by thiol-functionalized superparamagnetic carbon nanotubes. Chem Eng J 2012;210:45–52.
[77] Fayazi M, et al. Synthesis and application of novel ion-imprinted polymer coated magnetic multi-walled carbon nanotubes for selective solid phase extraction of lead(II) ions. Mater Sci Eng C 2016;60:365–73.
[78] Taghizadeh M, et al. Solid phase extraction of heavy metal ions based on a novel functionalized magnetic multi-walled carbon nanotube composite with the aid of experimental design methodology. Microchim Acta 2014;181(5–6):597–605.
[79] Fahimirad B, Asghari A, Rajabi M. A novel nanoadsorbent consisting of covalently functionalized melamine onto MWCNT/$Fe_3O_4$ nanoparticles for efficient microextraction of highly adverse metal ions from organic and inorganic vegetables: optimization by multivariate analysis. J Mol Liq 2018;252:383–91.
[80] Moazzen M, et al. Magnetic solid-phase extraction based on magnetic multi-walled carbon nanotubes for the determination of polycyclic aromatic hydrocarbons in grilled meat samples. Talanta 2013;115:957–65.
[81] Tang WW, et al. Simultaneous adsorption of atrazine and Cu (II) from wastewater by magnetic multi-walled carbon nanotube. Chem Eng J 2012;211:470–8.
[82] Behbahani M, et al. Monitoring of trace amounts of heavy metals in different food and water samples by flame atomic absorption spectrophotometer after preconcentration by amine-functionalized graphene nanosheet. Environ Monit Assess 2014;186(11): 7245–57.
[83] Wang YK, et al. Graphene-based solid-phase extraction combined with flame atomic absorption spectrometry for a sensitive determination of trace amounts of lead in environmental water and vegetable samples. Anal Chim Acta 2012;716:112–8.
[84] Sun JP, et al. One-step synthesis of magnetic graphene oxide nanocomposite and its application in magnetic solid phase extraction of heavy metal ions from biological samples. Talanta 2015;132:557–63.
[85] Kazemi E, Dadfarnia S, Haji Shabani AM. Dispersive solid phase microextraction with magnetic graphene oxide as the sorbent for separation and preconcentration of ultra-trace amounts of gold ions. Talanta 2015;141:273–8.

[86] Guo FY, et al. Adsorption behavior of Cr(VI) from aqueous solution onto magnetic graphene oxide functionalized with 1,2-diaminocyclohexanetetraacetic acid. RSC Adv 2015;5(56):45384–92.

[87] Cui LM, et al. EDTA functionalized magnetic graphene oxide for removal of Pb(II), Hg(II) and Cu(II) in water treatment: adsorption mechanism and separation property. Chem Eng J 2015;281:1–10.

[88] Zhao DL, et al. Facile preparation of amino functionalized graphene oxide decorated with $Fe_3O_4$ nanoparticles for the adsorption of Cr(VI). Appl Surf Sci 2016;384:1–9.

[89] Ma YX, et al. Synthesis of magnetic graphene oxide grafted polymaleicamide dendrimer nanohybrids for adsorption of Pb(II) in aqueous solution. J Hazard Mater 2017;340: 407–16.

[90] Wang L, et al. Anionic polypeptide poly($\gamma$-glutamic acid)-functionalized magnetic $Fe_3O_4$-GO-(o-MWCNTs) hybrid nanocomposite for high-efficiency removal of Cd(II), Cu(II) and Ni(II) heavy metal ions. Chem Eng J 2018;346:38–49.

[91] Jalilian N, et al. Extraction and determination of trace amounts of gold(III), palladium(II), platinum(II) and silver(I) with the aid of a magnetic nanosorbent made from $Fe_3O_4$-decorated and silica-coated graphene oxide modified with a polypyrrole-polythiophene copolymer. Microchim Acta 2017;184(7):2191–200.

[92] Rofouei MK, et al. A bucky gel consisting of $Fe_3O_4$ nanoparticles, graphene oxide and ionic liquid as an efficient sorbent for extraction of heavy metal ions from water prior to their determination by ICP-OES. Microchim Acta 2017;184(9):3425–32.

[93] Neyestani MR, et al. A magnetized graphene oxide modified with 2-mercaptobenzothiazole as a selective nanosorbent for magnetic solid phase extraction of gold(III), palladium(II) and silver(I). Microchim Acta 2017;184(8):2871–9.

[94] Chen J, et al. Magnetic solid-phase extraction for the removal of mercury from water with ternary hydrosulphonyl-based deep eutectic solvent modified magnetic graphene oxide. Talanta 2018;188:454–62.

[95] Su SW, et al. Determination of trace/ultratrace rare earth elements in environmental samples by ICP-MS after magnetic solid phase extraction with $Fe_3O_4$@$SiO_2$@polyaniline-graphene oxide composite. Talanta 2014;119:458–66.

[96] Zhang RZ, Su P, Yang Y. Microwave-assisted preparation of magnetic nanoparticles modified with graphene oxide for the extraction and analysis of phenolic compounds. J Sep Sci 2014;37(22):3339–46.

[97] Lee J, Kim J, Hyeon T. Recent progress in the synthesis of porous carbon materials. Adv Mater 2006;18(16):2073–94.

[98] Maya F, et al. Magnetic solid-phase extraction using metal-organic frameworks (MOFs) and their derived carbons. Trac Trends Anal Chem 2017;90:142–52.

[99] Dutta S, Bhaumik A, Wu KCW. Hierarchically porous carbon derived from polymers and biomass: effect of interconnected pores on energy applications. Energy Environ Sci 2014; 7(11):3574–92.

[100] Lu AH, et al. Highly stable carbon-protected cobalt nanoparticles and graphite shells. Chem Commun 2005;(1):98–100.

[101] Habila MA, et al. Carbon-coated $Fe_3O_4$ nanoparticles with surface amido groups for magnetic solid phase extraction of Cr(III), Co(II), Cd(II), Zn(II) and Pb(II) prior to their quantitation by ICP-MS. Microchim Acta 2017;184(8):2645–51.

[102] Peng CY, et al. Magnetic sulfur-doped porous carbon for preconcentration of trace mercury in environmental water prior to ICP-MS detection. Analyst 2017;142(23): 4570–9.

[103] Wen T, et al. Magnetic porous carbonaceous material produced from tea waste for efficient removal of As(V), Cr(VI), humic acid, and dyes. ACS Sustain Chem Eng 2017; 5(5):4371–80.
[104] Chen AB, et al. Solid-solid grinding/templating route to magnetically separable nitrogen-doped mesoporous carbon for the removal of $Cu^{2+}$ ions. J Hazard Mater 2014;279: 280–8.
[105] Yang GD, et al. Simultaneous removal of lead and phenol contamination from water by nitrogen-functionalized magnetic ordered mesoporous carbon. Chem Eng J 2015;259: 854–64.
[106] Huang LJ, et al. Facile fabrication of N-doped magnetic porous carbon for highly efficient mercury removal. ACS Sustain Chem Eng 2018;6(8):10191–9.
[107] Liu MX, et al. Magnetically separated and N, S co-doped mesoporous carbon microspheres for the removal of mercury ions. Chin Chem Lett 2016;27(5):795–800.
[108] Huang LJ, et al. Highly efficient magnetic nitrogen-doped porous carbon prepared by one-step carbonization strategy for $Hg^{2+}$ removal from water. ACS Appl Mater Interfaces 2017;9(3):2550–9.
[109] Liao Q, et al. Highly cuboid-shaped heterobimetallic metal-organic frameworks derived from porous Co/ZnO/C microrods with improved electromagnetic wave absorption capabilities. ACS Appl Mater Interfaces 2018;10(34):29136–44.
[110] Guo WL, et al. Synthesis and application of 8-hydroxyquinoline modified magnetic mesoporous carbon for adsorption of multivariate metal ions from aqueous solutions. J Ind Eng Chem 2015;21:340–9.
[111] Samadi A, Amjadi M. Magnetic $Fe_3O_4$@C nanoparticles modified with 1-(2-thiazolylazo)-2-naphthol as a novel solid-phase extraction sorbent for preconcentration of copper (II). Microchim Acta 2015;182(1–2):257–64.
[112] Wu YW, et al. $Fe_3O_4$@$ZrO_2$ nanoparticles magnetic solid phase extraction coupled with flame atomic absorption spectrometry for chromium(III) speciation in environmental and biological samples. Appl Surf Sci 2012;258(18):6772–6.
[113] Zhang N, Peng HY, Hu B. Light-induced pH change and its application to solid phase extraction of trace heavy metals by high-magnetization $Fe_3O_4$@$SiO_2$@$TiO_2$ nanoparticles followed by inductively coupled plasma mass spectrometry detection. Talanta 2012;94(0):278–83.
[114] Habila MA, et al. Synthesis and application of $Fe_3O_4$@$SiO_2$@$TiO_2$ for photocatalytic decomposition of organic matrix simultaneously with magnetic solid phase extraction of heavy metals prior to ICP-MS analysis. Talanta 2016;154:539–47.
[115] Abdolmohammad-Zadeh H, Talleb Z. Speciation of As(III)/As(V) in water samples by a magnetic solid phase extraction based on $Fe_3O_4$/Mg-Al layered double hydroxide nanohybrid followed by chemiluminescence detection. Talanta 2014;128:147–55.
[116] Nyaba L, Matong JM, Nomngongo PN. Nanoparticles consisting of magnetite and $Al_2O_3$ for ligandless ultrasound-assisted dispersive solid phase microextraction of Sb, Mo and V prior to their determination by ICP-OES. Microchim Acta 2016;183(4):1289–97.
[117] Li J, et al. Magnetic polydopamine decorated with Mg-Al LDH nanoflakes as a novel bio-based adsorbent for simultaneous removal of potentially toxic metals and anionic dyes. J Mater Chem 2016;4(5):1737–46.
[118] Lou ZM, et al. Enhanced removal of As(III)/(V) from water by simultaneously supported and stabilized Fe-Mn binary oxide nanohybrids. Chem Eng J 2017;322:710–21.
[119] Zhang GS, et al. Arsenate uptake and arsenite simultaneous sorption and oxidation by Fe-Mn binary oxides: influence of Mn/Fe ratio, pH, $Ca^{2+}$, and humic acid. J Colloid Interface Sci 2012;366(1):141–6.

[120] Shan C, Tong MP. Efficient removal of trace arsenite through oxidation and adsorption by magnetic nanoparticles modified with Fe-Mn binary oxide. Water Res 2013;47(10): 3411–21.

[121] Xiang B, et al. 3D hierarchical flower-like nickel ferrite/manganese dioxide toward lead (II) removal from aqueous water. J Hazard Mater 2017;325:178–88.

[122] Sun TY, et al. Efficient As(III) removal by magnetic CuO-Fe(3)O(4) nanoparticles through photo-oxidation and adsorption under light irradiation. J Colloid Interface Sci 2017;495:168–77.

[123] Lopez-Garcia I, et al. Speciation of very low amounts of antimony in waters using magnetic core-modified silver nanoparticles and electrothermal atomic absorption spectrometry. Talanta 2017;162:309–15.

[124] Hsu KC, et al. Microfluidic desorption-free magnetic solid phase extraction of $Hg^{2+}$ from biological samples using cysteine-coated gold-magnetite core-shell nanoparticles prior to its quantitation by ICP-MS. Talanta 2017;162:523–9.

[125] Yan P, et al. Fast preconcentration of trace rare earth elements from environmental samples by di(2-ethylhexyl)phosphoric acid grafted magnetic nanoparticles followed by inductively coupled plasma mass spectrometry detection. Spectrochim Acta B Atom Spectrosc 2017;136:73–80.

[126] Karimi MA, et al. Solid phase extraction of trace amounts of silver(I) using dithizone-immobilized alumina-coated magnetite nanoparticles prior to determination by flame atomic absorption spectrometry. Int J Environ Anal Chem 2012;92(12):1325–40.

[127] Tavallali H, Deilamy Rad G, Peykarimah P. Preconcentration and speciation of Cr(III) and Cr(VI) in water and soil samples by spectrometric detection via use of nanosized alumina-coated magnetite solid phase. Environ Monit Assess 2013;185(9):7723–38.

[128] Munonde TS, Maxakato NW, Nomngongo PN. Preconcentration and speciation of chromium species using ICP-OES after ultrasound-assisted magnetic solid phase extraction with an amino-modified magnetic nanocomposite prepared from $Fe_3O_4$, $MnO_2$ and $Al_2O_3$. Microchim Acta 2017;184(4):1223–32.

[129] Indira TK, Lakshml PK. Magnetic nanoparticles – a review. Int J Pharm Sci Nanotechnol 2010;3(3):1035–42.

[130] Lohe MR, et al. Heating and separation using nanomagnet-functionalized metal-organic frameworks. Chem Commun 2011;47(11):3075–7.

[131] Arai T, Kawasaki N, Kanoh H. Magnetically separable Cu-carboxylate MOF catalyst for the Henry reaction. Synlett 2012;(10):1549–53.

[132] Zhang T, et al. Synthesis of $Fe_3O_4$@ZIF-8 magnetic core-shell microspheres and their potential application in a capillary microreactor. Chem Eng J 2013;228:398–404.

[133] Ke F, et al. Facile fabrication of magnetic metal-organic framework nanocomposites for potential targeted drug delivery. J Mater Chem 2011;21(11):3843–8.

[134] Zhao M, Deng C, Zhang X. The design and synthesis of a hydrophilic core-shell-shell structured magnetic metal-organic framework as a novel immobilized metal ion affinity platform for phosphoproteome research. Chem Commun 2014;50(47):6228–31.

[135] Bagheri A, et al. Synthesis and characterization of magnetic metal-organic framework (MOF) as a novel sorbent, and its optimization by experimental design methodology for determination of palladium in environmental samples. Talanta 2012;99:132–9.

[136] Wang Y, et al. Preparation of a functionalized magnetic metal-organic framework sorbent for the extraction of lead prior to electrothermal atomic absorption spectrometer analysis. J Mater Chem 2013;1(31):8782–9.

[137] Ricco R, et al. Lead(II) uptake by aluminium based magnetic framework composites (MFCs) in water. J Mater Chem 2015;3(39):19822–31.

[138] Doherty CM, et al. Magnetic framework composites for polycyclic aromatic hydrocarbon sequestration. J Mater Chem 2012;22(23):11470−4.
[139] Huo S-H, Yan X-P. Facile magnetization of metal-organic framework MIL-101 for magnetic solid-phase extraction of polycyclic aromatic hydrocarbons in environmental water samples. Analyst 2012;137(15):3445−51.
[140] Zhao X, et al. Synthesis of magnetic metal-organic framework (MOF) for efficient removal of organic dyes from water. Sci Rep 2015;5.
[141] Lu G, et al. Imparting functionality to a metal-organic framework material by controlled nanoparticle encapsulation. Nat Chem 2012;4(4):310−6.
[142] Zhao M, Zhang X, Deng C. Rational synthesis of novel recyclable $Fe_3O_4$@MOF nanocomposites for enzymatic digestion. Chem Commun 2015;51(38):8116−9.
[143] Munuera C, et al. The controlled growth of oriented metal-organic frameworks on functionalized surfaces as followed by scanning force microscopy. Phys Chem Chem Phys 2008;10(48):7257−61.
[144] Shekhah O, et al. Controlling interpenetration in metal-organic frameworks by liquid-phase epitaxy. Nat Mater 2009;8(6):481−4.
[145] Shekhah O, et al. Layer-by-layer growth of oriented metal organic polymers on a functionalized organic surface. Langmuir 2007;23(14):7440−2.
[146] Shekhah O, et al. Step-by-step route for the synthesis of metal-organic frameworks. J Am Chem Soc 2007;129(49):15118−+.
[147] Makiura R, et al. Surface nano-architecture of a metal-organic framework. Nat Mater 2010;9(7):565−71.
[148] Ke F, et al. $Fe_3O_4$@MOF core-shell magnetic microspheres with a designable metal-organic framework shell. J Mater Chem 2012;22(19):9497−500.
[149] Xu Y, et al. Magnetization of a Cu(II)-1,3,5-benzenetricarboxylate metal-organic framework for efficient solid-phase extraction of Congo Red. Microchim Acta 2015; 182(13−14):2313−20.
[150] He Y, et al. Metal organic frameworks combining $CoFe_2O_4$ magnetic nanoparticles as highly efficient SERS sensing platform for ultrasensitive detection of N-terminal pro-brain natriuretic peptide. ACS Appl Mater Interfaces 2016;8(12):7683−90.
[151] Mandel K, et al. Smart optical composite materials: dispersions of metal-organic Framework@Superparamagnetic microrods for switchable isotropic-anisotropic optical properties. ACS Nano 2017;11(1):779−87.
[152] Rocio-Bautista P, et al. A magnetic-based dispersive micro-solid-phase extraction method using the metal-organic framework HKUST-1 and ultra-high-performance liquid chromatography with fluorescence detection for determining polycyclic aromatic hydrocarbons in waters and fruit tea infusions. J Chromatogr A 2016;1436:42−50.
[153] Saikia M, Bhuyan D, Saikia L. Facile synthesis of $Fe_3O_4$ nanoparticles on metal organic framework MIL-101(Cr): characterization and catalytic activity. New J Chem 2015; 39(1):64−7.
[154] Jabbari V, et al. Green synthesis of magnetic MOF@GO and MOF@CNT hybrid nanocomposites with high adsorption capacity towards organic pollutants. Chem Eng J 2016;304:774−83.
[155] Panda R, Rahut S, Basu JK. Preparation of a $Fe_2O_3$/MIL-53(Fe) composite by partial thermal decomposition of MIL-53(Fe) nanorods and their photocatalytic activity. RSC Adv 2016;6(84):80981−5.
[156] Sethi K, Sharma S, Roy I. Nanoscale iron carboxylate metal organic frameworks as drug carriers for magnetically aided intracellular delivery. RSC Adv 2016;6(80).

[157] Fan S, et al. In situ-induced synthesis of magnetic Cu-$CuFe_2O_4$@HKUST-1 heterostructures with enhanced catalytic performance for selective aerobic benzylic C-H oxidation. ACS Catal 2017;7(1):243−9.
[158] Tolmacheva VV, et al. Facile synthesis of magnetic hypercrosslinked polystyrene and its application in the magnetic solid-phase extraction of sulfonamides from water and milk samples before their HPLC determination. Talanta 2016;152:203−10.
[159] Zhang W, et al. Microwave-enhanced synthesis of magnetic porous covalent triazine-based framework composites for fast separation of organic dye from aqueous solution. J Hazard Mater 2011;186(2−3):984−90.
[160] Ren JY, et al. Magnetic covalent triazine-based frameworks as magnetic solid-phase extraction adsorbents for sensitive determination of perfluorinated compounds in environmental water samples. Anal Bioanal Chem 2018;410(6):1657−65.
[161] He SJ, et al. Facile synthesis of magnetic covalent organic framework with three-dimensional bouquet-like structure for enhanced extraction of organic targets. ACS Appl Mater Interfaces 2017;9(3):2959−65.
[162] Huang LJ, et al. Thiol-functionalized magnetic porous organic polymers for highly efficient removal of mercury. Ind Eng Chem Res 2017;56(46):13696−703.
[163] Zhou LJ, Hu YL, Li GK. Conjugated microporous polymers with built-in magnetic nanoparticles for excellent enrichment of trace hydroxylated polycyclic aromatic hydrocarbons in human urine. Anal Chem 2016;88(13):6930−8.
[164] Rengaraj A, et al. Covalent triazine polymer-$Fe_3O_4$ nanocomposite for strontium ion removal from seawater. Ind Eng Chem Res 2017;56(17):4984−92.
[165] Wang J, et al. Porphyrin based porous organic polymer modified with $Fe_3O_4$ nanoparticles as an efficient adsorbent for the enrichment of benzoylurea insecticides. Mikrochim Acta 2017;185(1):36.
[166] Yan ZJ, et al. Magnetic covalent triazine framework for rapid extraction of phthalate esters in plastic packaging materials followed by gas chromatography-flame ionization detection. J Chromatogr A 2017;1525:32−41.
[167] Huang X, et al. Design and synthesis of core-shell $Fe_3O_4$@PTMT composite magnetic microspheres for adsorption of heavy metals from high salinity wastewater. Chemosphere 2018;206:513−21.
[168] Zhao L, et al. Recent advances of mesoporous materials in sample preparation. J Chromatogr A 2012;1228:193−204.
[169] Li GL, et al. Effective heavy metal removal from aqueous systems by thiol functionalized magnetic mesoporous silica. J Hazard Mater 2011;192(1):277−83.
[170] Yuan Q, et al. Effect of large pore size of multifunctional mesoporous microsphere on removal of heavy metal ions. J Hazard Mater 2013;254:157−65.
[171] Zhao YG, et al. Amidoxime-functionalized magnetic mesoporous silica for selective sorption of U(VI). RSC Adv 2014;4(62):32710−7.
[172] Zhang N, et al. Fast and selective magnetic solid phase extraction of trace Cd, Mn and Pb in environmental and biological samples and their determination by ICP-MS. Microchim Acta 2011;175(1−2):121−8.
[173] Zhang XY, et al. Nitrogen rich core-shell magnetic mesoporous silica as an effective adsorbent for removal of silver nanoparticles from water. J Hazard Mater 2017;337:1−9.
[174] Xu D, et al. Selective enrichment of glycopeptides/phosphopeptides using $Fe_3O_4$@Au-$B(OH)_2$@m$TiO_2$ core-shell microspheres. Talanta 2017;166:154−61.
[175] Li LL, et al. Adsorption property of Cr(VI) on magnetic mesoporous titanium dioxide-graphene oxide core-shell microspheres. New J Chem 2014;38(12):6008−16.

[176] Zhao YJ, et al. Magnetic mesoporous $TiO_2$ microspheres for sustainable arsenate removal from acidic environments. Inorg Chem Frontiers 2018;5(9):2132–9.
[177] Li XS, et al. Preparation of titanium-grafted magnetic mesoporous silica for the enrichment of endogenous serum phosphopeptides. J Chromatogr A 2013;1315:61–9.
[178] Sorribes-Soriano A, et al. Magnetic molecularly imprinted polymers for the selective determination of cocaine by ion mobility spectrometry. J Chromatogr A 2018;1545: 22–31.
[179] Ming WN, et al. Magnetic molecularly imprinted polymers for the fluorescent detection of trace 17 beta-estradiol in environmental water. Sensor Actuator B Chem 2017;238: 1309–15.
[180] Li D, et al. Efficient preparation of surface imprinted magnetic nanoparticles using poly (2-anilinoethanol) as imprinting coating for the selective recognition of glycoprotein. Talanta 2018;184:316–24.
[181] Luo X, et al. Novel Cu (II) magnetic ion imprinted materials prepared by surface imprinted technique combined with a sol-gel process. J Hazard Mater 2011;192(3): 949–55.
[182] Luo XB, et al. A magnetic copper(II)-imprinted polymer for the selective enrichment of trace copper(II) ions in environmental water. Microchim Acta 2012;179(3–4):283–9.
[183] Cen SB, et al. Application of magnetic $Cd^{2+}$ ion-imprinted mesoporous organosilica nanocomposites for mineral wastewater treatment. RSC Adv 2017;7(13):7996–8003.
[184] Wang JJ, Wei J. Selective and simultaneous removal of dibenzothiophene and 4-methyldibenzothiophene using double-template molecularly imprinted polymers on the surface of magnetic mesoporous silica. J Mater Chem 2017;5(9):4651–9.
[185] Wei SL, et al. Preparation of magnetic Pb(II) and Cd(II) ion-imprinted microspheres and their application in determining the Pb(II) and Cd(II) contents of environmental and food samples. RSC Adv 2014;4(56):29715–23.
[186] Bagheri AR, et al. Dummy molecularly imprinted polymers based on a green synthesis strategy for magnetic solid-phase extraction of acrylamide in food samples. Talanta 2019; 195:390–400.
[187] Luo XB, et al. Synthesis of magnetic ion-imprinted fluorescent CdTe quantum dots by chemical etching and their visualization application for selective removal of Cd(II) from water. Colloid Surf Physicochem Eng Asp 2014;462:186–93.
[188] Li H, Wang L. Highly selective detection of polycyclic aromatic hydrocarbons using multifunctional magnetic–luminescent molecularly imprinted polymers. ACS Appl Mater Interfaces 2013;5(21):10502–9.
[189] de Faria HD, et al. New advances in restricted access materials for sample preparation: a review. Anal Chim Acta 2017;959:43–65.
[190] Ye L, et al. Restricted-access nanoparticles for magnetic solid-phase extraction of steroid hormones from environmental and biological samples. J Chromatogr A 2012;1244: 46–54.
[191] Yan P, et al. Restricted accessed nanoparticles for direct magnetic solid phase extraction of trace metal ions from human fluids followed by inductively coupled plasma mass spectrometry detection. Analyst 2015;140(12):4298–306.
[192] Liu XD, et al. Restricted access magnetic core-mesoporous shell microspheres with C8-modified interior pore-walls for the determination of diazepam in rat plasma by LC-MS. Talanta 2013;106:321–7.
[193] Lv YK, et al. Layer-by-layer fabrication of restricted access media-molecularly imprinted magnetic microspheres for magnetic dispersion microextraction of bisphenol A from milk samples. New J Chem 2015;39(3):1792–9.

# Metal-organic frameworks 10

*Gongke Li, Ling Xia, Jianwei Dong, Yanlong Chen, Yanxia Li*
School of Chemistry, Sun Yat-sen University, Guangzhou, Guangdong, PR China

## 10.1 Introduction

Porous metal-organic frameworks (MOFs), are a type of adsorption materials, formed from inorganic metals and multidentate organic ligands through coordination bonds, as a crystalline network [1,2]. The combination of inorganic and organic units in their construction results in favorable properties, such as high porosity and high specific surface area for use as selective adsorbents. The porosity of MOFs may reach 90% of their volume with specific surface areas up to over 7000 $m^2/g$ [3,4]. Another feature of MOFs is the designable structure. First, various inorganic metal and organic ligands can be chosen for producing MOFs for specific applications [5]. The pore size can be adjusted by varying the metal and ligands, as well as by controlling the reaction conditions [6]. In addition, MOFs can be further functionalized to meet the requirements of specific applications. By post-synthetic modification, functional groups with different properties can be introduced [7,8]. These characteristics allow MOFs to be utilized as a sorbent for solid-phase extraction.

MOFs are typically formed by self-assembly of metal ions and ligands in a one-pot reaction [9]. During the reaction, elevated temperature and pressures speed up product formation at the expense of single crystals growth. Rate-controlled reactions under mild conditions are preferred to obtain crystalline materials. Among the various possible synthetic methods, solvothermal/hydrothermal synthesis is the most widely used. In this method, ligands are mixed with metal ions or clusters in a solvent, encapsulated in a Teflon-lined bomb, and heated to a defined temperature [10]. Coordination reactions occur under conditions of high temperature and pressure. This method has several advantages, including perfect crystalline formation, simple operation, and low energy consumption [11]. Most of the well known MOFs, for example, the isoreticular metal-organic framework (IRMOF) series, materials of Institute Lavoisier (MIL) series, zeolitic imidazolate framework (ZIF) series, and University of Oslo (UiO) series, are synthesized by this method [12].

Alternative synthetic methods to prepare MOFs for SPE include the microwave/ultrasonic assisted method and the electrochemical method. Microwave/ultrasonic assisted synthesis method takes advantage of field assisted technology to accelerate coordination reactions reducing reaction times to minutes, instead of several days, typical of solvothermal methods [13]. For electrochemical synthesis, a metal plate is immersed in a solution of the organic ligands and used as an anode in an electrochemical cell. Application of a voltage results in the formation of a layer of the MOF on the cathode with reaction times typically less than 1 h [14].

**Solid-Phase Extraction. https://doi.org/10.1016/B978-0-12-816906-3.00010-8**

Since the first application of MOFs in analytical chemistry by Yan's group [15], their applications have grown to include sampling, solid-phase extraction, gas chromatography, and high-performance liquid chromatography. For solid-phase extraction, four series of MOFs have been used as sorbets, the IRMOF series, MIL series, ZIF series, and UiO series. Although high-throughput synthesis strategies for MOFs catch much attention [16], the amount typically used as sorbents for SPE has remained small. This favors their use in miniaturized SPE formats, including micro-solid-phase extraction (μSPE), solid-phase microextraction (SPME), dispersive solid-phase extraction (DSPE), and magnetic solid-phase extraction (MSPE), Fig. 10.1.

## 10.2 IRMOF in solid-phase extraction

The isoreticular metal-organic framework (IRMOF) series, are self-assembled from an inorganic unit $[Zn_4O]^{6+}$ and various aromatic carboxylic acid ligands to form octahedral crystalline structures [17]. The IRMOF series materials have the same topological structure but various pore sizes created by the use of different dicarboxylic acid ligands. Having a designable uniform structure with high porosity, they are suitable sorbents for SPE. Li's group described the use of an MOF-5 bar for direct extraction of volatile sulfides from plants [18]. However, the lack of water stability of IRMOF series prevents their application as a sorbent for extraction from aqueous samples. To enhance the stability of IRMOF in an aqueous environment, combining MOFs with hydrophobic "shield" materials is a common approach [19]. For example, carbon nanotubes(CNTs) hybrid MOF-199 was successfully applied for the noninvasive analysis of trace ethylene, methanol, and ethanol from fruit samples with relatively high humidity [20]. Carbonization can also be used to enhance the water stability of IRMOF series material and expand its application in an aqueous sample [21].

Due to the moisture-sensitive nature of IRMOF series materials, they are usually composited with other substrate materials to form the sorbent for SPE. As shown in

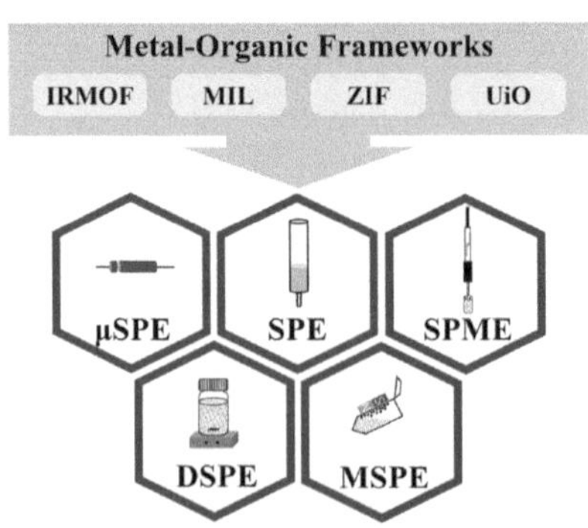


**Figure 10.1** Scheme of metal-organic frameworks application in solid-phase extraction.

**Table 10.1** Application of IRMOFs in SPE.

| MOFs-based sorbent | Analysis methods | Analyte | Sample matrix | References |
|---|---|---|---|---|
| MOF-5, MOF-199 | VA-D-SPME-HPLC | Parabens | Water, urine, cosmetics | [22] |
| MOF-5/PAN | SPE-HPLC | Drugs | Urine | [23] |
| Cu/MOF-5 | HS-SPME-GC/MS | Volatile sulfides | Plants | [18] |
| $Fe_3O_4$/MOF-5 | MSPE-LC-MS/MS or GC/MS | PAHs, pesticide | Soil, food, plant | [24] |
| $Fe_3O_4$/MOF-5 | MSPE-HPLC | Pesticides | Water | [25] |
| Fe/$SiO_2$/MOF-5 | MSPE-HPLC | PAHs | Water | [26] |
| $Fe_3O_4$/MOF-5-C | MSPE-HPLC | Chlorophenols | Mushroom | [21] |
| $Fe_3O_4$/IRMOF-3 | MSPE-AAS | Metal ions | Water | [27] |
| $Fe_3O_4$/SiO2/MOF-177 | MSPE-GC/MS | Phenols | Water | [28] |
| MOF-199/CNTs | HS-SPME-GC/MS | Ethylene, alcohols | Fruit | [20] |
| GO/MOF-199 | HS-SPME-GC | Pesticides | Water, soil | [29] |
| $Fe_3O_4$/MOF-199 | M-D-μSPE-UHPLC | PAHs | Waters, fruit-tea | [30] |
| $Fe_3O_4$/MOF-199 | MSPE-HPLC-MS/MS | Insecticides | Water | [31] |
| $Fe_3O_4$/MOF-199 | MSPE-AAS | Metal ions | Fish | [32] |
| $Fe_3O_4$/MOF-199 | MSPE-AAS | Metal ions | Food | [33] |
| MOF-199/$Fe_3O_4$/GA-MIP | UA-M-D-μSPE-UV-Vis | Gallic acid | Urine, plasma, water | [34] |
| Grafted $Fe_3O_4$/MOF-235 | MSPE-AAS | Metal ions | Water | [35] |

Table 10.1, MOF-5 [18,22–26], IRMOF-3 [27], MOF-177 [28], MOF-199 [20,29–34], and MOF-235 [35] have been applied in SPE. Magnetic SPE (MSPE) is the preferred extraction mode of IRMOF-based SPE. After extraction procedure, various analysis and detection methods, including gas chromatography (GC), liquid chromatography (LC), high-performance liquid chromatography (HPLC), and atomic absorption spectrum (AAS) can be selected according to target analytes and sample matrix.

## 10.2.1 MOF-5-based sorbent for SPE

### 10.2.1.1 Directly used as sorbent substrate

MOF-5 is the most famous MOF first reported by Yaghi's group [1]. It can be used for gas extraction, separation, and storage. However, when used in sample preparation, its powder form is inconvenient. To address the problems in sample processing, Hu and coworkers proposed a facile method for immobilization of MOF-5 on porous copper support and its application in the noninvasive sampling of volatile organic sulfur compounds from plants [18]. The sorbent preparation and extraction procedures are illustrated in Fig. 10.2 and summarized below.

#### 10.2.1.1.1 Preparation of Cu/MOF-5 bar

In situ solvothermal synthesis: foamed copper support was introduced into *N,N′*-dimethylformamide (DMF) solution containing $Zn(NO_3)_2 \cdot 6H_2O$ and terephthalic acid. Crystal growth in an autoclave at 120°C for 24 h. Activation: under reduced pressure at 250°C for 24 h.

#### 10.2.1.1.2 Headspace microextraction procedure

MOF-5 bar was exposed to the headspace of a hexane sample solution for 30 min.

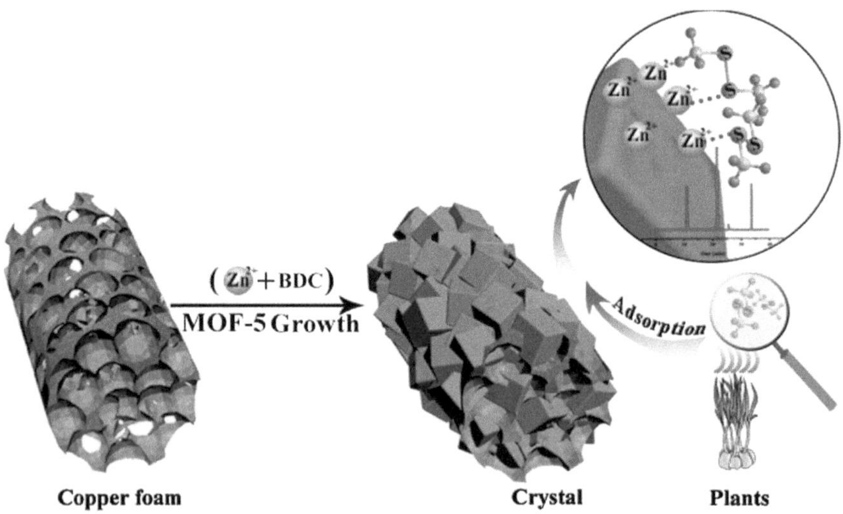


**Figure 10.2** The scheme of in situ solvothermal growth of Cu/MOF-5 bar for extraction of volatile sulfides from plants.
This figure is reprinted with permission from Hu YL, Lian HX, Zhou LJ, Li GK. In situ solvothermal growth of metal-organic framework-5 supported on porous copper foam for noninvasive sampling of plant volatile sulfides. Anal Chem 2015;87(1):406–12. Copyright (2015) American Chemical Society.

After extraction, MOF-5 bar was inserted into thermal desorption inlet immediately for thermal desorption.

##### 10.2.1.1.3 Application

The Cu/MOF-5 bar is a robust enrichment medium benefiting both from the large surface area of MOFs and the three-dimensional rigid skeleton of microporous Cu network. This MOF-5-based sorbent is an excellent adsorbent for trapping volatile organic sulfur compounds from Chinese chive and garlic sprout and for the preconcentration of other plant volatiles.

#### *10.2.1.2 MOF-5 composite materials for extraction*

The magnetic composite format is the preferred approach to enhance the stability and process capability of the IRMOF series materials for extraction from aqueous solutions. The chemical bonding between the magnetic nanoparticles and MOFs endows the composite material with higher chemical stability, durability, and improved reproducibility. Li's group reported a facile and efficient strategy for fabrication of hybrid magnetic MOF-5 nanoparticles and their application in trace enrichment from complex samples [24]. The sorbent preparation and extraction procedures are illustrated in Fig. 10.3 and summarized below.

##### 10.2.1.2.1 Preparation of $Fe_3O_4$/MOF-5 composite

Amino-functionalized $Fe_3O_4$ preparation: $Fe_3O_4$ nanoparticles were synthesized by a microwave-assisted method [36]. Tetraethylorthosilicate was added to the mixture of $Fe_3O_4$ nanoparticles in a mixture of ethanol and ammonia maintained at 40°C for 24 h. After drying, the activated nanoparticles were reacted with 3-aminopropyltriethoxysilane (APTES) with continuous agitation for 7 h at room temperature to form the amino-functionalized $Fe_3O_4$ nanoparticles. The amino-functionalized $Fe_3O_4$ nanoparticles were introduced into a DMF solution of $Zn(OAc)_2 \cdot 2H_2O$ and terephthalic acid. The composite material was formed in an autoclave at 120°C for 10 h.

##### 10.2.1.2.2 Extraction procedure

The $Fe_3O_4$/MOF-5 composite was dispersed into n-hexane sample solution with ultrasonication. After extraction, $Fe_3O_4$/MOF-5 composite was isolated with a magnet and the analytes eluted with acetonitrile containing 1% formic acid with ultrasound assistance. The supernatant was then evaporated under nitrogen for protection. Finally, the analytes were dissolved in a suitable solvent for analysis.

##### 10.2.1.2.3 Application

The chemical bonding between the $Fe_3O_4$ and MOF-5 allows the hybrid material to be reused without measurable loss of performance and affords high enrichment. These advantages were demonstrated by the enrichment of trace PAHs in soil, seaweed, and fish tissue prior to GC/MS analysis and for gibberellic acids in plants prior to LC-MS/MS analysis.

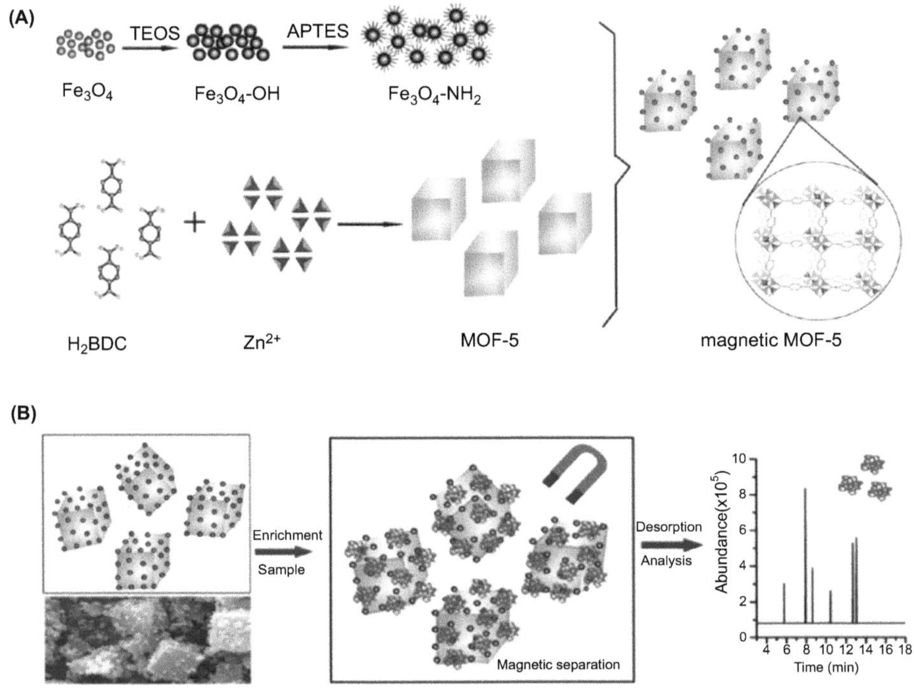


**Figure 10.3** The scheme of $Fe_3O_4$/MOF-5 composite preparation (A) and sample analysis procedures (B).
This figure is reprinted with permission from Hu YL, Huang ZL, Liao J, Li GK. Chemical bonding approach for fabrication of hybrid magnetic metal-organic framework-5: high efficient adsorbents for magnetic enrichment of trace analytes. Anal Chem 2013;85(14):6885−93 Copyright (2013) American Chemical Society.

### 10.2.2 MOF-199-based sorbent for SPE

MOF-199 is another typical IRMOF series material. Comparing with MOF-5, its higher porosity and larger pore size facilitate the adsorption of larger molecules. Its application for SPE remains limited by its moisture sensitivity. A simple and mild multilayer interparticle linking strategy was used by Li's group for the preparation of a carbon nanotube (CNTs) hybrid MOF-199 [20]. With the assistance of the hydrophobic "shield" provided by the CNTs, MOF-199 was used for extraction under high humidity conditions. The sorbent preparation and extraction procedures are illustrated in Fig. 10.4 and summarized below.

#### 10.2.2.1 Preparation of MOF-199/CNTs coated fibers

A $Cu(CH_3COO)_2$ solution was added to the CNTs dispersed in DMF/EtOH containing 1,3,5-tricarboxybenzene and stirred for 8 h. The composite material was dried at 120°C overnight and sieved to a particle size less than 106 μm. Clean silica fibers

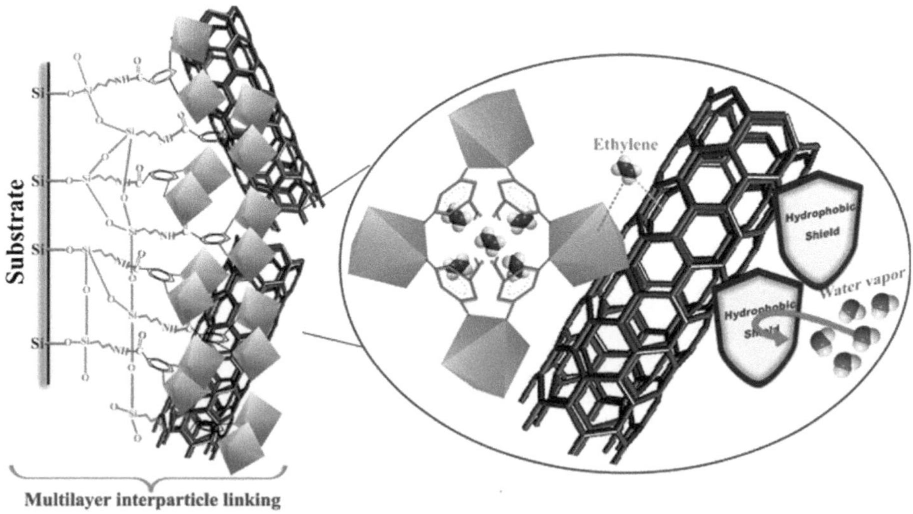


**Figure 10.4** The schematic working principle of MOF-199-based extraction with the protection of CNTs.
This figure is reprinted with permission from Zhang ZM, Huang YC, Ding WW, Li GK. Multilayer interparticle linking hybrid MOF-199 for noninvasive enrichment and analysis of plant hormone ethylene. Anal Chem 2014;86(7):3533–3540. Copyright (2014) American Chemical Society.

were immersed into APTES for 3 min, dipped into the MOF-199/CNTs composite ethanol solution, and then dried at 70°C for 1 h.

#### *10.2.2.2 Solid-phase microextraction procedure*

Weighed fruits were placed in a 500 mL sealed vial and purged with nitrogen. MOF-199/CNTs coated fibers were then exposed to the vial headspace for 20 min at room temperature and transferred to GC inlet for analysis.

#### *10.2.2.3 Application*

The multilayer interparticle linking strategy is a promising approach for the preparation of MOF-199/CNTs coating for extracting trace ethylene from fruits. Hybridization of hydrophobic materials will expand the application of MOFs to samples where moisture-resistance is important.

## 10.3 MIL in solid-phase extraction

Materials of Institute Lavoisier (MIL) series are an important type of MOFs material with high porosity and large surface areas. MIL series are constructed from various transition metals and dicarboxylic acid ligands, including succinic acid and glutaric acid [37]. Moreover, trivalent metals such as chromium, vanadium, aluminum, and iron can be used for their synthesis in combination with terephthalic acid and tribenzoic acid [38]. Unlike the IRMOF series, the MIL series materials are stable in water, allowing their use in humid and aqueous environments. Hu and coworkers reported a MIL-101(Cr) packed column for the online analysis of naproxen and its metabolite in urine [39]. Through magnetization, the procedural steps for MIL-based SPE can be simplified. A magnetic MIL-100(Cr) composite was used for the analysis malachite green from water and fish samples by Maya and coworkers [40]. Furthermore, sensitivity, selectivity, and biological compatibility of MIL-based sorbent can be enhanced by adopting other modification approaches.

As shown in Table 10.2, MIL-53 [41–45], MIL-68 [46], MIL-88 [47,48], MIL-96 [49], MIL-100 [40,50–57], and MIL-101 [39,58–85] have been used in SPE, and of these, MIL-101 is the most widely used. Magnetic solid-phase extraction (MSPE) is the preferred extraction format due to its easy operation. Some online strategies have been described to facilitate automation of sample processing [39,44,66].

## 10.3.1 MIL-100-based sorbent for SPE

MIL-100 is stable in water and strongly desired for extraction from aqueous samples. Moreover, the pore size of MIL-100 is relatively large and may be used to trap target compounds of high molecular weight. In addition, by magnetization, magnetic MIL-100-based SPE provides an easy operating mode for extraction. However, its hydrophobicity reduces sample interactions in aqueous solution. This drawback was solved by the automated method reported by Maya and coworkers [40]. The sorbent preparation and SPE procedures are summarized below:

### 10.3.1.1 Preparation of $Fe_3O_4$/MIL-100 composite

Chromium trioxide, trimesic acid, and hydrofluoric acid were added to an autoclave and reacted at 220°C for 96 h. $FeCl_2 \cdot 4H_2O$ and $FeCl_3$ were added to MIL-100(Cr) solution and stirred under nitrogen at room temperature for 2 h.

### 10.3.1.2 Automated magnetic dispersive micro-solid-phase extraction (automatic M-D-μSPE) extraction procedure

The extraction procedure is illustrated in Fig. 10.5. Step 1, loading magnet and $Fe_3O_4$/MIL-100(Cr) sorbent. Step 2, loading sample solution. Step 3, extraction in the stop flow mode. Step 4, elimination of the sample matrix. Step 5, introducing eluent. Step 6, transfer eluent to a spectrophotometric detector.

### 10.3.1.3 Application

This in-syringe M-D-μSPE strategy provides adequate dispersion and simplifies the extraction process using MIL-100(Cr) as a sorbent for extraction of malachite green from water and fish samples with direct spectrophotometric detection. Application of

**Table 10.2** Application of MIL in SPE.

| MOFs-based sorbent | Analysis methods | Analyte | Sample matrix | References |
|---|---|---|---|---|
| MIL-53(Al) | D-μSPE-UPLC-MS/MS | Hormones | Water, urine | [41] |
| MIL-53(Al) | VA-D-μSPE-HPLC | Phenols | Water | [42] |
| MIL-53(Al)-polymer | SPE-HPLC | Drugs | Water, urine | [43] |
| MIL-53(Al)-polymer | Online SPE-HPLC | Hormones | Urine | [44] |
| MIL-53(Fe)−C | MSPE-HPLC | EDCs | Food | [45] |
| PEEK/MIL-68(Al) | SBSE-HPLC-MS/MS | Parabens | Cosmetics, plasma | [46] |
| MIL-88(Fe) | SPME-GC-MS | PCBs | Water, soil | [47] |
| MIL-88(Fe)/GO | SPME-GC | Plasticizers | Vegetable oil | [48] |
| MIL-96(Al) | SPME-GC-MS | Disinfection byproducts | Water | [49] |
| $Fe_3O_4$/MIL-100(Cr) | Automatic M-D-μSPE-Vis | Dye | Water, fish | [40] |
| $Fe_3O_4$/MIL-100(Cr) | MSPE-GC−MS/MS | PCBs | Water | [50] |
| MIL-100(Fe)−C | MSPE-GC | PAHs | Water | [51] |
| Ionic liquid-MIL-100(Fe) | VA-D-μSPE-GC | PAHs | Water, food | [52] |
| $Fe_3O_4$/MIL-100(Fe) | MSPE-HPLC | Insect repellent | Water | [53] |
| $Fe_3O_4$/MIL-100(Fe) | MSPE-HPLC | Additive | Toothpastes | [54] |
| $Fe_3O_4$/MIL-100(Fe) | MSPE-LC-MS/MS | Peptides | Serum | [55] |
| $Fe_3O_4$/MIL-100(Fe)/GO | MSPE-fluorescence | Drugs | Food | [56] |
| Grafted $Fe_3O_4$/MIL-100(Fe) | M-D-μSPE-HPLC | Hormones | Water, urine | [57] |
| MIL-101(Cr) | SPE-DART-MS | Herbicides | Water | [58] |
| MIL-101(Cr) | SPE-UPLC-MS/MS | Drugs | Water | [59] |

*Continued*

**Table 10.2** Application of MIL in SPE.—cont'd

| MOFs-based sorbent | Analysis methods | Analyte | Sample matrix | References |
|---|---|---|---|---|
| MIL-101(Cr) | DSPE-GC-MS | EDCs | Water | [60] |
| MIL-101(Cr) | VA-D-SPE-UPLC-MS/MS | Drugs | Plasma | [61] |
| MIL-101(Cr) | Online SPE-HPLC | Drugs | Urine | [39] |
| MIL-101(Cr)-polymer | SPME-CEC | Penicillin | Water | [62] |
| MIL-101(Cr)/PVA | VA-D-SPE-HPLC-MS/MS | Drugs | Water | [63] |
| MIL-101(Cr)/RGO | SPE-HPLC/MALDI-TOF-MS | Drugs, proteins | Water | [64] |
| PDMS/MIL-101(Cr) | SBSE-GC | Pesticides | Water | [65] |
| $TiO_2$/MIL-101(Cr) | Online SPE-GC-MS | Volatile pollutants | Air | [66] |
| $Fe_3O_4$/MIL-101(Cr) | MSPE-GC | Pesticides | Water, tea | [67] |
| $Fe_3O_4$/MIL-101(Cr) | MSPE-GC | Pesticides | Water | [68] |
| $Fe_3O_4$/MIL-101(Cr) | M-D-SPE-GC-MS | Plasticizers | Water, plasma | [69] |
| $Fe_3O_4$/MIL-101(Cr) | MSPE-HPLC | Hormones | Water | [70] |
| $Fe_3O_4$/MIL-101(Cr) | MSPE-HPLC | Dyes | Food | [71] |
| $Fe_3O_4$/MIL-101(Cr) | MSPE-UPLC-MS/MS | Drugs | Water | [72] |
| $Fe_3O_4$/MIL-101(Cr) | MSPE-MALDI-TOF-MS | Proteins | Cell | [73] |
| Grafted $Fe_3O_4$/MIL-101(Cr) | MSPE-IC | Azide | Sartan drugs | [74] |
| Grafted $Fe_3O_4$/MIL-101(Cr) | MSPE-AAS | Metal ions | Water | [75] |
| Grafted $Fe_3O_4$/MIL-101(Cr) | MSPE-AAS | Metal ions | Water, food | [76] |
| $Fe_3O_4$/$SiO_2$/MIL-101(Cr) | MSPE-HPLC | PHAs | Water | [77] |
| $Fe_3O_4$/$SiO_2$/MIL-101(Cr) | MSPE-HPLC | Pesticides | Water | [78] |
| $Fe_3O_4$/$SiO_2$-GO/MIL-101(Cr) | MSPE-HPLC | Herbicides | Food | [79] |

**Table 10.2** Application of MIL in SPE.—cont'd

| MOFs-based sorbent | Analysis methods | Analyte | Sample matrix | References |
|---|---|---|---|---|
| $Fe_3O_4$/MIL-101(Fe) | MSPE-GC | Pesticides | Hair, urine | [80] |
| $Fe_3O_4$/MIL-101(Fe) | MSPE-AAS | Heavy metal ions | Plants | [81] |
| $Fe_3O_4$/MIL-101(Fe) | MSPE-AAS | Metal ions | Seafood | [82] |
| $Fe_3O_4$/MIL-101(Fe) | MSPE-AAS | Metal ions | Fish | [83] |
| $Fe_3O_4$/MIL-101(Fe) | MSPE-HPLC | Fungicides | Water | [84] |
| $Fe_3O_4$/MIL-101(Fe) | MSPE-HPLC | Herbicides | Water, vegetable | [85] |

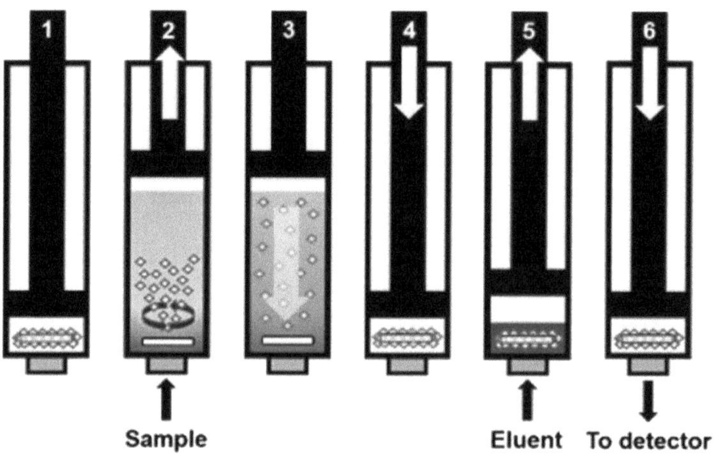


**Figure 10.5** The scheme of automatic M-D-μSPE procedure.
This figure is reprinted with permission from Maya F, Palomino Cabello C, Estela JM, Cerda V, Palomino TG. Automatic in-syringe dispersive micro-solid phase extraction using magnetic metal-organic frameworks. Anal Chem 2015;87(15):7545–7549. Copyright (2015) American Chemical Society.

this method can be extended to other water stable MOFs, as well as other target analytes with suitable detectors.

### *10.3.2 MIL-101-based sorbent for SPE*

MIL-101 has an even larger pore size than MIL-100 with retention of its water/solvent stability. MIL-101(Cr) was used by Li's group for the extraction of naproxen [39]. Fig. 10.6 demonstrates that MI-101(Cr) is a more efficient sorbent for the extraction of naproxen than $C_{18}$-bonded silica and multiwalled nanotubes (MWNTs). The

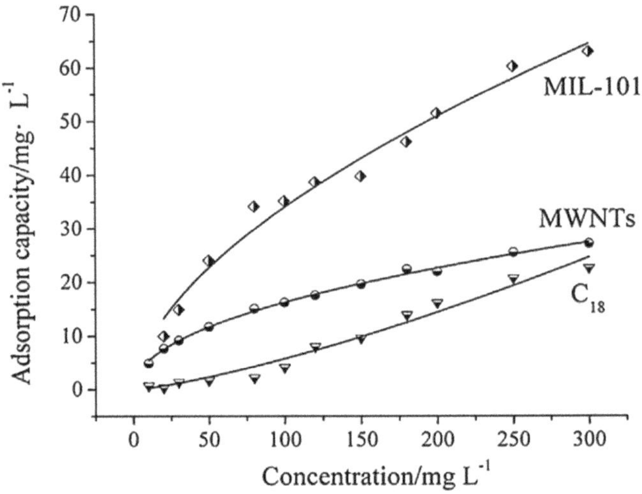


**Figure 10.6** Adsorption isotherms for naproxen on different SPE sorbents at 30°C.
This figure is reprinted with permission from Hu YL, Song CY, Liao J, Huang ZL, Li GK. Water stable metal-organic framework packed microcolumn for online sorptive extraction and direct analysis of naproxen and its metabolite from urine sample. J Chromatogr A 2013;1294:17–24 Copyright (2013) Elsevier.

preparation of the sorbent microcolumn for online extraction by micro-solid-phase extraction coupled to liquid chromatography (μSPE-HPLC) is shown in Fig. 10.7.

### 10.3.2.1 Preparation of MIL-101(Cr)-based microcolumn

To prepare the MIL-101(Cr) $Cr(NO_3)_3 \cdot 9H_2O$, terephthalic acid, and hydrofluoric acid were mixed and heated in an autoclave at 220°C for 8 h. The MOF was activated by heating at 150°C for 24 h under reduced pressure. The MIL-101(Cr) was dispersed in the slurry with ultrasonic ethanol assistance and immediately pumped into a polyether ether ketone (PEEK) tube.

### 10.3.2.2 The online micro-solid-phase extraction procedure

The extraction procedure is illustrated in Fig. 10.7. Step 1 preconditioning of the sorbent cartridge. Step 2, loading the sample solution. Step 3, desorption of naproxen from the microcolumn to the separation column. Step 4, flow rate adjustment for chromatographic separation.

### 10.3.2.3 Applications

This MIL-101(Cr)-based online μSPE-HPLC system opens up a new strategy for the application of MOFs-based material as sorbents for analysis of polar compounds from an aqueous matrix.

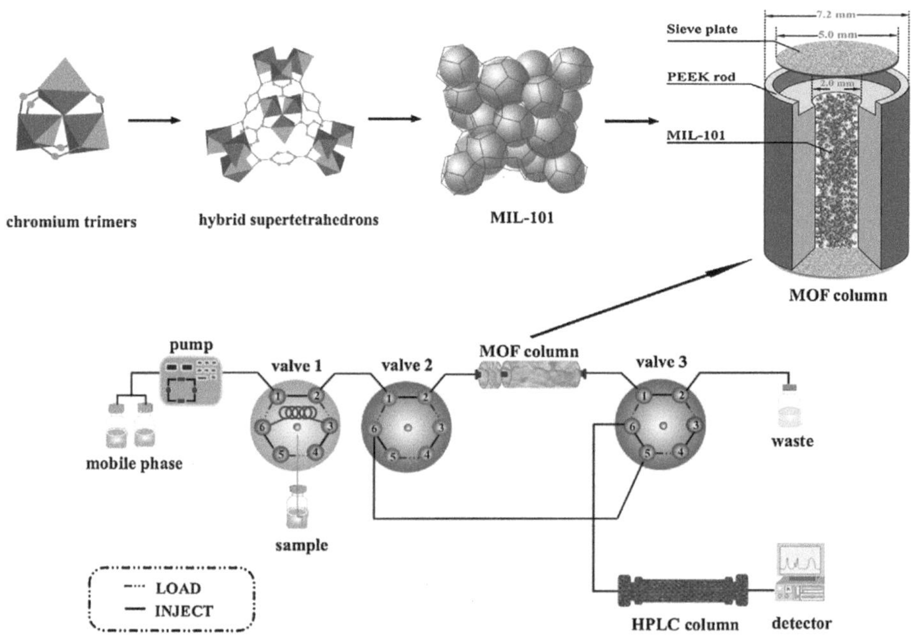


**Figure 10.7** The scheme of fabrication of the MIL-101(Cr) packed microcolumn and the online μSPE-HPLC system [43].
This figure is reprinted with permission from Hu YL, Song CY, Liao J, Huang ZL, Li GK. Water stable metal-organic framework packed microcolumn for online sorptive extraction and direct analysis of naproxen and its metabolite from urine sample. J Chromatogr A 2013;1294:17–24 Copyright (2013) Elsevier.

## 10.4 ZIF in solid-phase extraction

Zeolitic imidazolate framework (ZIF), are water and solvent stable MOFs, synthesized from Zn(II) or Co(II) and imidazole ligands [86]. Their zeolite-like tetrahedron structure and chemical stability in boiling alkaline solution and organic solvents distinguish these materials [87]. These features facilitate their chemical modification with a wide range of reagents. Thus, the ZIF series materials have been widely exploited for SPE [88–97]. Yan's group reported several applications of ZIF series materials, as well as modified ZIF-based sorbent for SPE [89,91,96]. Applications of ZIF-based SPE are summarized in Table 10.3.

### 10.4.1 ZIF-8-based sorbent for SPE

Chang and coworkers used ZIF-8-based sorbent and molecular sieves to for the analysis of n-alkanes from human serum and fuel samples, Fig. 10.8 [91]. The extraction procedure is summarized below.

**Table 10.3** Application of ZIF in SPE.

| MOFs-based sorbent | Analysis methods | Analyte | Sample matrix | References |
|---|---|---|---|---|
| ZIF-7, ZIF-11 | SPE-HPLC | PAHs | Water | [88] |
| ZIF-8 | Online SPE-HPLC | Drugs | Water, milk | [89] |
| ZIF-8 | μSPE-GC-MS | PAHs | Water | [90] |
| ZIF-8 | SPME-GC-MS | Alkanes | Fuel, serum | [91] |
| Cellulose/ZIF-8 | SPE-HPLC | PAHs | Water | [92] |
| $Fe_3O_4$/ZIF-8 | MSPE-AFS | Metal | Water, urine | [93] |
| ZIF-67-C | M-D-SPE-HPLC | Insecticides | Water, vegetable | [94] |
| ZIF-67-C | MSPE-HPLC | Herbicides | Plants | [95] |
| ZIF-90 | SPME-GC | Endocrine disruptors | Water, soil | [96] |
| $Fe_3O_4$/ZIF-90/trypsin | MSPE-MALDI-TOF-MS | Proteins, peptides | Cell | [97] |

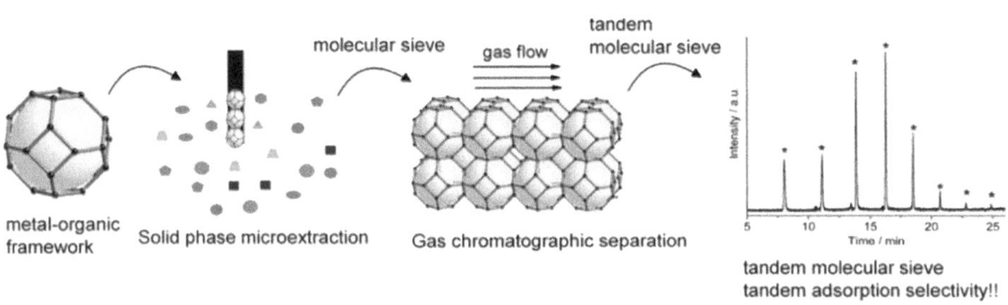


**Figure 10.8** The scheme of ZIF-8-based tandem molecular sieves for trace analytes in a complex sample.
This figure is reprinted with permission from Chang N, Gu ZY, Wang HF, Yan XP, Metal-organic-framework-based tandem molecular sieves as a dual platform for selective microextraction and high-resolution gas chromatographic separation of n-alkanes in complex matrixes. Anal Chem 2011;83(18):7094−101. Copyright (2011) American Chemical Society.

### *10.4.1.1 Preparation of ZIF-8 coated SPME fibers*

$Zn(NO_3)_2 \cdot 6H_2O$ was added to 2-methylimidazole in methanol and mixed at room temperature for 1h. Clean stainless steel wires were dipped into a ZIF-8 DMF solution for 20 s, and dried in a nitrogen atmosphere at 250°C for 10 min in the GC injector. Ten coating cycles were applied to achieve the final SPME fiber.

#### 10.4.1.2 Headspace solid-phase microextraction procedure

Fuel or human serum samples were introduced into a 100 μL sealed vial and sonicated at 25°C for 5 min. The ZIF-8 coated fiber was exposed to the vial headspace for 20 min and transferred to the GC inlet for analysis.

#### 10.4.1.3 Application

By coupling a ZIF-8-based sorbent for SPME with ZIF-8-based stationary phases for GC separation suitable enrichment and selectivity for the analysis of n-alkanes in human serum and fuel samples was achieved. This strategy can be adopted for other applications of MOFs materials.

## 10.5 UiO in solid-phase extraction

University of Oslo (UiO) series are zirconium-based novel MOFs materials [98]. Due to their outstanding physicochemical stability, the UiO has attracted considerable attention as an extraction sorbent [99—103]. Xia and coworkers used UiO-66 for the extraction of herbicides from vegetables [99]. Moreover, after suitable modification, UiO-66-based sorbent has been used for proteome research [101,102]. As shown in Table 10.4, functionalized UiO series materials are useful sorbents for the extraction of target compounds from biological samples.

### 10.5.1 UiO-66-based sorbent for SPE

Zhou and coworkers synthesized titanium immobilized magnetic UiO-66 material for the extraction of low phosphopeptides from human serum and milk samples [101]. The UiO-66-based sorbent is not only suitable for metal-oxide affinity chromatography (MOAC) for the extraction of monophosphorylated peptides but also as an immobilized metal affinity chromatography (IMAC) material for polyphosphorylated peptides. Beyond enhanced sensitivity and selectivity, sample pretreatment is simplified by UiO-66-based sorbents, as summarized below and illustrated in Fig. 10.9.

#### 10.5.1.1 Preparation of UiO-66-based sorbent

$Fe_3O_4$ nanoparticles were synthesized by a hydrothermal reaction and then dispersed into dopamine hydrochloride to form a $Fe_3O_4$ and polydopamine ($Fe_3O_4$/PDA) composite material. $Fe_3O_4$/PDA composite was added to a DMF solution containing $ZrCl_4$ and 2-aminoterephthalic acid and mixed at 120°C for 2 h. Three reaction cycles were used for the synthesis. The activated $Fe_3O_4$/UiO-66 composite was dispersed in a solution of adenosine triphosphate (ATP) and vibrated for 2 h. $Ti(SO_4)_2$ was then added to the ATP grafted $Fe_3O_4$/UiO-66 composite solution to immobilize the titanium (IV) ions.

**Table 10.4** Application of UiO in SPE.

| MOFs-based sorbent | Analysis methods | Analyte | Sample matrix | References |
|---|---|---|---|---|
| UiO-66 | D-μSPE- HPLC | Herbicides | Vegetable | [99] |
| Grafted $Fe_3O_4$/UiO-66 | M-D-SPE-GC-MS | PCBs | Soil | [100] |
| Grafted $Fe_3O_4$/UiO-66 | MSPE-MALDI-TOF-MS | Proteins, peptides | Serum, milk | [101] |
| Grafted $Fe_3O_4$/UiO-66/Au | MSPE-MALDI-TOF-MS | Peptides | Serum | [102] |
| $Fe_3O_4$/$SiO_2$/UiO-66 | MSPE-HPLC-MS/MS | Neurotoxin | Shellfish | [103] |

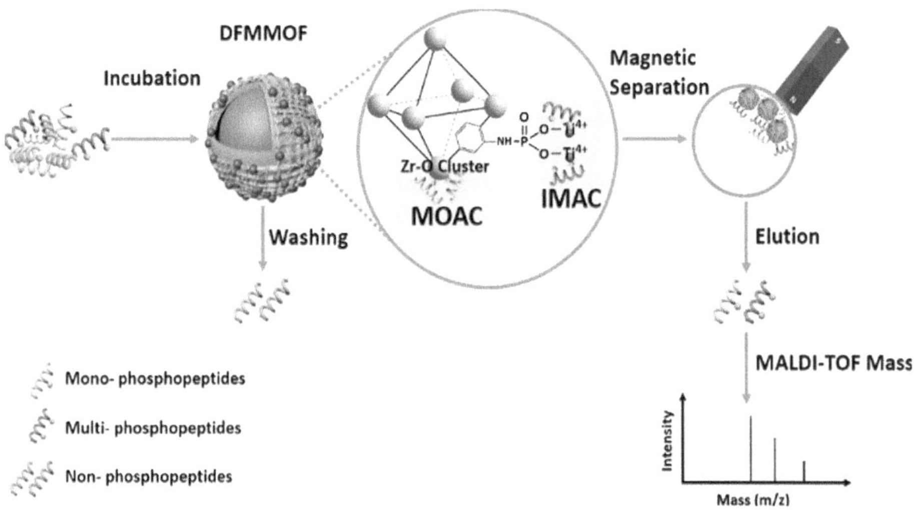


**Figure 10.9** The scheme of UiO-66-based dual-functionalized sorbent for analysis trace phosphorylated peptides from complex sample.
This figure is reprinted with permission from Zhou JQ, Liang YL, He XW, Chen LX, Zhang YK. Dual-functionalized magnetic metal-organic framework for highly specific enrichment of phosphopeptides. Chem Eng 2017;5(12):11413–21. Copyright (2017) American Chemical Society.

## *10.5.1.2 Magnetic solid-phase extraction*

The UiO-66-based sorbent was added to a solution of human serum and incubated at room temperature for 30 min. After extraction, the UiO-66-based sorbent was isolated using a magnet and phosphopeptides eluted with 5% ammonia solution by shaking for 15 min.

#### 10.5.1.3 Application

After modification, UiO-66-based sorbents exhibit several advantages, including biological compatibility and simple sample processing procedures [104].

## 10.6 Conclusions

MOFs are uniform structured porous materials suitable for solid-phase extraction. The application of MOFs materials for SPE needs to consider a wide range of sample properties to achieve the desired extraction performance. The water stability of MOFs-based sorbent is the first concern. Some MOFs can be used directly with aqueous samples, such as the MIL series, ZIF series and UiO series but not the IRMOF series sorbents. Postsynthesis modification with hydrophobic groups, composite materials with water-repellent components and carbonization are alternative approaches to improve water stability of moisture-sensitive MOFs. Beyond water stability, the selectivity of MOFs-based sorbents for the extraction of target analytes is another concern. The adsorption of analytes by many MOFs affords only low selectivity. To enhance the specific extraction of target compounds postsynthesis modifications enable MOFs-based sorbent to be modified for SPE. Packing monolithic columns with MOFs for μSPE and preparing magnetic MOFs composites for MSPE are the favored approaches to simplify processing procedures and for automation of sample processing.

## List of abbreviations

| | |
|---|---|
| **AAS** | Atomic absorption spectrum |
| **AFS** | Atomic fluorescence spectrometry |
| **APTES** | 3-aminopropyl triethoxysilane |
| **ATP** | Adenosine triphosphate |
| **CNTs** | Carbon nanotubes |
| **DMF** | N, N′- dimethylformamide |
| **EDCs** | endocrine disrupting compounds |
| **DFMMOF** | Dual-functionalized magnetic metal-organic framework |
| **DSPE** | Dispersive solid-phase extraction |
| **GAs** | Gibberellic acid |
| **GC** | Gas chromatography |
| **HPLC** | High-performance liquid chromatography |
| **IMAC** | Immobilized metal affinity chromatography |
| **IC** | Ion chromatography |
| **IRMOF** | Isoreticular metal-organic framework |
| **LC** | Liquid chromatography |
| **μSPE** | Micro-solid-phase extraction |
| **M-D-μSPE** | Magnetic dispersive micro-solid-phase extraction |
| **MIL** | Materials of Institute Lavoisier |

| | |
|---|---|
| **MOAC** | Metal-oxide affinity chromatography |
| **MOFs** | Metal-organic frameworks |
| **MALDI-TOF MS** | Matrix-assisted laser desorption ionization time-of-flight mass spectrometry |
| **MSPE** | Magnetic solid-phase extraction |
| **PCBs** | Polychlorinated biphenyls |
| **MWNTs** | Multiwalled nanotube |
| **PEEK** | Polyetheretherketone |
| **PHAs** | Polycyclic aromatic hydrocarbons |
| **SBSE** | Btir bar sorptive extraction |
| **SPME** | Solid-phase microextraction |
| **UPLC** | Ultra-performance liquid chromatography |
| **UiO** | University of Oslo |
| **ZIF** | Zeolitic imidazolate framework |

# References

[1] Furukawa H, Cordova KE, O'Keeffe M, Yaghi OM. The chemistry and applications of metal-organic frameworks. Science 2013;341(6149):974.

[2] Li JR, Sculley J, Zhou HC. Metal−organic frameworks for separations. Chem Rev 2012; 112(2):869−932.

[3] Furukawa H, Go YB, Ko N, Park YK, Uribe-Romo FJ, Kim J, et al. Isoreticular expansion of metal-organic frameworks with triangular and square building units and the lowest calculated density for porous crystals. Inorg Chem 2011;50(18):9147−52.

[4] Wang YH, Rui M, Lu GH. Recent applications of metal−organic frameworks in sample pretreatment. J Sep Sci 2018;41:180−94.

[5] Yaghi OM, O'Keeffe M, Ockwig NW, Chae HK, Eddaoudi M, Kim J. Reticular synthesis and the design of new materials. Nature 2003;423:705.

[6] Kitagawa S, Matsuda R. Chemistry of coordination space of porous coordination polymers. Coord Chem Rev 2007;251(21−24):2490−509.

[7] Song YF, Cronin L. Postsynthetic covalent modification of metal-organic framework (MOF) materials. Angew Chem Int Ed 2008;47:4635−7.

[8] Li B, Gui B, Hu G, Yuan D, Wang C. Postsynthetic modification of an alkyne-tagged zirconium metal−organic framework via a "click" reaction. Inorg Chem 2015;54(11): 5139−41.

[9] Li CQ, Qiu WG, Long W, Deng F, Bai GM, Zhang GZ, et al. Synthesis of porphyrin@ MOFs type catalysts through "one-pot" self-assembly. J Mol Catal A Chem 2014;393: 166−70.

[10] Qi Y, Luo F, Che YX, Zheng JM. Hydrothermal synthesis of metal-organic frameworks based on aromatic polycarboxylate and flexible bis(imidazole) ligands. Cryst Growth Des 2008;8(2):606−11.

[11] Guan L, Luo GH, Wang Y. Hydrothermal synthesis, structure and properties of a 3D pillar-layered metal-organic framework based on amino-arenedisulfonate ligand. Chin J Struct Chem 2018;37(11):1795−804.

[12] Li HL, Eddaoudi MM, O'Keeffe M, Yaghi OM. Design and synthesis of an exceptionally stable and highly porous metal-organic framework. Nature 1999;402(6759):276−9.

[13] Barton HF, Davis AK, Lee DT, Parsons GN. Solvothermal synthesis of MIL-96 and UiO-66-NH2 on atomic layer deposited metal oxide coatings on fiber mats. J Vis Exp 2018; 136:1−10.
[14] Nicolò C, Van Assche T, Boudewijns T, Denayer J, Binnemans K, Vos De D, et al. High pressure, high temperature electrochemical synthesis of metal−organic frameworks: films of mil-100 (fe) and hkust-1 in different morphologies. J Mater Chem A 2013;1(19):5827.
[15] Zhou YY, Yan XP, Kim KN, Wang SW, Liu MG. Exploration of coordination polymer as sorbent for flow injection solid-phase extraction on-line coupled with high performance liquid chromatography for determination of polycyclic aromatic hydrocarbons in environmental materials. J Chromatogr A 2006;1116:172−8.
[16] Volkringer C, Loiseau T, Guillou N, Férey G, Haouas M, Taulelle F, et al. High-throughput aided synthesis of the porous metal-organic framework-type aluminum pyromellitate, MIL-121, with extra carboxylic acid functionalization. Inorg Chem 2010; 49(21):9852−62.
[17] Prestipino C, Regli L, Vitillo J, Bonino F, Damin A, Lamberti C, et al. Local structure of framework Cu (II) in HKUST-1 metallorganic framework: spectroscopic characterization upon activation and interaction with adsorbates. Chem Mater 2006;18:1337−46.
[18] Hu YL, Lian HX, Zhou LJ, Li GK. In situ solvothermal growth of metal-organic framework-5 supported on porous copper foam for noninvasive sampling of plant volatile sulfides. Anal Chem 2015;87(1):406−12.
[19] Yang J, Grzech A, Mulder FM, Dingemans TJ. Methyl modified MOF-5: a water stable hydrogen storage material. Chem Commun 2011;47:5244−6.
[20] Zhang ZM, Huang YC, Ding WW, Li GK. Multilayer interparticle linking hybrid MOF-199 for noninvasive enrichment and analysis of plant hormone ethylene. Anal Chem 2014;86(7):3533−40.
[21] Hao L, Liu XL, Wang JT, Wang C, Wu QH, Wang Z. Metal-organic framework derived magnetic nanoporous carbon as an adsorbent for the magnetic solid-phase extraction of chlorophenols from mushroom sample. Chin Chem Lett 2016;27(5):783−8.
[22] Rocío-Bautista P, Martínez-Benito C, Pino V, Pasán J, Ayala JH, Ruiz-Pérez C, Afonso AM. The metal-organic framework HKUST-1 as efficient sorbent in a vortex-assisted dispersive micro solid-phase extraction of parabens from environmental waters, cosmetic creams, and human urine. Talanta 2015;139:13−20.
[23] Asiabi M, Mehdinia A, Jabbari A. Preparation of water stable methyl-modified metal−organic framework-5/polyacrylonitrile composite nanofibers via electrospinning and their application for solid-phase extraction of two estrogenic drugs in urine samples. J Chromatogr A 2015;1426:24−32.
[24] Hu YL, Huang ZL, Liao J, Li GK. Chemical bonding approach for fabrication of hybrid magnetic metal-organic framework-5: high efficient adsorbents for magnetic enrichment of trace analytes. Anal Chem 2013;85(14):6885−93.
[25] Ma JP, Wu GG, Li S, Tan WQ, Wang XY, Li JH, Chen LX. Magnetic solid-phase extraction of heterocyclic pesticides in environmental water samples using metal-organic frameworks coupled to high performance liquid chromatography determination. J Chromatogr A 2018;1553:57−66.
[26] Zhou QX, Lei M, Li J, Liu YL, Zhao KF, Zhao DC. Magnetic solid phase extraction of N- and S-containing polycyclic aromatic hydrocarbons at ppb levels by using a zerovalent iron nanoscale material modified with a metal organic framework of type Fe@MOF-5, and their determination by HPLC. Microchim Acta 2017;184(4):1029−36.

[27] Wang Y, Xie J, Wu YC, Hu XY. A magnetic metal-organic framework as a new sorbent for solid-phase extraction of copper (II), and its determination by electrothermal AAS. Microchim Acta 2014;181(9−10):949−56.
[28] Wang GH, Lei YQ, Song HC. Evaluation of $Fe_3O_4$@$SiO_2$-MOF-177 as an advantageous adsorbent for magnetic solid-phase extraction of phenols in environmental water samples. Anal Methods 2014;6:7842−7.
[29] Zhang SL, Du Z, Li GK. Metal-organic framework-199/graphite oxide hybrid composites coated solid-phase microextraction fibers coupled with gas chromatography for determination of organochlorine pesticides from complicated samples. Talanta 2013;115: 32−9.
[30] Rocío-Bautista P, Pino V, Ayala JH, Pasán J, Ruiz-Pérez C, Afonso AM. A magnetic-based dispersive micro-solid-phase extraction method using the metal-organic framework HKUST-1 and ultra-high-performance liquid chromatography with fluorescence detection for determining polycyclic aromatic hydrocarbons in waters and fruit tea infusions. J Chromatogr A 2016;1436:42−50.
[31] Cao XL, Liu GY, She YX, Jiang ZJ, Jin F, Jin MJ, Du PF, Zhao FN, Zhang YX, Wang J. Preparation of magnetic metal organic framework composites for the extraction of neonicotinoid insecticides from environmental water samples. RSC Adv 2016;6(114): 113144−51.
[32] Tadjarodi A, Abbaszadeh A. A magnetic nanocomposite prepared from chelator-modified magnetite ($Fe_3O_4$) and HKUST-1 (MOF-199) for separation and preconcentration of mercury (II). Microchim Acta 2016;183(4):1391−9.
[33] Ghorbani-Kalhor E. A metal-organic framework nanocomposite made from functionalized magnetite nanoparticles and HKUST-1(MOF-199) for preconcentration of Cd (II), Pb (II), and Ni (II). Microchim Acta 2016;183(9):2639−47.
[34] Asfaram A, Ghaedi M, Dashtian K. Rapid ultrasound-assisted magnetic microextraction of gallic acid from urine, plasma and water samples by HKUST-1-MOF-$Fe_3O_4$-GA-MIP-NPs: UV-Vis detection and optimization study. Ultrason Sonochem 2017;34:561−70.
[35] Moradi SE, Shabani AH, Dadfarnia S, Emami S. Sulfonated metal organic framework loaded on iron oxide nanoparticles as a new sorbent for the magnetic solid phase extraction of cadmium from environmental water samples. Anal Methods 2016;8(33): 6337−46.
[36] Hong RY, Pan TT, Li HZ. Microwave synthesis of magnetic $Fe_3O_4$ nanoparticles used as a precursor of nanocomposites and ferrofluids. J Magn Magn Mater 2006;303:60−8.
[37] Zhou X, Huang W, Shi J, Zhao Z, Xia Q, Li Y, et al. A novel MOF/graphene oxide composite GrO@Mil-101 with high adsorption capacity for acetone. J Mater Chem A 2014;2:4722−30.
[38] Yoon JW, Seo YK, Hwang YK, Chang JS, Leclerc H, Wuttke S, et al. Controlled reducibility of a metal-organic framework with coordinatively unsaturated sites for preferential gas sorption. Angew Chem Int Ed 2010;49:5949−52.
[39] Hu YL, Song CY, Liao J, Huang ZL, Li GK. Water stable metal-organic framework packed microcolumn for online sorptive extraction and direct analysis of naproxen and its metabolite from urine sample. J Chromatogr A 2013;1294:17−24.
[40] Maya F, Palomino Cabello C, Estela JM, Cerda V, Palomino TG. Automatic in-syringe dispersive micro-solid phase extraction using magnetic metal-organic frameworks. Anal Chem 2015;87(15):7545−9.
[41] Gao GH, Li SJ, Li S, Wang YD, Zhao P, Zhang XY, Hou XH. A combination of computational-experimental study on metal-organic frameworks MIL-53(Al) as sorbent for simultaneous determination of estrogens and glucocorticoids in water and urine

samples by dispersive micro-solid-phase extraction coupled to UPLC-MS/MS. Talanta 2018;180:358–67.
[42] Boontongto T, Siriwong K, Burakham R. Amine-functionalized metal-organic framework as a new sorbent for vortex-assisted dispersive micro-solid phase extraction of phenol residues in water samples prior to HPLC analysis: experimental and computational studies. Chromatographia 2018;81(5):735–47.
[43] Lyu DY, Yang CX, Yan XP. Fabrication of aluminum terephthalate metal-organic framework incorporated polymer monolith for the microextraction of non-steroidal anti-inflammatory drugs in water and urine samples. J Chromatogr A 2015;1393:1–7.
[44] Luo XL, Li GK, Hu YF. In-tube solid-phase microextraction based on $NH_2$-MIL-53(Al)-polymer monolithic column for online coupling with high-performance liquid chromatography for directly sensitive analysis of estrogens in human urine. Talanta 2017;165:377–83.
[45] Liu XL, Feng T, Wang CH, Hao L, Wang C, Wu QH, Wang Z. A metal-organic framework-derived nanoporous carbon/iron composite for enrichment of endocrine disrupting compounds from fruit juices and milk samples. Anal Methods 2016;8(17):3528–35.
[46] Wang CL, Zhou W, Liao XY, Wang XM, Chen ZL. Covalent immobilization of metal organic frameworks onto chemical resistant poly (ether ether ketone) jacket for stir bar extraction. Anal Chim Acta 2018;1025:124–33.
[47] Wu YY, Yang CX, Yan XP. Fabrication of metal-organic framework MIL-88B films on stainless steel fibers for solid-phase microextraction of polychlorinated biphenyls. J Chromatogr A 2014;1334:1–8.
[48] Zhang SH, Yang Q, Li Z, Wang WJ, Zang XH, Wang C, Wang Z. Solid phase microextraction of phthalic acid esters from vegetable oils using iron (III)-based metal-organic framework/graphene oxide coating. Food Chem 2018;263:258–64.
[49] Zhang BB, Xu GJ, Li L, Wang XL, Li N, Zhao RS, Lin JM. Facile fabrication of MIL-96 as coating fiber for solid-phase microextraction of trihalomethanes and halonitromethanes in water samples. Chem Eng J 2018;350:240–7.
[50] Chen XF, Ding N, Zang H, Yeung H, Zhao RS, Cheng C, Liu JH, Chan TD. $Fe_3O_4$@MOF core-shell magnetic microspheres for magnetic solid-phase extraction of polychlorinated biphenyls from environmental water samples. J Chromatogr A 2013;1304:241–5.
[51] Huo SH, An HY, Yu J, Mao XF, Zhang Z, Bai L, Huang YF, Zhou PX. Pyrolytic in situ magnetization of metal-organic framework MIL-100 for magnetic solid-phase extraction. J Chromatogr A 2017;1517:18–25.
[52] Nasrollahpour A, Moradi SE, Baniamerian MJ. Vortex-assisted dispersive solid-phase microextraction using ionic liquid-modified metal-organic frameworks of PAHs from environmental water, vegetable, and fruit juice samples. Food Anal Methods 2017;10(8):2815–26.
[53] Ma XW, Feng F, Yang Y, Dang XP, Huang JL, Chen HX. Magnetic solid-phase extraction of N, N-diethyl-m-toluamide from baby toilet water prior to its HPLC-UV detection. J Chromatogr Sci 2017;55(6):662–8.
[54] Yang Y, Ma XW, Feng F, Dang XP, Huang JL, Chen HX. Magnetic solid-phase extraction of triclosan using core-shell $Fe_3O_4$@MIL-100 magnetic nanoparticles, and its determination by HPLC with UV detection. Microchim Acta 2016;183(8):2467–72.
[55] Xiong ZC, Ji YS, Fang CL, Zhang QQ, Zhang LY, Ye ML, Zhang WB, Zou HF. Facile preparation of core-shell magnetic metal-organic framework nanospheres for the selective enrichment of endogenous peptides. Chem Eur J 2014|;20(24):7389–95.

[56] Li ZQ, Qi MY, Tu CY, Wang WP, Chen JR, Wang AJ. Magnetic metal-organic framework/graphene oxide-based solid-phase extraction combined with spectrofluorimetry for the determination of enrofloxacin in milk sample. Food Anal Methods 2017;10(12):4094−103.

[57] Khezeli T, Daneshfar A. Dispersive micro-solid-phase extraction of dopamine, epinephrine and norepinephrine from biological samples based on green deep eutectic solvents and $Fe_3O_4$@MIL-100(Fe) core-shell nanoparticles grafted with pyrocatechol. RSC Adv 2015;5(80):65264−73.

[58] Li XJ, Xing JW, Chang CL, Wang X, Bai Y, Yan XP, Liu HW. Solid-phase extraction with the metal-organic framework MIL-101(Cr) combined with direct analysis in real time mass spectrometry for the fast analysis of triazine herbicides. J Sep Sci 2014;37(12): 1489−95.

[59] Dai XP, Jia XN, Zhao P, Wang T, Wang J, Huang PT, He L, Hou XH. A combined experimental/computational study on metal-organic framework MIL-101(Cr) as a SPE sorbent for the determination of sulphonamides in environmental water samples coupling with UPLC-MS/MS. Talanta 2016;154:581−8.

[60] Huang ZZ, Lee JK. Performance of metal-organic frame work MIL-101 after surfactant modification in the extraction of endocrine disrupting chemicals from environmental water samples. Talanta 2015;143:366−73.

[61] Qi C, Cai QQ, Zhao P, Jia XN, Lu N, He L, Hou XH. The metal-organic framework MIL-101(Cr) as efficient adsorbent in a vortex-assisted dispersive solid-phase extraction of imatinib mesylate in rat plasma coupled with ultra-performance liquid chromatography/mass spectrometry: application to a pharmacokinetic study. J Chromatogr A 2016;1449: 30−8.

[62] Lin CL, Lirio S, Chen YT, Lin CH, Huang HY. A novel hybrid metal-organic framework-polymeric monolith for solid-phase microextraction. Chem Eur J 2014;20(12):3317−21.

[63] Wang YD, Zhang Y, Cui JN, Li S, Yuan M, Wang T, Hu Q, Hou XH. Fabrication and characterization of metal organic frameworks/polyvinyl alcohol cryogel and their application in extraction of non-steroidal anti-inflammatory drugs in water samples. Anal Chim Acta 2018;1022:45−52.

[64] Zhang XQ, Liang QL, Han Q, Wan W, Ding MY. Metal-organic frameworks@graphene hybrid aerogels for solid-phase extraction of non-steroidal anti-inflammatory drugs and selective enrichment of proteins. Analyst 2016;141(13):4219−26.

[65] Xiao ZW, He M, Chen BB, Hu B. Polydimethylsiloxane/metal-organic frameworks coated stir bar sorptive extraction coupled to gas chromatography-flame photometric detection for the determination of organophosphorus pesticides in environmental water samples. Talanta 2016;156−157:126−33.

[66] Hu YL, Huang ZL, Zhou LJ, Wang DM, Li GK. Synthesis of nanoscale titania embedded in MIL-101 for the adsorption and degradation of volatile pollutants with thermal desorption gas chromatography and mass spectrometry detection. J Sep Sci 2014;37(12): 1482−8.

[67] Lu N, He X, Wang T, Liu SY, Hou XH. Magnetic solid-phase extraction using MIL-101(Cr)-based composite combined with dispersive liquid-liquid microextraction based on solidification of a floating organic droplet for the determination of pyrethroids in environmental water and tea samples. Microchem J 2018;137:449−55.

[68] He X, Yang W, Li SJ, Liu Y, Hu BC, Wang T, Hou XH. An amino-functionalized magnetic framework composite of type $Fe_3O_4$-$NH_2$@MIL-101(Cr) for extraction of pyrethroids coupled with GC-ECD. Microchim Acta 2018;185(2):125−33.

[69] Dargahi R, Ebrahimzadeh H, Asgharinezhad AA, Hashemzadeh A, Amini MM. Dispersive magnetic solid-phase extraction of phthalate esters from water samples and

human plasma based on a nanosorbent composed of MIL-101(Cr) metal-organic framework and magnetite nanoparticles before their determination by GC-MS. J Sep Sci 2018;41(4):948–57.

[70] Huang YF, Liu M, Wang YQ, Li Y, Zhang JM, Huo SH. Hydrothermal synthesis of functionalized magnetic MIL-101 for magnetic enrichment of estrogens in environmental water samples. RSC Adv 2013;6(19):15362–9.

[71] Shi XR, Chen XL, Hao YL, Li L, Xu HJ, Wang MM. Magnetic metal-organic frameworks for fast and efficient solid-phase extraction of six Sudan dyes in tomato sauce. J Chromatogr B 2018;1086:146–52.

[72] Wang T, Liu SY, Gao GH, Zhao P, Lu N, Lun XW, Hou XH. Magnetic solid phase extraction of non-steroidal anti-inflammatory drugs from water samples using a metal organic framework of type $Fe_3O_4$/MIL-101(Cr), and their quantitation by UPLC-MS/MS. Microchim Acta 2017;184(8):2981–90.

[73] Wei JP, Qiao B, Song WJ, Chen T, li F, Li BZ, Wang J, Han Y, Huang YF, Zhou ZJ. Synthesis of magnetic framework composites for the discrimination of *Escherichia coli* at the strain level. Anal Chim Acta 2015;868:36–44.

[74] Zhang SN, Han PP, Xia Y. +Facile extraction of azide in sartan drugs using magnetized anion-exchange metal-organic frameworks prior to ion chromatography. J Chromatogr A 2017;1514:29–35.

[75] Rezabeyk S, Manoochehri M. Speciation analysis of Tl(I) and Tl (III) after magnetic solid phase extraction using a magnetite nanoparticle composite modified with aminodibenzo-18-crown-6 functionalized MIL-101(Cr). Microchim Acta 2018;185(8):365–73.

[76] Kalantari H, Manoochehri M. A nanocomposite consisting of MIL-101(Cr) and functionalized magnetite nanoparticles for extraction and determination of selenium (IV) and selenium (VI). Microchim Acta 2018;185(3):196–204.

[77] Huo SH, Yan XP. Facile magnetization of metal–organic framework MIL-101 for magnetic solid-phase extraction of polycyclic aromatic hydrocarbons in environmental water samples. Analyst 2012;137(15):3445–51.

[78] Ma JP, Yao ZD, Hou LW, Lu WH, Yang QP, Li JH, Chen LX. Metal organic frameworks (MOFs) for magnetic solid-phase extraction of pyrazole/pyrrole pesticides in environmental water samples followed by HPLC-DAD determination. Talanta 2016;161:686–92.

[79] Liang L, Wang XH, Sun Y, Ma PY, Li XP, Piao HL, Jiang YX, Song DQ. Magnetic solid-phase extraction of triazine herbicides from rice using metal-organic framework MIL-101(Cr) functionalized magnetic particles. Talanta 2018;179:512–9.

[80] Zhang SL, Jiao Z, Yao WX. A simple solvothermal process for fabrication of a metal-organic framework with an iron oxide enclosure for the determination of organophosphorus pesticides in biological samples. J Chromatogr A 2014;1371:74–81.

[81] Babazadeh M, Khanmiria RH, Abolhasania J, Kalhora EG, Hassanpour A. Solid phase extraction of heavy metal ions from agricultural samples with the aid of a novel functionalized magnetic metal-organic framework. RSC Adv 2015;5(26):19884–92.

[82] Ghorbani-Kalhor E, Hosseinzadeh-Khanmiri R, Abolhasani J, Babazadeh M, Hassanpour A. Determination of mercury (II) ions in seafood samples after extraction and preconcentration by a novel functionalized magnetic metal-organic framework nanocomposite. J Sep Sci 2015;38(7):1179–86.

[83] Babazadeh M, Khanmiri RH, Abolhasani J, Ghorbani-Kalhor E, Hassanpour A. Synthesis and application of a novel functionalized magnetic metal-organic framework sorbent for determination of heavy metal ions in fish samples. Bull Chem Soc Jpn 2015;88(6):871–9.

[84] Huang YF, Liu QH, Li K, Li Y, Chang N. Magnetic iron (III)-based framework composites for the magnetic solid-phase extraction of fungicides from environmental water samples. J Sep Sci 2018;41(5):1129−37.

[85] Deng YL, Zhang RQ, Li D, Sun P, Su P, Yang Y. Preparation of iron-based MIL-101 functionalized polydopamine@$Fe_3O_4$ magnetic composites for extracting sulfonylurea herbicides from environmental water and vegetable samples. J Sep Sci 2018;41(9): 2046−55.

[86] Liao YT, Dutta S, Chien CH, Hu CC, Shieh FK, Lin CH, et al. Synthesis of mixed-ligand zeolitic imidazolate framework (ZIF-8-90) for $CO_2$ adsorption. J Inorg Organomet Polym 2015;25(2):251−8.

[87] Li Y, Liu J, Chen C, Zhang X, Chen J. Preparation of NiCoP hollow quasi-polyhedra and their electrocatalytic properties for hydrogen evolution in alkaline solution. ACS Appl Mater Interfaces 2017;9(7):5982−91.

[88] Hu HP, Liu SQ, Chen CY, Wang JP, Zou Y, Lin LH, Yao SZ. Two novel zeolitic imidazolate frameworks (ZIFs) as sorbents for solid-phase extraction (SPE) of polycyclic aromatic hydrocarbons (PAHs) in environmental water samples. Analyst 2014;139(22): 5818−26.

[89] Yang XQ, Yang CX, Yan XP. Zeolite imidazolate framework-8 as sorbent for on-line solid-phase extraction coupled with high-performance liquid chromatography for the determination of tetracyclines in water and milk samples. J Chromatogr A 2013;1304: 28−33.

[90] Ge DD, Lee HK. Water stability of zeolite imidazolate framework 8 and application to porous membrane-protected micro-solid-phase extraction of polycyclic aromatic hydrocarbons from environmental water samples. J Chromatogr A 2011;1218(47):8490−5.

[91] Chang N, Gu ZY, Wang HF, Yan XP. Metal-organic-framework-based tandem molecular sieves as a dual platform for selective microextraction and high-resolution gas chromatographic separation of n-alkanes in complex matrixes. Anal Chem 2011;83(18): 7094−101.

[92] Liang XT, Liu SQ, Zhu R, Xiao LX, Yao SZ. Highly sensitive analysis of polycyclic aromatic hydrocarbons in environmental water with porous cellulose/zeolitic imidazolate framework-8 composite microspheres as a novel adsorbent coupled with high-performance liquid chromatography. J Sep Sci 2016;39(14):2806−14.

[93] Zou Z, Wang S, Jia J, Xu F, Long Z, Hou X. Ultrasensitive determination of inorganic arsenic by hydride generation-atomic fluorescence spectrometry using $Fe_3O_4$@ ZIF-8 nanoparticles for preconcentration. Microchem J 2016;124:578−83.

[94] Hao L, Wang C, Wu QH, Li Z, Zang XH, Wang Z. Metal-organic framework derived magnetic nanoporous carbon: novel adsorbent for magnetic solid-phase extraction. Anal Chem 2014;86(24):12199−205.

[95] Liu X, Wang C, Wu Q, Wang Z. Metal-organic framework-templated synthesis of magnetic nanoporous carbon as an efficient absorbent for enrichment of phenylurea herbicides. Anal Chim Acta 2015;870:67−74.

[96] Yu LQ, Yan XP. Covalent bonding of zeolitic imidazolate framework-90 to functionalized silica fibers for solid-phase microextraction. Chem Commun 2013;49(21):2142−4.

[97] Zhai R, Yuan YF, Jiao FL, Hao FR, Fang X, Zhang YJ, Qian XH. Facile synthesis of magnetic metal organic frameworks for highly efficient proteolytic digestion used in mass spectrometry-based proteomics. Anal Chim Acta 2017;994:19−28.

[98] Christel K, Georg N, Irena S, Volodymyr B, Stefan K, et al. Proline functionalized UiO-67 and UiO-68 type metal-organic frameworks showing reversed diastereoselectivity in aldol addition reactions. Chem Mater 2016;28(8):2573−80.

[99] Xia L, Liu LJ, Xu XL, Zhu FF, Wang XL, Zhang KY, Yang XC, You JM. Determination of chlorophenoxy acids herbicides by using zirconium based metal-organic framework as special sorbent for dispersive micro solid phase extraction and high performance liquid chromatography. New J Chem 2017;41(6):2241–8.

[101] Zhou JQ, Liang YL, He XW, Chen LX, Zhang YK. Dual-functionalized magnetic metal-organic framework for highly specific enrichment of phosphopeptides. Chem Eng 2017; 5(12):11413–21.

[102] Liu QJ, Deng CH, Sun NR. Hydrophilic tripeptide-functionalized magnetic metal-organic frameworks for the highly efficient enrichment of N-linked glycopeptides. Nanoscale 2018;10(25):12149–55.

[100] Lin SC, Gan N, Cao YT, Chen YJ, Jiang QL. Selective dispersive solid phase extraction-chromatography tandem mass spectrometry based on aptamer-functionalized UiO-66-$NH_2$ for determination of polychlorinated biphenyls. J Chromatogr A 2016;1446:34–40.

[103] Zhang WM, Yan ZM, Gao J, Tong P, Liu W, Zhang L. Metal-organic framework UiO-66 modified magnetite@silica core-shell magnetic microspheres for magnetic solid-phase extraction of domoic acid from shellfish samples. J Chromatogr A 2015;1400(31):10–8.

[104] Ma W, Li XJ, Bai Y, Liu HW. Applications of metal-organic frameworks as advanced sorbents in biomacromolecules sample preparation. Trends Anal Chem 2018;109: 154–62.

# Electrospun nanofibers

# 11

*Habib Bagheri, Omid Rezvani, Shakiba Zeinali, Sara Asgari, Tahereh Golzari Aqda, Faranak Manshaei*
Environmental and Bio-Analytical Laboratories, Department of Chemistry, Sharif University of Technology, Tehran, Iran

## 11.1 Introduction

There has been remarkable progress in material science and the development of new types of natural and synthetic structures with different properties over the last few decades. These materials have evolved from structures based on small molecules to macromolecules and polymers. Today, emphasis is placed on materials at the nanoscale [1]. A more generalized description of nanotechnology was established by the national nanotechnology initiative, which defines nanotechnology as the manipulation of matter with at least one dimension sized from 1 to 100 nm. This definition reflects the fact that quantum mechanical effects are more significant at this quantum-realm scale. Two main approaches are used in nanotechnology. In the "bottom-up" approach, materials and devices are built from molecular components which assemble themselves chemically by principles of molecular recognition [2]. In the "top-down" approach, nanoobjects are constructed from larger entities without atomic-level control [3]. The ever-growing library of nanostructures with variable sizes, shapes, and compositions has opened up many opportunities for the design of functional materials [4]. Controlling material structure at the molecular level has been enabled by innovative nanoscale manipulations of surface area, surface functionality, and porosity [5]. These developments are responsible for several enhancements in the field of separation and filtration [6]. Thus, these profound investigations have contributed to the synthesis and advancement of various forms of nanomaterials such as nanofibers, nanoparticles, nanotubes, and nanowires. Nanofibers, owing to their large surface area, high aspect ratio, flexibility, and other desirable features have garnered significant interests from both academia and industry. Several methods have been utilized for producing nanofibers, including electrospinning, melt or solution blowing, phase separation, self-assembly, and template synthesis. Among them, electrospinning is the most versatile technique for the synthesis of nanofibers from different materials, that is, polymers, ceramics, and metals. It is interesting that the history of the electrospinning process goes back to the observation of water behavior under the influence of an electric field [7]. Cooley [8] and Morton [9] were the first to describe electrospinning for fiber production.

**Solid-Phase Extraction. https://doi.org/10.1016/B978-0-12-816906-3.00011-X**

## 11.2 Electrospinning process

In electrospinning, a rather high voltage is applied to a polymeric solution in order to induce electrical charges within the fluid. When these charges reach a threshold value, a fluid jet will eject from the droplet at the tip of a needle resulting in the formation of a Taylor cone [10]. The produced jet dries during its flight toward a collecting electrode while simultaneously elongated by a whipping process caused by electrostatic repulsion at small bends in the fibers [11]. The elongation and thinning of the fibers resulting from this bending instability lead to the formation of uniform fibers with submicrometer-scale diameters [12]. Modification of the spinneret or the type of polymeric solution can generate miscellaneous fiber-fabricating protocols with unique structures and properties [13−16]. Electrospun fibers can adopt a porous or core-shell morphology depending on the type of materials being spun as well as the evaporation rates and solvent miscibility. For multiple spinning fluids, the general criteria for the creation of fibers, depends upon the spinnability of the outer solution. Accordingly, electrospinning is performed in three different formats: coaxial, emulsion, and melt electrospinning. A coaxial setup for a dual-solution system, allowing the inner feed to meet the outer one at the tip of the spinneret is shown in Fig. 11.1 [17,18]. If the solutions are immiscible then a core-shell structure is usually observed. Emulsions can also be used to create core-shell or composite fibers without modification of the spinneret [14]. However, these fibers are usually more difficult to produce compared with coaxial spinning due to the greater number of variables which need to be taken into account.

Electrospinning of polymer melts eliminates the need for volatile solvents typically used in solution electrospinning. This strategy allows semicrystalline polymer fibers such as PE, PET, and PP to be conveniently fabricated, which seem otherwise impossible or very difficult using solution spinning [19]. The setup is very similar to that employed for conventional electrospinning and includes a syringe or spinneret, a high voltage supply, and a collector. The polymer melt is usually produced by either resistance heating, circulating fluids, air heating, or lasers. However, electrospinning is

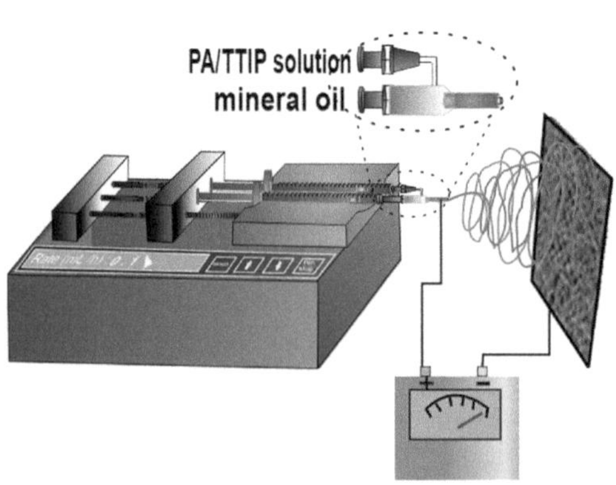


**Figure 11.1** A coaxial electrospinning nozzle for fabricating core-shell nanofibers [18].

usually carried out using a polymeric solution. Similar to most fabricating processes, there are many parameters which influence the morphology of the electrospun fibers, for example, beaded or porous fibers. Generally, the influential parameters for conventional electrospinning are associated with the applied voltage, temperature, collector, and ambient conditions [20,21]. With the understanding of these parameters, it is possible to devise setups for fibrous structures of various forms and arrangements. It is also possible to create nanofibers with different morphologies as the influential parameters are varied.

### 11.2.1 Polymer solution

The properties of the polymer solution including the molecular weight [20], solution viscosity [22], surface tension [22], solution conductivity [23], and dielectric effect of solvent have the most significant influence on the electrospinning process and the resultant fibers' morphology. The surface tension is particularly influential for the formation of beads along the fiber length. The viscosity of the solution and its electrical properties determine the extent of elongation. This will in turn have an effect on the diameter of the electrospun fibers. Thus if the conductivity of the solution is increased, more charges can be carried by the electrospinning jet. The conductivity of the solution can be enhanced by the addition of ions. As previously mentioned, bead formation will occur if the solution is not fully stretched. Therefore, when a small amount of salt or polyelectrolyte is added to the solution, the increased charges carried by the solution will increase the stretching of the solution.

### 11.2.2 Processing conditions

The electrospinning jet is influenced by various processing factors, including the applied voltage [24], the feed-rate [25], temperature of the solution, type of collector, diameter of the needle, and distance between the needle tip and collector [26]. These parameters are quite effective for altering the fiber morphology. The high voltage will firstly induce the electrical charges on the solution at the tip of the needle, creating electrostatic coulombic forces on the surface of the Taylor Cone. As a result, a competitive condition between the repulsion forces and surface tension is generated leading to a critical point under which electrostatic forces overcome the solution surface tension. Generally, both high negative or positive voltages ($>6$ kV) is necessary for fabricating fibers deposited at the collector.

The solution feed rate is a key element of the electrospinning process. When the feed rate is raised, there is an increase in fiber diameters as well as the number of beads. The solution temperature not only helps to facilitate the evaporation of the solvent, but also reduces the solution viscosity of the polymer solution. For instance, when polyurethane is electrospun at a higher temperature, the fibers have a more uniform diameter. This may be due to the lower viscosity of the solution and greater solubility of the polymer in the solvent which allows more stretched fibers to be formed. The internal diameter of the needle or the pipette orifice also has a characteristic effect on the electrospinning process. By selecting a needle with a smaller internal diameter, the clogging as well as the amount of beads within the fiber structure are reduced. The decrease in the internal diameter of the orifice leads to the production of electrospun fibers with reduced diameters.

### 11.2.3 Ambient parameters

The effect of the surrounding area of the electrospinning jet is still poorly understood. Any interaction between the surrounding and the polymeric solution may have an effect on the electrospun fiber morphology. High humidity [27], for example, was found to cause the formation of pores on the surface of the fibers. Since electrospinning is influenced by external electric fields, any change in its environment will also affect the formation process. At high humidity, it is likely that water condenses on the surface of the fibers. As a result, this may have an influence on the fiber morphology, especially for polymers which are dissolved in volatile solvents. The air composition is also influential in the electrospinning process. Different gases behave differently in high electrostatic fields. For example, helium will break down in high electrostatic fields making fiber fabrication more laborious. Generally, the surrounding pressure affects the resulting fibers. When the pressure is below atmospheric pressure, the polymer solution in the syringe will have a greater tendency to flow out of the needle causing unstable jet initiation.

## 11.3 Characterization of electrospun nanofibers

Solid phase extraction is a surface process, in which analytes are retained on the surface of extractive media. It means that each surface feature can influence the extraction efficiency. After preparation of electrospun sorbents, their aspects should be characterized. As explained earlier, some chemical, instrumental, and ambient parameters influence the final shape of submicron fibers, including porosity, surface area, fiber shape, and continuity. In addition, some of the instrumental and ambient parameters are uncontrollable or difficult to control. Consequently, the extraction performance of the fibers depends on their morphologies and chemical structures which are mostly unpredictable [28]. The fibers are often characterized on the basis of physical and chemical methods.

### 11.3.1 Physical fiber parameters

There are several surface relevant parameters including fiber diameters and their size distribution, porosity and surface roughness, which greatly affect extraction properties.

#### 11.3.1.1 Fiber diameter and size distribution

The simplest method for evaluating fiber morphology is light microscopy. Although, the method needs no sample preparation and is relatively straightforward, resolution is low and magnification is limited. Other imaging methods, including scanning electron microscopy (SEM) and transmission electron microscopy (TEM), are based on the use of an electron beam as incident source. In SEM devices an electron beam is projected onto the sample surface and the reflected electrons are collected. The most common collected signals are secondary electrons (SE) from excited atoms. The number of these electrons depends on the surface morphology. The collection of these electrons by the detector provides a morphological image which is quite useful to screen the surface. The alternative signal is from back-scattered electrons (BSE) reflected from the surface and depends on the atomic number. Thus, heavier atoms reflect more electrons toward the collecting electrode located above the surface. The more heavy atoms present in the fiber structure, the brighter the final image. This effect is mostly applied to imaging metal composite fibers to reveal the distribution of metallic particles throughout the fiber structure. Conventional SEM systems use thermoionic guns as an electron source, which requires heating an electron source such as tungsten for electron production. For these systems low brightness and thermal drifting during imaging are the main problems. To address this issue, a new generation of SEM instruments was introduced utilizing an electrical potential gradient source to overcome the work function and facilitate electrons release. This system is called field emission SEM (FE-SEM) and produces images with higher resolution. For instance, Ma et al. prepared a core-shell poly(vinylpyrrolidone) (PVP)-Ag NP electrospun fibers in which PVP acted as a reducing agent for production of Ag NPs from $AgNO_3$ during electrospinning. This concept was confirmed by observing the SEM images (Fig. 11.2A) [29].

It is also possible to calculate the fiber diameter distribution using SEM images [31]. There are two general pathways in this regard. The first method is to take several images from different parts of the fibers, and subsequently calculate fiber diameters with the aid of SEM software or Photoshop. A histogram of frequency versus fiber diameters is finally established. The drawback of this method is the bias of the results by inappropriate sampling. The other method calculates a fiber diameter distribution by the use of software, such as ImageJ. Coutinho et al. synthesized cellulose nanofibers in ionic liquid at room temperature. They used SEM images (Fig. 11.2B) to calculate fiber diameters and demonstrated the production of nanosized fibers [30]. This was a difficult task using other methods.

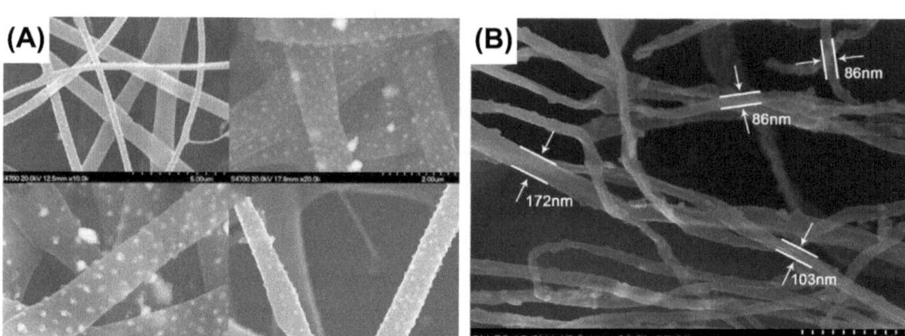


**Figure 11.2** SEM images (A) of deposited Ag NPs on electrospun PVP surface [29] (B) used for the fiber diameter measurement [30].

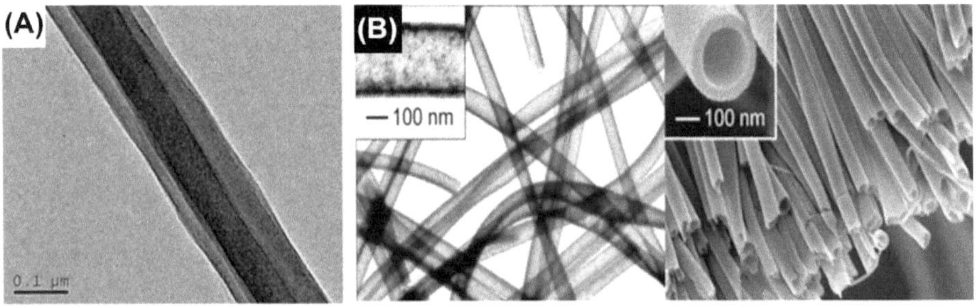


**Figure 11.3** (A) TEM image recorded from the core-shell polyethylene oxide/polylactic acid [32] and (B) hollow $TiO_2$/PVP [33].

TEM imaging is analogous to SEM in that both use electron beam. However, in TEM instruments the electron beam is transmitted from the sample being and collected by a detector located beneath the sample stage at the opposite side of electrons beam pathway. TEM images are particularly useful for imaging the electrospun hollow and core-shell fibers [31]. Xu et al. fabricated a core/shell polyethylene oxide/poly(lactic acid) structure. The TEM images (Fig. 11.3A) confirmed the core/shell structure and diameter of the fibers [32]. Also, Xia et al. prepared electrospun $TiO_2$/PVP hollow fibers. The core consisted of heavy mineral oil which was subsequently removed by immersion in isooctane. The surface smoothness and hollow structure of the fibers were characterized by TEM (Fig. 11.3B) [33].

### *11.3.1.2 Porosity and topography*

Porosity is a measure of the void (i.e., "empty") spaces in a material, and is considered as the fraction of void volumes compared with the total volume. Since in solid phase extraction techniques, the analytes are mostly adsorbed by the surface of the extracting phase, the porosity or surface area is of great importance. As mentioned earlier, higher surface porosity leads to higher surface areas and better extraction efficiency.

There are different methods for calculating surface porosity. The conventional method is mercury porosimetry, which pushes mercury into holes and pores of the material by pressure. Since mercury does not usually wet the surface, it is the only candidate to be applied in this method. Finally the applied pressure and volume of mercury are converted to surface porosity with mathematical calculations. This method is not common because of experimental difficulties and the toxicity of mercury [34]. The alternative is the Brunauer—Emmett—Teller (BET) method. This method is based on the adsorption of gases on a solid surface. Gases should be inert to avoid chemical interaction with the sample. An equation correlates gas pressure to the amount of adsorbed gas from which the specific surface area is obtained [35].

Xia et al. described a new method of electrospinning aimed at increasing the specific surface area of the fibers. They used a liquid nitrogen bath before the collector to induce thermal phase separation. The surface areas of the resulting fibers were threefold higher than theoretical values [36]. Yang et al. prepared porous carbon nanofibers with two different polymer mixtures. The electrospun nanofibers were thermally treated by which one of the polymers was removed and other was carbonized to achieve hollow and highly porous carbon fibers. Different ratios of the two polymers provided sufficient porosity with surface areas from 800 to 940 $m^2\ g^{-1}$. These data were obtained by the BET technique [37].

Surface topology is another important characteristic of solid extraction phases which provides information on surface smoothness and roughness. Scanning probe microscopy produces a 3D image of surface topology. The atomic force microscope (AFM) is the most common technique, since it is applicable to a wide variety of materials with no need for sample preparation. The interaction between the needle tip and surface atoms leads to an image from which the smoothness/roughness of the surface can be predicted. The main advantage of this technique is the fact that the surface morphology remains unchanged during imaging.

Xu et al. provided an AFM map and surface topography of polyethylene oxide/chitosan to illustrate the structural evolution of the nanofibers during electrospinning. They demonstrated that fiber formation is not just a simple drying and thinning process but consists of some surface topologically different steps, including columnar, platelike, bumpy, and other surface features (Fig. 11.4) [38].

There are other less common methods for characterizing the mechanical properties of fibers such as the vibrating sample magnetometer for magnetic properties [39], air permeability testing [40], conductivity measurement [41], and tensile testing [42]. However these properties are not of great significance for the use of electrospun fibers as solid phase extraction media. Thus, these methods are usually neglected. Enthusiastic readers are referred to Ref. [28] for further information.

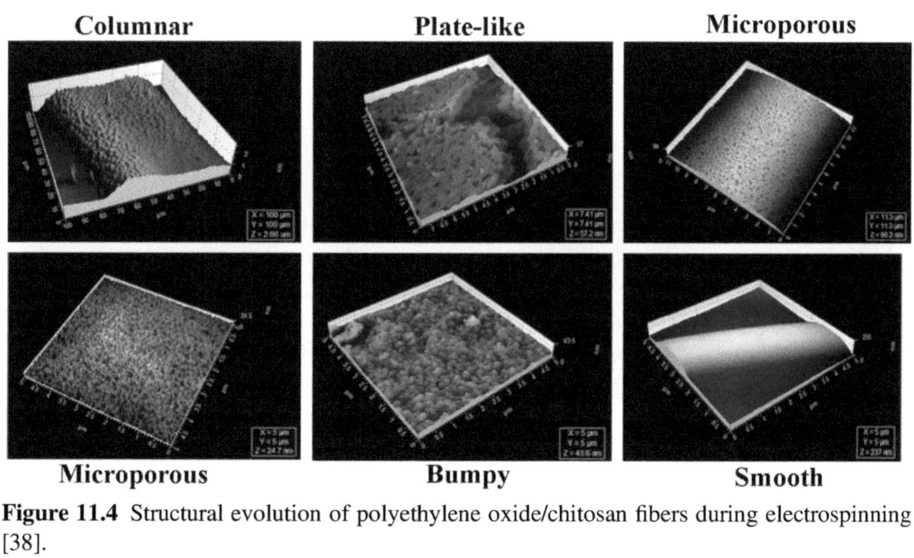


**Figure 11.4** Structural evolution of polyethylene oxide/chitosan fibers during electrospinning [38].

### 11.3.2 Chemical fiber parameters

Chemistry of electrospun sorbents plays a vital role in its extraction capability, particularly wettability, polarity, and fiber strength. As a result, the chemical characterization of electrospun nanofibers is a crucial step in their preparation.

#### 11.3.2.1 Elemental analysis

Elemental analysis can be exploited to understand the basic nature of fibers. Elemental analysis, also known as carbon hydrogen nitrogen sulfur (CHNS) analysis, is a destructive method of choice for fibers with organic backbones. It can determine the percentage of carbon, hydrogen, nitrogen, and sulfur by combustion of nanofibers and subsequent analysis of the gases produced. Ngial et al. prepared electrospun cellulose nanofibers modified with oxolane-2,5-dione for cadmium and lead removal. They calculated the percentage of ligand modification from CHNS analysis [43].

For inorganic fibers, energy dispersive spectroscopy (EDS) is typically used. This method is based on the measurement of the energy and intensity of X-ray photons resulting from the interaction of charged particles (electron or proton) and the fiber. The particles are focused on the surface and the X-rays produced, collected, and energy analyzed. The data can be utilized for quantitative or qualitative analysis of inorganic fibers. In most cases, SEM instruments capable of providing EDS data are used. EDS is nondestructive and produces point-specific elemental analysis. Xia et al. prepared fine composite fibers from a mixture of poly(urethane), poly(methyl methacrylate), and an organic lanthanide complex. The EDS spectra revealed the heterogeneous distribution of the starting materials along the fibers [44].

There are other X-ray-based methods such as X-ray photoelectron spectroscopy (XPS) measuring the kinetic energy and the number of ejected electrons, as a result of the interaction of an X-ray beam with the fiber. The method is surface sensitive and provides information for the outer layers of the surface ($\sim$10 nm). The information includes the structural formula and electronic and chemical state at the part per 1000 range. The X-ray-based methods are appropriate for metals or metal oxides composite fibers [45].

#### 11.3.2.2 Chemical bonding

To study the chemical bonding of nanofibers, Fourier transform infrared spectroscopy (FTIR) is an appropriate nondestructive technique. Interaction of the infrared beam excites the vibrational states of functional groups. The transmission is recorded as a function of wavenumber. Each functional group absorbs at a characteristic wavenumber that can be assigned to a chemical bond in the nanofibers. Electrospun nanofibers can be simply laid on a KBr disk for analysis. For surface-modified nanofibers, attenuated total reflectance (ATR)-based FTIR is implemented. This method is surface sensitive and the signal intensity is increased by extending the light pathway. Ma et al. functionalized the surface of electrospun poly(sulphone) with methyl methacrylate and a cerium salt. They used ATR-FTIR to verify the grafting of poly(methyl methacrylate) to the poly(sulphone) surface [46]. Polarized FT-IR, in which IR beams are aligned in different axis with a polarizer, can provide information about the orientation of molecules, especially when the nanofibers are spatially oriented. Kim et al. prepared randomly deposited and partially aligned electrospun nylon-6 nanofibers by using a rotating collector with different linear velocities. They showed that polarized FTIR data could distinguish between random and aligned fibers [47].

Nuclear magnetic resonance (NMR) spectroscopy provides structural information according to the concept of the chemical shift. The only problem is that electrospun fibers should be dissolved in a deuterated solvent, which makes this method destructive and expensive. Solid-state NMR is also possible for obtaining data without solvent dissolution. Ngila et al. prepared electrospun cellulose acetate fibers and were hydrolyzed confirming the deacetylation process by solid-state NMR [43].

### 11.3.3 Thermal stability

In some cases the thermal stability of electrospun nanofibers would be of great importance especially during extraction processes where thermal desorption or high-temperature extraction is used. Thermogravimetric analysis (TGA) can be used to evaluate the thermal stability of nanofibers. In this method, the sample is exposed to increasing temperature at a known rate, and the amount of weight loss is recorded. Ajji et al. used TGA to select amino-functionalized reduced-graphene oxide (Am-rGO) substrates among other GO derivatives for the preparation of fiber composites. The TGA data (Fig. 11.5A) demonstrated the higher thermal stability of Am-rGO compared with GO and r-GO [48].

Raman spectroscopy can be used to differentiate the difference in amorphous content between the initial precursor and final electrospun product. Using this approach Yang et al. studied the morphology of poly(acrylonitrile) fibers containing graphite crystallites [49]. The left peak in Fig. 11.5B is due to the disordered portion and the right peak the ordered state of the graphitic crystallites. The intensity and ratio of the peaks are related to the heat treatment applied to the sample.

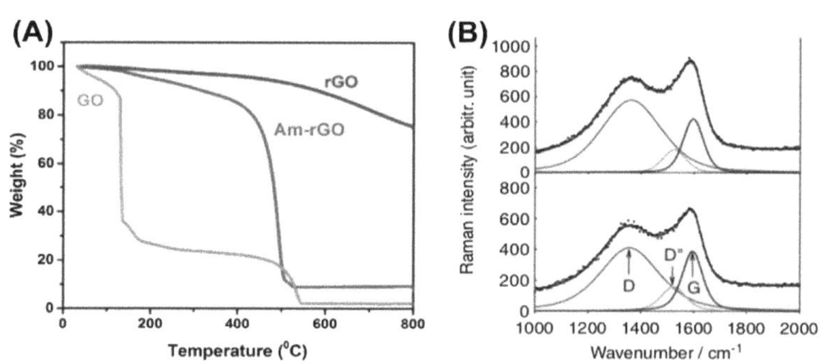


**Figure 11.5** (A).TGA curves of rGO, Am-rGO, GO [48]. (B). Raman spectra of PAN-based carbon nanofibers as a function of heat treatment temperature [49].

## 11.4 Electrospun nanofibers types

In recent decades a wide variety of electrospun nanofibers have been synthesized. The electrospinning method (e.g., nozzle or collector format) together with variation in synthetic methods has contributed to the development of nanofibers with unique properties for specific purposes. Some of the recent applications in the field of extraction science are discussed in more detail below.

### *11.4.1 Molecularly imprinted polymers (MIPs)*

Selectivity remains a concern for the developments of target compound extraction using solid phases. Recent decades have witnessed the development of various strategies for building selectivity into the solid phase extraction process. Molecular imprinting approaches are one of the more successful strategies. In this case a monomer and cross-linking agent are mixed in the presence of the target compound to achieve specific molecular recognition. Subsequent removal of the target compound by an appropriate solvent leads to an extraction phase with cavities matched to the shape of the target compound with complementary specific intermolecular interactions. The high specific surface area and low-cost production of electrospun fibers is attractive for the preparation of MIP-based adsorbents. However, a cross-linked polymeric network cannot be electrospun easily due to its insolubility. There are two approaches for producing electrospun MIP nanofibers: imprinting the electrospun fiber or encapsulation of MIP nanoparticles (NPs) into the electrospun nanofibers. In the direct method, an appropriate functional polymer and template molecule are blended and used for electrospinning. The template is then removed and leaves imprinted sites in the electrospun fibers [50]. Using the alternative approach, MIP-NPs are easily encapsulated inside the electrospun nanofibers [51–53]. The recognition sites in MIP-NPs remain intact even after encapsulation and can be used for trace analysis of different template compounds from complex samples [51,54]. Also, the centrifugation step commonly used to separate particles from solution, can be omitted for nanofiber-based adsorbents. The electrospun chiral imprinted membrane for (−)-cinchonidine showed the same target affinity for the immobilized microspheres and the free form [55]. Chronakis and coworkers used poly(ethylene terephthalate) (PET) as the supporting nanofibers matrix for encapsulation of 17 β-estradiol and theophylline imprinted nanoparticles [52]. Yoshimatsu et al. used MIP-NPs in a PET support through electrospinning for the solid phase extraction of propranolol. It was demonstrated that the binding sites in the nanofiber composites remained intact for the chiral-selective extraction of propanolol [51]. The same authors proposed a simple approach employing direct generation of recognition sites during the electrospinning process for the extraction of 2,4-dichlorophenoxyacetic acid (2,4-D) [50]. The synthesized nanofibers provide functional groups that interacted with the template compound during electrospinning. These results demonstrated the successful preparation of robust MIP nanofibers that can selectively rebind the target molecule. Poly(styrene)-based MIP nanofibers were synthesized for atrazine as a template via a noncovalent approach [56]. To compare the selectivity of the MIP-based nanofibers, a comparison was made with nonimprinted polymer (NIP) nanofibers for the extraction of atrazines from river water.

MIP-nanofiber membranes have been used for the selective enrichment of Rhodamine B (RhB) [57], bisphenol A, basic tebuconazole [58] and (−)-cinchonidine [55]. The molecularly imprinted membranes were produced by the electrospinning of RhB MIP microspheres and poly(ethylene terephthalate) as the matrix polymer. The electrospun imprinted polymer membranes demonstrated a higher affinity for the target compounds than molecularly imprinted particles. In addition, the imprinted nanofiber membranes had higher selectivity for the target compounds than nonimprinted analogs [57].

Electrospun molecularly imprinted polymer membranes were also prepared with two types of MIP-NPs as multianalyte selective membranes for the simultaneous isolation of trace amounts of bisphenol A and basic tebuconazole in vegetables and fruit juices [58]. MIP electrospun nanofibers of poly(styrene) were employed as a coating for solid phase microextraction (SPME) for the extraction of parabens from environmental waters [59]. More recently, the MIP methodology was used to produce an extraction material via sol-gel electrospinning. In this process, poly(amide)/tetraethoxyorthosilane (TEOS) solution was electrospun followed by a hydrolysis and condensation step using thermal treatment. After aging the organic backbone was removed by external heating. The imprinted-silica nanofibers were used for the extraction of atrazine in real samples by the μ-solid phase extraction technique [60].

## 11.4.2 Core-shell and hollow fibers

Core-shell nanofibers and hollow nanotubes were developed by coaxial electrospinning. The main applications of coelectrospinning consist of encapsulation of different biologically active compounds; formation of nanotubes, cell scaffolds, and drug release [61]. Core-shell nanofibers can also be produced using a single nozzle setup without complex coannular nozzles. Accordingly, Bazilevsky et al. synthesized core-shell nanofibers using an emulsion of two polymer solutions, PMMA/poly(acrylonitrile) (PAN) in *N,N*-dimethylformamide (DMF). To this end, PMMA/DMF and PAN/DMF were mixed and left for 1 day. During this period, PMMA droplets are dispersed in the PAN matrix. Precipitation of PMMA droplets in the PAN solution and its entrapment in the Taylor cone leads to core-shell formation during electrospinning. The use of a single nozzle system relies on the configuration of the large-particle emulsion. An emulsion of PMMA/DMF droplets in PAN/DMF gave multicore-shell nanofibers by electrospinning [14]. Core-shell nanofibers have also been synthesized by a two-step process. Tin oxide ($SnO_2$) was produced by electrospinning and zinc oxide (ZnO) was then deposited by atomic layer deposition (ALD) on the $SnO_2$ nanofibers as a shell layer. These nanofibers have potential for use as gas sensors for oxygen and nitrogen dioxide [62]. Similarly, $TiO_2$−ZnO core-shell nanofibers were synthesized by conventional electrospinning and ALD for sensing oxygen [63]. Bagheri et al. synthesized electrospun core-shell

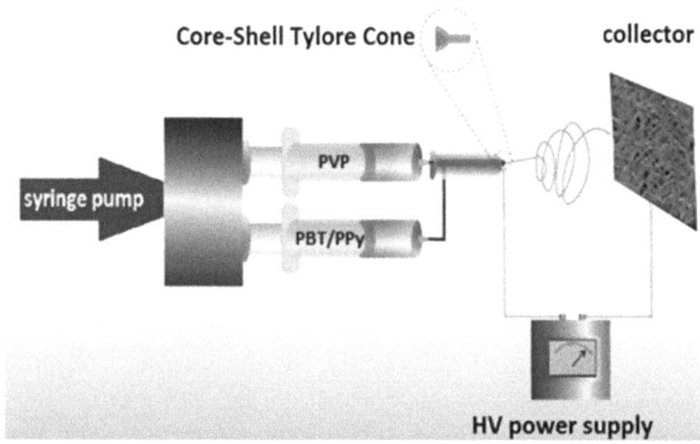


**Figure 11.6** Schematic diagram of the homemade core-shell electrospinning setup, PBT/PPy hollow nanofiber obtained from core-shell PVP-PBT/PPy nanofibers [17].

nanofibers of poly(vinylpyrrolidone)-poly(butylene terephthalate)/poly(pyrrole) (PVP—PBT/PPy) as a sorbent for μ-solid phase extraction (μ-SPE) of triazine herbicides from aqueous samples and wheat grains (Fig. 11.6). In order to remove the PVP, core-shell PVP-PBT/PPy nanofibers were submerged in water. The higher extraction capability of the hollow nanofibers compared with conventional fibers confirmed their suitability as an adsorbent for solid phase extraction [17].

A poly(pyrrole) functionalized core-shell electrospun nanofiber mat was used in disk-format for extraction of trace polar analytes from environmental water samples. Acid yellow 9, acid orange 7, and metanil yellow with different numbers of sulfonate functional groups were investigated as model analytes. Intercalation of multifunctional PPy in the PA nanofibrous mat led to significantly higher extraction capability for the organic dyes [64]. Tian et al. developed a simple method for fabrication of electrospun PPy hollow fibers with different surface morphologies. First, poly(caprolactone) (PCL) fibers were prepared by electrospinning, and used as a template for in situ polymerization of PPy as a shell layer. Then the PCL was removed and the hollow fibers used as sorbents for SPE of polar compounds. The PPy hollow fibers with a high specific surface area were used in the packed-fiber SPE (PFSPE) format for extraction of two neuroendocrine markers of behavioral disorders (5-hydroxyindole-3-acetic acid and homovanillic acid) [65].

### 11.4.3 Polymeric nanofibers

The application of electrospun nanofibers as sorbent materials in sample preparation dates back to 2007 [66]. In 2005 Shin et al. reported the use of poly(styrene) nanofibers as a filter to separate water from a water-oil emulsion [67]. This led to various attempts at packing nanofibers into different formats as membranes and packed tips for the extraction of organic compounds from different matrices [68]. PA has been used most widely for extraction purposes. In 2010, PA fibers were packed in a disk in a homemade device for extraction of estrogens from environmental water samples.

In this study, a comparison between a commercial nylon microporous membrane and octadecylsiloxane-bonded silica (ODS) cartridge was performed [69]. Also, electrospun PA nanofibers were applied for the extraction of phthalate esters by SPE and their performance compared with octadecylsiloxane-bonded silica cartridges. It was demonstrated that nanofibers had a good potential as an efficient SPE sorbent [70]. Docetaxel was extracted by the PA nanofibers (diameters 400–800 nm) from rabbit plasma [71]. Also, three types of nylon 6 nanofibers were prepared with different surface densities (5.04, 3.90, and 0.75 g $m^{-2}$) and used for solid phase extraction of parabens, steroids, flavonoids, and pesticides. These nanofibers were produced by needleless electrospinning. Their extraction performance was similar to Oasis HLB particle-packed cartridges [72]. In 2014, an online μ–SPE setup for extraction of clodinafop propargyl was developed by employing a cartridge packed with PA nanofibers as the HPLC sample loop located on a six-port valve. The online extraction was performed by pumping the sample through the cartridge followed by analytes desorption into the HPLC column (Fig. 11.7) [73].

In another study, PA nanofibers were used in headspace solid phase microextraction (HS-SPME) of chlorophenols from aqueous samples using GC–MS. The concentration of polymer in the electrospinning solution had a crucial role in the suitability of the polymer for this analysis [74]. Table 11.1 lists other polymers which have been used to produce electrospun nanofibers.

### 11.4.4 Copolymers

There are two general approaches for surface modification of electrospun polymeric sorbents. One is to synthesize a hydrophilic/hydrophobic copolymer using at least one monomer containing polar functional groups. An alternative route is to modify

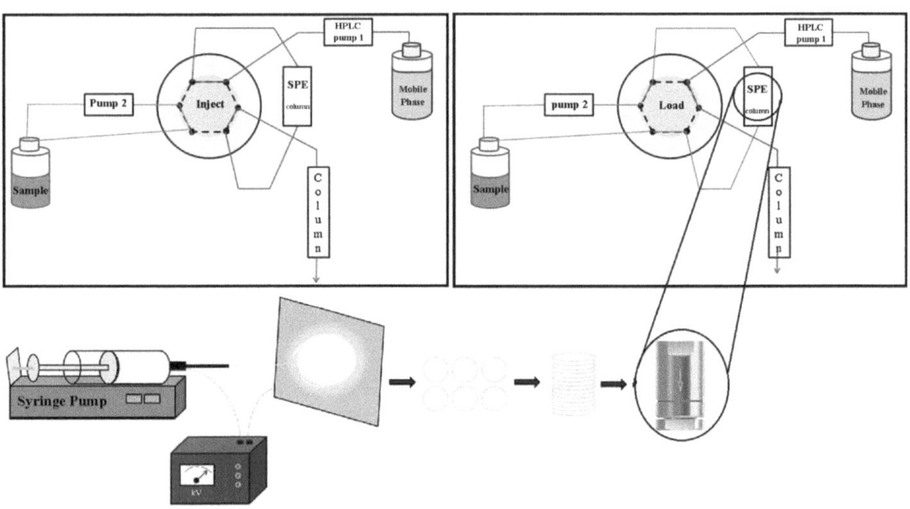


**Figure 11.7** The online μ-SPE of clodinafop propargyl using PA nanofibers [73].

**Table 11.1** Summary of applications of electrospun nanofibers.

| Polymer | Extraction technique | Instrumental technique | Analytes | Matrix | Reference |
|---|---|---|---|---|---|
| PS | SPME | GC-MS | Pesticides | Honey | [75] |
| PS | PFSPE | HPLC-UV | Trazodone | Plasma | [66] |
| PS | DLLME | HPLC-UV | Aldehydes | water | [76] |
| PS | Semimicro SPE | HPLC-DAD | Steroid compounds | Plasma & water | [68] |
| PS | PFSPE | LC-MS/MS | Diethylstilbestrol, hexestrol, and dienestrol | Milk | [77] |
| PS | SPE | UV−Vis spectroscopy at $\lambda_{max} = 638$ nm | Disulfine Blue | water | [78] |
| PS | SPE | HPLC-DAD | Microcystins | Water | [79] |
| PAN | PT-SPE | HPLC-UV | Nitroaromatic compounds | wastewater | [80] |
| Silk | SPME | GC-FID | Isopropyl alcohol | water | [81] |
| PET | SPME | UV | Chromium(VI) | water | [82] |
| PET | SPME | GC-MS | PAHs | water | [83] |
| PI | TFME | GC-MS | Phenols | water | [84] |
| PU | SPME | GC-MS | chlorobenzenes | Water | [85] |

a hydrophobic substrate by imparting polar moieties into the polymer structure [86]. Poly(styrene), due to its hydrophobic nature, is a suitable substrate for the extraction of moderate and nonpolar compounds but suffers from poor wettability [87]. Addition of surface polar groups can obviate this issue. Styrene-co-methacrylic acid copolymer has been elestrospun to enhance the extraction of aromatic hydrocarbons from water samples [88], as well as the determination of vitamins E and A in plasma samples [89]. Electrospun styrene-co-styrene sodium sulfonate copolymer was used for the extraction of several hormones from water samples. This copolymer demonstrated higher extraction efficiency than poly(styrene) fibers [90]. Ifegwu et al. prepared a styrene-co-methacrylic acid, styrene-co-p-sodium styrene sulfonate, and styrene-co-acrylamide polymers for SPE of 1-hydroxypyrene in urine, with a higher yield than styrene-co-acrylamide polymers [91].

### 11.4.5 Carbon fibers

Carbon fibers (CFs) have been developed for a wide range of applications [92]. CNFs can be fabricated by catalytic vapor deposition [93] and electrospinning, especially PAN-based CNFs [94,95]. To fabricate PAN-based CNFs, PAN nanofibers are prepared as a precursor for CNFs production. The properties of the CNFs are controlled by the selected polymer solution and the processing parameters. Once the PAN nanofibers are prepared, heat treatment is applied to the nanofibers [95]. This consists of a two-step stabilization and carbonization process [96]. Stabilization is usually carried out in air at temperatures between 200 and 300°C and carbonization is conducted in an inert atmosphere at 800–2800°C [97,98]. The applicability of activated CNFs was investigated for preconcentration of organophosphorus pesticides (OPPs) by LC-UV [99]. Zewe et al. dissolved SU-8 2100 in cyclopentanone to prepare a polymer solution for electrospinning subsequently used as SPME fibers. The carbon nanofibers-based coating was prepared by pyrolyzing the SU-8 nanofibers. The extraction characteristics of the SU-8 and pyrolyzed electrospun-coated wires were investigated for nonpolar (benzene, toluene, ethylbenzene, and o-xylene) and polar (phenol, 4-chlorophenol, and 4-nitrophenol) compounds for headspace extraction. The extraction efficiencies of the electrospun fibers–coated wires were compared with PDMS and PDMS/DVB fibers for the extraction of BTEX pollutants. The electrospun fibers prepared at temperatures between 600 and 800°C demonstrated enhanced extraction efficiencies compared to the commercial fibers. The extraction performance of the electrospun fibers for SPME was evaluated for the extraction of phenols and provided improved performance compared with commercial poly(acrylate) SPME fibers [100].

### 11.4.6 Inorganic fibers

Inorganic nanofibers can be synthesized by electrospinning via different synthetic routes (Fig. 11.8B). One approach is to add an inorganic precursor, typically a metal halide or a metal alkoxide, to a spinnable polymer solution. These precursors can be chemically transformed to the corresponding metal oxide via sol-gel reactions.

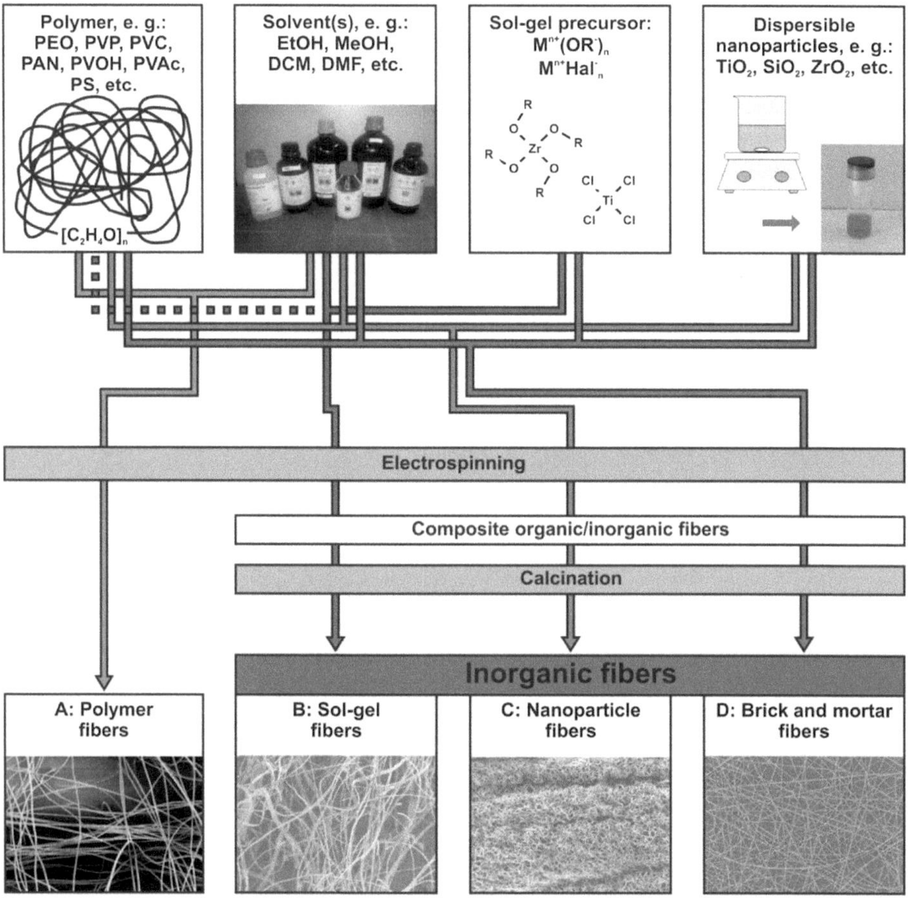


**Figure 11.8** Synthetic concepts for the preparation of electrospun fibers: A: preparation of polymeric fibers; B: preparation of inorganic fibers via the sol-gel route (with or without the addition of organic polymers); C: preparation of inorganic fibers via the nanoparticle approach; D: preparation of inorganic fibers via the sol-gel method and nanoparticle approach [107].

A thermal treatment is then used to remove the polymer by combustion. Sol-gel solutions may become very viscous during gelation of the precursors. During the hydrolysis reaction, a network of metal-oxygen bonds is established, resulting in gel formation (a highly viscous inorganic cross-linked polymer). Depending on the level of condensation, the viscosity of the sol-gel solution changes during the course of the reaction, which can lead to clogging of the electrospinning needle [101]. Another method is based on the use of inorganic nanoparticles rather than sol-gel precursors (Fig. 11.8C). In this strategy, a viscous solution of an organic polymer is combined with a dispersion of inorganic NPs. The difficulties with this approach are, firstly, the availability of suitable dispersible inorganic NPs, and secondly, the preparation of solutions with a certain "threshold" concentration of NPs. This nanoparticle-approach has been little investigated compared with the sol-gel approach and is categorized as part of a more general colloidal electrospinning concept, which is mostly used to prepare composite fibers [102]. An advantage of the NP-approach is that materials which cannot be prepared via sol-gel chemistry become applicable. In this case, a synthetic concept, which uses a spinning solution containing NPs and a small amount of sol-gel precursor as "molecular glue," can be applied (Fig. 11.8D) [103]. Titanium dioxide nanofibers (TDNFs) were prepared by this technique. $TiO_2$/poly(vinyl pyrrolidone) nanofibers were prepared from (PVP)/pluronic123 composite nanofibers by calcining at 450°C in air to remove the organic constituents and produce TDNFs. SPE with titanium dioxide nanofibers was used for the preconcentration of thallium from tea components, such as poly(phenols), soluble sugars, catechin, caffeine, and tea pigments [104].

The fabrication of ceramic nanofibers by electrospinning consists of three main steps. The first is the preparation of a homogenous spinnable polymer solution containing the inorganic precursor. The second step is electrospinning of the solution under controlled conditions, and the final step is calcination of the fibers in a furnace at elevated temperatures to remove the polymer. For this approach, poly(vinylpyrrolidone) is commonly used as the organic component of the sol during the electrospinning process. To improve the adsorption capacity, four different transition metals (Co, Mn, Ni, and Fe) are added to the sol for titanium-based nanofibers. These fibers are calcined at 500°C to remove the PVP. Ceramic nanofiber sheets containing the transition metals, Fe-Mn, Fe-Ni, Fe-Co, and Fe-Mn-Co-Ni were prepared. The electrospun fibers were used for the online extraction of naproxen and clobetasol from plasma and urine by online μ-SPE. All ceramic fibers exhibited acceptable efficiencies for the extraction of the target compounds largely independent of the metal incorporated into the fibers [105]. Ordered mesoporous silica fibers (OMSF) can be prepared by electrospinning combined with pseudomorphic synthesis. Initially, amorphous silica fibers (ASF) are fabricated via electrospinning of a PVA/$SiO_2$ composite followed by calcination to remove the organic component. Then the ASF was transformed into an ordered mesoporous phase by pseudomorphic synthesis. The surface area and total pore volume was dramatically increased in comparison with the ASF precursor. The OMSF was used for the preconcentration of endogenous peptides by a lab-in-syringe approach based on hydrophobic interactions and a size-exclusion mechanism [106].

### 11.4.7 Composites

Composite materials consist of two or more phases of different chemical composition, with mechanical and physical properties different from the original constituents. For instance, high strength and resistance to mechanical and thermal shock, and corrosion are the characteristics of many composites. The continuous phase in composites is called the matrix while the other phase is known as the dispersed phase. The usual classification of composite materials based on the type of matrix includes: (i) polymer matrix composites (PMCs); (ii) metal matrix composites (MMCs); (iii) ceramic matrix composites (CMCs); and (iv) carbon matrix composites (CAMCs). PMCs are the most widely used composites [108]. Nanofiber composites exhibit larger surface area-to-volume ratios and higher adsorption capacity, making them more suitable as extraction media compared with powder composites [109]. Composite nanofibers are synthesized by two techniques, known as the template method and electrospinning of a composite solution polymer. In the first method, the electrospun polymer fibers as a mat are immersed into a solution containing composite particles which adsorb onto the fiber surface. It is a time-consuming process and the amounts of entrapped materials are uncontrollable. The second method is based on the combination of polymer with composite which reduces the preparation time and results in spun fibers with a defined composition. The latter is widely used as a rapid fabrication process for composite nanofibers [110].

### *11.4.7.1 Polymer matrix composite nanofibers (PMCNs)*

PMCNs are the most common subset of composite materials designed to improve the properties of polymeric nanofibers and to introduce desired functionalities into the polymer matrix. These nanofibers have been used for the extraction of pesticides, industrial chemicals, polycyclic aromatic hydrocarbons, pharmaceuticals, heavy metals, and toxic compounds. Metal-organic frameworks (MOFs), graphene, inorganic oxide nanoparticles, and polymers have been used as dispersed phases in PMCNs. MOFs are compounds consisting of metal ions or clusters coordinated to organic ligands to form one-, two-, or three-dimensional (3D) structures. They are amenable to fine tuning of their pore structure appropriate for selective extraction. High back pressure is a challenging when MOF particles are used in SPE cartridges. An effective solution is to incorporate the MOF particles into electrospun nanofibers. Yan et al. investigated the applicability of the electrospun UiO-66/poly(acrylonitrile) nanofibers as MOF-polymer composite nanofibers for pipette tip-SPE (PT-SPE) of phytohormones in vegetable samples [111]. High extraction efficiency and sensitivity along with excellent reproducibility were obtained for the phytohormones. Spider web-like chitosan/MIL-68(Al) composite nanofibers were developed for the analysis of trace levels of Pb(II) and Cd(II) after SPE using inductively coupled plasma optical emission spectrometry [112]. Water-stable methyl-modified MOF-5/poly(acrylonitrile) composite nanofibers were used for the solid phase extraction of estrogens from urine [113]. The water solubility of MOFs is a general challenge for SPE. MOF-5 and several other MOFs exhibit limited solubility when exposed to water and moisture. The incorporation of hydrophobic functional groups, such as methyl, into the structure of MOF-5, modifies its water solubility. In addition, the combination of modified MOF-5 with poly(acrylonitrile) as a nanofiber composite increases $\pi-\pi$ interactions, hydrogen bonding, and hydrophobic interactions between the estrogenic drugs and the sorbent. Another limitation of MOFs is their separation from the substrate during thin film microextraction (TFME). To solve this problem, Xu et al. synthesized a poly(styrene)/MOF-199 electrospun nanofiber composite and investigated its use for the extraction of aldehydes from urine [114].

Graphene (G) is a single layer of carbon atoms arranged in a hexagonal lattice. Its excellent mechanical, thermal, electrical properties and ultrahigh specific surface area make it an interesting material for sample preparation. However, it has some limitation, such as: (i) leakage and blocking when the nanosized graphene is used as a sorbent in SPE and (ii) the tendency to aggregate leading to a decrease in the accessible surface area. These drawbacks were addressed by Xu et al. by incorporation of graphene into PS nanofibers as a thin film sorbent [115]. Acceptable recovery and low LODs demonstrated the suitability of the nanofiber composite for the isolation of aldehydes from complex breath samples.

Inorganic oxide NPs are another dispersed phase, which generate nanocomposite structures with nanoscale morphology. The effect of inorganic oxide nanoparticles on the extraction efficiency of electrospun poly(ethylene terephthalate) composites was investigated by Bagheri et al. [116]. Four types of nanoparticles ($Fe_3O_4$, $SiO_2$, CoO, and NiO) along with a PET polymer were electrospun and evaluated as SPME fiber coatings. The electrospun $SiO_2$—PET nanocomposite showed superior performance for the extraction of aromatic compounds. The porous structure, high specific surface area, and hydrophobicity of the fiber are probably the major factors contributing to its enhanced performance. Silica supported $Fe_3O_4$ nanoparticles ($Fe_3O_4$-$SiO_2$) were also used as the dispersed phase for the synthesis of an electrospun nanocomposite based on magnetic nanoparticles-poly(butylene terephthalate) (MNPs-PBT) [117]. The extraction efficiency was studied by online μ-SPE-LC-UV (Fig. 11.9A) for the analysis of antiinflammatory and loop diuretics as model analytes. High porosity and paramagnetic properties of the MNPs-PBT nanofibers enhanced the extraction efficiency of the selected drugs. By immersing the diamagnetic compounds into paramagnetic media, the analytes are more willing to accumulate in the areas in which the minimum magnetic field is applied. In another study, Feng et al. developed a novel sorbent by immobilizing oxidized carbon nanotubes (OCNTs) in PS and used the PS/OCNTs film as TFME adsorbent for matrix-assisted laser desorption/ionization time-of-flight mass spectrometry (MALDI—TOF—MS) (Fig. 11.9B) [118]. The PS/OCNTs film demonstrated a

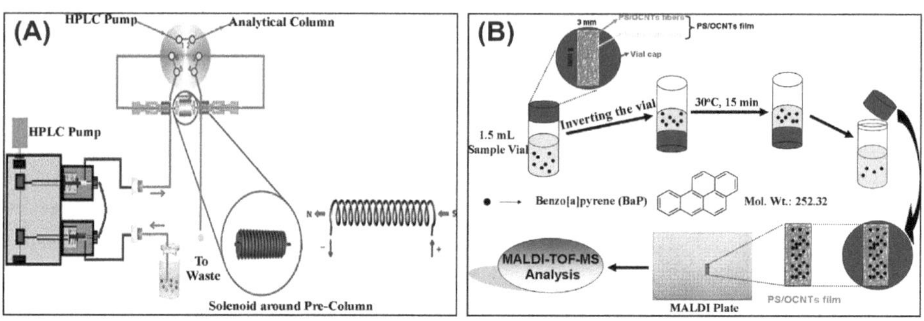


**Figure 11.9** (A) The schematic diagram of the magnetic field assisted online μ-SPE-LC-UV [117] (B) the Schematic diagram of the TFME technique coupled with MALDI-TOF-MS analysis using an electrospun PS/OCNTs film [118].

favorable adsorption capacity for benzo[a]pyrene and 1-hydroxypyrene from urine. In addition, the PS/OCNTs film was an excellent substrate for MALDI.

Various polymers can also be dispersed into the polymer matrix to produce composites with diverse properties, also called hybrid materials. Many reports describe the use of polymer-polymer composite nanofibers as sorbent s for various SPE and SPME formats. For instance, a poly(aniline)-nylon-6 (PANI-N6) nanofiber sheet was employed for headspace adsorptive microextraction of aromatic compounds (such as chlorobenzenes) from different environmental water samples by Bagheri et al. [119]. Hydrophilic nanofibers synthesized by the in-situ formation of Ag NPs on poly(dopamine) coated polystyrene electrospun fibers (PS@PDA-Ag) were utilized as μ-SPE sorbent for the online determination of PAHs in human urine [120]. The large surface area-to-volume ratio, high porosity, and resistance to matrix fouling are the attractive properties of this nanofiber sorbent. A PS/G nanofiber membrane was immersed in a dopamine solution for autopolymerization [121]. Then, the core-sheath conjugated electrospun PS/G@PDA nanofibers were used for the extraction of aldehyde metabolites in human urine. Ebrahimzadeh et al. also reported a reasonable approach to enhance the extraction capability of electrospun nanofibers [122]. A composite solution of poly(4-nitroaniline) (P4-NA) and PVA was electrospun and the porosity of the nanofibers increased by dissolving away the PVA fraction in hot water. The resulting nanofibers exhibited satisfactory performance for the extraction of organophosphorus pesticides from aqueous solution by SPME. Satínský et al. described an innovative approach for the preparation of nanofiber composites using a combination of melt-blown and electrospinning [123]. The combination of two different procedures created a stable 3D porous structure suitable for online SPE-LC system at high pressures. The extraction efficiency of the poly(caprolactone) and poly(vinylidene fluoride) composite nanofibers (PCL/PVDF) was compared with the 2D mat nanofibers. The 3D-based structure was shown to be suitable for the online extraction of ochratoxin A from beer samples.

#### 11.4.7.2 Ceramic matrix composite nanofibers (CMCNs)

The crystalline structure and strong atomic bonds in ceramic nanofibers impart excellent properties such as hardness, compressive strength, thermal and chemical stability, and corrosion resistance. However, the brittle nature, low toughness, poor tensile strength, and lack of flexibility limit the application of the pure ceramic nanofibers. To overcome these issues, ceramic composite nanofibers have been developed with improved surface area and porosity [124]. Bagheri et al. evaluated Ti-Fe-Mn, Ti-Fe-Ni, Ti-Fe-Co and Ti-Fe-Mn-Co-Ni ceramic nanofibers as sorbent for the extraction of naproxen and clobetasol using a μ-SPE setup online with HPLC [105]. The porous structure and high aspect ratio of the ceramic composite nanofibers indicated their potential for use in the cleanup and preconcentration of naproxen and clobetasol from urine and blood plasma samples.

### 11.4.7.3 Carbon matrix composite nanofibers (CAMCNs)

The large surface area and high chemical stability of CAMCNs are ideal features for their applications as sorbents. However, their application as extractive media has not been explored much. The applicability of electrospun CAMCNs as an extractive phase for the preconcentration of aniline compounds was reported by Basheer et al. [125]. The carbon nanofibers were obtained from carbon soot by combusting natural oil. These were mixed with PVA and electrospun to synthesize the CAMCNs mat. The extraction ability of the CAMCNs membrane in μ-SPE format was evaluated by HPLC.

## 11.4.8 Three-dimensional (3D) electrospun nanofibers

Three-dimensional electrospun nanofibrous scaffolds have many potential applications in different fields such as tissue engineering, solar cells, filters, and energy storage. 3D nanofibers can be synthesized by combining liquid-assisted collection in the electrospinning process, called wet electrospinning, without resorting to harsh chemical or experimental conditions. The nanofiber mats are expected to have relatively larger surface areas and higher porosity compared with 2D mats [126]. The first application of 3D electrospun nanofibers for the extraction of chlorobenzenes (CBs) was reported, recently [127]. Fig. 11.10 illustrates the wet electrospinning setup for the synthesis of 3D nanofibrous scaffolds and microoriented extraction setup, schematically. The comparison of 3D scaffolds and 2D nanofibers mats as an extractive phase in needle trap microextraction (NTME) of CBs demonstrates their superiority over conventional nanofibers. The induction of the third dimension has a surprising effect on extraction efficiency as the enhancement of the extraction yield is considerable. The interesting features of 3D nanofibrous scaffolds and promising results verifies their suitability as extractive phases in different extraction methods such as SPME, NTME, and SPE.

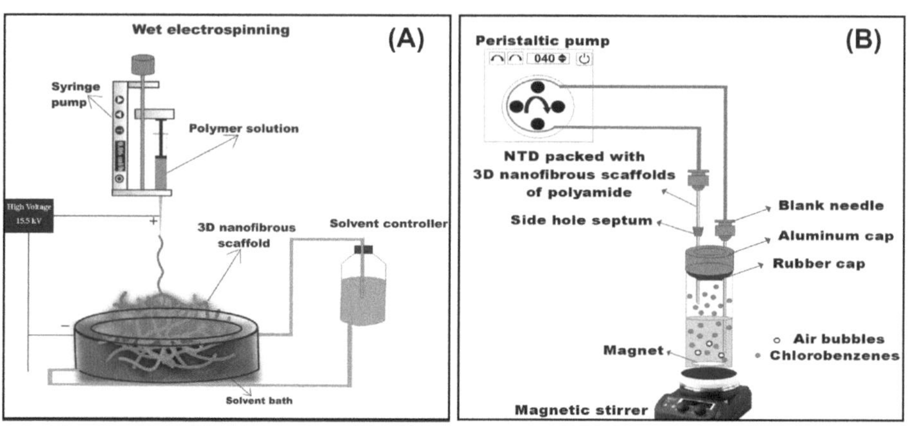


**Figure 11.10** (A) The schematic diagrams of wet electrospinning used for preparation of 3D polyamide scaffolds and (B) microoriented extraction setup [127].

# References


[1] Feynman RP. There's plenty of room at the bottom. In: Anthony JGH, editor. Feynman and computation: perseus books; 1999. p. 63–76.

[2] Kralj S, Makovec D. Magnetic assembly of superparamagnetic iron oxide nanoparticle clusters into nanochains and nanobundles. ACS Nano 2015;9(10):9700–7.

[3] Eijkel J CT, van den Berg A. The promise of nanotechnology for separation devices – from a top-down approach to nature-inspired separation devices. Electrophoresis 2006; 27(3):677–85.

[4] Pelaz B, Jaber S, de Aberasturi DJ, Wulf V, Aida T, de la Fuente JM, et al. The state of nanoparticle-based nanoscience and biotechnology: progress, promises, and challenges. ACS Nano 2012;6(10):8468–83.

[5] Richardson JJ, Björnmalm M, Caruso F. Technology-driven layer-by-layer assembly of nanofilms. Science 2015;348(6233).

[6] Reyes-Gallardo EM, Lucena R, Cárdenas S. Electrospun nanofibers as sorptive phases in microextraction. Trac Trends Anal Chem 2016;84:3–11.

[7] Gray S. A letter concerning the electricity of water, from Mr. Stephen Gray to Cromwell Mortimer. MD Secr R S Philos Trans 1731;37:227–60.

[8] Cooley JF. Apparatus for electrically dispersing fluids. 1902. p. 692631. US Patent.

[9] Morton WJ. Method of dispersing fluids. 1902. p. 705691. US Patent.

[10] Bhardwaj N, Kundu SC. Electrospinning: a fascinating fiber fabrication technique. Biotechnol Adv 2010;28(3):325–47.

[11] Bagheri H, Roostaie A, Daliri R. An electrospun magnetic nanocomposite for a facile micro-scaled analysis approach. Anal Method 2014;6(15):5838–46.

[12] Li D, Xia Y. Electrospinning of nanofibers: reinventing the wheel? Adv Mater 2004; 16(14):1151–70.

[13] Sinha-Ray S, Pelot DD, Zhou ZP, Rahman A, Wu XF, Yarin AL. Encapsulation of self-healing materials by coelectrospinning, emulsion electrospinning, solution blowing and intercalation. J Mater Chem 2012;22(18):9138–46.

[14] Bazilevsky AV, Yarin AL, Megaridis CM. Co-electrospinning of core–shell fibers using a single-nozzle technique. Langmuir 2007;23(5):2311–4.

[15] Xu X, Zhuang X, Chen X, Wang X, Yang L, Jing X. Preparation of core-sheath composite nanofibers by emulsion electrospinning. Macromol Rapid Commun 2006; 27(19):1637–42.

[16] Nagy ZK, Balogh A, Drávavölgyi G, Ferguson J, Pataki H, Vajna B, et al. Solvent-free melt electrospinning for preparation of fast dissolving drug delivery system and comparison with solvent-based electrospun and melt extruded systems. J Pharm Sci 2013; 102(2):508–17.

[17] Bagheri H, Rezvani O, Banihashemi S. Core–shell electrospun polybutylene terephthalate/polypyrrole hollow nanofibers for micro-solid phase extraction. J Chromatogr A 2016;1434:19–28.

[18] Rezvani O, Hedeshi MH, Bagheri H. Polyamide/titania hollow nanofibers prepared by core–shell electrospinning as a microextractive phase in a fabricated sandwiched format microfluidic device. J Chromatogr A 2017;1528:1–9.

[19] Hutmacher DW, Dalton PD. Melt electrospinning. Chem Asian J 2011;6(1):44–56.

[20] Gupta P, Elkins C, Long TE, Wilkes GL. Electrospinning of linear homopolymers of poly(methyl methacrylate): exploring relationships between fiber formation, viscosity, molecular weight and concentration in a good solvent. Polymer 2005;46(13):4799–810.

[21] Chong EJ, Phan TT, Lim IJ, Zhang YZ, Bay BH, Ramakrishna S, et al. Evaluation of electrospun PCL/gelatin nanofibrous scaffold for wound healing and layered dermal reconstitution. Acta Biomater 2007;3(3):321–30.
[22] Zhang C, Yuan X, Wu L, Han Y, Sheng J. Study on morphology of electrospun poly(-vinyl alcohol) mats. Eur Polym J 2005;41(3):423–32.
[23] Jiang H, Fang D, Hsiao BS, Chu B, Chen W. Optimization and characterization of dextran membranes prepared by electrospinning. Biomacromolecules 2004;5(2):326–33.
[24] Kim B, Park H, Lee S-H, Sigmund WM. Poly(acrylic acid) nanofibers by electrospinning. Mater Lett 2005;59(7):829–32.
[25] Sill TJ, von Recum HA. Electrospinning: applications in drug delivery and tissue engineering. Biomater 2008;29(13):1989–2006.
[26] Zhao Z, Li J, Yuan X, Li X, Zhang Y, Sheng J. Preparation and properties of electrospun poly(vinylidene fluoride) membranes. J Appl Polym Sci 2005;97(2):466–74.
[27] Mit-uppatham C, Nithitanakul M, Supaphol P. Ultrafine electrospun polyamide-6 fibers: effect of solution conditions on morphology and average fiber diameter. Macromol Chem Phys 2004;205(17):2327–38.
[28] Zhu J, Ge Y, Jasper S, Zhang X. 9 – physical characterization of electrospun nanofibers. In: Afshari M, editor. Electrospun nanofibers. Woodhead Publishing; 2017. p. 207–38.
[29] Jiang C, Nie J, Ma G. A polymer/metal core–shell nanofiber membrane by electrospinning with an electric field, and its application for catalyst support. RSC Adv 2016; 6(27):22996–3007.
[30] Freire MG, Teles ARR, Ferreira RAS, Carlos LD, Lopes-da-Silva JA, Coutinho JAP. Electrospun nanosized cellulose fibers using ionic liquids at room temperature. Green Chem 2011;13(11):3173–80.
[31] Merritt SR, Exner AA, Lee Z, von Recum HA. Electrospinning and imaging. Adv Eng Mater 2011;14(5):B266–78.
[32] Wang Y, Xu L. Preparation and characterization of porous core-shell fibers for slow release of tea polyphenols. Polymers 2018;10(2):144.
[33] Li D, Xia Y. Direct fabrication of composite and ceramic hollow nanofibers by electrospinning. Nano Lett 2004;4(5):933–8.
[34] Ni QQ, Jin XD, Xia H, Liu F. 7 – electrospinning, processing and characterization of polymer-based nano-composite fibers. In: Zhang D, editor. Advances in filament yarn spinning of textiles and polymers. Woodhead Publishing; 2014. p. 128–48.
[35] Im JS, Park S-J, Kim TJ, Kim YH, Lee Y-S. The study of controlling pore size on electrospun carbon nanofibers for hydrogen adsorption. J Colloid Interface Sci 2008; 318(1):42–9.
[36] McCann JT, Marquez M, Xia Y. Highly porous fibers by electrospinning into a cryogenic liquid. J Am Chem Soc 2006;128(5):1436–7.
[37] Kim C, Jeong YI, Ngoc BTN, Yang KS, Kojima M, Kim YA, et al. Synthesis and characterization of porous carbon nanofibers with hollow cores through the thermal treatment of electrospun copolymeric nanofiber webs. Small 2006;3(1):91–5.
[38] Xu P, Li W, Zhou H, Pan F, Xing H, Liu H. Investigations of the structural evolution of electrospun nanofibers using atomic force microscopy. RSC Adv 2012;2(29):11104–10.
[39] Ju Y-W, Park J-H, Jung H-R, Cho S-J, Lee W-J. Fabrication and characterization of cobalt ferrite ($CoFe_2O_4$) nanofibers by electrospinning. Mater Sci Eng, B 2008;147(1): 7–12.
[40] Gibson PW, Schreuder-Gibson HL, Rivin D. Electrospun fiber mats: transport properties. AIChE J 2004;45(1):190–5.

[41] Chronakis IS, Grapenson S, Jakob A. Conductive polypyrrole nanofibers via electrospinning: electrical and morphological properties. Polymer 2006;47(5):1597–603.
[42] Wong S-C, Baji A, Leng S. Effect of fiber diameter on tensile properties of electrospun poly(ε-caprolactone). Polymer 2008;49(21):4713–22.
[43] Stephen M, Catherine N, Brenda M, Andrew K, Leslie P, Corrine G. Oxolane-2,5-dione modified electrospun cellulose nanofibers for heavy metals adsorption. J Hazard Mater 2011;192(2):922–7.
[44] Xi P, Zhao T, Xia L, Shu D, Ma M, Cheng B. Fabrication and characterization of dual-functional ultrafine composite fibers with phase-change energy storage and luminescence properties. Sci Rep 2017;7:40390.
[45] Deitzel JM, Kosik W, McKnight SH, Beck Tan NC, DeSimone JM, Crette S. Electrospinning of polymer nanofibers with specific surface chemistry. Polymer 2002;43(3):1025–9.
[46] Ma Z, Kotaki M, Ramakrishna S. Surface modified nonwoven polysulphone (PSU) fiber mesh by electrospinning: a novel affinity membrane. J Membr Sci 2006;272(1):179–87.
[47] Lee K-H, Kim K-W, Pesapane A, Kim H-Y, Rabolt JF. Polarized FT-IR study of macroscopically oriented electrospun nylon-6 nanofibers. Macromolecules 2008;41(4):1494–8.
[48] Abdali H, Ajji A. Preparation of electrospun nanocomposite nanofibers of polyaniline/poly(methyl methacrylate) with amino-functionalized graphene. Polymers 2017;9(9):453.
[49] Kim C, Park S-H, Cho J-I, Lee D-Y, Park T-J, Lee W-J, et al. Raman spectroscopic evaluation of polyacrylonitrile-based carbon nanofibers prepared by electrospinning. J Raman Spectrosc 2004;35(11):928–33.
[50] Chronakis IS, Milosevic B, Frenot A, Ye L. Generation of molecular recognition sites in electrospun polymer nanofibers via molecular imprinting. Macromolecules 2006;39(1):357–61.
[51] Yoshimatsu K, Ye L, Lindberg J, Chronakis IS. Selective molecular adsorption using electrospun nanofiber affinity membranes. Biosens Bioelectron 2008;23(7):1208–15.
[52] Chronakis IS, Jakob A, Hagström B, Ye L. Encapsulation and selective recognition of molecularly imprinted theophylline and 17â-estradiol nanoparticles within electrospun polymer nanofibers. Langmuir 2006;22(21):8960–5.
[53] Piperno S, Tse Sum Bui B, Haupt K, Gheber LA. Immobilization of molecularly imprinted polymer nanoparticles in electrospun poly (vinyl alcohol) nanofibers. Langmuir 2011;27(5):1547–50.
[54] Tonglairoum P, Chaijaroenluk W, Rojanarata T, Ngawhirunpat T, Akkaramongkolporn P, Opanasopit P. Development and characterization of propranolol selective molecular imprinted polymer composite electrospun nanofiber membrane. AAPS Pharm Sci Tech 2013;14(2):838–46.
[55] Büttiker R, Ebert J, Hinderling C, Adlhart C. Membranes for specific adsorption: immobilizing molecularly imprinted polymer microspheres using electrospun nanofibers. CHIMIA Int J Chem 2011;65(3):182–6.
[56] Ruggieri F, D'Archivio A, Di Camillo D, Lozzi L, Maggi M, Mercorio R, et al. Development of molecularly imprinted polymeric nanofibers by electrospinning and applications to pesticide adsorption. J Sep Sci 2015;38(8):1402–10.
[57] Li L, Liu H, Lei X, Zhai Y. Electrospun nanofiber membranes containing molecularly imprinted polymer (MIP) for rhodamine B (RhB). Adv Chem Eng Sci 2012;2(02):266.

[58] Wu Y-t, Zhang Y-h, Zhang M, Liu F, Wan Y-c, Huang Z, et al. Selective and simultaneous determination of trace bisphenol A and tebuconazole in vegetable and juice samples by membrane-based molecularly imprinted solid-phase extraction and HPLC. Food Chemistry 2014;164:527–35.
[59] Demirkurt M, Olcer Y, Demir M, Eroglu A. Electrospun polystyrene fibers knitted around imprinted acrylate microspheres as sorbent for paraben derivatives. Anal Chim Acta 2018;1014:1–9.
[60] Asgari S, Bagheri H, Es-haghi A. Imprinted silica nanofiber formation via sol–gel-electrospinning for selective micro solid phase extraction. New J Chem 2018;42: 13864–72.
[61] Yarin A. Coaxial electrospinning and emulsion electrospinning of core–shell fibers. Polym Adv Technol 2011;22(3):310–7.
[62] Choi S-W, Park JY, Kim SS. Synthesis of $SnO_2$–ZnO core–shell nanofibers via a novel two-step process and their gas sensing properties. Nanotechnology 2009;20(46):465603.
[63] Park JY, Choi SW, Lee JW, Lee C, Kim SS. Synthesis and gas sensing properties of $TiO_2$–ZnO core-shell nanofibers. J Am Ceram Soc 2009;92(11):2551–4.
[64] Qi F, Li X, Yang B, Rong F, Xu Q. Disks solid phase extraction based polypyrrole functionalized core–shell nanofibers mat. Talanta 2015;144:129–35.
[65] Tian T, Deng J, Xie Z, Zhao Y, Feng Z, Kang X, et al. Polypyrrole hollow fiber for solid phase extraction. Analyst 2012;137(8):1846–52.
[66] Kang X, Pan C, Xu Q, Yao Y, Wang Y, Qi D, et al. The investigation of electrospun polymer nanofibers as a solid-phase extraction sorbent for the determination of trazodone in human plasma. Anal Chim Acta 2007;587(1):75–81.
[67] Shin C, Chase GG, Reneker DH. Recycled expanded polystyrene nanofibers applied in filter media. Colloid Surf Physicochem Eng Asp 2005;262(1–3):211–5.
[68] Chigome S, Darko G, Buttner U, Torto N. Semi-micro solid phase extraction with electrospun polystyrene fiber disks. Anal Method 2010;2(6):623–6.
[69] Xu Q, Wu S-Y, Wang M, Yin X-Y, Wen Z-Y, Ge W-N, et al. Electrospun nylon6 nanofibrous membrane as SPE adsorbent for the enrichment and determination of three estrogens in environmental water samples. Chromatographia 2010;71(5–6):487–92.
[70] Xu Q, Yin X, Wu S, Wang M, Wen Z, Gu Z. Determination of phthalate esters in water samples using Nylon6 nanofibers mat-based solid-phase extraction coupled to liquid chromatography. Microchim Acta 2010;168(3–4):267–75.
[71] Xu Q, Zhang N, Yin X, Wang M, Shen Y, Xu S, et al. Development and validation of a nylon6 nanofibers mat-based SPE coupled with HPLC method for the determination of docetaxel in rabbit plasma and its application to the relative bioavailability study. J Chromatogr B 2010;878(26):2403–8.
[72] Háková M, Raabová H, Havlíková LC, Chocholouš P, Chvojka J, Šatínský D. Testing of nylon 6 nanofibers with different surface densities as sorbents for solid phase extraction and their selectivity comparison with commercial sorbent. Talanta 2018;181:326–32.
[73] Bagheri H, Asgari S, Piri-Moghadam H. On-line micro solid-phase extraction of clodinafop propargyl from water, soil and wheat samples using electrospun polyamide nanofibers. Chromatographia 2014;77(9–10):723–8.
[74] Bagheri H, Aghakhani A, Baghernejad M, Akbarinejad A. Novel polyamide-based nanofibers prepared by electrospinning technique for headspace solid-phase microextraction of phenol and chlorophenols from environmental samples. Anal Chim Acta 2012;716:34–9.

[75] Zali S, Jalali F, Es-Haghi A, Shamsipur M. Electrospun nanostructured polystyrene as a new coating material for solid-phase microextraction: application to separation of multipesticides from honey samples. J Chromatogr B 2015;1002:387–93.
[76] Liu F, Song D, Huang X, Xu H. Electrospun polystyrene nanofibers as a novel adsorbent to transfer an organic phase from an aqueous phase. J Sep Sci 2016;39(7):1326–30.
[77] Hu W-Y, Kang X-J, Zhang C, Yang J, Ling R, Liu E-H, et al. Packed-fiber solid-phase extraction coupled with high performance liquid chromatography–tandem mass spectrometry for determination of diethylstilbestrol, hexestrol, and dienestrol residues in milk products. J Chromatogr B 2014;957:7–13.
[78] Hashemifard N, Shariati S. Electrospun polystyrene nanofiber as an adsorbent for solid-phase extraction of disulfine blue from aqueous samples. Arabian J Sci Eng 2016;41(7): 2487–92.
[79] Wei H, Yang F, Wang Y, Zhou Y, Yan Y, Liang G, et al. Electrospun polymer nanofibres as solid-phase extraction sorbents for extraction and quantification of microcystins. Environ Technol 2015;36(21):2796–802.
[80] Tavengwa NT, Nyamukamba P, Cukrowska E, Chimuka L. Miniaturized pipette-tip-based electrospun polyacrylonitrile nanofibers for the micro-solid-phase extraction of nitro-based explosive compounds. J Sep Sci 2016;39(24):4819–27.
[81] Müller V, Cestari M, Palácio SM, de Campos SD, Muniz EC, de Campos EA. Silk fibroin nanofibers electrospun on glass fiber as a potential device for solid phase microextraction. J Appl Polym Sci 2015;132(13).
[82] Sereshti H, Amini F, Najarzadekan H. Electrospun polyethylene terephthalate (PET) nanofibers as a new adsorbent for micro-solid phase extraction of chromium (VI) in environmental water samples. RSC Adv 2015;5(108):89195–203.
[83] Bagheri H, Akbarinejad A, Aghakhani A. A highly thermal-resistant electrospun-based polyetherimide nanofibers coating for solid-phase microextraction. Anal Bioanal Chem 2014;406(8):2141–9.
[84] Li S, Wu D, Yan X, Guan Y. Acetone-activated polyimide electrospun nanofiber membrane for thin-film microextraction and thermal desorption-gas chromatography–mass spectrometric analysis of phenols in environmental water. J Chromatogr A 2015; 1411:1–8.
[85] Bagheri H, Aghakhani A. Novel nanofiber coatings prepared by electrospinning technique for headspace solid-phase microextraction of chlorobenzenes from environmental samples. Anal Method 2011;3(6):1284–9.
[86] Fontanals N, Galià M, Marcé RM, Borrull F. Solid-phase extraction of polar compounds with a hydrophilic copolymeric sorbent. J Chromatogr A 2004;1030(1–2):63–8.
[87] Kang XJ, Chen LQ, Zhang YY, Liu YW, Gu ZZ. Performance of electrospun nanofibers for SPE of drugs from aqueous solutions. J Sep Sci 2008;31(18):3272–8.
[88] Qi D, Kang X, Chen L, Zhang Y, Wei H, Gu Z. Electrospun polymer nanofibers as a solid-phase extraction sorbent for the determination of trace pollutants in environmental water. Anal Bioanal Chem 2008;390(3):929–38.
[89] Liu Z, Kang X, Fang F. Solid phase extraction with electrospun nanofibers for determination of retinol and α-tocopherol in plasma. Microchim Acta 2010;168(1–2):59–64.
[90] Zhang Y, Kang X, Chen L, Pan C, Yao Y, Gu Z-Z. Fiber-packed SPE tips based on electrospun fibers. Anal Bioanal Chem 2008;391(6):2189–97.
[91] Ifegwu OC, Anyakora C, Chigome S, Torto N. Electrospun nanofiber sorbents for the pre-concentration of urinary 1-hydroxypyrene. J Anal Sci Tech 2015;6(1):14.

[92] Lu W, Zu M, Byun JH, Kim BS, Chou TW. State of the art of carbon nanotube fibers: opportunities and challenges. Adv Mater 2012;24(14):1805–33.
[93] Tibbetts GG, Lake ML, Strong KL, Rice BP. A review of the fabrication and properties of vapor-grown carbon nanofiber/polymer composites. Compos Sci Technol 2007;67(7–8): 1709–18.
[94] Qin X-H, Wan Y-Q, He J-H, Zhang J, Yu J-Y, Wang S-Y. Effect of LiCl on electrospinning of PAN polymer solution: theoretical analysis and experimental verification. Polymer 2004;45(18):6409–13.
[95] Kalayci VE, Patra PK, Kim YK, Ugbolue SC, Warner SB. Charge consequences in electrospun polyacrylonitrile (PAN) nanofibers. Polymer 2005;46(18):7191–200.
[96] Feng L, Xie N, Zhong J. Carbon nanofibers and their composites: a review of synthesizing, properties and applications. Materials 2014;7(5):3919–45.
[97] Zussman E, Chen X, Ding W, Calabri L, Dikin D, Quintana J, et al. Mechanical and structural characterization of electrospun PAN-derived carbon nanofibers. Carbon 2005; 43(10):2175–85.
[98] Wu M, Wang Q, Li K, Wu Y, Liu H. Optimization of stabilization conditions for electrospun polyacrylonitrile nanofibers. Polym Degrad Stabil 2012;97(8):1511–9.
[99] Maddah B, Soltaninezhad M, Adib K, Hasanzadeh M. Activated carbon nanofiber produced from electrospun PAN nanofiber as a solid phase extraction sorbent for the preconcentration of organophosphorus pesticides. Separ Sci Technol 2017;52(4):700–11.
[100] Zewe JW, Steach JK, Olesik SV. Electrospun fibers for solid-phase microextraction. Anal Chem 2010;82(12):5341–8.
[101] Ramaseshan R, Sundarrajan S, Jose R, Ramakrishna S. Nanostructured ceramics by electrospinning. J Appl Phys 2007;102(11):7.
[102] Wessel C, Ostermann R, Dersch R, Smarsly BM. Formation of inorganic nanofibers from preformed $TiO_2$ nanoparticles via electrospinning. J Phys Chem C 2010;115(2):362–72.
[103] Horzum N, Muñoz-Espí R, Glasser G, Demir MM, Landfester K, Crespy D. Hierarchically structured metal oxide/silica nanofibers by colloid electrospinning. ACS Appl Mater Interfaces 2012;4(11):6338–45.
[104] Chen S, Yan J, Li J, Zhang Y, Lu D. Solid phase extraction with titanium dioxide nanofibers combined with dispersive liquid-liquid microextraction for speciation of thallium prior to electrothermal vaporization ICP-MS. Microchim Acta 2017;184(8): 2797–803.
[105] Bagheri H, Piri-Moghadam H, Rastegar S, Taheri N. Electrospun titania sol–gel-based ceramic composite nanofibers for online micro-solid-phase extraction with high-performance liquid chromatography. J Sep Sci 2014;37(15):1982–8.
[106] Zhu G-T, Li X-S, Fu X-M, Wu J-Y, Yuan B-F, Feng Y-Q. Electrospinning-based synthesis of highly ordered mesoporous silica fiber for lab-in-syringe enrichment of plasma peptides. Chem Commun 2012;48(80):9980–2.
[107] Wessel C. Nanostructured mesoporous materials via electrospinning: principle concepts in the preparation of oxide nanofibers from different building blocks. 2016.
[108] Kutz M. Mechanical engineers' handbook. In: Materials and engineering mechanics, vol. 1. John Wiley & Sons; 2015.
[109] Asiabi M, Mehdinia A, Jabbari A. Electrospun biocompatible Chitosan/MIL-101 (Fe) composite nanofibers for solid-phase extraction of Δ 9-tetrahydrocannabinol in whole blood samples using Box-Behnken experimental design. J Chromatogr A 2017;1479: 71–80.

[110] Sawicka KM, Gouma P. Electrospun composite nanofibers for functional applications. J Nanoparticle Res 2006;8(6):769–81.

[111] Yan Z, Wu M, Hu B, Yao M, Zhang L, Lu Q, et al. Electrospun UiO-66/polyacrylonitrile nanofibers as efficient sorbent for pipette tip solid phase extraction of phytohormones in vegetable samples. J Chromatogr A 2018;1542:19–27.

[112] Asiabi M, Mehdinia A, Jabbari A. Spider-web-like chitosan/MIL-68 (Al) composite nanofibers for high-efficient solid phase extraction of Pb (II) and Cd (II). Microchim Acta 2017;184(11):4495–501.

[113] Asiabi M, Mehdinia A, Jabbari A. Preparation of water stable methyl-modified metal–organic framework-5/polyacrylonitrile composite nanofibers via electrospinning and their application for solid-phase extraction of two estrogenic drugs in urine samples. J Chromatogr A 2015;1426:24–32.

[114] Liu F, Xu H. Development of a novel polystyrene/metal-organic framework-199 electrospun nanofiber adsorbent for thin film microextraction of aldehydes in human urine. Talanta 2017;162:261–7.

[115] Huang J, Deng H, Song D, Xu H. Electrospun polystyrene/graphene nanofiber film as a novel adsorbent of thin film microextraction for extraction of aldehydes in human exhaled breath condensates. Anal Chim Acta 2015;878:102–8.

[116] Bagheri H, Roostaie A. Roles of inorganic oxide nanoparticles on extraction efficiency of electrospun polyethylene terephthalate nanocomposite as an unbreakable fiber coating. J Chromatogr A 2015;1375:8–16.

[117] Bagheri H, Khanipour P, Asgari S. Magnetic field assisted μ-solid phase extraction of anti-inflammatory and loop diuretic drugs by modified polybutylene terephthalate nanofibers. Anal Chim Acta 2016;934:88–97.

[118] He X-M, Zhu G-T, Yin J, Zhao Q, Yuan B-F, Feng Y-Q. Electrospun polystyrene/oxidized carbon nanotubes film as both sorbent for thin film microextraction and matrix for matrix-assisted laser desorption/ionization time-of-flight mass spectrometry. J Chromatogr A 2014;1351:29–36.

[119] Bagheri H, Aghakhani A. Polyaniline-nylon-6 electrospun nanofibers for headspace adsorptive microextraction. Anal Chim Acta 2012;713:63–9.

[120] Zhang H, Xu H. Electrospun nanofibers-based online micro-solid phase extraction for the determination of monohydroxy polycyclic aromatic hydrocarbons in human urine. J Chromatogr A 2017;1521:27–35.

[121] Zhang H, Hu S, Song D, Xu H. Polydopamine-sheathed electrospun nanofiber as adsorbent for determination of aldehydes metabolites in human urine. Anal Chim Acta 2016;943:74–81.

[122] Mehrani Z, Ebrahimzadeh H, Aliakbar AR, Asgharinezhad AA. A poly (4-nitroaniline)/poly (vinyl alcohol) electrospun nanofiber as an efficient nanosorbent for solid phase microextraction of diazinon and chlorpyrifos from water and juice samples. Microchim Acta 2018;185(8):384.

[123] Háková M, Havlíková LC, Chvojka J, Erben J, Solich P, Švec F, et al. A comparison study of nanofiber, microfiber, and new composite nano/microfiber polymers used as sorbents for on-line solid phase extraction in chromatography system. Anal Chim Acta 2018;1023:44–52.

[124] Rana D, Ramalingam M. Ceramic nanofiber composites. In: Nanofiber composites for biomedical applications. Elsevier; 2017. p. 33–54.

[125] Vadukumpully S, Basheer C, Jeng CS, Valiyaveettil S. Carbon nanofibers extracted from soot as a sorbent for the determination of aromatic amines from wastewater effluent samples. J Chromatogr A 2011;1218(23):3581−7.
[126] Sun B, Long Y, Zhang H, Li M, Duvail J, Jiang X, et al. Advances in three-dimensional nanofibrous macrostructures via electrospinning. Prog Polym Sci 2014;39(5):862−90.
[127] Bagheri H, Manshaei F, Rezvani O. Three-dimensional nanofiber scaffolds are superior to two-dimensional mats in micro-oriented extraction of chlorobenzenes. Microchim Acta 2018;185(7):322.

# Particle loaded membranes

**12**

*M.C. Díaz-Liñán, A.I. López-Lorente, R. Lucena, S. Cárdenas*
Departamento de Química Analítica, Instituto Universitario de Investigación en Nanoquímica (IUNAN), Edificio Marie Curie (anexo), Campus de Rabanales, Universidad de Córdoba, Córdoba, Spain

## 12.1 Introduction

Membranes have been extensively used at the industrial scale as, in general, they act as physical barriers which, based on permeability or size cut-off criteria allow the transfer of gas, liquid, or particulate material between two, usually miscible, phases. Therefore, the membrane can participate as an active or passive element with a common role in facilitating the contact of both phases at the membrane interface.

Membranes are employed in sample treatment, either for analyte preconcentration or sample cleanup. Moreover, they can be used in the microextraction context to prevent damage or blockage of the sorbent phase when dirty samples are processed. A paradigmatic example is hollow fiber protected liquid phase microextraction [1]. No doubt, the most interesting application is the use of membranes for analyte enrichment. In this case, target compounds migrate from the sample matrix to the membrane by a chemical, pressure, or electrical field gradient [2]. The so-called hollow fiber liquid phase microextraction (HF-LPME) is identified as a reference technique in microextraction [3]. It can be used in the two-phase (2D) or three-phase (3D) modes. 2D-HF-LPME uses two immiscible phases, with the extractant located in the lumen of the fiber. In 3D-HF-LPME the sample and the extractant are miscible (generally aqueous) while the pores of the HF are filled with an organic solvent or solution. The analytes are first extracted into the organic phase, and then transferred to the aqueous solution inside the HF. A chemical mechanism, usually a pH gradient, is used to avoid the reextraction of the analytes to the external aqueous matrix.

However, the advantages of using planar formats, mainly related to the higher surface area that can be exposed to the sample, have gained recognition in recent years. Therefore, to increase both the variety of chemistries available for extraction and the membrane capacity and selectivity, there is great interest in developing membranes incorporating nanoparticles and nanostructures. The potential of these materials in solid-phase microextraction is well-known [4]. These outstanding properties, however, are dramatically decreased if the nanoscale is lost. This is common for hydrophobic nanoparticles, especially carbon-based nanoparticles.

To avoid this inconvenience, carbon nanoparticles are either dispersed in a micellar media or immobilized on a solid surface. The addition of nanoparticles into a membrane is a synergic combination as, on the one hand aggregation is avoided and, on the other, new capabilities including higher surface area are conferred to the

**Solid-Phase Extraction. https://doi.org/10.1016/B978-0-12-816906-3.00012-1**

membrane. A clear precursor of this concept is disks-based solid-phase extraction, where the sorbent phase is placed in a planar element to minimize channeling during sample processing steps. In addition, higher flow rates can be used, thus allowing the preconcentration of large sample volumes in a reasonable time. If a low eluent volume is used, the enrichment factor that can be achieved leads to improved method sensitivity. There are many membrane formats which can be loaded with (nano)particles, namely planar and hollow fiber membranes, and membranes fabricated via electrospinning and thin films, among others.

This chapter provides an overview of the use of particle-loaded membranes in sample treatment. The first section describes the classical approach using membrane formats with a hollow fiber or planar geometry. In the second part, the potential of thin film microextraction as an extension of the particle-loaded membrane concept is described.

## 12.2 Membranes modified with nanoparticles

### *12.2.1 Hollow fiber membranes*

Polypropylene hollow fibers (HF) provide an excellent support for the confinement of the extraction solvent in either 2D or 3D formats. However, considering the small dimensions of HF microextraction units, and thus the small volume of extraction solvents, in the low microliter range, the efficiency of the process can be jeopardized. If an HF is intended to be used in micro solid phase extraction, a similar limitation would appear unless highly efficient sorbents, such as nanostructured solids, are used.

Carbon nanoparticles in general, and carbon nanotubes (CNTs) in particular, are by far the preferred nanomaterials for the construction of so-called reinforced HF on account of their outstanding sorbent capacity. These nanoparticles can be held in the lumen or immobilized into the pores of the HF, which eliminates the aggregation typically observed in other micro solid phase extraction formats, such as dispersive extraction techniques. The location of the CNTs inside the HF usually requires the preparation of a composite material using sol-gel technology. Eshaghy et al. prepared an organic/inorganic polymer containing oxidized multiwalled carbon nanotubes (o-MWNTs) in which orthosilicate was used as precursor and HCl as catalyst, for the in situ gelation inside the HF [5]. The reinforced-HF was immersed directly into the sample solution for the extraction, and the porosity of the HF allows fast resorption of the analytes. The fiber also prevents large molecules from reaching and diminishing the adsorption capacity of the extraction phase for the analytes. The fiber was discarded after each extraction. A simplification of this procedure was proposed by Song et al. by replacing the sol-gel approach with a surfactant hexadecyltrimethylammonium bromide (CTAB) [6]. No interference was observed for the extraction of carbamate pesticides from fruits [6] or strychnine and brucine in clinical samples [7].

If CNTs are dispersed in a liquid phase, the sealing of the fiber end is required to avoid material losses [8]. As an alternative, the dispersion of the CNTs either in 1-octanol [9] or water [10] can be added to HF, and external energy (ultrasounds or pressure) used to force the nanomaterial into the wall pores of the HF. In this case, a sort of CNTs-supported liquid membrane is obtained. The role of the CNTs is to increase the surface area of the extraction phase and also to increase the partition constant, especially if a 3D configuration is used.

Concerning other carbon nanoparticles, Jiménez-Soto et al. evaluated the use of oxidized single-walled carbon nanohorns (o-SWNHs) as the sorbent phase for 2D reinforced HF microextraction [11]. SWNHs are conical-ended carbon nanoparticles of smaller dimensions than CNTs capable of generating stable aggregates with a dahlia shape. These aggregates are still in the nanometric range (less than 100 nm diameter) and therefore maintain the extraction capacity associated with the nanoworld. The o-SWNHs were immobilized on the pores of the HF. An inserted stainless steel wire maintains the rigidity of the fiber allowing its direct immersion in the sample. Triazine herbicides were determined in waters with a sensitivity comparable to, or better than, other microextraction approaches designed for the same analytical problem. It was possible to reuse the fiber up to 30 times. Graphene (G), or its oxidized form (GO), exhibits an enhanced adsorption capacity for many compounds. Graphene was used as a dispersion in 1-octanol in the lumen of an HF membrane for the extraction of carbamate pesticides with better detection limits than those provided when CNTs were used [12]. GO is more suitable for the extraction of inorganic ions and was used for the extraction of heavy metals from waters by a hollow fiber membrane containing a composite adsorbent prepared from GO nanoparticles and silica. The hydrophilic interactions (electrostatic and coordination) that can be established between the oxygenated groups and the metals were identified as the driving force of the extraction [13].

### 12.2.2 Planar membranes

This section is focused on planar membranes modified with (nano)particles. Most of the functionalized membranes described within this section are commercially available and based on different polymers (e.g., nylon, poly(ethylene terephthalate) (PET), poly(vinylidene fluoride) (PVDF), among others), in which the particles have been incorporated via different routes, for example, dip coating, drop-casting directly on the membrane or after a membrane functionalization, that is, amino, carboxylic, etc. Different types of (nano)particles can be loaded onto the membranes, such as metal nanoparticles (Au, Ag, Pd, etc.) as well as some semiconductor nanoparticles ($SiO_2$, $Al_2O_3$, etc.) In addition, carbon nanomaterials have also been used as membrane coatings.

The incorporation of (nano)particles into membranes can respond to two general needs: (i) improvement of the adsorption/filtration process; and (ii) assist in the subsequent elution and/or detection procedure. For example, nanoparticles can provide interaction sites due to their affinity toward the analyte. In this regard, nylon membranes impregnated with metallic nanoparticles, namely Au and Ag, have been used for the solid phase extraction of Hg by filtration of the water through the membrane [14]. The large surface area of the nanoparticles is responsible for the high-efficiency adsorption of Hg into the membrane. Also, Pd nanoparticles have been deposited by dip coating on pore walls of carboxylic and amino group functionalized

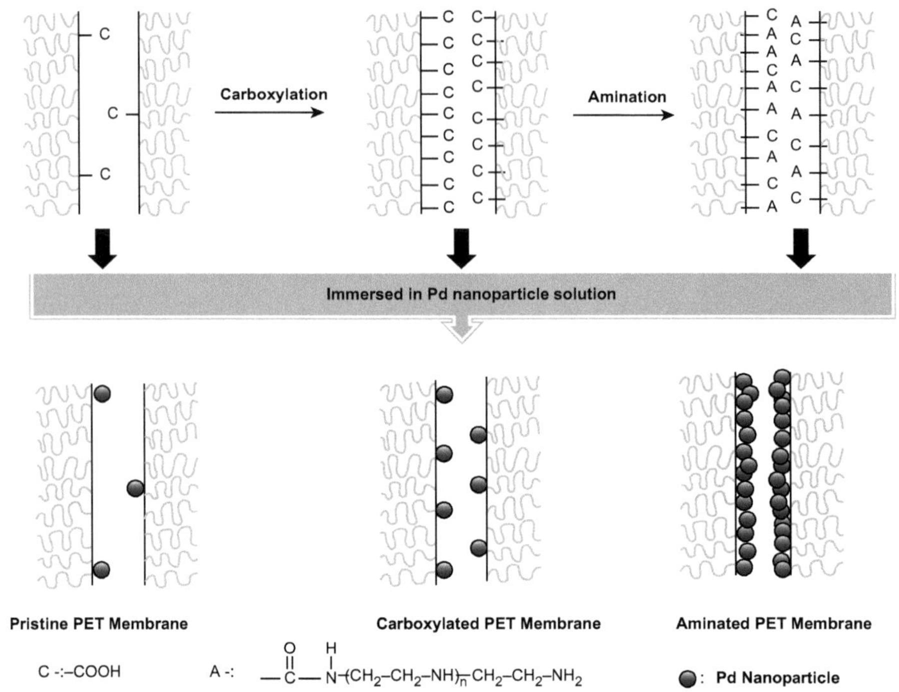


**Figure 12.1** Scheme of the functionalization process of the poly(ethylene terephthalate) membrane with carboxylic and amino groups and their effect on the immobilization of Pd NPs. Reproduced from Awasthi K, Choudhury S, Komber H, Simon F, Formanek P, Sharma A, et al. Functionalization of track-etched poly (ethylene terephthalate) membranes as a selective filter for hydrogen purification. Int J Hydrogen Energy 2014;39:9356–9365 with permission from Elsevier.

poly(ethylene terephthalate) membranes [15]. The functionalization was critical for the subsequent immobilization of the Pd nanoparticles (Fig. 12.1). These membranes were used in hydrogen separation and sensing. Furthermore, nanosilica extracted from rice straw was combined with a polyether-polyamide block copolymer (PEBA) to form a nanocomposite polymeric membrane for the separation of $CO_2$ in gas streams [16]. Huang et al. prepared PVDF/$SiO_2$ hybrid membranes by a sol-gel process for the preconcentration of fennel oil from herbal water extracts [17]. The presence of $SiO_2$ nanoparticles combined with PVDF resulted in an improvement of the thermal and mechanical properties of the membrane. Zhang et al. [18] developed a micro-solid phase extraction method for food colorants based on the impregnation of a nylon filter membrane with $Al_2O_3$ NPs. The performance of the $Al_2O_3$ NPs was compared with $SiO_2$ and $TiO_2$ NPs and conventional sorbents such as octadecylsiloxane-bonded silica for the extraction of tartrazine and Sunset yellow in food. The membrane impregnated with $Al_2O_3$ NPs exhibited the best extraction efficiency. In addition, solid phase micromembrane tip extraction using 1-butyl-3-methylimidazole iron nanoparticles enclosed within a membrane followed by chiral liquid chromatography was used to evaluate the stereoselective interactions of profens with proteins [19], as well as for the analysis of vitamin B in human plasma [20].

Wang et al. prepared a "pH-paper-like" chip for the enrichment and detection of organic pollutants[21]. The chip was composed of a PVDF membrane with a polydopamine (PDA) film and AuNPs. Each component of the chip membranes has a specific role, the PVDF membrane and PDA facilitating the adsorption of organic analytes via hydrophobic interactions and $\pi$-$\pi$ stacking, while AuNPs facilitated the subsequent quantification of the analytes via surface-assisted laser desorption/ionization time-of-flight mass spectrometry.

Metallic NPs are used extensively as a substrate in surface enhanced Raman spectroscopy (SERS) due to their capability to enhance the Raman signal. Typical substrates are based on rigid solid materials such as silicon, glass, etc.; however, recently SERS active membranes based on flexible materials have also been described. For example, liquid-crystal polymer (LCP) textile fibers decorated with AgNPs supported by a polyamide filter were used for the extraction and detection of the pesticide thiram [22]. This is an example of the role of metal nanoparticles on the subsequent detection step. Nevertheless, the NPs can play a dual role by supplying additional interaction sites for adsorption of the analytes, for example, via the binding affinity of gold toward sulfur or silver toward amine groups. Granger et al. developed a SERS-based point-of-need diagnostic test for alphafetoprotein (AFP) screening using solid phase microextraction membranes with functionalized AuNPs as Raman enhancer and selective tag for capturing AFP [23]. This field is very extensive and a more detailed description would go beyond the scope of this chapter [24]. However, particle-loaded membranes are expected to have a significant impact on the coupling of extraction with Raman and related detection methods. As well as optical sensors, metallic NPs loaded membranes are used in electrochemical devices, for example, AuNPs formed in situ in a polymer inclusion membrane were used for the electrode immobilization of anti-*Salmonella* antibodies [25].

Dispersions of MWCNTs and single layer graphene (SLG) nanoparticles have been incorporated into cellulose triacetate polymer matrix membranes [26]. The membranes are prepared via drop-casting of a cellulose triacetate (CTA) polymer matrix and carbon nanomaterials dispersed in dichloromethane on a flat glass surface. Carbon nanomaterials enable the extraction of aromatic compounds, for example, polycyclic aromatic hydrocarbons in sewage pond water samples.

One of the challenges to overcome in membrane-based extraction is the specific recognition properties of the membranes as well as their permeation performance. This led to the development of molecularly imprinted membranes (MIMs), which combine molecularly imprinted polymers with membrane separation techniques. Altintas et al. reported the fabrication of MIMs as filters coupled to SPE for the evaluation of water purification processes [27]. MIMs were formed by molecularly imprinted polymer nanoparticles (MIPNPs) toward pharmaceuticals, which were applied onto plasma-treated PVDF membranes (see Fig. 12.2). In addition, the synthesis of molecularly imprinted nanocomposite membranes (MINCMs) has also been described [28]. Nanocomposite membranes were prepared by infiltrating the functional monomer modified with the NPs onto a polydopamine-modified regenerated

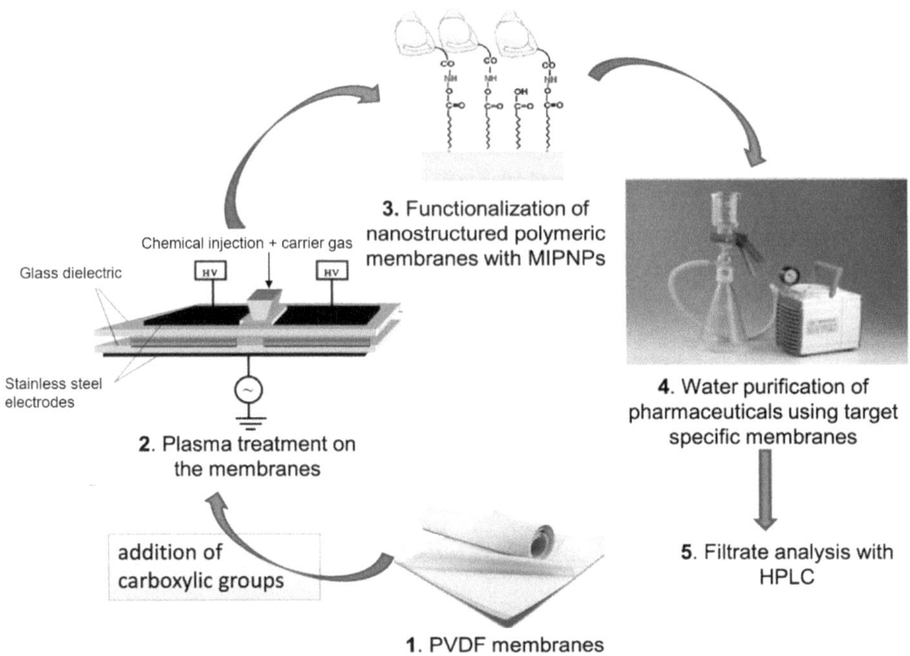


**Figure 12.2** Development of functionalized nanostructured polymeric membranes with MIPNPs from PVDF membranes for water purification of pharmaceuticals and subsequent HPLC analysis.
Reproduced from Altintas Z, Chianella I, Ponte GD, Paulussen S, Gaeta S, Tothill IE. Development of functionalized nanostructured polymeric membranes for water purification. Chem Eng J 2016;300:358−366 with permission from Elsevier.

cellulose membrane, followed by an imprinting procedure. Imprinted PDA@$SiO_2$ MINCMs for the selective extraction of m-cresol was reported in which the high surface-to-volume ratio and, thus, the large surface area of the $SiO_2$ nanoparticles played an important role in the extraction of the target analyte [29].

## 12.3 Particle-loaded membranes in a thin film microextraction format

Thin film microextraction (TFME) is a miniaturized sample preparation technique where a flat film of a high surface area-to-volume ratio is used as the extraction phase [30]. Originally proposed in 2001 [31], as an alternative approach to in-fiber SPME, it has continued to evolve with an expanding number of applications. Among the proposed formats, this chapter is focused only on thin films modified with nanoparticles in a membrane configuration generally prepared by dip coating and electrospinning techniques. Readers interested in different coating technologies, theoretical aspects, and applications are referred to more detailed reviews [30,32].

### 12.3.1 Nanoparticle-loaded membranes synthesized via electrospinning

Electrospinning is a simple, versatile, and widely used technique to synthesize fibrous materials with a variety of diameters, morphologies, and compositions. Briefly, the electrospinning technique consists of a viscous polymeric solution to which a high voltage is applied, after which the charged polymer ejects from the solution toward the collector wire as a fibrous filament [33]. This phenomenon is based on the repulsive electrostatic force that overcomes the surface tension of the polymer when the high voltage is applied. Different electrospinning procedures, for example, magnetic field–assisted electrospinning, melt electrospinning, or bilayer electrospinning, have been described for the synthesis of nanofibers [34].

An advantage of electrospinning is that the resulting material can be deposited uniformly on a flat surface, thus forming a thin film suitable as an extraction material for TFME or SPME, among others. Moreover, the resulting fibers possess a high surface area, porous structure, and flexibility, desirable for extraction techniques [35]. Sometimes, the electrospinning process is followed by secondary steps, such as reactions for surface modification and other procedures.

The incorporation of nanomaterials into electrospun fibers affords different characteristic properties compared with the starting materials; in particular, the surface area-to-volume ratio related to the adsorption capacity and in specific cases the adsorption selectivity. For instance, among magnetic nanoparticles, $Fe_3O_4$ nanoparticles have acquired special attention in the electrochemical field due to their unique properties, such as excellent conformation stability, better contact between a biocatalyst and its substrate, ease of preparation, and large surface area, among others [36].

The incorporation of nanoparticles into the fiber structure can be carried out by different procedures. Fig. 12.3 illustrates two methods employed to immobilize nanoparticles in electrospun nanofibers, that is, dispersing the nanoparticles into the precursor solution of the polymer or through a chemical procedure in which the fibers are modified to immobilize the nanoparticles into its structure. Both approaches will be discussed in this section.

#### 12.3.1.1 One pot synthesis of electrospun fibers with embedded nanoparticles

This approach is the fastest and easiest way to incorporate nanomaterials into electrospun fibers. It consists of the addition of the nanoparticles into the solution of monomers, prior to polymerization. When the nanoparticles are dispersed in the precursor solution of the polymer, they are easily incorporated to the fibers compared to the chemical modification of the fibers with the nanoparticles after electrospinning [34]. Bagheri et al. synthesized electrospun poly(butylene terephthalate) (PBT) nanofibers with embedded magnetic nanoparticles for the isolation of triazines [37]. The inorganic nanoparticles increased the surface area and porosity of the sorbent, thus leading to higher extraction efficiency. Furthermore, since these nanoparticles have magnetic properties, the nanofibers

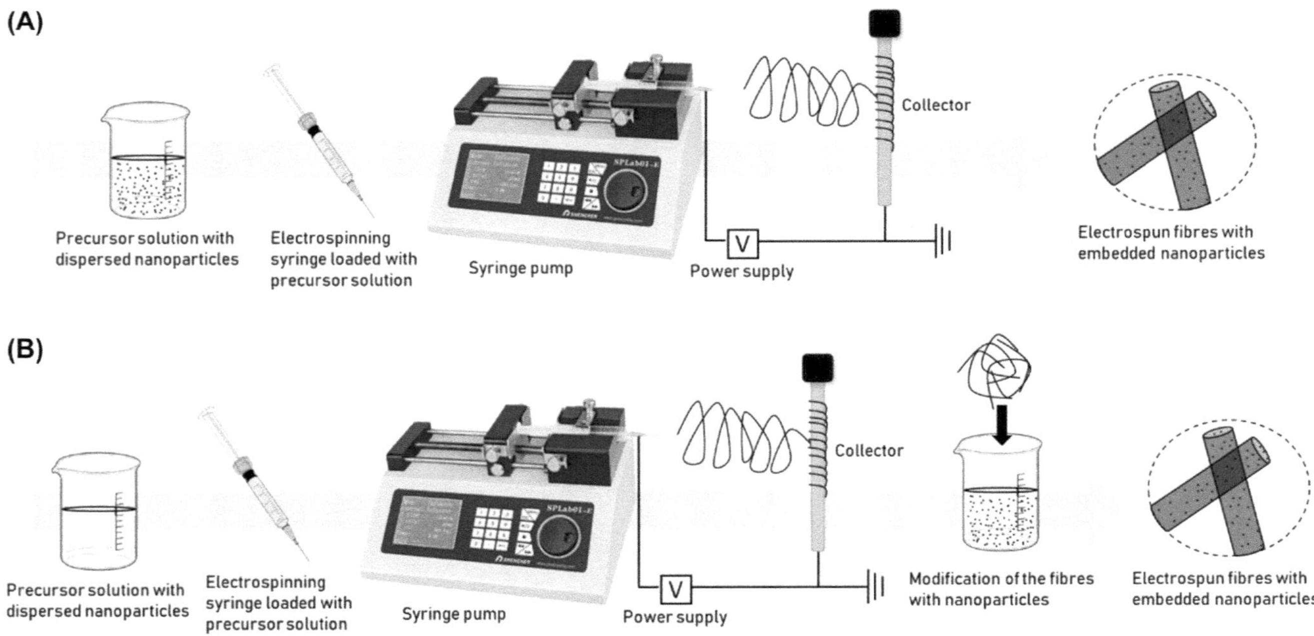


**Figure 12.3** Scheme of the alternatives employed to synthesize electrospun fibers with embedded nanoparticles (A) by dispersing the nanoparticles into the precursor solution and (B) by chemical modification of the electrospun fibers and immobilization of nanoparticles.

could be easily isolated from solution by an external magnet once the microextraction is complete.

Although electrospinning provides long synthetic fibers, these can be organized into a membrane format, especially useful for TFME. Liu and Xu obtained electrospun polystyrene/metal-organic frameworks-199 (PS/MOF-199) nanofiber film for TFME of aldehydes in human urine [38]. For the collection of the nanofiber, aluminum foil was employed. MOFs properties, such as tunable pore size, high surface area, and good thermal stability, enhanced the extraction efficiency of the PS/MOF-199 nanofiber film, compared with poly(styrene) films and an MOF-199 coated mesh. Furthermore, no loss of MOF-199 nanoparticles was found during the microextraction procedure, which demonstrates the enhanced mechanical stability of the material. Aldehydes were selected as model analytes, since they are considered to be potential biomarkers of lung cancer. Thus, this method was successfully applied for the determination of aldehydes in urine samples of eight healthy subjects and seven lung cancer patients. Wu et al. synthesized nanofibrous membranes with multianalyte selectivity by encapsulating two types of molecularly imprinted polymer nanoparticles (MIP-NPs) into electrospun poly(vinyl alcohol) nanofibers for the simultaneous extraction of bisphenol A and tebuconazole in vegetable and juice samples [39]. Nanofibrous molecularly imprinted membranes (nano-MIMs) retain the advantages of MIP-NPs, such as high molecular selectivity, high binding capacity, and fast binding kinetics, attributed to the large surface-to-volume ratio and the high number of recognition sites at the surface of the core-shell material. In addition, due to their geometric features MIP-NPs provide better accessibility to recognition sites for the analyte as well as lower mass-transfer resistance [40]. The combination of different MIP-NPs resulted in the selective extraction of both analytes from complex food samples.

#### 12.3.1.2 Chemical modification to immobilize nanoparticles in electrospun fibers

He et al. prepared polyaniline coated silica nanofibers (PANI/$SiO_2$) by combining electrospinning with in situ polymerization for the extraction of fluoroquinolones from honey [41]. After electrospinning of the silica nanofibers, aniline monomers were polymerized on the surface of the electrospun fibers. The extraction procedure was carried out by packing PANI/$SiO_2$ nanofibers in a syringe. When combining silica nanofibers with PANI nanoparticles, the aggregation problems of PANI particles are eliminated, thus obtaining a uniform coating of PANI nanoparticles on the surface of the silica fibers. Morillo et al. developed PAN nanofibers to which superparamagnetic iron oxide nanoparticles (SPION) were fixed after hydrolysis of the electrospun fibers [42]. Nonhydrolyzed PAN fibers resulted in a low fixation of nanoparticles compared with the hydrolyzed fibers. The PAN-modified fibers were used for the removal of As(V) from wastewater. Liu et al. prepared a composite membrane from poly(vinyl alcohol)/poly(acrylic acid)/carboxyl-functionalized graphene oxide (GO) modified with silver nanoparticles (PVA/PAA/GO-COOH@AgNPs) for the catalytic degradation of the dye methylene blue [43]. Electrospun fibers containing graphene oxide were impregnated with silver nanoparticles by treatment with a silver nitrate solution containing ascorbic acid. The PVA/PAA/GO-COOH@AgNPs nanocomposite showed a significantly enhanced performance for the elimination of model dyes.

#### 12.3.1.2.1 Nanoparticles in paper-based coated sorptive phases

The use of paper, a cellulose-based support, to fabricate thin film microextraction units is a low-cost and reliable alternative. Due to its porosity, it is typically employed for filtration purposes. The raw paper is, in general, incompatible with its immersion in liquid phases. However, if it is covered by a polymeric layer, a flexible and resistant device is obtained. As a general procedure, a piece of paper is immersed in a solution of the extraction phase (polymer), and after a number of dipping cycles, the polymeric layer is generated after solvent evaporation. Rios-Gómez et al. demonstrated that changing the dipping direction ensures a homogeneous distribution of the polymer (in this case polystyrene) which resulted in improved reproducibility for microextraction [44]. In this case, the polystyrene film was placed in a 100 μL pipette tip for the extraction of methadone from water. Zinc oxide nanorods have been synthesized onto paper by repeated immersion into a precursor solution. The functionalized extractant phase was used for the isolation of phenylurea herbicides from water [45].

The mechanical stability of the coating is the main challenge to overcome when NPs are used in the fabrication of paper-based sorptive phases. The nanomaterial must remain in the coating during the whole extraction process to obtain reproducible extractions. The inclusion of NPs in a polymeric network can be obtained by a simple solvent exchange [46,47]. Ríos-Gómez et al. synthesized a paper-based phase with photocatalytic activity by this approach [48]. A segment of paper is immersed in a dispersion consisting of a polyamide dissolved in formic acid to which $TiO_2$ NPs are subsequently added. Evaporation of the solvent induces the precipitation of the polyamide embedding the NPs in the polymeric network. The polymer entraps the NPs increasing their mechanical stability and enhances the sorption capacity of the paper-based membrane toward the target contaminants which are subsequently degraded by the photocatalytic action of the $TiO_2$ nanoparticles.

The use of a protective polymeric network is an efficient strategy to synthesize NPs-based coatings. However, Ríos-Gómez et al. also investigated the synthesis of coatings consisting only of NPs [49]. The capacity of SWNHs to form stable and ordered aggregates (dahlia) paved the way to this development. For this purpose, the paper is immersed in a dispersion of SWNHs in chloroform. When the solvent evaporates, a mechanically stable coating of SWNHs is obtained.

## 12.4 Conclusions

Membranes have been used extensively in the analytical chemistry for the extraction and preconcentration of target compounds from different samples. Membrane modification by loading (nano)particles on commercial membranes (both hollow-fiber or planar formats), paper, or electrospun supports, effectively expands their range of applications. In fact, the use of planar configurations increases the surface available for interaction with the analytes which clearly enhances the efficiency of the (micro) extraction step. Nanoparticles can be used for the extraction as well as for environmental remediation if nanoparticles with catalytic activity are embedded in the polymeric network. Electrospun membranes are easily doped with NPs and the appearance of new fabrication approaches (like paper-based sorptive phases) will increase interest and drive further applications of particle-loaded membranes in coming years.

## Acknowledgments

Financial support for the Spanish Ministry of Economy and Competitiveness (MINECO) is gratefully acknowledged (Grant number CTQ2017-83175R).

## References

[1] Pedersen-Bjergaard S, Rasmussen KE. Liquid–liquid–liquid microextraction for sample preparation of biological fluids prior to capillary electrophoresis. Anal Chem 1999;71: 2650–6.

[2] Patnaik P. Handbook of environmental analysis. CRC Press; 2010. https://doi.org/10.1201/b10505. Epub ahead of print 21 May 2010.

[3] Gjelstad A. Three-phase hollow fiber liquid-phase microextraction and parallel artificial liquid membrane extraction. TrAC Trends Anal Chem 2019;113:25–31.

[4] Lucena R, Simonet BM, Cárdenas S, Valcárcel M. Potential of nanoparticles in sample preparation. J Chromatogr A 2011;1218:620–37.

[5] Es'haghi Z, Rezaeifar Z, Rounaghi G-H, Nezhadi ZA, Golsefidi MA. Synthesis and application of a novel solid-phase microextraction adsorbent: hollow fiber supported carbon nanotube reinforced sol–gel for determination of phenobarbital. Anal Chim Acta 2011;689:122–8.

[6] Song X-Y, Shi Y-P, Chen J. Carbon nanotubes-reinforced hollow fibre solid-phase microextraction coupled with high performance liquid chromatography for the determination of carbamate pesticides in apples. Food Chem 2013;139:246–52.

[7] Song X-Y, Shi Y-P, Chen J. Carbon nanotubes reinforced hollow fiber solid phase microextraction for the determination of strychnine and brucine in urine. Talanta 2013;116: 188–94.

[8] Zhao G, Wang C, Wu Q, Wang Z. Determination of carbamate pesticides in water and fruit samples using carbon nanotube reinforced hollow fiber liquid-phase microextraction followed by high performance liquid chromatography. Anal Methods 2011;3: 1410.

[9] Wu C, Liu Y, Wu Q, Wang C, Wang Z. Combined use of liquid–liquid microextraction and carbon nanotube reinforced hollow fiber microporous membrane solid-phase microextraction for the determination of triazine herbicides in water and milk samples by high-performance liquid chromatography. Food Anal Methods 2012;5:540–50.

[10] Hylton K, Chen Y, Mitra S. Carbon nanotube mediated microscale membrane extraction. J Chromatogr A 2008;1211:43–8.

[11] Jiménez-Soto JM, Cárdenas S, Valcárcel M. Oxidized single-walled carbon nanohorns as sorbent for porous hollow fiber direct immersion solid-phase microextraction for the determination of triazines in waters. Anal Bioanal Chem 2013;405:2661–9.

[12] Ma X, Wang J, Wu Q, Wang C, Wang Z. Extraction of carbamate pesticides in fruit samples by graphene reinforced hollow fibre liquid microextraction followed by high performance liquid chromatographic detection. Food Chem 2014;157:119–24.

[13] Su S, Chen B, He M, Hu B. Graphene oxide–silica composite coating hollow fiber solid phase microextraction online coupled with inductively coupled plasma mass spectrometry for the determination of trace heavy metals in environmental water samples. Talanta 2014; 123:1–9.

[14] Panichev N, Kalumba MM, Mandiwana KL. Solid phase extraction of trace amount of mercury from natural waters on silver and gold nanoparticles. Anal Chim Acta 2014;813: 56–62.

[15] Awasthi K, Choudhury S, Komber H, Simon F, Formanek P, Sharma A, et al. Functionalization of track-etched poly (ethylene terephthalate) membranes as a selective filter for hydrogen purification. Int J Hydrogen Energy 2014;39:9356–65.

[16] Bhattacharya M, Mandal MK. Synthesis of rice straw extracted nano-silica-composite membrane for $CO_2$ separation. J Clean Prod 2018;186:241–52.

[17] Huang X, Zhang J, Wang W, Liu Y, Zhang Z, Li L, et al. Effects of PVDF/$SiO_2$ hybrid ultrafiltration membranes by sol–gel method for the concentration of fennel oil in herbal water extract. RSC Adv 2015;5:18258–66.

[18] Zhang Z, Wang L, Liu X, Zhang D, Zhang L, Li Q. Nano-$Al_2O_3$ -based micro solid-phase filter membrane extraction for simultaneous determination of tartrazine and sunset yellow in food. RSC Adv 2015;5:86445–52.

[19] Ali I, Suhail M, Alothman ZA, Badjah AY. Stereoselective interactions of profen stereomers with human plasma proteins using nano solid phase micro membrane tip extraction and chiral liquid chromatography. Separ Purif Technol 2018;197:336–44.

[20] Ali I, Kulsum U, AL-Othman ZA, Alwarthan A, Saleem K. Functionalized nanoparticles based solid-phase membrane micro-tip extraction and high-performance liquid chromatography analyses of vitamin B complex in human plasma. J Sep Sci 2016;39:2678–88.

[21] Wang S, Niu H, Cai Y, Cao D. Multifunctional Au NPs-polydopamine-polyvinylidene fluoride membrane chips as probe for enrichment and rapid detection of organic contaminants. Talanta 2018;181:340–5.

[22] Fateixa S, Raposo M, Nogueira HIS, Trindade T. A general strategy to prepare SERS active filter membranes for extraction and detection of pesticides in water. Talanta 2018; 182:558–66.

[23] Granger JH, Skuratovsky A, Porter MD, Scaife CL, Shea JE, Li Q, et al. Coupling solid-phase microextractions and surface-enhanced Raman scattering: towards a point-of-need tool for hepatic cancer screening. Anal Methods 2017;9:4641–6.

[24] Fedick PW, Bills BJ, Manicke NE, Cooks RG. Forensic sampling and analysis from a single substrate: surface-enhanced Raman spectroscopy followed by paper spray mass spectrometry. Anal Chem 2017;89:10973–9.

[25] Silva NFD, Magalhães JMCS, Barroso MF, Oliva-Teles T, Freire C, Delerue-Matos C. In situ formation of gold nanoparticles in polymer inclusion membrane: application as platform in a label-free potentiometric immunosensor for *Salmonella typhimurium* detection. Talanta 2019;194:134–42.

[26] Mukhtar NH, See HH. Carbonaceous nanomaterials immobilised mixed matrix membrane microextraction for the determination of polycyclic aromatic hydrocarbons in sewage pond water samples. Anal Chim Acta 2016;931:57–63.

[27] Altintas Z, Chianella I, Ponte GD, Paulussen S, Gaeta S, Tothill IE. Development of functionalized nanostructured polymeric membranes for water purification. Chem Eng J 2016;300:358–66.

[28] Wu Y, Liu X, Meng M, Lv P, Yan M, Wei X, et al. Bio-inspired adhesion: fabrication of molecularly imprinted nanocomposite membranes by developing a hybrid organic–inorganic nanoparticles composite structure. J Membr Sci 2015;490:169–78.
[29] Wu Y, Cui J, Meng M, Yan M, Yan Y, Li C. Bio-inspired adhesion: fabrication and evaluation of molecularly imprinted nanocomposite membranes by developing a "bio-glue" imprinted methodology. RSC Adv 2015;5:46146–57.
[30] Jiang R, Pawliszyn J. Thin-film microextraction offers another geometry for solid-phase microextraction. TrAC Trends Anal Chem 2012;39:245–53.
[31] Wilcockson JB, Gobas FAPC. Thin-film solid-phase extraction to measure fugacities of organic chemicals with low volatility in biological samples. Environ Sci Technol 2001;35: 1425–31.
[32] Olcer YA, Tascon M, Eroglu AE, Boyacı E. Thin film microextraction: towards faster and more sensitive microextraction. TrAC Trends Anal Chem 2019;113:93–101.
[33] Bagheri H, Roostaie A. Electrospun modified silica-polyamide nanocomposite as a novel fiber coating. J Chromatogr A 2014;1324:11–20.
[34] Reyes-Gallardo EM, Lucena R, Cárdenas S. Electrospun nanofibers as sorptive phases in microextraction. TrAC Trends Anal Chem 2016;84:3–11.
[35] Huang J, Deng H, Song D, Xu H. Electrospun polystyrene/graphene nanofiber film as a novel adsorbent of thin film microextraction for extraction of aldehydes in human exhaled breath condensates. Anal Chim Acta 2015;878:102–8.
[36] Bagheri H, Roostaie A. Roles of inorganic oxide nanoparticles on extraction efficiency of electrospun polyethylene terephthalate nanocomposite as an unbreakable fiber coating. J Chromatogr A 2015;1375:8–16.
[37] Bagheri H, Najafi Mobara M, Roostaie A, Baktash MY. Electrospun magnetic polybutylene terephthalate nanofibers for thin film microextraction. J Sep Sci 2017;40: 3857–65.
[38] Liu F, Xu H. Development of a novel polystyrene/metal-organic framework-199 electrospun nanofiber adsorbent for thin film microextraction of aldehydes in human urine. Talanta 2017;162:261–7.
[39] Wu Y, Zhang Y, Zhang M, Liu F, Wan Y, Huang Z, et al. Selective and simultaneous determination of trace bisphenol A and tebuconazole in vegetable and juice samples by membrane-based molecularly imprinted solid-phase extraction and HPLC. Food Chem 2014;164:527–35.
[40] Wackerlig J, Lieberzeit PA. Molecularly imprinted polymer nanoparticles in chemical sensing – synthesis, characterisation and application. Sensor Actuator B Chem 2015;207: 144–57.
[41] He X-M, Zhu G-T, Zheng H-B, Xu S-N, Yuan B-F, Feng Y-Q. Facile synthesis of polyaniline-coated $SiO_2$ nanofiber and its application in enrichment of fluoroquinolones from honey samples. Talanta 2015;140:29–35.
[42] Morillo D, Faccini M, Amantia D, Pérez G, García MA, Valiente M, et al. Superparamagnetic iron oxide nanoparticle-loaded polyacrylonitrile nanofibers with enhanced arsenate removal performance. Environ Sci Nano 2016;3:1165–73.
[43] Liu Y, Hou C, Jiao T, Song J, Zhang X, Xing R, et al. Self-assembled AgNP-containing nanocomposites constructed by electrospinning as efficient dye photocatalyst materials for wastewater treatment. Nanomaterials 2018;8:35.
[44] Ríos-Gómez J, Lucena R, Cárdenas S. Paper supported polystyrene membranes for thin film microextraction. Microchem J 2017;133:90–5.
[45] Saraji M, Mehrafza N. Phenyl carbamate functionalized zinc oxide nanorods for paper-based thin film microextraction. RSC Adv 2017;7:50210–5.

[46] Reyes-Gallardo EM, Lucena R, Cárdenas S, Valcárcel M. Magnetic nanoparticles-nylon 6 composite for the dispersive micro solid phase extraction of selected polycyclic aromatic hydrocarbons from water samples. J Chromatogr A 2014;1345:43–9.
[47] Reyes-Gallardo EM, Lucena R, Cárdenas S. Silica nanoparticles–nylon 6 composites: synthesis, characterization and potential use as sorbent. RSC Adv 2017;7:2308–14.
[48] Ríos-Gómez J, Ferrer-Monteagudo B, López-Lorente ÁI, Lucena R, Luque R, Cárdenas S. Efficient combined sorption/photobleaching of dyes promoted by cellulose/titania-based nanocomposite films. J Clean Prod 2018;194:167–73.
[49] Ríos-Gómez J, Fresco-Cala B, García-Valverde M, Lucena R, Cárdenas S. Carbon nanohorn suprastructures on a paper support as a sorptive phase. Molecules 2018;23:1252.

# Fabric phase sorptive extraction: a new genration, green sample preparation approach

**13**

*Abuzar Kabir, Kenneth G. Furton*
International Forensic Research Institute, Department of Chemistry and Biochemistry, Florida International University, Miami, FL, United States

## 13.1 Introduction

Although sample preparation has gained appreciable interests among separation scientists in recent years, the developments in analytical instruments has generally occurred at a faster pace than the progress in sample preparation [1]. The notable advances in modern analytical instrument are yet to be fully realized due to the absence of a universal sample preparation technique that can efficiently prepare the sample for instrumental analysis regardless of its origin, complexity, and the final chromatographic technique used for the separation, identification, and quantification of the target compounds. The importance of a universal and green sample preparation technique cannot be overstated. In fact, the success of an analytical method, developed for the analysis of target analytes at their trace and ultratrace level concentration present in complex sample matrices including environmental, pharmaceutical, food, and biological samples largely depends on the efficiency of the sample preparation technique used. The primary objectives of a sample preparation technique include: (a) selectively isolate and preconcentrate the target analytes from the complex sample matrix; (b) eliminate/minimize the matrix interferents from the sample; and (c) exchange the solvent so that the sample is compatible with the downstream chromatographic technique [2]. A clean, matrix interferent—free sample not only ensures easy separation, identification, and quantification of the target compounds but also safeguards the performance of the chromatographic system.

Among many classical sample preparation techniques, liquid-liquid extraction (LLE) and solid phase extraction (SPE) are the most prevalent techniques [3—6]. However, a lack of selectivity, inability to discriminate the target compounds from matrix interferents, formation of hard to break emulsions, and use of large volumes of toxic and hazardous organic solvents are the prime concern for LLE. For many contemporary applications solid-phase extraction (SPE) is considered the gold standard sample preparation technique. SPE, however, is a laborious, time-consuming, and multistep sample preparation technique that demands clean, particle-free samples and often requires solvent evaporation and sample reconstitution in a suitable solvent, resulting in potential analyte loss. The limited number of polar sorbents for SPE is also a major

**Solid-Phase Extraction. https://doi.org/10.1016/B978-0-12-816906-3.00013-3**

problem that hinders the selective extraction of numerous polar analytes and metabolites, especially from biological samples. In addition, protein precipitation is inevitable for biological samples prior to SPE that imposes an additional burden on the method. SPE is an exhaustive extraction technique that requires passage of the sample solution through a compact bed of sorbent particles dispensed either in the form of a cartridge or a disk. Due to the compactness of the sorbent particles, either positive or negative pressure must be applied to facilitate the steady flow through the sorbent bed and limits its field deployment for sampling and sample preparation. During the forced passage of the sample solution through the sorbent bed, unwanted matrix interferents also interact with the sorbent and become extracted. To eliminate the matrix interferents, a washing step is often employed after the extraction that may incur some analyte loss as well. SPE sorbents are primarily silica-based chemically modified sorbents with some polymeric phases such as hydrophilic-lipophilic balanced phase (HLB) also available [7].

To mitigate the shortcomings of SPE, solid phase microextraction (SPME) was introduced by Pawliszyn and coworkers in 1987 [8] as a solvent-free/solvent-minimized sampling and sample preparation technique. Numerous advantages of SPME compared to both SPE and LLE have promoted this technique toward high popularity in a short time. Unlike SPE, solid phase microextraction is an equilibrium driven sampling technique that utilizes predominantly organic polymer and particles blended with organic polymer as the sorbent, immobilized on the SPME fibers (fused silica fiber/stableflex fiber/metal fiber) by a free radical cross-linking reaction. Due to the lack of chemical bonding between the sorbent and the SPME fiber, the extraction phase is not generally stable to direct exposure to organic solvents. Most common SPME sorbents include polydimethylsiloxane (PDMS), polyacrylate (PA), polyethylene glycol (PEG), carboxen-polydimethyl siloxane (CAR-PDMS), divinylbenzene-polydimethylsiloxane (DVB-PDMS), and carbopack Z/PDMS [9]. For extracting the target analytes, SPME fiber coated with the polymeric sorbent is immersed into the headspace of the aqueous sample (HS-SPME) or directly immersed into the aqueous sample (DI-SPME). The analytes partition between the SPME sorbent and the aqueous solution based on the partition coefficient and the process continues until the equilibrium is reached. SPME is a selective extraction process and therefore does not require any washing step after the extraction. However, due to the viscous nature of the polymeric coating, the sample in SPME should be free of particles, debris, and proteins to avoid irreversible adsorption to the sorbent surface, leading to fouling of the sorbent coating. In addition, high viscosity of the polymeric sorbent in SPME often retards the analyte mass transfer into the sorbents and consequently prolongs the extraction time. SPME also suffers from poor extraction sensitivity due to the low phase volume (0.030–0.612 μL) [10]. Subsequently, a number of new microextraction techniques emerged with higher sorbent loading such as intube SPME [11], stir bar sorptive extraction (SBSE) [12], microextraction by packed sorbent (MEPS) [13], rotating-disk sorptive extraction (RDSE) [14], and thin film microextraction (TFME) [15].

Sorbent-based microextraction techniques are governed by two principle criteria: (a) thermodynamics and (b) kinetics [16]. The thermodynamic properties of the sorbent determine the maximum amount of analytes that can be extracted by unit mass of sorbent under a given set of extraction conditions. Since higher sorbent loading allows accumulation of a higher mass of analyte under equilibrium extraction conditions, sorbent loading should be maximized. However, the substrate type and geometry determine the maximum sorbent loading capacity for a microextraction 2device [17]. On the other hand, kinetics controls the rate of extraction and therefore the time needed to reach the extraction equilibrium. The faster the extraction equilibrium, the shorter is the sample preparation time and the higher is the throughput of the analytical laboratory. It is obvious that the extraction kinetics is directly related to the primary contact surface area (PCSA) of the device, which is defined as the surface area of the extraction device available for direct interaction with the analytes during the extraction process. The augmentation of PCSA also offers higher sorbent loading without changing the coating thickness and allows accumulating more analytes without saturating the sorption membrane. The sorbent coating technology is the key to improving the thermodynamic properties of sorbent materials. Classical organic/inorganic polymers such as poly(dimethylsiloxane) (PDMS), poly(ethylene glycol) (PEG) are highly viscous in nature. Mass transfer in these polymeric coatings is frustratingly slow, leading to prolonged extraction times. As such, increasing sorbent loading in microextraction device without changing the material properties does not significantly change the method sensitivity. As a result, stir bar sorptive extraction (SBSE) has failed to improve sensitivity significantly even though it has ~250 times higher PDMS sorbent loading than SPME fibers [18].

A closer look on SPE and SPME sorbent/devices reveals that both the sample preparation techniques use an inert substrate (silica particles for SPE and fused silica/metal/composite fiber for SPME) and an organic ligand (for SPE) or an inorganic/organic polymer/blended polymer (for SPME) as the sorbent. SPE primarily extracts the target analytes via a flow through system. On the other hand, SPME primarily extracts analytes from either headspace or by direct immersion. Elution of the analytes in SPE is carried out using a suitable organic solvent while for SPME fibers, thermal desorption is more common. Although, the primary objectives of both SPE and SPME are the same, their approach in achieving these objectives is profoundly different in (a) extraction mechanism; (b) sorbent material chemistry; and (c) analyte elution/desorption process. As such, the goal for the development of a new sample preparation technique should be devoted to the unification of both SPE and SPME so that the best of both the techniques can be beneficially exploited.

Some other shortcomings of SPE and SPME include (a) inherently low thermal, solvent, and chemical stability of the sorbents; (b) limited selectivity of the sorbents; (c) slow mass transfer of analytes due to the high viscosity of the polymeric sorbents; (d) absence of cation-exchanger, anion-exchanger, and mixed mode sorbents for SPME that limits the SPME applications to only neutral analytes unless matrix pH adjustment is used.

Taking all the shortcomings of SPE and SPME into consideration, fabric phase sorptive extraction (FPSE) has been developed as a universal sample preparation technique that innovatively combines the extraction mechanism of SPE and SPME into a single sample preparation platform [19–21]. FPSE has eloquently addressed most of the shortcomings of SPE and SPME and combines almost all the features of both techniques. FPSE utilizes a section of fabric (cellulose, polyester, or fiberglass) as the substrate with a sol-gel sorbent coating on its surface. The sorbent coatings are chemically bonded to the substrate for high pH and sorbent stability. When sol-gel sorbents are coated on fiberglass, they offer high thermal stability that allows thermal desorption of the extracted analytes via instantaneous heat shock (Thermal Desorption Unit). To extract the target analytes, FPSE membranes can be inserted directly into the sample matrix such as whole blood, urine, whole milk, environmental water without any sample pretreatment. The extraction rate is often enhanced by convection using a Teflon coated magnetic stir bar. After the extraction, the analytes are eluted in a small volume of organic solvent in a small back-extraction tube. Sponge-like porous architecture of the sol-gel sorbents allows fast diffusion of the aqueous sample for rapid analyte extraction as well as the organic solvent for quick elution of the analytes. Selection of a suitable elution solvent allows injecting an aliquot of the sample into gas chromatography (GC), liquid chromatography (LC), capillary electrophoresis (CE), and inductively coupled plasma-mass spectrometry (ICP-MS) to maximize the analytical information.

The current chapter describes the rationale of the invention of FPSE, building blocks of FPSE membranes, development and characterization of FPSE membranes, extraction principles exploited in FPSE, FPSE method development process, FPSE protocol, and different implementations of FPSE. Applications of FPSE are also briefly discussed.

## 13.2 Building blocks of fabric phase sorptive extraction membranes and their role in extraction

Fabric phase sorptive extraction membranes are built on the following building blocks: (1) a fabric substrate; (2) a sol-gel inorganic precursor/organically modified inorganic precursor; (3) a sol-gel active inorganic/inorganic polymer. Occasionally, high surface area carbonaceous particles such as carboxen, graphene, carbon nanotubes are also used in combination with inorganic/organic polymers to enhance the selectivity of the extracting sorbents. Fig. 13.1 presents a schematic representation of sol-gel sorbent-coated FPSE membrane.

The description of different building blocks of FPSE membranes are presented below:

(i) *Fabric substrate*:

Fabric substrate plays a key role in building the FPSE membrane. Unlike other extraction and microextraction techniques, the substrate in FPSE is not merely a host for the sorbent but also actively contributes to the overall selectivity of the FPSE membrane via hydrophilic/hydrophobic interactions. Among many potential candidates, cellulose, polyester, and fiberglass fabrics are primarily used as the substrates for FPSE membranes. All these fabrics possess sol-gel active functional groups on which the sol-gel sorbent networks are chemically bonded during the sol-gel

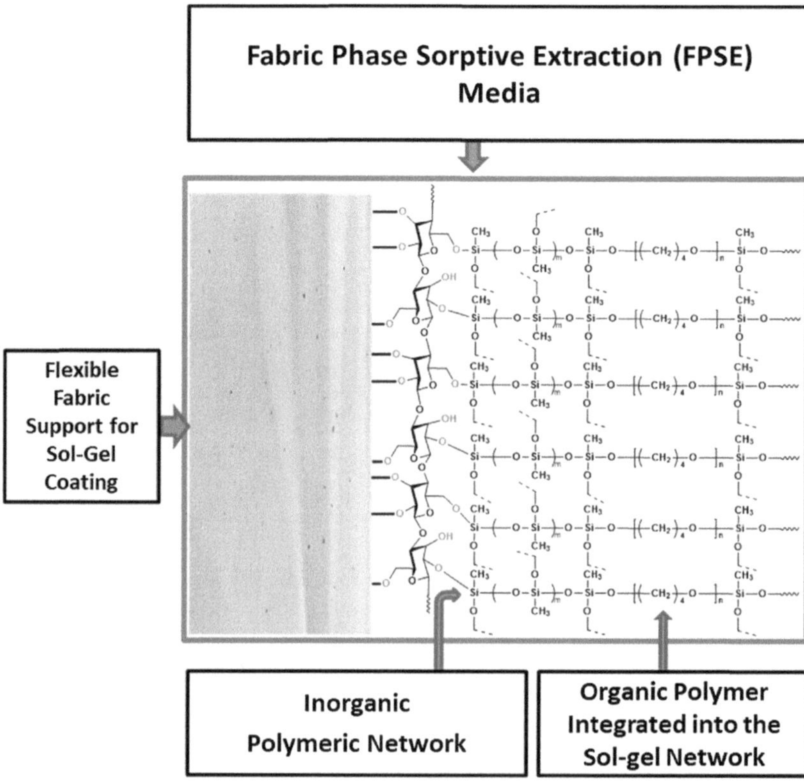


**Figure 13.1** Schematic representation of FPSE membrane and its different building blocks.

coating process. Since the sorbent loading on the FPSE membrane depends on the concentration of sol-gel active functional groups, the amount of sorbent loading per unit area on cellulose fabric is substantially higher than polyester fabric. It is important to note that the sol-gel sorbent networks chemically bind to the sol-gel active functional group, leaving the majority of the fabric substrate surface uncovered and available for interaction with target compounds. Due to the porosity of the fabric, FPSE membranes mimic a solid phase extraction disk. During the extraction process, aqueous sample permeates through the FPSE membrane and facilitates rapid and near exhaustive extraction. Fig. 13.2 presents the chemical structures of different FPSE substrates.

**(ii)** *Inorganic/organically modified sol-gel precursor:*

Inorganic/organically modified sol-gel precursor plays an important role in the sol-gel sorbent coating process. It creates the 3D networks of sol-gel sorbent by randomly incorporating the inorganic/organic polymer into the networks. It also acts as a linker to bind the sol-gel sorbent networks with the fabric substrate. When organically modified inorganic sol-gel precursors are used, the organic pendant moiety

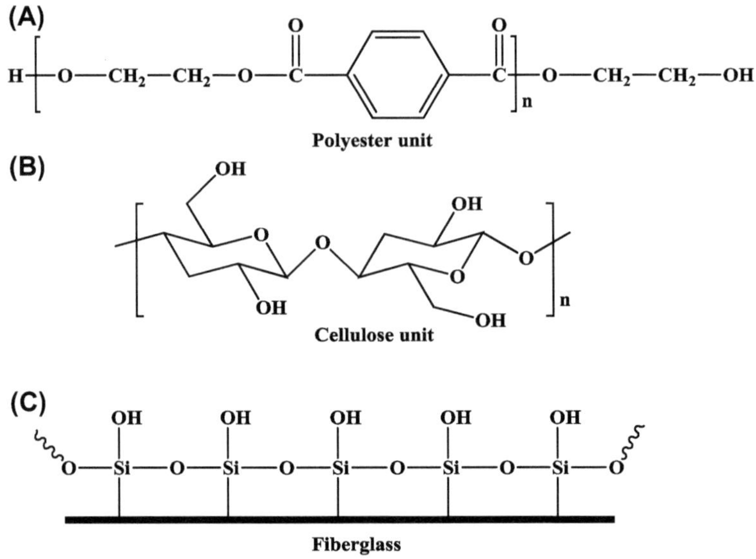


**Figure 13.2** Chemical structures of (A) a polyester unit; (B) a cellulose unit; and (C) fiberglass fabric.

actively contributes to the overall selectivity of the FPSE membrane. As such, the polarity and selectivity of the FPSE membrane can be fine tuned by a judicious selection of the sol-gel precursor. Among a large number of available sol-gel precursors, tetramethoxysilane (TMOS), tetraethoxysilane (TEOS), methyl trimethoxysilane (MTMOS), phenyltrimethoxysilane (PTMOS), 3-aminopropyltrimethoxysilane (3-APTMOS), octadecyltrimethoxysilane (C18-TMOS), octyltrimethoxysilane (C8-TMOS), titanium isoproxide, zirconium isopropoxide, teramethoxygermane are noteworthy.

**(iii)** *Sol-gel active inorganic/organic polymer:*

Sol-gel active inorganic polymers such as poly(dimethylsiloxane) (PDMS), poly(-dimethyldiphenylsiloxane), or organic polymers such as poly(ethylene glycol), poly(-tetrahydrofuran) are randomly integrated into the sol-gel networks via sol-gel synthesis and act as the primary source of the selectivity and extraction affinity toward the target analyte(s). The polymer used in the sol-gel sorbent coating process provides different intermolecular interactions, which are attributed to their structure and broadly differ from one polymer to the other.

## 13.3 Preparation of sol-gel sorbent coated fabric phase sorptive extraction membranes

Due to the strong covalent bonding between the sorbent and the fabric substrate, FPSE membranes demonstrate excellent thermal, solvent, and chemical stability. As such, sol-gel sorbent coated FPSE membranes can be exposed to any organic solvent, strong acidic or basic environment (pH 1-12), and high temperature (~350 °C) for thermal desorption of the extracted analytes (only sol-gel sorbents coated on fiberglass). Incorporation of hundreds of organic polymers/ligands into a large number of inorganic polymeric networks opens the floodgates for thousands of novel sorbents with different selectivity, surface morphology, and other advanced material properties [22–25].

Preparation of sol-gel sorbent coated fabric phase sorptive extraction membrane involves a number of sequential steps including: (1) selection and pretreatment of fabric phase sorptive extraction membrane; (2) design and preparation of the sol solution for sol-gel sorbent coating on the substrate; (3) sol-gel sorbent coating process using immersion coating technology; (4) aging, conditioning, and cleaning of sol-gel sorbent coated fabric phase sorptive extraction membrane; and (5) cutting the FPSE membrane into an appropriate size.

### 13.3.1 Selection and pretreatment of fabric phase sorptive extraction membrane

In order to qualify as a substrate for FPSE, the fabric should meet the following criteria: (a) presence of abundant surface hydroxyl groups; and (b) surface permeability so that aqueous solution can easily permeate through the FPSE membrane during analyte extraction and organic solvent during elution/back-extraction. Many commercially available fabrics meet these criteria and among them cellulose, polyester and fiberglass, nylon, polyamide fabric are noteworthy. The commercial fabric often retains residual finishing chemicals on their surface used to provide glossiness and protect the fiber from accumulating dust particles. These residual chemicals must be removed prior to the sol-gel sorbent coating on the fabric surface. In addition, the surface hydroxyl groups should be activated. To clean the fabric and activate surface hydroxyl groups, a fabric pretreatment regimen was developed. The detailed pretreatment process is described elsewhere [2,21].

### 13.3.2 Design and preparation of sol solution for sol-gel sorbent coating on the substrate

The design of the sol solution primarily depends on the nature of the target analytes and the sample matrix. Primary focus is on the selection of sol-gel precursor(s) and the organic polymer. If the analytes are hydrogen bond donors or acceptors, the organic polymer should possess hydrogen bond acceptor or donor groups for complementary interactions with the analytes.

### 13.3.3 Sol-gel sorbent coating process using immersion coating technology

Sol-gel sorbent coating process involves the following sequential reactions:

**(i)** Catalytic hydrolysis of sol-gel precursor or organically modified sol-gel precursor;

**(ii)** Polycondensation of hydrolyzed precursor, resulting in a growing sol-gel network;
**(iii)** Random incorporation of sol-gel active polymer into the growing sol-gel network on the fabric substrate surface via polycondensation.

Sol-gel reactions are generally carried out under ambient conditions. Typically, methyltrimethoxysilane (MTMS) is used as the sol-gel precursor and trifluoroacetic acid (TFA) (95%, 5% $H_2O$) as the sol-gel catalyst. During the hydrolysis, the three methoxy groups of MTMS are hydrolyzed to form hydroxyl groups and subsequently begin forming a 3D network via polycondensation. During the formation of 3D network, sol-gel active organic polymer randomly enters the network. Finally, the sol-gel sorbent network chemically bonds to the fabric substrate through the fabric hydroxyl groups via polycondensation. Fig. 13.3 illustrates the different chemical reactions involved in the sol-gel gel sorbent coating process.

The sol-gel sorbents are highly porous and possess a sponge-like porous architecture. Due to the high porosity of sol-gel sorbent and the permeability of the fabric substrate, aqueous sample travels through the FPSE membrane thousands of times during the extraction process and ensures exhaustive extraction of the target compounds in a relatively short time.

Sol-gel sorbent coating on fabric substrate is generally carried out at room temperature in a 2 oz reaction vessel for 6-8 h. Subsequently, the sol solution is drained from the reaction vessel, the sol-gel sorbent coated FPSE membrane is dried in air for 1h, and then transferred to a home-built thermal conditioning device.

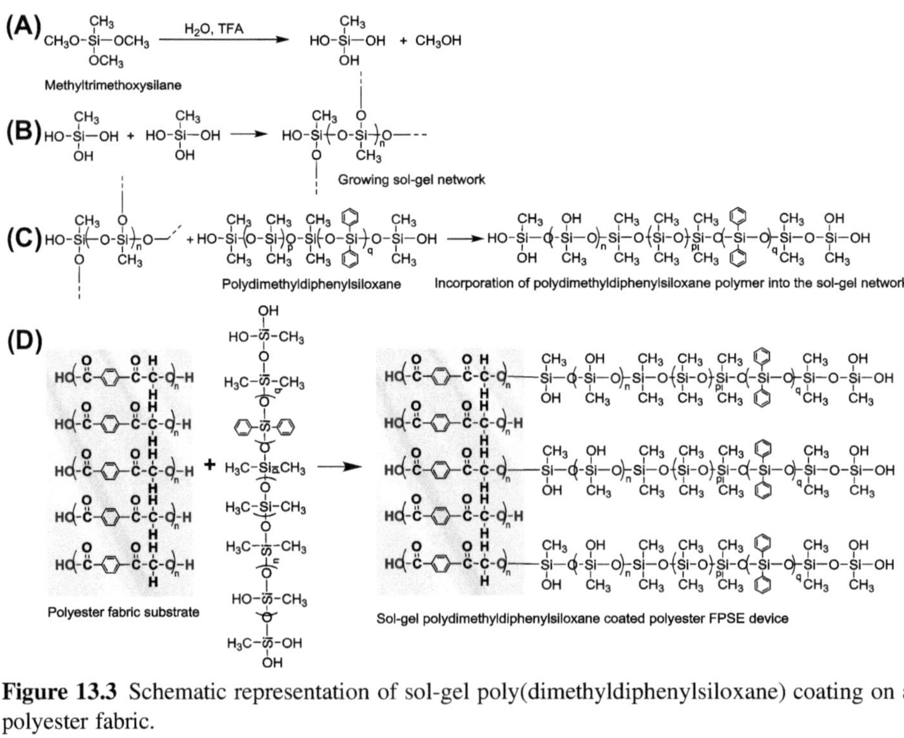


**Figure 13.3** Schematic representation of sol-gel poly(dimethyldiphenylsiloxane) coating on a polyester fabric.

### 13.3.4 Aging, conditioning, and cleaning of sol-gel sorbent-coated fabric phase sorptive extraction membrane

The objective of aging and conditioning of the sol-gel sorbent-coated FPSE membrane is to ensure that the sol-gel reaction goes to the completion. The aging and conditioning is carried out inside a homemade conditioning unit at 50°C under continuous flow of helium gas for 24 h. The conditioning is followed by cleaning the sol-gel sorbent-coated FPSE membrane with methanol/methylene chloride (50:50 v/v). During the cleaning process, any unreacted precursor, polymer, reaction intermediates are removed. After cleaning, the FPSE membrane is conditioned again for 12 h at 50°C.

### 13.3.5 Cutting the FPSE membrane into appropriate size

Unlike other sample preparation techniques, one of the major advantages of FPSE is its variable size that can be tailored to specific applications. For example, due to the small volume of typical biological samples such as whole blood and plasma, FPSE membrane disks of 1 $cm^2$ are used. For environmental samples, an FPSE membrane of 2.5 $cm^2$ is typically used. The sol-gel sorbent-coated FPSE membranes are cut into sizes using a homemade cutting device [26,27].

## 13.4 Sol-gel sorbents for fabric phase sorptive extraction

Among all extraction and microextraction techniques, FPSE has developed the highest number of sorbents, from nonpolar to medium polar to polar to cation exchanger to anion exchanger to zwitterionic to mixed mode chemistries. FPSE also is the first microextraction technique that offers anion exchanger, cation exchanger, zwitterionic and mixed mode sorbents. Table 13.1 provides a list of fabric phase sorptive extraction sorbent chemistries with other pertinent information.

## 13.5 Characterization of fabric phase sorptive extraction membranes

Sol-gel sorbent-coated fabric phase sorptive extraction membranes are often characterized by a number of techniques including scanning electron microscopy (SEM), atomic force microscopy (AFM), thermogravimetric analysis (TGA), Fourier transform-infrared spectroscopy (FT-IR), nitrogen adsorption isotherm, elemental analysis, energy dispersion spectroscopy (EDS). SEM provides details of the surface morphology of the fabric substrate, before and after the sol-gel sorbent coating. FT-IR spectra provide information on the chemical integration of different sol solution ingredients into the sol-gel sorbent network as well as integration of sol-gel sorbent network with the fabric substrate.

**Table 13.1** List of major fabric phase sorptive extraction sorbents.

| FPSE sorbent | Sorbent type | Sorbent loading (mg/cm$^2$) |
|---|---|---|
| 1. Sol-gel polydimethylsiloxane | Nonpolar | 2.30 |
| 2. Sol-gel polydimethyldiphenylsiloxane | Nonpolar | 1.93 |
| 3. Sol-gel ppolytetrahydrofuran | Medium polar | 3.96 |
| 4. Sol-gel polyethylene glycol-polypropylene glycol-polyethylene glycol | Medium polar | 5.68 |
| 5. Sol-gel polypropylene glycol-polyethylene glycol-polypropylene glycol | Medium polar | 5.25 |
| 6. Sol-gel methacrylate | Medium polar | 4.51 |
| 7. Sol-gel C4 | Medium polar | 1.90 |
| 8. Sol-gel C8 | Medium polar | 2.33 |
| 9. Sol-gel C12 | Nonpolar | 3.54 |
| 10. Sol-gel C18 | Nonpolar | 4.88 |
| 11. Sol-gel polycaprolactone diol | Polar | 3.46 |
| 12. Sol-gel polycaprolactone triol | Polar | N/A |
| 13. Sol-gel polycaprolactone-polytetrahydrofuran-polycaprolactone | Medium polar | 4.64 |
| 14. Sol-gel polycaprolactone-polydimethylsiloxane-polycaprolactone | Medium polar | 6.14 |
| 15. Sol-gel UCON | Polar | N/A |
| 16. Sol-gel graphene | Nonpolar | N/A |
| 17. Sol-gel multiwalled carbon nanotubes | Nonpolar | N/A |
| 18. Sol-gel sucrose | Polar | N/A |
| 19. Sol-gel sucralose | Polar | N/A |
| 20. Sol-gel chitosan | Polar | N/A |
| 21. Sol-gel carbowax 20M | Polar | 8.64 |
| 22. Sol-gel polyethylene glycol 10,000 | Polar | 6.36 |
| 23. Sol-gel polyethylene glycol 300 | Polar | 4.45 |

**Table 13.1** List of major fabric phase sorptive extraction sorbents.—cont'd

| FPSE sorbent | Sorbent type | Sorbent loading (mg/cm$^2$) |
|---|---|---|
| 24. Sol-gel cation exchanger | Cation exchanger | N/A |
| 25. Sol-gel anion exchanger | Anion exchanger | N/A |
| 26. Sol-gel zwitterionic sorbent | Cation and anion exchanger | N/A |
| 27. Sol-gel mixed mode sorbent | Medium polar, cation exchanger; medium polar, anion exchanger | N/A |
| 28. Sol-gel zwitterionic-mixed mode sorbent | Medium polar, cation exchanger, anion exchanger | N/A |

## 13.6 Working principle of fabric phase sorptive extraction

Solid sorbent-based sample preparation techniques are primarily classified into two major classes based on their extraction mechanism: (a) exhaustive extraction, as in solid phase extraction (SPE); and (b) equilibrium extraction, as in solid phase microextraction (SPME). Exhaustive extraction techniques employ extraction of the analytes during percolation of the sample through the sorbent bed. However, during this process many unwanted compounds from the sample matrix may also become adsorbed. Removal of these materials is necessary prior to chromatographic analysis and is often accomplished by incorporating a washing step into the workflow prior to eluting the analytes. The washing step may result in substantial analyte loss and should be avoided if possible.

Equilibrium extraction techniques, such as SPME, require positioning the device inside the sample matrix (direct immersion extraction) or in the headspace of the sample (headspace extraction) and the mass transfer of analytes continues until the equilibrium is reached. Due to the viscous nature of the extracting phases, a prolonged extraction time is required to achieve equilibrium. In addition, low sorbent mass (especially in SPME) often fails to accumulate a high mass of target analytes during the extraction, resulting in a lower overall method sensitivity.

Fabric phase sorptive extraction, on the other hand, integrates both extraction mechanisms. When the FPSE membrane is immersed into the sample during extraction, it mimics direct immersion solid phase microextraction. The coated fabric substrate is inherently permeable. In a stirred solution, the sample rapidly permeates through the porous bed of the FPSE membrane mimicking an SPE disk. This permeation occurs thousands of times during the extraction period. Analytes are extracted by the FPSE membrane almost exhaustively without matrix interferents due to the unique selectivity of the sorbent.

## 13.7 Fabric phase sorptive extraction protocol

Fabric phase sorptive extraction has profoundly simplified the overall sample preparation workflow. It has successfully eliminated all the sample pretreatment steps such as filtration, centrifugation, protein precipitation (for biological samples), and defatting (milk samples). It also eliminates post sample treatment such as solvent evaporation and sample reconstitution. The simplification of the sample preparation process not only saves time, effort, money, and organic solvent usage, but also improves the overall quality of the analytical data. Fig. 13.4 illustrates the different steps involved in fabric phase sorptive extraction.

The fabric phase sorptive extraction protocol is described below:

**Step 1.** Cleaning of FPSE Membrane

**(i)** Prior to the extraction, immerse all the FPSE membranes in 2 mL of acetonitrile:methanol (50:50 v/v) for 5 min (no shaking/vortexing/stirring is needed). Membranes are manipulated using clean tweezers.
**(ii)** Rinse the FPSE membranes in 2 mL DI water to remove residual organic solvents.
**(iii)** Now the FPSE membranes are ready for extraction.

**Step 2.** Extraction (10-40 min):

**(i)** Transfer a representative aliquot of the sample to a 10/20 mL SPME vial.
**(ii)** Insert a clean Teflon-coated magnetic stir bar into the vial.
**(iii)** Insert a clean FPSE membrane into the vial.

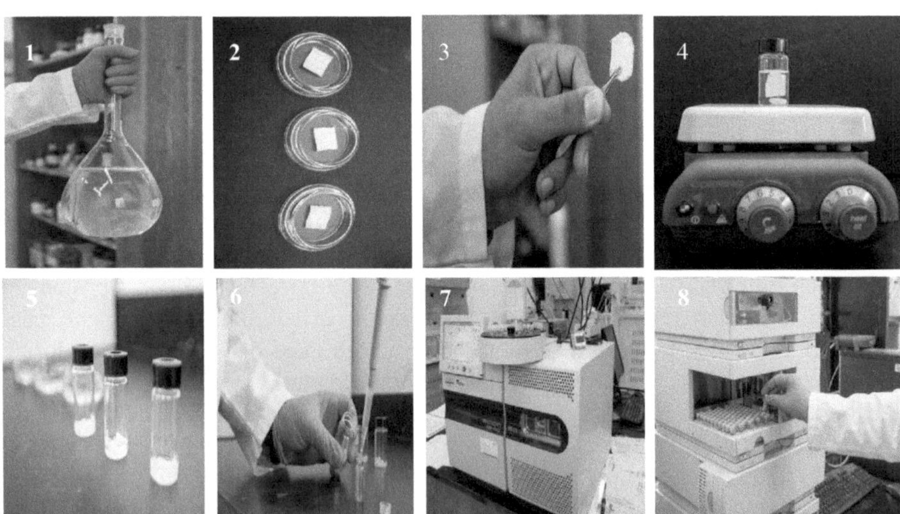


**Figure 13.4** Steps in fabric phase sorptive extraction (FPSE): (1) unfiltered, dirty environmental sample; (2) FPSE media on watch glass; (3) an FPSE medium on a tweezer; (4) fabric phase sorptive extraction; (5) back extraction in organic solvent; (6) transferring preconcentrated eluent into sample vial; (7) analysis in GC; (8) analysis in HPLC [2].

**(iv)** Set up magnetic stirrer to medium agitation to ensure adequate analyte convection.
**(v)** Extract for 10-40 min depending on the type of target analyte(s).
**(vi)** Remove the FPSE membrane from the sample and dry it by gently pressing against a paper towel.

**Step 3.** Elution/Back extraction (5-10 min):

**(i)** Use a deactivated glass vial/Eppendorf tube for analyte desorption.
**(ii)** Fold the FPSE membrane and insert it into the desorption vial.
**(iii)** Add 300-500 μL of solvent into the vial.
**(iv)** Allow 5-10 min for complete back extraction (unnecessity to vortex, centrifuge, or sonicate).
**(v)** Centrifuge the back-extraction solution if it looks cloudy or contains particulates.
**(vi)** Pipette the back-extraction solvent/eluent into a GC/LC sample vial for chromatographic analysis.

**Step 4.** Cleaning FPSE membrane and storing for future use:

**(i)** After the elution/back extraction, immerse the FPSE membrane in 2 mL acetonitrile:methanol (50:50 v/v) for 5 min (no shaking/vortexing/stirring is necessary).
**(ii)** Dry the FPSE membrane to drive off the solvents by putting it on a watch glass for 5-10 min.
**(iii)** Store in an airtight glass container for future use.

## 13.8 Fabric phase sorptive extraction method development

A number of factors that affect the FPSE selectivity and extraction efficiency should be optimized. These factors include: (a) FPSE sorbent chemistry; (b) sample volume; (c) stirring speed; (d) extraction time; (e) addition of salt; (f) elution solvent type, volume, and elution time; and (g) extraction temperature.

The selection of an appropriate sorbent is generally the most important step that determines the overall sensitivity of the method. In order to simplify the sorbent selection process, a second order mathematical model has been developed for each of the FPSE sorbent that allows calculating absolute recovery (%) of an analyte using its log Kow value. These models are valid for log Kow values between 0.3 and 5.07. Based on the log Kow values of the analytes, three to four potential sorbent candidates can usually be identified for systematic method development experiments. Fig. 13.5 illustrates this process for a sol-gel polycaprolactone-polydimethylsiloxane-polycaprolactone coated FPSE membrane and the corresponding absolute recovery versus log Kow correlation model.

Table 13.2 presents the absolute recovery calculation equations for representative FPSE sorbents.

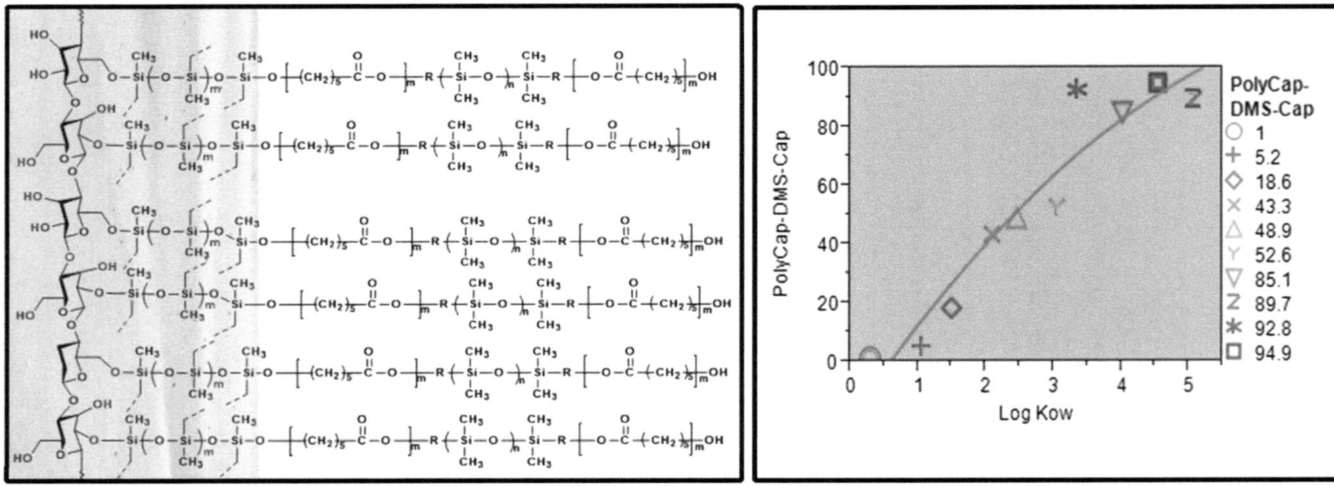


**Figure 13.5** Schematic representation of sol-gel silica polycaprolactone-polydimethylsiloxane-polycaprolactone coated FPSE membrane (left) and the associated absolute recovery versus log Kow correlation model (right).

**Table 13.2** Absolute recovery calculation equations for representative FPSE sorbents.

| Sorbent (substrate) | Absolute recovery equation |
|---|---|
| Si-CW20M (cellulose) | R, % = 4.298 + 22.823*log Kow − 3.134*(log Kow − 2.737)$^2$ |
| Si-PEG1000 (cellulose) | R, % = −11.535 + 20.950*log Kow − 0.402*(log Kow − 2.737)$^2$ |
| Si-PEG300 (cellulose) | R, % = +14.759 + 16.310*log Kow − 5.550*(Log Kow − 2.737)$^2$ |
| Si-CN-CW20M (cellulose) | R, % = −24.393 + 23.940*log Kow + 1.247*(log Kow − 2.737)$^2$ |
| Si-PPG-PEG-PPG (cellulose) | R, % = −3.649 + 21.546*log Kow −2.879*(Log Kow − 2.737)$^2$ |
| Si-PEG-PPG-PEG (cellulose) | R, % = −7.680 + 23.069*log Kow −1.726*(Log Kow − 2.737)$^2$ |
| Si-PTHF (cellulose) | R, % = +12.401 + 17.849*log Kow + 17.849*(Log Kow − 2.737)$^2$ |
| Si-PTHF (fiber glass) | R, % = −28.442 + 20.950*log Kow + 3.327*(Log Kow − 2.737)$^2$ |
| Si−C18 (cellulose) | R, % = −2.275 + 20.816*log Kow − 4.148*(Log Kow − 2.737)$^2$ |
| Si−C8 (cellulose) | R, % = −3.393 + 21.261*log Kow − 3.772*(log Kow − 2.737)$^2$ |
| Si-PDPS (cellulose) | R, % = −10.300 + 17.450*log Kow − 0.288*(log Kow − 2.737)$^2$ |
| Si-PDMDPS (polyester) | R, % = −9.185 + 17.816*log Kow − 1.966*(log Kow − 2.737)$^2$ |
| Si-PDMDPS (cellulose) | R, % = −19.602 + 15.454*log Kow − 1.622*(log Kow − 2.737)$^2$ |

Once the most suitable sorbent chemistry is identified, a systematic method development approach can be followed to optimize the other extraction parameters. A detailed approach is presented elsewhere [2,28−30].

## 13.9 Advantages of fabric phase sorptive extraction over conventional sorbent-based sample preparation techniques

Fabric phase sorptive extraction has substantially simplified sample preparation process by eliminating all sample pretreatment and post-treatment steps typically involved in conventional sample preparation techniques. A schematic comparison between SPE, SBSE, and FPSE illustrates the operational advantages of FPSE over other major sample preparation techniques (Fig. 13.6 and Table 13.3) [28].

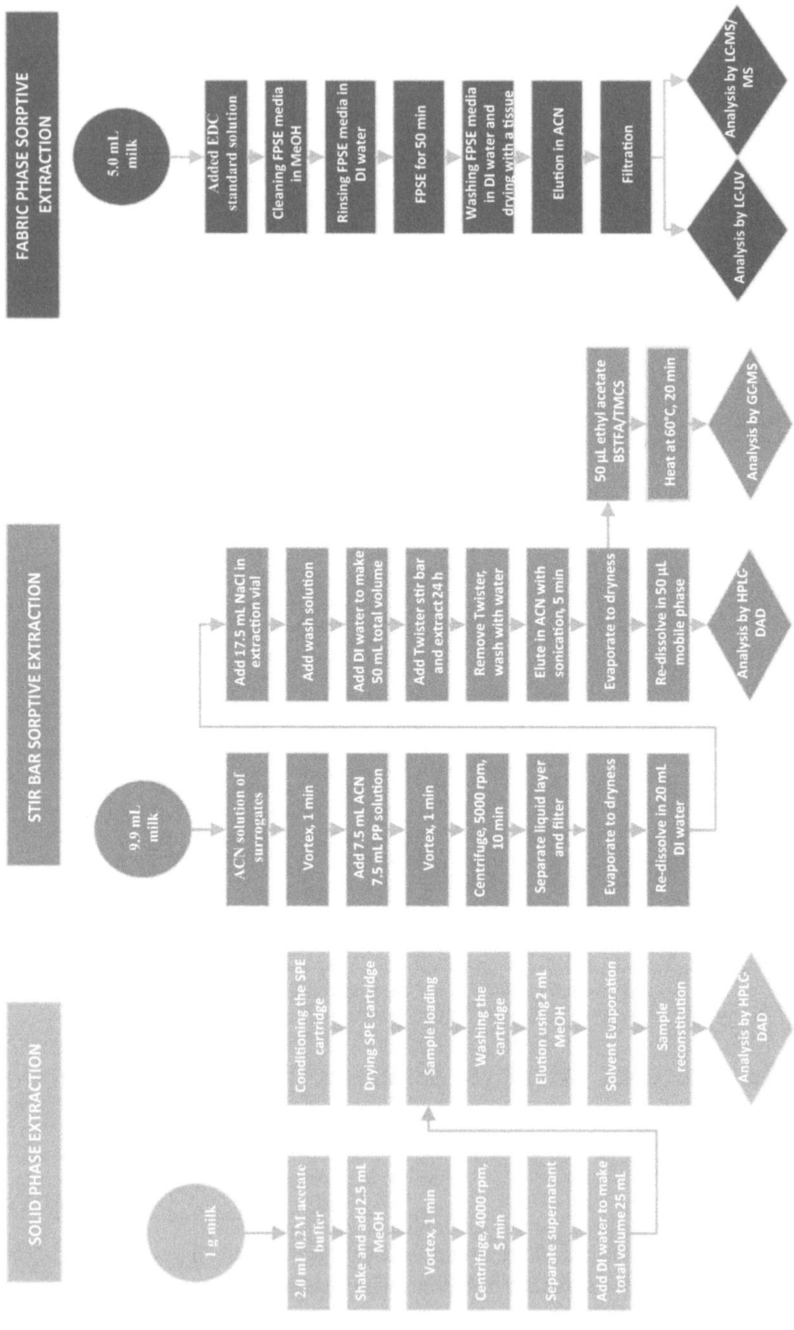


**Figure 13.6** Comparison between solid phase extraction, stir bar sorptive extraction, and fabric phase sorptive extraction [28].

**Table 13.3** Comparison between fabric phase sorptive extraction (FPSE), solid phase microextraction (SPME), and solid phase extraction (SPE).

| | FPSE | SPME | SPE |
|---|---|---|---|
| 1. Substrate geometry | FPSE utilizes a small piece of cotton fabric as the substrate that holds the sol-gel-derived hybrid organic-inorganic polymeric sorbent responsible for analyte extraction from different matrices. | A small segment of 100 μm i.d. fused silica rod mounted on the tip of a stainless steel tube holds the polymeric sorbent material. Easy to use, but prone to needle bending and fiber breakage if the operator is not properly trained. | Small, spherical silica particles are used to support the coating of sorbent materials on their surfaces. The coated silica particles are then packed in different geometrical formats including cartridges, disks, etc. The sorbent bed often exerts enough resistance to flow aqueous solution containing the target analyte(s) requiring application of either positive or negative pressure in the extraction process. |
| 2. Accessible surface area | Provides flexible, open, flat, and easily accessible surface for sorbent-analyte interaction. Typically, a 5 $cm^2$ piece of fabric is used as the extraction phase. Two sides of the fabric offer approximately 10 $cm^2$ directly accessible sorbent coated surface. | Due to the small surface area of the substrate, the sorbent-analyte contact surface area is approximately $<0.1$ $cm^2$. | Directly accessible surface area depends on the geometry of the SPE unit. |
| 3. Extraction principle | Can be used in equilibrium extraction mode as well as in exhaustive extraction mode. | Utilizes equilibrium extraction mode based on the distribution of the analyte between the donor phase (water) and the acceptor phase (SPME fiber coating). | Utilizes exhaustive extraction mode. |

*Continued*

**Table 13.3** Comparison between fabric phase sorptive extraction (FPSE), solid phase microextraction (SPME), and solid phase extraction (SPE).—cont'd

| | FPSE | SPME | SPE |
|---|---|---|---|
| 4. Solvent volume | Needs only a small volume of organic solvent for elution/back extraction of the extracted analyte(s). | Does not need any solvent for the gas chromatography–based analysis as the SPME fiber can be directly introduced into the GC inlet after the extraction. A small volume of organic solvent is required for eluting extracted analytes when LC is the analytical instrument of choice. | Requires high volume of organic solvent(s) to accomplish quantitative recovery of the analyte from the SPE bed after the extraction. |
| 5. Solvent type | Any organic solvent or solvent mixture can be used as the sorbent coatings are chemically bonded to the surface of the fabric. As such, different solvent or mixture thereof can be employed for back extraction depending on the analytical instrument (gas chromatography (GC), liquid chromatography (LC)). Utilization of proper solvent for back extraction ensures no analyte carry-over that may hinder future extractions. | For GC-based analysis, no solvent is needed. For LC-based analysis, a limited number of solvents can be used for analyte desorption as the organic solvent may dissolve the coating material, leading to significant extraction sensitivity loss in future extractions. Limited choices of organic solvents pose risk for analyte carry-over. | Polymers/ligands are physically immobilized on the spherical substrate (typically silica particles). As such, only limited numbers of organic solvents can be used for eluting extracted analytes from SPE bed. Many strong organic solvents wash away the polymeric coating from the bed and therefore are not suitable for SPE. |

| | | | |
|---|---|---|---|
| 6. Coating process | Hybrid organic-inorganic polymer coatings are generated on the surface of the fabric with the help of sol-gel chemistry. At the end, a strong covalent bond is formed between the hybrid material and the fabric substrate. | The extracting sorbents are immobilized on a piece of fused silica rod or metal surface by physical immobilization process followed by free radical cross-linking reaction. | The polymer coatings are generally formed by physical immobilization process followed by cross-linking reaction. New bonded SPE sorbents utilize silane chemistry to bind organic ligands to silica particles. |
| 7. Sorbent chemistry | Endless numbers of coatings can be created with tunable selectivity, controlled porosity, and adjustable coating thickness using well-understood inorganic precursor and organic polymer chemistry. | A limited number of commercially available sorbent types pose serious limitation to the healthy growth of the technique among scientist community. | Although there are a relatively large number of sorbents commercially available, its multistep procedure, dependence on high volume organic solvent, high resistance against the sample flow, inevitable analyte reconstitution step are challenging issues for the analyst. |
| 8. Chemical stability | Due to the strong covalent bonding between the hybrid coating and the fabric substrate, the coating demonstrates high chemical stability and can be exposed to high acidic and basic environment without compromising the integrity of the hybrid coating. Demonstrate pH stability from 1 to 13. | Due to the absence of chemical bonding between the polymeric coating and the substrate surface, the polymer cannot be exposed to any matrix with pH value beyond 2-11. | Due to the absence of chemical bonding between the polymeric coating and the substrate surface, the polymer cannot be exposed to any matrix with pH value beyond 2-8. However, new bonded phases demonstrate pH stability from 2 to 9. |

*Continued*

**Table 13.3** Comparison between fabric phase sorptive extraction (FPSE), solid phase microextraction (SPME), and solid phase extraction (SPE).—cont'd

| | FPSE | SPME | SPE |
|---|---|---|---|
| 9. Complexity of sample matrix | As the total surface area of the coated fabric is easily accessible for the analyte due to its open geometry, complex sample matrix containing particles, debris, and other nonchemical interferences can be used without any modification (any modification may lead to analyte loss) and therefore the extractant would be true representation of the sample matrix. | Cannot handle complex sample matrix. When the sample matrix is overly complicated, headspace extraction is recommended. The SPME fiber gets easily contaminated if it is exposed directly to the complex sample matrix. | Requires a sample cleaning step (e.g., filtration) before the sample containing the analyte(s) are passed through the SPE bed. Presence of any particulate material in the sample matrix potentially clogs the pores of the bed leading to destroying the SPE bed. Filtration prior to SPE may cause significant analyte loss from the sample matrix. |
| 10. Operational simplicity | The FPSE involves limited numbers of operational steps and hence does not require any special operator training. | SPME fibers are very fragile. Any scraping on the coating may lead to loss of the sorbent volume and consequently loss in extraction sensitivity. | SPE involves multistep operation, and hence time consuming and laborious. |

| | | | |
|---|---|---|---|
| 11. Forensic significance | As FPSE provides a relatively high volume of extractant, the extractant can be injected into the analytical instrument several times. In case, there is a legal challenge against the laboratory report, the result can be easily verified by injecting the preserved extractant maintained under proper chain of custody. | Although SPME is a nondestructive technique, the same sample matrix can be used for multiple extractions. However, the target analyte(s) may interact with the matrix and the overall profile may keep changing over the period of its storage. As a result, the profile may be different if the SPME is conducted after a tangible pause. | SPE is a destructive extraction procedure. As it requires high volume of organic solvent for eluting retained analyte(s) from the SPE bed, the eluent is evaporated first to drive the solvent off from the analyte followed by reconstituting with a known and small volume of organic solvent. Evaporation of the organic solvent may cause some analyte loss if the target analyte is highly volatile. |
| 12. Field application | Very effective for field applications. Extraction and back extraction can be carried out in situ and the extractant can be shipped in a small HPLC/GC sample vial overnight to the analytical lab. There is no need to open the sealed vial during the analysis as the instrument needle can puncture the vial septa to collect aliquot of the extractant for the injection. | Although a field compatible SPME kit is commercially available, there is no guarantee that the extracted analyte(s) will always remain into the SPME fiber coating prior to the instrumental analysis. As soon as the SPME fiber closure is opened for instrumental analysis, there is a potential risk of losing some of the extracted analytes. | Not suitable for field investigation. Often, samples are collected from the field and are brought to the analytical lab for sample preparation and instrumental analysis (GC/LC). |
| 13. Operational cost | As the FPSE does not require any specialized instrument for the extraction and back extraction, operation cost for the sample preparation per sample is nominal. | SPME fibers are very expensive. As such, cost of sample preparation per sample is quite high. | Generally, SPE cartridge/disk is built for single use. As a result, sample preparation cost per sample is high even though SPE cartridge/cylinder is relatively inexpensive. |

*Continued*

**Table 13.3** Comparison between fabric phase sorptive extraction (FPSE), solid phase microextraction (SPME), and solid phase extraction (SPE).—cont'd

| | FPSE | SPME | SPE |
|---|---|---|---|
| 14. Extraction sensitivity | Due to the exceedingly high sorbent- analyte contact surface compared to SPME and SPE, FPSE has been found to be at least 3 orders of magnitude more sensitive than SPME. | A small volume of sorbent often does not yield high extraction sensitivity, particularly for polar analytes. | In exhaustive extraction, analytes are retained and collected quantitatively. However, high volume of elution solvent reduces the preconcentration effect unless the solvent is evaporated and the analyte is reconstituted in a smaller volume of solvent. |
| 15. Enhancement of analyte preconcentration factor | When the objective of the analytical scheme is to achieve highest possible preconcentration factor (especially applicable to ultratrace analysis), multiple fabrics can be employed in independent extraction followed by simultaneous back extraction in the same back-extraction solvent. We have successfully applied this principle for extracting highly polar nitroaromatic explosives at parts per quadrillion (ppq) level concentrations. This attribute of FPSE would be highly beneficial for the analysis of emerging and persistent environmental pollutants. | Current state of the analytical instrument does not support extraction on multiple SPME fibers and simultaneous introduction into the analytical instrument. | Uses of multiple SPE cartridges are possible but extracted analytes must be eluted individually. If the extractants are combined, the analytes will be diluted due to the increased total solvent volume. |

| | | | |
|---|---|---|---|
| 16. Selectivity manipulation | FPSE allows user to use fabric coated with polymers of different polarity. As a result, if the objective of the analytical scheme is to extract analytes with different polarity, multiple fabrics coated with different polarity can be employed. At the end, all extracted analytes can be back extracted in the same back-extraction solvent. | SPME has different fibers to handle only either nonpolar, medium polar, or highly polar analytes. Current state of SPME does not support simultaneous extraction of nonpolar, medium polar, and highly polar analytes from a given matrix. | One of the major shortcomings of SPE is its inability to extract nonpolar, medium polar, and polar analytes simultaneously from a given matrix. |
| 17. Multiple use | FPSE can be used multiple times without compromising extraction sensitivity. As such, the operational cost as well as data quality can be significantly enhanced when FPSE is used as the sample preparation technique. | SPME fibers can be used multiple times when extraction is carried out from the headspace of the enclosed sample matrix. Exposing the fiber directly to the aqueous sample matrix is not generally recommended as particulates, debris, matrix interferents can easily get irreversibly absorbed on the surface of the sorbent coating. | SPE tubes and cartridges are generally used for single extraction application. |

## 13.10 Different implementations of fabric phase sorptive extraction

In the classical implementation of fabric phase sorptive extraction, a piece of sol-gel sorbent coated FPSE membrane is immersed into the aqueous sample and a Teflon coated magnetic stir bar is used to agitate the solution. During magnetic stirring, the FPSE membrane rotates in a random fashion, but may occasionally become static on the sidewall of the extraction vial or becomes coiled, resulting in prolonged extraction times. Due to the randomness of these processes, extraction reproducibility is also impacted. To ensure that the entire primary contact surface area of the FPSE membrane remain accessible during extraction, a number of new implementations have been developed in recent years including (1) stir fabric phase sorptive extraction (stir FPSE); (2) stir bar-fabric phase sorptive extraction (stir bar-FPSE); (3) dynamic fabric phase sorptive extraction (DFPSE); and (4) magnet integrated fabric phase sorptive extraction (MI-FPSE).

### *13.10.1 Stir fabric phase sorptive extraction*

Stir fabric phase sorptive extraction (stir FPSE) integrates FPSE membrane extraction with a magnetic stirring mechanism so that the integrated device spins on a magnetic stirrer during extraction. The rotation per min of stir FPSE can be precisely controlled [31]. The device was fabricated using a section of a polypropylene SPE cartridge, FPSE membrane, a magnetic iron wire, and a section of a pipette tip. The new device has demonstrated improved extraction efficiency for the determination of triazine herbicides in environmental water. Huang et al. described a modified form of stir FPSE known as magnetic stir fabric phase sorptive extraction (magnetic stir-FPSE) for the extraction of brominated fire retardants from environmental water, Fig. 13.7 [32]. Extraction was carried out for 15 min at 400 rpm. The device can be reused many times.

### *13.10.2 Stir bar-fabric phase sorptive extraction (stir bar-FPSE)*

Huang et al. developed a stir bar-FPSE, Fig. 13.7, that allows stirring the FPSE membrane during the whole extraction process. Stir bar-FPSE was created by cutting FPSE membranes into the shape of a house, clamped, and fixed with a stir bar. The device requires only 10 min to reach extraction equilibrium. Stir bar-FPSE is recommended for large sample volumes [32].

### *13.10.3 Magnet integrated fabric phase sorptive extraction (MI-FPSE)*

Magnet integrated fabric phase extraction (MI-FPSE) is the newest implementation of FPSE that integrates a Teflon coated bar magnet into an FPSE membrane. The MI-FPSE device is created by stitching two FPSE disks together with a pocket at the center

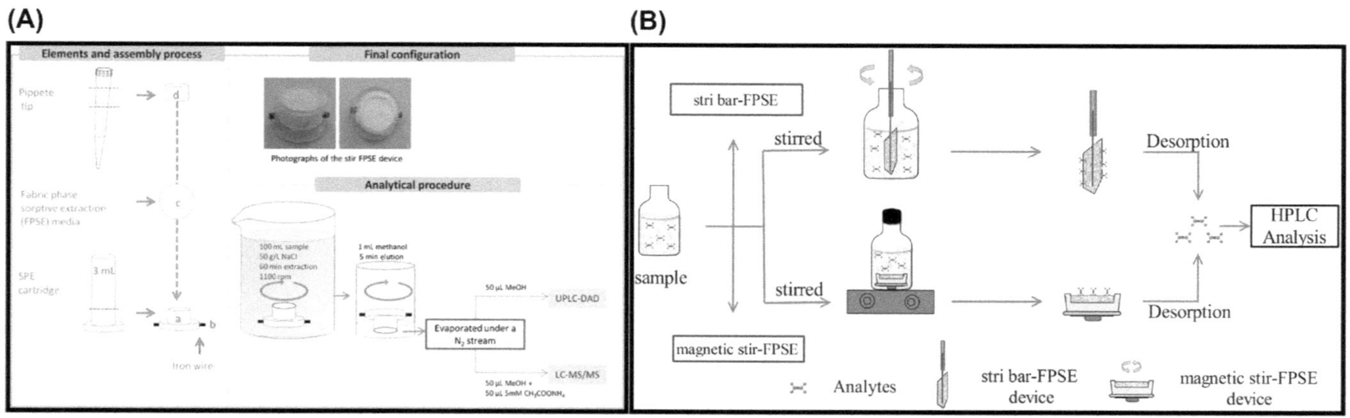


**Figure 13.7** Schematic representation of (A) Stir fabric phase sorptive extraction [32]; (B) Stir bar-fabric phase sorptive extraction [32].

to insert the bar magnet into the device. The magnet can be introduced into the device (during extraction) and can be removed from the device (during elution). Due to the continuous circular motion during extraction, MI-FPSE reduces the equilibrium extraction time with improved reproducibility.

### 13.10.4 Dynamic fabric phase sorptive extraction

Dynamic fabric phase sorptive extraction (DFPSE) mimics solid phase extraction in a disk format. The FPSE membranes can be cut into 47 cm disks and loaded onto the fritted disk of a standard filtration assembly for extraction. Lakade et al. [35] compared the extraction efficiency of FPSE and DFPSE for the extraction of pharmaceutical and personal care products from environmental water samples and observed similar recoveries. However, DFPSE substantially reduces the overall sample preparation time without compromising the extraction efficiency.

### 13.10.5 Fabric phase sorptive extraction interfacing with ion mobility spectrometry

Ion mobility spectrometry (IMS) has become popular in recent years due to its suitability for field deployment, rapid onsite analysis, and portability. However, sampling and sample preparation remain challenging tasks that limit some applications. T. Sun et al. [33] combined IMS with FPSE for sample preparation to mitigate some of these problems. A sol-gel poly(dimethylsiloxane) (sol-gel PDMS) coated fiberglass FPSE membrane was used for sampling polycyclic aromatic hydrocarbons (PAHs) in environmental water (Fig. 13.8). Subsequently, the FPSE membrane was introduced into the IMS inlet for thermal desorption of the extracted analytes. The detection limits for phenanthrene, benzo[a] anthracene, and benzo[a] pyrene were found as 5, 8, and 10 $ngmL^{-1}$, respectively. FPSE-IMS opens up new possibilities for rapid water quality monitoring in the field.

### 13.10.6 Automated online fabric phase sorptive extraction

Automated online fabric phase sorptive extraction was developed by the Anthemedis research group [34,36]. A simple online flow injection (FI) fabric phase sorptive extraction (FPSE) device was built to interface the extraction device to flame atomic absorption spectrometry (FAAS). A sol-gel PDMS sorbent coated polyester FPSE membrane was packed into a microcolumn as the extraction phase for the determination of Pb(II) and Cd(II) in energy and soft drink samples. The new device offered excellent detection limits of 1.6 $\mu gL^{-1}$ and 0.3 $\mu gL^{-1}$ for Pb(II) and Cd(II), respectively.

The same device packed with sol-gel poly(caprolactone-dimethylsiloxane-caprolactone) polyester FPSE membrane was used to monitor the concentration of Cu(II), Ni(II), Zn(II), Pb(II), and Cd(II) in human urine. The precision of the device was $<3.5\%$ for all metals. The detection limits were 0.15, 0.41, 1.62, 0.49 $\mu gL^{-1}$ for Cu(II), Ni(II), Pb(II), and Cd(II), respectively [36] (Fig. 13.8).

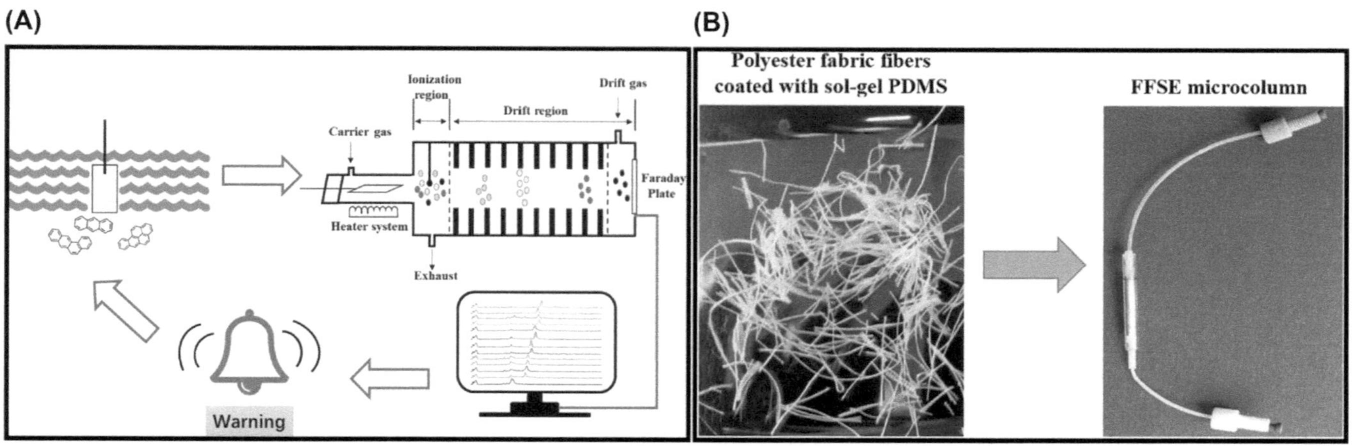


**Figure 13.8** Schematic representation of (A) fabric phase sorptive extraction-ion mobility spectrometry [33]; (B) preparation of microcolumn for FI-FPSE-FAAS [34].

## 13.11 Applications of fabric phase sorptive extraction

FPSE has been used to extract a wide range of samples including environmental water [21,32,33,35,37–44], food [28,45–53], biological samples [26,27,49,54,55], and air [56]. It has also been used for the extraction of metals from different sample matrices [34,36,57–59]. The applications of FPSE in different sample matrices have been reviewed elsewhere [29,30,60,61].

Among all these applications, FPSE has made a profound impact in the analysis of biological samples, especially whole blood. Traditional methods of analysis involve conversion to plasma or serum followed by protein precipitation and centrifugation to remove proteins and other cellular debris prior to further sample preparation for analysis. These processes may incur significant analyte losses. After analyte extraction, elution is carried out with usually relatively large volumes of organic solvent frequently requiring solvent evaporation and sample reconstitution steps, resulting in further analyte loss. FPSE allows extraction of analytes directly from whole blood. The viscosity of the whole blood can be reduced by diluting with deionized water. After the extraction, the FPSE membrane is exposed to a small volume of organic solvent. Elution of the analytes and protein precipitation occur simultaneously at this step. The solvent quantitatively scavenges all the analytes from the FPSE membrane and cellular materials. Finally, the eluant is centrifuged to remove particulate materials prior to chromatographic analysis. Since only a small volume of organic solvent is required for back extraction of analytes, no solvent evaporation or sample reconstitution is needed.

## 13.12 Conclusions

Fabric phase sorptive extraction has emerged as a useful sample preparation technology, which can satisfy the ever-increasing demands of modern analytical laboratories, thanks to its numerous positive attributes compared to conventional sorptive extraction and microextraction techniques. These include a high primary contact surface area, exploitation of the surface chemistry of different fabric substrates, chemically bonded sorbents created via sol-gel sorbent coating technology, spongelike porous architecture of sol-gel sorbents, availability of a large number of sorbents with different selectivity, rapid extraction equilibrium, substantial simplification of the sample preparation workflow with elimination of sample pretreatment and posttreatment steps, field deployability, as well as other favorable attributes are expected to maintain interest in current developments. The high pH stability of FPSE sorbents would simplify analysis of organic acids and bases without major efforts [62].

## References

[1] Kabir A, Furton KG, Malik A. Innovations in sol-gel microextraction phases for solvent-free sample preparation in analytical chemistry. Trac Trends Anal Chem 2013;45:197–218.

[2] Kabir A, Mesa R, Jurmain J, Furton KG. Fabric phase sorptive extraction explained. Sep 2017;4:21.

[3] Wang H, Yan H, Qiu M, Qiao J, Yang G. Determination of dicofol in aquatic products using molecularly imprinted solid-phase extraction coupled with GC-ECD detection. Talanta 2011;85:2100–5.
[4] Varga R, Somogyvári I, Eke Z, Torkos K. Determination of antihypertensive and anti-ulcer agents from surface water with solid-phase extraction–liquid chromatography–electrospray ionization tandem mass spectrometry. Talanta 2011;83:1447–54.
[5] Ribeiro C, Tiritan ME, Rocha E, Rocha MJ. Development and validation of a HPLC-DAD method for determination of several endocrine disrupting compounds in estuarine water. J Liq Chromatogr Relat Technol 2007;30:2729–46.
[6] Abdulra'uf LB, Chai MK, Tan GH. Applications of solid-phase microextraction for the analysis of pesticide residues in fruits and vegetables: a review. J AOAC Int 2012;95: 1272–90.
[7] Andrade-Eiroa A, Canle M, Leroy-Cancellieri V, Cerda V. Solid-phase extraction of organic compounds: a critical review (part I). Trac Trends Anal Chem 2016;80:641–54.
[8] Pawliszyn J, Liu S. Sample introduction for capillary gas chromatography with laser desorption and optical fibers. Anal Chem 1987;59:1475–8.
[9] Pawliszyn J, Lord HL. Handbook of sample preparation. Wiley; 2012.
[10] Pawliszyn J. Handbook of solid phase microextraction. Elsevier; 2011.
[11] Eisert R, Pawliszyn J. Automated in-tube solid-phase microextraction coupled to high-performance liquid chromatography. Anal Chem 1997;69:3140–7.
[12] Baltussen E, Sandra P, David F, Cramers C. Stir bar sorptive extraction (SBSE), a novel extraction technique for aqueous samples: theory and principles. J Microcolumn Sep 1999; 11:737–47.
[13] Abdel-Rehim M. Recent advances in microextraction by packed sorbent for bioanalysis. J Chromatogr A 2010;1217:2569–80.
[14] Richter P, Leiva C, Choque C, Giordano A, Sepulveda B. Rotating-disk sorptive extraction of nonylphenol from water samples. J Chromatogr A 2009;1216:8598–602.
[15] Bruheim I, Liu XC, Pawliszyn J. Thin-film microextraction. Anal Chem 2003;75: 1002–10.
[16] Cardenas S, Lucena R. Recent advances in extraction and stirring integrated techniques. Sep 2017;4:6.
[17] Piri-Moghadam H, Alam MN, Pawliszyn J. Review of geometries and coating materials in solid phase microextraction: opportunities, limitations, and future perspectives. Anal Chim Acta 2017;984:42–65.
[18] Gilart N, Marce RM, Borrull F, Fontanals N. New coatings for stir-bar sorptive extraction of polar emerging organic contaminants. Trac Trends Anal Chem 2014;54:11–23.
[19] Kabir A, Furton KG. Fabric phase sorptive extractors. United States Patents and Trademark Office. 2016.
[20] Kabir A, Furton KG. Fabric phase sorptive extractors (fpse). Google Patents. 2014.
[21] Kumar R, Gaurav H, Malik AK, Kabir A, Furton KG. Efficient analysis of selected estrogens using fabric phase sorptive extraction and high performance liquid chromatography-fluorescence detection. J Chromatogr A 2014;1359:16–25.
[22] Amiri A. Solid-phase microextraction-based sol–gel technique. Trac Trends Anal Chem 2016;75:57–74.
[23] Fumes BH, Silva MR, Andrade FN, Nazario CED, Lancas FM. Recent advances and future trends in new materials for sample preparation. Trac Trends Anal Chem 2015;71: 9–25.
[24] Aziz-Zanjani MO, Mehdinia A. A review on procedures for the preparation of coatings for solid phase microextraction. Microchim Acta 2014;181:1169–90.

[25] Bagheri H, Piri-Moghadam H. Recent advances in capillary microextraction. Trac Trends Anal Chem 2015;73:64–80.
[26] Locatelli M, Kabir A, Innosa D, Lopatriello T, Furton KG. A fabric phase sorptive extraction-high performance liquid chromatography-photo diode array detection method for the determination of twelve azole antimicrobial drug residues in human plasma and urine. J Chromatogr B 2017;1040:192–8.
[27] Kabir A, Furton KG, Tinari N, Grossi L, Innosa D, Macerola D, Tartaglia A, Di Donato V, D'Ovidio C, Locatelli M. Fabric phase sorptive extraction-high performance liquid chromatography-photo diode array detection method for simultaneous monitoring of three inflammatory bowel disease treatment drugs in whole blood, plasma and urine. J Chromatogr B 2018;1084:53–63.
[28] Mesa R, Kabir A, Samanidou V, Furton KG. Simultaneous determination of selected estrogenic endocrine disrupting chemicals and bisphenol A residues in whole milk using fabric phase sorptive extraction coupled to HPLC-UV detection and LC-MS/MS. J Sep Sci 2019;42:598–608.
[29] Kazantzi V, Anthemidis A. Fabric sol–gel phase sorptive extraction technique: a review. Sep 2017;4:20.
[30] Zilfidou E, Kabir A, Furton K, Samanidou V. Fabric phase sorptive extraction: current state of the art and future perspectives. Sep 2018;5:40.
[31] Roldan-Pijuan M, Lucena R, Cardenas S, Valcarcel M, Kabir A, Furton KG. Stir fabric phase sorptive extraction for the determination of triazine herbicides in environmental waters by liquid chromatography. J Chromatogr A 2015;1376:35–45.
[32] Huang G, Dong S, Zhang M, Zhang H, Huang T. Fabric phase sorptive extraction: two practical sample pretreatment techniques for brominated flame retardants in water. Water Res 2016;101:547–54.
[33] Sun T, Wang D, Tang Y, Xing X, Zhuang J, Cheng J, Du Z. Fabric-phase sorptive extraction coupled with ion mobility spectrometry for on-site rapid detection of PAHs in aquatic environment. Talanta 2019;195:109–16.
[34] Kazantzi V, Kabir A, Furton KG, Anthemidis A. Fabric fiber sorbent extraction for on-line toxic metal determination by atomic absorption spectrometry: determination of lead and cadmium in energy and soft drinks. Microchem J 2018;137:285–91.
[35] Lakade SS, Borrull F, Furton KG, Kabir A, Marce RM, Fontanals N. Dynamic fabric phase sorptive extraction for a group of pharmaceuticals and personal care products from environmental waters. J Chromatogr A 2015;1456:19–26.
[36] Kazantzi V, Samanidou V, Kabir A, Furton K, Anthemidis A. On-line fabric disk sorptive extraction via a flow preconcentration platform coupled with atomic absorption spectrometry for the determination of essential and toxic elements in biological samples. Sep 2018;5:34.
[37] Kumar R, Gaurav, Kabir A, Furton KG, Malik AK. Development of a fabric phase sorptive extraction with high-performance liquid chromatography and ultraviolet detection method for the analysis of alkyl phenols in environmental samples. J Sep Sci 2015;38:3228–38.
[38] Lakade SS, Borrull F, Furton KG, Kabir A, Fontanals N, Maria Marcé R. Comparative study of different fabric phase sorptive extraction sorbents to determine emerging contaminants from environmental water using liquid chromatography-tandem mass spectrometry. Talanta 2015;144:1342–51.
[39] Montesdeoca-Esponda S, Sosa-Ferrera Z, Kabir A, Furton KG, Santana-Rodriguez JJ. Fabric phase sorptive extraction followed by UHPLC-MS/MS for the analysis of benzotriazole UV stabilizers in sewage samples. Anal Bioanal Chem 2015;407:8137–50.

[40] Racamonde I, Rodil R, Quintana JB, Sieira BJ, Kabir A, Furton KG, Cela R. Fabric phase sorptive extraction: a new sorptive microextraction technique for the determination of non-steroidal anti-inflammatory drugs from environmental water samples. Anal Chim Acta 2015;865:22–30.

[41] García-Guerra RB, Montesdeoca-Esponda S, Sosa-Ferrera Z, Kabir A, Furton KG, Santana-Rodríguez JJ. Rapid monitoring of residual UV-stabilizers in seawater samples from beaches using fabric phase sorptive extraction and UHPLC-MS/MS. Chemosphere 2016;164:201–7.

[42] Guedes-Alonso R, Ciofi L, Sosa-Ferrera Z, Juan Santana-Rodriguez J, del Bubba M, Kabir A, Furton KG. Determination of androgens and progestogens in environmental and biological samples using fabric phase sorptive extraction coupled to ultra-high performance liquid chromatography tandem mass spectrometry. J Chromatogr A 2016;1437:116–26.

[43] Saini S, Kabir A, Rao A, Malik A, Furton K. A novel protocol to monitor trace levels of selected polycyclic aromatic hydrocarbons in environmental water using fabric phase sorptive extraction followed by high performance liquid chromatography-fluorescence detection. Sep 2017;4:22.

[44] Santana-Viera S, Guedes-Alonso R, Sosa-Ferrera Z, Santana-Rodríguez JJ, Kabir A, Furton KG. Optimization and application of fabric phase sorptive extraction coupled to ultra-high performance liquid chromatography tandem mass spectrometry for the determination of cytostatic drug residues in environmental waters. J Chromatogr A 2017;1529:39–49.

[45] Samanidou V, Galanopoulos L-D, Kabir A, Furton KG. Fast extraction of amphenicols residues from raw milk using novel fabric phase sorptive extraction followed by high-performance liquid chromatography-diode array detection. Anal Chim Acta 2015;855:41–50.

[46] Aznar M, Alfaro P, Nerin C, Kabir A, Furton KG. Fabric phase sorptive extraction: an innovative sample preparation approach applied to the analysis of specific migration from food packaging. Anal Chim Acta 2016;936:97–107.

[47] Karageorgou E, Manousi N, Samanidou V, Kabir A, Furton KG. Fabric phase sorptive extraction for the fast isolation of sulfonamides residues from raw milk followed by high performance liquid chromatography with ultraviolet detection. Food Chem 2016;196:428–36.

[48] Aznar M, Úbeda S, Nerin C, Kabir A, Furton KG. Fabric phase sorptive extraction as a reliable tool for rapid screening and detection of freshness markers in oranges. J Chromatogr A 2017;1500:32–42.

[49] Samanidou V, Filippou O, Marinou E, Kabir A, Furton KG. Sol–gel-graphene-based fabric-phase sorptive extraction for cow and human breast milk sample cleanup for screening bisphenol A and residual dental restorative material before analysis by HPLC with diode array detection. J Sep Sci 2017;40:2612–9.

[50] Samanidou V, Michaelidou K, Kabir A, Furton KG. Fabric phase sorptive extraction of selected penicillin antibiotic residues from intact milk followed by high performance liquid chromatography with diode array detection. Food Chem 2017;224:131–8.

[51] Pérez-Mayán L, Rodríguez I, Ramil M, Kabir A, Furton KG, Cela R. Fabric phase sorptive extraction followed by ultra-performance liquid chromatography-tandem mass spectrometry for the determination of fungicides and insecticides in wine. J Chromatogr A 2019;1584:13–23.

[52] Yang M, Gu Y, Wu X, Xi X, Yang X, Zhou W, Zeng H, Zhang S, Lu R, Gao H, Li J. Rapid analysis of fungicides in tea infusions using ionic liquid immobilized fabric phase sorptive

extraction with the assistance of surfactant fungicides analysis using IL-FPSE assisted with surfactant. Food Chem 2018;239:797–805.
[53] Kaur R, Kaur R, Rani S, Malik AK, Kabir A, Furton KG. Application of fabric phase sorptive extraction with gas chromatography and mass spectrometry for the determination of organophosphorus pesticides in selected vegetable samples. J Sep Sci 2019;42:862–70.
[54] Samanidou V, Kaltzi I, Kabir A, Furton KG. Simplifying sample preparation using fabric phase sorptive extraction technique for the determination of benzodiazepines in blood serum by high-performance liquid chromatography. Biomed Chromatogr 2016;30: 829–36.
[55] Locatelli M, Tinari N, Grassadonia A, Tartaglia A, Macerola D, Piccolantonio S, Sperandio E, D'Ovidio C, Carradori S, Ulusoy HI, Furton KG, Kabir A. FPSE-HPLC-DAD method for the quantification of anticancer drugs in human whole blood, plasma, and urine. J Chromatogr B 2018;1095:204–13.
[56] Alcudia-León MC, Lucena R, Cárdenas S, Valcárcel M, Kabir A, Furton KG. Integrated sampling and analysis unit for the determination of sexual pheromones in environmental air using fabric phase sorptive extraction and headspace-gas chromatography–mass spectrometry. J Chromatogr A 2017;1488:17–25.
[57] Heena, Kaur R, Rani S, Malik AK, Kabir A, Furton KG. Determination of cobalt(II), nickel(II) and palladium(II) Ions via fabric phase sorptive extraction in combination with high-performance liquid chromatography-UV detection. Separ Sci Technol 2017;52: 81–90.
[58] Anthemidis A, Kazantzi V, Samanidou V, Kabir A, Furton KG. An automated flow injection system for metal determination by flame atomic absorption spectrometry involving on-line fabric disk sorptive extraction technique. Talanta 2016;156–157:64–70.
[59] Rekhi H, Kaur R, Rani S, Malik AK, Kabir A, Furton KG. Direct rapid determination of trace aluminum in various water samples with quercetin by reverse phase high-performance liquid chromatography based on fabric phase sorptive extraction technique. J Chromatogr Sci 2018;56:452–60.
[60] Montesdeoca-Esponda S, Guedes-Alonso R, Santana-Viera S, Sosa-Ferrera Z, Santana-Rodríguez J. Applications of fabric phase sorptive extraction to the determination of micropollutants in liquid samples. Sep 2018;5:35.
[61] Kabir A, Locatelli M, Ulusoy H. Recent trends in microextraction techniques employed in analytical and bioanalytical sample preparation. Sep 2017;4:36.
[62] Qiu XZ, Wu Y, Chen DZ, Lai LY, Liang Y, Guo HS. Fabrication of novel stir bar sorptive extraction coating based on magnetic molecularly imprinted polymer through atom transfer radical polymerization for trace analysis of estrogens in milk. J Nanosci Nanotechnol 2016; 16:12374–81.

# In-tube solid-phase microextraction

# 14

*Y. Moliner-Martinez, A. Ballester-Caudet, J. Verdú-Andrés, R. Herráez-Hernández, C. Molins-Legua, P. Campíns-Falcó*
Miniaturization and Total Analysis Methods (MINTOTA) Research Group, Departament de Química Analítica, Facultat de Química. Universitat of València, Valencia, Spain

## 14.1 Introduction

Sample preparation is still a critical step in many analyses. It is estimated that about 65% of the total of samples analyzed require three or more operations in order to obtain a suitable instrumental response. In addition, sample preparation often involves the consumption of large amounts of solvents, reagents, materials, and energy. Therefore, it is a goal to decrease the time required for this step of the analysis and to achieve it in a sustainable manner. In this sense, in-tube solid-phase microextraction (IT-SPME) is an attractive technique. The potential of IT-SPME arises mainly from the online coupling to liquid chromatography (LC) combining the advantages of SPME with the online separation and detection of target compounds. Initially, IT-SPME was coupled with conventional high-performance liquid chromatography (HPLC). More recently, research aimed at developing the coupling of IT-SPME with ultrahigh performance liquid chromatography (UHPLC), capillary liquid chromatography (Cap-LC), and Nano-liquid chromatography (Nano-LC) are worthy of mention [1]. Although the sample dilution ratio for Cap-LC or Nano-LC is lower than that achieved in conventional HPLC, sensitivity for several applications is unsatisfactory mainly due to the low sample volume utilization. Moreover, in many cases, samples cannot be directly analyzed and require a clean-up step. The miniaturization of analytical systems in order to achieve exhaustive and real information from samples, in as short as possible time while taking account of environmental and economic issues, is one of the main goals in modern analytical chemistry.

The database ISI Web of Science with the keyword "in-tube solid-phase microextraction" indicated around 300 entries at the end of 2018, with an H index of 58 (representing about 20% of the published papers) and nearly 10,000 citations. The first report was published in 1997, by R. Eisert and J. Pawliszyn [2] and the number of citations has increased linearly since then with more than 800 citations in 2018. Table 14.1 summarizes the topics covered in the 27 review articles and the three books published up to the end of 2018. The trends in online IT-SPME are mainly focused on three scenarios as derived from the topics shown in Table 14.1: (1) the preparation of new extractive phases including the possibility of applying several

**Solid-Phase Extraction. https://doi.org/10.1016/B978-0-12-816906-3.00014-5**

**Table 14.1** Covered topics of the in-tube solid-phase microextraction in the 27 review articles and three books published. Database web of science (access January 2019).

| Covered topic | References |
|---|---|
| Current advances about configurations, optimization, and applications of the technique | [3] |
| Characterization of metallic nanoparticles using plasmonic measurements | [4] |
| Current status and future trends on automated techniques | [5] |
| Recent advances of online coupling of sample preparation techniques | [6] |
| The role of graphene-based sorbents | [7] |
| Different configurations of carbon nanotubes reinforced solid-phase microextraction techniques. | [8] |
| Trends of the technique: configurations, extractive phases, online applications, and figures of merit. | [9] |
| Strengths and weaknesses of the technique | [10] |
| Advances of the technique in configurations, chromatographic coupling, and capillary coatings | [1] |
| Online analysis of residues and contaminants in food matrices | [11] |
| Organic micropollutants in environmental water samples | [12] |
| Superparamagnetic silica nanoparticles for extraction of triazines | [13] |
| Capillary coating materials to determine drugs and biomarkers in biological samples | [14] |
| Advances in column switching sample preparation in bioanalysis | [15] |
| Steroid hormones in biological and environmental samples | [16] |
| Veterinary residues in food matrices by porous monolith microextraction-based techniques | [17] |
| Advances in biomedical analysis | [18] |
| Monolith as a new sample preparation material | [19] |
| The roles of ionic liquids in sorptive microextraction techniques | [20] |
| In-tube solid-phase microextraction coupled by in valve mode to capillary LC-DAD | [21] |
| Developments and applications in drug analysis | [22] |
| Sorptive microextraction for liquid-chromatographic determination of drugs in urine | [23] |
| New tendencies in liquid chromatography for controlling PAHs pollution | [24] |
| Developments and applications of capillary microextraction techniques | [25] |
| Application of solid-phase microextraction in analytical toxicology | [26] |

**Table 14.1** Covered topics of the in-tube solid-phase microextraction in the 27 review articles and three books published. Database web of science (access January 2019).—cont'd

| Covered topic | References |
|---|---|
| Analysis of herbicides in water by online in-tube solid-phase microextraction coupled with liquid chromatography-mass spectrometry | [27] |
| Pharmaceutical and biomedical analysis | [28] |
| Analysis of drugs in biological samples | [29] |
| Drug analysis | [30] |
| Automated sample preparation using in-tube solid-phase microextraction | [31] |

external forces (magnetic, electric, thermic), (2) the coupling of IT-SPME with new chromatographic modalities such as miniaturized liquid chromatography, and (3) new uses and utilities.

## 14.2 Theoretical considerations

Microextraction refers to the volume or mass of the extracting phase, which can be placed on different formats such as fibers, filters, capillary coating as in IT-SPME or other surfaces, or can even be suspended in the sample. If the extracting phase is a solid or semi-solid polymeric material, the technique is called *solid-phase microextraction* (SPME). IT-SPME is based typically on the use of a fused-silica capillary tube packed or coated on its inner surface with an extractive phase (see Fig. 14.1). When the sample is passed through the capillary, the analytes are extracted and concentrated by adsorption/absorption onto or into the internal coating of the capillary. Then, the

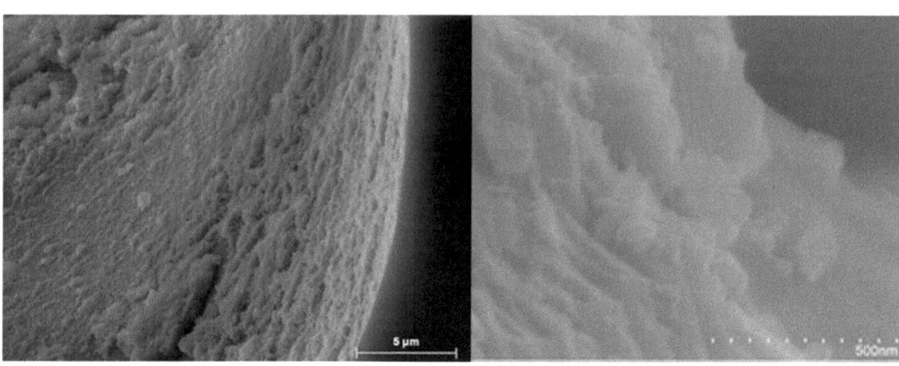


**Figure 14.1** SEM micrographs of the surface of a segment of commercial gas chromatography column of TRB-5 (0.32 mm i.d. and 3 μm of phase thickness). obtained with a HITACHI-S4800 operating at 20 kV. For measuring samples, Au/Pd coating was required.

extracted analytes are desorbed by filling the capillary with a solvent, which can be collected for further processing or can be transferred to the analytical instrument (static desorption), or by means of mobile phase (dynamic desorption). The static mode is mainly applied when analytes have strong interactions with the extraction phase and cannot be desorbed easily by the mobile phase or for applications as an off-line technique. The volume or amount of sample processed is highly dependent on the extracting phase, and generally, the extraction of the analytes is not exhaustive [9].

Fig. 14.1 shows the SEM micrograph of the surface of a segment of a gas chromatography column TRB-5 (coated with 95% dimethyl-5% diphenylpolysiloxane, 0.32 mm i.d., and 3 μm film thickness). A homogeneous coating is formed inside the capillary column. Its porosity facilitates diffusion of the analyte inside the extractive phase contributing to the extraction efficiency. Among the parameters affecting the extraction, one of the most important is the nature of the extractive phase, since it mainly determines the extraction efficiency. Undoubtedly, other experimental parameters, such as processed sample volume, pH, phase thickness, length and diameter of the IT-SPME capillary, and washing solvent, must be optimized in order to achieve the greatest benefits from this technique. It should be noted that sensitivity is influenced not only by the above parameters but also by the coupling with a suitable chromatographic system. Although this technique has been used in combination with gas chromatography (GC), liquid chromatography (LC) is by far the most used system in IT-SPME-based assays (Table 14.1).

Typically, commercial GC columns (Fig. 14.1) have been used to perform IT-SPME, and the sorbents mainly used can be divided into (1) Si-based extractive phases and (2) C-based extractive phases. 77% of the studies are carried out with sorbents [9]. The main limitation of conventionally coated capillary surfaces is the low extraction efficiency. This can be attributed to the low sorbent loading capacity, stability, and in some cases, the long extraction times required due to the slow diffusion of the analytes from the sample to the capillary coating. Therefore, continuous effort to find more efficient capillary coatings and explore new extractive phases is a major topic of IT-SPME (see Table 14.1). In [9] it is shown that different capillary lengths (between 2 and 80 cm) and internal diameters (i.d.) between 0.025 and 0.8 mm have been proposed for IT-SPME, which influence the time for online sample treatment between 3 and 105 min. The optimum i.d. depends on the coupling with the LC mode, whereas column length is not a critical parameter. The IT-SPME capillary i.d. for LC coupling needs to be smaller in the order nano, capillary, and conventional LC. There are two main modes to carry out this coupling, known as the draw/eject and flow-through modes, discussed in Section 14.4.1.

Clean-up, preconcentration, separation, and detection of analytes can be carried out online by the coupling IT-SPME-LC. Fig. 14.2 shows an example of the optimization of preconcentration time and elution solvent of the analytes in the IT-SPME technique. In order to study the IT-SPME retention of a sample, in this case, a mixture of two analytes (4-tert-octylphenol and technical nonylphenol), the samples were processed without a separation column. Acetonitrile-water (85:15, v/v) was

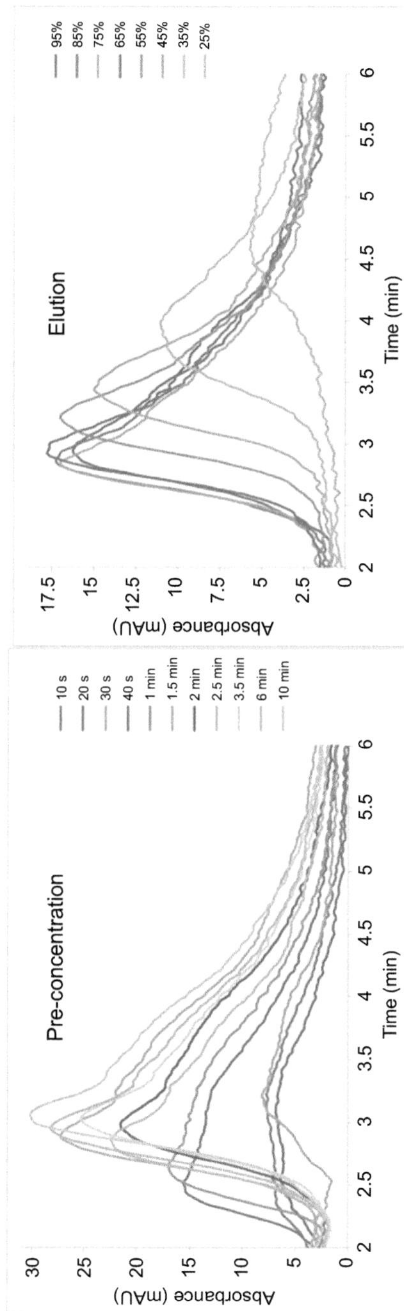


**Figure 14.2** Preconcentration and elution studies of the IT-SPME step of a mixture of 4-tert-Octylphenol and Technical Nonylphenol at 100 μg/L level. For more details see text.

used as the mobile phase. The sample volume was 8 μL of an aqueous solution of the target compounds. Samples were passed several times through the IT-SPME capillary for extraction at 50 μL/min and afterward, the flow was sent to 2.5 μL/min for online UV detection. The retention time of the mixture was between 2 and 3 min, a function of the preconcentration time, as can be seen in Fig. 14.2. The sample is observed as a broad peak in Fig. 14.2 because no analytical column was employed and the IT-SPME capillary coated with TEOS-MTEOS-$SiO_2$NPs (15 cm, i.d. 0.075 mm) prepared as described in [32] is open (not totally packed). After IT-SPME preconcentration, different mixtures of acetonitrile/water were evaluated as mobile-phases for elution of the samples, also without a separation column. Preconcentration times greater than 3 min achieved the best results for extraction and mobile phases containing more than 55% (v/v) acetonitrile for elution (see Fig. 14.2).

In addition, IT-SPME can be considered an environmentally friendly technique. The sustainability of an analytical method is governed by minimizing the use of toxic solvents, reducing solvent waste, and utilizing energy-efficient and cost-effective methodologies, while maintaining the reliability of the performance parameters, such as sensitivity, precision, and accuracy [33,34]. Previously our group [35,36] demonstrated that IT-SPME-CapLC-DAD achieves miniaturization of the sample pretreatment step, reduces the analysis time and wastes, with favorable costs, and thus, improvement of the analytical and environmental performance.

## 14.3 Preparation of designed capillary columns

The evolution of IT-SPME is directly linked to the development of new extractive phases. In recent years, extractive phases based on new nanomaterials have proven to be of considerable interest. Another key development is the use of physical stimuli, such as electrical or magnetic fields in advanced nanomaterials to improve the features of IT-SPME.

The practical use of these new extractive phases is not only related to improvements of the extraction efficiency but also the stability and robustness of the extractive phases deposited inside of the capillary columns. This is directly dependent on the synthetic route followed to fabricate the capillary column. IT-SPME can be classified in four modalities on the basis of how the extraction phase is localized inside the capillary column: (1) Surface coated capillary columns, (2) Monolithic capillary columns, (3) Fiber packed capillary columns (fiber-IT-SPME) and (4) Packed sorbent capillary columns (packed IT-SPME). Fig. 14.3 shows a schematic representation of the frontal view of a capillary column or tube used for each modality.

In each modality, the fabrication depends on the support capillary column in which the new extractive phase will be deposited, the synthetic route, and on the physical-chemical properties of the extraction phase, which leads to the different coating modalities. Fig. 14.4 shows a schematic diagram of the different modalities used to

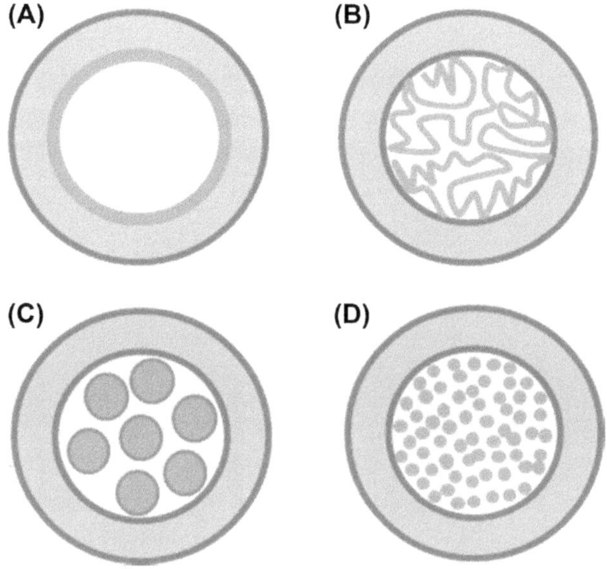


**Figure 14.3** Schematic representation of capillary columns (A) surface modified column, (B) Monolithic capillary column, (C) Fiber packed capillary column and (D) Sorbent packed capillary column.

fabricate capillary columns, the different synthetic routes developed for each modality, and the support capillary columns employed in each case.

Surface coated capillary columns are the most widely used devices to perform IT-SPME. The extractive phase is applied as a layer on the inner wall of the capillary column synthesized by one of several procedures, mainly: (1) polymerization, (2) electrodeposition and (3) direct surface modification (see Fig. 14.4).

### 14.3.1 Polymerization

Polymerization, especially thermal polymerization, is one of the main procedures to immobilize a layer of extraction phase on the inner surface of the capillary column to perform IT-SPME. The synthesis of these capillary columns depends on the support; with fused-silica capillary columns being most widely used. Their fabrication starts with the activation of silanol groups by means of NaOH and/or HCl followed by a drying step with $N_2$. Then, the polymeric coating is directly deposited on the surface of the capillary column. Poly(ethylene glycol) (PEG)/tetraethylorthosilicate (TEOS)/ trimethoxymethylsilane (MTEOS)-based extraction phases can be directly deposited on the surface of the capillary column. For this purpose, the column is filled with an optimized mixture of the compounds, and the capillary ends sealed with silicone septa. After which, the coating is deposited by a thermal treatment that depends on the selection of the functional monomers [9,32].

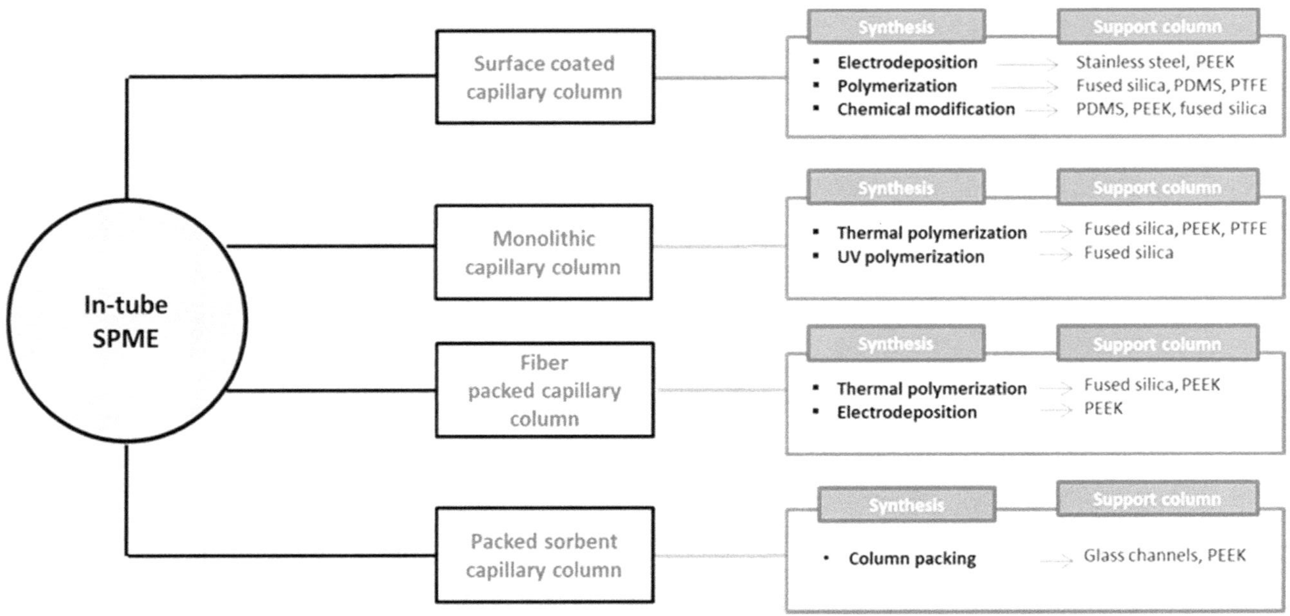


**Figure 14.4** Diagram of the different modalities to fabricate capillary columns, different synthetic routes developed for each modality and the support columns.

Magnetic coatings have also been proposed by our group [37,38], in order to take advantage of the magnetic fields to improve the extraction efficiency. In this case, the coating is synthesized by using $Fe_3O_4$ NPs (CTAB) in the polymerization mixture (PEG and acetic acid). This solution was mixed with an $Fe_3O_4$-CTAB water dispersion and TEOS, which polymerizes to give rise to a silica matrix with embedded $Fe_3O_4$-CTAB micelles. This mixture was stirred until it became a homogeneous gel, which was injected into the fused-silica capillary. The ends of the capillary were sealed and placed in an oven for 20 h at 43°C. Then, the temperature was raised to 120°C at 0.5°C/min and kept constant for 2 h, followed by water and methanol washes. After drying, the capillary was heated at 330°C for 25 h [37,38].

A similar strategy was used to fabricate a boronate-decorated polyethyleneimine (PEI) grafted coating. In this case, after the treatment with PEG and TEOS, polyethyleneimine can be immobilized and modified by boronate groups. This synthesis includes surface modification with epoxy groups, graft with PEI, and functionalization with 4-formylphenylboronic acid (4-FPBA) [39]. Another possibility is to modify the activated surface of a fused-silica capillary column by treatment with vinyltrimethoxysilane before filling the capillary column with the polymeric mixture. This strategy was used to fabricate capillary coatings based on imprinted polymers or ionic liquid [40,41].

In addition, polytetrafluoroethylene (PTFE) columns have also been proposed in recent years. In this case, the surface of the tube is conditioned with methanol, water, and dried with $N_2$. After which, diphenylamine (DPA) is used to introduce $-NH_2$ groups on the surface of the capillary column. Finally, a mixture of dialdehyde and chitosan is used to synthesize the coating after thermal treatment (80°C for 24 h). This extraction phase was used for the extraction of two liver biomarkers (hexanal and 2-butanone) in human blood [42]. Table 14.2 summarizes the new capillary coatings and the general conditions employed for their synthesis.

### 14.3.2 Electrodeposition

Electrodeposition is used to deposit conducting polymers on the inner wall of the capillary column. These coatings are used for electrochemically controlled IT-SPME (EC-IT-SPME). These coatings are utilized to improve the extraction efficiency by taking advantage of the multifunctional properties of the conducting polymers, such as hydrophobicity, acid-base character, $\pi$-$\pi$ interactions, interactions with polar groups, ion-exchange properties, and electroactivity [43]. The deposition of the conducting polymer coating is mainly carried out by electrochemical synthesis using an aqueous solution of electrolytes. In general, the extraction mechanism is based on the ion exchange properties of the extraction phases in the presence of an electric field. The application of voltage induces the reduction or oxidation of the polymer creating charged sites. These charged sites interact with cations or anions in the sample, enhancing the extraction efficiency. The analytes are desorbed by applying an opposite voltage to that used for extraction.

Typically, the support used is a stainless steel capillary column internally coated with polypyrrole (PPY) or its derivatives, or PPY functionalized with different

**Table 14.2** Summary of conditions to develop different surface coated IT-SPME capillary columns.

| | Support capillary | | | | | |
|---|---|---|---|---|---|---|
| | **TEOS/MTEOS** | **PEG-TEOS $Fe_3O_4$(CTAB)** | **MIP-RAM** | **PIL** | **Boronate decorated PEI** | **PDA/DAS/CHI** |
| Capillary support | Fused silica | Fused silica | Fused silica | Fused silica | Fused silica | PTFE |
| Activation | NaOH/HCl | NaOH/HCl | NaOH/HCl | NaOH/HCl | NaOH/HCl | MeOH |
| Previous modification | – | – | 3-MAPS | VTMS | PEG | DPA |
| Polymeric mixture | $SiO_2$NPs-PEG TEOS/MTEOS ($NH_4OH$) | $Fe_3O_4$ (CTAB)-PEG-TEOS (HAc) | MIP EGMA AIBN | IL-monomer[a] AIBN | GTPE/PEI 4-FPBA | DAS/CHI |
| Thermal treatment | 40°C 3h 120°C 15 h | 43°C 20 h 120°C (0,5°C/min) 2h 33°C25 h | 50°C 4h | 50°C 24 h | 65°C 5h | 80°C 24h Drying 4′°C 3h |

[a] 1-vinyl-3-hexylimidazolium chloride and 1-vinyl-3-hexadecylimidazolium

compounds. Recently, molecularly imprinted polymers (MIPs) [44], deep eutectic solvents [45], or graphene oxide [46], among others, have been used to improve the extraction properties of these conducting polymers.

The synthesis of these conducting coatings is carried out in the following way. The stainless steel capillary column is initially cleaned with solvents such as acetone and water. After drying, the capillary column is filled with the coating solution for the electrodeposition. The stainless steel tube is used as a working electrode; and Pt and Ag/AgCl electrodes or copper wire are employed as the counter and reference electrodes, respectively. Finally, the electrodeposition is performed by passing the solution through the tube and applying voltage for the optimum time. Voltages from 0.9 to 3 V have been applied and deposition times from 5 to 30 min used in the most recent procedures. Cyclic voltamperometry was also used with potentials between $-1.0$ and $+1.0$ V during cycles [45]. The IT-SPME device can be used after a washing step (methanol-water) and drying with $N_2$.

### 14.3.3 Direct chemical modification

Direct chemical modification is carried out by activation of the inner wall of the capillary column with functional groups in order to covalently bond compounds that can enhance the extraction capability of the capillary columns. With this aim, fused silica, PDMS and PEEK capillary columns have been chemically modified. The synthetic route is similar in all cases: activation of the capillary inner surface, replenishment of the capillary column, thermal treatment and, if necessary, a washing step.

Fused-silica capillary columns are activated with NaOH and or HCl treatment. The introduction of amino groups can be performed by using APTS (3-aminopropyl(trimethoxysilane)) for bonding with aldehyde groups via glutaraldehyde. By using this route, several nanoparticles or nanostructures were immobilized on the inner capillary surface. This is the case of PS/IgG nanoparticles and PGMA NPs. Thermal treatment was only employed in the case of PGMA NPs [47]. PDMS capillary columns have also been functionalized with carbon nanotubes by this procedure [32].

PEEK is also used as a substrate for chemical modification. The procedure starts with an activation of the inner surface with methanol, followed by a dry cycle. Then, PDA was deposited on the surface. In this case, the formation of PDA is based on the oxidation of catechol under alkaline condition and further oxidation and polymerized to form a highly crosslinked homopolymer. Then, ZIF-8 metal-organic framework was grown on the surface of the capillary column [48].

### 14.3.4 IT-SPME monolithic capillary columns

Nowadays, monolithic capillary columns are one of the most active areas of IT-SPME research. These extraction phases possess several advantages such as the high permeability, rapid mass transfer, high stability, high loading capacity, and the availability of a wide variety of surface chemistry [49]. The possibility of fabricating hydrophilic extraction phases for polar analytes is an added advantage of these capillary columns.

**Table 14.3** Experimental data to fabricate monolithic capillary columns for in-tube SPME.

| | Support capillary | | |
|---|---|---|---|
| | **Fused silica** | **PEEK** | **PTFE** |
| Activation | NaOH/HCl<br>Water<br>$N_2$ stream | MeOH<br>$N_2$ stream | MeOH<br>$N_2$ stream |
| Previous capillary functionalization | γ-MPS<br>γ-MAPS<br>VTMS | Dopamine<br>γ-MAPS | – |
| Polymeric mixture | Mixture: monomer, crosslinker, initiator, porogenic solvent | | |
| Polymerization | Thermal treatment<br>UV-treatment | Thermal treatment | Thermal treatment |
| Washing and drying | Methanol/acetonitrile | | |

*γ-MAPS,* γ -(trimethoxysilyl)propyl methacrylate; *γ-MPS,* 3-(methacryloxy)propyl-trimethoxysilane; *VTMS,* vinyltrimethoxysilane.

Thermal polymerization and UV-polymerization are the most commonly used procedures to synthesize monolithic capillary columns. The success of the synthesis depends not only on monolith formation but also on the anchorage of the monolith to the capillary column. Therefore, the fabrication of the monolithic capillary column can be divided into five steps: activation, capillary column surface modification, injection of the polymeric mixture into the capillary column, thermal or UV polymerization, and washing and drying.

Table 14.3 summarizes the main experimental approaches for the fabrication of monolithic capillary columns for IT-SPME. In fused-silica capillary columns, the first step is to activate the silanol groups of the capillary wall, afterward, the capillary walls are modified with different reagents (γ-MPS; γ-MAPS or VTMS) in order to achieve the covalent binding of the polymeric material to the inner capillary wall [50–52]. PEEK and PTFE capillary columns are activated by rinsing with methanol followed by functionalization with DPA or MAPS [49].

The monolithic capillary column can be prepared by using different polymerization mixtures. Monomers commonly used are choline chloride, butylmethacrylate (BMA), acrylamide (AM), divinylbenzene (DVB), methacrylic acid (MA). Ethylene glycol dimethacrylate (EGDMA) and 2,2-azobis(2-methylpropionitrile) (AIBN) as crosslinking agent and initiator, respectively, are also widely used. However, PEG [51], sodium hyaluronate-functionalized urea-formaldehyde monoliths has also been proposed [52]. Finally, porogenic solvents such as isopropanol, butanediol, toluene, methanol, and isooctane are needed to obtain the final monolithic structure. Moreover, additional materials, such as graphene oxide [53], magnetic nanoparticles [50], and molecular organic frameworks [54] can be employed to improve the extraction capability of conventional monolithic capillary columns.

Thermal treatment depends on the polymerization mixture used to fabricate the monolithic columns, typically temperatures up to 80°C with times from 3 to 24 h are used. In the case of UV-treatment, the capillary column is subjected to UV radiation for shorter times, i.e., 25 min.

### 14.3.5 Fiber IT-SPME

Fiber IT-SPME is based on the use of SPME fibers located inside the extraction capillary column. The main advantages of this modality are the reduction of peak broadening by minimizing the dead volume. In this case, the extraction phase is the fiber, and the capillary column only serves as support [55]. Commercially available fibers have been proposed for fiber-IT-SPME; however, recent advances are based on the modification of different fibers with nanomaterials in order to improve the extraction efficiency. This improvement is related to a large number of fibers that can be placed inside the capillary column. The synthetic route to prepare these fibers depends on the fiber material. Common procedures include (1) activation, (2) preparation of the coating solution and (3) fiber treatment to deposit the coating on the surface of the fiber. Copper wires in copper tubes, basalt fibers, and carbon fibers in PEEK tubes are the most recent supports used for fiber IT-SPME.

Copper fibers are activated by etching with nitric acid to increase the surface area and remove the oxide layer. Free radical polymerization reactions and crosslinked poly(ionic liquids) coated onto the copper fiber are used as the extraction phase [56]. Basalt fibers are first cleaned by sonication in water and ethanol. They are subsequently activated by oxidation (e.g., $H_2O_2$) followed by treatment with $H_2SO_4$ and finally drying with nitrogen. Afterward, different coatings can be deposited onto the fiber surface. A PANI coating formed via in situ hydrothermal treatment was used to extract UV filters from water samples [57]. The same basalt fibers were coated with silica aerogels by polymerization of the aldehyde groups of p-phthalaldehyde and amino groups of 3- aminopropyl(trimethoxysilane) with the hydrolysis of the methoxy groups and dehydration of silanol groups on the fiber to form Si—O—Si bonds. After the thermal treatment, the aerogel coated basalt fibers were packed inside a PEEK tube [58]. A melamine-formaldehyde coating has also been prepared on basalt fibers. The synthesis, in this case, was performed by introducing the fiber in a solution of melamine, sodium hydroxide, and formaldehyde at acid pH. The aerogel was formed by thermal treatment at 80° for 2h forming a wet gel, with the final coating obtained by heating the fiber at 40°, 50°, and 60°C for 12 h [59].

Carbon fibers have been used as extraction phase for fiber-IT-SPME after treatment with $HNO_3$ and packing into a PEEK tube [60]. More recently, a procedure to immobilize MOFs on the surface of carbon fibers was proposed. This synthesis is based on the growth of zeolitic imidazolate frameworks (ZIFs) on an electrochemically deposited ZnO layer on the carbon fibers [61]. Table 14.4 summarizes the thermal treatment of various fibers coated with different extraction phases; however, the optimization of the synthetic parameters is a key point in order to ensure the durability and stability of the extraction phase.

**Table 14.4** Summary of thermal treatments used for the functionalization of different fiber for fiber in-tube SPME.

| Fiber | Coating | Thermal treatment |
|---|---|---|
| Copper fibers | ILs | 70°C/24 h |
| Basalt fibers | PANI | 130°C/3h (functionalization)<br>50°C/10h (drying) |
| | Silica aerogel | 40°C/20h (−CHO−$NH_2$ polymerization)<br>80°C/24 h (aerogel formation) |
| | Melamine-formaldehyde aerogel | 80°C/2h (sol-gel reaction)<br>40, 50, 60°C/12h (aerogel formation) |
| Carbon fibers | ZIFs | 60°C/30 min(nucleation sites)<br>120°C/48h (crystal growth) |

### 14.3.6 Packed IT-SPME

For packed IT-SPME, the extraction phase is packed inside the capillary columns. Recent applications used PEEK capillary columns for C18 packing [62] and glass substrates for $TiO_2$ packing [63]. In both cases, slurry packing was employed. Once the extraction phase is deposited inside the capillary column, it is capped with frits. In the case of IT-SPME microchips, monolithic frits must be prepared by UV-polymerization of vinyl-functionalized microchannels. This modality is especially attractive for the further development of IT-SPME microchips.

## 14.4 Online coupling to liquid chromatography

In general terms, IT-SPME can be performed by two main operational modes, draw/eject and flow-through IT-SPME, each with advantages and limitations [1,9]. In the draw/eject mode, the sample is passed through the capillary repeatedly by reversing the flow direction of the sample (draw/eject) by means of a programmable autosampler; after a number of cycles, the analytes are desorbed and injected onto the separation column. In flow-through systems, the sample is passed through the extraction capillary once without flow reversion. The analytes are then desorbed from the extraction phase by filling the capillary with a desorption solvent (static desorption) before transferring them to the chromatographic column, or by flushing the capillary by mobile phase (dynamic desorption). In flow-through systems, the extraction capillary is connected directly or indirectly to the separation column, so that the extracted compounds are desorbed and then sent to the separation column by means of a switching arrangement. Different combinations of pumps, switching valves, or external accessories (for creating magnetic or electric fields, for controlling temperature, etc.) can be used to implement flow-through IT-SPME. Because of the inherent versatility of switching arrangements, IT-SPME flow-through systems are currently more popular.

### 14.4.1 Draw-eject mode IT-SPME

Draw/eject IT-SPME requires a programmable sample injector. The extraction capillary is placed between the needle and the injection loop (or the metering pump) of the autosampler, as depicted in Fig. 14.5. When the sample is aspirated from the sample vial, it comes into contact with the extraction capillary, and a fraction of the target compounds migrate from the sample to the extraction phase. The maximum amount extracted is reached under equilibrium conditions. In order to accelerate this process, aliquots of the samples are aspirated and ejected repeatedly, so that after a number of cycles, the distribution equilibrium for a given sample composition is reached. Finally, the analytes are desorbed and introduced into the separation column by means of the autosampler injection valve. Additional solvents, such as water or buffers, can be flushed through the extraction capillary to eliminate matrix components before the desorption stage, or to clean-up and condition the extraction capillary before loading the next sample. Besides the extraction phase, the main parameters to be optimized are the number of draw/eject cycles and the sample flow rate, the type, and volume of desorption solvent, and the composition and volume of clean-up solvents, if applicable.

An important advantage of this mode of IT-SPME is that it is compatible with low sample volumes; typically only a few hundred μL are used. Furthermore, it is easy to implement, as it only requires a programmable autosampler [64]. However, because of the high number of draw/eject cycles required to increase the extraction yield (>20−25) the extraction time can be significantly longer than the time required for the chromatographic separation. The draw/eject IT-SPME approach has been primarily used for the analysis of biological samples (see Table 14.5).

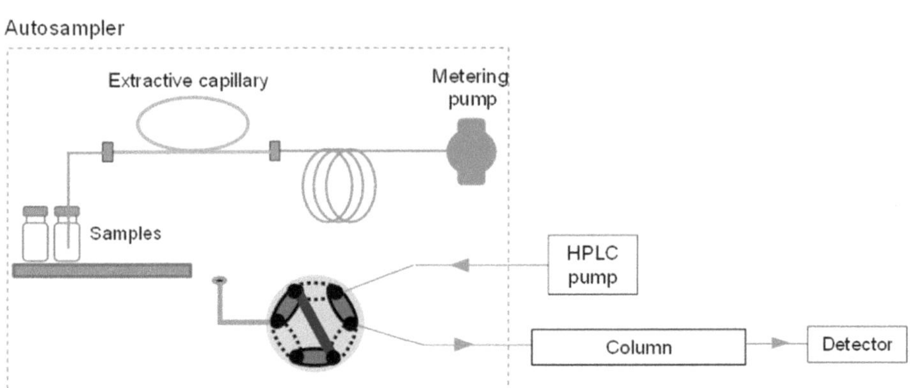


**Figure 14.5** Schematic representation of a draw/eject IT-SPME system.

**Table 14.5** Summary of selected applications of IT-SPME during the 2017–February 2019 period.

| IT-SPME mode | Analytes | Matrix | Extractive capillary | Separation/ detection | LOD, recovery (%) | RSD (%) | References |
|---|---|---|---|---|---|---|---|
| **Food analysis** | | | | | | | |
| Flow-through | Trans fatty acids (as monounsaturated fatty acid methyl esters) | French fries | Silver nanoparticles-coated monolithic column (20 cm, PTFE tube 750 μm i.d.) | HPLC-UV | 5.2 μg/kg<br>86.6%–96.1% | <5.2% | [65] |
| Flow-through | Trans fatty acids (as monounsaturated fatty acid methyl esters) | Instant coffee | Poly (octadecyl methacrylate-co-ethylene dimethacrylate) monolith in fused-silica capillary (15 cm, 250 μm i.d.) | HPLC-DAD | 3.0–7.1 μg/kg<br>83.0%–106.4% | 3.2%–4.7% | [66] |
| Draw/eject | *cis*-diol-containing polyphenols (catechin, chlorogenic acid, caffeic acid and epicatechin; ferulic acid as reference compound) | Litchi juice, green tea and haw juice | Boronate-decorated polyethyleneimine-grafted porous layer open tubular (BPPLOT) capillary (Fused-silica capillaries 0.25 mm ID) | HPLC-UV | 0.5–4 ng<br>80.0%–100% for litchi juice<br>82.8%–101% for green tea<br>80.9%–102% for haw juice<br><3% for non-cis-diol (ferulic acid) | 5.9%–7.2% (intraday) and 6.0%–8.6% (interday) for litchi juice<br>5.3%–7.3% (intraday) and 5.8%–8.4% (interday) for green tea<br>4.8%–7.2% (intraday) and 5.0%–8.0% (interday) for haw juice | [39] |

| | | | | | | | |
|---|---|---|---|---|---|---|---|
| Flow-through | Phthalates (benzyl butyl phthalate, di-n-butyl phthalate, di-n-pentyl phthalate, dicyclohexyl phthalate) | Plastic box and bottle | Basalt fibers grafted with poly(ILs) coating (1-dodecyl-3-vinylimidazolium bromide). A PEEK tube was used as the tubular matrix (30 cm, 0.75 mm i.d., 1.6 mm o.d.) | HPLC-DAD | 0.01–0.05 μg/L 92.2%–119.5%, | 0.5%–1.9% (intraday) 3.3%–4.6% (interday) | [67] |
| Flow-through | Mycotoxins (aflatoxin B1, zearalenone and sterigmatocystin) | Rice grains | Poly (methacrylic acid-co-divinyl-benzene) [poly (MAA-co-DVB)] monolithic column synthesized inside a fused silica capillary (20 cm, 0.25 mm i.d.) | HPLC-DAD | 0.69–2.03 μg/kg 78.0%–102.8% | 2.96%–4.74% | [68] |
| Flow-through | Mycotoxins (ochratoxin A) | Wine | PEEK tube (50 mm, 1/16 in e.d. 0.02 in i.d.) was filled with Luna C18 particles (dp = 10 μm; pore size, 100 Å). The column tubing was capped with a frit at both ends and connected to a PEEK tubing (150 mm, 1/16 in e.d. 0.005 in i.d.) | HPLC–MS/MS | 0.02 μg/L 61.4%–72.8% | <5.1% (intraday) <6.1% (interday) | [62] |

Continued

**Table 14.5** Summary of selected applications of IT-SPME during the 2017–February 2019 period.—cont'd

| IT-SPME mode | Analytes | Matrix | Extractive capillary | Separation/ detection | LOD, recovery (%) | RSD (%) | References |
|---|---|---|---|---|---|---|---|
| Flow-through | Melamine | Milk formulas | Sodium hyaluronate-functionalized urea-formaldehyde monolithic column in PTFE tube (20 cm, 750 μm i.d.) | HPLC-DAD | 0.2 μg/L<br>87.3%–96.7% | 1.9%<br>7.2% (tube-to-tube) | [52] |
| **Environmental analysis** | | | | | | | |
| Flow-through | 8 polycyclic aromatic hydrocarbons | snow water and river water | Natural cotton fiber directly packed into a PEEK tube (20 cm, 0.75 mm i.d.) | HPLC-DAD | 0.02–0.05 μg/L<br>80%–95% (2 μg/L)<br>81%–94% (5 μg/L) | 0.8%–3.6% (same tube)<br>1.7%–5.2% (tube-to-tube) | [69] |
| Flow-through | 8 polycyclic aromatic hydrocarbons | Cigarette smoke and soil | Diamond nanoparticles coated onto stainless steel wires filled into a polyetheretherketone (30 cm, 0.75 mm id, 1.6 mm od) | HPLC-DAD | (0.005–0.020 μg/L<br>89%–101% (cigarette smoke)<br>70%–96% (soil) | <2.4% (intraday)<br><8.4% (interday) | [70] |
| Flow-through | 8 polycyclic aromatic hydrocarbons | Water | Melamine-formaldehyde aerogel coating prepared on basalt fibers placed in a polyetheretherketone tube (0.75 mm i.d., 1.6 mm o.d.) | HPLC-DAD | 0.01–0.05 μg/L<br>81%–117% (2 μg/L)<br>86%–111% (5 μg/L) | 0.48%–2.04% (same tube)<br>0.88%–9.17% (tube-to-tube) | [59] |

| | | | | | | | |
|---|---|---|---|---|---|---|---|
| Flow-through | 10 polycyclic aromatic hydrocarbons | Water and milk | Poly (9-vinylanthracene-*co*-ethylene dimethacrylate) monolith (20 cm fused silica capillary column, 0.32 mm i.d.) | HPLC-FLD | 0.017–0.23 ng/L (water)<br>0.10–2.36 ng/L (milk)<br>78.5%–118% (wáter)<br>75.5%–119% (milk) | <9.8% (intraday)<br><11% (interday) | [71] |
| Flow-through | 8 polycyclic aromatic hydrocarbons | Tap and rain water | Co-Al bimetallic hydroxide nanocomposites coated fibers packed in a PEEK tube (20 cm, 0.75 mm i.d., 1.6 mm o.d.) | HPLC-DAD | <0.02 μg/L<br>83%–121% | <4.0% (same tube)<br><10.9% (tube-to-tube) | [72] |
| Flow-through | 8 polycyclic aromatic hydrocarbons | Groundwater and wastewater | Ionic liquids (1-methyl-3-(3-trimethoxysilyl propyl)imidazolium chloride) was prepared and chemically bonded onto basalt fibers packed in PEEK tube (40 cm, 0.75 mm i.d., 1.6 mm o.d.) | HPLC-DAD | 0.03–0.05 μg/L<br>82.1%–115.3% (5 μg/L)<br>80.4%–116.8% (10 μg/L) | 0.8%–2.8% (intraday)<br>1.3%–3.2% (interday) | [73] |
| Flow-through | 8 polycyclic aromatic hydrocarbons | Tap water and rain water | Basalt fibers functionalized with gold nanoparticles packed in a PEEK tube (20 cm, 0.75 mm i.d., 1.6 mm o.d.) | HPLC-DAD | 0.003–0.015 μg/L<br>89%–117% (5 μg/L)<br>92%–109% (10 μg/L) | 1.0%–5.2% (same tube)<br><14.7% (tube-to-tube) | [74] |

*Continued*

**Table 14.5** Summary of selected applications of IT-SPME during the 2017–February 2019 period.—cont'd

| IT-SPME mode | Analytes | Matrix | Extractive capillary | Separation/ detection | LOD, recovery (%) | RSD (%) | References |
|---|---|---|---|---|---|---|---|
| Flow-through | 10 polycyclic aromatic hydrocarbons | Soil and coal ash | Packed poly(ionic liquids) (1-vinyl-3-octylimidazolium bromide and 1,1′-(1,6-hexanediyl)bis(1-vinylimidazolium) dibromide)-coated stainless-steel wires (0.18 mm d) into a PEEK tube (35 cm, 0.50 mm id, 1.6 mm od) | HPLC-DAD | 0.010–0.020 μg/L<br>85.1%–118.9% (5 μg/L)<br>88.6%–113.5% (10 μg/L) | 0.2%–1.8% (same tube)<br>3.8%–19% (tube-to-tube) | [75] |
| Flow-through | 10 polycyclic aromatic hydrocarbons | Industrial sewage (chemical plant and coking plant wastewaters) | Carbon fibers functionalized with graphene oxide by electrophoretic deposition and filled (30 mg) into a PEEK tube (30 cm, 0.50 mm i.d., 1.59 mm o.d.) | HPLC-DAD | 0.001–0.004 μg/L<br>90.2%–112% | 2.1%–8.7% (intraday)<br>4.2%–9.0% (interday)<br>4.8%–16.7% (tube-to-tube) | [76] |
| Flow-through | Triazoles (triadimenol (TR), triadimefon (TRI) and hexaconazole (HE)) | Environmental waters (river, lake and sewage water samples) | Poly (4-vinyl pyridine co-ethylene dimethacrylate) monolith (21 cm fused silica capillary, 0.32 mm i.d.) | HPLC-DAD | 0.031 (TR), 0.022 (TRI) and 0.014 (HE) μg/L.<br>78.9%–106% | 0.20%–9.9% (same tube)<br>7.0%–8.0% (tube-to-tube) | [77] |

| | | | | | | | |
|---|---|---|---|---|---|---|---|
| Flow-through (offline) | Caffeine | Domestic wastewater samples | Ten modified stainless-steel wires (9 cm × 200 μm) in a tube (9 cm × 2.75 mm id) with a layer of polyaniline conductive polymer coated with Zn/Al layered double hydroxide | GC-MS | 0.14 μg/L<br>72.1%–89.5% | 5.7% (intraday)<br>8.3% (interday) | [78] |
| Flow-through | Estrogens (bisphenol A, 17α-ethylestradiol, estrone, diethylstilbestrol and hexestrol) | Tap water and mask water | basalt fibers coated with nano-calcium carbonate packed into a PEEK tube (30 cm, 0.75 mm i.d., 1.6 mm o.d.) | HPLC-DAD | 0.05–0.30 μg/L<br>93%–120% (1 μg/L)<br>87%–105% (10 μg/L) | 0.6%–1.9% (intraday)<br>4.1% (interday) | [79] |
| Flow-through | Estrogens (bisphenol A, 17α-ethylestradiol, estrone, diethylstilbestrol and hexestrol) | Sewage and emollient water | Basalt fibers organically modified with silica aerogel (phthalaldehyde and 3-aminopropyl) and filled into a PEEK tube (25 cm, 0.75 mm i.d., 1.6 mm o.d.) | HPLC-DAD | 0.01–0.05 μg/L<br>82.1%–109.4% (1 μg/L)<br>80.4%–104.5% (10 μg/L) | <4.5% (intraday)<br><6.7% (interday) | [58] |
| Flow-through | Silver nanoparticles | Compost, mountain, orchard and urban soils | Polydimethylsiloxane (PDMS) with 35% of polydiphenylsiloxane | CapLC-DAD | 0.009 nM | – | [80] |

*Continued*

**Table 14.5** Summary of selected applications of IT-SPME during the 2017–February 2019 period.—cont'd

| IT-SPME mode | Analytes | Matrix | Extractive capillary | Separation/ detection | LOD, recovery (%) | RSD (%) | References |
|---|---|---|---|---|---|---|---|
| Flow-through | PAHs, Alcohols, aldehydes, ketones, Amides and phenols | Wastewater, sea-water and swimming pool water | Sol-gel [bis(hydroxyethyl) amine] terminated polydimethylsiloxane - bis(trimethoxysily lpropyl) urea (BHEA-BPU) immobilized to the inner surface of a fused silica capillary (40 cm, 0.32 mm I.D., 0.25 mm I.D.) | HPLC-DAD | 5.21–11.90 μg/L 87.5%-114.8 | <10% | [81] |
| Flow-through | Herbicides (iodosulfuron-methyl-sodium, tribenuron-methyl, tritosulfuron, triflusulfuron-methyl, metribuzin, irgarol, bifenox, cypermethrin) | Water and extracts from agricultural soil. | Organic modified coating (tetraethyl orthosilicate (TEOS)-trimethoxyethylsilane (MTEOS)) reinforced with $SiO_2$ NPs in fused silica capillaries (30 cm, 0.32 mm o. d. and 75 μm i.d.) | CapLC-DAD | ≤1.0 μg/L (7.5 μg/L for cypermethrin) 84%–143% | 1%–9% | [82] |
| Flow-through | Aldehydes (propanal, butanal and pentanal) | Tap water and water put into a newly renovated house for three days | Silk fibers (0.10 g) were placed into a PEEK tube (30 cm, 750 μm i.d., 0.16 cm o.d.). Aldehydes were derivatized with 2,4-dinitrophenylhydrazine | HPLC-DAD | 0.01 μg/L (propanal and pentanal) 0.005 μg/L (butanal) 85%–106% (0.5 μg/L) 95%–102% (1 μg/L) | 2.32%–3.96% (intraday) 4.66%–6.36% (interday) 7.23%–10.22% (tube-to-tube) | [83] |

| | | | | | | | |
|---|---|---|---|---|---|---|---|
| Flow-through | 13 diesters and 2 monoesters of phthalic acid | Particulate matter ($PM_{10}$) in air | Divinylbenzene with 4-vinylpyridine (70 cm, 0.32 mm i.d., 10 μm) | HPLC-DAD | 0.09–0.52 $ng/m^3$ (method detection limit)<br>89%–98% (<80% for the four most polar compounds) | 4.2%–12.7% (interday) | [84] |
| Flow-through | Polar triazines and their degradation products (terbutylazine-2-hydroxy, terbutylazine-desethyl, atrazine, propazine, terbutylazine, Irgarol, atrazine-desisopropyl) | Water and recovered struvite from wastewater treatment plants | Polar-coated capillary based on tetramethyl orthosilicate (TEOS) and trimethoxymethy lsilane (MTEOS) containing $SiO_2$ nanoparticles (NPs) in fused silica capillary (20 cm, 0.075 mm i.d.) | NanoLC-DAD | 0.025–0.5 μg/L<br>82%–115% (struvite)<br>90%–106% (water) | <9.1% (interday)<br><11% (tube-to-tube) | [32] |
| **Biological and forensic analysis** | | | | | | | |
| Flow-through (offline) | Phosphorylated peptides (bovine α-casein, bovine β-casein, and bovine serum albumin (BSA)) | Digested chicken egg white | monolithic column based on graphene oxide–trimethyl-2-methacroyloxyethyl ammonium chloride-titania (GO-META-$TiO_2$) in fused-silica capillaries (15 cm, 0.10 mm I.D. × 0.36 mm O.D.) | MALDI-TOF-MS | 30 fmol | | [53] |

*Continued*

**Table 14.5** Summary of selected applications of IT-SPME during the 2017–February 2019 period.—cont'd

| IT-SPME mode | Analytes | Matrix | Extractive capillary | Separation/ detection | LOD, recovery (%) | RSD (%) | References |
|---|---|---|---|---|---|---|---|
| Flow-through (offline) | Phosphopeptides | Digested chicken egg white | $TiO_2$- nanoparticle-packed microchannel-array glass microchips (TMA-microchips).<br>Two cycles of photolithographic and chemical wet etching were used for the preparation of the SPME array channels and capillary connecting channels on a 2.3 mm thick 40 × 60 mm2 glass substrate layered with chromium and photoresist. | MALDI-TOF-MS | 40 fmol | | [63] |
| Flow-through | Endocannabinoids (anandamide, AEA, and 2-arachidonoyl glycerol, 2 A G) | Plasma samples | Polymeric ionic liquid ([$VC_{16}$IM][Br]/ [$(VIM)_2C_{10}$]2[Br]) based capillary in a fused silica capillary column (11.0 cm × 0.53 mm-id) | UHPLC-MS/MS | LLOQ:<br>0.10 μg/L (AEA)<br>0.05 μg/L ('2 A G)<br>Accuracy:<br>−19.6%–13.2% | 1.6%–14.0% | [41] |

| | | | | | | | |
|---|---|---|---|---|---|---|---|
| Flow-through | Cannabinoids (Delta(9)-tetrahydro cannabinol (THC), cannabidiol (CBD), and cannabinol (CBN)) | Different kind of items (plastic bags, office paper, aluminum foil, cotton cloths, and hand skin). | TBR-5 column (95% polydimethyl siloxane, 5% polydiphenylsiloxane) (15 cm, 0.320 mm id, 3 μm coating thickness) | nanoLC-DAD | 5 μg/L (2 μg/L for THC)<br>91% (CBN)<br>88% (CBD)<br>128% (THC) | 4% (homogenous solution)<br>23% (heterogeneous sample) | [85] |
| Flow-through | Antipsychotics (chlopromazine, clozapine, quetiapine, olanzapine) and their metabolites | Plasma samples from schizophrenic patients | Organic poly(butyl methacrylate-co-ethylene glycol dimethacrylate) monolith synthesized on the internal surface of a fused silica capillary (10 cm × 0.53 mm-id) | UHPLC-MS/MS | LLOQ:<br>10 μg/L<br>Accuracy:<br>−19.4%−19.9% (intra-assay)<br>−18.9%−19.3% (inter-assay) | 0.7−14.2 | [52] |
| Flow-through | Antipsychotic drugs (perphenazine, PPZ, and chlorpromazine, CPZ) | Urine and plasma | Nanostructured Cu−Cr−Al ternary layered double hydroxide/polythiophene coating was prepared on the inner surface of a stainless-steel tube (10 cm, 1.6 mm i.d.) and on the surfaces of eight stainless-steel wires (10 cm, 0.15 mm o.d.) | HPLC-UV | 0.1 μg/L (PPZ)<br>0.07 μg/L (CPZ)<br>94.0%−106.0% | <4.6% (intraday)<br><5.1% (interday)<br>3.0%−5.5% and 4.3%−6.3% (tube-to-tube) | [86] |

*Continued*

**Table 14.5** Summary of selected applications of IT-SPME during the 2017–February 2019 period.—cont'd

| IT-SPME mode | Analytes | Matrix | Extractive capillary | Separation/ detection | LOD, recovery (%) | RSD (%) | References |
|---|---|---|---|---|---|---|---|
| Flow-through | Non-steroidal anti-inflammatory drugs (NSAIDs): Ketoprofen, flurbiprofen and diclofenac sodium | Plasma and lake water | Deep eutectic solvent (DES) composed of choline chloride and itaconic acid was synthesized and applied to preparation of polymer monolith inside polydopamine-functionalized PEEK tube (10 cm, 1/16 in. o.d., 0.02 in. i.d.) | HPLC-UV | 0.01 μg/L (ketoprofen and flurbiprofen)<br>0.05 μg/L (diclofenac sodium)<br>87.3%–113.7% | 1.89%–2.45% (intra-batches)<br>3.70%–4.32% (inter-batches) | [49] |
| Draw/ eject | Nicotine and cotinine | Hair | Carboxen 1006 PLOT (60 cm, 0.32 mm i.d., film thickness 17 μm) | HPLC-MS/MS | 0.45 ng/L (nicotine)<br>0.13 ng/L (cotidine)<br>87%–96.6% | 1.62%–3.37% (nicotine)<br>2.35%–5.97% (cotidine) | [87] |
| Draw/ eject | Biomarkers for the evaluation of oxidative stress (8-isoprostane, 8-IP, 8-hydroxy-2′-deoxyguanosine, 8-OHdG, and 3-nitro-L-tyrosine, 3-NT) | Human urine | Carboxen 1006 PLOT (60 cm, 0.32 mm i.d., film thickness 17 μm) | HPLC-MS/MS | 3.4 (8-IP), 12.6 (8-OHdG) and 21.5 (3-NT) ng/L<br>63.1%–100% | 1.1%–5.0% (intraday)<br>2.3%–6.5% (interday) | [88] |

| | | | | | | | |
|---|---|---|---|---|---|---|---|
| Flow-through | Horseradish peroxidase (glycoprotein) | Mixture of bovine serum albumin and horseradish peroxidase.<br>Plasma. | poly(vinylphenylboronic acid−ethylene glycol dimethacrylate) monolithic material incorporated with graphene oxide synthesized inside a PEEK tube (10 cm, 1.58 mm o.d., 0.53 mm i.d.) | HPLC-UV | 0.01 mg/L<br>95.6% | 1.8% (intra-batches)<br>2.7% (inter-batches) | [89] |
| Flow-through | Meropenem | Endotracheal tubes | Polydimethylsiloxane (PDMS) with 35% of polydiphenylsiloxane (60 cm, 0.32 mm i.d. 3 μm thickness) | CapLC-DAD | 0.003 mg/L<br>94%−103% | 1% (intraday)<br>4% (interday) | [90] |
| Flow-through | Six formylated nucleosides (5-formyl-20-deoxycytidine (5-fodC), 5-formylcytidine (5forC), 5-formyl-20-deoxyuridine (5-fodU), 5-formyluridine (5-forU), 20-O-methyl-5-formylcytidine (5forCm) and 20-O-methyl-5-formyluridine (5-forUm)) | DNA and RNA of cultured human cells and multiple mammalian tissues. | Poly (methacrylic acid−ethylene glycol dimethacrylate) monolithic capillary in silica capillary (5 cm, 0.25 mm i.d., 0.36 mm o.d.) | UPLC-ESI-MS/MS | 0.03−0.18 fmol<br><14% (relative error) | <13.2% (intraday)<br><14.3% (interday) | [91] |

*Continued*

**Table 14.5** Summary of selected applications of IT-SPME during the 2017–February 2019 period.—cont'd

| IT-SPME mode | Analytes | Matrix | Extractive capillary | Separation/ detection | LOD, recovery (%) | RSD (%) | References |
|---|---|---|---|---|---|---|---|
| Flow-through (offline) | $\beta_2$-microglobin and cystatin C | – | poly(glycidyl methacrylate) nanoparticles-coated capillary was prepared in a fused-silica capillary (60 cm × 0.25 mm ID). Oriented antibody immobilized capillary was prepared immobilizing on it the rabbit anti-human cystatin C antibody. | Particle-enhanced turbidimetric immunoassay | 1 μg/L ($\beta_2$-microglobin)<br>0.5 μg/L (cystatin C)<br>96.8%–103.6% | 3.6%–3.8% (intra-batches)<br>6.4%–7.8% (inter-batches) | [47] |
| Flow-through | Positively charged quaternary alkaloids (jatrorrhizine, palmatine and berberine) | Cortex Phellodendri herb and rat plasma | PEEK tubes (10 cm, 1/16 in. o.d., 250 μm i.d.) were stepwise functionalized with polydopamine and graphene oxide. | HPLC-MS/MS | 0.1 ng/L<br>94.7%–102.6% | 0.47%–2.46% (intraday)<br>1.78%–7.80% (interday)<br>3.0%–5.5%<br>4.7%–6.2% (tube-to-tube) | [92] |
| Flow-through (offline) | Organic iodine (3-Iodo-L-tyrosine, MIT, 3,5-diiodo-L-tyrosine, DIT, 3,3′,5-triiodo-L-thyronine, T3), levothyroxine T4, 3-iodobenzoic acid and 3-Iodophenol) | Urine and human milk | DB-wax (10 cm, 0.25 mm i.d., 0.25 μm thickness) | Nanospray high-resolution mass spectrometry | At least 10 μg/L (objective rapid identification of unknown organic iodine) | – | [93] |

| | | | | | | | |
|---|---|---|---|---|---|---|---|
| Flow-through | Estrogens (estriol, 17 beta-estradiol, estrone, ethinyl estradiol and progesterone) | Urine | $NH_2$-MIL-53(Al) metal–organic framework incorporated in poly(styrene-divinylbenzene-methacrylic acid) monolith in fused silica capillary (20 cm, 0.53 mm i.d.) | HPLC-UV-Fl | 2.0–40 ng/L<br>75.1%–120% | 1.1%–4.2% (intra-batches)<br>0.5%–7.4% (inter-batches) | [54] |
| **Magnetic and electrochemically controlled** | | | | | | | |
| Flow-through | Parabens (isopropylparaben, IPP, propylparaben, PP, isobutylparaben, IBP, and butylparaben, BP) | Lake water, river water and grape juice | Octyl methacrylate monolithic capillary column doped with magnetic nanoparticles, $Fe_3O_4$@ $SiO_2$@γ-MAPS into fused silica capillary column (20 cm, i.d. 0.32 mm). 70 gs magnetic field applied for adsorption and −50 gs for desorption. | HPLC-UV | 0.010–0.026 μg/L (water)<br>0.019–0.051 μg/L (grape juice)<br>87.9%–108% (water)<br>86.9%–114% (grape juice) | 0.9%–8.3% | [94] |

*Continued*

**Table 14.5** Summary of selected applications of IT-SPME during the 2017–February 2019 period.—cont'd

| IT-SPME mode | Analytes | Matrix | Extractive capillary | Separation/ detection | LOD, recovery (%) | RSD (%) | References |
|---|---|---|---|---|---|---|---|
| Flow-through | Organic ultraviolet filters (2-hydroxy-4-methoxy benzophenon, BP-3, 3-(4-methylbenzylidene) camphor, 4-MBC, 2-ethylhexyl 4-methoxycinnamate, EHMC, octyl dimethyl-p-amino-benzoate, OD-PABA, and octyl salicylate, OS) | Lake water, river water and wastewater | Ionic liquid-based monolithic capillary column doped with magnetic nanoparticles ($Fe_3O_4$@$SiO_2$@γ-MAPS) in fused-silica capillary (20 cm, 320 μm i.d.). 20 gs magnetic field applied for adsorption and −20 gs for desorption. | HPLC-DAD | 0.04–0.26 μg/L<br>70.5%–119% | <10% (intra-batches)<br>2.6%–4.8% (inter-batches) | [95] |
| Flow-through | Hypertension drug (losartan) | Urine and plasma | Nanocomposite prepared from polypyrrole and deep eutectic solvent (choline chloride and urea) by electrochemical deposition on the inner walls of a stainless steel capillary (10 cm, 0.75 mm o.d.) that serves as a working electrode (anode), and a platinum electrode as the cathode. | HPLC-UV | 50–500 ng/L<br>95%–108.2% (urine)<br>91.8%–110.0% (plasma) | 1.9%–3.9% (intraday)<br>2.4%–4.6% (interday | [45] |

## 14.4.2 Flow-through IT-SPME

A wide variety of arrangements have been used for online flow-through IT-SPME. In the simplest configuration, only the extraction capillary is needed in addition to standard LC equipment. In these systems, the extraction capillary is used to replace the inner loop of the injection valve (in-valve IT-SPME). The analytes are retained by the extraction phase during loading of the sample with the valve in the load position; when the valve is switched to the inject position, the mobile-phase desorbs and transfers the extracted compounds to the chromatographic column for separation and detection. A schematic diagram of this configuration is shown in Fig. 14.6A.

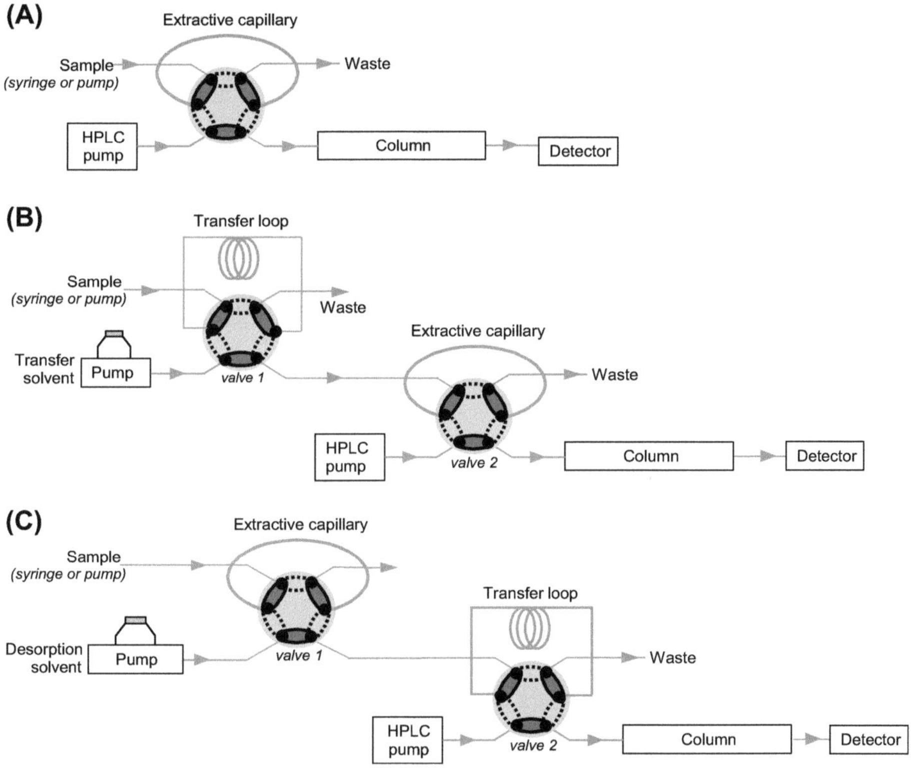


**Figure 14.6** Schematic representation of possible setups used for online flow-through IT-SPME. (A) System with one valve (*continuous line*, sample loading; *dotted line*, sample transfer and chromatographic separation and detection). (B) System with two valves and a loop for sample injection (Valve 1: *continuous line*, filling of the loop with the sample; *dotted line*, transfer of the sample plug to the extractive capillary. Valve 2: *continuous line*, filling of the extractive capillary with the sample plug; *dotted line*, chromatographic separation and detection). (C) System with two valves and a loop for the injection of the analytes after IT-SPME (Valve 1: *continuous line*, filling of the extractive capillary with the sample for extraction; *dotted line*, desorption and transfer of the analytes. Valve 2: *continuous line*, filling of the injection loop; *dotted line*, chromatographic separation and detection).

At the flow rates used to load samples, the distribution equilibrium between the sample and the extractive phase is not obtained, and only a fraction of the analyte concentration is retained when the sample is passed through the extraction capillary. This can be compensated by loading a large sample volume. This increases the amount of analyte transferred to the extraction phase in proportion to the sample volume, provided that the loading capacity of the extraction capillary is not exceeded.

Sample loading can be done with either a precision syringe or by using an additional pump. In the latter case, a high-pressure pump is typically used to deliver the sample, which is housed in the solvent compartment. Peristaltic pumps have also been used for the same purpose. The parameters to be optimized are the sampling time and the flow rate. Sampling volumes up to 30−50 mL can be loaded on some extraction capillaries at the expense of long sampling times. Excessively high sample flow rates are not suitable, as they may reduce the lifetime of the extraction capillary [72]. Although relatively large sample volumes can be loaded manually using multiple injections, the use of a pump is preferable when large sample volumes are to be processed.

A more precise control of the amount of sample transferred to the extraction capillary is possible using a loop of the required volume connected to an additional valve, as shown in Fig. 14.6B. A second pump is also necessary to transfer the sample. Loops with volumes up to 6 mL have been used [54]. The sample is loaded, manually or with a pump, with the first valve in the load position; next, the valve is switched so that the loop contents of is transferred to the extraction capillary with the second valve in the load position using a transfer solvent. Once the sample plug has been completely transferred to the extraction capillary, the second valve is rotated to the inject position to desorb the analytes from the extraction capillary with mobile phase and transferred to the separation column. Simultaneous sample clean-up is possible if the second valve is switched after a delay time, using the transfer solvent also for sample clean-up. If a binary pump or quaternary pump is used for the transfer step, different solvents can be selected for desorption, clean-up and capillary conditioning once the second valve has been switched to its original position; these operations can be performed during the separation of the previous sample. In the above configuration, as well as the sample volume, the eluent composition and flow-rate for transfer of the analytes have to be optimized; the valve switching times must also be controlled to achieve reproducible results.

Alternatively, the analytes can be desorbed from the extraction capillary and transferred to a loop; the loop is subsequently inserted into the mobile phase stream so that only the fraction that fills the loop enters the separation column. A possible scheme for this configuration is shown in Fig. 14.6C. An interesting feature of the system such as that shown in Fig. 14.6C is that there is no direct contact between the extraction capillary and the chromatographic flow-scheme, which facilitates performing the extraction and separation steps under different conditions. For example, these systems can be used with peristaltic pumps that allow sampling and desorption at high flow rates (up to 5 mL/min) [96]. As a result, sample volumes up to 30−80 mL can be processed in a reasonable time [77,96−99].

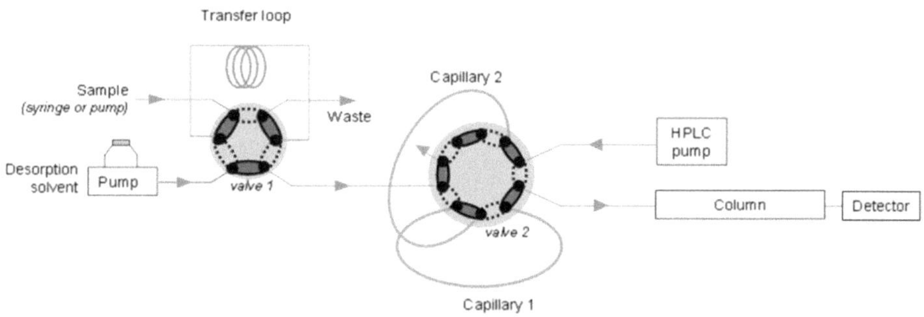


**Figure 14.7** Schematic representation of the system used for IT-SPME with two extractive capillaries coupled in series via a 10-port switching valve. Valve 1: *continuous line*, filling of the injection loop with the sample; *dotted line*, desorption and transfer of the sample. Valve 2: *dotted line*, filling sequentially capillaries one and two; *continuous line*, chromatographic separation and detection.

The installation of additional valves and pumps or other elements leads to more versatile systems, of which those combining two (or more) extraction capillaries are of special interest. This is because capillaries with different sorbents can be connected in series to enhance the extraction of samples containing analytes of different characteristics. An example of this type of systems using a 10-port switching valve is shown in Fig. 14.7.

For relatively simple matrices, such as water, a configuration, such as Fig. 14.6A, is suitable for many applications. Systems such as Fig. 14.6B allow some online clean-up of the analytes, which facilitates the analysis of complex samples with MS detection [91]. More versatile systems can be set up with systems such as Fig. 14.6C, which also facilitates coupling to MS. This configuration is also preferable to minimize damage to the extraction capillary which is operated at different pressures during the extraction and separation steps, for example, if large sample volumes or UHPLC are to be used.

So far IT-SPME has been used primarily prior to chromatographic separation. However, examples of applications have been reported with other instrumental techniques, such as capillary electrophoresis [100], capillary electrochromatography [101], or even atomic absorption spectrometry (AAS) [102]. The latter approach, reported by Yamini et al. illustrates the applicability of IT-SPME for the extraction of inorganic species, in this case, inorganic selenium from water samples. The authors used a setup similar to Fig. 14.6C with a peristaltic pump connected to the second valve to deliver the reagents necessary to convert the analyte into the hydride form for AAS detection.

### 14.4.3 Miniaturized LC systems

Miniaturized LC systems, such as capillary- and nano-LC, offer significant advantages such as enhanced sensitivity and easy hyphenation with MS detectors [1,9,36]. The coupling of IT-SPME to CapLC can be done with minor modifications compared with conventional-scale LC, as demonstrated by the number of applications mainly in the environmental and biomedical fields, using, either segments of GC column or capillaries with coatings, specifically synthesized for the target compounds (see Table 14.1).

Recently, the coupling of IT-SPME to Nano-LC was described [32,103]. Because of the low mobile-phase flow rates used in Nano-LC, capillaries with low internal diameters (0.1–0.075 mm i.d.) are necessary. As the number of extraction columns compatible with these systems is rather limited, new sorbents with improved extraction capabilities are needed [32,103]. As an example, Fig. 14.8 compares the signals for a mixture of three compounds of different polarity: bisphenol A, 4-n-nonylphenol and diethylhexyl phthalate using the configuration shown in Fig. 14.6A for the online coupling of IT-SPME to CapLC or to NanoLC. A 15 cm segment of a capillary column coated with TEOS-MTEOS-$SiO_2$NPs (0.32 mm i.d. for CapLC and 0.075 mm i.d. for NanoLC) was used as the extraction phase [32]. The IT-SPME capillary columns were connected to the six-port valve and used as an external loop (Fig. 14. 6A). Aliquots of 100 μL sample volumes were manually passed through the system by means of a glass precision syringe. Fig. 14.8 shows that the sensitivity achieved by NanoLC is at least 10 times higher than that obtained by CapLC, meanwhile the efficiency and the resolution of the two analytical columns were similar.

The chromatogram obtained by using the configuration of Fig. 14.7 for a mixture of phenols with similar polarities: 4-tert-octylphenol, technical nonylphenol and 4-n-nonylphenol is shown in Fig. 14.9. The same conditions and separation column previously indicated for NanoLC were used. For IT-SPME, two TEOS-MTEOS-SiO2NPs

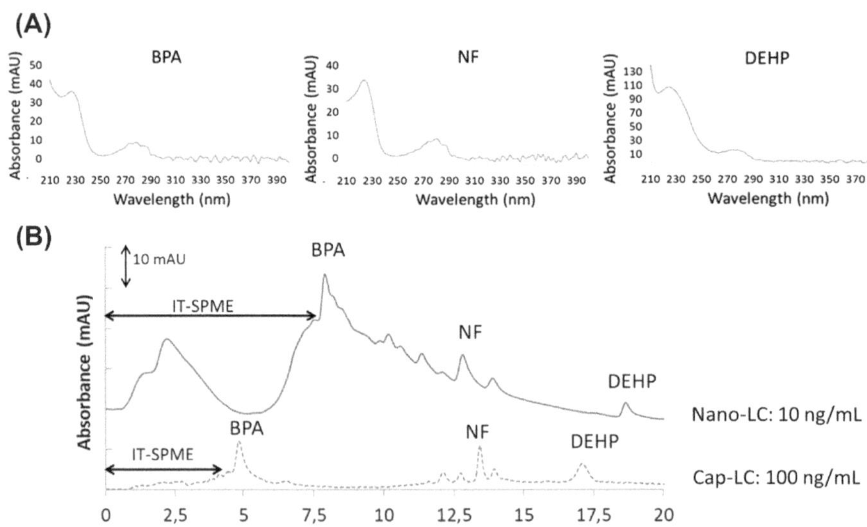


**Figure 14.8** Bysphenol A (BPA), 4-n-Nonylphenol (NF) and Diethyl hexyl phtalate (DEHP): (A) spectrum registered at each chromatographic peak for identifying the compounds and (B) chromatograhs for Nano-LC and Cap-LC, the registers included the time for online IT-SPME. For the configuration of the system see Fig. 14.6A, for more explanation see text.

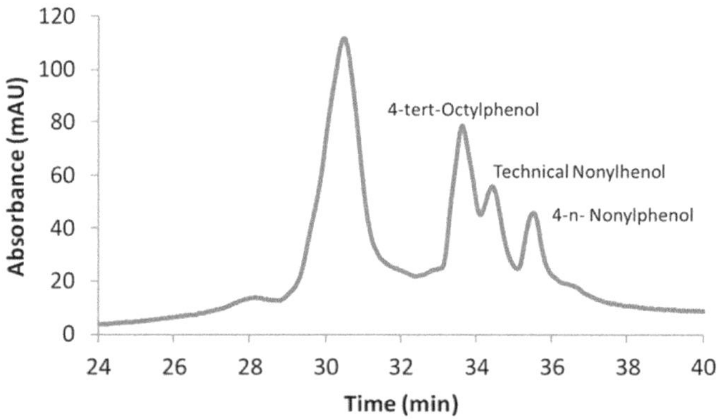


**Figure 14.9** Chromatograms obtained for a mixture of phenols with similar polarities: 4-tert-octylphenol (25 ng/mL), Technical Nonylphenol (25 ng/mL) and 4-n-Nonylphenol (50 ng/mL). For the configuration of the system see Fig. 14.7, for more explanation see text.

(0.075 mm i.d.) capillaries were employed in series to improve resolution. The configuration of Fig. 14.6A did not achieve the required resolution, and only two peaks were observed.

From another perspective, a new format of miniaturized IT-SPME was proposed by Guo et al. based on the immobilization of the extraction sorbent ($TiO_2$ nanoparticles) in the channels of a microchip [63]. The device was successfully applied to the selective isolation of phosphopeptides from protein digests (10 μL) using pressure as the driving force. Once isolated and purified, the peptides were desorbed and the extracts analyzed by MALDI-TOF-MS.

## 14.5 Off-line development

Whereas most reports describe the online coupling of IT-SPME to usually a separation system a few utilized off-line IT-SPME for sample preparation. Most utilized pressure systems, such as pumps to pass the sample through the capillary tube or capillary diffusion for in situ devices. All were combined with instrumental detection. The main advantage of off-line IT-SPME is that once the target compounds are extracted, clean-up operations can be carried out to improve the compatibility of the extract with the chromatographic or detection systems, for example, to minimize ion suppression for MS or to derivatize the extracted analytes.

Wu et al. [42] proposed the use of polydopamine/dialdehyde starch/chitosan composite coating for IT-SPME and in situ derivatization for the analysis of cancer biomarkers in human blood. This coating material was synthetized and immobilized on the inner wall of polytretrafluoroethylene (PTFE) tube. Due to the dimensions of the capillary tube (6 cm 1.6 mm o.d., and 0.5 mm i.d.) a syringe pump was needed to drive the solutions through the extraction tube. The sampling time was 11 min, with LODs around 1.5 nM for hexanal and 2-butanone. Recoveries were 70% and 91%, respectively. The pretreated sample was collected and analyzed after by HPLC.

For the extraction of triazines from environmental water samples, a magnetism-enhanced monolithic (octylmethacrylate-*co*-ethylene glycol dimethacrylate) fused-silica capillary column doped with magnetic particles was used. The capillary was connected to a pressure pump, and 900 μL of the sample was processed [104]. Satisfactory extraction efficiency for the target analytes (between 64.8% and 99.7%) was obtained.

A fiber in-tube solid-phase microextraction device with an electrochemically deposited hydroxide layer was proposed for the extraction of caffeine in wastewater samples [105]. In this configuration, both the fibers and tube were engaged in the extraction process. Ten stainless steel wires (9 cm × 200 μm) as fiber supports and a stainless steel tube (9 cm x 2.75 mm i.d.) were used to construct the fiber in-tube SPME device. A peristaltic pump was used to pass the sample solution through the device. The extracts were analyzed by GC-MS.

Molecularly imprinted polymer fibers typically possess limited mechanic stability and short lifetimes for multiple uses. To overcome this problem, in-tube SPME was proposed. Off-line extraction of 4-nitrophenol in a hypodermic needle connected to a MIP-capillary tube and used a syringe pump for liquid delivery [106]. Two mL of sample was passed through conditioned tubes with a flow rate of 2 mL/h, 0.5 mL of water (pH 3) was used for washing the tube, and 50 μL (ACN containing 0%.1% of acetic acid) to elute the analyte. The extracts were analyzed by HPLC-UV.

Other geometrical configurations for miniaturized in-tube SPME have emerged to process small sample volumes and for rapid analysis. Several approaches have been developed based on IT-SPME devices combined with analytical instruments. A biocompatible in-tube SMPE device based on a medical grade spinal needle with an electromechanically deposited layer of nanostructured polypyrrole (PPy), pore size less than 80 nm, was developed for the analysis of whole blood and untreated urine with MS detection [107]. The extractions were performed under static conditions by pulling 2 μL of the sample inside the spinal needle.

Ying et al. [47] used an immunoaffinity sorbent for in-tube SMPE in combination with a particle-enhanced turbidimetric immunoassay (PETIA). A combination of poly(glycidyl methacrylate) (PGMA) and oriented or nonoriented immobilized antibody (AI) on the internal surface of a fused-silica capillary for extraction of antigens were compared. The antigens β2MG and CysC were extracted by flushing the sample solution through the antibody immobilized capillary and left to incubate overnight at 37°C. The oriented AI capillaries provided higher sensitivity and extraction capacity.

The integration of capillaries to microchannels with almost zero dead volume was proposed in [108]. Based on this strategy, rapid fabrication of $TiO_2$ nanoparticle packed microchannels array glass microchips (TMA-microchips) for in tube solid-phase microextraction coupled to a matrix-assisted laser desorption/ionization time of flight mass spectrometer (MALDI-TOF-MS) for detection of the selective enrichment of phosphopeptides from a protein digest were reported [63]. Later, these authors reported a novel monolithic column based on graphene oxide−trimethyl-2-methacroyloxyethylammonium chloride-titania (GO-META-$TiO_2$) for the in-tube solid-phase microextraction of phosphopeptides coupled to MALDI-TOF-MS [53]. In this case, 10 μL of sample solution was introduced into the monolithic column. Following phosphopeptide enrichment, the column was washed with 10 μL of ultrapure water. Finally, the phosphopeptides were eluted with a few μL of 1% (v/v) ammonia in water.

## 14.6 Applications

Table 14.5 summarizes selected applications of IT-SPME for the period 2017–February 2019. The applications are grouped under food, environmental, biological, and forensic, as well as magnetic and electrochemically controlled analyses. In each instance, a trend toward the evaluation of new sorbent phases to increase selectivity or analyte recovery is observed. The most used support is fused-silica capillaries or poly(ether ether ketone) (PEEK) tubes, internally coated with various sorbents or as monoliths. These new extraction phases contain, in some cases, nanoparticles or nanomaterials, such as silver [65], diamond [70], gold [74], $CaCO_3$ [78], $SiO_2$ [32,82], $TiO_2$ [63], $Fe_3O_4@SiO_2@\gamma$-MAPS [53,63], Co-Al bimetallic hydroxide [53], graphene oxide [76,89,92], ternary layered double hydroxide [86], poly(glycidyl methacrylate) [47], or ionic liquids [41,67,73,75]. In other cases, natural materials in the form of fibers, with or without derivatization, such as basalt fibers [58,67,73,74,79], cotton [69], steel wires [70,75,78,86], carbon [76], and silk [83] inserted inside the capillary, were used.

In the field of food analysis, the analytes studied were *trans*-fatty acids [65,66], polyphenols [39], phthalates [67], mycotoxins [62,68], or melamine [52] in matrices, such as French fries, instant coffee, fruit juices, rice grains, wine, or milk.

In environmental analysis, polycyclic aromatic hydrocarbons are the most studied contaminants [59,69–76], as well as triazoles [77], caffeine [78], estrogens [58,79], silver nanoparticles [80], herbicides [32,82], aldehydes [83], and esters of phthalic acid [84]. The most studied matrix is environmental waters, but also soil, coal ash, air particulate matter or struvite from wastewater treatment plants.

In biological and forensic analysis, phosphopeptides [53,63], cannabinoids [41,85], antipsychotic drugs [52,86], non-steroidal anti-inflammatory drugs [49], meropenem [90], estrogens [54], nicotine and cotinine [87], biomarkers for the evaluation of oxidative stress [88], glycoproteins [89], formylated nucleosides [91], quaternary alkaloids [92], and organic iodine [93] have been studied. Biological fluids are the most studied matrix (urine, plasma, milk), as well as hair or clinical devices (endotracheal tubes).

## 14.7 Conclusions

IT-SPME is an attractive option for modernizing sample preparation procedures since this it integrates the clean-up, extraction, and pre-concentration into a single process. The most important parameter for method development is the selection of the extraction phase. Other experimental parameters, such as the sample volume, pH, phase thickness, length, and diameter of the extraction capillary, and washing solvent, must be optimized to achieve the greatest benefits.

IT-SPME can be classified in four modalities based on the extraction phase: (1) surface coated capillary columns, (2) monolithic capillary columns, (3) fiber packed capillary columns and (4) packed sorbent capillary columns. The fabrication depends on the support capillary column (typically fused-silica capillaries or poly(ether ether ketone tubes)). Various extraction phases are formed typically by polymerization, electrodeposition, or direct surface modification. Nowadays, monolithic capillary columns are the most active areas of research for IT-SPME.

## Acknowledgements

The authors are grateful to MINECO-AEI of Spain and EU-FEDER (project CTQ2017-90082-P) and the Generalitat Valenciana (PROMETEO 2016/109) for the financial support received. A.B.-C. expresses his gratitude to the GV for her PROMETEO post-doctoral grant.

## References

[1] Moliner-Martinez Y, Herráez-Hernández R, Verdú-Andrés J, Molins-Legua C, Campíns-Falcó P. Trac Trends Anal Chem 2015;71:205–13.
[2] Eisert R, Pawliszyn J. Anal Chem 1997;69:3140–7.
[3] MEC Queiroz MEC, ID de Souza ID, C Marchioni C. Trac Trends Anal Chem 2019;111: 261–78.
[4] Gonzalez-Fuenzalida RA, Moliner-Martinez Y, Molins-Legua C, Campíns-Falcó P. Anal Chim Acta 2019;1045:23–41.
[5] Maciel EVS, de Toffoli AL, Lancas FM. J Sep Sci 2019;42:258–72.
[6] Liang YS, Zhou T. J Sep Sci 2019;42:226–42.
[7] De Toffoli AL, Maciel EVS, Fumes BH, Lancas FM. J Sep Sci 2018;41:288–302.
[8] Song XY, Chen J, Shi YP. Trac Trends Anal Chem 2017;86:263–75.
[9] Serra-Mora P, Moliner-Martinez Y, Molins-Legua C, Herráez-Hernández R, Verdú-Andrés J, Campíns-Falcó P, Ibanez E, Cifuentes A. Wilson and Wilson's comprehensive analytical chemistry. In: Barceló D, editor. Green extraction techniques: principles, advances and applications, vol. 76. Amsterdam: Elsevier; 2017. p. 427–61.
[10] Fernandez-Amado M, Prieto-Blanco MC, Lopez-Mahia P, Muniategui-Lorenzo S, Prada-Rodriguez D. Anal Chim Acta 2016;906:41–57.
[11] Barreiro JC, Luiz AL, Maciel SCF, Maciel EVS, Lancas FM. J Sep Sci 2015;38: 1721–32.
[12] Padron MET, Afonso-Olivares C, Sosa-Ferrera Z, Santana-Rodriguez JJ. Molecules 2014;19:10320–49.
[13] Gonzalez-Fuenzalida RA, Moliner-Martinez Y, Prima-Garcia H, Ribera A, Campins-Falco P, Zaragoza RJ. Nanomaterials 2014;4:242–55.
[14] Queiroz MEC, Melo LP. Anal Chim Acta 2014;826:1–11.
[15] Kataoka H, Saito K. Bioanalysis 2012;4:809–32.

[16] Aufartova J, Mahugo-Santana C, Sosa-Ferrera Z, Santana-Rodriguez JJ, Novakova L, Solich P. Anal Chim Acta 2011;704:33–46.
[17] Wei F, Feng YQ. Anal Meth 2011;3:1246–56.
[18] Kataoka H, Saito K. J Pharm Biomed Anal 2011;54:926–50.
[19] Namera A, Nakamoto A, Saito T, Miyazaki S. J Sep Sci 2011;34:901–24.
[20] Aguilera-Herrador E, Lucena R, Cardenas S, Valcarcel M. Trac Trends Anal Chem 2010; 29:602–16.
[21] Campíns-Falcó P, Verdú-Andrés J, Sevillano-Cabeza A, Herráez-Hernández R, Molins-Legua C, Moliner-Martinez Y. J Chromatogr A 2010;1217:2695–702.
[22] Kataoka H. Anal Bioanal Chem. 2010;396:339–64.
[23] Cruz-Vera M, Lucena R, Cardenas S, Valcarcel M. Trac Trends Anal Chem 2009;28: 1164–73.
[24] Herraez-Hernandez R, Campíns-Falcó P, Verdú-Andrés J, Sevillano-Cabeza A, Ferrando-Climent L. In: Haines PA, Hendrickson MD, editors. Polymer science and technology series. Polycyclic aromatic hydrocarbons: pollution, health effects and chemistry. New York: Nova Science Publishers INC; 2009. p. 201–20 [chapter 7].
[25] Kataoka H, Ishizaki A, Nonaka Y, Saito K. Anal Chim Acta 2009;65:8–29.
[26] Pragst F. Anal Bioanal Chem 2007;388:1393–414.
[27] Kataoka H, Mitani K, Takino M. In: Vidal JLM, Frenich AG, editors. Methods in biotechnology book series. Pesticide protocols, vol. 19. Totowa, NJ: Humana Press Springer; 2005. p. 365–82.
[28] Kataoka H. Curr Pharmaceut Anal 2005;1:65–84.
[29] Queiroz MEC, Lancas FM. Quim Nova 2005;28:880–6.
[30] Kumazawa T, Lee XP, Sato K, Suzuki O. Anal Chim Acta 2003;492:49–67.
[31] Kataoka H. Anal Bioanal Chem 2002;373:31–45.
[32] Serra-Mora P, Jornet-Martínez N, Moliner-Martínez Y, Campíns-Falcó P. J Chromatogr A 2017;1513:51–8.
[33] Galuszka A, Konieczka P, Migaszewski ZM, Namiesnik J. Trac Trends Anal Chem 2012; 37:61–72.
[34] Plotka-Wasylka J. Talanta 2018;181:204–9.
[35] Pla-Tolós J, Serra-Mora P, Hakobyan L, Molins-Legua C, Moliner-Martinez Y, Campíns-Falcó P. Sci Total Environ 2016;560–570:611–8.
[36] Jornet-Martínez N, Bocanegra-Rodríguez S, González-Fuenzalida RA, Molins-Legua C, Campíns-Falcó P. In: Grumezescu AM, Holban AM, editors. Processing and sustainability of beverages, vol. 2. Amsterdam: Elsevier; 2019. p. 275–317 [chapter 9].
[37] Moliner-Martínez Y, Prima H, Ribera A, Coronado E, Campins-Falcó P. Anal Chem 2012;84:7233–40.
[38] Moliner-Martinez Y, Vitta Y, Prima-Garcia H, Gonzalez-Fuenzalida R, Ribera A, Campins-Falco P. Anal Bioanal Chem 2014;406:2211–5.
[39] Ying LL, Wnag DY, Yang HP, Deng XY, Peng C, Zheng C, et al. J Chromatogr A 2018; 1544:23–32.
[40] Souza ID, Melo LP, Jardim ICSF, Monteiro JCS, Nakano AMS, Queiroz MEC. Anal Chim Acta 2016;932:49–59.
[41] Souza ID, Hantao LW, Queiroz MEC. Anal Chim Acta 2019;1045:108–16.
[42] Wu S, Cai C, Cheng J, Cheng M, Zhou H, Deng J. Anal Chim Acta 2016;935:113–20.
[43] Ahmadi SH, Manbohi A, Heydat KT. Anal Chim Acta 2015;853:335–41.
[44] Asiabi H, Yamini Y, Seidi S, Ghahramanifard F. Anal Chim Acta 2016;913:76–85.
[45] Asiabi H, Yamini Y, Shamsayei M, Mehraban JA. Microchim Acta 2018;185:169–77.
[46] Shamsayei M, Yamini Y, Asiabi H. J Chromatogr A 2016;1475:8–17.

[47] Ying LL, Ma YC, Xu B, Wang XH, Dong LY, Wang DM, et al. J Chromatogr A 2017; 1509:1−8.
[48] Zhang J, Zhang W, Bao T, Chen Z. J Chromatogr A 2015;1388:9−16.
[49] Wang R, Li W, Chen Z. Anal Chim Acta 2018;1018:111−8.
[50] Mei M, Huang X, Luo Q, Yuan D. Anal Chem 2016;88:1900−7.
[51] Wang T, Chen Y, Ma J, Qian Q, Jin Z, Zhang L, et al. Anal Chem 2016;88:1535−41.
[52] (a) Beloti LGM, Miranda LFC, Queiroz MEC. Molecules 2019;24:310−23.
(b) Wang J, Jiang N, Cai Z, Li W, Li J, Lin X, et al. J Chromatogr A 2017;1515:54−61.
[53] Zhao S, Wang S, Yan Y, Wang L, Guo G, Wang X. Talanta 2019;192:360−7.
[54] Luo X, Li G, Hu Y. Talanta 2017;165:377−83.
[55] Sun M, Feng J, Bu Y, Luo C. J Chromatogr A 2015;1408:41−8.
[56] Sun M, Feng J, Bu Y, Luo C. J Chromatogr A 2016;1458:1−8.
[57] Bu Y, Feng J, Wang X, Tian Y, Sun M, Luo C. J Chromatogr A 2017;1483:48−53.
[58] Bu Y, Feng J, Tian Y, Wang X, Sun M, Luo C. J Chromatogr A 2017;1517:203−8.
[59] Feng J, Wang X, Tian Y, Luo C, Sun M. J Chromatogr A 2018;1577:8−14.
[60] Feng J, Sun M, Bu Y, Luo C. Talanta 2016;148:313−20.
[61] Ling X, Chen Z. Talanta 2019;192:142−6.
[62] Andrade MA, Lanças FM. J Chromatogr A 2017;1493:41−8.
[63] Wang S, Wang X, Wang L, Pu Q, Du W. Anal Chim Acta 2018;1018:70−7.
[64] Cháfer-Pericás C, Campíns-Falcó P, Prieto-Blanco MC. Anal Chim Acta 2008;610: 268−73.
[65] Jiang N, Wang JB, Li WB, Xiao JH, Li JH, Lin XC, et al. J Chromatogr A 2019;1585: 19−26.
[66] Wu FL, Wang JB, Zhao Q, Jiang N, Lin XC, Xie ZH, Li JH, Zhang QQ. J Chromatogr B 2017;1040:214−21.
[67] Feng JJ, Wang XQ, Tian Y, Luo CN, Sun M. J Sep Sci 2018;41:3267−74.
[68] Wu FL, Xu CS, Jiang N, Wang JB, Ding CF. Talanta 2018;178:1−8.
[69] Feng JJ, Han S, Ji X, Li C, Wang X, Tian Y, Sun M. J Sep Sci 2019;42:1051−7.
[70] Ji XP, Feng JJ, Wang XQ, Tian Y, Li CY, Luo C, Sun M. J Sep Sci 2018;41:4480−7.
[71] Pang JL, Yuan DX, Huang XJ. J Chromatogr A 2018;1571:29−37.
[72] Wang X, Feng J, Tian Y, Luo C, Sun M. J Chromatogr A 2018;1550:1−7.
[73] Feng JJ, Mao HJ, Wang XQ, Tian Y, Luo CN, Sun M. J Sep Sci 2018;41:1839−46.
[74] Feng JJ, Tian Y, Wang XQ, Luo CN, Sun M. J Sep Sci 2018;41:1149−55.
[75] Feng JJ, Wang XQ, Tian Y, Luo CN, Sun M. J Sep Sci 2017;40:4773−9.
[76] Feng JJ, Wang XQ, Tian Y, Bu YA, Luo CN, Sun M. J Chromatogr A 2017;1517: 209−14.
[77] Pang J, Mei M, Yuan D, Huang X. Talanta 2018;184:411−7.
[78] Lashgari M, Yamini Y. J Sep Sci 2018;41:2393−400.
[79] Wang XQ, Feng JJ, Feng JQ, Tian Y, Luo CN, Sun M. Anal Meth 2018;10:2234−41.
[80] Gonzalez-Fuenzalida RA, Sanjuan-Navarro L, Moliner-Martinez Y, Campins-Falco P. Sci Total Environ 2018;630:1226−36.
[81] Jillani SMS, Alhooshani K. J Chromatogr A 2018;1543:14−22.
[82] Serra-Mora P, Rodriguez-Palma CE, Verdu-Andres J, Herraez-Hernandez R, Campins-Falco P. Sep 2018;5:10.
[83] Wang XQ, Pan L, Feng JJ, Tian Y, Luo CN, Sun M. J Chromatogr A 2017;1522:16−22.
[84] Fernandez-Amado M, Prieto-Blanco MC, Lopez-Mahia P, Muniategui-Lorenzo S, Prada-Rodriguez D. J Chromatogr A 2017;1520:35−47.
[85] Jornet-Martínez N, Ortega-Sierra A, Verdú-Andrés J, Herráez-Hernández R, Campíns-Falcó P. Molecules 2018;23:2359.

[86] Shamsayei M, Yamini Y, Asiabi H. J Sep Sci 2018;41(18):3598−606.
[87] Inukai T, Kaji S, Kataoka H. J Pharm Biomed Anal 2018;156:272−7.
[88] Saito A, Hamano M, Kataoka H. J Sep Sci 2018;41:2743−9.
[89] Wang R, Chen ZL. J Sep Sci 2018;41:2767−73.
[90] Hakobyan L, Pla-Tolos J, Moliner-Martinez Y, Molins-Legua C, Ruiz-Ramos J, Gordon M, Ramirez-Galleymore P, Campins-Falco P. J Pharm Biomed Anal 2018;151: 170−7.
[91] Jiang HP, Liu T, Guo N, Yu L, Yuan BF, Feng YQ. Anal Chim Acta 2017;981:1−10.
[92] Wang CL, Zhou W, Liao XY, Zhang WP, Chen ZL. Microchim Acta 2017;184:2715−21.
[93] Chang Q, Peng YE, Yun LF, Zhu QX, Hu SH, Shuai Q. Anal Chem 2017;89:4147−52.
[94] Mei M, Pang JL, Huang XJ. Anal Meth 2018;10:1977−85.
[95] Mei M, Huang XJ. J Chromatogr A 2017;1525:1−9.
[96] Alidoust M, Seidi S, Rouhollahi A, Shanehsaz M. Microchim Acta 2017;184:2473−81.
[97] Asiabi H, Yamini Y, Shamsayei M. Talanta 2018;185:80−8.
[98] Shamsayei M, Yamini Y, Asiabi H. J Chromatogr A 2018;1569:70−8.
[99] Shamsayei M, Yamini Y, Asiabi H, Safari M. Microchim Acta 2018;185:192.
[100] Xia Z, Gan T, Chen H, Lv R, Wei W, Yang F. J Sep Sci 2010;33:3221−30.
[101] Zheng MM, Lin B, Feng YQ. J Chromatogr A 2007;1164:48−55.
[102] Asiabi H, Yamini Y, Seidi S, Shamsayei M, Safari M, Rezaei F. Anal Chim Acta 2016; 922:37−47.
[103] González-Fuenzalida RA, López-García E, Moliner-Martínez Y, Campíns-Falcó P. J Chromatogr A 2016;1432:17−25.
[104] Mei M, Huang X, Ynag X, Luo Q. Anal Chim Acta 2016;937:69−79.
[105] Sajid M, Basheer C. Trac Trends Anal Chem 2016;75:174−82.
[106] Zarejousheghani M, Moder M, Borsdorf H. Anal Chim Acta 2013;708:48−55.
[107] Moghadam HP, Lendo S, Pawliszyn J. Anal Chem 2016;88:12188−95.
[108] Wang W, Gu C, Lunch KB, Lu JJ, Zhang Z, Pu Q, et al. Anal Chem 2014;86:1958−64.

# Needle extraction device

15

*Ikuo Ueta* [1], *Yoshihiro Saito* [2]
[1]Department of Applied Chemistry, University of Yamanashi, Kofu, Yamanashi, Japan;
[2]Department of Applied Chemistry and Life Science, Toyohashi University of Technology, Toyohashi, Aichi, Japan

## 15.1 Introduction

A needle extraction device (also known as needle trap or as a type of in-tube extraction) is an apparatus utilized for the sample preparation of volatile organic compounds (VOCs) for gas chromatographic (GC) analysis developed in the early 2000s [1−4]. A needle extraction device typically includes a solid extraction medium or coating material contained within a stainless-steel needle. A needle-shaped extraction device containing Within the needle a polymeric liquid phase coating, for instance, poly(dimethylsiloxane) (PDMS) is called a solid phase dynamic extraction (SPDE) device. When the objective of the procedure is the extraction of organic compounds from gas samples, the extraction needle is attached to a gas sampling pump or syringe, and the gas sample is collected via the tip of the extraction needle. Subsequently, the needle is inserted into the injection port of a gas chromatograph, and the extracted analytes are thermally desorbed into the gas phase. Since the extracted analytes are quickly desorbed from the extraction needle into a conventional GC injection port, the performance of the method does not require the use of a thermal desorption instrument or the implementation of a cryofocusing process.

An extraction needle device can be used to perform the extraction of VOCs from either gas or aqueous samples. Since sample preparation using a needle extraction device is typically performed via active sampling, the sensitivity of the ensuing analysis depends on the sample volume up to the corresponding breakthrough amount of the analytes from the needle device. Although to guarantee the reliability of the analytical data the extraction medium must have sufficient extraction power to collect the required amount of target compounds, a concern to be kept in mind is that the desorption of the extracted analytes into the injection port of the GC apparatus achieved by heating must be rapid and quantitative. Therefore, when using a needle extraction device, an appropriate choice of the adsorbent needs to be made based on the volatility and/or polarity of the target analyte(s). Notably, the needle extraction device can be reusable, which renders the utilization of this device cost-effective for VOC analysis. In fact, the extraction needle device has been employed for the analysis of a wide range of VOCs.

**Solid-Phase Extraction. https://doi.org/10.1016/B978-0-12-816906-3.00015-7**

## 15.2 Construction of a needle extraction device

When manufacturing a needle extraction device, stainless-steel needles, usually ranging in size between 22 (0.7 mm o.d. and 0.5 mm i.d.) and 20 gauge (0.9 mm o.d. and 0.7 mm i.d.), and in length between 50 and 100 mm, are typically employed [3,4]. The needle of the device has a side hole or a tip hole used to collect gas samples and to allow the extracted analytes to be desorbed into a GC injection port. A porous particulate extraction medium, for example, activated carbon (AC) or polymer beads, is packed inside the needle to act as the adsorbent (see the illustration in Fig. 15.1A). As to the particle size of the adsorbent, particles ranging in diameter from 200 to 850 μm (corresponding to 80 to 20 mesh) are typically used for gas or aqueous matrices. Choosing an adequate adsorbent is of course very important to achieve the quantitative extraction and desorption of the analytes [5]. Quartz wool or heat-resistant synthetic fiber is used to fix the particulate adsorbent. Fibrous extraction media of outer diameters of about 10 μm have also been used to pack extraction needles (Fig. 15.1B).

The most advantageous feature of needle extraction devices is the fact that they can be inserted directly into the injection port of conventional GC instruments. Therefore, analytical approaches that rely on needle extraction devices do not require expensive, specially designed desorption instruments. In fact, a quite simple desorption process can be realized using just a syringe containing desorption gas. In particular, small volumes (typically 0.5−1.0 mL) of nitrogen or helium are employed as desorption gases. Notably, desorption can also be achieved in some cases with a flow of GC carrier gas. Importantly, if the analyte is effectively desorbed during the injection process, high sensitivity can be obtained, even when the needle extraction device has a limited sampling volume.

## 15.3 Determination of VOC in gas samples

### 15.3.1 Extraction/desorption of VOCs

When aiming to collect gaseous samples, a gas sampling pump, a manual gas sampler, or a syringe is connected to a needle extraction device, and the gaseous sample is collected through the tip of the needle (Fig. 15.2). The sampling speed depends on

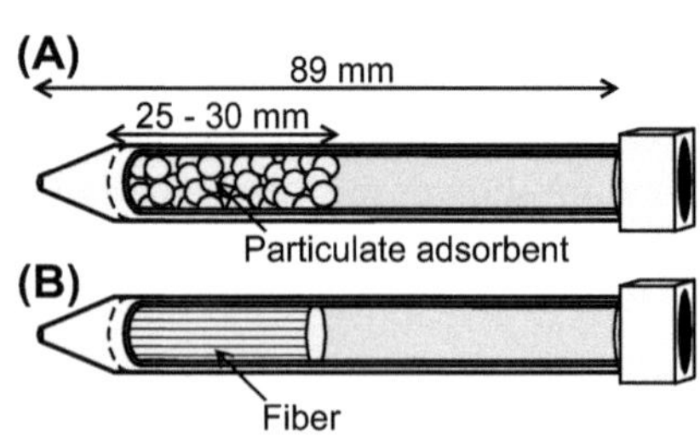


**Figure 15.1** Illustration of a needle extraction device. (A) Device packed with a particulate adsorbent. (B) Device packed with fiber.

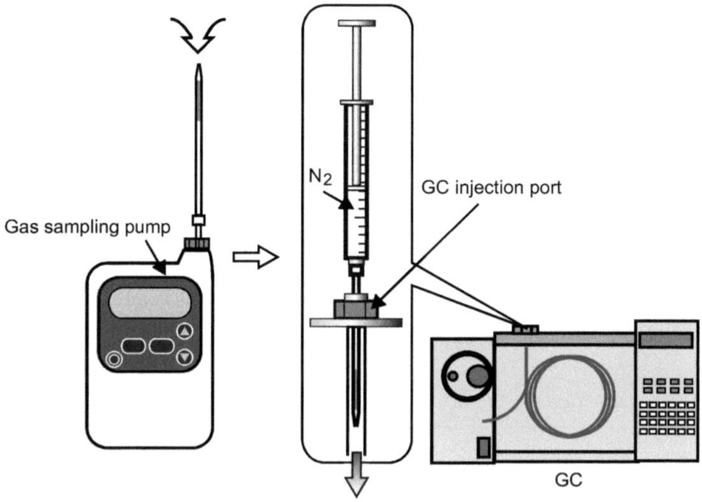


**Figure 15.2** Illustrations of the extraction and desorption processes for gaseous volatile organic compounds using a needle extraction device.

the particle size of the adsorbent and its packing length. Typically, the gas sampling speed using a needle extraction device is designed to range between a few mL/min and 10 mL/min. During sample collection, target VOCs are adsorbed onto the adsorbent packed inside the stainless-steel needle.

An adsorbent based on activated carbon (AC) showed greater extraction efficiency for a group of VOCs than polymer-based adsorbents, particularly for very volatile organic compounds (VVOCs) [6]. On the other hand, compounds of relatively high molecular weight (low-volatility organic compounds) were not successfully desorbed from a porous AC-based adsorbent, due to strong adsorption interactions between adsorbent and analyte. Therefore, use of a multibed-type adsorbent consisting of a polymer-based adsorbent, like divinylbenzene (DVB), and an AC-based adsorbent is a promising approach for the analysis of a wide range of VOCs [5,7,8]. In fact, compounds of relatively low molecular weight (i.e., highly volatile compounds) easily pass through the section of the polymer-based adsorbent, which is packed in the tip side of the needle extraction device; these compounds are instead adsorbed onto the AC-based adsorbent packed at the top of the adsorbent bed. By contrast, compounds of relatively low volatility are adsorbed on the polymer-based adsorbent. Importantly, both classes of compounds were successfully desorbed from the adsorbent by heating [7]. One of the examples is illustrated in Fig. 15.3, where the peak areas of high-volatility (top) and low-volatility (bottom) compounds at different gas sampling volumes using a DVB-packed extraction needle and a DVB/AC-packed multibed-type extraction needle are reported. As can be seen from the data in Fig. 15.3, highly volatile compounds are passed through the DVB-packed needle at large sample volumes, but they were adsorbed onto the AC layer.

In Table 15.1 a selection of suitable adsorbents for target VOCs is summarized [7]. Evidence indicates that volatile VOCs break through polymer-based adsorbents,

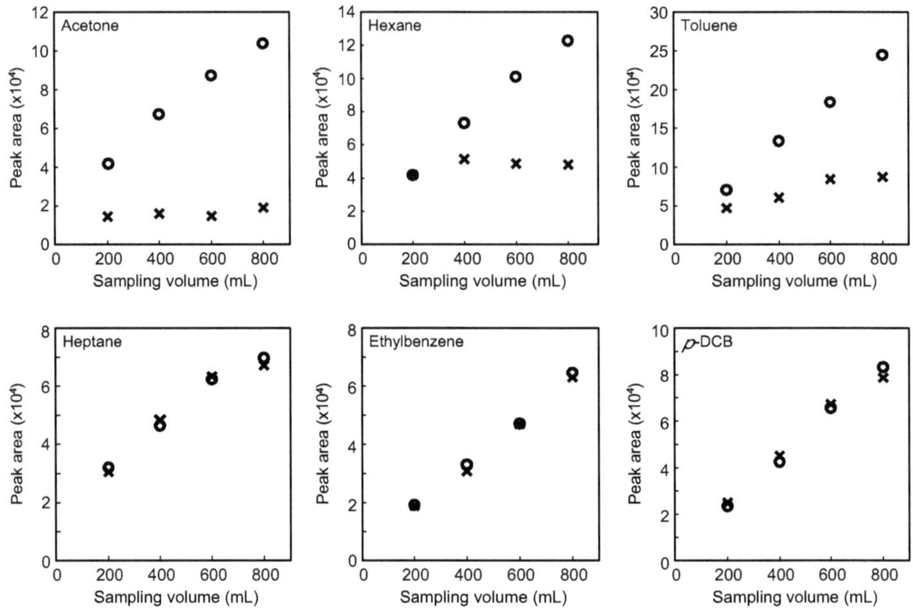


**Figure 15.3** Sampling volume profiles of divinylbenzene/activated carbon-packed needles (*circles*) and divinylbenzene-packed needles (*crosses*) for relatively high-volatility compounds (*top*) and low-volatility compounds (*bottom*).

**Table 15.1** Extraction and desorption Performances of various adsorbents of volatile organic compounds (VOCs).

| Adsorbent | Maximum temperature (°C) | Extraction of high-volatility VOCs[a] | Desorption of low-volatility VOCs[b] |
|---|---|---|---|
| MA-EGDMA | 200 | Fair | Poor |
| AC | 400 | Excellent | Poor |
| Tenax TA | 350 | Poor | Good |
| Tenax TA/AC | 350 | Collapse of Tenax TA | |
| DVB | 280 | Poor | Excellent |
| DVB/AC | 280 | Excellent | Excellent |

*AC*: activated carbon; *DVB*: divinylbenzene; *MA-EGDMA*: methacrylic acid-ethyleneglycol-dimethacrylate.
[a]Acetone, hexane, and toluene.
[b]Ethylbenzene, *p*-dichlorobenzene, and tetradecane.

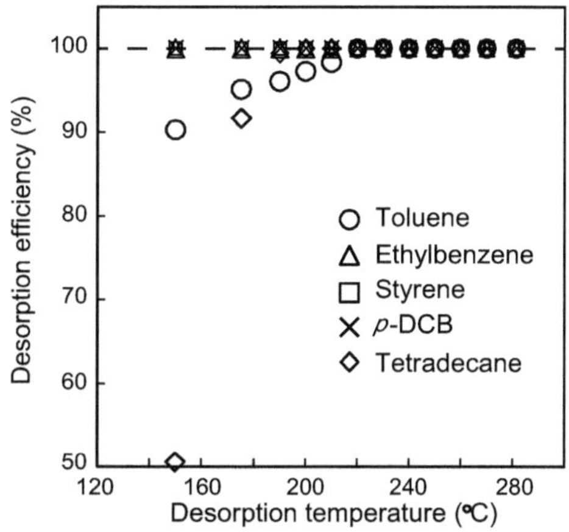


**Figure 15.4** Desorption efficiency of volatile organic compounds adsorbed onto a divinylbenzene/activated carbon-packed extraction needle.

whereas the recovery of low-volatility VOCs from AC-based adsorbents is poor. Notably, the extracted analytes are desorbed by heating the adsorbent (typically by thermal desorption performed in the injection port of the GC apparatus); therefore, the desorption temperature (GC injector temperature) is the most important parameter for the quantitative desorption of the extracted analytes from the extraction needle. As can be seen from the data reported in Fig. 15.4, the desorption efficiency of the extracted VOCs increases with the desorption temperature [7]. However, the upper temperature limit for polymeric adsorbents like styrene–divinylbenzene, ethylvinylbenzene–divinylbenzene, and methacrylic acid–ethyleneglycol-dimethacrylate (MA-EGDMA) ranges between 200 and 250°C. AC-based adsorbents, on the other hand, are typically used at higher temperatures [9]. By optimizing the desorption conditions, the extracted analytes are quantitatively desorbed from the needle extraction device with a desorption ratio of more than 99% in one injection. Notably, split injection is typically employed to prevent peak splitting and broadening. The desorption time is generally less than 1 min, but desorption times under 10 s can also be achieved with particular needle extraction devices. The extracted VOCs can be stored in the extraction needle at room temperature. For this purpose, a polytetrafluoroethylene (PTFE) cap and plug are attached to both ends of the needle, and a quantitative recovery of the extracted compounds after storage for up to 3 days has been reported for typical VOCs [4,7]. When utilizing the needle extraction device for environmental analyses, the percent recovery of the extracted VOCs stored in the needle is a significant parameter.

A number of different methods exist to desorb VOCs from the needle extraction device: (1) use of a gas-tight syringe containing a desorption gas ($N_2$ or air) [4,7], (2) use of the flow of a GC carrier gas [3], (3) use of the expansion of internal air [9], and (4) use of an autosampler [8,10]. Heating the adsorbent section by a separate heater was also reported as a VOC desorption strategy [11].

### 15.3.2 Use of needle extraction devices for gas analysis

Needle extraction devices have been applied in several fields for the determination of VOCs in gas samples. One of the most important examplesis the evaluation of the quality of indoor air in households and in the work environment. A needle extraction device packed with DVB and AC particles was used for the determination of a wide variety of VOCs in indoor air samples, with a limit of quantification (LOQ) of less than 1 ng/L reported for all target analytes, using mass spectrometry (MS) as the detection method [7]. In this study, toluene-$d_8$ was used as an internal standard. An extraction needle device was also employed for the determination of the concentrations of VOCs in the indoor air of school facilities [12]. Time-weighted average sampling using needle extraction devices was reported for passive sampling [3,13]. A nanoporous silica aerogel was utilized as adsorbent in a needle extraction device for the extraction of volatile aldehydes [14]. A needle extraction device coupled with a barrier discharge ionization detector (BID), which was recently developed as a universal GC detector that relies on the ionization of analytes by helium plasma, was reported for the determination of gaseous compounds of low molecular weight. In particular, an AC-packed needle was utilized for the extraction of formic and acetic acid from indoor air samples [15]; on the other hand, a carboxylic acid-coated terephthalic acid particle-packed needle was developed to extract volatile amines like ammonia, methylamine, and trimethylamine from air samples [16]. These low-molecular-weight compounds provide no or a low response when a flame ionization detector or an MS detector is used. Detection of gasoline and kerosene at the scene of fires using needle extraction devices was also reported [17]. These accelerants were successfully extracted from fire debris in a simulated fire. An MA-EGDMA polymer-beads-packed needle was used for the evaluation of third-hand smoking by detecting tobacco-related VOCs extracted from several types of fabrics and from the breath of smokers [18].

A variety of VOCs are present in the human breath, and the usefulness of some VOCs found in breath as biomarkers has been investigated for the rapid and noninvasive diagnosis of several diseases. A multibed-type extraction needle device containing Carboxen 1000, Carbopack X, and Tenax was used for the in vivo sampling of animal (pig) breath [19]. The same method was also employed for the determination of VOCs in the breath of healthy volunteers by comprehensive two-dimensional GC/time-of-flight MS [20]. In these studies, the extraction needle was connected to an automated syringe, and the extracted analytes were automatically desorbed using an autosampler. The concentrations of VOCs extracted from the breath of patients with heart failure were determined using a needle packed with DVB, Carbopack X, and Carboxen 1000 [21]; the concentrations of VOCs extracted from human breath and the gas generated from lung cancer cell lines were determined using a needle extraction device containing graphene coated with electrodeposited polyaniline [22]. The concentration of acetone in the breath of patients with type-2 diabetes was determined with an MA-EGDMA copolymer beads-packed needle. In this case, the breath sample was collected in a gas sampling bag, and a 50-mL aliquot of the sample was extracted using the above packed needle device [23]. In Fig. 15.5 the time-dependent profiles of the acetone concentration extracted from the breath and urine of a healthy volunteer

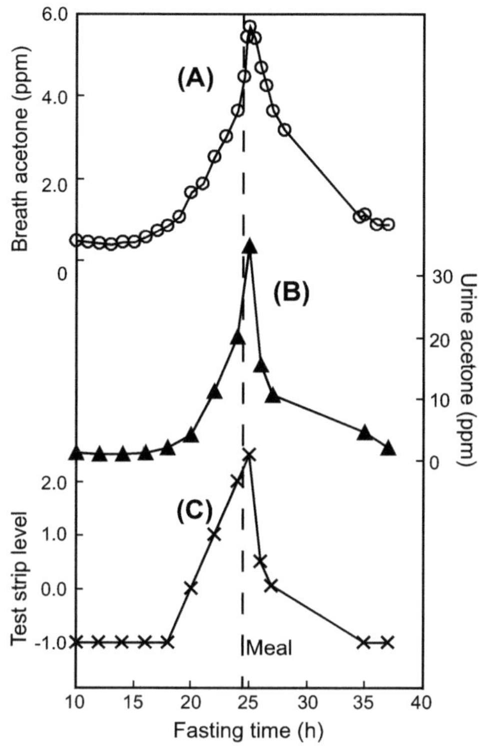


**Figure 15.5** Time-dependent variations in the level of acetone in the breath (A) and urine (B) of a healthy volunteer kept in fasting conditions and of the level of ketone bodies (C) in the urine of the same volunteer. A meal was made available to the volunteer when 24 h had passed since the previous meal.

under fasting conditions and also the ketone bodies extracted from the volunteer's urine are shown. These results clearly indicate the existence of a correlation between breath acetone and urine acetone levels; they also suggest that this approach could be used for the basis of a screening method for diabetes. In other studies, breath acetone and isoprene concentrations were determined with an extraction needle device comprising a Carbopack X and carbon molecular sieve (CMS)-packed needle [6,24].

Procedures based on the on-fiber derivatization/extraction of volatile compounds were reported that rely on the use of a fiber-packed needle extraction device. This approach allowed the detection of formaldehyde and acetaldehyde [25], ethylene oxide [26], and volatile amines [27]. A heat-resistant synthetic fiber, Zylon, was coated with PDMS as sorbent [28]. In this approach, the solution of the derivatization reagent is initially loaded onto the fiber-packed section of the needle. During the subsequent collection of the gas sample, the analytes react with the derivatization reagent, and the derivatized compounds are adsorbed onto the fiber. These compounds are then desorbed by a small amount of solvent passed through the fiber-packed needle during injection into the gas chromatograph.

# 15.4 Determination of VOCs in water samples

## 15.4.1 Extraction of VOCs

Water samples may contain several VOCs, some of which have adverse effects on human health, while others are used as flavoring agents in beverages and food items. If the water sample contains nonvolatile organic or inorganic compounds, a sample cleanup step is required prior to GC analysis. Purge and trap (PT) extraction (dynamic headspace extraction) is a popular sample preparation method for the analysis of VOCs in water samples. In the typical PT method, the sample is placed in a sealed vessel and the VOCs purged from the water into the headspace of the sealed vessel using a purging gas, typically helium or nitrogen. The VOCs are then transferred to a trap and extracted onto an adsorbent. Generally, implementation of the PT method affords higher sensitivity than for static headspace extraction methods [29]. In PT-based methods, the needle extraction device is exposed to the headspace of the sealed vessel, and the gas phase is dynamically collected through the needle using a gas sampling pump or a syringe. Fig. 15.6 illustrates a typical PT-based procedure for the extraction of VOCs from water using a needle extraction device. During the sampling of the headspace, the purge gas is continuously bubbled through the water by decompression in the vessel. Notably, PT-based extraction can also be effected by introducing a purge

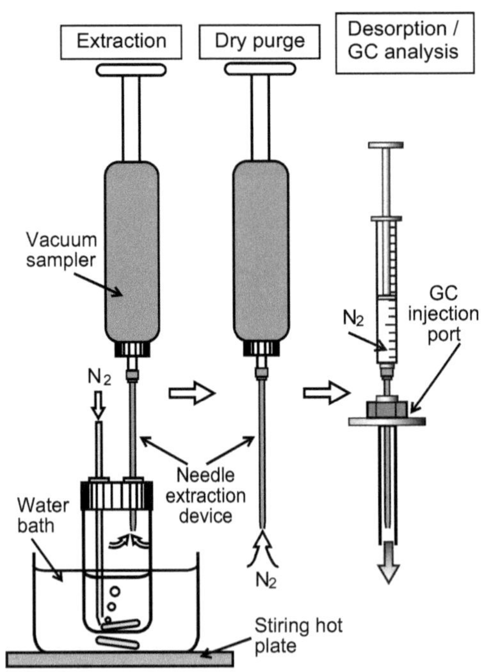


**Figure 15.6** Illustration of the purge and trap extraction of volatile organic compounds present in a water sample using a needle extraction device.

gas at a constant flow rate [30,31]. An inorganic salt (usually NaCl) can be added to the water sample to increase the efficiency with which target VOCs are purged into the headspace of the sealed vessel, a strategy particularly useful for relatively hydrophilic compounds. Adjusting the pH of the sample can be used to improve the purging efficiency of weak acids and bases. Following extraction of the headspace, dry nitrogen gas is used to remove excess water from the adsorbent (dry purge). In fact, too much moisture on the adsorbent results in low repeatability of the procedure and a low peak response, due to water being vaporized in the GC injection port. A volume of nitrogen gas of 50–100 mL is typically employed for the dry purge.

When the sample solution does not include nonvolatiles, the target analytes can be trapped onto the adsorbent's surface by forcing the sample solution through the extraction needle [32–34]. Static headspace sampling using a needle has also been reported, where the effects of the solution temperature, salt addition, and headspace gas sampling volume on the sensitivity of the analytes in static sampling were confirmed [35].

## 15.4.2 Use of needle extraction devices for water analysis

The concentrations of benzene, toluene, ethylbenzene, and xylene in water samples were determined using a Porapak Q-packed [32] and Carbopack X-packed extraction needles [30,35]. Furthermore, multibed-type extraction needle devices were shown to be suitable for the extraction of a wide range of VOCs from water samples [36]. A multibed-type extraction needle packed with DVB, AC, and CMS was developed for the extraction of 23 different VOCs from water samples [37]. In this study, the LOQs for the VOCs ranged from 0.03 to 0.6 μg/L; all values were less than 1/10 of the reference value established for drinking water by the WHO, indicating the suitability of the technique for drinking water analysis.

The chromatograms for the analysis of VOC in a standard water sample and tap water are shown in Fig. 15.7. In both cases, fluorobenzene was used as the internal standard. This method was implemented for the analysis of musty odor compounds of 2-methylisoborneol and geosmin present in trace amounts in tap water. Although typically the analyte peak area increases with the purge temperature, due to an increase in the efficiency of VOC purging from the water sample, in the PT-based extraction using a needle extraction device, the purge temperature is limited to approximately 40°C, because at higher temperatures the extraction of the headspace gas is rendered difficult by condensation of large amounts of water vapor. In this method, the LOD for 2-methylisoborneol and geosmin were 1.0 and 0.5 ng/L, respectively [38]. The concentrations of flavoring compounds like limonene and monoterpene alcohols in fruit juice were determined using extraction needle devices containing a Carbopack-X-packed extraction needle [39]. In general, detecting hydrophilic compounds by a PT-based method is difficult, because the purging of these compounds from water is a low-efficiency process. Another class of compounds that are difficult to extract by this method are low-molecular-weight compounds, due to their high volatility. The determination of the concentration of hydrophilic and low-molecular-weight compounds, like methanol, acetaldehyde, and acetone, has been performed using a

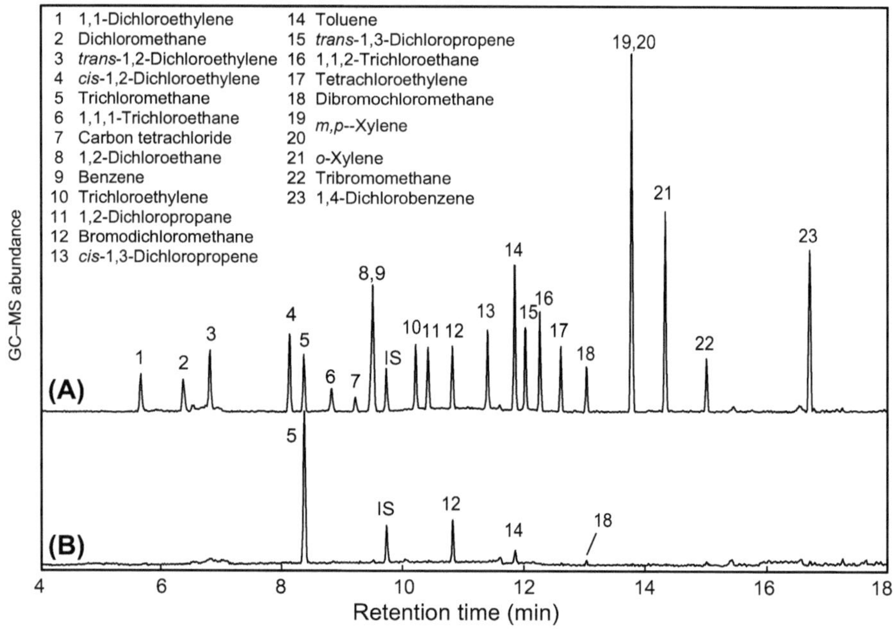


**Figure 15.7** Determination of volatile organic compounds (VOCs) contained in a water sample performed with a needle extraction device and purge and trap-based gas chromatography/mass spectrometry. (A) Results of an analysis performed on a standard water sample containing 23 VOCs. (B) Results of an analysis performed on a tap water sample.

needle extraction device packed with CMS particles [40]. In this case, the LOQs for methanol, acetaldehyde, and acetone were 75, 75, and 0.5 μg/L, respectively. As an application of the method, trace amounts of methanol, acetaldehyde, and ethanol were identified in grape juice samples (Fig. 15.8). Trace amounts of methanol and acetaldehyde were also found in orange juice samples.

An extraction needle packed with Tenax TA and Carboxen 1000 was utilized for the extraction of VOC from water [41]. Analytical methods combining PT-based needle extraction with GC-BID were reported for the determination of formic acid and acetic acid [42]. In this case, the pH of the water sample was adjusted to below 2 by addition of $H_2SO_4$ to increase the purge efficiency. The PT procedure coupled with a needle extraction device was also employed for the determination of ammonia in water [43]. To determine the concentration of aqueous ammonia or of the ammonium ion, the Nessler method or the indophenol method is widely used; notably, however, when implementing these methods, reaction conditions, such as reaction temperature, pH, and time, must be strictly controlled. The PT-based needle extraction followed by GC-BID analysis proved to be a simple and rapid procedure to determine the concentration of aqueous ammonia. The sample pH was adjusted to above 12 by addition of a NaOH solution. The determination of polycyclic aromatic hydrocarbons

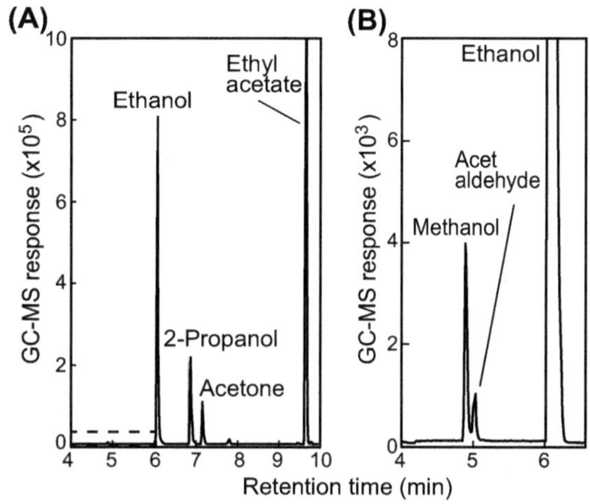


**Figure 15.8** Determination of very volatile organic compounds (A) in a grape juice sample using a needle extraction device and purge and trap—gas chromatography/mass spectrometry and (B) its enlarged view.

by a carbon nanotube powder-packed extraction needle [44] and the determination of chlorobenzenes by a modified nanoporous silica aerogel-packed needles [45,46] have also been reported. In addition, the determination of several VOCs using a PDMS/AC-coated needle (SPDE needle) was reported [47,48]. In PDMS-coated needles, the analytes are adsorbed on the PDMS layer based on their equilibrium distribution between the PDMS layer and the headspace gas. For the extraction of low-volatility organic compounds, an extraction method relying on the analyte equilibrium distribution between the adsorbent and gas phase is suitable. A needle extraction device coated with sol-gel titania was used for the extraction of the morphine derivatives desomorphine and desocodeine from water samples [49].

## 15.5 Conclusions

Needle extraction devices allow the simple, rapid, and sensitive analyses of various VOCs at low concentrations in gaseous and aqueous samples. In these analyses, the VOCs can be quantitatively extracted and desorbed by selecting the appropriate adsorbent for the target compounds. These extraction devices have been employed in several contexts, including environmental, food and beverage, forensic, pharmaceutical, and biological analyses. Needle extraction technique can be employed in combination with an autosampler, enabling high-throughput in the routine analysis of VOCs. Notably, recent applications of needle extraction devices are described in a well-written review article [50].

# References


[1] Lipinski J. Automated solid phase dynamic extraction – extraction of organics using a wall coated syringe needle. Fresenius J Anal Chem 2001;369:57–62.

[2] Koziel JA, Odziemkowski M, Pawliszyn J. Sampling and analysis of airborne particulate matter and aerosols using in-needle trap and SPME fiber devices. Anal Chem 2001;73: 47–54.

[3] Wang A, Fang F, Pawliszyn J. Sampling and determination of volatile organic compounds with needle trap devices. J Chromatogr A 2005;1072:127–35.

[4] Saito Y, Ueta I, Kotera K, Ogawa M, Wada H, Jinno K. In-needle extraction device designed for gas chromatographic analysis of volatile organic compounds. J Chromatogr A 2006;1106:190–5.

[5] Trefz P, Kischkel S, Hein D, James ES, Schubert JK, Miekisch W. Needle trap micro-extraction for VOC analysis: effects of packing materials and desorption parameters. J Chromatogr A 2012;1219:29–38.

[6] Ueta I, Samsudin EL, Mizuguchi A, Takeuchi H, Shinki T, Kawakubo S, et al. Double-bed-type extraction needle packed with activated-carbon-based sorbents for very volatile organic compounds. J Pharm Biomed Anal 2014;88:423–8.

[7] Ueta I, Mizuguchi A, Fujimura K, Kawakubo S, Saito Y. Novel sample preparation technique with needle-type micro-extraction device for volatile organic compounds in indoor air samples. Anal Chim Acta 2012;746:77–83.

[8] Laaks J, Jochmann MA, Schilling B, Schmidt TC. Optimization strategies of in-tube extraction (ITEX) methods. Anal Bioanal Chem 2015;407:6827–38.

[9] Eom IY, Pawliszyn J. Simple sample transfer technique by internally expanded desorptive flow for needle trap devices. J Sep Sci 2008;31:2283–7.

[10] Trefz P, Rösner L, Hein D, Schubert JK, Miekisch W. Evaluation of needle trap micro-extraction and automatic alveolar sampling for point-of-care breath analysis. Anal Bioanal Chem 2013;405:3105–15.

[11] Jochmann MA, Yuan X, Schilling B, Schmidt TC. In-tube extraction for enrichment of volatile organic hydrocarbons from aqueous samples. J Chromatogr A 2008;1179:96–105.

[12] Inoue M, Mizuguchi A, Ueta I, Takahashi K, Saito Y. Rapid on-site air sampling with a needle extraction device for evaluating the indoor air environment in school facilities. Anal Sci 2013;29:519–25.

[13] Cheng WH, Zhan W, Pawliszyn J. Extraction of gaseous VOCs using passive needle trap samplers. Aerosol Air Qual Res 2011;11:387–92.

[14] Azari MR, Barkhordari A, Zendehdel R, Heidari M. A novel needle trap device with nanoporous silica aerogel packed for sampling and analysis of volatile aldehyde compounds in air. Microchem J 2017;134:270–6.

[15] Ueta I, Nakamura Y, Fujimura K, Kawakubo S, Saito Y. Determination of gaseous formic and acetic acids by a needle-type extraction device coupled to a gas chromatography-barrier discharge ionization detector. Chromatographia 2017;80:151–6.

[16] Ueta I, Nakamura Y, Fujikawa H, Fujimura K, Saito Y. Determination of volatile amines using needle-type extraction coupled with gas chromatography-barrier discharge ionization detection. Chromatographia 2019;82:317–23.

[17] Ueta I, Saito Y, Teraoka K, Matsuura H, Fujimura K, Jinno K. Novel fire investigation technique using needle extraction in gas chromatography. Anal Sci 2010;26:1127–32.

[18] Ueta I, Saito Y, Teraoka K, Miura T, Jinno K. Determination of volatile organic compounds for a systematic evaluation of third-hand smoking. Anal Sci 2010;26:569–74.

[19] Mieth M, Kischkel S, Schubert JK, Hein D, Miekisch W. Multibed needle trap devices for on site sampling and preconcentration of volatile breath biomarkers. Anal Chem 2009;81: 5851–7.

[20] Mieth M, Schubert JK, Gröger T, Sabel B, Kischkel S, Fuchs P, et al. Automated needle trap heart-cut GC/MS and needle trap comprehensive two-dimensional GC/TOF-MS for breath gas analysis in the clinical environment. Anal Chem 2010;82:2541–51.

[21] Biagini D, Lomonaco T, Ghimenti S, Bellagambi FG, Onor M, Scali MC, et al. Determination of volatile organic compounds in exhaled breath of heart failure patients by needle trap micro-extraction coupled with gas chromatography-tandem mass spectrometry. J Breath Res 2017;11:047110.

[22] Li Y, Li JH, Xu H. Graphene/polyaniline electrodeposited needle trap device for the determination of volatile organic compounds in human exhaled breath vapor and A549 cell. RSC Adv 2017;7:11959.

[23] Ueta I, Saito Y, Hosoe M, Okamoto M, Ohkita H, Shirai S, et al. Breath acetone analysis with miniaturized sample preparation device: in-needle preconcentration and subsequent determination by gas chromatography–mass spectroscopy. J Chromatogr B 2009;877: 2551–6.

[24] Ueta I, Mizuguchi A, Okamoto M, Sakamaki H, Hosoe M, Ishiguro M, et al. Determination of breath isoprene and acetone concentration with a needle-type extraction device in gas chromatography–mass spectrometry. Clin Chim Acta 2014;430:156–9.

[25] Saito Y, Ueta I, Ogawa M, Jinno K. Simultaneous derivatization/preconcentration of volatile aldehydes with a miniaturized fiber-packed sample preparation device designed for gas chromatographic analysis. Anal Bioanal Chem 2006;386:725–32.

[26] Ueta I, Saito Y, Ghani NBA, Ogawa M, Yogo K, Abe A, et al. Rapid determination of ethylene oxide with fiber-packed sample preparation needle. J Chromatogr A 2009;1216: 2848–53.

[27] Nakagami K, Tazawa T, Sumiya O, Ueta I, Saito Y. Simultaneous derivatization and extraction of volatile amines with fiber-packed needle and subsequent analysis in gas chromatography. Chromatography 2018;39:75–81.

[28] Saito Y, Ueta I, Ogawa M, Abe A, Yogo K, Shirai S, et al. Fiber-packed needle-type sample preparation device designed for gas chromatographic analysis. Anal Bioanal Chem 2009;393:861–9.

[29] Kremser A, Jochmann MA, Schmidt TC. Systematic comparison of static and dynamic headspace sampling techniques for gas chromatography. Anal Bioanal Chem 2016;408: 6567–79.

[30] Jurdáková H, Kubinec R, Jurčišinová M, Krkošová Z, Blaško J, Ostrovský I, et al. Gas chromatography analysis of benzene, toluene, ethylbenzene and xylenes using newly designed needle trap device in aqueous samples. J Chromatogr A 2008;1194:161–4.

[31] Mesarchaki E, Yassaa N, Hein D, Lutterbeck HE, Zindler C, Williams J. A novel method for the measurement of VOCs in seawater using needle trap devices and GC–MS. Mar Chem 2014;159:1–8.

[32] Kubinec R, Berezkin VG, Górová R, Addová G, Mračnová H, Soják L. Needle concentrator for gas chromatographic determination of BTEX in aqueous samples. J Chromatogr B 2004;800:295–301.

[33] Ogawa M, Saito Y, Ueta I, Jinno K. Fiber-packed needle for dynamic extraction of aromatic compounds. Anal Bioanal Chem 2007;388:619–25.

[34] Pietrzyńska M, Voelkel A. Optimization of the in-needle extraction device for the direct flow of the liquid sample through the sorbent layer. Talanta 2014;129:392–7.

[35] Crom JD, Claeys S, Godayol A, Alonso M, Anticó E, Sanchez JM. Sorbent-packed needle microextraction trap for benzene, toluene, ethylbenzene, and xylenes determination in aqueous samples. J Sep Sci 2010;33:2833–40.

[36] Laaks J, Jochmann MA, Schilling B, Schmidt TC. In-tube extraction of volatile organic compounds from aqueous samples: an economical alternative to purge and trap enrichment. Anal Chem 2010;82:7641–8.

[37] Ueta I, Razak NA, Mizuguchi A, Kawakubo S, Saito Y, Jinno K. Needle-type extraction device for the purge and trap analysis of 23 volatile organic compounds in tap water. J Chromatogr A 2013;1317:211–6.

[38] Ueta I, Mitsumori T, Kawakubo S, Saito Y. Determination of musty odor compounds in water by gas chromatography-mass spectrometry with a needle-type sample preparation device. Anal Sci 2014;30:979–83.

[39] Ueta I, Mitsumori T, Suzuki Y, Kawakubo S, Saito Y. Purge-and-trap analysis of flavor compounds in aqueous samples by a needle-type extraction device. Chromatography 2015; 36:99–104.

[40] Ueta I, Mitsumori T, Suzuki Y, Kawakubo S, Saito Y. Determination of very volatile organic compounds in water samples by purge and trap analysis with a needle-type extraction device. J Chromatogr A 2015;1397:27–31.

[41] Alonso M, Cerdan L, Godayol A, Anticó E, Sanchez JM. Headspace needle-trap analysis of priority volatile organic compounds from aqueous samples: application to the analysis of natural and waste waters. J Chromatogr A 2011;1218:8131–9.

[42] Ueta I, Nakamura Y, Kawakubo S, Saito Y. Determination of aqueous formic and acetic acids by purge-and-trap analysis with a needle-type extraction device and gas chromatography barrier discharge ionization detector. Anal Sci 2018;34:201–5.

[43] Ueta I, Fujikawa H, Fujimura K, Saito Y. Purge-and-trap determination of ammonia in water samples using needle-type extraction coupled with gas chromatography-barrier discharge ionization detection. Anal Sci 2019;35:759–62.

[44] Bagheri H, Ayazi Z, Aghakhani A. A novel needle trap sorbent based on carbon nanotube-sol–gel for microextraction of polycyclic aromatic hydrocarbons from aquatic media. Anal Chim Acta 2011;683:212–20.

[45] Roostaie A, Bargozin H, Mohammadiazar S, Ehteshami S. Nanoporous silica aerogel modified by triethylchlorosilane as a new sorbent for the needle-trap extraction. Sep Sci plus 2018;1:76–82.

[46] Roostaie A, Mohammadiazar S, Bargozin H, Ehteshami S. A modified nanoporous silica aerogel as a new sorbent for needle trap extraction of chlorobenzenes from water samples. Chromatographia 2018;81:649–55.

[47] Malherbe S, Watts V, Nieuwoudt HH, Bauer FF, Toit MD. Analysis of volatile profiles of fermenting grape must by headspace solid-phase dynamic extraction coupled with gas chromatography-mass spectrometry (HS-SPDE GC-MS): novel application to investigate problem fermentations. J Agric Food Chem 2009;57:5161–6.

[48] Bicchi C, Cordero C, Liberto E, Rubiolo P, Sgorbini B. Automated headspace solid-phase dynamic extraction to analyse the volatile fraction of food matrices. J Chromatogr A 2004; 1024:217–26.

[49] Su CJ, Srimurugan S, Chen C, Shu HC. Sol-gel titania-coated needles for solid phase dynamic extraction-GC/MS analysis of desomorphine and desocodeine. Anal Sci 2011;27: 1107–13.

[50] Kedziora-Koch K, Wasiak W. Needle-based extraction techniques with protected sorbent as powerful sample preparation tools to gas chromatographic analysis: trends in application. J Chromatogr A 2018;1565:1–18.

# Micro-solid-phase extraction

16

*Nyi Nyi Naing* [1], *Sze Chieh Tan* [2,3], *Hian Kee Lee* [1,2,3,4]
[1]National University of Singapore Environmental Research Institute, National University of Singapore, Singapore, Singapore; [2]NUS Graduate School for Integrative Sciences and Engineering, National University of Singapore, Singapore, Singapore; [3]Department of Chemistry, National University of Singapore, Singapore, Singapore; [4]Tropical Marine Science Institute, National University of Singapore, Singapore, Singapore

## 16.1 Introduction

Sampling and sample preparation represent precursory steps of an analytical procedure that are subsequently followed by detection and quantification, and data collection and evaluation. Sampling, particularly in environmental analysis, is important since the representativeness and integrity of the sample need to be assured. Sample preparation, however, is even more critical whether or not the sample matrices are simple or complex to ensure that the analytes are satisfactorily isolated, cleaned up, and enriched in the final extract, and are in a state that is compatible with the intended analytical technique [1].

Diverse sample preparation approaches have been developed for aqueous samples (the main matrix considered in this chapter). Particularly, in the past two-and-a-half decades, scientists have given strong emphasis to the miniaturization of these processes [2–5]. The incentive for the development of miniaturized procedures is the desire to minimize, if not eliminate, the following disadvantages of conventional-scale methods: consumption of large amounts of organic solvents with the concomitant generation of wastes, processing of large volumes of aqueous samples, and excessive number of operational steps [2].

A typical feature of such miniaturized procedures is their solventless or solvent-minimized character. In the former, best exemplified by solid-phase microextraction (SPME), a small amount of sorbent material in the form of a coating is presented to a liquid sample or its headspace. Solvent-minimized procedures can be classified broadly as liquid-phase microextraction (LPME) [6], and include the following variants, single-drop microextraction [7], hollow fiber-liquid-phase microextraction (HF-LPME) [8], and dispersive liquid-liquid microextraction (DLLME) [9,10], among others. While the present article is largely on sorbent-based sample preparation, the procedures, as shall be seen later, can also be synergistically combined with various LPME approaches to improve overall extraction efficiencies.

It was the use of a sorbent, calcium carbonate, to extract plant pigments by Mikhail Tswett [11,12] that set off the chromatography field, in the early part of the 20th century. Essentially, the procedure originated by Tswett is now referred to as solid-phase extraction (SPE) which was commercially introduced in the mid-1970s [13,14].

**Solid-Phase Extraction. https://doi.org/10.1016/B978-0-12-816906-3.00016-9**

SPE remains one of the most widely-used sample preparation procedures today for processing liquid (mainly aqueous) matrices.

SPME and stir-bar sorptive extraction are miniaturized progressions of sorbent-based extraction devices, with both introduced in the late 1980s/early 1990s. Their formats are different from conventional SPE, however, with coatings as the extractive phase, rather than fine particles or powder. The most commonly utilized SPE devices are in the form of cylindrical cartridges packed with sorbents (although disk sorbent-impregnated membranes are also available).

Over the past 15–20 years, there have been extensive studies involving the miniaturization of SPE with retention of the traditional constructs, that is, fixed sorbent bed (cartridge- or disk-based) SPE and loose media (dispersive SPE (DSPE)) formats. DSPE is a significantly different implementation of SPE from cartridge- or disk-based systems. In this procedure, loose sorbent is simply dispersed in the sample for extraction. The spent sample and the sorbent with adsorbed analytes are then physically separated (by filtration, centrifugation, or magnetism (if the sorbent is magnetized), etc.) [15], before a solvent is used to elute the analytes for analysis using a suitable technique [16]. The generic term, *micro-SPE* (μ-SPE) has been used in many such SPE studies in reference to the trend toward miniaturization, with the "μ" appellation meant to indicate reduction in device dimensions, reduced amount of sorbents, and/or use of microsized or nanosized materials (particularly the latter with their advantageously large surface area and exceptional physicochemical properties).

Numerous contributions to the sample preparation field have involved the in-house synthesis, characterization, and application of novel materials as sorbents for analyte extraction [17–19]. In most cases, these new materials and their structural functionalities are considered for applications in which their selectivity and stability are exploited, to provide higher extraction efficiency for specific classes of compounds [11]. These sorbents are usually claimed to overcome some of the limitations of commercial ones that are generally more universal and therefore lacking in selectivity [2,20–23]. Nevertheless, new materials appear on the market frequently, and can also be evaluated for their applicability in sample preparation. For example, one recent μ-SPE work [24] reported the use of buckypaper, an aggregate of intertwined and entangled multiwalled carbon nanotubes. The material was in the shape of a three-dimensional disk; it was enclosed within a mesh which also contained a small magnetic stir-bar to provide a self-contained microextraction unit.

Since DSPE is easy to use and implement, the evaluation of the properties and performance of in-house prepared sorbents is most conveniently examined using this approach. The term *dispersive μ-SPE* was coined to refer to DSPE using a small amount (microgram or milligram quantities) of a sorbent of, for example, nanoparticles or their functionalized counterparts [16]. The dispersion approach enables the sorbent to interact more rapidly and uniformly with the target analytes (compared to the fixed-bed format), leading to faster equilibrium being attained. This improves extraction efficiency and extraction time. Additionally, sample agitation by ultrasonication [25] or vortex [26] can further facilitate the mass transfer kinetics pertaining to the extraction.

There are several commercially available miniaturized SPE devices (primarily based on their reduced dimensions) although they are not specifically designated as "μ-SPE," even though they certainly fit the criteria. These are microextraction by packed sorbent (MEPS), pipette-tip SPE (PT-SPE), disposable pipette extraction (DPX), and extraction by needle trap device (NTD). (It should be stated that researchers can synthesize and use their own materials and devices for these techniques.) For the sake of completeness in covering miniaturized modes of SPE, these procedures are discussed briefly, with examples of applications.

In MEPS [27,28], about 1 mg of the sorbent is packed inside a syringe (with volumes ranging between 100 and 150 μL), or in a cartridge that is inserted between the syringe barrel and the needle. In contrast with conventional SPE, sample and elution solvents are forced through the sorbent via the needle end. Sorbents available for MEPS are the same as those for conventional SPE. Similar to MEPS is PT-SPE in which a pipette tip is partially filled with sorbent (between 1 and 5 mg). The asymmetric pipette tip-shaped configuration is considered beneficial, with the wider end useful for sample loading and solvent elution. The narrow diameter tip allows a smaller amount of sorbent to be used while still providing efficient extraction. PT-SPE applications include those reporting the extraction of fluoroquinolones [29] and sulfonamides [30] in which molecularly imprinted microspheres and graphene, respectively, are used as sorbents. The commercial PT-SPE products are given names such as ZipTip, NuTip, and MonoTip $C_{18}$ [31,32].

DPX [33] is a combination of PT-SPE and DSPE: the sorbent is loosely confined with a section of the pipette space flanked by two filters. Like MEPS, the sample is drawn in through the bottom of the device. Air is then aspirated, generating bubbles to mix the samples and sorbent more effectively. After discarding the spent sample, the elution solvent and air are sequentially aspirated to desorb the analytes. As is common with the other procedures, DPX consumes only small amounts of sorbent and solvent [34,35], relative to SPE. Apart from the sorbents supplied by the manufacturer, in-house sorbents can also be applied [36]. Like MEPS, DPX can be automated on commercial autosampler systems.

Finally, the last of the commercialized miniaturized SPE procedures to be mentioned here is NTD extraction. The latter resembles an MEPS device except that the sorbent is housed within the syringe needle itself [37,38]. Operationswise, there is little difference among the MEPS, DPX, and NTD methods.

Most of the device-based procedures described above have similar physical configurations to conventional cartridge SPE in that the sorbents are contained within plastic, glass, or metallic structures; the only difference is that the physical dimensions are smaller. In 2006, a rather unusual SPE format, leveraging to some extent on the advantages of DSPE, was reported. Termed as "μ-SPE" [39], the extraction is conducted using a device that comprises a porous membrane envelope containing a small amount of sorbent. As in DSPE, this μ-SPE device is simply added to the sample in which it moves freely in the vortex of the stirred solution during extraction. After extraction, the device is conveniently retrieved and placed in a small-volume tube, and eluted with a solvent to release the analytes. In contrast to other miniaturized SPE formats, membrane-based μ-SPE provides protection of the sorbent from challengingly complex matrices by virtue of the envelope; further cleanup is thus unnecessary. In this regard, μ-SPE lends itself to the processing of a variety of aqueous matrices, including biological and environmental samples [40]. Like DSPE methods in general, another attractive feature is that sorbents, outside of what are available commercially, can be designed and tailored to the requirements of particular analytes. μ-SPE devices can be also cleaned and reused. This, together with the minimal requirements of sample, sorbent, and solvent, mean that the procedure subscribes to green chemistry principles, like all other microextraction techniques.

There have been only a limited number of reviews on membrane-based μ-SPE. Thus, the remainder of this article is devoted to this procedure in which there is a growing interest, accompanied by an increasing number of publications utilizing this technique. The chapter discusses the basic features, essential extraction parameters, variations and improved modes, and representative and recent applications of this μ-SPE format in various matrices.

### 16.1.1 Features of membrane-based μ-SPE

To make a μ-SPE device, a few milligrams of sorbent are enclosed inside a porous membrane bag which is made by superimposing two sheets of porous polymer membrane and heat-sealing the edges [39] (Fig. 16.1). Before use, the device is rinsed in a suitable solvent and conditioned under a vital process and was carried out by ultrasonication. It is then kept in the same solvent until ready for use.

μ-SPE is a sorption-based, nonexhaustive, and time-dependent extraction procedure. Extraction is considered complete when the analyte concentration has reached a distribution equilibrium between the sample matrix and the sorbent. The analyte adsorbs onto the sorbent phase, and hence, not only is the surface area important but also the total amount of the extraction phase. The extraction efficiency of the μ-SPE is therefore enhanced principally with respect to the following dynamic

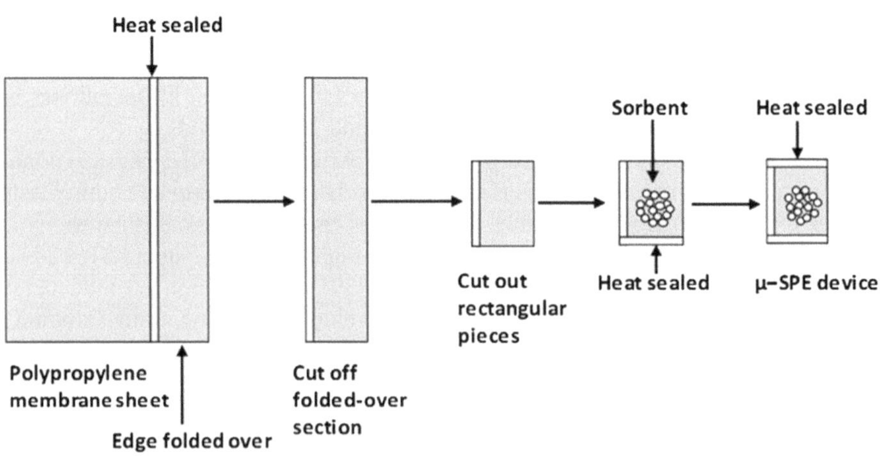


**Figure 16.1** Preparation of a μ-SPE device.

parameters to be optimized for specific classes of analytes: amount (usually in the milligram range) of sorbent used, duration of extraction, effect of desorption solvent, desorption time, volume of desorption solvent, and volume of sample solution. Extraction of relatively more polar analytes especially needs crucial supplementary optimization relating to the effects of sample ionic strength and pH.

The unique feature of μ-SPE is that extraction, cleanup, and preconcentration of target compounds in various matrices are carried out in a single step (Fig. 16.2). Among its proponents, this procedure is considered to be a portable, robust, durable, and cost-effective microextraction technique. Furthermore, a single bag can be used for multiple times (for example, up to 30 times) for some applications [40].

## 16.2 Sorbents used in μ-SPE

The type of sorbent used in μ-SPE is essential for the extraction of specific analytes or classes of analytes. Particularly with respect to DSPE, a wide variety of commercial and inhouse-synthesized sorbent materials can be utilized in μ-SPE, with the choice of sorbent predominantly dictated by the nature of the target analytes. Due to the interactions and dynamic partitioning of analytes between the sorbent and the matrix, large surface area, high adsorption capacity, acceptable selectivity, good stability, durability, and desirable adsorption kinetics are some of the main characteristics of an ideal sorbent [41,42]. These features are governed by properties of the sorbent such as the pore size, pore size distribution, specific surface area, as well as its surface chemistry [42]. Besides these, the sorbent should possess excellent solid state flow properties, such that it is "non-sticky" to ensure accurate and reproducible measurement of its mass [43]. These properties also facilitate the transfer and packing of the sorbent into the membrane envelopes without undesirable aggregation. In addition, it is also crucial for the sorbent to be nonadhesive to the membrane so that it can easily move toward the bottom of the membrane envelope, and not stick to the walls close to the opening. Particles interfering with the heat sealing of the open edge will eventually compromise the integrity of the envelope. This can cause tearing, or damage the seal if stronger agitation (stirring or ultrasonication) is used during the extraction [43].

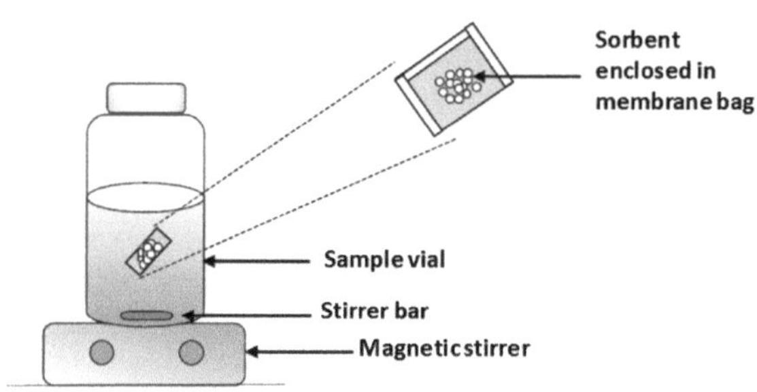


**Figure 16.2** Schematic of a μ-SPE system.

### 16.2.1 Conventional sorbents

An extensive range of sorbents of various properties are commercially available that may be used for DSPE and therefore μ-SPE. Generally, the polarity of the more widely used commercial sorbents decreases in this order: activated alumina > Haye-Sep B > Haye-Sep A ≈ Porapak R > ethylsilane ($C_2$) modified silica > octylsilane ($C_8$) modified silica > octadecylsilane ($C_{18}$) modified silica > activated carbon (AC) ≈ Carbograph (CG) ≈ multiwalled carbon nanotubes (MWCNTs) [44–46]. In most cases, polar analytes show a greater affinity to polar sorbents, based on the principle of "like dissolves like," in comparison to their nonpolar counterparts. For instance, $C_2$ was found to be more effective in extracting polar estrogenic compounds from cyst fluids as compared to $C_8$, $C_{18}$, AC, CG, Haye-Sep A, and Haye-Sep B [45]. Despite the high complexity of cyst fluid samples, the protection afforded by the porous propylene membrane in μ-SPE prevented the interferences posed by particulates, proteins, and humic substances, ultimately giving rise to low limits of detection (LODs) and excellent relative recoveries, when extracts were analyzed by gas chromatography-mass spectrometry (GC-MS). In another study, 2,4-dinitrophenylhydrazine (2,4-DNPH)-coated $C_2$ provided an approach for the simultaneous derivatization and extraction of highly volatile and polar aldehydes from rainwater [44]. The superior extraction efficiency exhibited by $C_2$-based material over other sorbents was attributed to its moderate polarity which allowed it to better interact with the polar derivatizing reagent and the aldehydes.

On the other hand, $C_{18}$ is more efficient at extracting relatively nonpolar analytes as expected. In a study, $C_{18}$ was chosen over MWCNTs and AC, in the stir-bar supported μ-SPE of polychlorinated biphenyl (PCB) congeners in human serum samples [46]. This is likely to be a result of the strong hydrophobic interactions between the analytes and $C_{18}$. Moreover, nonpolar analytes tend to adsorb overly strongly onto carbon-based materials, rendering desorption a difficult task. Consequently, the analytes could not be effectively recovered for subsequent analysis. Similarly, $C_{18}$ has also been applied as sorbent for the extraction of carbamazepine in surface waters [47], carbamate pesticides in soil samples [48], and acidic drugs in wastewater [49]. Very recently, 2,4-DNPH-coated $C_{18}$ was utilized for the extraction of glyoxal and methylglyoxal from atmospheric particulate matter (particles of size < 2.5 μm) [50] (after the particulates collected on filters were first taken into aqueous solution). Due to the low polarity of the analytes, the compounds were best extracted using $C_{18}$ in comparison to graphene oxide (GO), reduced-graphene oxide (r-GO), AC, and MWCNTs. Only 3 minutes of vortex mixing were required to attain the optimal extraction efficiency. Oxygen-containing functional groups of GO and r-GO made these materials more polar than $C_{18}$, and despite being hydrophobic, AC primarily interact with substances of low polarity and high molecular weights while MWCNTs preferentially interacted with organic molecules containing aromatic rings through $\pi$ interactions. Thus, the universal $C_{18}$-based material was demonstrated to be a more effective sorbent than the much newer ones like the graphenes and CNTs. The work also highlighted the versatility of μ-SPE, in this case, the translation of an extraction approach originally developed for water analysis, to air pollution research.

In attempts to identify sorbents of the most suitable polarity for particular analytes, mixtures of various commercial sorbents have been evaluated. Extraction efficiencies of nine combinations of polar/nonpolar sorbents consisting of $C_2$, $C_8$, $C_{18}$, Haye-Sep A, Haye-Sep B, and Porapak R were evaluated for the extraction of persistent organic pollutants from human ovarian cancer tissues [51]. Considering the nature of the analytes, the moderate-to-high hydrophobic mixture of Haye-Sep A/$C_{18}$ was found to be most efficient. Under optimized conditions, the μ-SPE-GC-MS method achieved low LODs.

For relatively polar compounds such as parabens, a porous sorbent with intermediate polarity, such as Haye-Sep A, was more effective [52]. Despite being an exhaustive extraction technique, the recoveries of parabens achieved using microwave-assisted solvent extraction (MASE)-SPE with $C_{18}$ sorbent and high-performance liquid chromatography-ultraviolet detection (HPLC-UV) were lower than those of MASE-μ-SPE (Haye-Sep A), an equilibrium-based technique, due to the high preconcentration that could be achieved. The LODs achieved using MASE-μ-SPE-HPLC-UV were also low. A summary of μ-SPE studies using various commercial sorbents, along with relevant analytical information, is presented in Table 16.1.

### 16.2.2 Carbon-based sorbents

An allotrope of carbon, graphene is a two-dimensional atomic crystal consisting of a monolayer of $sp^2$ hybridized carbon atoms packed into a dense honeycomb structure. Graphene's unique planar hexagonal nanostructure and high theoretical specific area (2630 $m^2/g$) gives it high sorption capacity. Given its large double-sided delocalized $\pi$-electron system, graphene is particularly effective for the adsorption of aromatic compounds via strong $\pi$ interactions. In addition, the material is characterized by its low cost, good thermal and chemical stability, and high mechanical strength. However, the easy aggregation of graphene sheets, due to the high hydrophobicity of the material, severely reduces the surface available for adsorption. The dispersibility of graphene in water, can be improved by functionalization, for example, by sulfonyl (-$SO_3H$) groups. The negatively charged $-\ SO_3^-$ induces electrostatic repulsion that prevents aggregation. When used as a μ-SPE sorbent for benzenoid compounds, sulfonated graphene showed improved extraction capability compared with $C_8$ and $C_{18}$, due to its larger surface area and planar structure [53]. Under optimized conditions, the μ-SPE method achieved low LODs using GC-MS. More recently, graphene was also intercalated into a layered double hydroxide for the stir-bar supported μ-SPE of 15 organochlorine pesticides (OCPs) from urine samples [54].

GO, the oxidized counterpart of graphene, contains functional groups such as epoxide, phenol, hydroxyl, and carboxylic groups which can participate in covalent, electrostatic or hydrogen bonding interactions with the analytes [55,56]. These functionalities make GO more hydrophilic than graphene itself, allowing for stable dispersion in water without greatly altering its original properties. In a study, GO was revealed to extract parabens by μ-SPE more effectively from water and vinegar samples compared with graphene, MWCNTs, and $C_{18}$ [55]. This could be attributed to the strong hydrophobic, hydrogen bonding and $\pi$-stacking interactions that could

**Table 16.1** Analytical aspects of conventional sorbents used in membrane-based μ-SPE.

| Analyte | Sorbent | Matrix | Analytical method | Analytical performance | Reference |
|---|---|---|---|---|---|
| Estrogens | $C_2$ | Cyst fluids | GC-MS | Recovery (%): 86–97 LOD: 9–22 ng $L^{-1}$ RSD (%): 4–11 | [45] |
| Aldehydes | 2,4-DNPH-coated $C_2$ | Rainwater | HPLC-UV | Recovery (%): 85.3–106.7 LOD: 27–150 ng $L^{-1}$ RSD (%): 7.9–12.0 | [44] |
| Aflatoxins | $C_8$ | Coffee and malt | LC-MS/MS | LOD: 0.12–0.76 ng $g^{-1}$ RSD (%): 0.22–11.23 | [87] |
| PCBs | $C_{18}$ | Human serum | GC-MS | Recovery (%): 85.2–102.7 LOD: 3–47 ng $L^{-1}$ RSD (%): 3.1–9.1 | [46] |
| Acidic drugs | $C_{18}$ | Wastewater | HPLC-UV | Recovery (%): 94–112. LOD: 30–80 ng $L^{-1}$ RSD (%): 6–8 | [49] |
| α–Dicarbonyl | 2,4-DNPH-coated $C_{18}$ | $PM_{2.5}$ particulate matter | LC-DAD | Recovery (%): 69–97 LOD: 140–150 ng $L^{-1}$ RSD (%): 1.39–6.56 | [50] |
| Carbamate pesticides | $C_{18}$ | Soil | HPLC-UV | Recovery (%): 60–110 LOD: 0.01–0.4 ng g-1 RSD (%): 4–11 | [48] |
| Carbamazepine | $C_{18}$ | Surface waters | LC-MS/MS | Recovery (%): 96.7–103.5 LOD: 0.5; 1.5 ng $L^{-1}$ | [47] |
| OCPs and PCBs | $C_{18}$ and Haye-Sep A | Human ovarian cancer tissues | GC-MS | LOD: 0.001–0.014 ng $g^{-1}$ RSD (%): 0.14–12.70 | [51] |
| Aromatic hydrocarbons and aldehydes | $C_{18}$, HLB, and MWCNTs | Saliva | GC-IMS | LOD: 0.18–0.31; 0.38–0.49 mg $L^{-1}$ RSD (%): 8.4–21.2 | [41] |
| Parabens | Haye-Sep A | Human ovarian cancer tissues | HPLC-UV | LOD: 0.005–0.0244 ng $g^{-1}$ RSD (%): 0.09–2.81 | [52] |
| Heterocyclic aromatic amines | Activated alumina | Seawater | HPLC-FLD | Recovery (%): 79.9–95.2 LOD: 4–26 ng $L^{-1}$ RSD (%): 1.3–7.5 | [88] |

*DAD*, Diode-array detection; *FLD*, Fluorescence detection; *GC*, Gas chromatography; *HPLC*, High-performance liquid chromatography; *IMS*, Ion-mobility spectrometry; *MS*, Mass spectrometry; *MS/MS*, Tandem mass spectrometry; *OCPs*, Organochlorine pesticides; *PCBs*, Polychlorinated biphenyls; *UV*, Ultraviolet detection be formed between GO and parabens. Moreover, without aggregation, the available surface adsorption sites of GO were not compromised. Together with the protection provided by the membrane bag, this μ-SPE-GC-MS method achieved lower LODs and good relative recovery ranges for genuine samples [55]. In another report, GO was functionalized with magnetic chitosan for the μ-SPE-GC-MS analysis of polycyclic aromatic hydrocarbons (PAHs) from river water [57]. The cross-linked chitosan, coupled with GO, provided a high active surface area with superior hydrophobic and π interactions with the analytes. The method achieved low LODs with good precision.

Carbon nanotubes have unique electronic, mechanical, and chemical properties. Based on the carbon layers of the walls, the materials may be classified as single-walled carbon nanotubes or MWCNTs. Given its tubular structure, high specific surface area and strong π interactions, MWCNTs were chosen as the sorbent in the first work introducing μ-SPE, for the extraction of the relatively hydrophobic organophosphorus pesticides (OPPs) from sewage sludge [39]. Later on, MWCNTs were also applied as sorbent for the μ-SPE of PAHs where the μ-SPE device was attached to the bottom of a portable battery-operated whisk stirrer [58]. The purpose of the handheld stirrer was to provide agitation to replace the normal laboratory magnetic stirrer or vortex, thereby making an onsite μ-SPE approach possible.

Recently, O-CMK-3, an oxidative surface-modified ordered mesoporous carbon, was introduced for the first time as a sorbent for the μ-SPE of eight *N*-nitrosamines, some of which are important and undesirable disinfection byproducts, from domestic wastewater and swimming pool water samples [59]. When compared with 10 other commercial carbonaceous sorbents, O-CMK-3 with GC tandem-mass spectrometry (GC-MS/MS), exhibited the best extraction and recoveries of *N*-nitrosamines [60]. This is because desorption of nonpolar analytes from carbonaceous materials is not an easy task which normally results in poor recoveries and high LODs, as discussed previously. By modifying the surfaces of CMK-3, a favorable hydrophilic/hydrophobic balance could be achieved on the inert carbonaceous sorbent, making recovery of the analytes more effective [59,60]. Coupled with its high specific surface area (999 $m^2/g$) and large pore size (3.84 nm), *N*-nitrosamines with a range of polarities could be reversibly extracted and subsequently desorbed for analysis. A collection of μ-SPE applications in which various carbonaceous sorbents are used is presented in Table 16.2.

### 16.2.3 Zeolites, silica, and metal-organic frameworks

Zeolites are crystalline microporous aluminosilicates with open framework structures and exchangeable cations. They contain cavities of specific size and shape which can be used for the selective entrapment of molecules. Linde Type L (LTL) is a zeolite with a one-dimensional 12-membered ring channel system. In a study, various sizes, lengths, and diameters of LTLs were exploited in their use as sorbents for the μ-SPE of ochratoxin A from coffee and cereal [61]. The results revealed that cylindrical LTL was most effective, due to its large number (1,068,000) of parallel channels and relatively long channel lengths (3.4 μm). This allowed the analytes to diffuse deeper and be trapped in the channels more effectively. LTL exhibited comparable

**Table 16.2** Analytical aspects of carbonaceous sorbents used in membrane-based μ-SPE.

| Analytes | Sorbent | Matrix | Analytical method | Analytical performance | Reference |
|---|---|---|---|---|---|
| PAHs | Sulfonated graphene | River water | GC-MS | Recovery (%): 81.6–113.5<br>LOD: 0.8–3.9 ng $L^{-1}$<br>RSD (%): 1.0–11.2 | [53] |
| OCPs | Layered double hydroxide/graphene | Human urine | GC-MS | Recovery (%): 84.2–100.2<br>LOD: 220–1380 ng $L^{-1}$<br>RSD (%): 2.7–9.5 | [54] |
| Parabens | GO | Water and vinegar | GC-MS | LOD: 5–50 ng $L^{-1}$<br>RSD (%): 5.4–8.9 | [55] |
| PAHs | Magnetic chitosan-functionalized GO | River water | GC-MS | Recovery (%): 67.5–106.9<br>LOD: 0.2–1.8 ng $L^{-1}$<br>RSD (%): 2.1–8.2 | [57] |
| Estrogens | SDS-loaded r-GO | Canal water | HPLC-UV | Recovery (%): 91–113<br>LOD: 0.24–0.52 ng $L^{-1}$<br>RSD (%): 2–5 | [56] |
| Chlorophenols | $Fe_3O_4$@rGO-g-$C_3N_4$ | Cosmetic | HPLC-UV | Recovery (%): 80.5–104.0<br>LOD: 0.2–0.3 ng $g^{-1}$<br>RSD (%): 5.6–9.6 | [89] |
| PAHs | Graphite fibers | Soil | GC-MS | LOD: 0.0017–0.0057 ng $g^{-1}$<br>RSD (%): 4.9–10.2 | [90] |
| OPPs | MWCNTs | Sewage sludge | GC-MS | LOD: 0.001–0.007 ng $g^{-1}$<br>RSD (%): 2–8 | [39] |
| PAHs | MWCNTs | River water | GC-MS | Recovery (%): 70.9–96.8<br>LOD: 4.2–46.5 ng $L^{-1}$<br>RSD (%): <12 | [91] |

| | | | | | |
|---|---|---|---|---|---|
| PAHs | MWCNTs | Lake, river, and farm water | HPLC-UV | Recovery (%): 77.3–104.1<br>LOD: 10–590 ng $L^{-1}$<br>RSD (%): 4.8–7.1 | [58] |
| Ultraviolet filters | MWCNTs | Tap, river, dam, spa, sewage treatment plant water | GC-MS | Recovery (%): 67–116<br>LOD: 3–18 ng $L^{-1}$<br>RSD (%): 4.7–13.8 | [92] |
| *N*-Nitrosamines | O-CMK-3 | Domestic wastewater and swimming pool water | GC-MS/MS | LOD: 5–283 ng $L^{-1}$<br>RSD (%): 3.09–6.22 | [59] |
| OCPs | Zinc oxide incorporated carbon foam | Milk | GC-MS | Recovery (%): 85.1–100.7<br>LOD: 190–1640 ng $L^{-1}$<br>RSD (%): 2.3–10.2 | [93] |
| PAHs | Zinc oxide incorporated carbon foam | Wastewater | GC-MS | Recovery (%): 91.8–102<br>LOD: 290–8400 ng $L^{-1}$<br>RSD (%): 3.8–10.9 | [94] |
| OPPs | AC | Vegetables and fruits | GC-MS | LOD: 0.07–0.23 ng $g^{-1}$<br>RSD (%): 1.5–8.7 | [95] |

*GO*, Graphene oxide; *MWCNTs*, Multiwalled carbon nanotubes; *OPPs*, Organophosphorus pesticides; *PAHs*, Polycyclic aromatic hydrocarbons extraction capability and excellent recovery over the commercial molecularly imprinted polymer (MIP) (consisting of crushed monolith and beads (AFFINIMIP OTA)). It was also superior to $C_8$, $C_{18}$, and $C_{30}$ phases.

Lately, mesoporous silica-based materials have garnered attention as extraction sorbents. The greatest advantage of mesoporous silica materials is their gross surface area, uniform pore size, and controlled surface chemistry [62]. Among these materials, MCM-41 is a well-known mesoporous molecular sieve that has been widely explored in selective adsorption and separation processes. The narrow pore size distribution, high surface area, and high pore volume of MCM-41 make it a promising extraction sorbent. In a μ-SPE study of perfluorinated carboxylic acids (PFCs) present in fish fillet, the performance of cetyltrimethylammonium bromide-templated MCM-41 (CTAB-MCM-41) was evaluated [63]. Subsequently, the performance of the material for extracting PFCs from rainwater [42] and human plasma [64], was evaluated. There has been a growing awareness of the hazards posed by PFCs in the environment, in view of their bioaccumulation in wildlife and humans. Thus, new extraction approaches in respect of this emerging class of contaminants are always welcomed.

Another class of ordered mesoporous silica, SBA-15, is a structurally well-ordered material possessing a regular two-dimensional hexagonal array of channels with narrow pore size distribution of 0.15–30 nm. Larger pore size and thicker pore walls than those of MCM-41 give it the stability and capability to capture larger and more complex molecules. In a recent report, SBA-15 was coated with positively charged sulfate-doped polyaniline for the extraction of parabens in wastewater and cosmetic products [62]. In addition to pore entrapment, hydrophobic and $\pi$ interactions, the prepared material was also able to form electrostatic interactions with the analytes during the ionic exchange of anionic parabens with sulfate anions bonded on polyaniline, therefore enhancing the adsorption of parabens onto the material.

Metal-organic frameworks (MOFs) are a new class of advanced hybrid crystalline materials formed via the self-assembly of metal ions and organic ligands. For nearly 20 years now, MOFs have generated a tremendous amount of interest. They have been the subject of a large number of studies to identify potential applications in many fields. Featured by their extremely large surface areas, pore volumes, high loading capacities, tuneable topologies, and ease of modification, the exploitation of these materials in analytical chemistry has similarly garnered a lot of attention. Zeolite imidazolate framework-8 (ZIF-8), a hydrophobic, water, chemical, and thermally stable material coupled with a high surface area (1810 $m^2/g$) is perhaps the most widely studied MOF in this respect [65]. When applied as a sorbent for the μ-SPE of PAHs in canal waters, ZIF-8 showed improved extraction efficiency in comparison to $C_8$ and $C_{18}$ [66]. Apart from the large pore sizes, ZIF-8 possesses Lewis acid sites from Zn, thus allowing selective interaction between these sites and the electron-rich analytes, which is not possible with $C_8$ and $C_{18}$. Also, ZIF-8 could be reused for 50 times without any significant changes in its polyhedral crystal structure. Besides nonpolar analytes, ZIF-8 was shown to be more effective for extracting polar compounds such as estrogens [65] and acidic drugs [67], than some commercial sorbents.

Another interesting MOF material that has been applied as sorbent in μ-SPE is the chromium terephthalate-based mesoporous MIL-101 [68]. Like ZIF-8, MIL-101 is also characterized by a high surface area (2600–3600 $m^2/g$), porosity, large pore windows (2.9 and 3.4 nm), numerous unsaturated metal sites, and excellent chemical, water, and solvent stability. In a study, MIL-101 was first dispersed in acetone, then filled into a hollow fiber device for the extraction of PCBs in river and wastewater [68] (Note that in this work, a tubular hollow fiber membrane was used to hold the sorbent, not an envelope made from a flat sheet membrane). The chemical stability of MIL-101 was evaluated by immersing it in pH 2 or pH 12 solutions for 2 hours. It was found that after the immersion, only a slight drop in extraction efficiency was observed, proving the chemical stability of MIL-101. Besides that, the μ-SPE device could be reused for up to 80 analyses without any observable damage and loss of extraction performance. Also, satisfactory recoveries were attained, ranging between 80.6% and 115.9%. Later, MIL-101 was also employed as μ-SPE sorbent for the extraction of OCPs in river water [69] and phthalate esters in drinking water [70], both in conjunction with GC-MS analysis. Adsorption of OCPs and phthalate esters onto MIL-101 was primarily dependent on pore encapsulation, and surface adsorption by coordination bonds, π interactions, and hydrophobic effects. A number of μ-SPE studies in which zeolites, silica, and MOFs are used as sorbents are listed in Table 16.3.

### 16.2.4 Biopolymeric materials, inclusion compounds, and molecularly imprinted polymers

The use of a selective extraction phase is usually preferred for the extraction of analytes from complex matrices as it helps to minimize interferences and coextraction of undesirable substances. Selectivity enhancement represents a great challenge in modern SPE. The best implementation of a selective sorbent is molecular imprinting. A molecularly imprinted polymer (MIP) is a specially prepared material with a specific recognition ability bestowed upon it by the involvement of a template molecule during synthesis. The functional monomers are intrinsically linked to the template covalently or noncovalently. The spatial arrangement between the template and monomers are fixed during the polymerization process. After the template is removed, a molecular recognition cavity is left behind, which allows for subsequent specific (an analyte identical to the original template model) or selective (group of analytes structurally analogous to the template molecules) extraction. Due to its specificity or selectivity, a particular MIP does not adsorb matrix interfering compounds. An example of an MIP-based μ-SPE application is given by the extraction of the flavonoids, hyperoside, and isoquercitrin, from rat plasma [71]. The method exhibited improved extraction efficiency compared with SPE using $C_{18}$ cartridges. MIP-μ-SPE with HPLC-UV analysis resulted in cleaner extracts, which translated into less matrix effects, clearer chromatograms, and subsequently, lower LODs. MIPs were also used for the μ-SPE of analytes in complex samples such as environmental waters [72,73], fruits [74], beverages and urine [75–78], and human plasma [77,79].

As in other aspects of chemistry, the consideration of environmentally sustainable natural materials is a significant trend in analytical chemistry. For instance, naturally abundant and nontoxic biopolymers such as chitosan have also been shown to be

**Table 16.3** Analytical aspects of zeolite, silica and MOF-based sorbents used in membrane-based μ-SPE.

| Analyte | Sorbent | Matrix | Analytical method | Analytical performance | Reference |
|---|---|---|---|---|---|
| Ochratoxin A | Cylinder zeolite LTL | Coffee and cereal | HPLC-FLD | LOD: 0.03–0.09 ng $g^{-1}$<br>RSD (%): 1.1–6.8 | [61] |
| Pharmaceutical drugs | Ureido-functionalized silica | Reservoir water | HPLC-DAD | Recovery (%): 65.6–123.0<br>LOD: 660–2360 ng $L^{-1}$<br>RSD (%): 2.1–12.6 | [43] |
| β-blockers | Rice husk silica/15% Fe | Hospital wastewater | LC-MS/MS | Recovery (%): 80.6–105.1<br>LOD: 4.0–6.4 ng $L^{-1}$<br>RSD (%): 2.34–8.30 | [96] |
| Parabens | SBA-15/polyaniline para-toluenesulfonic acid | Wastewater and cosmetic | HPLC-UV | Recovery (%): 82.2–108.3<br>LOD: 80–400 ng $L^{-1}$<br>RSD (%): 4.5–7.1 | [62] |
| PFCs | CTAB-MCM-41 | Fish fillet | LC-MS/MS | Recovery (%): 77.5–121.2<br>LOD: 0.97–2.70 ng $g^{-1}$<br>RSD (%): 5.4–13.5 | [63] |
| PFCs | CTAB-MCM-41 | Rain and river water | LC-MS/MS | Recovery (%): 64.7–127.0<br>LOD: 0.02–0.08 ng $L^{-1}$<br>RSD (%):1.9–10.5 | [42] |
| PFCs | CTAB-MCM-41 | Human plasma | LC-MS/MS | Recovery (%): 87.6–102.5<br>LOD: 21.23–65.07 ng $L^{-1}$<br>RSD (%): 5.68–15.22 | [64] |
| PAHs | ZIF-8 | Canal water | GC-MS | Recovery (%): 96.9–106.5<br>LOD: 2–12 ng $L^{-1}$<br>RSD (%): 2.1–8.5 | [66] |

| | | | | | |
|---|---|---|---|---|---|
| Estrogens | ZIF-8 | Sewage water | LC-MS/MS | Recovery (%) 85.6–108.2<br>LOD: 50–100 ng $L^{-1}$<br>RSD (%): 4.9–9.2 | [65] |
| Acidic drugs | ZIF-8 | Canal water | GC-MS | Recovery (%): 88.9–115.1<br>LOD: 10–40 ng $L^{-1}$<br>RSD (%):4.1–7.6 | [67] |
| PAHs | Mixed ZIF-8 | Tap water | GC-MS | Recovery (%): 88.1–111.0<br>LOD: 4–11 ng $L^{-1}$<br>RSD (%): 3.4–9.9 | [97] |
| PCBs | MIL-101 | River and wastewater | GC-MS/MS | LOD: 0.15–0.63 ng $L^{-1}$<br>RSD (%): 3.5–13.6 | [68] |
| OCPs | MIL-101 | River water | GC-MS | Recovery (%): 87.6–98.6<br>LOD: 2.5–16.0 ng $L^{-1}$<br>RSD (%): 4.2–11.0 | [69] |
| Phthalate esters | MIL-101 | Drinking water | GC-MS | LOD: 4–20 ng $L^{-1}$<br>RSD (%): 0.8–10.9 | [70] |
| PBDEs | Copper (II) isonicotinate | Soil | GC-μ-ECD | LOD: 0.026–0.066 ng $g^{-1}$<br>RSD (%): 1.3–10.1 | [98] |
| Bisphenol A and phthalate esters | Copper(II) isonicotinate | Atmospheric particulate matter | HPLC-UV | LOD: 2.5–8.5 ng $L^{-1}$<br>RSD (%): 3.6–9.4 | [99] |
| Tetracycline antibiotics | Copper(II) isonicotinate | Soil | HPLC-UV | LOD: 0.1–6.3 ng $g^{-1}$<br>RSD (%): 5.7–15.1 | [100] |
| Tetracycline antibiotics | Copper(II) isonicotinate | Pork, chicken, and clam meat | HPLC-UV | LOD: 7.4–16.3 ng $g^{-1}$<br>RSD (%): 5.7–9.3 | [101] |

*μ-ECD*, microcell electron capture detection; *PBDEs*, Polybrominated diphenyl ethers; *PFCs*, Perfluorinated carboxylic acids excellent sorbents (see also the following section). When applied to the extraction of volatile organic compounds such as benzene, toluene, ethylbenzene, xylenes, and styrene (BTEX-S), cross-linked chitosan demonstrated a better extraction capability than $C_8$ and aluminum oxide, and has comparable performance to that of $C_{18}$ [80]. This is because chitosan is endowed with a large number of amino and hydroxyl groups which allow it to form electrostatic, dipole, and hydrogen bonding interactions with organic compounds. Upon cross-linking, chitosan possesses greater surface porosity and higher nonpolar character, enabling hydrophobic interactions and pore entrapment of analytes on the surface of the material. In another study, cross-linked β-cyclodextrin was utilized as sorbent for the extraction of BTEX [81]. It was suggested that the β-cyclodextrin-BTEX interactions were mainly due to (1) interactions in the polymer cavities (formation of inclusion complexes), (2) interactions in the internal pores of polymeric pearls (beads), and (3) interactions on the surface (physisorption). However, due to the small surface area of β-cyclodextrin, formation of inclusion complexes with the target analytes was deemed to be the dominant mechanism over the physical adsorption of analytes [81]. The use of biopolymers, inclusion compounds, and MIPs as μ-SPE sorbents is illustrated by the listings in Table 16.4.

### 16.2.5 Miscellaneous sorbents

Given the creative efforts that go into the design and preparation of new materials for various purposes, it is not surprising that μ-SPE has also benefitted from these activities. Several examples are featured in this section.

Hydrazone-based ligands immobilized in a silica sol-gel matrix were synthesized and evaluated in the μ-SPE of biogenic amines in orange juice [82]. Among the eight ligands prepared, benzophenone-2,4-dinitrophenylhydrazone, the most hydrophobic derivative, with the largest number of benzene rings, was the most effective sorbent. It was able to participate in enhanced π interactions with the analytes, improving the overall extraction efficiency. It was also observed that in general, electrostatic interactions were insufficient to ensure good adsorption of biogenic amines [82]. Steric hindrances due to bulky groups on the ligands would also prevent effective extraction of the analytes.

In another study, nano-sized alumina ($Al_2O_3$) was functionalized with 3-(2-aminoethylamino) propyltrimethoxysilane (nano-$Al_2O_3$/AAPTMS) for the extraction of metal ions from gasoline [83]. Given the large surface areas, strong mechanical properties, high resistance to thermal degradation, and ability to chemisorb many substances, metal oxides such as $Al_2O_3$, titania, and zirconia are suitable sorbents for the extraction of metal ions. However, by itself, nano-$Al_2O_3$ exhibited relatively poor sorption efficiency for certain metal ions. As such, it was surface-modified with functional groups containing oxygen, nitrogen, and sulfur. By doing so, metal ions such as cobalt, chromium, manganese, nickel, and titanium could be adsorbed onto the surface of $Al_2O_3$ and attracted to the functional groups via chemical bonding, thus enhancing the extraction capability of the sorbent [83].

**Table 16.4** Analytical aspects of biopolymers, inclusion compounds, and MIPs used as sorbents in membrane-based μ-SPE.

| Analyte | Sorbent | Matrix | Analytical method | Analytical performance | Reference |
|---|---|---|---|---|---|
| Hyperoside and isoquercitrin | Hyperoside MIP | Rat plasma | HPLC-UV | LOD: 220–240 μg $L^{-1}$<br>RSD (%): 7.7–8.4 | [71] |
| Phenols | 2,4,6-Trichlorophenol MIP | River and raw sewage water | HPLC-UV | LOD: 560–4500 ng $L^{-1}$<br>RSD (%): 0.8–10.0 | [72] |
| Ochratoxin A | AFFINIMIP™ OTA | Coffee, grape juice and urine | HPLC-FLD | LOD: 0.06 ng $g^{-1}$; 20 ng $L^{-1}$<br>RSD (%): 0.2–4.8 | [75] |
| Sulfonamides | Sulfadimethoxine MIP | River and reservoir waters | HPLC-DAD | Recovery (%): 70–120<br>LOD: 200–3000 ng $L^{-1}$ | [73] |
| Cocaine and its metabolites | Cocaine MIP | Urine | LC-MS/MS | LOD: 49–500 ng $L^{-1}$<br>RSD (%): 2–10 | [76] |
| Cocaine and its metabolites | Cocaine MIP | Human plasma | LC-MS/MS | LOD: 61–870 ng $L^{-1}$<br>RSD (%): 1–12 | [79] |
| Cannabinoids | Δ9-THC-COOH MIP | Human plasma and urine | LC-MS/MS | LOD: 110–170 ng $L^{-1}$<br>RSD (%): 1–11 | [77] |
| Thiabendazole | Thiabendazole MIP | Orange and lemon | HPLC-DAD | LOD: 4–9 ng $g^{-1}$<br>RSD (%): <4 | [74] |
| Cannabinoids | JWH-015 MIP | Urine | LC-MS/MS | LOD: 32–750 ng $L^{-1}$<br>RSD (%): 1–8 | [78] |
| Triazine herbicides | Prometryn MWCNTs-MIP | River water, wastewater, and milk | HPLC-UV | LOD: 80–380 ng $L^{-1}$<br>RSD (%): 6.4–10.2 | [102] |

*Continued*

**Table 16.4** Analytical aspects of biopolymers, inclusion compounds, and MIPs used as sorbents in membrane-based μ-SPE.—cont'd

| Analyte | Sorbent | Matrix | Analytical method | Analytical performance | Reference |
|---|---|---|---|---|---|
| Sulfonamides | p-MAA-EDMA | Milk and chicken muscle samples | HPLC-UV | Recovery (%): 43.11–83.8<br>LOD: 380–620 ng $L^{-1}$; 4.52–15.63 ng $g^{-1}$,<br>RSD (%): 2.7–15.2 | [103] |
| BTEX-S | Chitosan microspheres | Canal water | GC-MS | Recovery (%): 59–97<br>LOD: 10–40 ng $L^{-1}$<br>RSD (%): 2–3 | [80] |
| BTEX | β-cyclodextrin | Drinking, rain, agricultural, and well water | GC-FID | Recovery (%): 64.5–101.3<br>LOD: 150–600 ng $L^{-1}$<br>RSD (%): 5.5–11.2 | [81] |
| Steroids | β-cyclodextrin | Urine | LC-MS/MS | LOD: 250–500 ng $L^{-1}$<br>RSD (%): 2.1–4.9 | [104] |

*p-MAA-EDMA*, Poly(methacrylic acid ethylene glycol dimethacrylate).

Although synthetic sorbents offer high affinity and can be designed to exhibit selectivity toward particular target compounds, their preparation procedures are often time-consuming and tedious, and consume considerable resources (solvents, reagents, etc). Besides, slight variation in the synthesis conditions may affect the activity of the sorbent. Consequently, apart from synthetic materials, natural, green, and easily accessible materials are often explored as extraction sorbents. To be sure, some processing or pretreatment of these biological materials are often needed. Nonetheless, there is also a growing trend in their application as sorbents, as an example, powdered seed from *Moringa oleifera*, a plant used as a traditional herbal medicine, was utilized as a sorbent for the μ-SPE of phthalate esters from milk [84]. The fine powder from the seeds, which are naturally enriched with a variety of functional groups, has elevated protein content (38%) and a highly interconnected fibrous structure. Its extraction capability was proven to be superior to that of $C_{18}$, and even other biosorbents such as banana peel and rice husk [84]. The use of natural and green materials as extraction sorbents is advantageous as such sorbents are easy to use, reuse, and dispose of; they are also of low cost and readily available. Also, unlike synthetic sorbents, the nature and properties of biosorbents are less likely to differ across batches, hence allowing for more consistent extraction outcomes. Additionally, unlike sorbents synthesized in-house, larger amounts (kilogram quantities) can be prepared at a time. More analytical details of the above applications are given in Table 16.5.

## 16.3 Enhanced μ-SPE: Combination with other microextraction procedures

Undoubtedly, effective isolation and enrichment of analytes are two of the major objectives in all extraction techniques. Several reports have described alternative approaches for μ-SPE in that it was either combined with other extraction techniques, or some improvements were made directly to the procedure itself. This section gives two accounts of these enhanced μ-SPE options.

The first example is a combined microextraction procedure involving μ-SPE as the first step to determine phthalate esters in river water [85]. The second step was DLLME in which a low-density solvent (LDS) was used. The analytes were initially extracted and preconcentrated by vortex-assisted (VA)-μ-SPE, and then desorbed by ultrasonication into acetonitrile. The latter served as the dispersive solvent in the subsequent DLLME step, which preconcentrated the analytes further. The development of this combined approach (VA-μ-SPE−LDS-DLLME) was inspired by the desire to reduce the total extraction time and enhance analyte preconcentration, and concurrently, to handle more complex aqueous samples. The μ-SPE step, of course, dealt with the latter challenge. Taking advantage of the small size of a μ-SPE device, the portability of the procedure was evaluated, indicating the possibility of onsite application.

Another combined technique utilizing ionic liquid-based DLLME and μ-SPE (IL-DLLME-μ-SPE), followed by HPLC-UV, is for the determination of tricyclic

**Table 16.5** Analytical aspects of miscellaneous sorbents used in membrane-based μ-SPE.

| Analyte | Sorbent | Matrix | Analytical method | Analytical performance | Reference |
|---|---|---|---|---|---|
| Biogenic amines | Benzophenone 2,4-dinitrophenylhydrazone | Orange juice | HPLC-UV | LOD: 3.82–31.30 ng $L^{-1}$<br>RSD (%): ≤11.3 | [82] |
| Co, Cr, Mn, Ni, and Ti | Nano-$Al_2O_3$/AAPTMS | Gasoline | ICP-MS | LOD: 2.3–11.5 ng $L^{-1}$<br>RSD (%): 1.1–4.5 | [83] |
| Phthalate esters | Seed powder of *Moringa oleifera* | Milk | GC-MS | Recovery (%): 77–103<br>LOD: 10–1200 ng $L^{-1}$<br>RSD (%): 3.6–10.2 | [84] |

*ICP-MS*, Inductively coupled plasma mass spectrometry; *Nano-$Al_2O_3$/AAPTMS*, Nanometer-sized alumina functionalized with [3-(2-aminoethylamino) propyl] trimethoxysilane antidepressants (TCAs) in water [86]. Sonication-assisted IL-DLLME was carried out prior to the addition of the μ-SPE device containing ZIF-4 as sorbent. Then, VA-μ-SPE was conducted to isolate the TCAs. The normally tedious steps associated with VA-DLLME, such as centrifugation or refrigeration of solvent using a conical-bottomed tube were avoided in this procedure. Good analytical performance with satisfactory relative recoveries, LODs and RSDs, were achieved. More detailed information on the combined techniques including those described above is given in Table 16.6.

## 16.4 Conclusion and future trends

The advent of solid-phase extraction (SPE) as a commercial product over 40 years ago introduced a convenient and efficient approach to process liquid samples, prior to analysis, using solid materials as sorbents. This sample preparation procedure has certainly proven its value. The original cartridge-, and later, disk-based, SPE subsequently laid the groundwork for the development of microscale or miniaturized devices. Micro-SPE (μ-SPE) is largely an outcome of this progression. Some μ-SPE developments eventually became commercialized (disposable pipette extraction, microextraction by packed sorbent, etc.). Currently, there is a lot of effort being expended in designing and preparing new materials for SPE applications, with most of them being used and reported usually independently of any commercial intentions or motives. The starting point is usually dispersive SPE, since, in comparison to cartridge- and disk-based SPE, it is the most easily implementable mode, and is the most expedient when evaluating new sorbents. At the same time, new miniaturized inhouse devices or contrivances have been, and are still being, devised to exploit the properties of new materials, and perhaps also in order to avoid issues concerning the infringement of proprietary configurations. Membrane-based μ-SPE, in which a polymeric membrane envelope serves to protect the enclosed sorbent from challenging matrices, is an example of such innovation.

In this review, the focus was on membrane-based μ-SPE, primarily covering the various types of sorbent materials used, and the applications of the procedure, normally in combination with gas and/or liquid chromatography. Membrane-based μ-SPE, it is believed, represents the forerunner of what the future of SPE (or μ-SPE) may be like. It is miniaturized and thus portable (with high possibility of onsite operations, as has been demonstrated preliminarily in some reports). Uniquely, membrane-based μ-SPE is protective of its sorbent, making it amenable to handle complex environmental and biological liquid matrices. Concomitantly, the membrane preserves the sorbent better, and therefore permits repeated use of the device. With its protective function, this μ-SPE mode can also be applied in a combined microextraction method advantageously if the other procedure lacks this feature. μ-SPE is suitable for use with a wide range of materials that are conventional (i.e., available commercially), current or newly introduced, and even those yet to be synthesized. It is also an environmentally friendly technique. In all likelihood, in the immediate

**Table 16.6** List of studies in which enhanced μ-SPE is applied.

| Analyte | Sorbent | Matrix | Extraction method | Analytical method | Analytical performance | Reference |
|---|---|---|---|---|---|---|
| Phthalate esters | MWCNTs | River water | μ-SPE-DLLME | GC-MS | Recovery (%): 85.8–104.1<br>LOD: 0.006–0.020 μg $L^{-1}$<br>RSD (%): 5.8–9.2 | [85] |
| TCAs | ZIF-4 | Environmental water | IL-DLLME-μ-SPE | HPLC-UV | Recovery (%): 94.3–114.7<br>LOD: 0.3–1.0 μg $L^{-1}$<br>RSD (%): 1.5–7.8 | [86] |
| OCPs | MWCNTs | Soil | μ-SPE-DLLME | GC-ECD | LOD: 0.11–1.07 ng $g^{-1}$<br>RSD (%): <14 | [105] |

*TCAs*, Tricyclic antidepressants.

future, advances in material chemistry will drive the further development of membrane-based (and also nonmembrane-based) μ-SPE, particularly in terms of the selectivity of the sorbents for analytes, and their stability and robustness. In the longer term, greater efforts to explore the possibility of automating some aspects of μ-SPE operations can be expected.

# References

[1] Costa R. Newly introduced sample preparation techniques: towards miniaturization. Crit Rev Anal Chem 2014;44:299–310.
[2] Huang Z, Lee HK. Materials-based approaches to minimizing solvent usage in analytical sample preparation. Trends Anal Chem 2012;39:228–44.
[3] Ramos L, Ramos J, Brinkman U. Miniaturization in sample treatment for environmental analysis. Anal Bioanal Chem 2004;381:119–40.
[4] Tankiewicz M, Fenik J, Biziuk M. Solventless and solvent-minimized sample preparation techniques for determining currently used pesticides in water samples: a review. Talanta 2011;86:8–22.
[5] Kataoka H. Recent developments and applications of microextraction techniques in drug analysis. Anal Bioanal Chem 2009;396:339–64.
[6] Abdulra'uf L, Sirhan A, Huat Tan G. Recent developments and applications of liquid phase microextraction in fruits and vegetables analysis. J Sep Sci 2012;35:3540–53.
[7] Jeannot M, Przyjazny A, Kokosa J. Single drop microextraction—development, applications and future trends. J Chromatogr A 2010;1217:2326–36.
[8] Lee J, Lee HK, Rasmussen K, Pedersen-Bjergaard S. Environmental and bioanalytical applications of hollow fiber membrane liquid-phase microextraction: a review. Anal Chim Acta 2008;624:253–68.
[9] Rezaee M, Yamini Y, Faraji M. Evolution of dispersive liquid–liquid microextraction method. J Chromatogr A 2010;1217:2342–57.
[10] Namera A, Nakamoto A, Saito T, Miyazaki S. Monolith as a new sample preparation material: recent devices and applications. J Sep Sci 2011;34:901–24.
[11] Pawliszyn J. Comprehensive sampling and sample preparation. Analytical techniques for scientists. 1st ed., vol. 3. elsevier; 2012.
[12] Ali I, Aboul-Enein H, Cazes J. A journey from Mikhail Tswett to nano-liquid chromatography. J Liq Chromatogr Relat Technol 2010;33:645–53.
[13] May WE, Chesler SN, Cram SP, Gump BH, Hertz HS, Enagonio DP, Dyszel SM. Chromatographic analysis of hydrocarbons in marine sediments and sea water. J Chromatogr Sci 1975;13:535–40.
[14] Little JN, Fallick GJ. New considerations in detector-application relationships. J Chromatogr A 1975;112:389–97.
[15] Khezeli T, Daneshfar A. Development of dispersive micro-solid phase extraction based on micro and nano sorbents. Trends Anal Chem 2017;89:99–118.
[16] Kocot K, Zawisza B, Marguí E, Queralt I, Hidalgo M, Sitko R. Dispersive micro solid-phase extraction using multiwalled carbon nanotubes combined with portable total-reflection X-ray fluorescence spectrometry for the determination of trace amounts of Pb and Cd in water samples. J Anal At Spectrom 2013;28:736–42.

[17] Fontanals N, Marcé R, Borrull F. New materials in sorptive extraction techniques for polar compounds. J Chromatogr A 2007;1152:14−31.
[18] Augusto F, Carasek E, Silva R, Rivellino S, Batista A, Martendal E. New sorbents for extraction and microextraction techniques. J Chromatogr A 2010;1217:2533−42.
[19] Scida K, Stege P, Haby G, Messina G, García C. Recent applications of carbon-based nanomaterials in analytical chemistry: critical review. Anal Chim Acta 2011;691:6−17.
[20] Poole C. New trends in solid-phase extraction. Trends Anal Chem 2003;22:362−73.
[21] Ouyang G, Pawliszyn J. SPME in environmental analysis. Anal Bioanal Chem 2006;386: 1059−73.
[22] Sánchez-Rojas F, Bosch-Ojeda C, Cano-Pavón J. A review of stir bar sorptive extraction. Chromatographia 2008;69:79−94.
[23] David F, Sandra P. Stir bar sorptive extraction for trace analysis. J Chromatogr A 2007; 1152:54−69.
[24] Tomai P, Martinelli A, Gasperi T, Bianchi M, Purcaro V, Teofili L, Papacci P, Cori MS, Vento G, Curini R, Fanali S, Gentili A. Corrigendum to "rotating-disc micro-solid phase extraction of F2-isoprostanes from maternal and cord plasma by using oxidized bucky-paper as sorbent membrane". J Chromatogr A 2019;1586:30−9.
[25] Dil EA, Ghaedi M, Asfaram A, Mehrabi F, Bazrafshan AA, Ghaedi AM. Trace determination of safranin O dye using ultrasound assisted dispersive solid-phase micro extraction: artificial neural network-genetic algorithm and response surface methodology. Ultrason Sonochem 2016;33:129−40.
[26] Salisaeng P, Arnnok P, Patdhanagul N, Burakham R. Vortex-assisted dispersive micro-solid phase extraction using CTAB-modified zeolite NaY sorbent coupled with HPLC for the determination of carbamate insecticides. J Agric Food Chem 2016;64: 2145−52.
[27] Moein MM, Abdel-Rehim A, Abdel-Rehim M. Microextraction by packed sorbent (MEPS). Trends Anal Chem 2015;67:34−44.
[28] Silva CL, Goncalves JL, Camara JS. A sensitive microextraction by packed sorbent-based methodology combined with ultra-high pressure liquid chromatography as a powerful technique for analysis of biologically active flavonols in wines. Anal Chim Acta 2012; 739:89−98.
[29] Liu S, Yan H, Wang M, Wang L. Water-compatible molecularly imprinted microspheres in pipette tip solid-phase extraction for simultaneous determination of five fluoroquinolones in eggs. J Agric Food Chem 2013;61:11974−80.
[30] Yan H, Sun N, Liu S, Row KH, Song Y. Miniaturized graphene-based pipette tip extraction coupled with liquid chromatography for the determination of sulfonamide residues in bovine milk. Food Chem 2014;158:239−44.
[31] Yan Z, Wu M, Hu B, Yao M, Zhang L, Lu Q, Pang J. Electrospun UiO-66/ polyacrylonitrile nanofibers as efficient sorbent for pipette tip solid phase extraction of phytohormones in vegetable samples. J Chromatogr A 2018;1542:19−27.
[32] Sun N, Han Y, Yan H, Song Y. A self-assembly pipette tip graphene solid-phase extraction coupled with liquid chromatography for the determination of three sulfonamides in environmental water. Anal Chim Acta 2014;810:25−31.
[33] htpp://www.gerstel.com/pdf/S00135-030-02_DPX_Flyer_en.pdf.
[34] Mafra G, Spudeit D, Brognoli R, Merib J, Carasek E. Expanding the applicability of cork as extraction phase for disposable pipette extraction in multiresidue analysis of pharmaceuticals in urine samples. J Chromatogr B 2018;1102−1103:159−66.

[35] Corazza G, Merib J, Magosso HA, Bittencourt OR, Carasek E. A hybrid material as a sorbent phase for the disposable pipette extraction technique enhances efficiency in the determination of phenolic endocrine-disrupting compounds. J Chromatogr A 2017;1513: 42−50.
[36] Pinto MAL, de Souza ID, Queiroz MEC. Determination of drugs in plasma samples by disposable pipette extraction with C18-BSA phase and liquid chromatography-tandem mass spectrometry. J Pharm Biomed Anal 2017;139:116−24.
[37] htpp://Shinwa-cpc.co.jp/en/products/needlex/.
[38] Lord H, Zhan W, Pawliszyn J. Fundamentals and applications of needle trap devices. Anal Chim Acta 2010;677:3−18.
[39] Basheer C, Alnedhary A, Rao B, Valliyaveettil S, Lee HK. Development and application of porous membrane-protected carbon nanotube micro-solid-phase extraction combined with gas chromatography/mass spectrometry. Anal Chem 2006;78:2853−8.
[40] Sajid M. Porous membrane protected micro-solid-phase extraction: a review of features, advancements and applications. Anal Chim Acta 2017;965:36−53.
[41] Criado-Garcia L, Arce L. Extraction of toxic compounds from saliva by magnetic-stirring- assisted micro-solid-phase extraction step followed by headspace-gas chromatography-ion mobility spectrometry. Anal Bioanal Chem 2016;408:6813−22.
[42] Lashgari M, Basheer C, Lee HK. Application of surfactant-templated ordered meso-porous material as sorbent in micro-solid phase extraction followed by liquid chromatography-triple quadrupole mass spectrometry for determination of perfluorinated carboxylic acids in aqueous media. Talanta 2015;141:200−6.
[43] Lim TH, Hu L, Yang C, He C, Lee HK. Membrane assisted micro-solid phase extraction of pharmaceuticals with amino and urea-grafted silica gel. J Chromatogr A 2013;1316: 8−14.
[44] Basheer C, Pavagadhi S, Yu H, Balasubramanian R, Lee HK. Determination of aldehydes in rainwater using micro-solid-phase extraction and high-performance liquid chromatography. J Chromatogr A 2010;1217:6366−72.
[45] Kanimozhi S, Basheer C, Narasimhan K, Liu L, Koh S, Xue F, Choolani M, Lee HK. Application of porous membrane protected micro-solid-phase-extraction combined with gas chromatography-mass spectrometry for the determination of estrogens in ovarian cyst fluid samples. Anal Chim Acta 2011;687:56−60.
[46] Sajid M, Basheer C. Stir-bar supported micro-solid-phase extraction for the determination of polychlorinated biphenyl congeners in serum samples. J Chromatogr A 2016;1455: 37−44.
[47] Teo HL, Wong L, Liu Q, Teo TL, Lee TK, Lee HK. Simple and accurate measurement of carbamazepine in surface water by use of porous membrane-protected micro-solid-phase extraction coupled with isotope dilution mass spectrometry. Anal Chim Acta 2016;912: 49−57.
[48] Basheer C, Alnedhary AA, Rao BS, Lee HK. Determination of carbamate pesticides using micro-solid-phase extraction combined with high-performance liquid chromatography. J Chromatogr A 2009;1216:211−6.
[49] Basheer C, Chong HG, Hii TM, Lee HK. Application of porous membrane-protected micro-solid-phase extraction combined with HPLC for the analysis of acidic drugs in wastewater. Anal Chem 2007;79:6845−50.
[50] Naing NN, Lee HK. Determination of glyoxal and methylglyoxal in atmospheric fine particulate matter by vortex-assisted micro-solid phase extraction and liquid chromatography-diode array detection. J Chromatogr A 2018;1573:42−7.

[51] Basheer C, Narasimhan K, Yin M, Zhao C, Choolani M, Lee HK. Application of micro-solid-phase extraction for the determination of persistent organic pollutants in tissue samples. J Chromatogr A 2008;1186:358−64.
[52] Sajid M, Basheer C, Narasimhan K, Choolani M, Lee HK. Application of microwave-assisted micro-solid-phase extraction for determination of parabens in human ovarian cancer tissues. J Chromatogr B 2015;1000:192−8.
[53] Zhang H, Low WP, Lee HK. Evaluation of sulfonated graphene sheets as sorbent for micro-solid-phase extraction combined with gas chromatography-mass spectrometry. J Chromatogr A 2012;1233:16−21.
[54] Sajid M, Basheer C, Daud M, Alsharaa A. Evaluation of layered double hydroxide/graphene hybrid as a sorbent in membrane-protected stir-bar supported micro-solid-phase extraction for determination of organochlorine pesticides in urine samples. J Chromatogr A 2017;1489:1−8.
[55] Wang L, Zang X, Wang C, Wang Z. Graphene oxide as a micro-solid-phase extraction sorbent for the enrichment of parabens from water and vinegar samples. J Sep Sci 2014; 37:1656−62.
[56] Naing NN, Li SFY, Lee HK. Evaluation of graphene-based sorbent in the determination of polar environmental contaminants in water by micro-solid phase extraction-high performance liquid chromatography. J Chromatogr A 2016;1427:29−36.
[57] Naing NN, Li SFY, Lee HK. Magnetic micro-solid-phase-extraction of polycyclic aromatic hydrocarbons in water. J Chromatogr A 2016;1440:23−30.
[58] Abidin NNZ, Sanagi MM, Ibrahim WAW, Endud S, Shukri DSM. Portable micro-solid phase extraction for the determination of polycyclic aromatic hydrocarbons in water samples. Anal Methods 2014;6:5512−8.
[59] Lashgari M, Yamini Y, Basheer C, Lee HK. Ordered mesoporous carbon as sorbent for the extraction of N-nitrosamines in wastewater and swimming pool water. J Chromatogr A 2017;1513:35−41.
[60] Lashgari M, Lee HK. Introducing surface-modified ordered mesoporous carbon as a promising sorbent for extraction of N-nitrosamines. J Colloid Interface Sci 2016;481: 39−46.
[61] Lee TP, Saad B, Ng EP, Salleh B. Zeolite Linde Type L as micro-solid phase extraction sorbent for the high performance liquid chromatography determination of ochratoxin A in coffee and cereal. J Chromatogr A 2012;1237:46−54.
[62] Ara KM, Pandidan S, Aliakbari A, Raofie F, Amini MM. Porous-membrane-protected polyaniline-coated SBA-15 nanocomposite micro-solid-phase extraction followed by high-performance liquid chromatography for the determination of parabens in cosmetic products and wastewater. J Sep Sci 2015;38:1213−24.
[63] Lashgari M, Lee HK. Determination of perfluorinated carboxylic acids in fish fillet by micro-solid phase extraction, followed by liquid chromatography-triple quadrupole mass spectrometry. J Chromatogr A 2014;1369:26−32.
[64] Lashgari M, Lee HK. Micro-solid phase extraction of perfluorinated carboxylic acids from human plasma. J Chromatogr A 2016;1432:7−16.
[65] Wang Y, Jin S, Wang Q, Lu G, Jiang J, Zhu D. Zeolitic imidazolate framework-8 as sorbent of micro-solid-phase extraction to determine estrogens in environmental water samples. J Chromatogr A 2013;1291:27−32.
[66] Ge D, Lee HK. Water stability of zeolite imidazolate framework 8 and application to porous membrane-protected micro-solid-phase extraction of polycyclic aromatic hydrocarbons from environmental water samples. J Chromatogr A 2011;1218:8490−5.

[67] Ge D, Lee HK. Zeolite imidazolate frameworks 8 as sorbent and its application to sonication-assisted emulsification microextraction combined with vortex-assisted porous membrane-protected micro-solid-phase extraction for fast analysis of acidic drugs in environmental water samples. J Chromatogr A 2012;1257:19–24.

[68] Zang H, Yuan J-P, Chen X-F, Liu C-A, Cheng C-G, Zhao R-S. Hollow fiber-protected metal–organic framework materials as micro-solid-phase extraction adsorbents for the determination of polychlorinated biphenyls in water samples by gas chromatography-tandem mass spectrometry. Anal Methods 2013;5.

[69] Huang Z, Lee HK. Micro-solid-phase extraction of organochlorine pesticides using porous metal-organic framework MIL-101 as sorbent. J Chromatogr A 2015;1401:9–16.

[70] Wang T, Wang J, Zhang C, Yang Z, Dai X, Cheng M, Hou X. Metal-organic framework MIL-101(Cr) as a sorbent of porous membrane-protected micro-solid-phase extraction for the analysis of six phthalate esters from drinking water: a combination of experimental and computational study. Analyst 2015;140:5308–16.

[71] Yin XY, Luo YM, Fu JJ, Zhong YQ, Liu QS. Determination of hyperoside and isoquercitrin in rat plasma by membrane-protected micro-solid-phase extraction with high-performance liquid chromatography. J Sep Sci 2012;35:384–91.

[72] Feng Q, Zhao L, Lin JM. Molecularly imprinted polymer as micro-solid phase extraction combined with high performance liquid chromatography to determine phenolic compounds in environmental water samples. Anal Chim Acta 2009;650:70–6.

[73] Diaz-Alvarez M, Barahona F, Turiel E, Martin-Esteban A. Supported liquid membrane-protected molecularly imprinted beads for micro-solid phase extraction of sulfonamides in environmental waters. J Chromatogr A 2014;1357:158–64.

[74] Diaz-Alvarez M, Martin-Esteban A. Hollow fiber membrane-protected molecularly imprinted microspheres for micro solid-phase extraction and clean-up of thiabendazole in citrus samples. J Chromatogr A 2018;1531:39–45.

[75] Lee TP, Saad B, Khayoon WS, Salleh B. Molecularly imprinted polymer as sorbent in micro-solid phase extraction of ochratoxin A in coffee, grape juice and urine. Talanta 2012;88:129–35.

[76] Sanchez-Gonzalez J, Tabernero MJ, Bermejo AM, Bermejo-Barrera P, Moreda-Pineiro A, Strano-Rossi S. Porous membrane-protected molecularly imprinted polymer micro-solid-phase extraction for analysis of urinary cocaine and its metabolites using liquid chromatography - tandem mass spectrometry. Anal Chim Acta 2015;898:50–9.

[77] Sanchez-Gonzalez J, Salgueiro-Fernandez R, Cabarcos P, Bermejo AM, Bermejo-Barrera P, Moreda-Pineiro A. Cannabinoids assessment in plasma and urine by high performance liquid chromatography-tandem mass spectrometry after molecularly imprinted polymer microsolid-phase extraction. Anal Bioanal Chem 2017;409:1207–20.

[78] Sanchez-Gonzalez J, Odoardi S, Bermejo AM, Bermejo-Barrera P, Romolo FS, Moreda-Pineiro A. Development of a micro-solid-phase extraction molecularly imprinted polymer technique for synthetic cannabinoids assessment in urine followed by liquid chromatography-tandem mass spectrometry. J Chromatogr A 2018;1550:8–20.

[79] Sanchez-Gonzalez J, Garcia-Carballal S, Cabarcos P, Tabernero MJ, Bermejo-Barrera P, Moreda-Pineiro A. Determination of cocaine and its metabolites in plasma by porous membrane-protected molecularly imprinted polymer micro-solid-phase extraction and liquid chromatography-tandem mass spectrometry. J Chromatogr A 2016;1451:15–22.

[80] Naing NN, Li SFY, Lee HK. Application of porous membrane-protected chitosan microspheres to determine benzene, toluene, ethylbenzene, xylenes and styrene in water. J Chromatogr A 2016;1448:42–8.

[81] Nojavan S, Yazdanpanah M. Micro-solid phase extraction of benzene, toluene, ethylbenzene and xylenes from aqueous solutions using water-insoluble beta-cyclodextrin polymer as sorbent. J Chromatogr A 2017;1525:51−9.
[82] Basheer C, Wong W, Makahleh A, Tameem AA, Salhin A, Saad B, Lee HK. Hydrazone-based ligands for micro-solid phase extraction-high performance liquid chromatographic determination of biogenic amines in orange juice. J Chromatogr A 2011;1218:4332−9.
[83] Nomngongo PN, Ngila JC. Functionalized nanometer-sized alumina supported micro-solid phase extraction coupled to inductively coupled plasma mass spectrometry for preconcentration and determination of trace metal ions in gasoline samples. RSC Adv 2014;4:46257−64.
[84] Sajid M, Basheer C, Alsharaa A, Narasimhan K, Buhmeida A, Al Qahtani M, Shaheen Al-Ahwal M. Development of natural sorbent based micro-solid-phase extraction for determination of phthalate esters in milk samples. Anal Chim Acta 2016;924:35−44.
[85] Guo L, Lee HK. Vortex-assisted micro-solid-phase extraction followed by low-density solvent based dispersive liquid−liquid microextraction for the fast and efficient determination of phthalate esters in river water samples. J Chromatogr A 2013;1300:24−30.
[86] Ge D, Lee HK. Ionic liquid based dispersive liquid−liquid microextraction coupled with micro-solid phase extraction of antidepressant drugs from environmental water samples. J Chromatogr A 2013;1317:217−22.
[87] Khayoon WS, Saad B, Salleh B, Manaf NH, Latiff AA. Micro-solid phase extraction with liquid chromatography-tandem mass spectrometry for the determination of aflatoxins in coffee and malt beverage. Food Chem 2014;147:287−94.
[88] Basheer C. Application of micro-solid-phase extraction for the on-site extraction of heterocyclic aromatic amines in seawater. J Sep Sci 2018;41:1610−7.
[89] Zang X, Chang Q, Liang W, Wu T, Wang C, Wang Z. Micro-solid phase extraction of chlorophenols using reduced graphene oxide functionalized with magnetic nanoparticles and graphitic carbon nitride as the adsorbent. Microchim Acta 2017;185:18.
[90] Xu L, Lee HK. Novel approach to microwave-assisted extraction and micro-solid-phase extraction from soil using graphite fibers as sorbent. J Chromatogr A 2008;1192:203−7.
[91] Guo L, Lee HK. Development of multiwalled carbon nanotubes based micro-solid-phase extraction for the determination of trace levels of sixteen polycyclic aromatic hydrocarbons in environmental water samples. J Chromatogr A 2011;1218:9321−7.
[92] Vosough M, Hassanbeigi Z, Salemi A. Determination of ultraviolet filter compounds in environmental water samples using membrane-protected micro-solid-phase extraction. J Sep Sci 2018;41:2401−10.
[93] Sajid M, Basheer C, Mansha M. Membrane protected micro-solid-phase extraction of organochlorine pesticides in milk samples using zinc oxide incorporated carbon foam as sorbent. J Chromatogr A 2016;1475:110−5.
[94] Jillani SMS, Sajid M, Alhooshani K. Evaluation of carbon foam as an adsorbent in stir-bar supported micro-solid-phase extraction coupled with gas chromatography−mass spectrometry for the determination of polyaromatic hydrocarbons in wastewater samples. Microchem J 2019;144:361−8.
[95] Wang Z, Zhao X, Xu X, Wu L, Su R, Zhao Y, Jiang C, Zhang H, Ma Q, Lu C, Dong D. An absorbing microwave micro-solid-phase extraction device used in non-polar solvent microwave-assisted extraction for the determination of organophosphorus pesticides. Anal Chim Acta 2013;760:60−8.
[96] Nsubuga H, Basheer C, Haider MB, Bakdash R. Sol-gel based biogenic silica composite as green nanosorbent for chemometric optimization of micro-solid-phase extraction of beta blockers. J Chromatogr A 2018;1554:16−27.

[97] Ge D, Lee HK. Sonication-assisted emulsification microextraction combined with vortex-assisted porous membrane-protected micro-solid-phase extraction using mixed zeolitic imidazolate frameworks 8 as sorbent. J Chromatogr A 2012;1263:1–6.
[98] Zhou YY, Zhang CY, Yan ZG, Li KJ, Wang L, Xie YB, Li FS, Liu Z, Yang J. The use of copper(II) isonicotinate-based micro-solid-phase extraction for the analysis of polybrominated diphenyl ethers in soils. Anal Chim Acta 2012;747:36–41.
[99] Jiao Z, Guo Z, Zhang S, Chen H, Xie H, Zeng S. Novel extraction for endocrine disruptors in atmospheric particulate matter. Anal Lett 2014;48:1355–66.
[100] Jiao Z, Guo Z, Zhang S, Chen H. Microwave-assisted micro-solid-phase extraction for analysis of tetracycline antibiotics in environmental samples. Int J Environ Anal Chem 2014;95:82–91.
[101] Jiao Z, Zhu D, Yao W. Combination of accelerated solvent extraction and micro-solid-phase extraction for determination of trace antibiotics in food samples. Food Anal Methods 2015;8:2163–8.
[102] Tan F, Deng M, Liu X, Zhao H, Li X, Quan X, Chen J. Evaluation of a novel micro-extraction technique for aqueous samples: porous membrane envelope filled with multiwalled carbon nanotubes coated with molecularly imprinted polymer. J Sep Sci 2011; 34:707–15.
[103] Huang J, Liu J, Zhang C, Wei J, Mei L, Yu S, Li G, Xu L. Determination of sulfonamides in food samples by membrane-protected micro-solid phase extraction coupled with high performance liquid chromatography. J Chromatogr A 2012;1219:66–74.
[104] Manaf NA, Saad B, Mohamed MH, Wilson LD, Latiff AA. Cyclodextrin based polymer sorbents for micro-solid phase extraction followed by liquid chromatography tandem mass spectrometry in determination of endogenous steroids. J Chromatogr A 2018;1543: 23–33.
[105] Mirikaram N, Salemi A, Vosough M. Optimization of membrane-protected micro-solid-phase extraction coupled with dispersive liquid-liquid microextraction for determination of organochlorine pesticides in soil media. Chromatographia 2017;80:157–64.

# Microextraction by packed sorbent (MEPS) and monolithic packed pipette tips for 96-well plates

17

*Mohammad Mahdi Moein* [1], *Abbi Abdel-Rehim* [2], *Mohamed Abdel-Rehim* [1]
[1]Department of Clinical Neuroscience, Center for Psychiatry Research, Karolinska Institutet and Stockholm County Council, Stockholm, Sweden; [2]Faculty of Science and Engineering, University of Manchester, Manchester, United Kingdom

## 17.1 Introduction

The major purpose of sample preparation is to prepare samples for the separation/detection part of the analysis. Sample preparation is a difficult and a relatively uncared for part of the systems for chemical analysis. Sample preparation has been performed by multistep procedures using methods such as conventional solid or liquid phase extraction. During recent years, sample preparation has attracted increasing attention, and the need for new improved methods has been largely recognized. Demands made by bioanalysis moved forward the development of more selective, sensitive, repeatable, cost-effective, and green analysis methods. Among modern trends in sample preparation, techniques which can connect online to analytical instruments and operate with low sample volumes have won favor. There is a critical need for faster and high-throughput methods. In the pharmaceutical industry, approaches such as combinatorial chemistry and parallel synthesis provide a means of production of large numbers of compounds to be used for drug-discovery experiments. Therapeutic drug monitoring of drug concentrations in biological fluids is established, since it helps to arrive at the optimal dose for therapeutic efficacy as well as to minimize toxicity. Therefore, there is a need for robust and sensitive high-throughput sample preparation with a short turnaround time. Therefore, recent developments in sample handling techniques are directed, from one side, toward miniaturization, automation, and online coupling of sample preparation units and detection systems.

Recently microextraction by packed sorbent (MEPS) and packed pipette tips have emerged as promising miniaturized SPE techniques used to simplify sample preparation of complex samples. This chapter provides a short overview of advances, advantages, and future prospects for MEPS and monolithic packed tips sample preparation methods.

**Solid-Phase Extraction. https://doi.org/10.1016/B978-0-12-816906-3.00017-0**

## 17.2 Microextraction by packed sorbent (MEPS)

MEPS is a scaled-down solid phase extraction (SPE) sample preparation technique utilizing low sample and organic solution volumes [1−3]. MEPS is a nimble, easy, and environmentally friendly method compared with conventional SPE. In MEPS a solid bed is packed between two frits in a 50−500 μL glass or plastic syringe, or in a cartridge format well known as BIN (Fig. 17.1) [3]. MEPS steps are similar to SPE contain the conditioning, loading, washing, and eluting the analyte which can be performed in short time with a few milliliters solution. In conventional SPE sorbent drying is also a common step, which can be neglected in MEPS, reducing the sample analysis time. The effective factors for MEPS were reported by Abdel-rehim [1]: briefly, in the first step the sorbent bed is conditioned with a proper solution; the sample is applied and the target compounds trapped by the sorbent by passing the sample several times through the sorbent bed; then suitable solvents for washing away matrix components (optional step) and elution of the analytes are applied. The selection of suitable solvent/solvents for the above steps is a crucial factor to not only increase the recovery but also reduce carryover. An advantage for MEPS compared with conventional SPE is that it can be used for over 100 consequences extraction if proper washing steps after each extraction are employed. MEPS is a flexible method capable of handling sample volumes from few microliters for bioanalysis to several millimetres for water samples [3]. MEPS reduces the sample preparation time due to the low volume of organic solvents required in the different procedural steps.

### *17.2.1 Effective factors on MEPS performance*

Several parameters can affect the MEPS efficiency, some of which are briefly listed here;

- Sample loading flow rate and time: sample loading flow rate and time are key steps to have maximum analyte absorption and increase the method repeatability. MEPS efficiency is affected by operator expertise in minimizing variations between experiments. To reduce this sort of errors eVol XR digital analytical syringe can be used on/semiline as shown in Fig. 17.1. This method can perform all MEPS steps with the same optimized conditions and increase reproducibility and reduce laboratory errors.
- Optimization of solutions: The selection of a proper type and volume of solutions for conditioning, loading, washing, and eluting steps is an important factor in MEPS performance. Proper conditioning increases the sorbent adsorption capability. In addition, proper washing solution can reduce interferences and increase method selectivity, especially when molecularly imprinted polymers are used as the MEPS sorbent. The elution solution is crucial to enhance the analyte recovery.
- Carryover is a well-known problem in MEPS, especially for biological matrixes; it needs to be minimized using proper washing solutions between each extraction. Typically a 2−5 times volume of cleaning solution to sample volume between each run can reduce the carryover and increase the sorbent lifetime. In addition, the amount of solid phase employed can also be a cause of carryover and should be optimized during the initial steps of method development. Based on our experiences 1−2 mg sorbents provide a better response and less carryover.

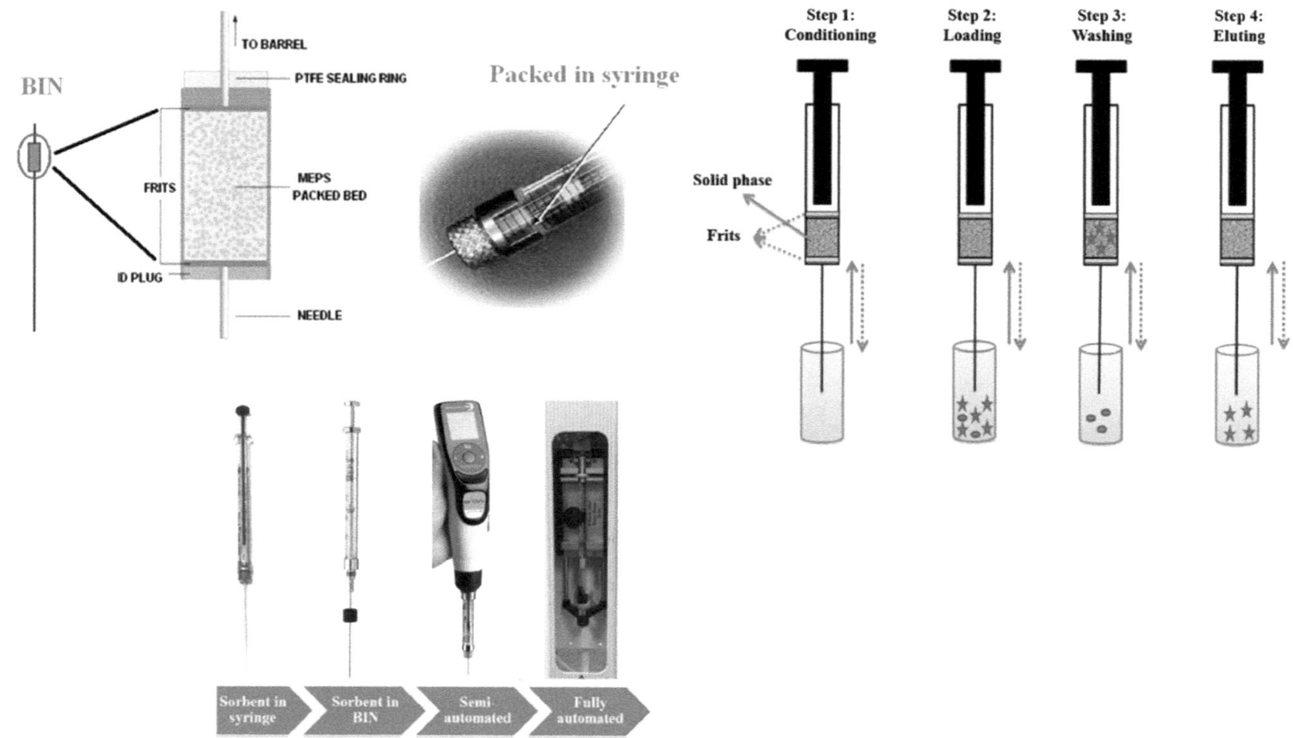


**Figure 17.1** MEPS tools sorts and performance steps.
Reused with permission Yanga L, Said R, Abdel-Rehim M. Sorbent, device, matrix and application in microextraction by packed sorbent (MEPS): a review. J Chromatogr B 2017;1043:33–43.

- Matrix effect (ME) is a vital defective factor in MEPS performance when analyzing samples electrospray-ionization mass (ESI-MS). ME can be reduced by controlling the MEPS sorbent particle size and selecting the proper solutions for the different extraction steps. In addition, sample dilution or a presample preparation method, such as plasma protein precipitation, can effectively reduce the ME.
- Sorbent adsorption capacity: The capacity and lifetime of an MEPS sorbent is also a factor in MEPS performance. Finding a sorbent with sufficient porosity and stability in harsh conditions is a challenge. This factor will be explained shortly in the following section.

### 17.2.2 MEPS sorbents

The selection of a proper sorbent is the most effectual factor in MEPS performance. Several types of commercial and noncommercial sorbents are used in MEPS. Commercial MEPS sorbents include silica-based (normal phase), C2, C8, C18, mixed mode, polystyrene-divinylbenzene (PS-DVB), and porous graphitic carbon as is shown in Fig. 17.2 [2]. Noncommercial sorbents include molecularly imprinted polymers (MIPs), nanomaterials, and graphitic-based materials [4]. In addition MEPS sorbents have been prepared in different formats such as particulate and monolithic [5] for different applications.

The sample matrix and the physiochemical properties of the analytes are the most important factors in selecting the MEPS sorbent [6]. Typically, for the extraction of polar compounds silica-based or mixed mode sorbents provide better performance in MEPS while poly(styrene-co-divinylbenzene) is more suitable for nonpolar compounds [2].

Finally the sorbent particle size can affect the adsorption capacity and permeability of the sorbent bed. In addition, nanomaterials used as solid phases in MEPS resulted in improved sensitivity and reproducibility [2].

### 17.2.3 MEPS off- and online application

MEPS can be combined off- and online with liquid chromatography (LC), gas chromatography (GC), capillary electrochromatography (CEC), and in combination with various sorts of mass spectrometry (MS) instruments, Table 17.1 [2]. MEPS can be a programmed fully automated method with the capability to introduce the analyte directly to analytical instruments without evaporating the elution solvent. This interesting capability enhances the extraction recovery in comparison with other microextraction methods such as SPME.

### 17.2.4 MEPS fields of application

For food analysis, MEPS on account of its simplicity and high recovery is a suitable sample preparation method for the analysis of various analytes in food samples. MEPS was used for the extraction of ethyl carbamate [7], furanic derivatives [8], phenolic constituents [9], flavonols [10], L-ascorbic acid [11], hydroxybenzoic and hydroxycinnamic acids [12] in wine and beverage samples. MEPS was also used

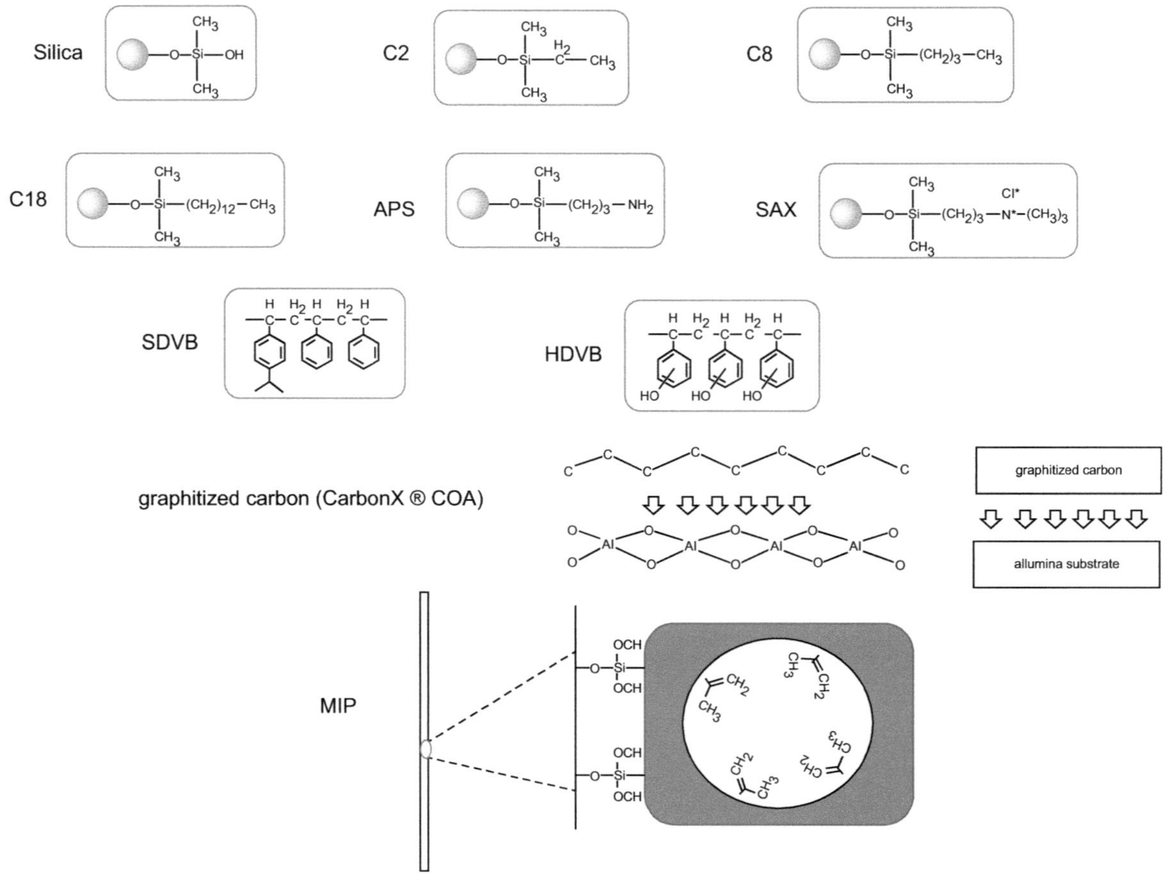


**Figure 17.2** MEPS common sorbents.
Reused with permission Yanga L, Said R, Abdel-Rehim M. Sorbent, device, matrix and application in microextraction by packed sorbent (MEPS): a review. J Chromatogr B 2017;1043:33–43.

**Table 17.1** Recent MEPS applications.

| Analyte | Sorbent | Matrix | Instrumentation | Ref |
|---|---|---|---|---|
| Sarcosine | MIPs | Human plasma | LC-MS/MS | [4] |
| Bupivacaine | C8, C18 | Human plasma | LC–MS/MS | [23] |
| Oxcarbazepine10,11-dihydro1-10-hydroxycarbamazepine-Trans-10,11-dihydro-10,11-dihydroxycarbamazepine3-hydroxycarbamazepine | C18 | Human plasma and saliva | HPLC– DAD | [24] |
| Buprenorphine, Norbuprenorphine, Methadone, Naloxone | C8 | Human plasma | HPLC-CD | [25] |
| Oxcarbazepine, Carbamazepine, Phenytoin, Alprazolam | C18 | Human plasma and urine | GC-MS | [26] |
| Immunosuppressive drugs | C8 | Whole blood | LC-MS/MS | [27] |
| Lidocaine, prilocaine, and ropivacaine | R. Graphene | Human plasma and saliva | LC-MS/MS | [28] |
| Busulfan | C8 | Human plasma | LC-MS/MS | [29] |
| Hippuric acid | MIPs | Human plasma and urine | LC-MS/MS | [30] |
| Pyrethroid metabolites | C18 | Urine sample | | [31] |
| Aromatic microbial metabolites (phenylcarboxylic acids) | C18 | Blood sample | | [32] |
| Organophosphate flame retardant | C2, C8, C18, Silica, and DVB | Environmental aqueous matrices | | [33] |
| Psychoactive substances | C8/SCX | Oral fluid | | [34] |

| | | | | |
|---|---|---|---|---|
| Zonisamide | C18 BIN | Human plasma | | [35] |
| Phthalate esters | Three-dimensional carbon nanotube/ carbon nanofiber-graphene nanostructure | Water samples | | [36] |
| Trans,trans-muconic acid | MIP | Urine samples | | [37] |
| Fluoxetine, Norfluoxetine, and Paroxetine | C8 | Plasma samples | | [38] |
| Dexamethasone disodium phosphate, and dexamethasone | C2, C8 and C18 | Aqueous humor of patients with uveitis | | [39] |
| Statins | C18 | Human plasma | | [40] |
| Furanic-derivatives formation | C8 | Wine samples | | [8] |
| Drugs of abuse | C8/SCX | Plasma sample | | [41] |
| THC and main metabolites | C8/SCX | Human plasma | | [42] |
| Volatile compounds | C8/C18 | Water samples | | [43] |
| Lamotrigine | C18 | Human plasma and saliva | | [44] |
| Multiple antiinflammatory drugs and fluoroquinolones | C18 | Human plasma and urine | | [45] |
| Lamotrigine | C18 | Rat plasma and brain samples | HPLC-DAD | [46] |
| Ochratoxin A | C18 | Wine sample | HPLC-FLD | [47] |
| Metoprolol enantiomers | C18 | Human plasma, saliva samples | LC-MS/MS | [48] |

*Continued*

**Table 17.1** Recent MEPS applications.—cont'd

| Analyte | Sorbent | Matrix | Instrumentation | Ref |
|---|---|---|---|---|
| Pesticides (tebuthiuron, carbofuran, atrazine, metribuzine, ametryn, and bifenthrin) | C18, C8, SCX, SAX, and SIL. | Sugarcane juice | GC-MS | [49] |
| Chlordiazepoxide; lorazepam; diazepam; oxazepam; medazepam | C18 | Alcoholic beverage | UPLC-MS/MS | [50] |
| Polyamines and related compounds | C18 | Urine sample | GC-MS | [51] |
| Polychlorinated biphenyls | C18 | Bovine serum | GC-MS | [52] |
| Phthalates | C18 | Cold drink and cosmetic samples | GC-MS | [53] |
| Ciprofloxacin and levofloxacin | C18 | Human sputum | HPLC-PAD | [54] |
| Ketamine and its major metabolite, norketamine | C8/SCX | Urine, plasma samples | GC-MS/MS | [55] |
| Remifentanil | C8 | Plasma | LC-MS/MS | [56] |
| Local anesthetics | MIPs | Plasma, urine | LC-MS/MS | [57] |
| Phenoxy acid herbicides | Mesoporous org.silica | Aqueous samples | GC-FID | [58] |
| Childhood-asthma biomarkers | R-AX sorbent | Urine samples | UHPLC-PDA | [59] |
| Levofloxacin | MIPs | Plasma samples | UHPLC-DAD | [60] |
| Beta-blocker | C18 | Urine samples | HPLC-Fluorescence | [61] |

for the extraction of ractopamine from porcine muscle [13]. In addition, MEPS was used for the evaluation of novel sorbents such as metal-organic frameworks and molecularly imprinted polymers for determination of triazine in corn samples. MEPS in combination with LC-DAD was used for the determination of sulfonamides in egg samples and for the analysis of different compounds in various matrices such as honey and soil. Details of the most recent MEPS applications are presented in Table 17.1.

### 17.2.5 MEPS challenges and future aspects

MEPS is a simple, speedy, straightforward, and green method but similar to all other sample preparation methods still suffers from some defects which need further investigation to solve. The most cited problems and some possible solutions are:

**(a)** The blockage of sorbent, especially for complex matrixes, is a commonplace MEPS problem. Sample dilution and deproteinization are the appropriate solutions to reduce these defects.
**(b)** Sorbents lifetime, stability, selectivity, and capacity are a further challenge in MEPS which needs more investigation by organic chemists, such as the synthesis of more efficient sorbents like nanomaterials.

## 17.3 Monolithic packed 96-tips

The monolithic packed tips, in a 96-format (Fig. 17.3), were first presented by our group at HPLC-2005 conference followed by several papers utilizing packed 96-tips

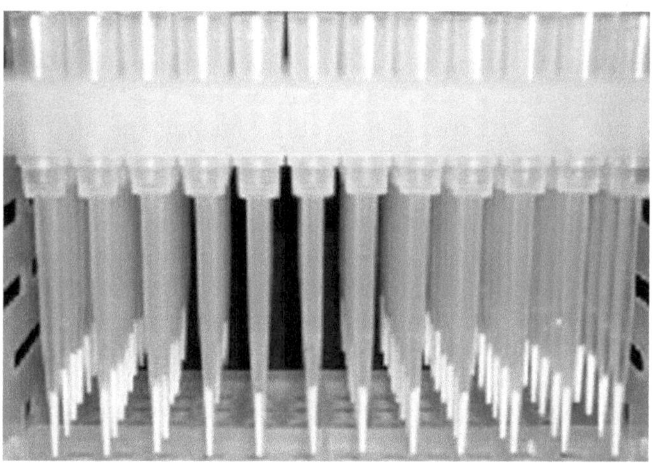

**Figure 17.3** UV-polymerized 96-tips packed with monolithic methacrylate polymer. Reused with permission Altun Z, Skoglund C, Abdel-Rehim M. Monolithic methacrylate packed 96-tips for high throughput bioanalysis. J Chromatogr A 2010; 1217:2581–2588.

for the extraction of drugs and biomarkers from biological fluids [14–20]. Other groups developed polymeric packed tips for single-use applications utilizing monoliths, monolithic silica, or monoliths modified with nanoparticles.

## 17.3.1 Preparation of monolithic packed tips

### 17.3.1.1 Treatment of pipette surface

The treatment of the surface of poly(propylene) pipette tips (PP) is an important issue to improve the interaction of the monolith to the pipette wall. There are different methods for surface modification to increase the surface energy to improve the substrate's wetting and adhesion capability. UV-irradiation using photoinitiators was used for the surface modification of PP tips. Alternatively, the PP tips can be modified by grafting with chains of ethylene glycol dimethacrylate (EDMA). The advantages of using the UV technique are low cost, simplicity, mild reaction conditions, and rapidity [15].

### 17.3.1.2 Manufacture of monolithic pipettes

Organic polymer monoliths are produced by the one-step polymerization of a mixture consisting of organic monomers, cross-linkers, porogenic solvents, and initiators. Within pipette tips, the mixture is typically sonicated, purged under nitrogen, filled into the tip itself, and in most cases polymerized using UV irradiation [14,21].

Briefly, a solution containing EGDMA (70%), BMA (30%), BPO (2% m/m with respect to monomers), 1,4-butanediol (25%), 1-propanol (65%), and water (10%) was vortexed for 10 min and purged with nitrogen for 10 min to remove oxygen. The type and composition of the porogenic mixture is the key factor for fine-tuning the final properties of the polymer monoliths. The pipette tips were filled to approximately 8 mm (6–7 μL) by capillary action and placed vertically inside the polymerization apparatus, a Spectrolinker XL-1500 UV Cross-linker. The polymerization was allowed to proceed for 60 min with the sharp end of the tip pointing downwards at a distance of 15 cm from the lamp, and then for 25 min with the sharp end pointing upwards at a distance of 5 cm from the lamp. After completion of the polymerization, the tips were removed, inspected under a microscope for bubbles, and washed with acetone to remove the porogenic solvents and other compounds remaining in the monolith. An important issue is the good mechanical stability of the monolithic plug and the monoliths should have good hydrodynamic properties for easy and uniform pumping (draw-eject) of sample through the bed. Therefore, prior to in situ polymerization, the surface of the PP tips was modified [18,21]. Briefly, pipette tips were filled to about 8 mm with methanol containing 5% (v/v) benzophenone. After purging with nitrogen to remove oxygen, the tips were placed inside the polymerization apparatus (mentioned above) and irradiated using 254 nm UV light for a certain time. Afterward the tips were washed with methanol and dried at 40°C for 10 min (Fig. 17.4).

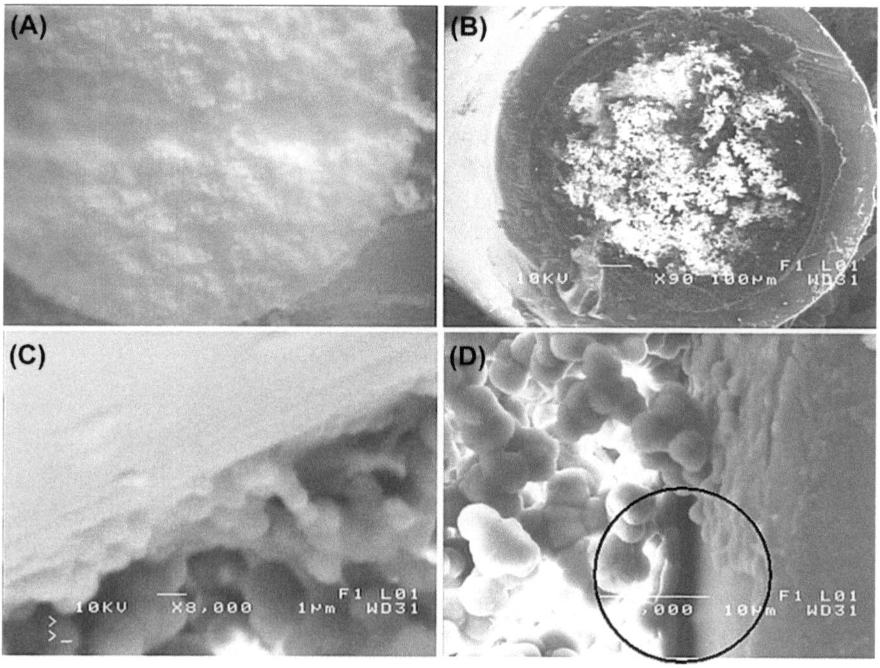


**Figure 17.4** (A–D) displays scanning electron microscopy (SEM) images of the monolithic structure inside a pipette tip indicating complete filling of the polymer across the tube. No significant differences and no voids between the polymer matrix and the tip can be seen. In (C), the SEM image of a monolith inside a pipette tip modified with MMA/EGDMA 1:1 with 5% BP shows binding between the monolith and the pipette wall. The monolith inside the tip surface modified with 5% BP in MeOH shows binding between monolith and wall in the upper part of the image (D). The lower part of the image shows a gap between the monolith and the pipette wall. The surface of the pipette wall in this part appears to be much smoother compared to the surface in the upper part of the image. Hence, the surface modified part seems to have obtained a more adhesive surface.
Reused with permission Altun Z, Skoglund C, Abdel-Rehim M. Monolithic methacrylate packed 96-tips for high throughput bioanalysis. J Chromatogr A 2010;1217:2581–2588.

## 17.3.2 Factors affecting the performance of packed 96-tips

There are a number of factors affecting the porous properties of the monoliths. The factors to be considered are not only all components included in the polymerization mixture but also the polymerization temperature and intensity of the UV light. Of these factors, the type and composition of the porogenic mixture seem to be the key factors for fine-tuning of the final properties of the monoliths. A commonly used porogenic mixture for the preparation of the monoliths is cyclohexanol and 1-dodecanol. In such a porogenic mixture the pore size increases with the concentration of 1-dodecanol in the polymeric mixture [18]. The polymerization temperature is another factor effecting pore size distribution of the monoliths. Usually, at higher temperature smaller pores are obtained. Several reports dealing with factors effecting surface properties of monoliths have been published [18].

### 17.3.3 Modified monoliths phases for packed pipette tips

In several reports nanoparticles were incorporated in polymer monoliths to increase the selectivity and extraction capacity. El Wael et al. modified a monolithic phase with gold nanoparticles for selective extraction of glycoproteins [21]. Hsieh et al. prepared a polymer monolith in a pipette tip incorporating titanium dioxide nanoparticles for the selective extraction of phosphorylated peptides prior to mass spectrometric analysis [22]. Nanostructured polymer monoliths were prepared by the encapsulation of selected nanoparticles in the monomer mixture prior to polymerization [21,22]. Additionally in situ monolithic molecularly imprinted polymer sol-gel packed tips were reported by Moein et al. for the extraction of L-tyrosine from human plasma and urine samples [20].

### 17.3.4 Monolithic packed tips applications

The extraction of metoprolol and pindolol from human plasma was carried out using packed 96-tips [17,18]. The accuracy varied from 94% to 114% for pindolol, and from 101% to 103% for metoprolol (n = 18). The between-day precisions (RSD) were 9%−13% for pindolol, and 9%−15% for metoprolol (n = 18). In Fig. 17.5,

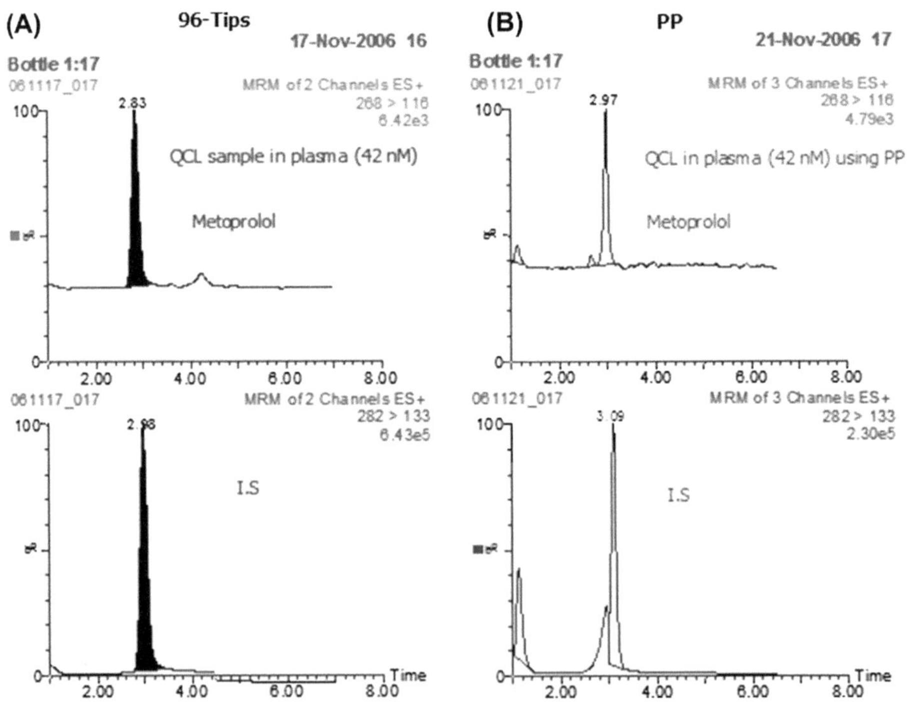


**Figure 17.5** Total ion chromatogram of the MRM analysis of metoprolol and internal standard from spiked human plasma sample (42 nmol/L): (A) treated by 96-tips, (B) protein precipitation was used.

Reused with permission Altun Z, Skoglund C, Abdel-Rehim M. Monolithic methacrylate packed 96-tips for high throughput bioanalysis. J Chromatogr A 2010;1217:2581−2588.

packed 96-tips extraction method was compared with protein precipitation for the sample preparation of metoprolol and pindolol from human plasma. Using monolithic sorbents to clean up the extracts eliminated ion suppression while high ion suppression was observed using the protein precipitation method [18]. In addition, monolithic methacrylate 96-tips were compared with commercial silica-based monolithic packed tips (OMIX C18). The precision was within the acceptable range for the two methods, but better accuracy was obtained for methacrylate monolithic tips [18].

Monolithic packed 96-tips were used for the extraction of local anaesthetics (lidocaine, ropivacaine, and bupivacaine) from human plasma [14,16]. The calibration range was 2–2000 nmol/L and the extraction recovery was 60%.

Monolithic packed tips were used for the extraction of the anticancer agent busulfan in human plasma, cyclophosphamide in mice blood, and roscovitine in human plasma [19]. Additionally the extraction of L-tyrosine from human plasma and urine samples utilizing packed pipette tips was reported [20].

## 17.4 Concluding remarks

Sample preparation for a long time has been a bottleneck in bioanalysis. During recent years innovations in sample preparation techniques have gone some way to address this problem. Further progress in the automated systems for sample preparation is expected to meet the need for high sample throughput or 24 h process monitoring. The main attractive features of automated sample preparation techniques include miniaturization and high throughput. MEPS and monolithic packed 96-tips techniques provide significant advantages such as speed and simplicity. A key feature of MEPS is that typical elution volumes are of a suitable size for direct injection into GC or LC systems. The sample preparation procedure using MEPS is automated, simple, and does not involve costly and time-consuming processes. In this chapter the progress, application, and future aspect of MEPS and monolithic packed tips such as green, fast, high-throughput, and sensitive sample preparation techniques were briefly reviewed.

## References

[1] Abdel-Rehim M. New trend in sample preparation: on-line microextraction in packed syringe for liquid and gas chromatography applications I. Determination of local anaesthetics in human plasma samples using gas chromatography–mass spectrometry. J Chromatogr B 2004;801:317–21.

[2] Moein MM, Abdel-Rehim A, Abdel-Rehim M. Microextraction by packed sorbent (MEPS). TrAC Trends Anal Chem 2015;67:34–44.

[3] Yanga L, Said R, Abdel-Rehim M. Sorbent, device, matrix and application in microextraction by packed sorbent (MEPS): a review. J Chromatogr B 2017;1043:33–43.

[4] Moein MM, Abdel-Rehim A, Abdel-Rehim M. On-line determination of sarcosine in biological fluids utilizing dummy molecularly imprinted polymers in microextraction by packed sorbent. J Sep Sci 2015;38(5):788–95.
[5] de Souza ID, Domingues DS, Queiroz MEC. Hybrid silica monolith for microextraction by packed sorbent to determine drugs from plasma samples by liquid chromatography–tandem mass spectrometry. Talanta 2015;140:166–75.
[6] Szultka M, Krzeminski R, Szeliga J, Jackowski M, Buszewski B. A new approach for antibiotic drugs determination in human plasma by liquid chromatography-mass spectrometry. J Chromatogr A 2013;1272:41–9.
[7] Leca JM, Pereira V, Pereira AC, Marques JC. Rapid and sensitive methodology for determination of ethyl carbamate in fortified wines using microextraction by packed sorbent and gas chromatography with mass spectrometric detection. Anal Chem Acta 2014; 811:29–35.
[8] Perestrelo R, Rodriguez E, Câmara JS. Impact of storage time and temperature on furanic derivatives formation in wines using microextraction by packed sorbent tandem with ultrahigh pressure liquid chromatography. LWT – Food Sci Tech 2017;76:40–7.
[9] Goncalves J, Mendes B, Silva CL, Camara JS. Development of a novel microextraction by packed sorbent-based approach followed by ultrahigh pressure liquid chromatography as a powerful technique for quantification phenolic constituents of biological interest in wines. J Chromatogr A 2012;1229:13–23.
[10] Silva CL, Goncalves JL, Câmara JS. A sensitive microextraction by packed sorbent-based methodology combined with ultra-high pressure liquid chromatography as a powerful technique for analysis of biologically active flavonols in wines. Anal Chim Acta 2012;739: 89–98.
[11] Adam M, Pavlíková P, Čížková A, Ventura K. Microextraction by packed sorbent (MEPS) as a suitable selective method for l-ascorbic acid determination in beverages. Food Chem 2012;135(3):1613–8.
[12] Gonçalves J, Silva CL, Castilho PC, Câmara JS. An attractive, sensitive and high-throughput strategy based on microextraction by packed sorbent followed by UHPLC-PDA analysis for quantification of hydroxybenzoic and hydroxycinnamic acids in wines. Microchem J 2013;106:129–38.
[13] Du W, Zhao G, Fu Q, Sun M, Zhou H, Chang C. Combined microextraction by packed sorbent and high-performance liquid chromatography-ultraviolet detection for rapid analysis of ractopamine in porcine muscle and urine samples. Food Chem 2014;145: 789–95.
[14] Altun Z, Blomberg L, Abdel-Rehim M. Increasing sample preparation throughput using monolithic methacrylate polymer as packing material for 96-tip robotic device. J Liq Chromatogr Relat Technol 2006;29(10):477–1489.
[15] Altun Z, Hjelmström A, Abdel-Rehim M, Blomberg L. Surface modified polypropylene pipette tips packed with a monolithic plug of adsorbent for high throughput sample preparation. J Sep Sci 2007;30:1964–72.
[16] Altun Z, Hjelmström A, Blomberg L, Abdel-Rehim M. Evaluation of monolithic packed 96-tips for solid-phase extraction of local anesthetics from human plasma for quantitation by liquid chromatography tandem mass spectrometry. J Liq Chromatogr Relat Technol 2008;31:743–75.
[17] Abdel-Rehim M, Persson C, Altun Z, Blomberg L. Evaluation of monolithic packed 96 tips and liquid chromatography–tandem mass spectrometry for extraction and quantification of pindolol and metoprolol in human plasma samples. J Chromatogr A 2008;1196–1197: 23–7.

[18] Altun Z, Skoglund C, Abdel-Rehim M. Monolithic methacrylate packed 96-tips for high throughput bioanalysis. J Chromatogr A 2010;1217:2581–8.

[19] Skoglund C, Bassyouni F, Abdel-Rehim M. Monolithic packed 96-tips set for high throughput sample preparation: determination of cyclophosphamide and busulfan in whole blood samples by monolithic packed 96-tips and LC-MS. Biomed Chromatogr 2013;27: 714–9.

[20] Moein MM, El-Beqqali A, Abdel-Rehim A, Jeppsson-Dadoun A, Abdel-Rehim M. Preparation of monolithic molecularly imprinted polymer sol–gelpacked tips for high throughput bioanalysis: extraction and quantification of l-tyrosine in human plasma and urine samplesutilizing liquid chromatography and tandem mass spectrometry. J Chromatogr B 2014;967:168–73.

[21] Alwael H, Connolly D, Clarke P, Thompson R, Twamley B, O'Connora B, Paull B. Pipette-tip selective extraction of glycoproteins with lectin modified gold nano-particles on a polymer monolithic phase. Analyst 2011;136(12):2619–28.

[22] Hsieh HC, Sheu C, Shi FK, Li T. Development of a titanium dioxide nanoparticle pipette-tip for the selective enrichment of phosphorylated peptides. J Chromatogr A 2007;1165: 128–35.

[23] Altun Z, Abdel-Rehim M. Study of the factors affecting the performance of micro-extraction by packed sorbent (MEPS) using liquid scintillation counter and liquid chromatography-tandem mass spectrometry. Anal Chim Acta 2008;630(2):116–23.

[24] Saracino MA, Tallarico K, Raggi MA. Liquid chromatographic analysis of oxcarbazepine and its metabolites in plasma and saliva after a novel microextraction by packed sorbent procedure. Anal Chim Acta 2010;661(2):222–8.

[25] Somaini L, Saracino MA, Marcheselli C, Zanchini S, Gerra G, Raggi MA. Combined liquid chromatography-coulometric detectionand microextraction by packed sorbent for the plasma analysis of long acting opioids in heroin addicted patients. Anal Chim Acta 2011;702(2):280–7.

[26] Rani S, Malik AK. A novel microextraction by packed sorbent-gas chromatography procedure for the simultaneous analysis of antiepileptic drugs in human plasma and urine. J Sep Sci 2010;35(21):2970–7.

[27] Said R, Pohank A, Abdel-Rehim M, Beck O. Determination of four immunosuppressive drugs in whole blood using MEPS and LC–MS/MS allowing automated sample work-up and analysis. J Chromatogr B 2012;897:42–9.

[28] Ahmadi M, Moein MM, Madrakian T, Afkhami A, Bahar S, Abdel-Rehim M. Reduced graphene oxide as an efficient sorbent in microextraction by packed sorbent: determination of local anesthetics in human plasma and saliva samples utilizing liquid chromatography-tandem mass spectrometry. J Chromatogr B 2018;1095:177–82.

[29] Abdel-Rehim M, Hassan Z, Skansem P, Hassan M. Simultaneous determination of busulphan in plasma samples by liquid chromatography-electrospray ionization mass spectrometry utilizing microextraction in packed syringe (MEPS) as on-line sample preparation method. J Liq Chromatogr Rel Tech 2007;30:3029–41.

[30] Moein MM, El-Beqqali A, Javanbakht M, Karimi M, Akbari-adergani B, Abdel-Rehim M. On-line detection of hippuric acid by microextraction with a molecularly-imprinted polysulfone membrane sorbent and liquid chromatography-tandem mass spectrometry 2014;1372:55–62.

[31] Klimowska A, Wielgomas B. Off-line microextraction by packed sorbent combined with on solid support derivatization and GC-MS: application for the analysis of five pyrethroid metabolites in urine sample. Talanta 2018;176:165–71.

[32] Pautova AK, Revelsky AI, Sobolev PD, Pautova AK. Microextraction of aromatic microbial metabolites by packed sorbent (MEPS) from model solutions followed by gas chromatography/mass spectrometry analysis of their Silyl derivatives. J Anal Chem 2017; 72:1426–33.

[33] Naccarato A, Elliani R, Sindona G, Tagarelli A. Multivariate optimization of a microextraction by packed sorbent-programmed temperature vaporization-gas chromatography–tandem mass spectrometry method for organophosphate flame retardant analysis in environmental aqueous matrices. Anal Bioanal Chem 2017;409(30): 7105–20.

[34] Fernandez P, Regenjo M, Fernandez AM, Carro AM, Ares AM, Fernández P, Regenjo M, Fernández AM, Carro AM, Lorenzo RA. A fast bioanalytical method based on microextraction by packed sorbent and UPLC–MS/MS for determining new psychoactive substances in oral fluid. Talanta 2017;174:454–61.

[35] Sarraguça D, Araujo M, Rodrigues ARTS, Rodrigues M. A novel HPLC method for the determination of zonisamide in human plasma using microextraction by packed sorbent optimised by experimental design. Anal Methods 2017;9:5910–9.

[36] Amiri A, Ghaemi F. Microextraction in packed syringe by using a three-dimensional carbon nanotube/carbon nanofiber–graphene nanostructure coupled to dispersive liquid-liquid microextraction for the determination of phthalate esters in water samples. Micro Chim Acta 2017;184:3851–8.

[37] Soleimani E, Bahrami A, Afkhami A, Shahna FG. Determination of urinary trans,trans-muconic acid using molecularly imprinted polymer in microextraction by packed sorbent followed by liquid chromatography with ultraviolet detection. J Chromatogr B 2017; 1067:65–71.

[38] Llerena P, Alves G, Llerena A, Falcão A. Therapeutic drug monitoring of fluoxetine, norfluoxetine and paroxetine: a new tool based on microextraction by packed sorbent coupled to liquid chromatography. J Anal Toxicol 2017;41(7):631–8.

[39] Bianchia F, Mattarozzi M, Riboni N, Mora P, Gandolfi SA, Careri M. A rapid microextraction by packed sorbent – liquid chromatography tandem mass spectrometry method for the determination of dexamethasone disodium phosphate and dexamethasone in aqueous humor of patients with uveitis. J Pharm Biomed Anal 2017;142:343–7.

[40] Ortega SN, Santos-Neto AJ, Lancas FM. Development and optimization of a fast method for the determination of statins in human plasma using microextraction by packed sorbent (MEPS) followed by ultra high-performance liquid chromatography-tandem mass spectrometry (UHPLC-MS/MS). Anal Methods 2017;9:3039–48.

[41] Fernández P, González M, Regenjo M, Ares AM, Fernández AM, Lorenzo RA, Carro AM. Analysis of drugs of abuse in human plasma using microextraction by packed sorbents and ultra-high-performance liquid chromatography. J Chromatogr A 2017;1485:8–19.

[42] Rosado T, Fernandes L, Barroso M, Gallardo E. Sensitive determination of THC and main metabolites in human plasma by means of microextraction in packed sorbent and gas chromatography–tandem mass spectrometry. J Chromatogr B 2017;1043:63–73.

[43] Ferreira AMC, Cordero BM, Pavón JLP. Use of microextraction by packed sorbent directly coupled to an electron ionization single quadrupole mass spectrometer as an alternative for non-separative determinations. J Chromatogr B 2017;1043:74–80.

[44] Ventura S, Rodrigues M, Pousinho S, Falcão A, Alves G. Determination of lamotrigine in human plasma and saliva using microextraction by packed sorbent and high performance liquid chromatography–diode array detection: an innovative bioanalytical tool for therapeutic drug monitoring. Microch J 2017;130:221–8.

[45] D'Angelo V, Tessari F, Bellagamba G, De Luca E, Cifelli R, Celia C, Primavera R, Di Francesco M, Paolino D, Di Marzio L, Locatelli M. Microextraction by packed sorbent and HPLC–PDA quantification of multiple anti-inflammatory drugs and fluoroquinolones in human plasma and urine. J Enzym Inhib Med Chem 2016;31:110–6.

[46] Ventura S, Rodrigues M, Pousinho S, Falcão A, Alves G. An easy-to-use liquid chromatography assay for the analysis of lamotrigine in rat plasma and brain samples using microextraction by packed sorbent: application to a pharmacokinetic study. J Chromatogr B 2016;1035:67–75.

[47] Savastano ML, Losito I, Pati S. Rapid and automatable determination of ochratoxin A in wine based on microextraction by packed sorbent followed by HPLC-FLD. Food Control 2016;68:391–8.

[48] Elmongy H, Ahmed H, Wahbi AA, Amini A, Colmsjö A, Abdel-Rehim M. Determination of metoprolol enantiomers in human plasma and saliva samples utilizing microextraction by packed sorbent and liquid chromatography-tandem mass spectrometry. Biomed Chromatogr 2016;30(8):1309–17.

[49] Fumes BH, Andrade FN, Neto ÁJd S, Lanças FM. Determination of pesticides in sugarcane juice employing microextraction by packed sorbent followed by gas chromatography and mass spectrometry. J Sep Sci 2016;39(14):2823–30.

[50] Magrini L, Cappiello A, Famiglini G, Palma P. Microextraction by packed sorbent (MEPS)-UHPLC-UV: a simple and efficient method for the determination of five benzodiazepines in an alcoholic beverage. J Pharm Biomed Anal 2016;125:48–53.

[51] Ferreira AMC, Cordero BM, Pozas ÁPC, Pavón JLP. Use of microextraction by packed sorbents and gas chromatography-mass spectrometry for the determination of polyamines and related compounds in urine. J Chromatogr A 2016;1444:32–41.

[52] Yang L, Han Q, Cao S, Yang J, Zhao J, Qin M, Ding M. Self-made microextraction by packed sorbent device for the cleanup of polychlorinated biphenyls from bovine serum. J Sep Sci 2016;39(8):1518–23.

[53] Kaur R, Heena, Kaur R, Rani S, Malik AK. Simple and rapid determination of phthalates using microextraction by packed sorbent and gas chromatography with mass spectrometry quantification in cold drink and cosmetic samples. J Sep Sci 2016;39(5):923–31.

[54] Locatelli M, Ciavarella MT, Paolino D, Celia C, Fiscarelli E, Ricciotti G, Pompilio A, Bonaventura GD, Grande R, Zengin G, Marzio LD. Determination of ciprofloxacin and levofloxacin in human sputum collected from cystic fibrosis patients using microextraction by packed sorbent-high performance liquid chromatography photodiode array detector. J Chromatogr A 2015;1419:58–66.

[55] Moreno I, Barroso M, Martinho A, Cruzc A, Gallardo E. Determination of ketamine and its major metabolite, norketamine, in urine and plasma samples using microextraction by packed sorbent and gas chromatography-tandem mass spectrometry. J Chromatogr B 2015; 1004:67–78.

[56] Said R, Pohank A, Andersson M, Beck O, Abdel-Rehim M. Determination of remifentanil in human plasma by liquidchromatography-tandem mass spectrometry utilizing micro extraction inpacked syringe (MEPS) as sample preparation. J Chromatogr B 2011;879: 815–8.

[57] Daryanavard SM, Jeppsson- Dadoun A, Andersson LA, Hashemi M, Colmsj A, Abdel-Rehim M. Molecularly imprinted polymer in microextraction by packed sorbent for the simultaneous determination of local anesthetics: lidocaine, ropivacaine,mepivacaine and bupivacaine in plasma and urine samples. Biomed Chromatogr 2013;27:1481–8.

[58] Mousavi ZK, Yamini Y, Karimi B, Seidi S, Khorasani M, Ghaemmaghami M, Vali H. Imidazolium-based mesoporous organosilicas with bridging organic groups for microextraction by packed sorbent of phenoxy acid herbicides, polycyclic aromatic hydrocarbons and chlorophenols. Microchim Acta 2019;186:239. https://doi.org/10.1007/s00604-019-3355-3.

[59] Berenguer PH, Camacho IC, Câmara R, Oliveira S, Câmara JS. Determination of potential childhood asthma biomarkers using a powerful methodology based on microextraction by packed sorbent combined with ultra-high pressure liquid chromatography. Eicosanoids as case study. J Chromatogr A 2019;1584:42–56.

[60] Meng J, Wang X. Microextraction by packed molecularly imprinted polymer combined ultra-high-performance liquid chromatography for the determination of levofloxacin in human plasma. J Chem 2019;2019:1–9. Article ID 4783432.

[61] Šatínský D, Sobek V, Lhotská I, Solich P. Micro-extraction by packed sorbent coupled on-line to a column-switching chromatography system – a case study on the determination of three beta-blockers in human urine.


## Further reading


[1] Jiang Y, Ma P, Li X, Piao H, Li D, Sun Y, Wang X, Song D. Application of metal-organic framework MIL-101(Cr) to microextraction in packed syringe for determination of triazine herbicides in corn samples by liquid chromatography-tandem mass spectrometry. J Chromatogr A 2018;157:36–41.

[2] Ayazi Z, Shekari Esfahlan F, Matin P. Graphene oxide reinforced polyamide nanocomposite coated on paper as a novel layered sorbent for microextraction by packed sorbent. Int J Environ Anal Chem 2018;98:1118–34.

[3] Maciel EVS, Fumes BH, de Toffoli AL, Lanças FM. Graphene particles supported on silica as sorbent for residue analysis of tetracyclines in milk employing microextraction by packed sorbent. Electrophoresis 2018;39:2047–55.

[4] Rocchi R, Simeoni MC, Montesano C, Vannutelli G, Curini R, Sergi M, Compagnone D. Analysis of new psychoactive substances in oral fluids by means of microextraction by packed sorbent followed by ultra-high performance liquid chromatography tandem mass spectrometry. Drug Test Anal 2018;10(5):865–73.

[5] Taghani A, Goudarzi N, Bagherian GA, Chamjangali MA, Amin AH. Application of nanoperlite as a new natural sorbent in the preconcentration of three organophosphorus pesticides by microextraction in packed syringe coupled with gas chromatography and mass spectrometry. J Sep Sci 2018;41(10):2245–52.

[6] Dhingra G, Bansal P, Dhingra N, Rani S, Malik AK. Development of a microextraction by packed sorbent with gas chromatography mass spectrometry method for quantification of nitroexplosives in aqueous and fluidic biological samples. J Sep Sci 2018;41(3):639–47.

[7] Silva CF, Borges KB, do Nascimento CS. Rational design of a molecularly imprinted polymer for dinotefuran: theoretical and experimental studies aimed at the development of an efficient adsorbent for microextraction by packed sorbent. Analyst 2018;143(1):141–9.

[8] Soleimani E, Bahrami A, Afkhami A, Shahna FG. Selective determination of mandelic acid in urine using molecularly imprinted polymer in microextraction by packed sorbent. Arch Toxicol 2018;92(1):213–22.

[9] El Beqqali A, Ahmadi M, Abdel-Rehim M. Determination of AZD6118 in dog plasma samples utilizing microextraction by packed sorbent and liquid chromatography-electrospray ionization tandem mass spectrometry. J Chromatogr A 2017;1043:20–4.

[10] Karimiyan H, Uheida A, Hadjmohammadi M, Moein MM, Abdel-Rehim M. Polyacrylonitrile / rapheme oxide nanofibers for packed sorbent microextraction of drugs and their metabolites from human plasma samples. Talanta 2019;201:474–9.

[11] Abdel-Rehim A, Abdel-Rehim M. Advantages of Saliva Sampling in Bioanalysis Using Microextraction by Packed Sorbent and Dried Saliva Spot with LC-MS-MS. LC GC Europe 2014;27(10):529–31.

[12] Abdel-Rehim A, Abdel-Rehim M. Evaluation of microextraction by packed sorbent and micro-liquid chromatography-tandem mass spectrometry as a green approach in bioanalysis. Biomedical Chromatography 2013;27(10). https://doi.org/10.1002/bmc.2839.

[13] Zohdi Z, Hashemi M, Uheida A, Moein MM, Abdel-Rehim M. Graphene Oxide Tablets for Sample Preparation of Drugs in Biological Fluids: Determination of Omeprazole in Human Saliva for Liquid Chromatography Tandem Mass Spectrometry. Molecules 2014; 24:1191.

# Stir-bar sorptive extraction

# 18

*Olatz Zuloaga* [1,2], *Nestor Etxebarria* [1,2], *Belén González-Gaya* [1,2], *Maitane Olivares* [1,2], *Ailette Prieto* [1,2], *Aresatz Usobiaga* [1,2]
[1]Department of Analytical Chemistry, Faculty of Science and Technology, University of the Basque Country (UPV/EHU), Basque Country, Spain; [2]Research Centre for Experimental Marine Biology and Biotechnology (PIE), University of the Basque Country (UPV/EHU), Basque Country, Spain

## 18.1 Theory of stir-bar sorptive extraction

Determination of organic and organometallic compounds at trace levels requires an extraction step in order to isolate and preconcentrate the target compounds from its matrix. Within the past decades a large effort has been made to reduce or eliminate solvent consumption and to simplify and automate the extraction step in analytical procedures [1]. In this context, stir-bar sorptive extraction (SBSE) was first described by Baltussen and coworkers in 1999 as a new, solventless, sample enrichment technique, which combines target analyte extraction and preconcentration [2]. It is based on the same principles as solid phase microextraction (SPME), while enhancing the effectiveness of poly(dimethylsiloxane) (PDMS) as a sorbent material. The new format had its origin in packed PDMS beds already validated for gas and liquid samples [3,4]. SBSE was commercialized by Gerstel GmbH & Co. KG (Mülheim and der Ruhr, Germany) under the trademark of Twister.

The theory for SPME and SBSE is based on the sorption of target compounds on the acceptor phase and, therefore, depends on the corresponding distribution constant ($Kd$), which is characteristic of each analyte [2,5]. If the solid support is PDMS, $K_{PDMS/W}$ is closely correlated with $K_{o/w}$ as stated in Eq. (18.1):

$$K_{o/w} \approx K_{PDMS/W} = \frac{C_{SBSE}}{C_W} = \frac{m_{SBSE}}{m_W} \times \frac{V_W}{V_{SBSE}} \tag{18.1}$$

where the ratio of the concentrations of the analyte in the SBSE bulk material ($C_{SBSE}$) and water ($C_W$) is related to the ratio of the mass of the analyte in the SBSE ($m_{SBSE}$) and water ($m_W$) and the ratio of the water (donor) and SBSE (acceptor) phase volumes ($V_W/V_{SBSE}$). Replacing the volume ratio by $\beta$, Eq. (18.1) can be expressed as:

$$\frac{K_{o/w}}{\beta} = \frac{m_{SBSE}}{m_W} = \frac{m_{SBSE}}{m_0 - m_{SBSE}} \tag{18.2}$$

Solid-Phase Extraction. https://doi.org/10.1016/B978-0-12-816906-3.00018-2

where $m_0$ is the initial amount of analyte present in the water sample. Rearranging Eq. (18.2) to obtain the method recovery ($R$, ratio between the mass of analyte in the acceptor phase and the initial mass) then:

$$R = \frac{m_{SBSE}}{m_0} = \frac{\left(\frac{K_{o/w}}{\beta}\right)}{1 + \left(\frac{K_{o/w}}{\beta}\right)} = \frac{K_{o/w}}{\beta + K_{o/w}} \tag{18.3}$$

According to Eq. (18.3), $\beta$ governs the design criteria for the sampling device and the analyte hydrophobicity ($K_{o/w}$) its recovery. Eq. (18.3) provides a good practical approach to predict recovery data in SBSE.

However, Eq. (18.3) is only valid when the acceptor phase is a liquid and equilibrium has been reached. Fig. 18.1 shows the theoretical extraction efficiency of SBSE and SPME for different $K_{o/w}$ values. As the amount of PDMS in SPME (approx. 0.5 μL) is much lower than for the stir-bar (approx. 25 μL), $\beta_{SPME} >>> \beta_{SBSE}$ and, therefore, the extraction efficiency for SBSE will always be higher, allowing almost quantitative extraction of more polar compounds by SBSE, hardly achievable with SPME. In this sense, the SBSE design is particularly more effective than SPME for the extraction of polar and medium polar compounds (log $K_{o/w} < 3$) [6], providing a further advantage for this kind of extraction.

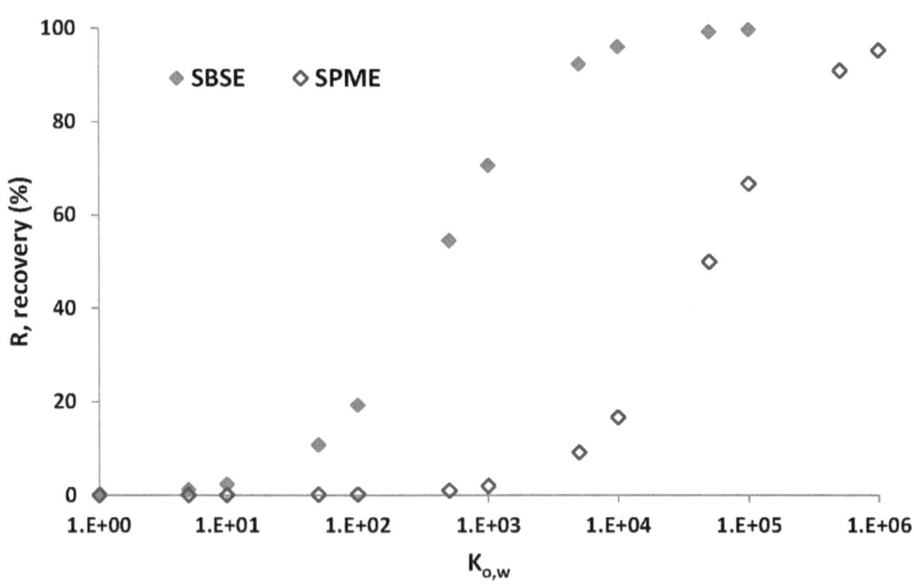


**Figure 18.1** Extraction efficiency of SPME and SBSE as a function of the octanol-water partition coefficient ($K_{o/w}$) of the extracted analyte.

The theoretical recoveries at equilibrium can be estimated for a given sample volume, selected stir-bar dimensions and target analyte using the KowWIN software (Syracuse Research Corp, Syracuse, New York) [1]. Working under equilibrium conditions guarantees a better precision for the extraction step, while small changes in the experimental variables imply a significant loss of precision for kinetic controlled conditions. Nevertheless, equilibrium extraction is not always feasible and compromise conditions must be established in this case.

## 18.2 Method development

For most applications (see Section 18.4) SBSE is typically coupled to separation and detection systems, and the relevant parameters affecting both extraction and desorption are optimized to obtain the most sensitive, accurate, and precise method. Additionally, depending on the physicochemical properties of the target compounds, particularly their volatility and polarity, SBSE can be coupled to gas or liquid chromatography. The choice of separation/detection system will largely depend on the goals of the analysis and the instrumentation available [7–11].

### *18.2.1 Selection of the extraction phase*

The selection of the stir-bar coating is the first step in SBSE method development. Currently, there are three commercially available coatings for SBSE: PDMS (typically 10–20 mm length and 0.5–1 mm thick), poly(acrylate) (PA), and ethylene glycol-silicone (EG-silicone). The last two were introduced by Gerstel in 2011 for the extraction of medium to high polarity compounds. Higher extraction efficiency is obtained for these new sorbents when extracting polar compounds compared with PDMS [7]. However, poorer mechanical stability and degradation of these polar coatings has also been reported. In fact, EG-silicone coatings degrade quickly and each stir-bar may only be used about 20 times compared with approximately 100 reuses for PDMS coating [7]. Due to the increased interest in the extraction of more polar or even ionic compounds, researchers have recently focused on the developments in this area [7,9].

The stir-bar coatings are typically prepared by two technologies: sol-gel processes and molecular imprinting technology. Through sol-gel technology, different chemical groups have been introduced into a PDMS network, including β-cyclodextrin (PDMS/β-CD), divinylbenzene (PDMS/DVB), and poly(vinyl alcohol) (PDMS/PVA) [9,11]. This method is also used for the manufacture of organic-inorganic hybrid materials, mainly metal oxides of silicon, titanium, and zirconium [7]. However, coatings prepared by sol-gel technology suffer from cracking of the polymer layer, as mentioned before. On the other hand, molecularly imprinted technology allows the creation of artificial ligands with a predetermined selectivity and specificity for target analytes. These materials are known as molecularly imprinted polymers (MIPs) and can be prepared in a variety of physical forms tailored to their intended applications [12].

The conventional approach is to synthesize MIPs in bulk. Alternative methods, such as MIP beads, membranes, *in situ* prepared monoliths, surface imprinting, and molecularly imprinted monolayers have been developed in recent years. In this sense, a series of coatings including hydrophobic, hydrophilic, and mix-mode extraction phases based on monoliths for SBSE have been prepared for the extraction of a variety of organic and inorganic compounds. These coatings are generally inexpensive, show high permeability and favorable mass-transfer characteristics [9].

Stir-bar conditioning is performed either thermally or by solvent conditioning [7,11]. The former can be carried out online or off-line with dedicated equipment. The Tube Conditioner (TC) commercialized by Gerstel is expensive and the use of a thermal desorption unit (TDU) coupled to a GC is time-consuming. Solvent conditioning might be the best approach but the solvent must be carefully selected to avoid damaging the coating (excessive swelling). Acetonitrile and mixtures of acetonitrile with dichloromethane (typically 1:1) are the most common solvents used for this purpose [7].

### 18.2.2 Selection of the sampling mode

SBSE method offers two general sampling modes: (1) direct immersion stir-bar sorptive extraction (SBSE) and (2) headspace sorptive extraction (HSSE) (Fig. 18.2). While SBSE is more suitable for gaseous or simple liquid samples, HSSE is preferred for the extraction of volatile or semivolatile compounds from complex liquid and solid samples.

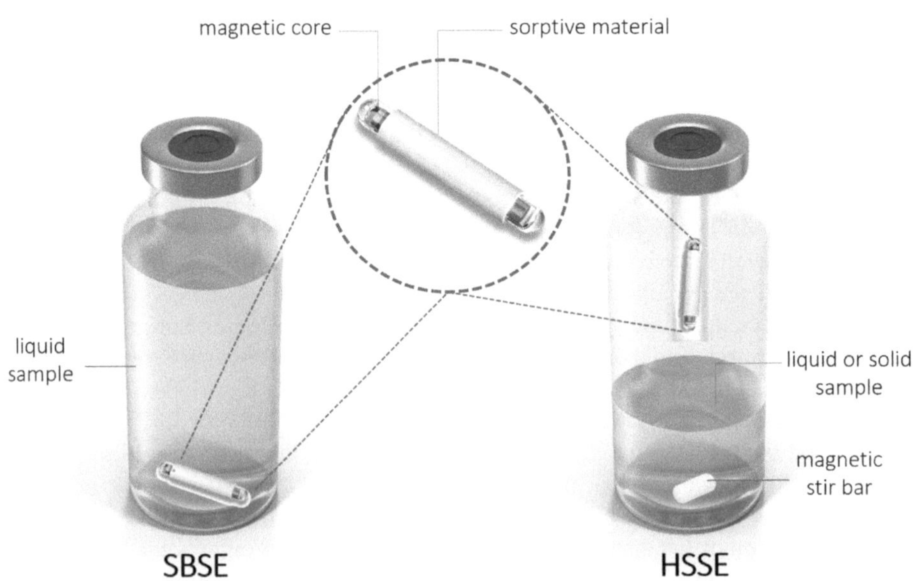


**Figure 18.2** Stir-bar sorptive extraction in the immersion (*left*) and headspace (*right*) sampling modes.

The simplest SBSE sampling mode is direct immersion which has been used for most applications. In this case, a polymer-coated stir-bar is placed inside a HS vial that contains an aqueous sample that is continuously stirred. When applied to solid samples, a previous solid-liquid extraction is generally performed and the extract subsequently treated as a liquid sample.

HSSE sampling is performed by suspending a polymer-coated stir-bar in a HS vial where the polymeric phase is in static contact with the vapor phase of a solid or liquid matrix; for this purpose, special devices to hold the stir-bar are available [1]. A magnetic stirrer is usually used to homogenize the sample, and to enhance the mass transfer from the bulk sample to the headspace.

The term "dual SBSE" was introduced by Ochiai et al. [13] for the multiresidue determination of pesticides in food matrices. In this case, two stir-bars were added to a twofold and a fivefold dilution of the same sample and, after the extraction, both stir-bars were simultaneously desorbed. Since then, the term dual SBSE has been applied to a duplicate extraction of the same sample under different extraction conditions, which can imply differences in the matrix (including extraction pH, amount of organic modifier or inert salts, for instance) or differences in the sampling mode and/or the nature or dimensions of the stir-bar used.

### 18.2.3 Optimization of extraction parameters

Addition of inert salts (e.g., NaCl, up to 20%) or an organic modifier (e.g., methanol, up to 20% v/v), or pH adjustment (between 2 and 11) for aqueous samples are typically employed to increase the transfer of the analytes to the extraction phase [7,9,11]. Inert electrolytes, such as NaCl cause a "salting-out effect" or "oil-effect" that improves the extraction efficiency of hydrophilic compounds (log $K_{o/w} < 3$) from aqueous samples [9]. However, for nonpolar compounds it decreases the mass-transfer rate owing to the higher water density [7]. Addition of an organic solvent, such as methanol, increases the water solubility of nonpolar compounds (log $K_{o/w} > 3$) and minimizes the adsorption of these compounds onto the glass walls. Finally, pH is adjusted for the extraction of weak acids and bases because only neutral (nonionic) species are extracted. Proper selection of the pH or addition of ion-pair reagents facilitates the extraction of acids and bases as their neutral forms by typical PDMS coatings [7]. For instance, the use of tetra-*n*-butylammonium salts or trifluoroacetic acid allows the extraction of anionic and cationic compounds, respectively, together with polar organic compounds from aqueous solution [14].

The total mass of analyte extracted from the sample depends on the phase ratio, see section 18.1. For typical sample volumes of 5 −200 mL, lower extraction efficiency is expected for large sample volumes [7]. However, large sample volumes are a reasonable choice because a higher analyte mass is extracted, providing lower limits of quantification [7,11]. In fact, the extracted mass increases in proportion to the sample mass. Similarly, the volume of the extraction phase also affects the extraction yield with higher recovery achieved for larger volumes of the extraction phase. For example, the sensitivity is increased by a factor of 50−250 when PDMS based stir-bars are used instead of PDMS-based SPME. Stir-bars with a coating of 25−125 μL of

PDMS (typically, 0.5−1 mm thick) are commercially available and theoretical recoveries of 100% can be reached at optimum extraction conditions for compounds with log $K_{o/w} > 3$ [8].

The extraction time is kinetically controlled and the optimum value is selected by determining the analyte recovery as a function of extraction time. The extraction can be performed under equilibrium or nonequilibrium conditions. The maximum recovery is obtained for equilibrium conditions, when recovery values are constant no matter how long the extraction time is extended [1]. Polar compounds with low log $K_{o/w}$ values require longer extraction times to attain equilibrium (i.e., the kinetic process is slow) compared with the time required for nonpolar compounds [1]. In this sense, the possibility to perform the extraction overnight and, hence, the enrichment of those compounds that need higher extraction times to reach steady state conditions, is affordable. When long extraction times are not practical, kinetic controlled extraction (i.e., extraction at fixed times lower than those required to reach equilibrium) can be used. Although this approach is feasible, lower repeatability and sensitivity are typically achieved [9].

Agitation speed and temperature also affect the extraction efficiency in SBSE. The efficiency is enhanced when vigorous mixing is used but excessive agitation speed can promote bubble formation, which reduces the extraction efficiency. Hence, agitation speeds between 700 and 1500 rpm are typically used [9]. Temperature has two opposing effects on SBSE: the equilibrium state is reached faster at higher temperatures but, at the same time, the water solubility of the analytes increases ($K_{o/w}$ decreases), leading to a lower extraction efficiency [7]. The effect of temperature on SBSE is not always evaluated and extractions are often performed at room temperature [7]. For HSSE, however, temperature is a key factor, since the vapor pressures of the target compounds depend on the temperature. Typical temperatures employed for HSSE range from 20°C to 90°C.

### 18.2.4 Selection of the desorption mode

After extraction, the stir-bar is normally rinsed with distilled water, dried using a paper tissue, and the analytes desorbed either thermally or by solvent before analysis. Most applications involve the use of thermal desorption (TD) when used in conjunction with GC since the online combination of the desorption and analysis steps guarantees good limits of detection (LODs) as the analytes trapped in the extraction phase are transferred completely into the GC system. The TD approach needs a glass tube and a dedicated unit, so called TDU, consisting of two programmable temperature vaporizers (PTVs) for the bake-out operation (up to 350°C). The first PTV is operated at temperatures between 150 and 300°C for up to 15 min. Due to the long desorption times necessary in SBSE, and so as to reduce peak broadening, the analytes are cryofocused in the second PTV at low temperatures (−150°C to 40°C range) using liquid $N_2$, followed by a vaporization process using rapid heating for the transfer of the analytes [11]. Consequently, TD is limited to thermally stable volatile and semivolatile compounds. Two systems are commercially available: the TDS-A classic TD system and a specially designed TDU (both from Gerstel).

Liquid desorption (LD) provides an alternative to TD for thermally labile compounds, when LC or capillary electrophoresis (CE) are used for the separation step, or when a TDU is not available for GC. In this case, the stir-bar is immersed in the desorption solvent with sonication or stirring for 10−20 min. The desorption solvent should be compatible with the extraction phase to avoid excessive swelling or incomplete wetting, it should ensure quantitative desorption of the target compounds, and require a small volume for the total immersion of the stir-bar [11]. The main limitation of LD is that only a portion of the analyte extracted is typically introduced into the corresponding analysis system with a reduction in sensitivity. When LD is followed by GC, large volume injection (LVI) can be utilized to improve sensitivity [113,114]. LVI with a PTV injector can be achieved by splitless injection, vapor overflow, and solvent-split injection. Of these, solvent-split injection is the most widely used for LVI applications. On-column LVI is superior for highly volatile and thermally labile compounds, whereas LVI-PTV is preferred for the analysis of extracts with a dirty matrix [8,113,114].

### 18.2.5 Derivatization

Polar organic compounds with hydroxyl (-OH), carboxylic (-COOH), amine (-$NH_2$), or thiol (-SH) functional groups are only poorly extracted by PDMS coated stir-bars due to their low log $K_{o/w}$ values. Derivatization of the target compounds increases their transfer to the extraction phase after, typically, a chemical reaction to form alkyl, aryl, acyl, or silyl derivatives of polar functional groups [1,115]. Derivatization can be carried out either off-line or online for SBSE, Fig. 18.3. *In situ* derivatization is common for liquid samples to enhance transfer to the extraction phase, as well as to improve the chromatographic separation and detection of the target compounds. *In situ* derivatization takes place in the aqueous phase either before or simultaneously with the extraction step, and the derivatized compounds are preconcentrated in the polymeric phase, in both the immersion and HS modes. Acetic acid anhydride (AAA) for phenolic groups, ethyl chloroformate (ECF) for carboxylic groups, *O*-(2,3,4,5,6-pentafluorobenzyl)hydroxylamine (PFBHA) for carbonyl groups, and sodium tetraethylborate ($NaBEt_4$) for organometallic compounds are widely used with aqueous samples for *in situ* derivatization

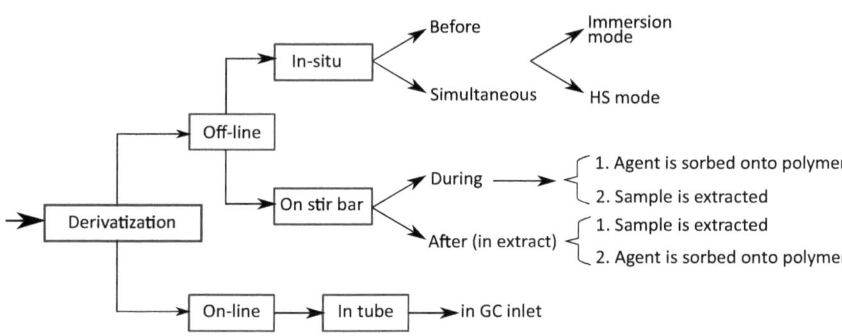


**Figure 18.3** Off-line and online derivatization in SBSE.

[8,10]. *In situ* derivatization requires that the reagents are water compatible, which is not the case for silylation reagents [1]. Off-line silylation reactions require an extra step after the extraction, with the possibility of analyte loss, contamination, and interference from water [116]. Off-line derivatization can also be performed on-polymer, also known as on-stir-bar derivatization. On-stir-bar derivatization can be performed during or after the extraction step. For the former, the derivatization reagent (i.e., AAA) is sorbed into the polymeric coating and, afterward, the stir-bar containing the reagent is exposed to the sample. For the latter, the analytes are first extracted by the stir-bar, which is then exposed to the derivatization reagent. This last step can be performed in-extract, after a back-extraction, or in-tube [10,117,118]. In-tube derivatization can be classified as an on-line process, where the derivatization reaction is performed simultaneously to the separation/detection step by adding few μLs of the reagent into the inlet of the GC together with the stir-bar. In-port derivatization occurs in the gas phase and, apart from simplifying the analytical procedure, avoids the use of large amounts of expensive and/or toxic reagents and solvents [8,119].

## 18.3 Calibration and method validation

Traditional calibration methods, such as external calibration, internal standard or standard addition, as well as less traditional matrix-matched calibration, can be used for quantification in SBSE, and the selection of the method is mainly determined by the complexity of the sample. External calibration in SBSE requires the preparation of several standard solutions in order to obtain the relationship between the analytical signal and the target compound concentration. Calibration solutions should be analyzed under the same extraction conditions used for the analysis of real samples. External calibration is useful when there is little variability in target sample composition or for those samples without strong matrix effects (e.g., tap water) [8].

Matrix effect in samples with high organic matter or suspended solid components (e.g., environmental samples, biological fluids, food samples) makes quantification more complicated. The simplest solution is often to dilute the sample until matrix effects are minimized but higher LODs will be obtained [7]. The addition of an internal standard (e.g., an isotopically labeled surrogate) to the sample at the beginning of the procedure is another strategy to compensate for matrix effects [7,11]. However, this approach may be limited if labeled homologs of the target compounds are not commercially available or are too expensive, and of course, if the labeled surrogates fail to address issues with the matrix. Then, matrix-matched calibration is an effective strategy [7]. Matrix-matched calibration standards are prepared by standard additions at three to six concentrations of target compounds to an analyte-free matrix similar to the sample, and the resulting calibration curve is used for the quantification of the target compounds in all real samples. However, sample variability may limit options for identifying suitable matching matrices. Actually, the only efficient way to overcome, or at least minimize, matrix effects is the use of standard addition calibration. In this case, the calibration curve is built by the standard addition of three to six concentrations of target compounds to sample aliquots of a constant volume/mass. Although this strategy can provide the level of accuracy required for trace analysis, it can be tedious for routine analysis when a large number of samples must be processed [11]. Besides, standard addition requires some indication of the amount of analyte in the sample for the selection of the correct amount of added analyte.

## 18.4 Application of stir-bar sorptive extraction

Environmental [1,7,8], food [1,7,120,121], and biomedical applications [1,7,8] of SBSE are reviewed in the cited sources. In this section, recent applications (restricted to last 10 years and selected from ScienceDirect according to their relevance) are compiled in Tables 18.1−18.3. Table 18.4 summarizes applications of SBSE as passive sampler. The main aim was to select a limited set of key applications that might be useful for potential users instead of a comprehensive list of applications.

### *18.4.1 Environmental analysis*

SBSE was originally used for extraction of environmental samples [1] and this remains the field with the largest number of applications. Although SBSE has been applied to both volatile and semivolatile organic compounds [1,8,10], it is mostly used to determine semivolatile low-polarity contaminants, including PAHs, organochlorine pesticides (OCPs), organophosphorus pesticides (OPPs), organonitrogen pesticides (ONPs), polychlorinated biphenyls (PCBs), polybrominated diphenyl ethers (PBDEs), and alkylphenols (APs) [1,7]. This is due to the good extraction efficiency obtained with PDMS coatings and the high sensitivity when TD coupled to GC is used. Recent applications include emerging pollutants, such as bisphenols, estrogens, pharmaceuticals, personal care products (PCPs), and organometallic compounds (see Table 18.1). From a matrix perspective, SBSE is mostly applied to water samples of different types and less frequently to solid matrices (e.g., soil, sediment, dust, or biota). For solid samples, a preliminary analyte extraction before SBSE is typically performed and only a few reports describe simultaneously solid-liquid extraction and preconcentration in the PDMS sorbent [25]. Although most applications still utilize commercially available sorbents, especially the PDMS polymer, a growing use of LD is observed, since LC is often better suited than GC for the analysis of emerging contaminants. In the case of derivatization, *in situ* is mostly applied and in-tube in combination with acetylation [19]. Dual-SBSE was used in certain cases where different modes of derivatization were necessary. For instance, Giráldez and coworkers [46] used dual-SBSE for the determination of selenite and selenoamino acids in clams after ultrasonic probe enzymatic hydrolysis. While selenoamino acids were *in situ* derivatized using isobutyl chloroformate, 4-chloro-1,2-phenylenediamine provided higher sensitivity for selenite. Aguirre and coworkers [25] studied both HSSE and SBSE for the determination of several musk fragrances in vegetables and amended soils with better sensitivity provided by SBSE.

**Table 18.1** Experimental conditions for the analysis of environmental samples by SBSE.

| Analyte | Matrix | Sample amount | Pretreatment | Mode | Modifiers | Derivatization | Phase/dimension | Time/ Temperature | Desorption | Analysis | LODs | Repeatability | Ref. |
|---|---|---|---|---|---|---|---|---|---|---|---|---|---|
| Benzothiazole | Untreated wastewater, river water | 18 mL | – | SBSE | pH = 2, 30% NaCl | – | PA-PEG/10 mm, 25 μL | 240 min/ 30°C | TD | GC-MS | 0.256 ng $mL^{-1}$ | 9.8% | [15] |
| Bisphenols, APs, and estrogens | Water and plastics | 15 mL (water) 0.1 g (plastics) | (1) Water: filtering (2) Cut, extract in 5 mL acetone for 12 h, evaporate, redisolve in 15 mL water | SBSE | – | – | Dual-template MIP coating/5 μm | 90 min | Ultrasonic LD in 150 μL of 10% acetic acid in water for 10 min | HPLC-DAD | 1.0–5.0 ng $mL^{-1}$ | <11% | [16] |
| Booster biocides | Seawater | 10 mL | pH = 3 | SBSE | pH = 7, 0.2 g $mL^{-1}$ KCl | – | PDMS/10 mm, 0.5 mm, 24 μL | 90 min/25°C | TD | GC-MS | 0.005–0.9 ng $mL^{-1}$ | 7%–15% | [17] |
| Chlorophenols, BPA, 4-t-OP, 4-NPs | Soil | 1 g | Ultrasonic extraction in 15 mL methanol for 30 min | SBSE | Dilute to 100 mL water | *In situ*: AAA | PDMS/20 mm, 0.5 mm | 6 h | TD | GC-MS | 0.2–1.7 ng $g^{-1}$ | <20% | [18] |
| EDCs | WWTP effluent | 10 mL | – | SBSE | * 20% methanol in the case of apolar compounds | * Phenols: *In situ* AAA, pH = 11+ in-tube BSTFA * Amines and acids: pH 5, ethyl chloroformate * Organotins: pH = 5 1% $NaBEt_4$ | PDMS | 60 min | TD | GC-MS | – | – | [19] |
| Emerging and priority contaminants | Sewage, surface seawater, river water, groundwater | 100 mL | | SBSE | pH = 7, 100 g $L^{-1}$, 10 g $L^{-1}$ $Na_2CO_3$ | *In situ*: AAA for 5 | PDMS/10 mm, 0.5 mm 24 μL | 5 h/20°C | Ultrasonic LD 200 μL ethyl acetate for 30 min | APGC-ToF-MS | <0.02–18 ng $L^{-1}$ | | [20] |
| Ergosterol | Soil | 1 g | Ultrasonication for 20 min at 25°C in 10 mL of 0.14 mol $L^{-1}$ KOH in methanol | SBSE | 10 mL extract, 7.0 mmol $L^{-1}$ HCl, 0.4 mol $L^{-1}$ | – | PDMS/20 mm, 1.0 mm | 4 min | LD in 2 mL acetonitrile | HPLC-UV | 0.08 μg $g^{-1}$ | 2.37%–3.71% | [21] |
| EU list of priority substances and US EPA 625 | Groundwater | 20 mL (two aliquots) | Dilution 100 times | Dual SBSE | (1) 6 g NaCl (2) 4 mL methanol | – | PDMS/10 mm, 0.5 mm 24 μL | 2h/24°C | TD | GC-MS | 0.5–431 ng $L^{-1}$ | <23% | [22] |
| Explosives | Spring, mineral, and underground water | Filtering | 30 mL | SBSE | 35% NaCl | – | PDMS-PEG/32 μL | 18 h/22°C | Sonication LD in 400 μL of methanol for 40 min | UHPLC-MS/MS | 0.003–0.7 ng $mL^{-1}$ | 4%–16% | [23] |
| Methylmercury, mercury (II) | River and tap water | 10 mL | – | SBSE | – | *In situ* alkylation with $NaBEt_4$ | PDMS/10 mm, 0.5 mm, 24 μL | 15 min/r.t. | TD | GC-MS | 2–5 ng $L^{-1}$ | 7.0%–8.9% | [24] |

| | | | | | | | | | | | | |
|---|---|---|---|---|---|---|---|---|---|---|---|---|
| Musk fragrances | Vegetables, amended soils | 0.5 g | – | SBSE | 9 mL water: methanol (80:20, v/v) | – | PDMS/20 mm, 0.5 mm | 180 min/ 40°C | TD | GC-MS | 0.01–0.8 ng $g^{-1}$ (vegetables), 0.01–1.1 ng $g^{-1}$ (amended soil) | 1%–23% | [25] |
| NSAIDs | Sewage and lake water | 20 mL | – | MPSBSE | pH = 3.0 | – | C18 | 60 min/50°C | LD in 150 μL methanol for 15 min | HPLC-UV | 7.89–9.52 ng $mL^{-1}$ (wastewater), 7.69–6.90 ng $mL^{-1}$ (pure water) | 4.1%–9.2% | [26] |
| OPPs | Lake and pond water | 10 mL | Filtering | SBSE | 5% NaCl | – | PDMS/MIL-101-Cr-$NH_2$ | 25 min | LD in 50 μL of acetone for 10 min | GC-FPD | 43–85 ng $L^{-1}$ | 6.1%–10.7% | [27] |
| Organotins | Seawater | 10 mL | – | HSSE | pH = 4.8 (HAc/Ac 0.2M buffer) | *In situ* ethylation with sodium tetraethylborate (2% 75 μL) | PDMS/10 mm, 0.5 mm | 3 h/90°C | TD | GC-MS | <10% | 1.7–7.0 ng $L^{-1}$ | [28] |
| PAHs | Wastewater | 50 mL | – | SBSE | 25% MeOH | – | PDMS/10 mm, 0.5 mm (24 μL) | 4 h | Ultrasonic LD in 1 mL ethyl acetate for 15 min | GC-MS/MS | 0.002–0.010 ng $mL^{-1}$ | 11%–28% | [29] |
| Parabens, triclosan, methyl triclosan | Soil, sediment, sludge | 0.5 g | – | SBSE | 5 mL $NaHCO_3$ 0.4 M | *In situ*: AAA | PDMS/10 mm, 0.5 mm, 24 μL | 60 min | TD | GC-MS | 80–1060 ng $kg^{-1}$ | <13% | [30] |
| PCBs | Fish | 5 g | Freeze-drying, ultrasonic extraction in hexane, filtrate, acidic silica clean-up | SBSE | 10 mL, pH = 7,0. 10% NaCl | – | MOF-5(Fe)/40 mg | 30 min | Ultrasonic LD in 2.5 mL of *n*-hexane for 3 min | GC-MS | 0.061–0.096 ng $g^{-1}$ | 3.2%–4.3% | [31] |
| PCBs | Fish | | Freeze-drying, ultrasonic extraction in hexane, filtrate, acidic silica clean-up | SBSE | Dilute in phosphate buffer | – | Apt-MOF/5 mm | 50 min/50°C | LD in 5 mL dichloromethane-pH 3 glycine HCl (1/10, v/v) for 20 min | GC-MS | 0.003–0.004 ng $mL^{-1}$ | 1.8%–2.6% | [32] |
| PCBs | Soil | 10 g | – | HSSE | 10 mL $H_2O$ | | Magnetic mesoporous titanium dioxide@covalent organic frameworks ($Fe_3O_4$@m$TiO_2$@COFs) composite | 30 min/50°C | TD | GC-MS | 0.3–0.6 ng $g^{-1}$ | 2.7%–4.9% | [33] |
| PEs | Seawater | 30 mL | – | SBSE | 5% NaCl, 10% methanol | – | PDMS/20 mm, 1 mm 150 μL | 2 h/25°C | Ultrasonic LD in 250 μL methanol:acetonitrile (4:1) for 50 min | GC-MS | 0.07–5.71 ng $mL^{-1}$ | 6.65%–11.04% | [34] |
| Pesticides | Surface water | 20 mL | Filtering | SBSE | 10% NaCl | – | PDMS/20 mm, 1.0 mm (126 μL) | 3 h/r.t. | Ultrasonic LD in 200 μL of methanol:acetonitrile (50:50, v/v)for 15 min | LC-MS/MS | 0.02–1 ng $mL^{-1a}$ | <17% | [35] |
| PFAAs | River and lake water | 5 mL | Filtering | SBSE | 5 mL methanol, pH = 3 | – | Poly(VI-EMDA) monolith/20 mm | 60 min | Ultrasonic LD in 300 mL methanol with 0.4% ammonia for 10 min | HPLC-ESI-MS/MS | 0.06–0.40 ng $L^{-1}$ | <13% | [36] |
| Phenols | Lake and river water | 10 mL | Filtering | SBSE | pH = 4, 30% NaCl | – | PDMS/CTFs/20 mm | 50 min | Ultrasonic LD in 50 μL of methanol:10 mmol $L^{-1}$ NaOH (8:2) for 25 min | HPLC-UV | 0.08–0.30 ng $mL^{-1}$ | 4.3%–9.4% | [37] |

*Continued*

**Table 18.1** Experimental conditions for the analysis of environmental samples by SBSE.—cont'd

| Analyte | Matrix | Sample amount | Pretreatment | Mode | Modifiers | Derivatization | Phase/dimension | Time/Temperature | Desorption | Analysis | LODs | Repeatability | Ref. |
|---|---|---|---|---|---|---|---|---|---|---|---|---|---|
| Phenols, NSAIDs, PCBs | River water and sediment | 0.5 g (sediment) 10 mL (water) | * River: filtration<br>* Sediment: dryed, groud, extracted in 5 mL acetonitrile, filter, dilute 10 times in water | SBSE | pH = 4 | – | PANi/MWCNTs-OH | 50 min/25°C | Ultrasonic LD in methanol-10 mmol $L^{-1}$ NaOH for (92:8, v/v) 30 min | HPLC-UV | 0.09–0.81 ng $mL^{-1}$ | 6.5%–11.8% | [38] |
| Phosphoric acid esters | Underground and river water | 2 mL | – | SBSE | pH = 2 | – | PDMS/10 mm, 0.5 mm, 24 μL | 60 min/25°C | TD | DART-Orbitrap-MS | 0.1 ng $mL^{-1}$ | 3%–4% | [39] |
| Polar and nonpolar emerging and priority pollutants | Environmental waters | 100 mL | Filtering | SBSE | 38 g NaCl, pH = 3 EG-Silicone and pH = 2 PDMS | – | PDMS/10 mm, 0.5 mm (24 μL)/ EG-Silicone/10 mm, 32 μL | 24 h | Ultrasonic LD in 500 μL methanol for 15 min | LC-MS/MS | 7–165 ng $L^{-1}$ (tap water) 3.1–45 ng $L^{-1}$ (surface water) | <20% | [40] |
| Polar pharmaceuticals | River water and WWTP effluent | 100 mL | Filtering | SBSE | pH = 3 | – | poly(MAA-co-DVB) monolithic sorbent | 4 h/25°C | LD in 5 mL methanol for 20 min | LC-MS/MS | <10 ng $L^{-1}$ | <15% | [41] |
| Polar PPCPs | River water and WWTP effluent and influent | 50 mL | pH = 3, filtering | SBSE | pH = 5, 10% NaCl | – | poly(VPD-co-DVB) monolithic sorbent/10 mm, 0.5 mm, 70 μL | 4 h/r.t. | LD in 5 mL methanol for 15 min | LC-MS/MS | 50–100 ng $L^{-1}$ | <17% | [42] |
| PPCPs | Wastewater | 50 mL | – | SBSE | pH = 7, 15% NaCl | – | Poly(PEGMA-co-PETRA monolith sorbent/12 mm, 1 mm, 225 μL | 60 min/25°C | LD by stirring in 5 mL methanol:acetonitrile (1:1, v/v) for 10 min | LC-MS/MS | 0.15–0.50 ng $mL^{-1}$ | <23% | [43] |
| Parabens | In-house dust | Sieving, PHWE, filtering, dilute in 100 mL | 100 mg | SBSE | – | *In situ*: AAA | PDMS/20 mm, 0.5 mm 48 μL | 3 h/r.t. | TD | GC-MS | 1.0–2.1 ng $g^{-1}$ | <10% | [44] |
| SCCPs | Surface water | 100 mL | – | SBSE | 20 mL methanol | – | PDMS/10 mm, 0.5 mm 24 μL | 16 h/24°C | TD | GC-MS/MS | 0.02 ng $mL^{-1}$ | 6.4%–7.7% | [45] |
| Selenite and selenoamino acids | Clams | 50 mg | Freeze-drying, ultrasonic probe assisted enzymatic hydrolysis | Dual SBSE | 100–500 μL extract | * Selenoamino acids: Isobutyl chloroformate<br>* Selenite: 4-chloro-1,2-phenylenediamine<br>* Selenite: $NaBEt_4$ | PDMS/20 mm, 0.5 mm or 10 mm, 0.5 mm | 20–60 min/ 23–25°C | TD | GC-MS | 0.008–0.180 ng | <13% | [46] |

| Selenomethionine, seleno-methyl-selenocysteine | Clams | 50 mg | Freeze-drying, ultrasonic assisted enzymatic digestion | Multi-shot SBSE | 70 g $L^{-1}$ NaCl, pH = 6 | *In situ*: Ethyl chloroformate | PDMS/10 mm, 0.5 mm, 24 μL | 20 min/23°C | TD | GC-MS | 0.9–1.3 ng | 8.1% | [47] |
|---|---|---|---|---|---|---|---|---|---|---|---|---|---|
| Steroid sex hormones | Wastewater, seawater, lake water | 50 mL | – | SBSE | pH = 11, 0.6% NaCl | – | SBSEM/30 mm, 1.0 mm | 20 min/r.t. | LD in 3.0 mL acetonitrile | HPLC-DAD | 0.10–0.41 ng $mL^{-1}$ | 0.73%–5.36% | [48] |
| Sulfonylurea herbicides | Soil, lake water | Soil: 10 g, water: 5 mL | (1) Water: filter (2) Sieving, sonication in 20 mL in acetone:water (3:1), evaporate acetone and dilute to 20 mL | SBSE | 5 mL, pH = 5 | – | Hybrid MOFs and monolith UIO-66(Zr)–$NH_2$ | 30 min | LD in 200 μL methanol: acetic acid (9:1, v/v) for 30 min | HPLC-UV | 0.04–0.84 ng $mL^{-1}$ | <13.8% | [49] |
| Synthetic musks | Influent, effluent, and river water | Filtering | 100 mL, dilution (5/100) for urban influent and effluent | SBSE | pH = 7 | – | PDMS/20 mm, 0.5 mm, 48 μL | 4 h/25°C | TD | GC-MS | 0.02–0.3 ng $L^{-1}$ | <14.8% | [50] |
| TBT | Seawater | 200 mL | Filtering | SBSE | 30 g NaCl | – | PDMS/20 mm, 0.5 mm | 24 h | Ultrasonic LD in 1.0 mL acetonitrile:formic acid (0.1%, v/v) for 39 min | LC-MS/MS | 0.8 ng $mL^{-1}$ | <16% | [51] |
| Triclocarban | Wastewater effluent | 10 mL | – | SBSE | – | – | PDMS/10 mm, 0.5 mm, 24 μL | 22 h | Sonic LD in 1.5 mL of methanol for 60 min | LC-MS/MS | 1 ng $L^{-1}$ | 2% | [52] |
| UV filters | Bathing water | 25 mL | Ultrasonic leaching for 15 min at pH 2.5 using glacial acetic acid, filtering | SBSDμE | – | – | Neodymium stir-bar magnetically coated with oleic acid-coated cobalt ferrite MNPs/6.4 mm, 3.1 mm, 25 mg | 30 min | TD | GC-MS | 13–148 ng $L^{-1}$ | <12% | [53] |
| UV filters | Bathing water | 25 mL | Ultrasonic leaching for 15 min at pH 2.5 using glacial acetic acid, filtering | SBSDμE | pH = 2.0 | - | $CoFe_2O_4$@$SiO_2$-nylon 6 composite/ 11 mm, 6 mm, 100 mg | 30 min/r.t. | LD in 2.5 mL HCl 1 M for 5 min | LC-UV | 1.6–2.9 ng $mL^{-1}$ | <13% | [54] |
| UV filters | Coastal sand samples | 1 g | – | SBSDμE | 25 mL water, pH = 2, 20% NaCl | – | $CoFe_2O_4$@oleic acid MNPs /6 mm, 3 mm, 15 mg | 10 min | LD in 250 μL ethanol | GC-MS | 0.01–0.55 ng $g^{-1}$ | <14% | [55] |
| UV filters | Seawater | 10 mL | Filtering | SBSE | pH = 6, 0.5 mL methanol | – | PDMS/10 mm, 0.5 mm, 24 μL | 180 min/r.t. | LD in 1 mL methanol for 30 min at r.t. | LC-(APCI) MS/MS | 8–31 ng $L^{-1}$ | 5%–13% | [56] |

*4-NP*, 4- nonylphenols; *4-t-OP, 4-tert-octylphenol*; *AAA*, acetic acid anhydride; *APGC-ToF-MS*, atmospheric pressure gas chromatography-time of flight-mass spectrometry; *APs*, alkylphenols; *Apt-MOF*, aptame-functionalized-metal-organic framework; *BPA*, bisphenol-A; *BSTFA*, *N,O*-bis(trimethylsilyl)trifluoroacetamide; *C18, octadecylsilyl; EDCs*, endocrine disrupting chemicals; *EG-Silicone*, ethylene glycol-silicone; *LDTD-APCI-MS/MS*, laser diode thermal desorption/atmospheric pressure chemical ionization coupled to tandem mass spectrometry; *MOF*, metal-organic framework; *$NaBEt_4$*, sodium tetraethylborate; *NSAIDs*, nonsteroidal antiinflammatory drugs; *MNPs*, magnetic nanoparticles; *MPSBSE*, membrane-protected-SBSE; *OPPs*, organophosphorous pesticides; *PANi/MWCNTs-OH*, polyaniline/hydroxyl multiwalled carbon nanotubes composite; *PDMS/CTFs*, polydimethylsiloxane, covalent triazine networks; *PDMS/MIL-101-Cr-$NH_2$*, polydimethylsiloxane/metal-organic framework; *PEs*, phthalate esters; *PFAAs*, perfluoroalkyl acids; *PHWE*, Pressurized-hot water extraction; *Poly(VI-EMDA) monolith*, poly(1-vinylimidazole-ethyleneglycol dimethacrylate); *poly(VPD-co-DVB)*, poly(N-vinylpyrrolidone-co-divinylbenzene); *PPCPs*, pharmaceutical and personal care products; *SBSDμE*, stir-bar sorptive–dispersive microextraction; *SBSEM*, poly(vinylpyridine-ethylene dimethacrylate) monolithic material; *poly(MAA-co-DVB)*, copolymer of methacrylic acid and divinylbenzene monolithic coating; *poly(PEGMA-co-PETRA)*, poly(poly(ethylene glycol) methacrylate-co-pentaerythritol triacrylate; *SCCPs*, short-chain chlorinated paraffins; *TBT*, tributyltin.

[a]Limit of quantification.

**Table 18.2** Experimental conditions for the analysis of food samples by SBSE.

| Analyte | Matrix | Samle amount | Pretreatment | Mode | Addition | Derivatisation | Phase/dimensions | Time (min)/ Temperature (°C) | Desorption | Analysis | LODs | Repeatability | Ref. |
|---|---|---|---|---|---|---|---|---|---|---|---|---|---|
| Acidic phytohormones | Cucumbers and green bean sprouts | 0.1 g | Cut, grind to fine powder, sieve, add 4 mL methanol and ultrasonicate 30 min at r.t., centrifuge, evaporate to dryness, reconstitute in 50 mL water | IP-SBSE | 10 mL extract, pH = 8, 0.4 mmol $L^{-1}$ CTAB | - | $C_{18}$ | 20 min | Ultrasonic LD in 100 μL of methanol for 20 min | HPLC-UV | 0.3–2.7 ng $mL^{-1}$ | 5.0%–8.3% | [57] |
| APs and PEs | Vegetables | 2 g | Probe ultrasonication for 1 min in 10 mL methanol, centrifugation, evaporation, redissolve in 1 mL methanol | SBSE | 10 mL 2% NaCl | - | PDMS/10 mm, 0.5 mm (24 μL) | 4 h | TD | GC-MS | 2.5–21 ng $L^{-1}$ | <9.4% | [58] |
| APs and PEs | Packaging material | 30 mg | Migration to 5 mL water for 10 days at room temperature | SBSE | Dilute 1 mL extract in 10 mL of water | - | PDMS/10 mm, 0.5 mm (24 μL) | 4 h | TD | GC-MS | 2.5–21 ng $L^{-1}$ | <9.4% | [58] |
| Aroma compounds | Fruit juice | 5 mL | - | HSSE | - | - | PDMS/10 mm, 0.5 mm | 120 min/rt and 40°C | TD | GC-MS | - | 1.7%–14.3% | [59] |
| Bisphenols | Canned beverages, filling liquids of canned vegetables | 2 mL | Degas and dilute 1:10 times the sample in water | SBSE | *In situ*: 200 μL of acetic anhydride and 300 mg $Na_2CO_3$<br>In tube: pH = 7, 10% NaCl | In situ acetylation<br>In-tube silylation | PDMS/10 mm, 0.5 mm (24 μL) | In situ: 3h<br>In tube: 4h | TD (3 mL of BSTFA added for in tube silylation) | GC-MS | In situ: 4.7–12.5 ng $L^{-1}$ | In situ: < 6% | [60] |
| Chlorophenols, chloroanisoles | Wine | 10 mL | - | SBSE | 1.0 g NaCl | - | EG-silicone (32 mm), PA (32 mm) | 2 h | TD | GC-MS | EG-silicone: 0.2–1.5 ng $L^{-1}$ | 4.7% | [61] |
| *Cis/trans* isomers of resveratrol, piceatannol, oxyresveratrol | Wine | 10 mL | 1:1 to 1:10 dilution of the sample | SBSE | *In situ* Acetylation | 210 μL acetic anhydride, 650 mg $Na_2HPO_4$ | PDMS/10 mm, 0.5 mm (24 μL) | 3 h | TD | GC-MS | 13.5–49.2 ng $L^{-1}$ | 5%–9% | [62] |
| Estrogens | Milk | 10 mL | 5 mL acetonitrile, shake 10 min (x3), Add 5 mL hexane, shake 15 min, evaporate extract | SBSE | Redisolve in 0.5 mL methanol and dilute in 10 mL hexane | - | Luminol MIP/CG | 20 min | LD in 5 mL methanol: acetic acid (99:1) for 20 min | HPLC-UV | 0.36–1.09 ng $mL^{-1}$ | <7.1% | [63] |
| Estrogens | Pork and chicken meat | 1.0 g | Trituration and freeze-drying, ultrasonication for 30 min in 10 mL acetonitrile, centrifugation, and filtering | SBSE | Dilute to 50 mL water with 25% NaCl, pH = 6.0–8.0 | - | m-TiMIF | 20 min | LD in 5 mL methanol: acetic acid (99:1) for 15 min | HPLC-PAD | 0.28–0.47 ng $mL^{-1}$ | 3.9%–5.8% | [64] |
| Flavor profile | Wine | 10 mL | | Dual-SBSE | 15% NaCl | | PDMS/10 mm, 0.5 mm (24 μL) | 300 min/60°C | TD | GC-MS | - | - | [65] |

| | | | | | | | | | | | | |
|---|---|---|---|---|---|---|---|---|---|---|---|---|
| Flavors | Fresh rapeseed oil | 1 mL | - | HSSE | - | - | PDMS/10 mm, 0.5 mm | 60 min/90°C | TD | GC-MS | 0.44–2.17 ng $mL^{-1}$ | 0.6%–4.4% | [66] |
| Fluoroquinolones | Chicken muscle and liver | 2 g | Grinding, freeze-drying, grinding to powder, add 20 mL water, vortex 1 min, ultrasonicate 15 min, centrifuge, transfer supernatant to a clean tube (x2), combine extracts | SBSE | 10 mL, pH = 3.0 (HCl) | - | GO/PEG/20 mm | 50 min | Ultrasonic LD in 100 μL 10 mmol NaOH for 10 min | HPLC-FLD | 0.0045–0.0079 ng $mL^{-1}$ | 4.6%–12.1% | [67] |
| Free medium-chain fatty acids | Beer | 10 mL | - | SBSE | - | - | PDMS/10 mm, 0.5 mm (24 μL) | 60 min/r.t. | LD in 200 μL of dichloromethane: hexane for 40 min | GC-FID | - | <6.7% | [68] |
| Furan | Coffee, jarred baby food | 0.5 g or 1 g | Add 5 mL water | SBSE | - | - | PDMS/20 mm, 1 mm or 10 mm, 1 mm | 1 h/r.t. | TD | GC-MS | 2 ng $g^{-1}$ | 2.37%–10.46% | [69] |
| Limonene-derived monoterpenes | Red wine | 90 mL | SPE preconcentration and elution in 1.5 mL methanol | SBSE | Dilute in 10 mL | - | PDMS/20 mm, 1 mm (126 μL) | 60 min/20°C | TD | GC-MS | 2–45 ng $L^{-1}$ | <18.8% | [70] |
| Medium and long-chain free fatty acids | Cooked ham exudates | 10 mL | | SBSE | pH = 2 | - | PDMS/10 mm, 0.5 mm (24 μL) | 15 h | TD | GC-MS | 2.68–30.21 ng $mL^{-1}$ | <4.4% | [71] |
| Melamine | Milk powder | 3.0 g | Ultrasonication in 60 mL methanol, centrifugation for 15 min | SBSE | 10 mL, pH = 8.5–9.5 | - | MIP-SB/30 mm | 50 min | Ultrasonic LD in 2 mL methanol:acetic acid (9:1, v/v) | HPLC-UV | 0.0127 ng $mL^{-1}$ | <8.3% | [72] |
| Methoxypyrazines | Wine | 5 mL | - | HSSE | pH = 6 (NaCit 0.5M buffer) 2 g NaCl | - | PDMS/10 mm, 0.5 mm | 2 h/rt | | | 0.25–0.5 ng $g^{-1}$ | <14% | [73] |
| Off-flavor compounds, 2-methylisoborneol and geosmin | Salmon fillet | 1.0 g | Thaw, grind | SBSE | 9 mL saturated NaCl solution | - | PDMS/10 mm, 0.5 mm (24 μL) | 2 h | TD | GC-MS | 0.3 ng $L^{-1}$ | <8% | [74] |
| Off-flavor profile | Wine | 35 mL | - | SBSE | pH = 10 ($Na_2CO_3$) | - | PDMS 10 mm, 10 mm (63 μL) | 1 h/r.t. | TD | GC-MS | 0.1–3.3 ng $L^{-1}$ | 24% | [75] |
| PAHs | Herbal infusion | 1.5 g | Dried leaves were poured into 50 mL of boiling water, filtered after 5 min, and cooled to room temperature to extract 10 mL | HSSE | 1.5 g NaCl | - | PDMS/10 mm, 0.5 mm | 4 h/88°C | TD | GC-MS | 9%–13.2% | 11–26 ng $L^{-1}$ | [76] |
| Preservatives | Beverages | - | Dilute 10 times | Dual-phase dual-SBSE | 10 mL, pH = 3 | - | APTES–OH–TSO and $C_{18}$-PDMS | 40 min | Ultrasonic LD in 200 μL methanol/20 mmol $L^{-1}$ $CH_3COONH_4$ (40:60, v/v) for 25 min | HPLC-UV | 0.6–2.7 ng $mL^{-1}$ | 4.6%–9.2% | [77] |

*Continued*

**Table 18.2** Experimental conditions for the analysis of food samples by SBSE.—cont'd

| Analyte | Matrix | Samle amount | Pretreatment | Mode | Addition | Derivatisation | Phase/dimensions | Time (min)/ Temperature (°C) | Desorption | Analysis | LODs | Repeatability | Ref. |
|---|---|---|---|---|---|---|---|---|---|---|---|---|---|
| Pyrethroids | Tea | 2.0 g | Ultrasonication for 15 min in 8 mL methanol | SBSE | 2 mL extract + 18 mL water | - | PDMS/10 mm, 0.5 mm (24 μL) | 1 h | TD | GC-MS | - | >9.7% | [78] |
| Seleno-amino acids, seleno-oligopeptides | Garlic samples | 0.5 g | Extraction in 10 mL water, 24 h, 50°C + ultrasonication 30 min, centrifugation, dilution to 25 mL water=> 10 mL only used for SBSE | SBSE | pH = 5 | - | PSP-$TiO_2$/20 mm | 20 min | Ultrasonication LD, 0.5 mol $L^{-1}$ ammonium citrate 150 μL, 20 min | HPLC-ICPMS | 50.2–185.5 ng $L^{-1}$ (as 77 Se) and 45.9–158.8 ng $L^{-1}$ (as 82 Se) | 4.9%–11.7% | [79] |
| Strobilurin fungicides | Fruits | 5 g | Probe ultrasonication in 2 mL ethanol 5 min | SBSE | Dilute the extract in 14 mL acetate buffer (0.04 M) and 5% NaCl | - | PDMS/10 mm, 0.5 mm (24 μL) | 20 min/45°C | LD in 100 μL acetonitrile/water (50:50), 15 min | LC-DAD | 0.3-n ng $g^{-1}$ | <11% | [80] |
| Sulfa drugs | Pork, liver, chicken | 5 g | Thinly slice, extract with 10 mL dichloromethane under 10 min ultrasonication (x2), evaporate, redissolve in 10 mL toluene | SBSE | - | - | Sulfamethazine-MIP 20 μm | 60 min | Ultrasonic LD, 150 μL methanol, 10 min | HPLC-UV | 0.20–0.72 ng $mL^{-1}$ | <6.7% | [81] |
| Sulfonamide antibacterials | Milk | 5.0 mL | 1:10 dilution in water | SBSE | 15% NaCl, pH = 5.0 | - | VIDB/30 mm, 1.0 mm | 60 min/r.t. | LD in 3.0 mL acetonitrile:water (9:1) for 40 min | HPLC-DAD | 1.30–7.90 ng $mL^{-1}$ | <10.9% | [82] |
| Sulfonamides | Milk and milk powder | 1.0 mL (milk), 0.5 g (milk powder) | 10 mL HCl (pH = 2) containing 15% NaCl, centrifugation for 10 min | SBSE | 4 mL, pH = 2 | - | $C_{18}$ | 10 min | Ultrasonic LD in 90 μL methanol for 10 min | HPLC-ESI-MS/MS | 0.9–10.5 ng $mL^{-1}$ (milk); 2.7–31.5 ng $g^{-1}$ (milk powder) | 7.3%–16.7% | [83] |
| Taints | Water or aqueous soft drinks | 5 mL | - | SBSE | pH = 2 | | PDMS/20 mm, 1 mm | 1 h | TD | GC-MS | 0.002–0.065 μg | 4.6%–47.4% | [84] |
| Taints | Water or aqueous soft drinks | 5 mL | - | HSSE | 2 g NaCl, pH = 2 | | PDMS/20 mm, 1 mm | 25 min/50°C | TD | GC-MS | 0.003–0.174 μg | 21.4%–69.6% | [84] |
| Terpenes | Honey | 2 g | - | HSSE | 2 mL HAc/Ac 0.2 M buffer 4% NaCl | - | PDMS/10 mm, 0.5 mm | 2 h/50°C | TD | GC-MS | 9%–17% | 0.007–0.032 ng $g^{-1}$ | [85] |
| Tetramethylene disulfotetramine | Food | 3–10 g | Solid matrices premixed in 10 g of water 1 min, stand for 1 h | SBSE | - | - | PDMS/10 mm, 0.5 mm (24 μL) | 60 min | TD | GC-MS | 0.2–2.1 ng $g^{-1}$ | 4.4%–9.8% | [86] |
| Volatile compounds | Sherry brandy | 35 mL | 1:1 dilution in water | SBSE | - | - | PDMS/10 mm, 0.5 mm (24 μL) | 100 min 25°C | TD | GC-MS | 0.101–246 ng $mL^{-1}$ | 4.97%–17.9% | [87] |

| | | | | | | | | | | | | | |
|---|---|---|---|---|---|---|---|---|---|---|---|---|---|
| Volatile compounds | Wine vinegars | 5 mL | - | HSSE | 1.67 g of NaCl | - | PDMS/10 mm, 0.5 mm | 60 min/62°C | TD | GC-MS | - | - | [88] |
| Volatile compounds | Buckwheat grain | 2 g | - | HSSE | 3 mL $H_2O$ | - | PDMS/20 mm, 1 mm | 30 min/50°C | TD | GC-MS | - | - | [89] |
| Volatile compounds | Wine | | - | HSSE | - | - | PDMS | 2 h/22°C | TD | GC-MS | <15% | - | [90] |
| Volatile compounds | Wine | 5 mL | - | HSSE | NaCl under saturation | - | PDMS/20 mm, 0.5 mm | 120 min/rt | TD | GC-MS | 3.5%–12% | 8–135 μg $L^{-1}$ | [91] |
| Volatile compounds | Wine | 7.5 mL | | HSSE | 30% NaCl | | PDMS/10 mm, 0.5 mm | 1 h/64°C | TD | GC-MS | - | - | [92] |
| Volatile compounds | Coffee | 2 g | | HSSE | | | PU/13 mm, 0.5 mm | 60 min/90°C | LC, 1.5 mL methanol (60 min sonication) | LVI-GC-MS | - | - | [93] |
| Volatile constituents | Grape juice, wine | 20 mL | - | SBSE | - | - | PDMS/10 mm, 0.5 mm (24 μL) | 2 h/r.t. | TD | GC-MS | 0.008–138 ng $mL^{-1}$ | 2.5%–37.8% | [94] |
| Volatile phenols | Alcoholic beverages | 4.0 mL | Dilute in 16 mL phosphate buffer (1M, pH = 7) | SBSE | - | - | EG-PDMS/10 mm, 0.5 mm | 3 h | TD | GC-MS | 0.10–1.39 ng $mL^{-1}$ | <5.8% | [95] |
| Volatile constituents | Sparkling wine | 7.5 mL | - | Sequential-SBSE | 2.25 g NaCl, | - | PDMS/10 mm, 0.5 mm (24 μL) EG-Silicone/ 10 mm, 32 μL | SBSE: 1 h/r.t. HSSE: 1 h/ 62°C | TD | GC-MS | - | - | [92] |

*APTES–OH–TSO*, 3-aminopropyltriethoxysilane-hydroxy-terminated silicone; *APs*, alkylphenols; *BSTFA*, *N,O*–bis(trimethylsilyl)trifluoroacetamide; *$C_{18}$-PDMS*, octadecylsilyl-polydimethylsiloxane; *CTBA*, cetyltrimethylammonium bromide; *EG-silicone*, ethylene glycol-silicone copolymer; *GO-PEG*, graphene oxide/polyethyleneglycol composite; *sulfamethazine-MIP*, sulfamethazine molecularly imprinted polymer; *IP-SBSE*, ion-pair SBSE; *VIDB*, poly(vinylimidazole–divinylbenzene) monolithic material; *MIP-SB*, molecularly imprinted-stir-bar; *m-TiMIF*, $Fe_3O_4$@ meso-/macroporous $TiO_2$ microspheres and molecularly imprinted film; *PEs*, phthalate esters; *SEM-MIP*, semicarbazide-molecularly imprinted polymer.

**Table 18.3** Experimental conditions for the analysis of biomedical samples by SBSE.

| Analyte | Matrix | Sample amount | Pretreatment | Mode | Addition | Derivatization | Phase/Dimension | Time/Temperature °C | Desorption | Analysis | LODs | Repeatability | Ref. |
|---|---|---|---|---|---|---|---|---|---|---|---|---|---|
| Amino acids | Rat's urine, cerebrospinal fluid | 2 mL | 3:1 dilution in ethanol | HF-SBSE | 20% NaCl, pH = 6 | Microwave assisted/ BSTFA | Silica microspheres/ 50 μm | 30 min/30°C | LD 1 mL BSTFA/$CCl_4$(1:9) | GC-MS | 3–17 ng $mL^{-1}$ | <12.4% | [96] |
| Carvedilol | Human serum | 5 mL |  | SBSE | pH = 8.5 | - | PA-EG | 45 min/50°C | Ultrasonic LD in 350 μL of methanol with [Omim][$BF_4$] | HPLC-UV | 0.3 ng $mL^{-1}$ | 3.2% | [97] |
| Diclofenac | Liquid formulations | 5 mL | - | SBSE | 15% NaCl, pH = 2.5 with HCl | - | PDMS/20 mm, 0.5 mm (47 μL) | 120 min/20°C | Magnetic stirring LD, 3 mL acetonitrile | HPLC-UV | 16.06 ng $mL^{-1}$ | <3.0% | [98] |
| Diclofenac | Pediatric urine | 1 mL | 1:10 dilution in water, only 5 mL diluted sample used for extraction | SBSE | 15% NaCl, pH = 2.5 with HCl | - | PDMS/20 mm, 0.5 mm (47 μL) | 120 min/20°C | Magnetic stirring LD, 3 mL acetonitrile, 40 min | HPLC-UV | 12.03 ng $mL^{-1}$ | <20% | [99] |
| Dimethyl trisulfide | Rabbit and mouse blood | 100 μL | 1:5 dilution in water, protein denaturalization in 1 mL $HNO_3$ | SBSE | - | - | PDMS/10 mm, 0.5 mm (24 μL) | 60 min | TD | GC-MS | 0.06 mM | <10% | [100] |
| Drugs of abuse | Urine | 3 mL | - | SBSE | 20% NaCl, pH = 11 | - | Titania–OH–TSO | 30 min | LD 60 μL (80:20 (v/v) methanol:phosphate buffer, pH = 1.5), 20 min | HPLC-UV | 2.3–9.1 ng $mL^{-1}$ | 7.3%–8.9% | [101] |
| EDCs | Human breast milk | 9.9 mL | 7.5 mL acetonitrile + 7.5 mL fat/protein precipitation solution, centrifugation, filtration, evaporation to near dryness, and dilute in 50 mL water | SBSE | 17.5 g NaCl | GC: BSTFA/ 1% TMCS (60:40, v/v) | PDMS/20 mm, 0.5 mm | 24 h | Ultrasonic LD, 1 mL acetonitrile, 5 min | GC-MS/MS UHPLC-MS/MS | GC: 0.1–1.5 ng $mL^{-1}$; LC:0.1–0.3 ng $mL^{-1}$ | <15% | [102] |
| Haloanisoles, halophenols | Drug products | 4 tablets | Extraction in 100 mL 0.1% formic acid in water, sonication 30 min | SBSE | - | - | PDMS/10 mm, 1 mm (63 μL) | 90 min | TD | GC-MS/MS | 1–100 pg/tablet 0.04–4 ng/L | <12% | [103] |
| Leachables | Protein-based parental drugs | 1 mL | 1:5 dilution in water | SBSE | - | - | PDMS/10 mm, 0.5 mm (24 μL) | 60 min/19°C | TD | GC-MS | 0.6–2.8 μg/mL | - | [104] |
| Losartan, valsartan | Human plasma | 1 mL | 1 mL acetonitrile, vortexing + centrifugation | SBSE | 25% NaCl, pH = 2 | - | Acrylate monolithic polymer/ 10 mm, 35 μL | 120 min/70°C | Ultrasonic LD, 250 μL methanol, 30 min | HPLC-DAD-ITMS | 7–27 ng $mL^{-1}$ | <8% | [105] |

| | | | | | | | | | | | | |
|---|---|---|---|---|---|---|---|---|---|---|---|---|
| OH-PAHs | Human urine | 5 mL | Enzymatic hydrolysis, 37°C overnight | SBSE | pH = 5.0 | - | PDMS/20 mm, 1 mm | 3 h | LD, 5 mL methanol | HPLC-MS/MS | 0.001–0.022 ng $mL^{-1}$ | 3.1%–13.0% | [106] |
| Propanolol | Urine | 10 mL | 1:10 dilution in water, pH = 9 | SBSE | pH = 9 | - | GO/MIP/ 2000 mm, 700 μm | 40 min | Ultrasonic LD, 100 μL methanol:10 mmol $L^{-1}$ NaOH (60/40, v/v), 20 min | HPLC-UV | 0.37 ng $mL^{-1}$ | 7.3% | [107] |
| Rifampicin | Plasma | 200 μL | Centrifugation, addition of ascorbic acid, storing at −20°C, 1/20 dilution with sodium acetate (pH = 5.0) | SBSE | - | - | PDMS/10 mm, 0.5 mm (24 μL) | 50 min/24°C | LD, acetonitrile, 20 min | HPLC-UV | 0.125 mg $L^{-1}$ | <10% | [108] |
| Seleno-amino acids, seleno-oligopeptides | Human urine | | Filter, dilution in 25 mL water (10 mL only used for SBSE) | SBSE | pH = 5 | - | PSP-$TiO_2$/20 mm | 20 min | Ultrasonication LD, 0.5 mol $L^{-1}$ ammonium citrate 150 μL, 20 min | HPLC-ICPMS | 50.2–185.5 ng $L^{-1}$ (as 77 Se) and 45.9–158.8 ng $L^{-1}$ (as 82 Se) | 4.9%–11.7% | [79] |
| Thymol phase I metabolites | Human urine | 875 μL | | HSSE | 100 μL of a sodium acetate buffer (pH 5; 0.01 M), 50 μL glucuronidase and 1000 μL $H_2O$ | - | PDMS/10 mm, 0.5 mm | 20 min/60°C | TD | GC-MS | - | - | [109] |
| Thyroxines and metabolites | Urine | 10 mL | - | SBSE | pH = 5.5 | - | PEG/OH-PDMS/ γ-MPTS/ 20 mm, 30 μm | 30 min | LD, 100 μL (80:20) methanol: 0.05 mol $L^{-1}$ $Na_2CO_3$, 15 min | HPLC-UV HPLC-ICPMS | UV: 0.6–2.2 ng $mL^{-1}$ ICPMS: 0.0071–0.0355 ng $mL^{-1}$ | UV: 4%–13.8% ICPMS: 3.7%–6.1% | [110] |
| Trimethoprim and sulfonamides | Urine, plasma, milk | 5 mL | Urine: Preextraction in 30 mL dichloromethane, 10 min, evaporate to dryness and Plasma/milk: 5 mL, add 1 mL $HClO_4$ 10% and centrifuge 5 min, collect supernatant and add 2 mL NaOH 1 M and 30 mL dichloromethane, evaporate to dryness an | SBSE | Redissolve in 10 mL toluene | - | Trimethoprim-MIP/15 cm, 21.5 μm | 45 min/r.t. | LD 150 μL acetonitrile/0.01% (v/v) phosphoric acid (13: 87, v/v), 2 min | HPLC-UV | 1.6 ng $mL^{-1}$ | 4.1%–8.0% | [111] |
| Volatile metabolites | Human airway cell cultures | 5 mL | The vial was placed into a thermostated water bath (37°C) for 60 min before extraction | HSSE | OPT media | - | PDMS/10 mm, 0.5 mm | 24 h/37°C | TD | GC-MS | - | - | [112] |

*BSTFA*, *N,O*–bis(trimethylsilyl)trifluroacetamide; *BSTFA/1% TMCS (60:40*, v/v), BSTFA/1% trimethylchlorosilane; *EDC*, endocrine disrupting chemicals; *GO/MIP*, graphene oxide/molecularly imprinted polymer; *HF-SBSE*, hollow fiber-SBSE; *PA-EG*, poly(methyl methacrylate/ethyleneglycol dimethacrylate); *OH-PAHs*, monohydroxylated polycyclic aromatic hydrocarbons; *[Omim][$BF_4$]*, 1-methyl-3-octylimidazolium tetrafluoroborate ionic liquid; *PEG/OH-PDMS/γ-MPTS*, polyethyleneglycol/hydroxyl polydimethylsiloxane/γ-mercaptopropyl trimethoxysilane; *Titania–OH–TSO*, titania-hydroxy-terminated silicone oil; *PSP-$TiO_2$*, partially sulfonated polystyrene-titania organic–inorganic hybrid; *Trimethoprim-MIP*, trimethoprim-molecularly imprinted polymer.

**Table 18.4** Applications of PDMS-SBSE as a passive sampler.

| Analytes | SBSE length x thickness (mm) | Sample | Deployment time | Analysis | Ref |
|---|---|---|---|---|---|
| 10 microcontaminants (DDT organophosphorous pesticides, hexachlorohexanes, phthalates, alkylphenols, musk fragrances) | 20 × 0.5 | Estuarine waters | 7−14 days | TD-GC-MS | [122] |
| 10 microcontaminants (DDT, organophosphorous pesticides, hexachlorohexanes, phthalates, alkylphenols, musk fragrances) | 20 × 0.5 | WWTP effluents | 7-14-28 days | TD-GC-MS | [123] |
| 16 pesticides (herbicides, fungicides, metabolites, and insecticides) | 20 × 1 | Freshwater | 1 month | SBSE-LD-LC-MS/MS | [124] |
| 18 pesticides (triazines, substituted ureas, triazoles, and organophosphate compounds) | 20 × 1 | Freshwater | 1 month | SBSE-LD-LC-MS/MS | [125] |
| 20 Pesticides (herbicides, fungicides, metabolites, and insecticides) | 20 × 1 | Freshwater | 7 days | SBSE-LD-LC-MS/MS | [126] |
| a-,g-and d-hexachlorocyclohexane (a-,g- and d −HCH), hexachlorobenzene (HCB), 2,4,4'-trichlorobiphenyl (PCB 28), 2,2,5,5' -tetrachlorobiphenyl (PCB 52), 2,2′,4,5,5′-pentachloro-biphenyl (PCB 101), and fluoranthene (FLU) | 10 × 0.5 (volume of about 24 μL) | Air | 7, 14, 21, and 28 days | TD-GC-MS | [127] |
| HCHs, PCBs | 20 × 0.5 | Air | <500 h | TD-GC-MS | [128] |
| Legacy contaminants (16 PAHs, 7 PCBs) | 10 × 0.5 | Lab application | <16 days | TD-GC-MS | [129] |
| Legacy contaminants (16 PAHs, 7 PCBs) | 10 × 0.5 | Lab application | <180 h | TD-GC-MS | [130] |

| | | | | | |
|---|---|---|---|---|---|
| *p*-dichlorobenzene (PDCB), naphthalene, and camphor (moth repellents) | 10 × 0.5 | Air (indoor, method development) | 1 day | GC-FID | [131] |
| PAHs, PCBs, HCHs | 14 × 0.5 | Air | <15 days | TD-GC-MS | [132] |
| PAHs | 20 × 0.5 | Air (indoor) | 2 h | GC/MS/MS | [133] |
| Pesticides (herbicides, fungicides, metabolites, and insecticides) | 20 × 1 | Freshwater | 1 month | SBSE-LD-LC-MS/MS | [134] |
| Phosphorus flame retardants (alkyl, chloroalkyl, and phenylphosphates) | 20 × 1 | Air | 6–16 days | TD-GC-MS | [135] |
| PEs | (24 μL) | Air (indoor and outdoor) | 1 day | TD-capillary GC-MS | [136] |

Among new coatings, magnetic nanoparticles (MNPs) can be highlighted, giving rise to stir-bar sorptive–dispersive microextraction (SBSDμE) [53–55]. MNPs are detached and dispersed into solution at high stirring rate and are magnetically retrieved onto the stir-bar when stirring ceases. SBSDμE allowed a shorter extraction time compared to conventional SBSE, a simpler postextraction treatment compared with dispersive-SPE (dSPE), and solvent-free injection of the entire extract. Monolithic sorbents [41] were used for the extraction of polar compounds, including PCPs [42,43], steroid sex hormones [82], and perfluoroalkyl acids (PFAAs) [36]. Metal-organic frameworks (MOFs) [27,31,32] or combinations of monoliths and MOFs were used for the extraction of PCBs, OPPs, and sulfonylurea herbicides, respectively. The MIPs [16] and composite materials [38] have been used to a lesser extent in environmental applications. Membrane protected SBSE (MPSBSE) was proposed by Mao and coworkers [26] to protect the sorbent phases from high molecular weight compounds for the analysis of environmental samples without filtration or centrifugation. A poly(tetrafluorethylene) (PTFE) membrane impregnated with methanol was used to protect octadecylsilyl ($C_{18}$) coated stir-bars.

A general trend in environmental analysis is the development of multiresidue methods. Aparicio et al. [40] studied PDMS and EG-silicone stir-bars for the determination of up to 48 polar and nonpolar emerging and priority pollutants in environmental waters. The best extraction efficiency for most polar and semipolar compounds at an acidic pH was observed for the EG-silicone coated stir-bars. Pintado-Herrera and coworkers [20] developed a multiresidue method for the determination of 102 priority and emerging organic pollutants (fragrances, UV filters, repellents, endocrine disruptors, biocides, PAHs, PCBs, and different types of pesticides) in environmental water samples. A single extraction onto a PDMS sorbent with *in situ* acetylation of several of the target compounds, followed by gas chromatography-time of flight-mass spectrometry (GC-ToF-MS) resulted in acceptable results for all compounds.

HSSE has been used to a lesser extent for the determination of semivolatile compounds such as PCBs in soil [33] and organotin compounds in seawater [28].

### 18.4.2 Food analysis

While SBSE was mainly applied to the analysis of minor constituents (e.g., volatiles, additives), trace compounds responsible for off-odours (e.g., aldehydes, haloanisoles), toxins, and pollutants in food [1,7,10], few applications were identified for the analysis of nutrients and major constituents [7], such as the determination of free medium-chain fatty acids in beer [68] (see Table 18.2). Although mostly applications are used to analyze liquid samples (e.g., wine, juices, and beverages), solid samples after an initial solid-liquid extraction can also be analyzed. Most of the applications included in Table 18.2 use commercially available PDMS stir-bars. For polar compounds a few reports adopted the derivatization approach [60,62]. For the analysis of bisphenols [60] *in situ* acetylation provided improved sensitivity compared with in-tube silylation. To avoid derivatization, polar stir-bars, such as EG-silicone [61,95] and PA [61], were used. For the determination of chlorophenols and chloroanisoles in wine, EG-silicone copolymer provided higher sensitivity than the PA sorbent. Home-made sorbents have also received attention for the extraction of polar compounds [57,67,77,79,83], with three of those applications using $C_{18}$-based sorbents [57,77,83]. When selective extraction was necessary, MIP-based sorbents were used for complex samples, such as pork and chicken [64,81] and powdered milk [72].

For the determination of flavor profiles, where a large number of analytes must be extracted, dual or sequential-SBSE was used [65,92] with one stir-bar immersed in the sample and the other suspended in the headspace. While Arbulu and coworkers [65] used PDMS both in immersion and HS extraction modes, Ubeda and coworkers [92] found that PDMS provided better results, in terms of the number of compounds detected and the extraction efficiency, during immersion, while SBSE EG-Silicone provided better compound coverage and extraction efficiency for alcohols and volatile phenols. Dual-SBSE was also used for the determination of preservatives in beverages [77] but in this case performing two consecutive immersion extractions with sorbents of complimentary polarity.

HSSE is extensively used in food analysis (see Table 18.2). Although most of the applications utilized TD, LD was used for the determination of volatile compounds in coffee using a poly(urethane)-coated stir-bar [93].

### 18.4.3 Biomedical analysis

SBSE has been used to a lesser extent in the biomedical field probably due to either the lack of need for preconcentration in pharmaceutical production and quality assurance (QA) or to the complexity of biological samples, which requires a more selective extraction technique [7]. Besides, the available sample volume for plasma, serum, bile, etc., is very low (from few μLs to few mLs) and, due to the polar nature of the analytes, PDMS does not yield good extraction efficiency and, consequently, in situ derivatization [1] and new coatings [137] are of greater importance in this application field. Besides, LD coupled to LC provides lower sensitivity than TD [7]. As can be seen from the applications compiled in Table 18.3, the number of reports describing new coatings, either more appropriate for polar compounds [79,97,101,105,110,138] or more selective [107,111], is equal to those using commercially available PDMS stir-bars [98,100,102–104,108]. Due to the general type of target compounds, LD coupled to LC is mostly used. In those applications where derivatization was applied for GC analysis, off-line silylation using BSTFA was most common [96,102]. Dilution of urine samples is a common practice [96,99,107], while treatment with acetonitrile is typical for plasma and milk samples [102,105,111]. Although SBSE is usually applied to aqueous samples or aqueous extracts of solid samples, some of the MIP sorbents used for preconcentration required redissolving the extract in an organic solvent, such as toluene [81,111] and, in this case, the aim of minimizing solvent consumption by using SBSE is lost.

### 18.4.4 As a passive sampler

Among the varied applications of PDMS polymers in the analytical sciences their use as passive samplers should be highlighted. Recalling the kinetic features of SBSE extraction presented before, and in more detail by Seethapathy and Górecki [139], the use of PDMS to collect targeted or nontargeted, nonpolar compounds from the environment without any further maintenance can be easily understood. This way, a time-weighted average concentration ($C_{TWA}$) of the compounds can be provided in contrast to snap sampling. Fluctuations that happen along the exposure or deployment time can be integrated over time. Though the background of passive sampling is out of the scope of this work, some recent reports that describe the background and applications of passive samplers are recommended [140–143].

PDMS-based samplers not only include SBSE (see Table 18.4) but rods and sheets, SPME fibers, and pellets as well, for monitoring nonpolar contaminants in water, sediment-pore water, and air samples. Bartolomé et al. [144] used PDMS fibers as passive samplers to measure the free concentration of PAHs in water unsaturated soils. In a different approach, Vetter et al. [145] used PDMS sheets to measure the temporal variation of several halogenated natural products in the Great Barrier Reef (Australia). Finally, PDMS strips are used to monitor personal exposure to semivolatile organic compounds (phthalates and organophosphate esters) in indoor air and to assess inhalation exposure [146].

The most comprehensive description of the use of SBSE as passive samplers is found in reports by Vrana et al. [129] on calibration and the applications described by Posada-Ureta et al. in WWTP effluents and estuarine media [122,123]. Other contributions are summarized in Table 18.4. Compared with other devices, SBSE has the advantage of easy direct coupling by TD for GC analysis. Additionally, it is easily coupled with a magnetic stirrer, and is of an intermediate mass, between that of SPME fibers and the much higher mass of the rods or sheets, allowing nondepleting extractions. The availability of stir-bars of different sizes (length and thickness) can address the problem of obtaining a high extraction efficiency as well as the kinetic features of the extraction itself.

The work of Vrana et al. [130] is the seminal work of SBSE as passive sampler, where they fully described MESCO-stir-bar (membrane enclosed sorptive coated samplers, see Fig. 18.4) to analyze legacy nonpolar contaminants (PAHs and PCBs). In addition to this work, the five reports of Assumani and coworkers [124–126,134,147] should also be highlighted, because they developed the full calibration, deployment, and quantification process of passive SBSEs for the analysis of a wide variety of pesticides typically found in surface waters from agricultural watersheds. Ureta-Posada and coworkers [122,123] compared the use of both free or naked stir-bars and MESCO-stir-bars, for the analysis of nonpolar emerging contaminants commonly found in WWTP effluents and in estuarine media of different salinities.

The last application of SBSE in water analysis, and in a way close to passive sampling, is the continuous flow integrative sampler (CFIS) described by Llorca et al. [18]. This immersible and automatic device includes a peristaltic pump that flushes continuously and at a fixed flow the water to a holder containing typically one or more stir-bars. Owing to this fact, it cannot be strictly considered as a passive sampler although the application is rather similar.

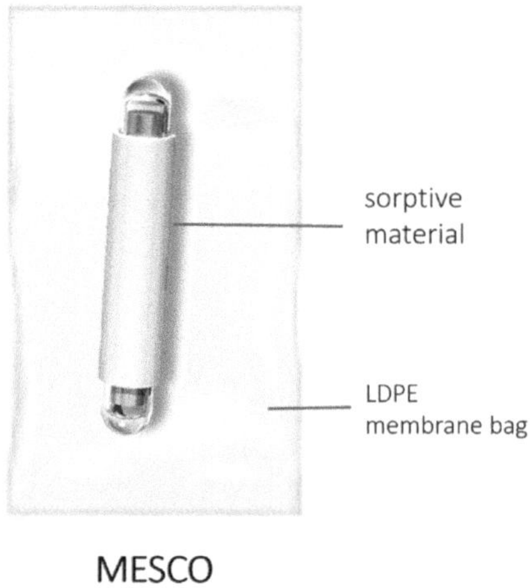


**Figure 18.4** Membrane enclose sorptive coated sampler (MESCO).

Passive sampling is also a useful approach in air analysis, especially in urban and indoor areas, and although stir-bars are well suited for headspace analysis, their use as passive air samplers is rather limited. In fact, the most recent application where SBSE is used as a passive sampler is to the monitoring of phosphorous flame retardants by Matsiko and coworkers [135]. The seminal works on passive air sampling are those of Wennrich and coworkers [132] and Paschke and Popp [128]. In both cases, they analyzed legacy organic contaminants (PAHs, PCBs, hexachlorocyclohexane) comparing different PDMS samplers and studying thoroughly the suitability of the method.

## 18.5 Conclusions

Based on the intended analysis, several steps are suggested to set up the method design, Fig. 18.5. First, the selection of the sorptive material is of major concern. The polymeric phase design has undergone a recent revolution with the development of new coatings (especially for polar and ionic compounds) and technologies, like sol-gel and molecular imprinting, allow the hybridization of new materials and the creation of 3D structures which enhance the selectivity and performance of the extraction phases. Sampling mode must also be directed to the type of sample and target analytes; immersion for gas and liquid samples, HSSE for volatile compounds from liquid or solid samples, and the newer method of "dual" SBSE covering a couple of extraction conditions (with different setup parameters) to ensure an optimal recovery of the analytes. Desorption is probably the most "technically" improved step of the extraction process. Online TDUs ensure the full recovery of analytes sorbed on the stir-bar, but limit its use to stable volatile and semivolatile compounds analyzed by GC. For LD, even further technical improvements have been conducted, as the use of PTV

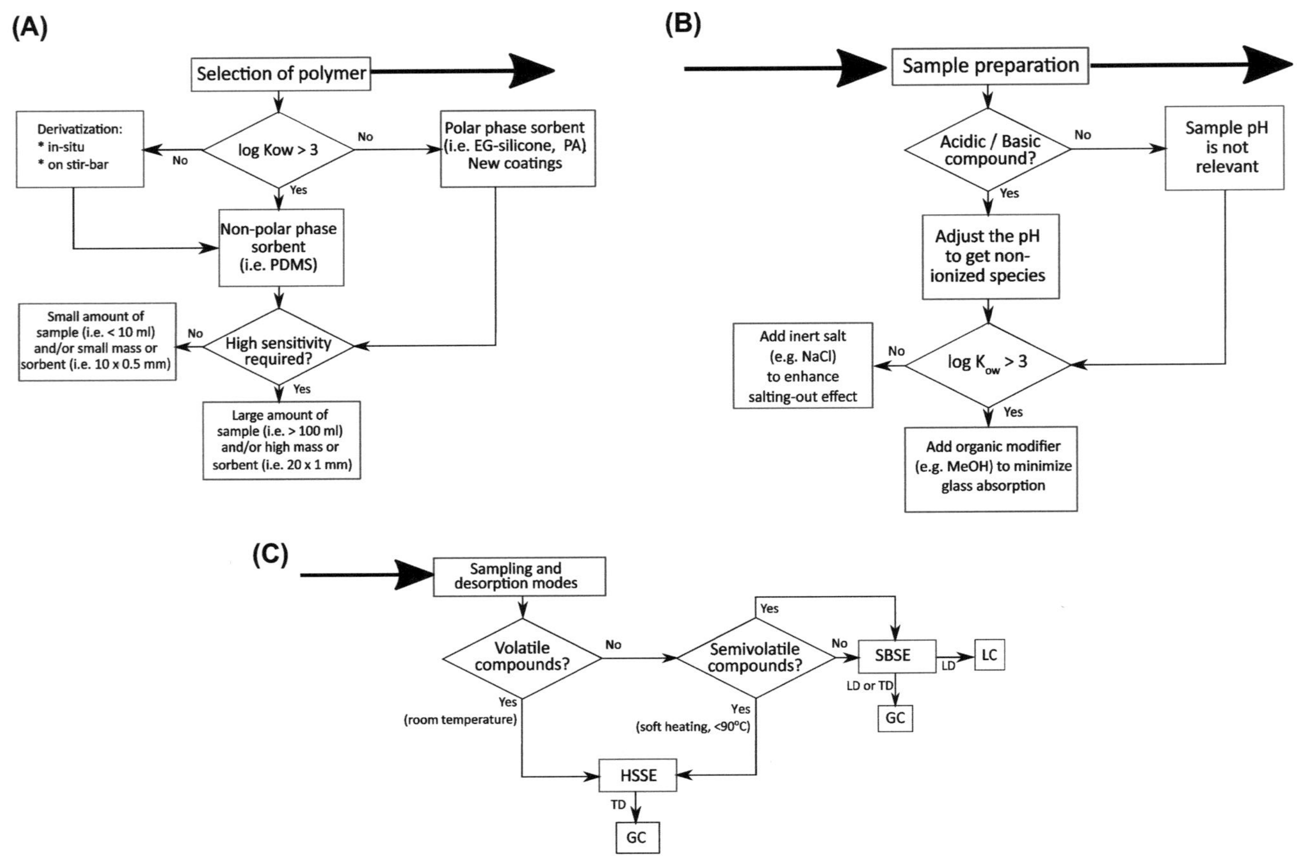


**Figure 18.5** Workflow for coating (A), sample preparation (B), and sampling and desorption mode (C) selection in SBSE.

injectors following desorption for the online introduction of larger volumes (approx. 100–1000 μL) for even thermally unstable analytes in the GC. LD can also be combined with LC, the latter being a more appropriate separation technique for polar and ionic compounds. However, when combined with LC, the major drawback of LD is that only a part of the final extract is introduced into the chromatographic system, decreasing method sensitivity. Derivatization should be considered for polar compounds, as it increases the extraction efficiency, particularly for PDMS sorbents. On-stir-bar derivatization is a quite specific practice that can be evaluated by off-line and innovative online approaches (see Fig. 18.3).

Calibration and validation methods have also been suggested in this chapter. Similar to most analytical procedures, external calibration as well as internal standard or matrix-matched calibration techniques can be evaluated, depending on the matrix effect and the nature of sample and analytes. Besides, nonequilibrium extractions, frequent with SBSE, should consider a time dependence for quantification. It is worth mentioning that due to the novelty of many coating materials, interlaboratory comparisons and further validations need to be performed to homogenize the use of stir-bars. Several applications can gain leverage of the SBSE performance, including environmental, food, and even biomedical analysis. However, the two formers are the most commonly found in the reviewed literature. The potential of new coatings for environmental analysis is also remarkable, with a promising future regarding the use of nanoparticles and newly designed sorbents for the analysis of polar and priority substances, but also for reducing the price and consumption of time and materials when performing environmental monitoring. Less modernization has been dedicated to food analysis, mostly focused in commercial stir-bars for the analysis of minor constituents for liquid or solid products quality control, even if few examples for polar compounds take the advantage of SBSE properties compared with other techniques. Still, some procedures for flavors and odors can be found, especially on HSSE mode, which points the positive evolution of this technique also in the alimentary sector. Similarly, development of biomedical applications is still ongoing, facing the difficulties of selectiveness, matrix effects with biological samples, and boost of LD-LC and GC online systems. Lastly, SBSE practice as passive samplers point out as well the future usage in long-term monitoring programs in air and water sampling, all of them profiting the new implementations of material science and chromatographic techniques.

## References

[1] David F, Sandra P. Stir bar sorptive extraction for trace analysis. J Chromatogr A June 8, 2007;1152(1):54–69.

[2] Baltussen E, Sandra P, David F, Cramers C. Stir bar sorptive extraction (SBSE), a novel extraction technique for aqueous samples: theory and principles. J Microcolumn Sep January 1, 1999;11(10):737–47.

[3] Baltussen E, Janssen H-G, Sandra P, Cramers CA. A new method for sorptive enrichment of gaseous samples: application in air analysis and natural gas characterization. J High Resolut Chromatogr July 1997;20(7):385–93.

[4] Baltussen E, Janssen H-G, Sandra P, Cramers CA. A novel type of liquid/liquid extraction for the preconcentration of organic micropollutants from aqueous samples: applications to the analysis of PAH's and OCP's in water. J High Resolut Chromatogr July 1, 1997; 20(7):395–9.

[5] Saraullo A, Martos PA, Pawliszyn J. Water analysis by solid phase microextraction based on physical chemical properties of the coating. Anal Chem June 1, 1997;69(11):1992–8.

[6] Nogueira JMF. Chapter fifteen – stir bar sorptive extraction. In: Ibáñez E, Cifuentes A, editors. Comprehensive analytical chemistry [internet]. Green extraction techniques, vol. 76. Elsevier; 2017. p. 463–81. Available from: http://www.sciencedirect.com/science/article/pii/S0166526X17300065.

[7] Camino-Sánchez FJ, Rodríguez-Gómez R, Zafra-Gómez A, Santos-Fandila A, Vílchez JL. Stir bar sorptive extraction: recent applications, limitations and future trends. Talanta December 1, 2014;130:388–99.

[8] Kawaguchi M, Ito R, Saito K, Nakazawa H. Novel stir bar sorptive extraction methods for environmental and biomedical analysis. J Pharm Biomed Anal February 24, 2006;40(3): 500–8.

[9] Nogueira JMF. Stir-bar sorptive extraction: 15 years making sample preparation more environment-friendly. Trac Trends Anal Chem September 1, 2015;71:214–23.

[10] Nogueira JMF. Novel sorption-based methodologies for static microextraction analysis: a review on SBSE and related techniques. Anal Chim Acta December 13, 2012;757:1–10.

[11] Prieto A, Basauri O, Rodil R, Usobiaga A, Fernández LA, Etxebarria N, et al. Stir-bar sorptive extraction: a view on method optimisation, novel applications, limitations and potential solutions. J Chromatogr A April 16, 2010;1217(16):2642–66.

[12] Vasapollo G, Sole RD, Mergola L, Lazzoi MR, Scardino A, Scorrano S, et al. Molecularly imprinted polymers: present and future prospective. Int J Mol Sci 2011;12(9).

[13] Ochiai N, Sasamoto K, Kanda H, Yamagami T, David F, Tienpont B, et al. Optimization of a multi-residue screening method for the determination of 85 pesticides in selected food matrices by stir bar sorptive extraction and thermal desorption GC-MS. J Sep Sci 2005; 28(9–10):1083–92.

[14] Villaverde-de-Sáa E, Racamonde I, Quintana JB, Rodil R, Cela R. Ion-pair sorptive extraction of perfluorinated compounds from water with low-cost polymeric materials: polyethersulfone vs polydimethylsiloxane. Anal Chim Acta August 31, 2012;740:50–7.

[15] Fries E. Determination of benzothiazole in untreated wastewater using polar-phase stir bar sorptive extraction and gas chromatography–mass spectrometry. Anal Chim Acta March 9, 2011;689(1):65–8.

[16] Xu Z, Yang Z, Liu Z. Development of dual-templates molecularly imprinted stir bar sorptive extraction and its application for the analysis of environmental estrogens in water and plastic samples. J Chromatogr A September 5, 2014;1358:52–9.

[17] Giráldez I, Chaguaceda E, Bujalance M, Morales E. Determination of five booster biocides in seawater by stir bar sorptive extraction–thermal desorption–gas chromatography–mass spectrometry. J Chromatogr A January 4, 2013;1271(1):17–26.

[18] Llorca-Pórcel J, Martínez-Parreño M, Martínez-Soriano E, Valor I. Analysis of chlorophenols, bisphenol-A, 4-tert-octylphenol and 4-nonylphenols in soil by means of ultrasonic solvent extraction and stir bar sorptive extraction with in situ derivatisation. J Chromatogr A August 7, 2009;1216(32):5955–61.

[19] Bicchi C, Schilirò T, Pignata C, Fea E, Cordero C, Canale F, et al. Analysis of environmental endocrine disrupting chemicals using the E-screen method and stir bar sorptive extraction in wastewater treatment plant effluents. Sci Total Environ March 1, 2009; 407(6):1842−51.

[20] Pintado-Herrera MG, González-Mazo E, Lara-Martín PA. Atmospheric pressure gas chromatography−time-of-flight-mass spectrometry (APGC−ToF-MS) for the determination of regulated and emerging contaminants in aqueous samples after stir bar sorptive extraction (SBSE). Anal Chim Acta December 3, 2014;851:1−13.

[21] Beni Á, Lajtha K, Kozma J, Fekete I. Application of a stir bar sorptive extraction sample preparation method with HPLC for soil fungal biomass determination in soils from a detrital manipulation study. J Microbiol Methods May 1, 2017;136:1−5.

[22] Tölgyessy P, Vrana B, Krascsenits Z. Development of a screening method for the analysis of organic pollutants in water using dual stir bar sorptive extraction-thermal desorption-gas chromatography−mass spectrometry. Talanta December 15, 2011;87:152−60.

[23] Schramm S, Vailhen D, Bridoux MC. Use of experimental design in the investigation of stir bar sorptive extraction followed by ultra-high-performance liquid chromatography−tandem mass spectrometry for the analysis of explosives in water samples. J Chromatogr A February 12, 2016;1433:24−33.

[24] Ito R, Kawaguchi M, Sakui N, Okanouchi N, Saito K, Seto Y, et al. Stir bar sorptive extraction with in situ derivatization and thermal desorption−gas chromatography−mass spectrometry for trace analysis of methylmercury and mercury(II) in water sample. Talanta February 15, 2009;77(4):1295−8.

[25] Aguirre J, Bizkarguenaga E, Iparraguirre A, Fernández LÁ, Zuloaga O, Prieto A. Development of stir-bar sorptive extraction−thermal desorption−gas chromatography−mass spectrometry for the analysis of musks in vegetables and amended soils. Anal Chim Acta February 17, 2014;812:74−82.

[26] Mao X, He M, Chen B, Hu B. Membrane protected C18 coated stir bar sorptive extraction combined with high performance liquid chromatography-ultraviolet detection for the determination of non-steroidal anti-inflammatory drugs in water samples. J Chromatogr A November 11, 2016;1472:27−34.

[27] Xiao Z, He M, Chen B, Hu B. Polydimethylsiloxane/metal-organic frameworks coated stir bar sorptive extraction coupled to gas chromatography-flame photometric detection for the determination of organophosphorus pesticides in environmental water samples. Talanta August 15, 2016;156−157:126−33.

[28] Cacho JI, Campillo N, Viñas P, Hernández-Córdoba M. Headspace sorptive extraction for the analysis of organotin compounds using thermal desorption and gas chromatography with mass spectrometry. J Chromatogr A March 1, 2013;1279:1−6.

[29] Barco-Bonilla N, Romero-González R, Plaza-Bolaños P, Fernández-Moreno JL, Garrido Frenich A, Martínez Vidal JL. Comprehensive analysis of polycyclic aromatic hydrocarbons in wastewater using stir bar sorptive extraction and gas chromatography coupled to tandem mass spectrometry. Anal Chim Acta May 5, 2011;693(1):62−71.

[30] Ferreira AMC, Möder M, Laespada MEF. Stir bar sorptive extraction of parabens, triclosan and methyl triclosan from soil, sediment and sludge with in situ derivatization and determination by gas chromatography−mass spectrometry. J Chromatogr A June 24, 2011;1218(25):3837−44.

[31] Lin S, Gan N, Qiao L, Zhang J, Cao Y, Chen Y. Magnetic metal-organic frameworks coated stir bar sorptive extraction coupled with GC−MS for determination of polychlorinated biphenyls in fish samples. Talanta November 1, 2015;144:1139−45.

[32] Lin S, Gan N, Zhang J, Qiao L, Chen Y, Cao Y. Aptamer-functionalized stir bar sorptive extraction coupled with gas chromatography–mass spectrometry for selective enrichment and determination of polychlorinated biphenyls in fish samples. Talanta March 1, 2016; 149:266–74.

[33] Wang Q, Wu H, Lv F, Cao Y, Zhou Y, Gan N. A headspace sorptive extraction method with magnetic mesoporous titanium dioxide@covalent organic frameworks composite coating for selective determination of trace polychlorinated biphenyls in soils. J Chromatogr A October 19, 2018;1572:1–8.

[34] Si Q, Li F, Gao C, Wang C, Wang Z, Zhao J. Detection of phthalate esters in seawater by stir bar sorptive extraction and gas chromatography–mass spectrometry. Mar Pollut Bull July 15, 2016;108(1):163–70.

[35] Margoum C, Guillemain C, Yang X, Coquery M. Stir bar sorptive extraction coupled to liquid chromatography-tandem mass spectrometry for the determination of pesticides in water samples: method validation and measurement uncertainty. Talanta November 15, 2013;116:1–7.

[36] Yao X, Zhou Z, He M, Chen B, Liang Y, Hu B. One-pot polymerization of monolith coated stir bar for high efficient sorptive extraction of perfluoroalkyl acids from environmental water samples followed by high performance liquid chromatography-electrospray tandem mass spectrometry detection. J Chromatogr A June 8, 2018;1553: 7–15.

[37] Zhong C, He M, Liao H, Chen B, Wang C, Hu B. Polydimethylsiloxane/covalent triazine frameworks coated stir bar sorptive extraction coupled with high performance liquid chromatography-ultraviolet detection for the determination of phenols in environmental water samples. J Chromatogr A April 8, 2016;1441:8–15.

[38] Hu C, He M, Chen B, Hu B. Simultaneous determination of polar and apolar compounds in environmental samples by a polyaniline/hydroxyl multi-walled carbon nanotubes composite-coated stir bar sorptive extraction coupled with high performance liquid chromatography. J Chromatogr A May 15, 2015;1394:36–45.

[39] Bridoux MC, Malandain H, Leprince F, Progent F, Machuron-Mandard X. Quantitative analysis of phosphoric acid esters in aqueous samples by isotope dilution stir-bar sorptive extraction combined with direct analysis in real time (DART)-Orbitrap mass spectrometry. Anal Chim Acta April 15, 2015;869:1–10.

[40] Aparicio I, Martín J, Santos JL, Malvar JL, Alonso E. Stir bar sorptive extraction and liquid chromatography–tandem mass spectrometry determination of polar and non-polar emerging and priority pollutants in environmental waters. J Chromatogr A June 2, 2017; 1500:43–52.

[41] Bratkowska D, Fontanals N, Cormack PAG, Borrull F, Marcé RM. Preparation of a polar monolithic stir bar based on methacrylic acid and divinylbenzene for the sorptive extraction of polar pharmaceuticals from complex water samples. J Chromatogr A February 17, 2012;1225:1–7.

[42] Bratkowska D, Marcé RM, Cormack PAG, Borrull F, Fontanals N. Development and application of a polar coating for stir bar sorptive extraction of emerging pollutants from environmental water samples. Anal Chim Acta November 7, 2011;706(1):135–42.

[43] Gilart N, Cormack PAG, Marcé RM, Borrull F, Fontanals N. Preparation of a polar monolithic coating for stir bar sorptive extraction of emerging contaminants from wastewaters. J Chromatogr A June 21, 2013;1295:42–7.

[44] Ramírez N, Marcé RM, Borrull F. Determination of parabens in house dust by pressurised hot water extraction followed by stir bar sorptive extraction and thermal desorption−gas chromatography−mass spectrometry. J Chromatogr A September 16, 2011;1218(37): 6226−31.
[45] Tölgyessy P, Nagyová S, Sládkovičová M. Determination of short chain chlorinated paraffins in water by stir bar sorptive extraction−thermal desorption−gas chromatography−triple quadrupole tandem mass spectrometry. J Chromatogr A April 21, 2017;1494:77−80.
[46] Giráldez I, Ruiz-Azcona P, Vidal A, Morales E. Speciation of selenite and selenoamino acids in biota samples by dual stir bar sorptive extraction-single desorption-capillary gas chromatography/mass spectrometry. Microchem J September 1, 2015;122:197−204.
[47] Mellano F, Bujalance M, Giráldez I, Ruiz-Azcona P, Sánchez-Rodas D, Morales E. Determination of selenomethionine and seleno-methyl-selenocysteine in biota by ultrasonic-assisted enzymatic digestion and multi-shot stir bar sorptive extraction−thermal desorption−gas chromatography−mass spectrometry. J Chromatogr A July 26, 2013;1300:151−8.
[48] Huang X, Lin J, Yuan D, Hu R. Determination of steroid sex hormones in wastewater by stir bar sorptive extraction based on poly(vinylpyridine-ethylene dimethacrylate) monolithic material and liquid chromatographic analysis. J Chromatogr A April 17, 2009; 1216(16):3508−11.
[49] Yang J-H, Cui C-X, Qu L-B, Chen J, Zhou X-M, Zhang Y-P. Preparation of a monolithic magnetic stir bar for the determination of sulfonylurea herbicides coupled with HPLC. Microchem J September 1, 2018;141:369−76.
[50] Ramírez N, Marcé RM, Borrull F. Development of a stir bar sorptive extraction and thermal desorption−gas chromatography−mass spectrometry method for determining synthetic musks in water samples. J Chromatogr A January 7, 2011;1218(1):156−61.
[51] Camino-Sánchez FJ, Zafra-Gómez A, Oliver-Rodríguez B, Ruiz-Naranjo I, Ruiz-García J, Vílchez JL. Validation of a method for the determination of tributyltin in seawater by stir bar sorptive extraction−liquid chromatography tandem mass spectrometry. J Chromatogr A November 9, 2012;1263:14−20.
[52] Klein DR, Flannelly DF, Schultz MM. Quantitative determination of triclocarban in wastewater effluent by stir bar sorptive extraction and liquid desorption−liquid chromatography−tandem mass spectrometry. J Chromatogr A March 12, 2010;1217(11): 1742−7.
[53] Benedé JL, Chisvert A, Giokas DL, Salvador A. Determination of ultraviolet filters in bathing waters by stir bar sorptive−dispersive microextraction coupled to thermal desorption−gas chromatography−mass spectrometry. Talanta January 15, 2016;147: 246−52.
[54] Benedé JL, Chisvert A, Giokas DL, Salvador A. Stir bar sorptive-dispersive microextraction mediated by magnetic nanoparticles−nylon 6 composite for the extraction of hydrophilic organic compounds in aqueous media. Anal Chim Acta July 5, 2016;926: 63−71.
[55] Benedé JL, Chisvert A, Moyano C, Giokas DL, Salvador A. Expanding the application of stir bar sorptive-dispersive microextraction approach to solid matrices: determination of ultraviolet filters in coastal sand samples. J Chromatogr A August 24, 2018;1564:25−33.

[56] Nguyen KTN, Scapolla C, Di Carro M, Magi E. Rapid and selective determination of UV filters in seawater by liquid chromatography–tandem mass spectrometry combined with stir bar sorptive extraction. Talanta October 15, 2011;85(5):2375–84.

[57] Wang W, He M, Chen B, Hu B. Simultaneous determination of acidic phytohormones in cucumbers and green bean sprouts by ion-pair stir bar sorptive extraction-high performance liquid chromatography. Talanta August 1, 2017;170:128–36.

[58] Cacho JI, Campillo N, Viñas P, Hernández-Córdoba M. Determination of alkylphenols and phthalate esters in vegetables and migration studies from their packages by means of stir bar sorptive extraction coupled to gas chromatography–mass spectrometry. J Chromatogr A June 8, 2012;1241:21–7.

[59] Barba C, Thomas-Danguin T, Guichard E. Comparison of stir bar sorptive extraction in the liquid and vapour phases, solvent-assisted flavour evaporation and headspace solid-phase microextraction for the (non)-targeted analysis of volatiles in fruit juice. LWT – Food Sci Technol (Lebensmittel-Wissenschaft -Technol) November 1, 2017;85:334–44.

[60] Cacho JI, Campillo N, Viñas P, Hernández-Córdoba M. Stir bar sorptive extraction coupled to gas chromatography–mass spectrometry for the determination of bisphenols in canned beverages and filling liquids of canned vegetables. J Chromatogr A July 20, 2012;1247:146–53.

[61] Cacho JI, Campillo N, Viñas P, Hernández-Córdoba M. Stir bar sorptive extraction polar coatings for the determination of chlorophenols and chloroanisoles in wines using gas chromatography and mass spectrometry. Talanta January 15, 2014;118:30–6.

[62] Cacho JI, Campillo N, Viñas P, Hernández-Córdoba M. Stir bar sorptive extraction with gas chromatography–mass spectrometry for the determination of resveratrol, piceatannol and oxyresveratrol isomers in wines. J Chromatogr A November 8, 2013;1315:21–7.

[63] Liu H, Qiao L, Gan N, Lin S, Cao Y, Hu F, et al. Electro-deposited poly-luminol molecularly imprinted polymer coating on carboxyl graphene for stir bar sorptive extraction of estrogens in milk. J Chromatogr B August 1, 2016;1027:50–6.

[64] Qiao L, Gan N, Wang J, Gao H, Hu F, Wang H, et al. Novel molecularly imprinted stir bar sorptive extraction based on an 8-electrode array for preconcentration of trace exogenous estrogens in meat. Anal Chim Acta January 1, 2015;853:342–50.

[65] Arbulu M, Sampedro MC, Sanchez-Ortega A, Gómez-Caballero A, Unceta N, Goicolea MA, et al. Characterisation of the flavour profile from *Graciano Vitis vinifera* wine variety by a novel dual stir bar sorptive extraction methodology coupled to thermal desorption and gas chromatography–mass spectrometry. Anal Chim Acta May 13, 2013; 777:41–8.

[66] Sghaier L, Vial J, Sassiat P, Thiebaut D, Watiez M, Breton S, et al. Analysis of target volatile compounds related to fishy off-flavor in heated rapeseed oil: a comparative study of different headspace techniques. Eur J Lipid Sci Technol 2016;118(6):906–18.

[67] Fan W, He M, Wu X, Chen B, Hu B. Graphene oxide/polyethyleneglycol composite coated stir bar for sorptive extraction of fluoroquinolones from chicken muscle and liver. J Chromatogr A October 30, 2015;1418:36–44.

[68] Horák T, Čulík J, Jurková M, Čejka P, Kellner V. Determination of free medium-chain fatty acids in beer by stir bar sorptive extraction. J Chromatogr A July 4, 2008; 1196–1197:96–9.

[69] Ridgway K, Lalljie SPD, Smith RM. The use of stir bar sorptive extraction—a potential alternative method for the determination of furan, evaluated using two example food matrices. Anal Chim Acta January 11, 2010;657(2):169–74.

[70] Picard M, Franc C, de Revel G, Marchand S. Dual solid-phase and stir bar sorptive extraction combined with gas chromatography-mass spectrometry analysis provides a suitable tool for assaying limonene-derived mint aroma compounds in red wine. Anal Chim Acta February 25, 2018;1001:168–78.

[71] Benet I, Ibañez C, Guàrdia MD, Solà J, Arnau J, Roura E. Optimisation of stir-bar sorptive extraction (SBSE), targeting medium and long-chain free fatty acids in cooked ham exudates. Food Chem October 15, 2015;185:75–83.

[72] Zhu L, Xu G, Wei F, Yang J, Hu Q. Determination of melamine in powdered milk by molecularly imprinted stir bar sorptive extraction coupled with HPLC. J Colloid Interface Sci September 15, 2015;454:8–13.

[73] Hjelmeland AK, Wylie PL, Ebeler SE. A comparison of sorptive extraction techniques coupled to a new quantitative, sensitive, high throughput GC–MS/MS method for methoxypyrazine analysis in wine. Talanta February 1, 2016;148:336–45.

[74] Ruan ED, Aalhus JL, Summerfelt ST, Davidson J, Swift B, Juárez M. Determination of off-flavor compounds, 2-methylisoborneol and geosmin, in salmon fillets using stir bar sorptive extraction–thermal desorption coupled with gas chromatography–mass spectrometry. J Chromatogr A December 20, 2013;1321:133–6.

[75] Franc C, David F, de Revel G. Multi-residue off-flavour profiling in wine using stir bar sorptive extraction–thermal desorption–gas chromatography–mass spectrometry. J Chromatogr A April 10, 2009;1216(15):3318–27.

[76] Cacho JI, Campillo N, Viñas P, Hernández-Córdoba M. Use of headspace sorptive extraction coupled to gas chromatography–mass spectrometry for the analysis of volatile polycyclic aromatic hydrocarbons in herbal infusions. J Chromatogr A August 22, 2014; 1356:38–44.

[77] Xu J, Chen B, He M, Hu B. Analysis of preservatives with different polarities in beverage samples by dual-phase dual stir bar sorptive extraction combined with high-performance liquid chromatography. J Chromatogr A February 22, 2013;1278:8–15.

[78] Li B, Zeng F, Dong Q, Cao Y, Fan H, Deng C. Rapid determination method for 12 pyrethroid pesticide residues in tea by stir bar sorptive extraction-thermal desorption-gas chromatography. Phys Procedia January 1, 2012;25:1776–80.

[79] Mao X, Hu B, He M, Chen B. High polar organic–inorganic hybrid coating stir bar sorptive extraction combined with high performance liquid chromatography–inductively coupled plasma mass spectrometry for the speciation of seleno-amino acids and seleno-oligopeptides in biological samples. J Chromatogr A September 21, 2012;1256:32–9.

[80] Campillo N, Viñas P, Aguinaga N, Férez G, Hernández-Córdoba M. Stir bar sorptive extraction coupled to liquid chromatography for the analysis of strobilurin fungicides in fruit samples. J Chromatogr A July 2, 2010;1217(27):4529–34.

[81] Xu Z, Song C, Hu Y, Li G. Molecularly imprinted stir bar sorptive extraction coupled with high performance liquid chromatography for trace analysis of sulfa drugs in complex samples. Talanta July 15, 2011;85(1):97–103.

[82] Huang X, Qiu N, Yuan D. Simple and sensitive monitoring of sulfonamide veterinary residues in milk by stir bar sorptive extraction based on monolithic material and high performance liquid chromatography analysis. J Chromatogr A November 13, 2009; 1216(46):8240–5.

[83] Yu C, Hu B. C18-coated stir bar sorptive extraction combined with high performance liquid chromatography–electrospray tandem mass spectrometry for the analysis of sulfonamides in milk and milk powder. Talanta February 15, 2012;90:77–84.

[84] Ridgway K, Lalljie SPD, Smith RM. An alternative method for analysis of food taints using stir bar sorptive extraction. Anal Chim Acta September 10, 2010;677(1):29–36.
[85] Cacho JI, Campillo N, Viñas P, Hernández-Córdoba M. Evaluation of three headspace sorptive extraction coatings for the determination of volatile terpenes in honey using gas chromatography–mass spectrometry. J Chromatogr A June 19, 2015;1399:18–24.
[86] De Jager LS, Perfetti GA, Diachenko GW. Stir bar sorptive extraction–gas chromatography–mass spectrometry analysis of tetramethylene disulfotetramine in food: method development and comparison to solid-phase microextraction. Anal Chim Acta March 9, 2009;635(2):162–6.
[87] Delgado R, Durán E, Castro R, Natera R, Barroso CG. Development of a stir bar sorptive extraction method coupled to gas chromatography-mass spectrometry for the analysis of volatile compounds in Sherry brandy. Anal Chim Acta July 5, 2010;672(1):130–6.
[88] Callejón RM, Torija MJ, Mas A, Morales ML, Troncoso AM. Changes of volatile compounds in wine vinegars during their elaboration in barrels made from different woods. Food Chem May 15, 2010;120(2):561–71.
[89] Prosen H, Kokalj M, Janeš D, Kreft S. Comparison of isolation methods for the determination of buckwheat volatile compounds. Food Chem July 1, 2010;121(1):298–306.
[90] Morales ML, Fierro-Risco J, Callejón RM, Paneque P. Monitoring volatile compounds production throughout fermentation by *Saccharomyces* and non-*Saccharomyces* strains using headspace sorptive extraction. J Food Sci Technol February 1, 2017;54(2):538–57.
[91] Hevia K, Castro R, Natera R, González-García JA, Barroso CG, Durán-Guerrero E. Optimization of head space sorptive extraction to determine volatile compounds from oak wood in fortified wines. Chromatographia June 1, 2016;79(11):763–71.
[92] Ubeda C, Callejón RM, Troncoso AM, Peña-Neira A, Morales ML. Volatile profile characterisation of chilean sparkling wines produced by traditional and Charmat methods via sequential stir bar sorptive extraction. Food Chem September 15, 2016;207:261–71.
[93] Rodrigues C, Portugal FCM, Nogueira JMF. Static headspace analysis using polyurethane phases – application to roasted coffee volatiles characterization. Talanta January 30, 2012;89:521–5.
[94] Caven-Quantrill DJ, Buglass AJ. Comparison of volatile constituents extracted from model grape juice and model wine by stir bar sorptive extraction–gas chromatography–mass spectrometry. J Chromatogr A February 18, 2011;1218(7):875–81.
[95] Zhou Q, Qian Y, Qian MC. Analysis of volatile phenols in alcoholic beverage by ethylene glycol-polydimethylsiloxane based stir bar sorptive extraction and gas chromatography–mass spectrometry. J Chromatogr A April 17, 2015;1390:22–7.
[96] Li J, Qi H-Y, Wang Y-B, Su Q, Wu S, Wu L. Hollow fiber–stir bar sorptive extraction and microwave assisted derivatization of amino acids in biological matrices. J Chromatogr A November 25, 2016;1474:32–9.
[97] Talebpour Z, Taraji M, Adib N. Stir bar sorptive extraction and high performance liquid chromatographic determination of carvedilol in human serum using two different polymeric phases and an ionic liquid as desorption solvent. J Chromatogr A May 4, 2012;1236:1–6.
[98] Kole PL, Millership J, McElnay JC. Stir bar sorptive extraction of diclofenac from liquid formulations: a proof of concept study. J Pharm Biomed Anal March 25, 2011;54(4):701–10.
[99] Kole PL, Millership J, McElnay JC. Determination of diclofenac from paediatric urine samples by stir bar sorptive extraction (SBSE)–HPLC–UV technique. Talanta September 30, 2011;85(4):1948–58.

[100] Manandhar E, Maslamani N, Petrikovics I, Rockwood GA, Logue BA. Determination of dimethyl trisulfide in rabbit blood using stir bar sorptive extraction gas chromatography-mass spectrometry. J Chromatogr A August 26, 2016;1461:10–7.

[101] Lan L, Hu B, Yu C. pH-resistant titania hybrid organic–inorganic coating for stir bar sorptive extraction of drugs of abuse in urine samples followed by high performance liquid chromatography–ultraviolet visible detection. J Chromatogr A November 5, 2010; 1217(45):7003–9.

[102] Rodríguez-Gómez R, Zafra-Gómez A, Camino-Sánchez FJ, Ballesteros O, Navalón A. Gas chromatography and ultra high performance liquid chromatography tandem mass spectrometry methods for the determination of selected endocrine disrupting chemicals in human breast milk after stir-bar sorptive extraction. J Chromatogr A July 4, 2014;1349: 69–79.

[103] Huang J-T, Alquier L, Kaisa JP, Reed G, Gilmor T, Vas G. Method development and validation for the determination of 2,4,6-tribromoanisole, 2,4,6-tribromophenol, 2,4,6-trichloroanisole, and 2,4,6-trichlorophenol in various drug products using stir bar sorptive extraction and gas chromatography–tandem mass spectrometry detection. J Chromatogr A November 2, 2012;1262:196–204.

[104] Scherer N, Marcseková K, Posset T, Winter G. Evaluation of stir-bar sorptive extraction coupled with thermal desorption GC–MS for the detection of leachables from polymer single use systems to drugs. J Pharm Biomed Anal April 15, 2018;152:66–73.

[105] Babarahimi V, Talebpour Z, Haghighi F, Adib N, Vahidi H. Validated determination of losartan and valsartan in human plasma by stir bar sorptive extraction based on acrylate monolithic polymer, liquid chromatographic analysis and experimental design methodology. J Pharm Biomed Anal May 10, 2018;153:204–13.

[106] Zhao G, Chen Y, Wang S, Yu J, Wang X, Xie F, et al. Simultaneous determination of 11 monohydroxylated PAHs in human urine by stir bar sorptive extraction and liquid chromatography/tandem mass spectrometry. Talanta November 15, 2013;116:822–6.

[107] Fan W, He M, You L, Zhu X, Chen B, Hu B. Water-compatible graphene oxide/molecularly imprinted polymer coated stir bar sorptive extraction of propranolol from urine samples followed by high performance liquid chromatography-ultraviolet detection. J Chromatogr A April 22, 2016;1443:1–9.

[108] Balbão MS, Bertucci C, Bergamaschi MM, Queiroz RHC, Malfará WR, Dreossi SAC, et al. Rifampicin determination in plasma by stir bar-sorptive extraction and liquid chromatography. J Pharm Biomed Anal April 6, 2010;51(5):1078–83.

[109] Thalhamer B, Buchberger W, Waser M. Identification of thymol phase I metabolites in human urine by headspace sorptive extraction combined with thermal desorption and gas chromatography mass spectrometry. J Pharm Biomed Anal August 25, 2011;56(1):64–9.

[110] Fan W, Mao X, He M, Chen B, Hu B. Stir bar sorptive extraction combined with high performance liquid chromatography-ultraviolet/inductively coupled plasma mass spectrometry for analysis of thyroxine in urine samples. J Chromatogr A November 29, 2013; 1318:49–57.

[111] Xu Z-G, Du Z, Hu Y-L, Hu Y-F, Pan Y-P, Li G-K. Preparation of trimethoprim molecularly imprinted stir bar sorptive extraction and its application for trace analysis of trimethoprim and sulfonamides in complex samples. Chin J Anal Chem July 1, 2012; 40(7):1002–10.

[112] Yamaguchi MS, McCartney MM, Linderholm AL, Ebeler SE, Schivo M, Davis CE. Headspace sorptive extraction-gas chromatography–mass spectrometry method to

measure volatile emissions from human airway cell cultures. J Chromatogr B July 15, 2018;1090:36–42.

[113] Li Y, Whitaker JS, McCarty CL. New advances in large-volume injection gas chromatography-mass spectrometry. J Liq Chromatogr Relat Technol 2009;32(11–12): 1644–71.

[114] Teske J, Engewald W. Methods for, and applications of, large-volume injection in capillary gas chromatography. Trac Trends Anal Chem October 2002;21(9–10):584–93.

[115] Moldoveanu S, David V. Chapter 10 – chemical reactions used in derivatizations. In: Moldoveanu S, David V, editors. Modern sample preparation for chromatography [internet]. Amsterdam: Elsevier; 2015. p. 333–91. Available from: http://www.sciencedirect.com/science/article/pii/B9780444543196000104.

[116] Docherty KS, Ziemann PJ. On-line, inlet-based trimethylsilyl derivatization for gas chromatography of mono- and dicarboxylic acids. J Chromatogr A July 6, 2001;921(2): 265–75.

[117] Pan L, Pawliszyn J. Derivatization/solid-phase Microextraction: new approach to polar analytes. Anal Chem January 1, 1997;69(2):196–205.

[118] Stashenko EE, Martínez JR. Derivatization and solid-phase microextraction. Trac Trends Anal Chem September 1, 2004;23(8):553–61.

[119] Bizkarguenaga E, Iparragirre A, Navarro P, Olivares M, Prieto A, Vallejo A, et al. In-port derivatization after sorptive extractions. J Chromatogr A June 28, 2013;1296:36–46.

[120] Kawaguchi M, Takatsu A, Ito R, Nakazawa H. Applications of stir-bar sorptive extraction to food analysis. Trac Trends Anal Chem April 1, 2013;45:280–93.

[121] David F. 4.21 – Application of stir-bar sorptive extraction in food analysis. In: Pawliszyn J, editor. Comprehensive sampling and sample preparation [internet]. Oxford: Academic Press; 2012. p. 473–93. Available from: http://www.sciencedirect.com/science/article/pii/B9780123813732001459.

[122] Posada-Ureta O, Olivares M, Zatón L, Delgado A, Prieto A, Vallejo A, et al. Uptake calibration of polymer-based passive samplers for monitoring priority and emerging organic non-polar pollutants in WWTP effluents. Anal Bioanal Chem 2016;408(12): 3165–75.

[123] Posada-Ureta O, Olivares M, Delgado A, Prieto A, Vallejo A, Irazola M, et al. Applicability of polydimethylsiloxane (PDMS) and polyethersulfone (PES) as passive samplers of more hydrophobic organic compounds in intertidal estuarine environments. Sci Total Environ 2017;578:392–8.

[124] Assoumani A, Margoum C, Chataing S, Guillemain C, Coquery M. Use of passive stir bar sorptive extraction as a simple integrative sampling technique of pesticides in freshwaters: determination of sampling rates and lag-phases. J Chromatogr A 2014;1333:1–8.

[125] Assoumani A, Coquery M, Liger L, Mazzella N, Margoum C. Field application of passive SBSE for the monitoring of pesticides in surface waters. Environ Sci Pollut Control Ser 2015;22(6):3997–4008.

[126] Assoumani A, Margoum C, Lombard A, Guillemain C, Coquery M. How do PDMS-coated stir bars used as passive samplers integrate concentration peaks of pesticides in freshwater? Environ Sci Pollut Control Ser 2017;24(8):6844–52.

[127] Popp P, Paschke H, Vrana B, Wennrich L, Paschke A. Chapter 5 Membrane-enclosed sorptive coating as integrative sampler for monitoring organic compounds in air. In: Comprehensive analytical chemistry [internet]. Elsevier; 2007. p. 107–23. Available from: http://linkinghub.elsevier.com/retrieve/pii/S0166526X06480053.

[128] Paschke H, Popp P. New passive samplers for chlorinated semivolatile organic pollutants in ambient air. Chemosphere February 1, 2005;58(7):855–63.

[129] Vrana B, Komancová L, Sobotka J. Calibration of a passive sampler based on stir bar sorptive extraction for the monitoring of hydrophobic organic pollutants in water. Talanta May 15, 2016;152:90–7.

[130] Vrana B, Popp P, Paschke A, Schüürmann G. Membrane-enclosed sorptive coating. An integrative passive sampler for monitoring organic contaminants in water. Anal Chem 2001;73(21):5191–200.

[131] De Coensel N, Desmet K, Górecki T, Sandra P. Determination of polydimethylsiloxane-air partition coefficients using headspace sorptive extraction. J Chromatogr A May 25, 2007;1150(1):183–9.

[132] Wennrich L, Popp P, Hafner C. Novel integrative passive samplers for the long-term monitoring of semivolatile organic air pollutants. J Environ Monit May 24, 2002;4(3): 371–6.

[133] Li X-M, Zhang Q-H, Wang P, Li Y-M, Jiang G-B. Determination of polycyclic aromatic hydrocarbons in air by stir bar sorptive extraction-thermal desorption-gas chromatography tandem mass spectrometry. Chin J Anal Chem November 1, 2011;39(11):1641–6.

[134] Assoumani A, Margoum C, Lassalle Y, Herbreteau B, Faure K, Coquery M, et al. Polydimethylsiloxane rods for the passive sampling of pesticides in surface waters. Water 2013;5(3):1366–79.

[135] Matsiko J, Li H, Wang P, Sun H, Zheng S, Wang D, et al. Stir bar sorptive extraction and thermal desorption-gas chromatography/mass spectrometry for determining phosphorus flame retardants in air samples. Anal Methods 2018;10(16):1918–27.

[136] Tienpont B. Determination of phthalates in environmental, food and biomatrices – an analytical challenge. PhD Thesis. 2014. p. 258.

[137] Gilart N, Marcé RM, Borrull F, Fontanals N. New coatings for stir-bar sorptive extraction of polar emerging organic contaminants. Trac Trends Anal Chem February 1, 2014;54: 11–23.

[138] Li X, Mei X, Xu L, Shen X, Zhu W, Hong J, et al. Development and application of novel clonazepam molecularly imprinted coatings for stir bar sorptive extraction. J Colloid Interface Sci April 15, 2016;468:183–91.

[139] Seethapathy S, Górecki T. Applications of polydimethylsiloxane in analytical chemistry: a review. Anal Chim Acta 2012;750:48–62.

[140] Difilippo EL, Eganhouse RP. Assessment of PDMS-water partition coefficients: implications for passive environmental sampling of hydrophobic organic compounds. Environ Sci Technol 2010;44(18):6917–25.

[141] Allan IJ, Harman C, Ranneklev SB, Thomas KV, Grung M. Passive sampling for target and nontarget analyses of moderately polar and nonpolar substances in water. Environ Toxicol Chem 2013;32(8):1718–26.

[142] Miège C, Mazzella N, Allan I, Dulio V, Smedes F, Tixier C, et al. Position paper on passive sampling techniques for the monitoring of contaminants in the aquatic environment – achievements to date and perspectives. Trends Environ Anal Chem November 2015;8:20–6.

[143] Booij K, Robinson CD, Burgess RM, Mayer P, Roberts CA, Ahrens L, et al. Passive sampling in regulatory chemical monitoring of nonpolar organic compounds in the aquatic environment. Environ Sci Technol January 5, 2016;50(1):3–17.

[144] Bartolomé N, Hilber I, Schulin R, Mayer P, Witt G, Reininghaus M, et al. Comparison of freely dissolved concentrations of PAHs in contaminated pot soils under saturated and unsaturated water conditions. Sci Total Environ December 2018;644:835–43.

[145] Vetter W, Kaserzon S, Gallen C, Knoll S, Gallen M, Hauler C, et al. Occurrence and concentrations of halogenated natural products derived from seven years of passive water sampling (2007–2013) at Normanby Island, Great Barrier Reef, Australia. Mar Pollut Bull December 1, 2018;137:81–90.
[146] Okeme JO, Yang C, Abdollahi A, Dhal S, Harris SA, Jantunen LM, et al. Passive air sampling of flame retardants and plasticizers in Canadian homes using PDMS, XAD-coated PDMS and PUF samplers. Environ Pollut 2018;239:109–17.
[147] Assoumani A, Lissalde S, Margoum C, Mazzella N, Coquery M. *In situ* application of stir bar sorptive extraction as a passive sampling technique for the monitoring of agricultural pesticides in surface waters. Sci Total Environ October 1, 2013;463–464:829–35.

# Matrix solid phase dispersion

**19**

*Rosa Ana Pérez, Beatriz Albero, José L. Tadeo*
Departamento de Medio Ambiente y Agronomía, Instituto Nacional de Investigación y Tecnología Agraria y Alimentaria (INIA), Madrid, Spain

## 19.1 Introduction

Classical procedures to extract target organic analytes from solid and semisolid samples (shaking and filtering and Soxhlet extraction methods) are time-consuming, laborious, difficult to automate, nonselective methods that use a large volume of organic solvents. To overcome these problems, several modern extraction techniques have been developed, such as pressurized liquid extraction (PLE), ultrasonic-assisted extraction (UAE), microwave-assisted extraction (MAE), matrix solid phase dispersion (MSPD) or the QuEChERS method (acronym for "quick, easy, cheap, effective, rugged and safe"). Thus, from the point of view of a multiresidue analysis, the efforts to carry out analyses in complex matrices at low levels have led to the development of new analytical techniques and the improvement of existing techniques.

In comparison with other modern extraction techniques that use high pressure, high temperature, or the application of supplementary energy (PLE, UAE, or MAE), MSPD carries out the extraction process under ambient conditions and does not need any special laboratory equipment. This extraction technique was developed by Barker et al. in 1989 [1] with the advantages over conventional techniques of needing small amounts of sample and solvent to carry out the extraction procedure with only a few simple steps. Since its introduction, it has been used in the isolation of a wide variety of organic compounds (contaminants or naturally occurring constituents) from a wide variety of complex matrices (environmental samples, plant and animal samples). This extraction technique contributes to the simplification of the sample preparation process, since it reduces the handling of samples and, in some cases, the additional purification. Thus, it can be considered a simple, generic, fast, cost-effective, environmentally friendly, and adaptable technique for a variety of applications [2,3].

The objective of this chapter is to present an overview of the MSPD methods developed to extract natural compounds and organic contaminants from environmental and foodstuff samples, focused on the last 5 years.

## 19.2 General procedure, traditional sorbents, and new materials

Barker et al. [1] firstly described MSPD as a new method based on the dispersion of tissues with an SPE column packing material octadecylsiloxane-bonded silica (C18).

**Solid-Phase Extraction. https://doi.org/10.1016/B978-0-12-816906-3.00019-4**

Thus, a semidry and easy to handle material was obtained that was used as a column packing from which several drugs were isolated. Briefly, the samples were prepared by adding the tissue (0.5 g) to 2.0 g of C18 placed in a glass mortar. Then, the sample was blended with a glass pestle to produce a semidry and homogenous appearing material that was added to a syringe barrel-column (10 mL) containing a frit and Cl8 sorbent (0.5 g) at the bottom and, finally, another portion of the sorbent (0.25 g) was added to the top of the column. The column was slightly tapped to remove the air pockets, before adding a suitable extraction solvent. In a simple way, MSPD is a simultaneous extraction and cleanup technique developed for solid or semisolid samples in which the sample is disintegrated in the presence of a solid support that produces complete sample disruption and interactions between the sample matrix and the sorbent, creating a new phase that is transferred to a column for the elution of the analytes. Accordingly, the main steps of MSPD are summarized in Fig. 19.1: blending, transferring to a column, compressing, and elution with solvent. In addition, a cosorbent can be added to the bottom of the extraction column to carry out an additional in-situ cleanup.

The blending with the dispersing material is generally carried out in a mortar using a glass or porcelain pestle to crush the sample. In this step, the sorbent acts as an abrasive that promotes the disruption of the initial physical structure of the sample matrix, dispersing it in small pieces, and, in addition, it may absorb the analytes or other sample components of interest. After complete homogenization, the sample is transferred to an empty column or solid phase cartridge, with a frit or paper filter at the bottom, which is empty or contains an additional cleanup material. Then, a new frit or filter paper is often placed on top of the sample before compression using a syringe plunger. The compressing step avoids the formation of channels or air bubbles. Finally, the elution of the analytes is carried out using a vacuum system.

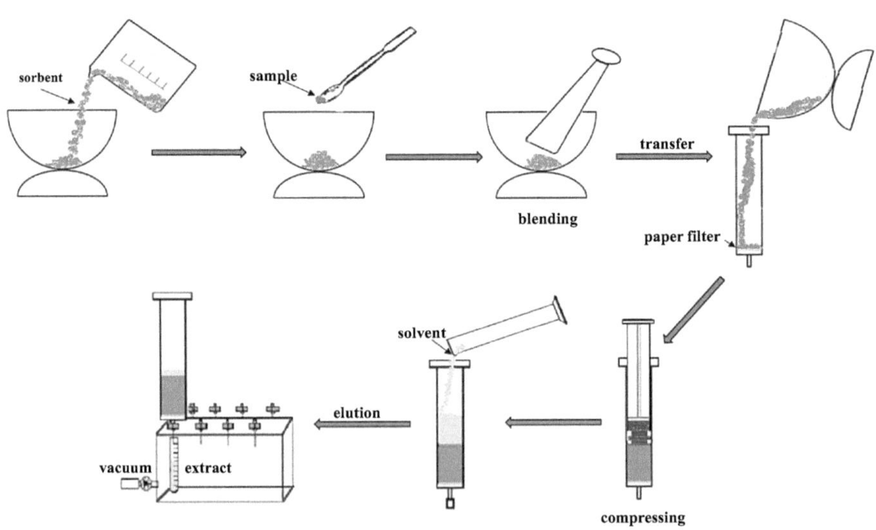


**Figure 19.1** Schematic diagram of the MSPD procedure.

Although it may seem that this extraction technique is very similar to SPE, there are clear differences between both techniques. Thus, in MSPD a complete disruption and dispersion of the sample in small particles is achieved, providing an improved surface for the subsequent extraction of the sample; while in SPE sample disruption should be carried out in an earlier step, because the sample must be liquid, and many of the sample components are discarded before SPE extraction. Furthermore, in SPE the sample is usually retained in the first few millimeters of the sorbent packed in the column; while, in MSPD, the sample is dispersed throughout the column. Finally, the physical and chemical interactions of the components of the system are greater in MSPD than in SPE.

In general, the sorbents used in MSPD are similar to those used in SPE in order to enhance selectivity and perform a simultaneous cleanup. Table 19.1 reports the composition and characteristics of traditional and new sorbent materials used in MSPD. The selection of the sorbent used to homogenize the sample depends on the nature of the material to be analyzed.

In recent years, the most commonly used sorbents in MSPD continue to be C18 and Florisil, a reverse-phase and a normal phase material, respectively. Nevertheless, MSPD

**Table 19.1** Composition and characteristics of the sorbents used in MSPD.

| Sorbents | Composition | Characteristic properties |
|---|---|---|
| Diatomite [2] | Diatomaceous earth | Inert support. It is a natural source of silica |
| Silica [2] | $SiO_2$ | For extraction of polar compounds |
| C18 [2] | Octadecyl silane bonded to silica | For reversed phase extraction of analytes with wide polarity range. High retentive properties for nonpolar compounds. Nonselective sorbent |
| C8 [2] | Octyl silane bonded to silica | Properties similar to C18, but with less holding capacity for non-polar compounds due to its shorter hydrocarbon chain |
| Alumina [2] | $Al_2O_3$ | High activity. Available in acidic, neutral and basic pH surface options. The surface-molecule interaction can be with the aluminum metal center, by hydrogen bonding with surface hydroxyl groups, or by ion exchange if the surface carries a charge |

*Continued*

**Table 19.1** Composition and characteristics of the sorbents used in MSPD.—cont'd

| **Sorbents** | **Composition** | **Characteristic properties** |
|---|---|---|
| Florisil [2] | Activated magnesium silicate | Highly polar material for adsorption extraction of polar compounds from nonpolar matrices |
| GCB [2] | Graphitized carbon black | Nonporous carbon that has a high attraction for organic polar and nonpolar compounds from both polar and nonpolar matrices |
| PSA [2] | Primary and secondary amine | Polymerically bonded, ethylenediamine-N-propyl phase that contains both primary and secondary amines. Strong affinity and high capacity for removing fatty acids, organic acids, and some polar pigments and sugars |
| Graphene [2] | Two-dimensional form of carbon | Ultrahigh specific surface area, superior chemical stability, excellent thermal stability and flexible nanosheet morphology |
| MWCNTs [2] | Multiwalled carbon nanotubes (unmodified or modified) are hollow, cylindrically shaped allotropes of carbon | Excellent adsorption properties due to a surface area extremely large and hydrophobic and peculiar structural features |
| MIPs [2] | Polymers around a template molecule from functionalized monomers | High selectivity and affinity for the target molecule used in the imprinting procedure |
| Magnetic sorbent (for example MGO@PIT) [3] | Poly(indole-thiophene)-coated magnetic graphene oxide | Easy separation. Simplicity, low extraction time and eliminates the column packing |
| PTS-MgO [41] | Magnesium oxide microspheres functionalized with phenyltrichlorosilane | Interaction between oxygen of MgO and $\pi$-electron system of the compounds. High purification efficiency to remove interfering substances (lipids, sulfur, pigments) |

*MIPs*, molecularly imprinted polymers; *MWCNTs*, multiwalled carbon nanotubes.

procedures using other sorbents (such as C8, silica, or alumina) and new sorbent materials [such us graphene, multiwalled carbon nanotubes (MWCNTs), molecularly imprinted polymers (MIPs), magnetic sorbents, and magnesium oxide microspheres functionalized with phenyltrichlorosilane (PTS-MgO)] have also been reported (see Sections 19.3 and 19.4). In general, a good compromise between a large surface area, good dispersion, and improved contact with the solvent are provided with sorbents with a particle size in the range of 40–100 μm. However, sorbents with a smaller particle size have also been used, although this leads to prolonged solvent elution times and possible blockage of the columns. Sorbents with small particle size are used in some of the MSPD procedures developed using new sorbents [4–7].

## 19.3 Extraction of natural components of plants and fruits by MSPD

At the beginning, MSPD was applied almost exclusively to the analysis of contaminants in food, and there were not many articles that addressed its application for the analysis of natural compounds in plant samples [8]. Nevertheless, in the past 5 years, numerous papers have focused on the characterization or determination of natural constituents (such as isoflavones, fatty acids, or polyphenols) in plants and fruits. Thus, the evaluation of four marker constituents in pomegranate [9], the characterization of the fatty acids profiles of four varieties of *Camellia* seeds [10], the main compounds of *Carthamus tinctorius* L [11], or isoflavones in nine *Medicago* species [12] have been reported by MSPD.

Flavonoids are an important group of dietary polyphenols commonly found in fruits and vegetables that are divided into the following subclasses: anthocyanins, flavanols, isoflavones, flavones, flavanones, and flavonols. These compounds have been studied in many fruits (mainly citrus) and pharmacological studies and clinical trials [13–15]. Several methods based on MSPD have been developed for the determination of flavonoids in plants and fruits [12–18]. However, there is no general MSPD procedure for the extraction of flavonoids, or for other natural components. Therefore, samples from 25 to 1000 mg, and sample-to-sorbent mass ratio from 1:0.5 to 1:4 w/w have been used. Silica and C18 were the sorbents most used based on their efficiency [12,14,16,18,19]; although other conventional sorbents, such as Florisil [13] or alumina (for the determination of vitamers of vitamin E in whole grain barley) [20] have also been employed. On the other hand, nonconventional solid supports, such as molecular sieve for flavonoids in orange fruit peel and plants [15,17], MIPs for quercetin in *Herba lysimachiae* [21], titanium dioxide nanoparticles for phospholipids in olive fruit and oil [22], graphene-encapsulated silica for polymethoxylated flavonoids in *Murraya paniculata* leaves [23], silica-supported ionic liquids for phenolic acids and flavonoids in raw propolis [24], microcrystalline cellulose for triterpenoic acids in loquat leaves [25], middle molecular-weight chitosan for phenols in olive fruit [26], or carbon molecular sieve for the extraction of polyphenols from pomegranate peel [27] have also been used as sorbent materials.

In general, the extraction of natural components by MSPD has usually been carried out without additional cleanup material. Methanol (MeOH) was the extraction solvent most frequently used, although a mixture MeOH:water was also used because the recovery of target compounds were remarkably increased with a more polar elution solvent [11,26]. Recently, ionic liquids, nonionic detergents, and limonene have been reported as adequate green solvents to extract bioactive compounds from fruit by MSPD [10,13,28,29].

## 19.4 MSPD methods for the extraction of contaminants from different matrices

The choice of the sorbent, the sample-to-sorbent ratio, the use of cosorbent for cleanup and the selection of a suitable solvent, or mixture of solvents are the main parameters that must be optimized in MSPD methods, often using a statistical design of experiments [4,30–33]. This extraction technique has been more frequently used in the extraction of contaminants from foodstuff than from environmental samples.

Anhydrous magnesium sulfate ($MgSO_4$) or anhydrous sodium sulfate ($Na_2SO_4$) are usually employed together with a sorbent in the matrix dispersion step, or used as cleanup cosorbent, in order to eliminate residual water, when the extracts are analyzed by gas chromatography (GC). In some cases, they are the only agent used in the extraction step, as cleanup and drying agent, without applying any other SPE sorbent [34,35]. A summary of MSPD methods for the extraction of contaminants from foodstuff and environmental samples is given in Tables 19.2 and 19.3, respectively.

The MSPD methods for the extraction of contaminants from foodstuff mainly used traditional sorbents, such as Florisil or C18 (see Table 19.2), although new sorbents such as a hybrid polymer with poly(vinylimidazole) grafted onto silica particles and MIPs were also used [6,36,37]. The synthesis of new sorbents with improved extraction properties is an easily identified trend in recent years.

Pesticides are the main contaminants extracted by MSPD from foodstuff samples, although emerging contaminants, such as preservatives, antibiotics, ultraviolet (UV) filters, and steroid hormones, have also been analyzed.

A multiresidue method for the determination of 163 pesticides (6 acaricides, 62 fungicides, 18 herbicides, and 77 insecticides) in herbs (chamomile, thyme, linden, lungwort, melissa, and peppermint) was developed by Lozowicka et al. [38]. In this study, two analytical methods, one based on MSPD and the other on liquid–solid extraction, were evaluated. Both methods were optimized considering different parameters (sample-to-sorbent mass ratio, extraction solvent, sorbents for cleanup, etc.). MSPD allowed a reduction in solvent consumption with a significant time saving, degradation of sample components was avoided, efficiency and selectivity of the extraction was improved, and an additional cleanup step before the chromatographic analysis was not required. The authors concluded that the developed method could be recommended for routine monitoring of pesticide in herbs. On the other hand, a method for the simultaneous determination of 16 pesticide residues in tea by MSPD

**Table 19.2** MSPD methods developed for the determination of organic contaminants in foodstuff.

| Compounds (number) | Sample | Sorbent | Cosorbent | Solvent | Determination | Recovery (%) | LOQ (ng/g) | References |
|---|---|---|---|---|---|---|---|---|
| OPPs (5) | Propolis (0.1 g) | $SiO_2$–PVI (1.9 g) | | ACN:DCM (25:75 v/v) (8 mL) | GC-MS | 81–122 | | [6] |
| Parabens (4) | Seafood (0.5 g) | Florisil: $Na_2SO_4$ (1g:0.5 g) | Silica gel (1.5 g) + C18 (9:1 w/w) | ACN (12 mL) | GC-MS | 93–112 | 0.2–1 | [30] |
| UV filters (5) | Fish (0.5 g) | Florisil (1 g) | C18 silica (1 g) | ACN (7 mL) | GC-MS/MS | 71–102 | <0.1 | [32] |
| Parabens (7) | Mollusks (0.5 g) | Silica (1.2 g) | C18 (3 g) | ACN (10 mL) | LC-MS/MS | 71–117 | <1.4 | [33] |
| Quinolones (8) | Fish meat (0.2 g) | MIP (0.05 g) | MIP (50 mg) | (99:1, v/v) ACN: Trifluoroacetic acid (4 mL) | HPLC-FLD | 64–103 | 0.06–0.2 (LOD) | [36] |
| Ketoprofen | Powder milk (50 mg) | MIS (25 mg) $Na_2SO_4$ (50 mg) | | ACN (1 mL) | LC/MS/MS | 89–97 | 2.5 | [37] |
| Pesticides (20) | Apples (5 g) | Florisil (8 g) | | EtAc (60 mL) | GC-MS | ≥90 | <5 | [40] |
| β-Lactam antibiotics (15) | Pork muscle (2 g) | Oasis HLB (3 g) | | ACN:water (10:90 v/v) + 0.2% $CH_2O_2$ | LC-MS/MS | >90 | 0.1–20.1 | [49] |

*Continued*

**Table 19.2** MSPD methods developed for the determination of organic contaminants in foodstuff.—cont'd

| Compounds (number) | Sample | Sorbent | Cosorbent | Solvent | Determination | Recovery (%) | LOQ (ng/g) | References |
|---|---|---|---|---|---|---|---|---|
| PBDEs, OH-PBDEs, MeO-PBDEs | Lettuce, (0.25 g) Carrot (0.5 g) | C18 (0.5 g) | $SiO_2$ (0.5 g) + acidified $SiO_2$ (1.75 g) | hexane:DCM (75:25, v/v) (10 mL) + DCM (30 mL) (OH-PBDEs) | GC-MS | 81–129 | 0.01–1.8 | [50] |
| OCPs (17) + PCBs (7) | Oil seeds (0.5 g) | 40% (w:w) $H_2SO_4$-silica (3.5 g) | Florisil (1.2 g) + C18 (100 mg) | hexane:DCM (7:3, v/v) (10 mL) | GC-ECD | 69–103 | 0.4–6.7 | [51] |
| OCPs (14), PCBs (7) | Vegetable oils (0.5 g) | $SiO_2$ ($H_2$ S-impregnated) (3.5 g) | $SiO_2$ gel (0.8 g) | hexane:DCM (70:30, v/v) (10 mL) | GC-ECD | 70–105 | 0.14–2.5 | [52] |
| Triazines (9) | Seaweeds (1 g) | C8 (2 g) | ENVI-CarbII (0.5 g)+ PSA (0.5 g) | EtAc:ACN (80:20, v/v) (25 mL) | HPLC-DAD | 80–92 | 4.1–7.3 | [53] |
| Pesticides (16) | Tea (0.5 g) | C18 (3 g) + Florisil (3 g) | GCB (50 mg)+sea sand (450 mg) //PSA (1.0 g)+sea sand (1.5g)// PVPP (750 mg) | EtAc (20 mL) + ACN (5 mL) | LC-MS/MS | 88–100 | 0.03–4.7 | [39] |

| | | | | | | | | |
|---|---|---|---|---|---|---|---|---|
| Triazines (9) | Mussels (0.5 g) | C18 (2 g) | EnviCarb-II/ SAX/PSA (0.5 g/0.5 g/ 0.5 g) | EtAc (20 mL) + ACN (5 mL) | HPLC−DAD | 79−99 | 100−180 | [54] |
| Pesticides (166) | Herbs (2 g) | Florisil (4 g) | $Na_2SO_4$ (5 g) + C18 (1 g) + silica gel (2.5 g) | acetone:MeOH (9:1, v/v) (25 mL) | GC-ECD GC-NPD | 70−116 | 5−40 | [38] |
| Steroid hormones (10) | Animal origin food (1 g) | Florisil (3 g) | | ACN:EtAc (4:1, v/v) (10 mL) | LC/MS/MS | 77−99 | 0.05−0.2 | [55] |

*ACN*, acetonitrile; *$CH_2O_2$*, formic acid; *DAD*, diode array detector; *DCM*, dichloromethane; *EtAc*, ethyl acetate; *MeOH*, methanol; *MeO-PBDEs*, methoxylated polybrominated diphenyl ethers; *MIP*, molecularly imprinted polymer.; *MIS*, mesoporous imprinted silica; *Oasis HLB*, copolymer of divinyl-benzene and N-vinyl-pyrrolidone; *OCPs*, organochlorine pesticides; *OH-PBDEs*, hydroxylated polybrominated diphenyl ethers; *OPPs*, organophosphorus pesticides; *PBDEs*, polybrominated diphenyl ethers; *PCBs*, polychlorinated biphenyls; *PSA*, primary and secondary amine; *PVPP*, polyvinylpolypyrrolidone; *$SiO_2$−PVI*, $SiO_2$/polyvinylimidazole hybrid polymer.

**Table 19.3** MSPD methods used for the determination of organic contaminants in environmental samples.

| Compounds (number) | Sample | Sorbent | Cosorbent | Solvent | Determination | Recovery (%) | LOQ (ng/g) | References |
|---|---|---|---|---|---|---|---|---|
| PAHs (7) | Soil (0.1 g) | PTS-MgO (0.1 g) | PTS-MgO (50 mg) | DCM (4 mL) | HPLC-FLD | 72.2−113 | 0.07−0.4 | [41] |
| PBDEs, OH-PBDEs, MeO-PBDEs | Soil (0.5 g) | C18-silica (0.5 g) | Silica (0.5 g) + acidified silica (1.75 g) | Hexane:DCM (75:25, v/v) (10 or 30 mL) | GC-MS | 81−129 | 0.01−1.8 | [50] |
| Chlorophenols (6) | Soil (5 g) | GAs (10 g) | $Na_2SO_4$ (50 mg) Florian soil (50 mg) | Acetone (1 mL) | HPLC-UV | 86−111 | 0.1−0.33 (μg/L) | [42] |
| Drugs (8), metabolite (1) | Sludge (0.5 g) | C18-silica (2 g) | Diatomaceous earth (1 g) | MeOH/ACN/formic acid (30:69:1 v/v/v) | LC-MS/MS | 82−124 | 2−10 | [56] |
| Triazines (9) | Marine sediment (1 g) | GCB (1 g) | | EtAc (20 mL) | HPLC-DAD | >85 | 22−37 | [44] |
| PPCPs (45) | Sludge (0.1 g) | C18-silica (0.4 g) | | MeOH (6 mL) and ACN + 5% oxalic acid (8/2, v/v) (10 mL) | LC-MS/MS | 50−107 | 0.12−5.5 | [43] |
| OPs (8) | Sludge (0.5 g) | C18-silica (2 g) | PSA (1 g) | ACN (15 mL) | LC-QTOF-MS | 69−123 | 2−50 | [57] |

*ACN*, acetonitrile; *C18*, silica modified with octadecyl groups; *DAD*, diode array detector; *DCM*, dichloromethane; *EtAc*, ethyl acetate; *FLD*, fluorescence detection; *GAs*, graphene aerogels; *MeO-PBDEs*, methoxylated polybrominated diphenyl ethers; *MS/MS*, tandem mass spectrometry; $Na_2SO_4$, anhydrous sodium sulfate; *OH-PBDEs*, hydroxylated polybrominated diphenyl ethers; *OPs*, Organophosphate compounds; *PBDEs*, polybrominated diphenyl ethers; *PPCPs*, pharmaceuticals and personal care products; *PSA*, primary and secondary amine; *PTS-MgO*, Magnesium oxide microspheres functionalized with phenyltrichlorosilane; *QTOF-MS*, quadrupole time-of-flight mass spectrometry; UV, ultraviolet.

and liquid chromatography-tandem mass spectrometry (LC-MS/MS) analysis was described by Cao et al. [39]. In this method, the sample was homogenized with a mixture of C18 and Florisil, with poly(vinylpyrrolidone), PSA, and GCB added to the bottom of the MSPD column as cleanup sorbents to remove coeluting matrix components. In this way, extraction and purification were achieved in a single step. MSPD was used for multiresidue analysis of pesticides in apple for the production of baby foods [40]. Florisil was used as sorbent with elution by 60 mL of ethyl acetate (EtAc) to extract the pesticides from the homogenized sample.

The extraction of UV filters and parabens preservatives in fish, mollusks, and seafood are some examples of MSPD methods described for the analysis of emerging contaminants [30,32,33]. The extraction was carried out using acetonitrile (ACN) as solvent. ACN extracts had the lowest dry residue with high recoveries, while recoveries with five other organic solvents were low or the extracts had high lipid content and/or colored polar species, which interfered in the chromatographic analysis.

The methods for the analysis of parabens were quite different, mainly due to the technique used for their determination. Owing to the polar nature of these compounds, derivatization is required before analysis by GC to reduce retention and improve sensitivity and resolution. Seafood samples were analyzed using MSPD and online acetylation for GC-MS analysis [30]. A portion of powdered seafood (0.5 g) was mixed with 0.5 g of $Na_2SO_4$ and dispersed with 1.0 g of Florisil using a vortex for 2 min, instead of using a mortar and a pestle. MSPD was combined with LC-MS/MS in the method developed by Villaverde-de-Sáa et al. [33] for the determination of parabens in mollusks. After optimization of the extraction parameters, the final conditions were the dispersion of 0.5 g of freeze-dried mollusk with 1.2 g of silica subsequently packed into a cartridge containing 3 g of C18, as online cleanup sorbent.

The extraction and cleanup of UV filters from fish samples were carried out in a simple step using Florisil and C18 (as dispersant and cleanup cosorbent, respectively) and ACN as the elution solvent [32]. First, a portion of powdered fish sample was mixed with the same amount of $Na_2SO_4$, and dispersed with Florisil using a mortar and pestle. This blend was transferred to a cartridge containing C18 at the bottom. With these conditions, less than 0.3% (w/w) of lipids were detected in the extract; however, a relatively high lipid content was observed when the dispersion was carried out with C18 and Florisil placed at the bottom of the column, or when the cleanup cosorbent (C18) was mixed together with the powdered fish sample, $Na_2SO_4$, and the dispersant (Florisil).

New sorbents with a great potential have been described for the selective extraction of target compounds by MSPD. Thus, imprinted mesoporous silica and class-selective dummy MIPs were synthesized and characterized as selective sorbents for MSPD extraction of ketoprofen from powdered milk and eight fluoroquinolones from fish tissue, respectively [36,37]. Surface molecular imprinting of mesoporous silica may overcome the poor site accessibility of MIPs to target compounds because the recognition sites are located on the surface of the solid support instead of being embedded in a highly cross-linked polymer. On the other hand, the preparation of MIPs using a dummy template can overcome the impact of template leakage on the accuracy of the method as the template is different from the target analyte.

MSPD was employed for the determination of organic contaminants in solid environmental samples: sediment, soil, or sludge (Table 19.3). MgO microspheres modified with phenyltrichlorosilane (PTS-MgO) were used for the selective extraction of seven dioxinlike polycyclic aromatic hydrocarbons (PAHs) from soil avoiding the coextraction of chlorinated compounds [41]. The homogenized mixture of soil and PTS-MgO was transferred to a prepackaged column containing an additional amount of PTS-MgO (50 mg) that acted as a cleanup cosorbent for further removal of interfering matrix components. Graphene aerogel was used for the determination of chlorophenols in soil due to its outstanding capacity to adsorb chlorophenols in comparison to C18 or single-walled carbon nanotubes owing to the strong π-π stacking interaction between the sorbent and the target analytes [42].

The multiresidue extraction of 45 pharmaceutical and personal care products (PPCPs) from sewage sludge by MSPD and LC-MS/MS was developed by Li et al. [43]. The best results were achieved using C18 as sorbent, a sample-sorbent ratio of 1:4 (w/w), and 6 mL MeOH and 10 mL ACN/5% oxalic acid (8/2, v/v) as elution solvents. Most of the PPCPs were eluted with MeOH but ACN/oxalic acid played an important role in the elution of quinolone and tetracycline antibiotics. In this method, the extraction was carried out without a cleanup cosorbent or a subsequent purification step. Recently, Rodríguez-Gonzalez et al. [44] developed a method based on MSPD for the determination of nine triazine herbicides in marine sediments. Three dispersing agents (C18, GCB, and diatomaceous earth) and four cleanup cosorbents (Florisil, GCB, PSA/SAX, and GCB/PSA) were evaluated. The best results were obtained with GCB as dispersing agent, without the need for an additional cleanup, using 20 mL of ethyl acetate as elution solvent.

### 19.4.1 Assisted MSPD extraction and combination with other extraction techniques

The combination of MSPD with other extraction techniques (such us UAE, MAE, dispersive liquid-liquid extraction, or SPE) to overcome the deficiencies of each technique, or to obtain the advantages of both, has been employed in some methods. Although most MSPD methods are carried out without additional energy sources, the extraction of analytes strongly sorbed to the matrix is facilitated by use of ultrasound or microwave sources, resulting in higher efficiency while reducing the solvent volume required. Ultrasonic radiation generates a disturbance when transmitted through a medium; and, in solid-liquid extractions, the mechanical effect of ultrasounds induces a greater penetration of the solvent into the matrix and the mechanical erosion of solids, including particle rupture, which improves mass transfer and enhances the extraction efficiency. UA-MSPD with an extraction column immersed in an ultrasonic bath was used in our laboratory for the multiresidue analysis of different organic contaminants in poultry manure, vegetables, and aquatic plants [34,35,45]. The effect of sonication on the extraction of analytes from vegetables was evaluated by preparing two sets of samples, extracting one set with sonication in an ultrasonic bath for 15 min and the other set without sonication [35]. Analysis of the extracts showed that the recoveries obtained with sonication were higher than those without sonication.

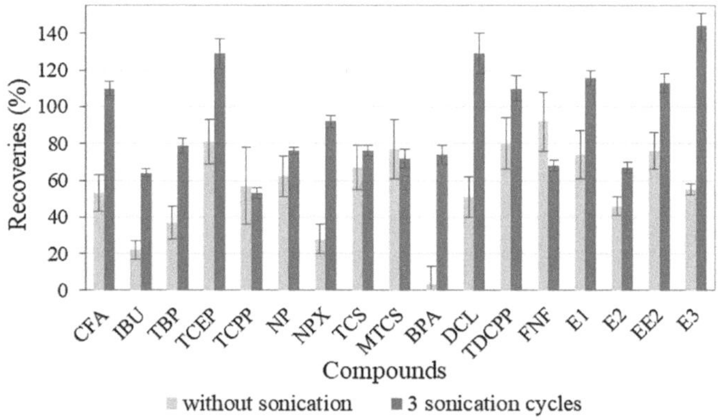


**Figure 19.2** Influence of sonication in the extraction of selected analytes from lettuce samples. Elution with EtAc-MeOH (9:1, v/v) + 3% $NH_4OH$ followed by elution with EtAc-MeOH (9:1, v/v) + 3% formic acid with or without sonication (3 × 15 min).
Adapted from data published by Albero B, Sanchez-Brunete C, Miguel E, Tadeo JL. Application of matrix solid-phase dispersion followed by GC-MS/MS to the analysis of emerging contaminants in vegetables. Food Chem. 2017;217:660–7, with permission.

As shown in Fig. 19.2, good extraction yields were obtained, with lower relative standard deviations (RSD), when three sonication cycles were employed.

In MAE, the temperature rapidly increases due to the interaction of microwave energy by dipole rotation and ionic conduction of solvent and sample molecules [23]. MAE provides several advantages, such as high recovery and low solvent consumption. Dynamic microwave-assisted extraction (DMAE) was coupled with MSPD for the extraction of six triazine herbicides from rice [46]. Zhang et al. [47] developed a method for the determination of chlorfenapyr and abamectin in rice. Elution of the target compounds was completed in 2.7 min and only 7 mL of solvent was required. In addition, the combination of MSPD with DMAE accomplished the extraction and cleanup processes simultaneously.

SPE and UAE are the most commonly employed extraction techniques used in combination with MSPD. Examples include the analysis of antimycotic drugs in digested sludge [31] and emerging contaminants in aquatic plants [45]. The multiresidue analysis of eight antimycotic drugs, belonging to three different chemical classes, in digested sludge used an MSPD extraction column connected in series with an SPE cartridge containing a cation-exchange sorbent. The target analytes were extracted with MeOH (10 mL) from the freeze-dried sludge sample (0.5 g) dispersed with C18 (2 g) plus a layer of PSA as cleanup sorbent (1 g). The extract flowed through the SPE cartridge, where the target compounds were trapped and separated from neutral interfering compounds. The MSPD extraction column was discarded and the target analytes eluted from the SPE cartridge with 10 mL of MeOH containing 0.5% (v/v) ammonia) for analysis by LC-MS/MS. The recoveries ranged from 70% to 118% with LOQs from 5 to 8 ng/g. A multiresidue method was developed for the determination of 31 emerging contaminants (pharmaceutical and personal care products, hormones, biocides, and flame retardants) in aquatic plants [45]. The target compounds were extracted by UA-MSPD followed by a cleanup step to minimize interference from chlorophyll that hindered MS quantification. SPE with C18 provided the best performance of four different sorbents evaluated for eliminating interferences. Recoveries of 70%−120% were obtained with low method detection limits (0.3−2.2 g/g wet weight).

Most of the reported MSPD methods are offline procedures, with the potential risk of sample loss and contamination during the extraction process. Although the online coupling of MSPD with various detection techniques is feasible, methods employing online system are scarcely reported. Gutierrez-Valencia et al. [48] developed an online MSPD-SPE-HPLC/FLD system for the analysis of four PAHs in bovine tissues. The MSPD extraction column was prepared by mixing C18 sorbent (200 mg) and 50 mg of bovine tissue to obtain a homogeneous mixture subsequently packed into a stainless steel cartridge (30 × 8.0 mm I.D.) containing a poly(tetrafluoroethylene) frit. The MSPD cartridge was closed with two steel nuts and connected to an MSPD-SPE-HPLC/FLD system. A C18 precolumn was used in the SPE preconcentration step. Recoveries from 96.4% to 98.8% were obtained.

MSPD can be combined with dispersive liquid-liquid microextraction (DLLME) as illustrated by the extraction of organochlorine pesticides (OCPs) from soil and parabens from breast milk [3,4]. Ghadiri et al. [4] used MWCNTs as sorbent for the extraction of OCPs. The soil sample (0.2 g) was mixed with 0.020 g of MWCNTs and 0.400 g of silica gel and extracted with dichloromethane. The resulting extract was preconcentrated using DLLME to improve sensitivity before chromatographic analysis. Fotouhi et al. [3] used MSPD with magnetic particles as sorbent for the analysis of three parabens in breast milk. The method combines magnetically assisted matrix solid phase dispersion (MA-MSPD), using poly(indole-thiophene) coated magnetic graphene oxide (MGO@PIT) as sorbent, followed by DLLME. Sodium sulfate, 550 mg (as drying and matrix dispersing agent) and 50 mg of MGO@PIT were added into 200 μL of milk. The mixture was gently blended to obtain a dry powder and the blend powder stirred into ultrapure water. The magnetic sorbent was separated using a magnet and the target compounds eluted with 1 mL of MeOH. Then, the DLLME was carried out by mixing 1-octanol as extraction solvent (90 μL) with the MeOH extract, as the dispersing solvent. This mixture was injected rapidly into 9.0 mL of ultrapure water in a glass vial, and immersed in an ultrasonic water bath. Ultrasonic assistance facilitated the formation of a cloudy solution of fine droplets of 1-octanol dispersed throughout the extract solution; which was centrifuged to separate the 1-octanol containing the target compounds as a thin layer on the upper surface of the aqueous phase. Finally, the analysis was carried out by LC-UV and LC-MS/MS.

## 19.5 Conclusion

MSPD is a well-established extraction technique that is speedy, cost-effective, environmentally friendly, and adaptable to a variety of sample types, without needing any special laboratory equipment. MSPD methods usually employ small amounts of sample, sorbents, and organic solvents in comparison with other extraction techniques. In general, MSPD is applicable for the isolation of compounds from a variety of solid and semisolid matrices. The development of new sorbents that are more selective, with larger surface area, good dispersion, and improved contact with the solvent will promote the development of new miniaturized MSPD methods. Miniaturization, together with automation of the process by the online coupling with chromatographic techniques, is driving recent advances in this extraction technique.

# References

[1] Barker SA, Long AR, Short CR. Isolation of drug residues from tissues by solid-phase dispersion. J Chromatogr A 1989;475:353–61.

[2] Capriotti AL, Cavaliere C, Foglia P, Samperi R, Stampachiacchiere S, Ventura S, Lagana A. Recent advances and developments in matrix solid-phase dispersion. Trac Trends Anal Chem 2015;71:186–93.

[3] Fotouhi M, Seidi S, Shanehsaz M, Naseri MT. Magnetically assisted matrix solid phase dispersion for extraction of parabens from breast milks. J Chromatogr A 2017;1504:17–26.

[4] Ghadiri A, Salemi A. Matrix solid-phase dispersion based on carbon nanotube coupled with dispersive liquid-liquid microextraction for determination of organochlorine pesticides in soil. J Chromatogr Sci 2017;55:578–85.

[5] Qiao FX, Du JJ. Rapid screening of clenbuterol hydrochloride in chicken samples by molecularly imprinted matrix solid-phase dispersion coupled with liquid chromatography. J Chromatogr B 2013;923:136–40.

[6] Medina-Dzul K, Carrera-Figueiras C, Perez-Padilla Y, Vilchis-Nestor RA, Lopez-Tellez G, Sanchez M, Munoz-Rodriguez D. SiO2/polyvinylimidazole hybrid polymer as a sorbent for extraction by matrix solid-phase dispersion (MSPD): synthesis, characterization, and evaluation. J Polym Res 2015;22:45.

[7] Llorent-Martinez EJ, Ortega-Barrales P, Fernandez-de Cordova ML, Ruiz-Medina A. Trends in flow-based analytical methods applied to pesticide detection: a review. Anal Chim Acta 2011;684:30–9.

[8] Karasova G, Lehotay J, Klodzinska E, Gadzala-Kopciuch R, Buszewski B. Comparison of several extraction methods for the isolation of benzoic acid derivatives from Melissa officinalis. J Liq Chromatogr Relat Technol 2006;29:1633–44.

[9] Kamal YT, Yusufoglu HS, Alam P, Foudah AI. Matrix solid-phase dispersion method coupled with high-performance liquid chromatography for the simultaneous quantification of selected compounds from freeze dry pomegranate juice. Acta Chromatogr 2018;30:153–7.

[10] Garcia-Jares C, Sanchez-Nande M, Lamas JP, Lores M. Profiling the fatty acids content of ornamental *Camellia* seeds cultivated in Galicia by an optimized matrix solid-phase dispersion extraction. Bioengineering 2017;4:87.

[11] Hong B, Wang Z, Xu T, Li C, Li W. Matrix solid-phase dispersion extraction followed by high performance liquid chromatography-diode array detection and ultra performance liquid chromatography-quadrupole-time of flight-mass spectrometer method for the determination of the main compounds from *Carthamus tinctorius* L. (Hong-hua). J Pharm Biomed Anal 2015;107:464–72.

[12] Visnevschi-Necrasov T, Barreira JCM, Cunha SC, Pereira G, Nunes E, Oliveira MBPP. Advances in isoflavone profile characterisation using matrix solid-phase dispersion coupled to HPLC/DAD in *Medicago* species. Phytochem Anal 2015;26:40–6.

[13] Xu J-J, Yang R, Ye L-H, Cao J, Cao W, Hu S-S, Peng L-Q. Application of ionic liquids for elution of bioactive flavonoid glycosides from lime fruit by miniaturized matrix solid-phase dispersion. Food Chem 2016;204:167–75.

[14] Barfi B, Asghari A, Rajabi M, Barfi A, Saeidi I. Simplified miniaturized ultrasound-assisted matrix solid phase dispersion extraction and high performance liquid chromatographic determination of seven flavonoids in citrus fruit juice and human fluid samples: hesperetin and naringenin as biomarkers. J Chromatogr A 2013;1311:30–40.

[15] Cao W, Hu S-S, Ye L-H, Cao J, Pang X-Q, Xu J-J. Trace matrix solid phase dispersion using a molecular sieve as the sorbent for the determination of flavonoids in fruit peels by ultra-performance liquid chromatography. Food Chem 2016;190:474–80.

[16] Aresta A, Cotugno P, Massari F, Zambonin C. Determination of isoflavones in soybean flour by matrix solid-phase dispersion extraction and liquid chromatography with UV-diode array detection. J Food Qual 2017. 5 pp.

[17] Cao W, Cao J, Ye L-H, Xu J-J, Hu S-S, Peng L-Q. Synthesis and application of mesoporous molecular sieve for miniaturized matrix solid-phase dispersion extraction of bioactive flavonoids from toothpaste, plant, and saliva. Electrophoresis 2015;36:2951–60.

[18] Ma S, Tu X, Dong J, Long P, Yang W, Miao X, Chen W, Wu Z. Soxhlet-assisted matrix solid phase dispersion to extract flavonoids from rape (*Brassica campestris*) bee pollen. J Chromatogr B 2015;1005:17–22.

[19] Liu X-G, Yang H, Cheng X-L, Liu L, Qin Y, Wang Q, Qi L-W, Li P. Direct analysis of 18 flavonol glycosides, aglycones and terpene trilactones in *Ginkgo biloba* tablets by matrix solid phase dispersion coupled with ultra-high performance liquid chromatography tandem triple quadrupole mass spectrometry. J Pharm Biomed Anal 2014;97:123–8.

[20] Tsochatzis ED, Tzimou-Tsitouridou R. Validated RP-HPLC method for simultaneous determination of tocopherols and tocotrienols in whole grain barley using matrix solid-phase dispersion. Food Anal Methods 2015;8:392–400.

[21] Hong Y, Chen L. Extraction of quercetin from Herba *Lysimachiae* by molecularly imprinted-matrix solid phase dispersion. J Chromatogr B 2013;941:38–44.

[22] Shen Q, Dong W, Yang M, Baibado JT, Wang Y, Alqouqa I, Cheung H-Y. Lipidomic study of olive fruit and oil using $TiO_2$ nanoparticle based matrix solid-phase dispersion and MALDI-TOF/MS. Food Res Int 2013;54:2054–61.

[23] Sun T, Li X, Yang J, Li L, Jin Y, Shi X. Graphene-encapsulated silica as matrix solid-phase dispersion extraction sorbents for the analysis of poly-methoxylated flavonoids in the leaves of *Murraya panaculata* (L.) Jack. J Sep Sci 2015;38:2132–9.

[24] Wang Z, Sun R, Wang Y, Li N, Lei L, Yang X, Yu A, Qiu F, Zhang H. Determination of phenolic acids and flavonoids in raw propolis by silica-supported ionic liquid-based matrix solid phase dispersion extraction high performance liquid chromatography-diode array detection. J Chromatogr B 2014;969:205–12.

[25] Cao J, Peng L-Q, Xu J-J. Microcrystalline cellulose based matrix solid phase dispersion microextration for isomeric triterpenoid acids in loquat leaves by ultrahigh-performance liquid chromatography and quadrupole time-of-flight mass spectrometry. J Chromatogr A 2016;1472:16–26.

[26] Peng L-Q, Li Q, Chang Y-x, An M, Yang R, Tan Z, Hao J, Cao J, Xu J-J, Hu S-S. Determination of natural phenols in olive fruits by chitosan assisted matrix solid-phase dispersion microextraction and ultrahigh performance liquid chromatography with quadrupole time-of-flight tandem mass spectrometry. J Chromatogr A 2016;1456:68–76.

[27] Du L-J, Huang J-P, Wang B, Wang C-H, Wang Q-Y, Hu Y-H, Yi L, Cao J, Peng L-Q, Chen Y-B, Zhang Q-D. Carbon molecular sieve based micro-matrix-solid-phase dispersion for the extraction of polyphenols in pomegranate peel by UHPLC-Q-TOF/MS. Electrophoresis 2018;39(17):2218–27.

[28] Du K, Li J, Bai Y, An M, Gao X-m, Chang Y-x. A green ionic liquid-based vortex-forced MSPD method for the simultaneous determination of 5-HMF and iridoid glycosides from *Fructus Corni* by ultra-high performance liquid chromatography. Food Chem 2018;244: 190–6.

[29] Du K, Li J, Tian F, Chang Y-x. Non-ionic detergent triton X-114 based vortex-synchronized matrix solid-phase dispersion method for the simultaneous determination of six compounds with various polarities from *Forsythiae Fructus* by ultra high-performance liquid chromatography. J Pharm Biomed Anal 2018;150:59–66.

[30] Djatinika R, Hsieh CC, Chen JM, Ding WH. Determination of paraben preservatives in seafood using matrix solid-phase dispersion and on-line acetylation gas chromatography-mass spectrometry. J Chromatogr B 2016;1036:93–9.

[31] Casado J, Castro G, Rodriguez I, Ramil M, Cela R. Selective extraction of antimycotic drugs from sludge samples using matrix solid-phase dispersion followed by on-line clean-up. Anal Bioanal Chem 2015;407:907–17.

[32] Tsai DY, Chen CL, Ding WH. Optimization of matrix solid-phase dispersion for the rapid determination of salicylate and benzophenone-type UV absorbing substances in marketed fish. Food Chem 2014;154:211–6.

[33] Villaverde-de-Saa E, Rodil R, Quintana JB, Cela R. Matrix solid-phase dispersion combined to liquid chromatography-tandem mass spectrometry for the determination of paraben preservatives in mollusks. J Chromatogr A 2016;1459:57–66.

[34] Aznar R, Albero B, Perez RA, Sanchez-Brunete C, Miguel E, Tadeo JL. Analysis of emerging organic contaminants in poultry manure by gas chromatography-tandem mass spectrometry. J Sep Sci 2018;41:940–7.

[35] Albero B, Sanchez-Brunete C, Miguel E, Tadeo JL. Application of matrix solid-phase dispersion followed by GC-MS/MS to the analysis of emerging contaminants in vegetables. Food Chem 2017;217:660–7.

[36] Sun XL, Wang JC, Li Y, Yang JJ, Jin J, Shah SM, Chen JP. Novel dummy molecularly imprinted polymers for matrix solid-phase dispersion extraction of eight fluoroquinolones from fish samples. J Chromatogr A 2014;1359:1–7.

[37] Ganan J, Morante-Zarcero S, Perez-Quintanilla D, Sierra I. Evaluation of mesoporous imprinted silicas as MSPD selective sorbents of ketoprofen in powder milk. Mater Lett 2017;197:5–7.

[38] Lozowicka B, Jankowska M, Rutkowska E, Hrynko I, Kaczynski P, Micinski J. The evaluation of a fast and simple pesticide multiresidue method in various herbs by gas chromatography. J Nat Med 2014;68:95–111.

[39] Cao YL, Tang H, Chen DZ, Li L. A novel method based on MSPD for simultaneous determination of 16 pesticide residues in tea by LC-MS/MS. J Chromatogr B 2015;998: 72–9.

[40] Purdesova A, Domotorova M. MSPD as sample preparation method for determination of selected pesticide residues in apples. Acta Chim Slov 2017;10:41–6.

[41] Tan DQ, Jin J, Li F, Sun XL, Dhanjai, Ni YW, Chen JP. Phenyltrichlorosilane-functionalized magnesium oxide microspheres: preparation, characterization and application for the selective extraction of dioxin-like polycyclic aromatic hydrocarbons in soils with matrix solid-phase dispersion. Anal Chim Acta 2017;956:14–23.

[42] Wang X, Du T, Wang J, Kou H, Du X, Lu X. Assessment of graphene aerogel matrix solid-phase dispersion as sample preparation for the determination of chlorophenols in soil. New J Chem 2018;42:6778–84.
[43] Li MY, Sun Q, Li Y, Lv M, Lin LF, Wu Y, Ashfaq M, Yu CP. Simultaneous analysis of 45 pharmaceuticals and personal care products in sludge by matrix solid-phase dispersion and liquid chromatography tandem mass spectrometry. Anal Bioanal Chem 2016;408: 4953–64.
[44] Rodriguez-Gonzalez N, Gonzalez-Castro MJ, Beceiro-Gonzalez E, Muniategui-Lorenzo S. Development of a matrix solid phase dispersion methodology for the determination of triazine herbicides in marine sediments. Microchem J 2017;133:137–43.
[45] Aznar R, Albero B, Sanchez-Brunete C, Miguel E, Martin-Girela I, Tadeo JL. Simultaneous determination of multiclass emerging contaminants in aquatic plants by ultrasound-assisted matrix solid-phase dispersion and GC-MS. Environ Sci Pollut Res 2017;24: 7911–20.
[46] Li D, Zhang Z, Li N, Wang K, Zang S, Jiang J, Yu A, Zhang H, Li X. Dispersant-assisted dynamic microwave extraction of triazine herbicides from rice. Anal Methods 2016;8: 3788–94.
[47] Zhang P, Ding J, Hou J, Zhao L, Chen YH, Ding L. Dynamic microwave assisted extraction coupled with matrix solid phase dispersion for the determination of chlorfenapyr and abamectin in rice by LC-MS/MS. Microchem J 2017;133:404–11.
[48] Gutierrez-Valencia TM, de Llasera MPG. On-line MSPD-SPE-HPLC/FLD analysis of polycyclic aromatic hydrocarbons in bovine tissues. Food Chem 2017;223:82–8.
[49] Huang Z, Pan XD, Huang BF, Xu JJ, Wang ML, Ren YP. Determination of 15 beta-lactam antibiotics in pork muscle by matrix solid-phase dispersion extraction (MSPD) and ultra-high pressure liquid chromatography tandem mass spectrometry. Food Control 2016;66: 145–50.
[50] Iparraguirre A, Rodil R, Quintana JB, Bizkarguenaga E, Prieto A, Zuloaga O, Cela R, Fernandez LA. Matrix solid-phase dispersion of polybrominated diphenyl ethers and their hydroxylated and methoxylated analogues in lettuce, carrot and soil. J Chromatogr A 2014; 1360:57–65.
[51] Zhan J, Li JD, Liu DH, Liu C, Yang GG, Zhou ZQ, Wang P. A simple method for the determination of organochlorine pollutants and the enantiomers in oil seeds based on matrix solid-phase dispersion. Food Chem 2016;194:319–24.
[52] Li J, Liu D, Wu T, Zhao W, Zhou Z, Wang P. A simplified procedure for the determination of organochlorine pesticides and polychlorobiphenyls in edible vegetable oils. Food Chem 2014;151:47–52.
[53] Rodriguez-Gonzalez N, Gonzalez-Castro MJ, Beceiro-Gonzalez E, Muniategui-Lorenzo S, Prada-Rodriguez D. Determination of triazine herbicides in seaweeds: development of a sample preparation method based on matrix solid phase dispersion and solid phase extraction clean-up. Talanta 2014;121:194–8.
[54] Rodriguez-Gonzalez N, Gonzalez-Castro MJ, Beceiro-Gonzalez E, Muniategui-Lorenzo S. Development of a matrix solid phase dispersion methodology for the determination of triazine herbicides in mussels. Food Chem 2015;173:391–6.
[55] Fan YB, Yin YM, Jiang WB, Chen YP, Yang JW, Wu J, Xie MX. Simultaneous determination of ten steroid hormones in animal origin food by matrix solid-phase dispersion and liquid chromatography-electrospray tandem mass spectrometry. Food Chem 2014; 142:170–7.

[56] Castro G, Carpinteiro I, Rodriguez I, Cela R. Determination of cardiovascular drugs in sewage sludge by matrix solid-phase dispersion and ultra-performance liquid chromatography tandem mass spectrometry. Anal Bioanal Chem 2018;410:6807–17.
[57] Celano R, Rodriguez I, Cela R, Rastrelli L, Piccinelli AL. Liquid chromatography quadrupole time-of-flight mass spectrometry quantification and screening of organophosphate compounds in sludge. Talanta 2014;118:312–20.

# Solid-phase analytical derivatizations

**20**

*Sanka N. Atapattu* [1], *Jack M. Rosenfeld* [2]
[1]CanAm Bioresearch Inc., Winnipeg, MB, Canada; [2]Department of Pathology and Molecular Medicine, McMaster University, Hamilton, ON, Canada

## 20.1 Introduction

Analytical method development typically consists of sampling, sample treatment, isolation of the target compounds, identification, quantification, and data handling. All steps carried out with samples before instrumental determination of the test compounds are considered to be part of the sample preparation step. Sample preparation is a critical step in analytical method development, as it enhances the reliability and accuracy of the analysis [1,2]. The main objectives of sample preparation are to reduce or eliminate matrix interference by separating the analyte of interest from matrix components, improve selectivity, enhance sensitivity, precision, and accuracy of analysis. Sample preparation is important, especially when analyzing biological and environmental samples in which target compounds are present at trace levels, and the sample matrix is usually not compatible with direct chromatographic analysis due to its complicated nature.

Liquid-liquid-extraction (LLE) is considered the classical sample preparation method which led to many analytical methods in the 1950s for the analysis of biological and environmental samples for organic analytes. Classical analytical methods based on LLE have many drawbacks; use of large quantities of toxic organic solvents, time consuming, labor intensive, low extraction efficiencies, and difficult to automate [3]. Current methods are miniaturized and more amenable to automations but laboratories analyzing large number of samples can still use substantial volumes of organic solvents [3]. Solid-phase extraction (SPE) based analytical methods were developed to overcome some of these drawbacks in LLE methods. In SPE, the analyte is partitioned between the aqueous liquid phase or gaseous phase of the sample onto a solid phase [4]. The analyte is sorbed to the solid phase and retained. In subsequent washings with organic solvents various components sorbed to the solid phase are desorbed.

While the established methods in LLE and SPE can isolate analytes and deliver them to the analytical instruments, some analytes still require modification via analytical derivatization (AD).This can increase sensitivity typically by 1–3 orders of magnitude compared with the native analyte as well as improving chromatographic separation [5–7]. Classically, this requires an extra step in the sample preparation that can be time consuming and uses additional chemical reagents to alter the structure of an analyte to produce a derivatized form more suitable for analysis.

Solid-Phase Extraction. https://doi.org/10.1016/B978-0-12-816906-3.00020-0

## 20.2 Solid-phase analytical derivatizations

Solid-phase analytical derivatization (SPAD) is a technique that combines the advantages of both SPE and AD [8−10]. This combination of techniques substantially reduces sample preparation time for methods that require the derivatization of analytes [11]. In SPAD, the analytes are converted to their derivative product on a solid phase or at the liquid-solid interface. Derivatized products of the analytes are isolated/concentrated on the solid phase and later desorbed by solvent or thermally for chromatographic analysis. In addition to simplifying sample preparation SPAD is readily amenable to automation.

SPAD provided conditions that can selectively derivatize various functional groups of an analyte. Organic acids fall into a broad class. Carboxylic acids can be selectively derivatized in the presence of phenols by reaction at neutral pH which does not ionize phenols and analytes with phenol functional groups can be derivatized in the presence of acids by selectively inhibiting derivatization of carboxylic acids at alkaline pH [10]. These reaction characteristics hold whether carboxylic acids are on different compounds or on the same compounds. In the case of carbonyls, SPAD enhances the reaction rate over solution chemistry so that both aldehydes and the usually slower reacting ketones are rapidly extracted/derivatized in one step. The derivatization incorporates chromophores, fluorophores, and electrophores for the purpose of detection. In the case of primary and secondary amines, SPAD functional groups that enhance sensitivity and/or lipophilicity [9,10]. Initially, SPAD was a batch process but transitioned into an automated microextraction/derivatization technique.

## 20.3 Solid-phase analytical derivatization of organic acids

Analysis of phenols and carboxylic acids on wall-coated open tubular columns (WCOT) by gas chromatography (GC) is somewhat compromised by loss of sensitivity, poor resolution, and peak tailing can result when accessible silanol groups of the fused-silica surface interact with phenols [12,13]. Conversion of phenols to ethers or carboxylic acids to esters alleviates these problems. Common derivatization methods for phenols and carboxylic acids are silylation, acylation, pentafluorobenzylation or injector port pyrolysis. Some of the reagents, solid-phase, and organic acid type combinations in SPAD methods reported in literature are listed in Tables 20.1 and 20.2.

Initially, such reactions were carried out in a solution which required extraction of the analyte and drying the isolate before derivatization. An alternative technique was extractive alkylation, which combined derivatization with liquid/liquid extraction but suffered from the same problems as LLE, such as formation of emulsions and use of toxic solvents.

SPAD provided a simple resolution to these problems. Most common SPAD methods are derivatizing on a highly crosslinked macroreticular resin with either

**Table 20.1** Phenol analysis by SPAD.

| Analyte | Reagent | Instrument | Solid phase |
|---|---|---|---|
| Lignin derived phenols | BSTFA [20] | GC-FID | SPME fiber |
| Surface water phenolic compounds | BSTFA [19] | GC-MS | SPME fiber |
| Alkylphenols, halogenated phenols | AA [15] | GC-MS | $C_{18}$ bonded silica |
| Alkylphenols and BPA | PFBBr [18] | GC-MS | Bond Elute PPL |
| Alkylphenols and halogenated phenols | PFBBr [14] | GC-MS | XAD-4 |
| Alkylphenols and BPA | PFPdy [25] | GC-MS | XAD-4 |
| Alkylphenols and halogenated phenols | PFPdy [16] | GC-MS | Oasis MAX |
| Phenolic compounds in air | AA [21] | GC-MS | PDMS SPME fiber |
| Alkylphenols and fatty acids in cow slurry | PDAM [22] | GC-MS | CAR/PDMS SPME fiber |
| Phenol, hydroquinone, catechol | HMDS [26] | GC-MS | SPME fiber |
| Phenols, acids and indoles in cow slurry | MTBSTFA [27] | GC-MS | SPME fiber |
| Estrogens | DSC [28] | LC-MS/MS | Oasis HLB column |
| 4-Hydroxynonenal, testosterone and epitestosterone | AA [24] | GC-MS | PDME coated stir bar |
| Alkyl phenols | BSFTA [17] | GC-MS | SPE column |
| Recreational drugs, THC, THCCOOH | MTBSTFA [29] | GC-MS | SPME on-fiber |
| THC, cannabidiol and cannabinol | MSTFA [23,30] | GC-MS | On-fiber |

*AA*, acetic anhydride; *DSC*, dansyl Chloride; *HMDS*, hexamethyldisilazane; *MSTFA*, *N*-methyl-*N*-trimethylsilyltrifluoroacetamide; *MTBSTFA*, N-(tert-butyldimethylsilyl)-N-methyltrifluoroacetamide; *PDAM*, 1-pirenyldiazomethane; *PFBBr*, pentafluobenzylbromide; *PFPdy*, pentafluoropyridine; *THC*, Δ9-tetrahydrocannabinol; *THCCOOH*, 11-nor-9-carboxy- Δ9-tetrahydrocannabinol

polystyrene divinyl benzene or methacrylate esters, strong anion exchange resin, SPME on-fiber coated with organic polymer, and stir bar polymer coatings.

Highly crosslinked macroreticular polymeric, such as resins in the XAD series provide a lipophilic surface that can be impregnated with reagent and can sorb ionized phenols from alkaline pH to a degree that is sufficient for rapid irreversible reaction. The resin is also stable at both acidic and alkaline pH. In SPAD, the solid phase is impregnated with a suitable derivatizing reagent, usually, before the sample is added.

At alkaline pH, phenolate anions are diffused from the bulk aqueous phase to the surface of the solid phase. High-molecular-weight phenolate anions are sorbed onto the solid surface due to strong hydrophobic interactions, but polar low-molecular-weight phenols, which form weak interactions with the solid phase are poorly adsorbed onto XAD solid phases. As a result, polar low-molecular-weight phenols, therefore, show low yields in SPAD when using crosslinked macroreticular polymeric resins. One approach to increase the yield of polar low-molecular-weight phenols is to use anionic surfactant or a phase transfer catalyst (PTC). The ionized phenolate forms a lipophilic ion pair with a counter-ion of the opposite charge. The lipophilic ion pair formation enhances the adsorption of analyte onto the crosslinked macroreticular polymeric resin coated with a derivatization reagent.

The effect of a PTC is complex [14]. While the use of a PTC would increase the yield for derivatization of hydrophilic analytes with lower log P values, a PTC, however, inhibits pentafluorobenzylation of high-molecular-weight analytes with higher log P values. In order to derivatize both lipophilic and hydrophilic analytes in one method, a two-step derivatization method can be used. In the first step, lipophilic analytes are derivatized without a PTC. In the second step, a PTC is added to increase the derivatizing yield of hydrophilic low-molecular-weight analytes. Although it is a two-step procedure, it is a "one pot" process. The effect of counter-ion on analytes with higher log P values provides specificity for analysis targeting low-molecular-weight compounds.

There are two ways to develop SPME on-fiber derivatization methods. One is to expose the SPME fiber for phenol adsorption in an aqueous media or air. The adsorbed phenol can then be derivatized exposing the fiber to the derivatizing reagent and finally thermal desorption of the SPME fiber in a GC injection port of analysis. The second approach is to coat the derivatizing reagent on the SPME fiber by immersing in a solution of the derivatizing reagent. Phenol analytes are simultaneously extracted and derivatized on-fiber and injected to a GC inlet for thermal desorption and analysis.

Rate adsorption of analyte onto the solid phase in SPAD methods can have different mechanisms depending on the type of solid phase used. In an SPAD method where an XAD phase is used, the shaking speed has an effect on bringing the analyte and reagent into contact. When an SPE solid phase is used in an SPAD method, the mobile phase flow rate is a critical parameter to bring the analyte and reagent coated in the solid phase into contact with each other.

In SPME on-fiber derivatization methods, where the fiber is immersed in a liquid, the rate of bringing the analyte and the reagent into contact with each other depend on two factors. If the liquid phase is not stirred, diffusion determines the adsorption of analyte onto the fiber, and thereby bringing the analyte and reagent into contact with each other, whereas when the liquid phase is stirred, convection and diffusion are the two factors that bring the analyte and reagent into contact with each other on an SPME fiber. In SPME head space on-fiber derivatization methods, diffusion determines the analyte and the reagent coming into contact with each other.

An alternative to SPME is SBSME in which, typically, a 0.5−1 mm thick polydimethylsiloxane coating of stir bar is the extraction phase. In SPAD methods, first the analytes are extracted from aqueous sample onto the stir bar coated stationary phase. Next, the stir bar coating is exposed to the derivatizing reagent for analyte derivatization. Finally, the derivatives are desorbed thermally or use a solvent for analysis. Also a derivatizing reagent coated stir bar could be used for simultaneous extraction and derivatization of phenols.

## 20.3.1 Phenols

Analytical methods for phenol analysis reported in literature include both GC and LC based methods. However, most common methods reported are GC-based methods equipped with mass spectrometry detectors using wall-coated open-tubular (WCOT) column. Hydroxyl functional groups in phenol analytes can interact with accessible silanol groups on the fused-silica surface of WCOT columns resulting in the tailing of peaks, reduced resolution, and sensitivity in detection. SPAD provides a simple solution to these issues in phenol analysis converting the hydroxyl group to an ether group in the derivatization step. Most common solid phases used in SPAD include crosslinked macroreticular polymeric resins in the XAD series, strong anion exchange resin, SPME fibers, stir bar polymer coatings, and SPE cartridges [9].

### 20.3.1.1 Anion exchange resin and solid-phase extraction cartridges

Use of anion exchange resins is a common practice to isolate phenolates from the sample matrix before derivatization. First, the phenolates are immobilized in the resin followed by a drying and derivatizing steps. Li and Lee [15] coated the ionic surfactant cetyltrimethylammonium bromide (CTAB) on a $C_{18}$ boned SPE cartridge. The surfactant CTAB is adsorbed on the $C_{18}$ bonded silica solid-phase acting as ion-exchange sites. Phenolates were immobilized by the surfactant molecules from aqueous samples. After the phenolates were adsorbed, the SPE cartridge was dried using a stream of nitrogen gas, and then derivatized using acetic anhydride on-column before GC/MS analysis. This method was validated in humic acid and surrogate sea water conditions.

This method can be modified using a strong anion exchange SPE cartridge instead of coating a surfactant on the solid phase. Kojima et al. [16] passed an alkaline solution of low-molecular-weight alkyl and halogenated phenols through a strong anion exchange Oasis MAX SPE cartridge for the analytes to get adsorbed. Next, the column was dried using a stream of nitrogen and derivatized on column passing a solution of the derivatizing reagent pentafluoropyridine (PFPdy). Jonsson et al. [17] utilized a Discovery DSC-18 SPE cartridge to adsorb lipophilic alkylphenols from fish bile. An ionic surfactant or a strong anion exchange cartridge was not required in their work as high-molecular-weight more lipophilic alkylphenols were readily adsorbed onto the solid phase due to strong hydrophobic interactions. After adsorption of alkylphenols the cartridge was dried using the stream of nitrogen gas and derivatized on-column with the derivatizing reagent BSTFA for analysis. Another approach is to impregnate the derivatizing reagent on a solid phase followed by sorption of the phenolate on to the surface for derivatization. Kuklenyik et al. [18] used tetrabutylammonium (TBA) cation as a PTC to transfer ionized alkylphenols of from alkalinized urine onto a solid phase impregnated with pentafluorobenzyl bromide (PFBBr). The phenolates reacted with the reagent to form the pentafluorobenzyl (PFB) ethers of the analytes.

### 20.3.1.2 SPME on-fiber derivatizations

Demands for reduced cost and time of analysis led to a focus on developing fully automated analytical methods. The instrumentation-exploited SPME fibers served as the solid phase for AD. Automation included the following steps: SPME fiber preconditioning: sample extraction: exposing the analytes that were loaded on the fiber to the derivatizing reagent for on -fiber derivatization; thermal desorption of derivatized analytes in the GC inlet; and finally, cleaning and postconditioning the fiber for the next sample analysis. Yuan and coworkers described an automated method for analysis of 11 chloro- and bromo-phenolic endocrine disruptive compounds (EDC) [19]. The reported method included: SPME fiber preconditioning by heating to 290°C; sorption of analyte in the headspace: exposure of sorbed analyte to BSTFA in a second head space mode extraction for on-fiber derivatization and finally thermal desorption of silylated EDC derivatives for GC-MS analysis. The proposed method was validated for 11 EDC compounds in surface water and reported LLOD values 8.8–42.9 ng/L.

Pinto et al. [20] focused on the need for green chemistry in their determination of lignin derived phenols in sediments. The authors coated BSTFA on a polyacrylate fiber SPME fiber. The solid sample was heated in a vial to vaporize the target analytes, which were sorbed and derivatized on the fiber. Derivatized phenols were analyzed by GC-FID.

Es-haghi et al. [21] used SPME on fiber derivatization method to detect phenol in air samples. After the SPME fiber was conditioned, it was exposed to a vial containing gaseous phenol for adsorption. The adsorbed phenol was derivatized exposing the fiber to the derivatizing reagent acetic anhydride in a 2 mL glass vial. The derivatization reaction took place for 1 min at 50°C. Finally the fiber was thermally desorbed onto a GC injection port for analysis immediately after extraction. Similar on-fiber derivatization approaches have been reported for the analysis of polar phenol compounds in urine and cow slurries.

Larreta et al. [22] reported an HS-SPME on fiber derivatization method for fatty acid and phenol analysis in cow slurries. In this method, the SPME fiber was first immersed in a solution of 1-Phenyldiazomethane, the derivatizing reagent. Next, the SPME fiber was inserted into a vial containing the sample. Phenols, indoles, and fatty acids were simultaneously extracted and derivatized on-fiber in headspace mode at for 15 min at 35°C. Finally, the derivatized analytes were thermally desorbed at 300°C for 9 min in a GC injection port before analysis.

Although SPME on-fiber derivatization has been widely used it suffers from two weaknesses: (i) the fragile fused silica and (ii) the outer surface of the stationary phase coating is exposed when extended through the syringe needle. To overcome these issues, the solid-phase dynamic extraction (SPDE) was developed. In this method, the stationary phase is coated inside a stainless steel needle. Musshoff and co-workers [23] extracted cannabinoids in HS-SPDE mode and derivatized the adsorbed analytes exposing the fiber to MSTFA. When SPDE and SPME were compared in cannabinoid, the analysis showed similar results.

Another alternative to SPME is SBSE. In SPAD methods similar to SMPE on-fiber derivatization, first, the analytes are extracted from aqueous sample to the stir bar coated stationary phase. Next, the stir bar coating is exposed to the derivatizing reagent for analyte derivatization. Finally, the derivatives are desorbed thermally or using a solvent for analysis. Sandra and co-workers [24] extracted testosterone and epitestosterone from human urine to a PDMS coated stir bar. After extraction, the stir bar was dried and exposed to acetic anhydride for derivatization. In the final step, the stir bar was placed in a glass desorption tube of an autosampler for GC analysis.

#### *20.3.1.3 Derivatizing on a poly styrene–divinylbenzene crosslinked macroreticular resins*

XAD phases have been extensively used in SPAD method development in a wide range of applications. Usually, before the sample is added, the solid phase is impregnated with a suitable derivatizing reagent. Analyte is diffused from the bulk aqueous phase to the surface of the solid phase. The derivatization reaction occurs on the surface of the solid phase. High-molecular-weight lipophilic analytes are readily adsorbed on to the solid surface due to strong hydrophobic interactions. However, polar low-molecular-weight analytes form weak interactions with the solid phase and are poorly adsorbed on to XAD solid phases. One solution is the use of a PTC; the ionized analyte forms a lipophilic ion pair with a counter-ion of the opposite charge [14]. As expected, the addition of a counter-ion increased the yields of hydrophilic phenols, carbonyls, and acids on XAD resins. However, counter-ions inhibited the pentafluorobenzylation of high-molecular-weight alkylphenols, chlorophenols, and carboxylic acids on the XAD-4 resin. This could be due to slow diffusion of bulky ion-pairs. To overcome these issues, a two-step derivatization method was reported [14]. In the first step of the method, the lipophilic analytes were derivatized without any counter-ions. After a reaction time of 20 min, TBA counter-ion was added to the reaction mixture and derivatized for further 15 min. This two-step derivatization method showed good yields for both hydrophilic and lipophilic alkylphenols, chlorophenols, and carboxylic acids. The method was validated in humic acid, surrogate sea water, and surrogate lake water conditions.

Lipophilic phenols have the ability to form strong hydrophobic interactions with solid supports. Thus, these compounds are readily adsorbed on solid phases. Some of these compounds posses more than one functional group, and therefore, require more than a derivatizing reagent. Analysis of hormones, cannabinoids, and biomarkers with the hydroxyl functional group is important in forensic, bio-medical, and environmental sample analysis.

The earlier SPAD work on cannabinoids was carried out in XAD-2 phases. First, the conditioned XAD-2 phase was impregnated with the derivatizing reagent. After adding the sample solution to the XAD-2 phase, the container was shaken for a specified period of time for the derivatization reaction to occur. Next, the supernatant was removed and the XAD-2 phase was dried and extracted with an organic solvent of

GC analysis. The derivatizing reagent PFBBr has been widely used in SPAD applications on XAD resins. The reagent PFBBr forms pentafluorobenzyl ethers with phenols and pentafluorobenzyl esters with carboxylic acids. This reagent has a high boiling point of 174°C making it difficult to remove the excess reagent. It is well known that PFBBr excess reagent degrade GC columns. An alternative is the use of PFPdy reagent instead of PFBBr [16,25]. The reagent PFPdy forms pentafluoropyridyl ethers with phenols. Atapattu et al. [25] reported an SPAD method for the determination of four endocrine disrupting chemicals, namely, 4-tert-butylphenol, 4-tert-octylphenol, bisphenol A, and 17β-estradiol using PFPdy as the derivatizing reagent. These lipophilic phenols did not require any PTC, as they form strong hydrophobic interactions with the XAD-4 resin. The method was validated with human urine samples.

Most of the recent SPAD work on cannabinoids was carried out with SPME on-fiber derivatization [10]. Cannabinoids can be easily extracted from biological and environmental samples in HS-SPME mode. After extraction, the fiber is exposed to a suitable derivatizing reagent for on-fiber derivatization. Finally, the fiber is thermally desorbed in a GC inlet for analysis. Cannabinoids in human hair and waste water samples have been analyzed successfully in this mode.

## 20.3.2 Carboxylic acids

Carboxylic acids are present in a wide range of environmental and biological samples. In most GC and LC methods reported, carboxylic acids are derivatized to esters. In GC methods, derivatization improves volatility and interaction with accessible silanol groups in WCOT columns. Objectives of derivatization in LC-based methods are to add a chromophere, fluorophore, or an electrophore in order to improve sensitivity and selectivity.

Pentafluobenzylbromide, *N*-methyl-*N*-trimethylsilyltrifluoroacetamide, and N-(tert-butyldimethylsilyl)-N-methyltrifluoroacetamide are the most commonly used analytical derivatizing reagents for analytes with carboxylic acid functionalgroups. Many different SPAD based methods have been reported in recent literature for derivatization ofcarboxylic acids after isolating and using strong anion exchange resins, derivatizing on a poly styrene–divinylbenzene crosslinked macroreticular resin, SPME on-fiber, hollow fiber membranes, metal surfaces, and GC injection syringes. Some of the reagents, solid-phase, and analyte combinations are listed in Table 20.2 in SPAD methods.

Many SPAD-based analytical methods have been developed to isolate, derivatize, and extract fatty acids in a single step on poly styrene–divinylbenzene crosslinked macroreticular resins and solid phases of SPE cartridges. In methods where XAD phases were utilized, the derivatizing reagent is impregnated on the solid phase and the aqueous sample was in contact with the solid phase. High-molecular-weight acids react at a faster rate than low-molecular-weight acids in this approach. Quaternary amines can be used as phase transfer catalysts to increase derivatize yields of low-molecular-weight carboxylic acids. The ion-pair formed brings the carboxylate onto the solid surface to react due to stronger intermolecular interactions with the solid phase. Although derivatization of hydrophilic analytes with low-log P values increase

**Table 20.2** Organic acid analysis by SPAD.

| Analyte | Reagent | Instrument | Solid phase |
|---|---|---|---|
| Volatile fatty acids in cow slurries | MTBSTFA [27] | GC-MS | SPME fiber |
| Recreational drugs in hair | AA [37] | GC-MS | SPME fiber |
| Organic acids | Methyl iodide [32] | GC-MS | Anion exchange resin |
| Fatty acids | PDAM [38] | GC-FID | SPME fiber |
| Fatty acids | Diazomethane [39] | GC-ECD, GC-MS | SPME fiber |
| Phytocannabinoids | MSTFA [40] | GC-MS | Buccal swabs |
| Acidic phytohormones | BTA [41] | UPLC-MS/MS | MSPE |
| Fatty acids and phenols | PFBBr [14] | GC-MS | XAD-4 |
| Peracetic acid in airborne samples | MTSO [42] | GC-MS | SPME fiber |
| Ursolic and oleanic acid | CPR [43] | LC-MS | Metal oxide |
| Volatile organic acids | BSTFA [36] | GC-MS | In syringe derivatization |
| Volatile fatty acids in cow slurries | PDAM [22] | GC-MS | SPME fiber |
| Fatty acids | PFBBr [44,45] | GC-ECD | XAD-2 |
| Metabolites in human plasma | Methoxyamine/ MSTFA [46] | GC-MS | Two ion exchange resins |
| Chlorinated acid herbicides | Methyl iodide [34] | GC-ECD JCA | Anion exchange resin |
| Organic acids | Methyl iodide [47] | GC-FID | Anion exchange resin |
| Antiinflammatory drugs | MTBSTFA [48] | GC-MS | SPME fiber |
| Acidic herbicides | BF3-MeOH/TMSD [49] | GC-ECD | C18/PS-DVB |
| Fatty acids | Methanol [50] | GC-MS | SPME fiber |
| Alkyphosphonic acids | BSTFA [35] | GC-MS | SPE anion exchange disk |
| Pharmaceutically active compounds | MTBSTFA [51] | GC-MS | Hollow fiber |

*Continued*

**Table 20.2** Organic acid analysis by SPAD.—cont'd

| Analyte | Reagent | Instrument | Solid phase |
|---|---|---|---|
| Volatile fatty acids | PFBBr [33] | GC-MS | Anion exchange resin |
| Cannabinoids | FMTPS [52] | MALDI-ESI | MALDI glass plate |
| Fatty acids and amino acids | MTBSTFA [31] | GC-MS | Metal surface |
| Pharmaceutical and personal care products | MTBSTFA [53] | GC-MS | SMPE fiber |

*BSTFA*, (bis(trimethylsilyl)-trifluoroacetamide; *BTA*, 3-bromoactonyltrimethylammonium bromide; *CPR*, 2′-carbonyl-piperazine rhodamine B; *FMTPS*, 2-fluoro-1-methylpyridinium ptolunesolfonate; *MSPE*, magnetic solid-phase extraction; *MTBSTFA*, N-methyl-N-(tert-butyldimethylsilyl)trifluoroacetamide; *MTSO*, methyl *p*-tolyl sulfoxide; *PDAM*, 1-pirenyldiazomethane; *PFBBr*, pentafluobenzylbromide; *TMSD*, trimethylsilyldiazomethane.

the yields using a PTC inhibited pentafluorobenzylation of high-molecular-weight analytes with higher log P values. To overcome these issues, a two-step derivatization procedure can be used, where lipophilic analytes are derivatized in the first step, and in the second step, a PTC is added to increase the yield of hydrophilic low-molecular-weight analytes.

SPAD based on-fiber SPME methods have been reported in literature. In the first approach, the SPME fiber impregnated with the reagent is immersed in the aqueous solution for simultaneous isolation, derivatization, and extraction. Alternatively, the analytes are first extracted from the aqueous solution, and then the fiber is exposed to the chemical reagent for on-fiber derivatization to an ester and then finally, thermally desorbed for analysis.

There are few reports on carboxylic acids having been derivatized of on a metal surface. A NASA-based research group reported an interesting SPAD variant, in which the organic and amino acids were derivatized on a derivatization cup made out of stainless steel and then analyzed by GC/MS [31]. The acids were derivatized with the reagent MTBSTFA at 300°C for several minutes. The reported method has potential applications in analyzing mineral samples in space missions in the future.

The use of strong anion exchange resin to isolate carboxylates is a common practice in Analytical Chemistry. The derivatization after isolating the carboxylates can have two approaches (i) derivatization on the anion exchange resin (SPAD) after drying the resin or (ii) derivatizing after eluting the carboxylate from resin. However, the analyte elution followed by derivatization is a tedious process and time consuming; thus, the SPAD approach is more widely applied. Chatfield et al. [32] isolated chlorinated acid herbicides on an anion exchange resin. The isolated carboxylates were derivatized using supercritical carbon dioxide containing alkylation reagent methyl iodide. Later derivatives were eluted in a supercritical fluid extraction (SFE) system as methyl esters. Cummins and Wells [33] reported a similar method to determine low-molecular-weight fatty acids using the derivatizing reagent pentafluorobenzyl bromide. Field and Monohan [34] demonstrated an "in-vial" elution technique for the determination of chlorophenoxy acid herbicides in aqueous samples. A strong anion-exchange solid-phase extraction disks were used to concentrate the chlorinated acid carboxylate. The chlorinated acids were eluted from the disk and derivatized with methyl iodide in a single step to their methyl esters with an in-vial elution technique. In later reports, Subramaniam et al. [35] used an anion exchange disk and in-vial elution technique to determine the alkylphosphonic acid nerve agent markers. In their work, a wide range of alkylphosphonic acids were derivatized using BSTFA as the chemical reagent.

Many SPAD-based methods have been developed derivatizing and extracting fatty acids in a single step on poly styrene–divinylbenzene crosslinked macroreticular resins and solid phases of SPE cartridges [9]. In methods where XAD phases were utilized, the derivatizing reagent was impregnated on the solid phase and the aqueous sample was made to contact with the solid phase. High-molecular-weight acids react at a faster rate than low-molecular-weight acids in this mode. In order to increase the reaction rates for low-molecular-weight analytes counter-ions are used. The ion-pair formed brings the carboxylate on to the solid surface to react due to stronger intermolecular interactions with the solid phase. However, when analyzing a mixture of low- and high-molecular-weight acids, this approach could cause difficulties. High-molecular-weight carboxylates form a bulky ion pair with counter-ions, as a the result hinders the diffusion process. To overcome this issue a two-step derivatization method was reported. First, high-molecular-weight acids were reacted without any counter-ions, and in the second step, low-molecular-weight acids were reacted with the aid of a catalyst.

Sha et al. [36] derivatized volatile organic acids in a syringe with BSFTA. The volatile organic acids were extracted from aqueous samples in a head space SDME mode. After extraction, the acids were in-syringe derivatized. This method was validated by analyzing tobacco samples. A NASA-based research group recently reported an interesting SPAD variant in which organic and amino acids were derivatized on a derivatization cup made out of stainless steel [31]. The acids were derivatized with the reagent MTBSTFA at 300°C for several minutes. The reported method has potential applications in analyzing mineral samples in space missions in the future.

## 20.4 Solid-phase analytical derivatization of aldehydes and ketones

Analysis of aldehydes and ketones are important in environmental chemistry and biomedical applications. Most of these compounds are found in such samples at trace levels challenging the existing analytical methods because their concentrations are lower than most instrument limits of detection, and in addition, some important carbonyls are highly water soluble (e.g., acetone and glucose). Recent research has focused on new sample preparation and derivatization methods to improve selectivity, sensitivity, and speed of analysis. SPAD based GC and LC methods have been reported in literature for aldehyde and ketone analysis.

Solid phases explored in SPAD methods for simultaneous derivatization and extraction include monolith frits, porous polypropylene frits with polydimethylsiloxane

(PDMS) coating, SPME fibers, microfluidic chips, magnetic solid-phase extraction, and XAD phases.

Aldehyde and ketone compounds present in the atmosphere are mainly a result of anthropogenic activities and natural processes. Airborne carbonyl compounds are usually derivatized with the reagent DNPH impregnated on a solid supporting SPAD-based methods. The formed 2, 4-dinitrophenylhydrazone derivatives of aldehyde and ketone compounds are extracted to a solvent and then analyzed by HPLC/UV.

There are two drawbacks in analyzing airborne aldehyde and ketone compounds using 2, 4-dinitrophenylhydrazone as a derivatizing reagent. One is, the ozone in the atmosphere reacts with 2, 4-dinitrophenylhydrazone derivatives interfering with airborne aldehyde and ketone analysis, and the other is, E- and Z-geometrical isomer presence of 2, 4-dinitrophenylhydrazone derivatives giving multiple peaks in HPLC analysis [54–58].

In order to avoid the ozone reacting with 2, 4-dinitrophenylhydrazone derivatives, research groups have explored having a two-bed cartridge system to trap the ozone [55]. The first cartridge bed is with an ozone scrubber impregnated silica particles and the second cartridge bed with DNPH impregnated silica particles. Ozone scrubbers 1, 2-bis-(4-pyridyl) ethylene, 1, 2-bis-(2-pyridyl) ethylene and hydroquinone have been explored by research groups. Hydroquinone and 1, 2-bis-(2-pyridyl) ethylene showed more promising results than 1, 2-bis-(4-pyridyl) ethylene. To convert the C=N double bond of the 2, 4-dinitrophenylhydrazone derivatives to give a C–N single bond the reagent 2-picoline borane has been explored successfully [56].

In most GC-based SPAD methods, the reagents pentafluorophenylhydrazine and O-(2, 3, 4, 5, 6-pentafluorobenzyl) hydroxylamine are impregnated on a solid phase [9,10]. Derivatized aldehydes and ketones are then desorbed and then injected to a GC for analysis. This GC approach is preferred over HPLC/UV based methods where DNPH impregnated solid materials are used. The approach of impregnating DNPH on a solid material and analyzing by HPLC/UV suffer from several disadvantages; limited HPLC peak capacity, overlap of peaks due to different isomers, and difficulties in analyzing high-molecular-weight carbonyl compounds. Methods based on impregnating PFPH on solid material and analyzing using GC/MS have been compared with classical DNPH coated solid sorbent HPLC/UV analysis methods. Results showed that GC/MS methods are superior than classical methods based on HPLC/UV especially for high-molecular-weight carbonyl compounds.

Use of dansyl hydrazine as a reagent has many advantages since it can act as a chromophore, a fluorophore, an electrophore or a preionized molecule improving sensitivity in mass spectrometry detection.

Magnetic solid-phase extraction (MSPE) has many advantages over conventional SPE methods as the magnetic or magnetizable sorbent can be dispersed in sample solution, and thereby increases the interfacial area between sorbent and sample [41]. After extracting the analytes, the magnetic sorbent can be separated from the sample solution by applying an external magnetic field. SPAD methods have been extended to MSPE. In a recently published research report, several acidic phytohormones were derivatized on $TiO_2$/magnetic hollow mesoporous silica spheres using

3-bromoactonyltrimethylammonium bromide as the derivatizing reagent [41]. After derivatizing the analytes, the magnetic sorbent was separated from the sample solution by applying a magnetic field and then analyzed the derivatized acidic phytohormones by UPLC-MS/MS (Table 20.3).

**Table 20.3** Aldehydes and ketones analysis by SPAD.

| Analyte | Reagent | Instrument | Solid phase |
|---|---|---|---|
| Carbonyls in vapor | PFBHA [59] | GC-MS | SPME fiber |
| Formaldehyde | Acetyacetone [60] | HPLC-DAD | Hollow fiber |
| 4HNE | PFBHA [61] | GC-MS | Stir bar |
| Hexanal, heptanal | DNPH [62] | HPLC-DAD | PDMS frit |
| Airborne carbonyls | PFBHA [63] | GC-MS | Tenax sorbent packed tube |
| Airborne carbonyls | PFPH [64] | GC-MS | Tenax sorbent packed tube |
| Hexanal, heptanal | DNPH [65] | HPLC-DAD | Monolith frit |
| Airborne carbonyls | DNPH [56] | HPLC-DAD | DNPH-coated silica particles |
| Carbonyls in cigarette smoke | DNPH [54] | HPLC-DAD | DNPH-coated silica particles |
| Ozone & airborne carbonyls | BPE & DNPH [55,57,58] | HPLC-UV | Impregnated silica particles |
| Hexanal, heptanal | DNPH [66] | HPLC-DAD | Monolith capillary column |
| Airborne carbonyls | DNPH [67] | HPLC-DAD | GC capillary column |
| Airborne carbonyls | PFPH/PFBHA [68] | GC-MS | Microfluidic chip |
| Aldehydes | PFBHA [69] | GC-MS | SPME fiber |
| Aldehydes | PFPH/PFBHA [70] | GC-FID | SPME fiber |
| Malondialdehyde | DANSYL hydrazine [71,72] | HPLC-fluorescence | XAD-2 |
| Butanone, 2-Pentanone Malondialdehyde | DNPH [73] | HPLC-UV | XAD-2 |

*4HNE*, 4-hydroxynonenal; *DNPH*, 2,4-dinitrophenylhydrazine; *PFBHA*, *O*-(2,3,4,5,6-pentafluorobenzyl)hydroxylamine; *BPE*, *trans*-1,2-bis-(4-pyridyl)ethylene; *PFPH*, pentafluorophenyl hydrazine.

## 20.5 Solid-phase analytical derivatization of amines

In amine compounds, the hydrogen atom is attached to the nitrogen atom, and therefore, the amine group has a partial positive charge due to the electro-negativity difference between the two atoms. The analysis of amine compounds by GC using WCOT columns is a difficult task as amines can interact with accessible silanol groups on the fused-silica surface and result in tailing of peaks, especially at low analyte concentrations and higher temperatures, thus lowering the sensitivity and selectivity. Derivatization of amines prior to gas chromatographic analysis on WCOT columns reduces the interaction with accessible silanol groups, thus improving the chromatography. Derivatization also improves the volatility, selectivity, sensitivity, and separation. Most amines found in environmental and biological samples lack strong chromophores except in the case of aromatic amines. Low molar absorptivity coefficients especially due to lack of chromophores and little or no electrochemical or fluorescence activity are challenges in amines analysis using HPLC methods. Therefore, analytical derivatization of amines are important before HPLC and GC analysis.

Pentafluorobenzoyl chloride, 9-fluorenylmethyl chloroformate, 2,3,4,5-pentafluorobenzaldehyde, propyl chloroformate, and trifluoroacetic anhydride are some of the common derivatizing reagents used in primary and secondary amine analysis in SPAD-based methods. Only few reports are found in literature where tertiary amines have been derivatized before analysis. Pentafluorobenzyl chloroformate, 9-fluorenylmethyl chloroformate and ethyl chloroformate are few reagents found in literature used in tertiary amines derivatization. Many solid-phase and chemical reagent combinations have been reported in SPAD-based applications in amine analysis. Most of these reported methods use an SMPE fiber to derivatize amines.

Analysis of organic ionic species is very important in metabolomics. A recent research report demonstrated the simultaneous derivatization of organic cations and anions in an SPAD based metabolite analysis method [46]. To analyze a wide range of metabolites, the authors combined a strong anion-exchange resin with a strong anion exchange resin. A wide range of amino acids and organic acids were derivatized first by methoximation, and then followed by trimethylsilylation. The developed method was validated analyzing amino acids and organic acids in human plasma.

Other SPAD methods reported in recent literature include in-tube derivatization, hollow fiber derivatization, simultaneous extraction, and derivatization on SPE cartridges, microfluidic on-chip derivatization, and in-syringe methods [9] (Table 20.4).

**Table 20.4** Amine analysis by SPAD.

| Analyte | Reagent | Instrument | Solid phase |
|---|---|---|---|
| Sarcosine | DMSCL [74] | HPLC-UV | Hollow fiber |
| Aliphatic amine | Isobutyl chloroformate [75] | GC-MS | SPME fiber |
| Aromatic amines | Allyl isothiocyanate [76] | GC-MS | SPME fiber |
| Aliphatic amines | PFBOC [77] | GC-MS | In syringe |
| Methylamine | FMOC [78] | HPLC-UV | SPME fiber |

**Table 20.4** Amine analysis by SPAD.—cont'd

| Analyte | Reagent | Instrument | Solid phase |
|---|---|---|---|
| Aliphatic amines | FMOC [79] | HPLC-UV | SPME fiber |
| Primary amines | PFBAY [80] | GC-MS/MS | SPME fiber |
| Trimethylamine | FMOC [81] | HPLC-UV | SPME fiber |
| Diamines | TFAA [82] | GC-MS | SPME fiber |
| Biogenic amines | Dansyl chloride [83] | HPLC-UV | Polypropylene hollow fiber |
| Dimethylamine | FMOC [84] | HPLC-UV | GC column |
| Aliphatic amines | FMOC [85] | HPLC-UV | GC column |
| Amphetamines | Fluorescamine [86] | CE-fluorescence | Oasis HLB |
| Aromatic amines | TFANA [87] | GC-MS | Oasis HLB |
| Amphetamines | PCF [88] | GC-MS | Monolithic silica spin column |

*DMSCL*, 4-dimethylarminoazobenzene-4-sulfonyl chloride; *DTAF*, dichlorotriazine fluorescein; *FMOC*, 9-fluorenylmethyl chloroformate; *PCF*, propyl chloroformate; *PFBAY*, 2,3,4,5-pentafluorobenzaldehyde; *PFBOC*, pentafluorobenzoyl chloride; *TFAA*, 1, 1, 1-trifluoroacetylacetone; *TFANA*, trifluoroacetic anhydride.

## 20.6 Conclusions

Since the first reported work on SPAD in 1986, SPAD has emerged over the years as an important sample preparation method for a wide range of analyte analysis. Analytical derivatization in SPAD methods expands the breath of analytical detectors and improves chromatographic separation and selectivity. Most chemical reagents currently used in analytical derivatizations are toxic and considered to be not environmentally friendly. It is worthwhile to research the use of nontoxic or less toxic derivatizing reagents; thus, SPAD methods would be more attractive as a green analytical chemistry sample preparation method.

## References

[1] Ferreira AM, Laespada ME, Pavon JL, Cordero BM. In situ aqueous derivatization as sample preparation technique for gas chromatographic determinations. J Chromatogr A 2013;1296:70–83.

[2] Nazario CE, Fumes BH, da Silva MR, Lancas FM. New materials for sample preparation techniques in bioanalysis. J Chromatogr B Anal Technol Biomed Life Sci 2017;1043: 81–95.

[3] Yamini Y, Rezazadeh M, Seidi S. Liquid-phase microextraction- the different principles and configurations. Trends Anal Chem 2019;112:264–72.

[4] Andrade-Eiroa A, Canle M, Lerory-Cancellieri V, Cerda V. Solid-phase extraction of organic compounds: a critical review.part ii. Trends Anal Chem 2016;80:655–67.
[5] Elahi M, Kamankesh M, Mohammadi A, Jazaer S. Acrylamide in cookie samples: analysis using an efficient co-derivatization coupled with sensitive microextraction method followed by gas chromatography-mass spectrometry. Food Anal Methods 2019;12(6): 1439–47. https://doi.org/10.1007/s12161-12019-01479-12167.
[6] Ohno K, Hasegawa T, Tamura T, Utsumi H. Proton affinitive derivatization for highly sensitive determination of testosterone and dihydrotestosterone in saliva samples by LC-ESI-MS/MS. Anal Sci 2018;34:1017–21.
[7] Nishikaze T, Tsumoto H, Sekiya S, Iwamoto S, Miura Y, Tanaka K. Differentiation of sialyl linkage isomers by one-pot sialic acid derivatization for mass spectrometry-based glycan profiling. Anal Chem 2017;89(4):2353–60.
[8] Rosenfeld JM. Solid-phase analytical derivatization: enhancement of sensitivity and selectivity of analysis. J Chromatogr A 1999;843(1–2):19–27.
[9] Atapattu SN, Rosenfeld JM. Solid phase analytical derivatization as a sample preparation method. J Chromatogr A 2013;1296:204–13.
[10] Atapattu SN, Rosenfeld JM. Micro scale analytical derivatizations on solid phase. Trends Anal Chem 2019;113:351–6.
[11] Wang X, Garcia CT, Gong G, Wishnok JS, Tannenbaum SR. Automated online solid-phase derivatization for sensitive quantification of endogenous S-nitrosoglutathione and rapid capture of other low-molecular-mass S-nitrosothiols. Anal Chem 2018;90(3): 1967–75.
[12] Atapattu SN, Poole CF. Selectivity equivalence of two poly(methylphenylsiloxane) open-tubular columns prepared with different deactivation techniques for gas chromatography. J Chromatogr A 2008;1185(2):305–9.
[13] Poole CF, Poole SK. Separation characteristics of wall-coated open-tubular columns for gas chromatography. J Chromatogr A 2008;1184(1–2):254–80.
[14] Atapattu SN, Fortuna JN, Rosenfeld JM. Effect of tetrabutylammonium cation on solid-phase analytical derivatization as a function of analyte lipophilicity. Chromatographia 2012;(75):47–54.
[15] Li N, Lee HK. Trace enrichment of phenolic compounds from aqueous samples by dynamic ion-exchange solid-phase extraction. Anal Chem 1997;69:5193–9.
[16] Kojima M, Tsunoi S, Tanaka M. High performance solid-phase analytical derivatization of phenols for gas chromatography-mass spectrometry. J Chromatogr A 2004;1042(1–2):1–7.
[17] Jonsson G, Cavcic A, Stokke TU, Beyer J, Sundt RC, Brede C. Solid-phase analytical derivatization of alkylphenols in fish bile for gas chromatography-mass spectrometry analysis. J Chromatogr A 2008;1183(1–2):6–14.
[18] Kuklenyik Z, Ekong J, Cutchins CD, Needham LL, Calafat AM. Simultaneous measurement of urinary bisphenol A and alkylphenols by automated solid-phase extractive derivatization gas chromatography/mass spectrometry. Anal Chem 2003;75(24):6820–5.
[19] Yuan SF, Liu ZH, Lian HX, Yang CT, Lin Q, Yin H, Lin Z, Dang Z. Fast trace determination of nine odorant and estrogenic chloro- and bromo-phenolic compounds in real water samples through automated solid-phase extraction coupled with liquid chromatography tandem mass spectrometry. Environ Sci Pollut Res Int 2018;25(4):3813–22.
[20] Pinto M, Frena M, Dos Santos Madureira LA. Solvent-free method for the determination of lignin-derived phenols in sediments. J Sep Sci 2017;40(9):2002–8.
[21] Es-haghi A, Baghernejad M, Bagheri H. In situ solid-phase microextraction and post on-fiber derivatization combined with gas chromatography-mass spectrometry for determination of phenol in occupational air. Anal Chim Acta 2012;742:17–21.

[22] Larreta J, Vallejo A, Bilbao U, Usobiaga A, Arana G, Zuloaga O. Headspace-solid-phase microextraction preconcentration of phenols, indoles and on-fibre derivatised volatile fatty acids in liquid and gas samples from cow slurries. J Sep Sci 2007;30(14):2293–304.

[23] Musshoff F, Lachenmeier DW, Kroener L, Madea B. Automated headspace solid-phase dynamic extraction for the determination of cannabinoids in hair samples. Forensic Sci Int 2003;133(1–2):32–8.

[24] Stopforth A, Grobbelaar CJ, Crouch AM, Sandra P. Quantification of testosterone and epitestosterone in human urine samples by stir bar sorptive extraction–thermal desorption–gas chromatography/mass spectrometry: application to HIV-positive urine samples. J Sep Sci 2007;30(2):257–65.

[25] Atapattu SN, Rosenfeld JM. Solid phase analytical derivatization of anthropogenic and natural phenolic estrogen mimics with pentafluoropyridine for gas chromatography-mass spectrometry. J Chromatogr A 2011;1218(51):9135–41.

[26] Lourenco ELB, Ferreira A, Pinto E, Yonamine M, Farsky SHP. On-fiber derivatization of SPME extracts of phenol, hydroquinone and catechol with GC-MS detection. Chromatographia 2006;(63):175–9.

[27] Larreta J, Usobiaga A, Etxebarria N, Arana G, Zuloaga O. Optimisation of the on-fibre derivatisation of volatile fatty acids in the simultaneous determination together with phenols and indoles in cow slurries. Anal Bioanal Chem 2007;389(5):1603–9.

[28] Salvador A, Moretton C, Piram A, Faure R. On-line solid-phase extraction with on-support derivatization for high-sensitivity liquid chromatography tandem mass spectrometry of estrogens in influent/effluent of wastewater treatment plants. J Chromatogr A 2007; 1145(1–2):102–9.

[29] Racamonde I, Villaverde-de-Saa E, Rodil R, Quintana JB, Cela R. Determination of Delta9-tetrahydrocannabinol and 11-nor-9-carboxy-Delta9-tetrahydrocannabinol in water samples by solid-phase microextraction with on-fiber derivatization and gas chromatography-mass spectrometry. J Chromatogr A 2012;1245:167–74.

[30] Lachenmeier DW, Kroener L, Musshoff F, Madea B. Determination of cannabinoids in hemp food products by use of headspace solid-phase microextraction and gas chromatography-mass spectrometry. Anal Bioanal Chem 2004;378(1):183–9.

[31] Stalport F, Glavin DP, Eigenbrode JL, Bish D, Blake D, Coll P, Szopa C, Buch A, McAdam A, Dworkin JP, et al. The influence of mineralogy on recovering organic acids from Mars analogue materials using the "one-pot" derivatization experiment on the Sample Analysis at Mars (SAM) instrument suite. Planet Space Sci 2012;67:1–13.

[32] Chatfield SN, Croft MY, Dang T, Murby EJ, Yu GYF, Wells RJ. Simultaneous extraction and methylation of acidic analytes adsorbed onto ion exchange resins using supercritical carbon dioxide containing methyl iodide. Anal Chem 1995;67:945–51.

[33] Cummins MT, Wells RJ. In situ derivatisation and extraction of volatile fatty acids entrapped on anion-exchange resin from aqueous solutions and urine as a test matrix using pentafluorobenzyl bromide in supercritical carbon dioxide. J Chromatogr B Biomed Sci Appl 1997;694(1):11–9.

[34] Field JA, Monohan K. Chlorinated acid herbicides in water by strong anion-exchange disk extraction and in-vial elution and derivatization. J Chromatogr A 1996;741:85–90.

[35] Subramaniam R, Astot C, Nilsson C, Ostin A. Combination of solid phase extraction and in vial solid phase derivatization using a strong anion exchange disk for the determination of nerve agent markers. J Chromatogr A 2009;1216(48):8452–9.

[36] Sha Y, Meng J, Zhang Y, Deng C, Wu D. Determination of volatile organic acids in tobacco by single-drop microextraction with in-syringe derivatization followed by GC-MS. J Sep Sci 2010;33(2):212–7.

[37] Merola G, Gentili S, Tagliaro F, Macchia T. Determination of different recreational drugs in hair by HS-SPME and GC/MS. Anal Bioanal Chem 2010;397(7):2987−95.
[38] Pan L, Adams M, Pawliszyn J. Determination of fatty acids using solid-phase microextraction. Anal Chem 1995;67:4396−403.
[39] Pan L, Pawliszyn J. Derivatization/solid-phase microextraction: new approach to polar analytes. Anal Chem 1997;69:196−205.
[40] Franklin T, Perry L, Shih W-C, Yu J. Detection of phytocannabinoids from buccal swabs by headspace solid phase microextraction − gas chromatography/mass spectrometry. Anal Methods 2018;10:942−6.
[41] Liu JF, Ding J, Yuan BF, Feng YQ. Magnetic solid phase extraction coupled with in situ derivatization for the highly sensitive determination of acidic phytohormones in rice leaves by UPLC-MS/MS. Analyst 2014;139(21):5605−13.
[42] Dugheri S, Bonari A, Pompilio I, Colpo M, Montalti M, Mucci N, Arcangeli G. Assessment of occupational exposure to gaseous peracetic acid. Int J Occup Med Environ Health 2018;31:527−35.
[43] Zheng Z, Zhao XE, Zhu S, Dang J, Qiao X, Qiu Z, Tao Y. Simultaneous determination of oleanolic acid and ursolic acid by in vivo microdialysis via UHPLC-MS/MS using magnetic dispersive solid phase extraction coupling with microwave-assisted derivatization and its application to a pharmacokinetic study of *Arctiumlappa* L. root extract in rats. J Agric Food Chem 2018;66(15):3975−82.
[44] Rosenfeld J, Mureika-Russell M, Yeroushalmi S. Solid-supported reagents for the simultaneous extraction and derivatization of carboxylic acids from aqueous matrices. Studies on optimization of reaction conditions. J Chromatogr 1986;358(1):137−46.
[45] Rosenfeld JM, Yeroushalmi S, Osei-Twum EY. Effect of solvent on solid-supported reactions on styrene/divinylbenzene copolymeric macroreticular resin. Anal Chem 1986; 58(14):3044−7.
[46] Takeo E, Sasano R, Shimma S, Bamba T, Fukusaki E. Solid-phase analytical derivatization for gas-chromatography-mass-spectrometry-based metabolomics. J Biosci Bioeng 2017; 124(6):700−6.
[47] Barden TJ, Croft MY, Murby EJ, Wells RJ. Gas chromatographic determination of organic acids from fruit juices by combined resin mediated methylation and extraction in supercritical carbon dioxide. J Chromatogr A 1997;785(1−2):251−61.
[48] Rodriguez I, Carpinteiro J, Quintana JB, Carro AM, Lorenzo RA, Cela R. Solid-phase microextraction with on-fiber derivatization for the analysis of anti-inflammatory drugs in water samples. J Chromatogr A 2004;1024(1−2):1−8.
[49] Yu LZ, Wells MJ. Establishing the feasibility of coupled solid-phase extraction-solid-phase derivatization for acidic herbicides. J Chromatogr A 2007;1143(1−2):16−25.
[50] Cha D, Cheng D, Liu M, Zeng Z, Hu X, Guan W. Analysis of fatty acids in sputum from patients with pulmonary tuberculosis using gas chromatography-mass spectrometry preceded by solid-phase microextraction and post-derivatization on the fiber. J Chromatogr A 2009;1216(9):1450−7.
[51] Guo L, Lee HK. One step solvent bar microextraction and derivatization followed by gas chromatography-mass spectrometry for the determination of pharmaceutically active compounds in drain water samples. J Chromatogr A 2012;1235:26−33.
[52] Beasley E, Francese S, Bassindale T. Detection and mapping of cannabinoids in single hair samples through rapid derivatization and matrix-assisted laser desorption ionization mass spectrometry. Anal Chem 2016;88(20):10328−34.

[53] Lopez-Serna R, Marin-de-Jesus D, Irusta-Mata R, Garcia-Encina PA, Lebrero R, Fdez-Polanco M, Munoz R. Multiresidue analytical method for pharmaceuticals and personal care products in sewage and sewage sludge by online direct immersion SPME on-fiber derivatization - GCMS. Talanta 2018;186:506–12.

[54] Uchiyama S, Inaba Y, Kunugita N. Determination of acrolein and other carbonyls in cigarette smoke using coupled silica cartridges impregnated with hydroquinone and 2,4-dinitrophenylhydrazine. J Chromatogr A 2010;1217(26):4383–8.

[55] Uchiyama S, Inaba Y, Kunugita N. Ozone removal in the collection of carbonyl compounds in air. J Chromatogr A 2012;1229:293–7.

[56] Uchiyama S, Inaba Y, Matsumoto M, Suzuki G. Reductive amination of aldehyde 2,4-dinitorophenylhydrazones using 2-picoline borane and high-performance liquid chromatographic analysis. Anal Chem 2009;81(1):485–9.

[57] Uchiyama S, Naito S, Matsumoto M, Inaba Y, Kunugita N. Improved measurement of ozone and carbonyls using a dual-bed sampling cartridge containing trans-1,2-bis(2-pyridyl)ethylene and 2,4-dinitrophenylhydrazine-impregnated silica. Anal Chem 2009; 81(15):6552–7.

[58] Uchiyama S, Otsubo Y. Simultaneous determination of ozone and carbonyls using trans-1,2-bis(4-pyridyl)ethylene as an ozone scrubber for 2,4-dinitrophenylhydrazine-impregnated silica cartridge. Anal Chem 2008;80(9):3285–90.

[59] Sala C, Medana C, Pellegrino R, Aigotti R, Bello FD, Bianchi G, Davoli E. Dynamic measurement of newly formed carbonyl compounds in vapors from electronic cigarettes. Eur J Mass Spectrom 2017;23(2):64–9.

[60] Ganjikhah M, Shariati S, Bozorgzadeh E. Preconcentration and spectrophotometric determination of trace amount of formaldehyde using hollow fiber liquid-phase microextraction based on derivatization by Hantzsch reaction. J Iran Chem Soc 2017;14:763–9.

[61] Stopforth A, Burger BV, Crouch AM, Sandra P. Urinalysis of 4-hydroxynonenal, a marker of oxidative stress, using stir bar sorptive extraction-thermal desorption-gas chromatography/mass spectrometry. J Chromatogr B Anal Technol Biomed Life Sci 2006;834(1–2): 134–40.

[62] Xu H, Wang S. A novel sorptive extraction method based on polydimethylsiloxane frit for determination of lung cancer biomarkers in human serum. Anal Chim Acta 2012;724: 61–6.

[63] Ho S, Yu JZ. Feasibility of collection and analysis of airborne carbonyls by on-sorbent derivatization and thermal desorption. Anal Chem 2002;74(6):1232–40.

[64] Pang X, Lewis AC, Hamilton JF. Determination of airborne carbonyls via pentafluorophenylhydrazine derivatisation by GC-MS and its comparison with HPLC method. Talanta 2011;85(1):406–14.

[65] Xu H, Wang S, Zhang G, Huang S, Song D, Zhou Y, Long G. A novel solid-phase microextraction method based on polymer monolith frit combining with high-performance liquid chromatography for determination of aldehydes in biological samples. Anal Chim Acta 2011;690(1):86–93.

[66] Zhang HJ, Huang JF, Lin B, Feng YQ. Polymer monolith microextraction with in situ derivatization and its application to high-performance liquid chromatography determination of hexanal and heptanal in plasma. J Chromatogr A 2007;1160(1–2):114–9.

[67] Prieto-Blanco MC, Lopez-Mahia P, Campins-Falco P. On-line analysis of carbonyl compounds with derivatization in aqueous extracts of atmospheric particulate PM10 by in-tube solid-phase microextraction coupled to capillary liquid chromatography. J Chromatogr A 2011;1218(30):4834–9.

[68] Pang X, Lewis AC, Rodenas-Garcia M. Microfluidic lab-on-a-chip derivatization for gaseous carbonyl analysis. J Chromatogr A 2013;1296:93−103.
[69] Beranek J, Kubatova A. Evaluation of solid-phase microextraction methods for determination of trace concentration aldehydes in aqueous solution. J Chromatogr A 2008; 1209(1−2):44−54.
[70] Wang Q, O'Reilly J, Pawliszyn J. Determination of low-molecular mass aldehydes by automated headspace solid-phase microextraction with in-fibre derivatisation. J Chromatogr A 2005;1071(1−2):147−54.
[71] Lord HL, Rosenfeld J, Volovich V, Kumbhare D, Parkinson B. Determination of malondialdehyde in human plasma by fully automated solid phase analytical derivatization. J Chromatogr B Anal Technol Biomed Life Sci 2009;877(13):1292−8.
[72] Lord HL, Rosenfeld J, Raha S, Hamadeh MJ. Automated derivatization and analysis of malondialdehyde using column switching sample preparation HPLC with fluorescence detection. J Sep Sci 2008;31(2):387−401.
[73] Rosenfeld J, Kim M, Rullo A. Development of an impregnated reagent and automation of solid-phase analytical derivatization for carbonyls: proof of principle. J Chromatogr Sci 2006;44:333−9.
[74] Huang Y, Huang X, Huang L, Liu Q, Lei Y, Yang L. Three-phase solvent bar liquid-phase microextraction combined with high-performance liquid chromatography to determine sarcosine in human urine. J Sep Sci 2018;41(15):3121−8.
[75] Huang J, Ou C, Lv F, Cao Y, Tang H, Zhou Y, Gan N. Determination of aliphatic amines in food by on-fiber derivatization solid-phase microextraction with a novel zeolitic imidazolate framework 8-coated stainless steel fiber. Talanta 2017;165:326−31.
[76] Sharma N, Jain A, Verma KK. Headspace solid-phase microextraction and on-fibre derivatization of primary aromatic amines for their determination by pyrolysis to aryl isothiocyanates and gas chromatography-mass spectrometry. Anal Methods 2011;3: 970−6.
[77] Singh DK, Sanghi SK, Gowri S, Chandra N, Sanghi SB. Determination of aliphatic amines by gas chromatography-mass spectrometry after in-syringe derivatization with pentafluorobenzoyl chloride. J Chromatogr A 2011;1218(33):5683−7.
[78] Herráez-Hernández R, Cháfer-Pericás C, Campíns-Falcó P. Analysis of methylamine by solid-phase microextraction and HPLC after on-fibre derivatization with 9-fluorenylmethyl chloroformate. Anal Chim Acta 2004;503:425−33.
[79] Herraez-Hernandez R, Chafer-Pericas C, Verdu-Andres J, Campins-Falco P. An evaluation of solid phase microextraction for aliphatic amines using derivatization with 9-fluorenylmethyl chloroformate and liquid chromatography. J Chromatogr A 2006; 1104(1−2):40−6.
[80] Llop A, Pocurull E, Borrull F. Automated on-fiber derivatization with headspace SPME-GC-MS-MS for the determination of primary amines in sewage sludge using pressurized hot water extraction. J Sep Sci 2011;34(13):1531−7.
[81] Chafer-Pericas C, Campins-Falco P, Herraez-Hernandez R. Comparative study of the determination of trimethylamine in water and air by combining liquid chromatography and solid-phase microextraction with on-fiber derivatization. Talanta 2006;69(3):716−23.
[82] Ali Awan M, Fleet I, Paul Thomas CL. Determination of biogenic diamines with a vaporisation derivatisation approach using solid-phase microextraction gas chromatography-mass spectrometry. Food Chem 2008;111(2):462−8.
[83] Saaid M, Saad B, Ali AS, Saleh MI, Basheer C, Lee HK. In situ derivatization hollow fibre liquid-phase microextraction for the determination of biogenic amines in food samples. J Chromatogr A 2009;1216(27):5165−70.

[84] Prieto-Blanco MC, Chafer-Pericas C, Lopez-Mahia P, Campins-Falco P. Automated on-line in-tube solid-phase microextraction-assisted derivatization coupled to liquid chromatography for quantifying residual dimethylamine in cationic polymers. J Chromatogr A 2008;1188(2):118–23.

[85] Campíns-Falcó P, Herráez-Hernández R, Verdú-Andrés J, Cháfer-Pericás C. On-line determination of aliphatic amines in water using in-tube solid-phase microextraction-assisted derivatisation in in-valve mode for processing large sample volumes in LC. Anal Bioanal Chem 2009;394:557–65.

[86] Gremaud L, Schappler J, Veuthey J-L, Rohrbasser C, Rudaz S. Analysis of amphetamine derivatives in plasma using capillary zone electrophoresis coupled with laser-induced fluorescence after derivatization on solid-phase extraction support. Chimia 2008;62:210–4.

[87] Brede C, Skjevrak I, Herikstad H. Determination of primary aromatic amines in water food simulant using solid-phase analytical derivatization followed by gas chromatography coupled with mass spectrometry. J Chromatogr A 2003;983(1–2):35–42.

[88] Nakamoto A, Nishida M, Saito T, Kishiyama I, Miyazaki S, Murakami K, Nagao M, Namura A. Monolithic silica spin column extraction and simultaneous derivatization of amphetamines and 3,4-methylenedioxyamphetamines in human urine for gas chromatographic-mass spectrometric detection. Anal Chim Acta 2010;661(1):42–6.

# Automated and high-throughput extraction approaches

# 21

*Naiyu Zheng*
Department of Bioanalytical Sciences, Bristol-Myers Squibb Company, Princeton, NJ, United States

## 21.1 Introduction

Solid-phase extraction (SPE) is a separation process using packed solid sorbent(s) and appropriate elution solvent(s) by which the target analyte(s) are retained by the sorbent and the interferences are washed off. Without automation, the SPE procedure can be tedious and time-consuming when a large batch sample extraction is required. Automation of the SPE procedure is essential to improve throughput so that a large number of samples can be processed in a single batch. The advantages of automated SPE include the standardization of the sample extraction procedure, increased assay throughput, improved assay reproducibility, and reduced overall cost. Automated and high-throughput SPE is widely used in several fields, such as the fractionation of natural products for drug discovery [1], GC-MS analysis of chemical contaminants in foods [2], high-throughput GC-MS/MS and UHPLC-MS/MS for the analysis of pesticides and environmental contaminants in meats and poultry [3], UHPLC-MS/MS analysis of aflatoxin M1 in milk and dairy products [4] and LC-MS/MS analysis of thiopental in human plasma [5]. However, the applications of automated and high-throughput SPE are mainly for sample extraction for LC-MS/MS bioanalysis of drugs or their metabolites in drug discovery and development [5–7].

In drug discovery and development, LC-MS/MS bioanalysis for drugs or metabolites in biological samples for toxicokinetic or pharmacokinetic evaluation is an important part of the USFDA and other regulatory authorities' requirements for new drug applications. In most cases, a large number of samples need to be analyzed with a fast turnaround to support preclinical and clinical studies, such as drug toxicity, dose escalating, and first-in-human studies [8–10]. Recent advances in ultra-high-performance liquid chromatography (UHPLC) separation, mass spectrometry detection, and data analysis software have significantly enhanced the speed of bioanalysis and improved data quality in quantitative LC-MS/MS bioanalysis. However, sample preparation remains the bottleneck for throughput because biological matrices contain significant amounts of endogenous compounds, which can interfere with the detection of target compounds or cause ion suppression in MS detection [11]. The purpose of the sample extraction is mainly to extract the analytes of interest with high extraction recovery and remove the matrix-effect causing phospholipids in order to improve the detection sensitivity and selectivity of analyte(s) at

**Solid-Phase Extraction. https://doi.org/10.1016/B978-0-12-816906-3.00021-2**

the level of lower limit of quantification (LLOQ), as well as the assay's robustness [11]. Traditionally, these unwanted matrix components are removed using protein precipitation (PPT), liquid–liquid extraction (LLE), or solid-phase extraction (SPE). The major advantage of using an SPE method is that the method can be easily automated with high-throughput using a robotic liquid handler (RLH) or an automated online SPE system. The diversity of available SPE sorbent chemistries makes it possible to apply a generic protocol to many applications without changing the automation platform. The focus of this chapter will be on automated and high-throughput SPE approaches.

## 21.2 Evolution of automated SPE approaches toward high-throughput sample extraction

A typical SPE procedure involves preconditioning, sample loading, washing (one or more times), and analyte elution (one or more times), (as shown in Fig. 21.1). SPE for a small number of samples can be done manually using a 12-position vacuum SPE manifold, as shown in Fig. 21.2A. The 12-position manifold allows the extraction of samples with different volumes using SPE cartridges of varied size depending on the application. However, manual SPE has limitations due to its low throughput. As shown in Fig. 21.1, all sample handling steps, including steps one to six require manual addition of solvents for preconditioning, washing, and elution, as well as sample transfer. In addition, to speed up the SPE extraction, the use of vacuum in steps 7–10 (Fig. 21.1) is required.

Due to many labor-intensive steps with manual SPE, parallel sample processing in a 96-well format (or 96-tube cluster format) using a 96-well SPE plate vacuum manifold or positive pressure SPE processor have become the most common approach (Fig. 21.2B). The operation of the 96-well plate vacuum manifold or positive pressure SPE processor is usually performed manually. In some cases, elution of the analytes from the SPE plate can be performed using a centrifuge.

Using a 96-well SPE plate, the liquid transfer steps can be significantly simplified and automated using an RLH, such as JANUS Mini RLH (PerkinElmer), Tomtec Quadra 4 (Tomtec, Inc.) [12] (Fig. 21.2C), or Hamilton's Microlab NIMBUS96 (Hamilton Company) [13]. Automated SPE using a 96-well SPE plate is also known as the automated off-line SPE method because the SPE is performed using a separate procedure from the sample analysis.

Today, the majority of SPE procedures for LC-MS/MS bioanalysis are performed using some degree of automation to improve the throughput. Based on the characteristics of each methodology, automated and high-throughput SPE approaches can be summarized into three approaches:

**(1)** *Automated off-line SPE approach*: This approach requires an RLH. The SPE extraction is performed on a 96-well format SPE plate [7,14], as shown in Fig. 21.2C–21.2D.

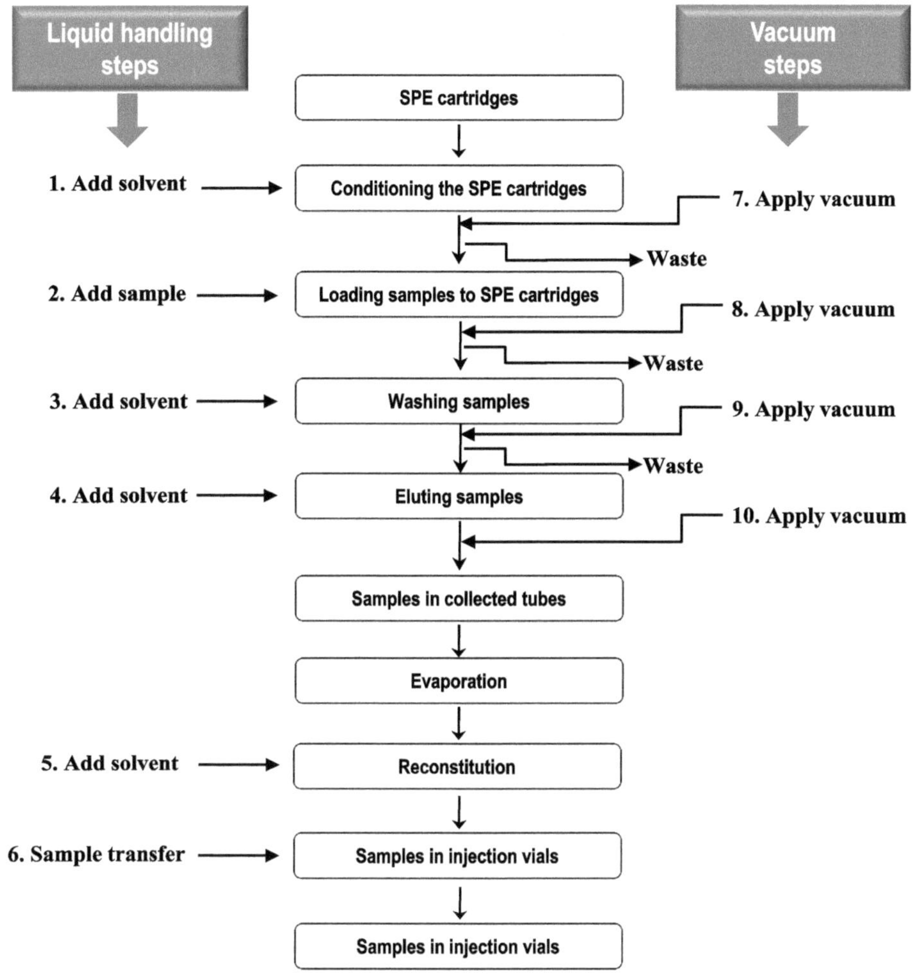


**Figure 21.1** Illustration of a typical manual SPE sample extraction process for bioanalysis using LC−MS/MS: in which steps one to six and steps 7−10 may require manual interventions [7].

**(2)** *Automated online SPE approach*: This approach requires an LC pump and an autosampler. The SPE extraction is performed using online LC columns or disposable SPE cartridges [4,15];

**(3)** *Dispersive pipette extraction (DPX) approach*: This approach requires an RLH. The SPE extraction is performed on disposable tips loaded with the SPE sorbent [16,17].

The comparison of these three approaches is shown in Table 21.1. The detailed methodology and application for each automated SPE method will be discussed in the following sections.

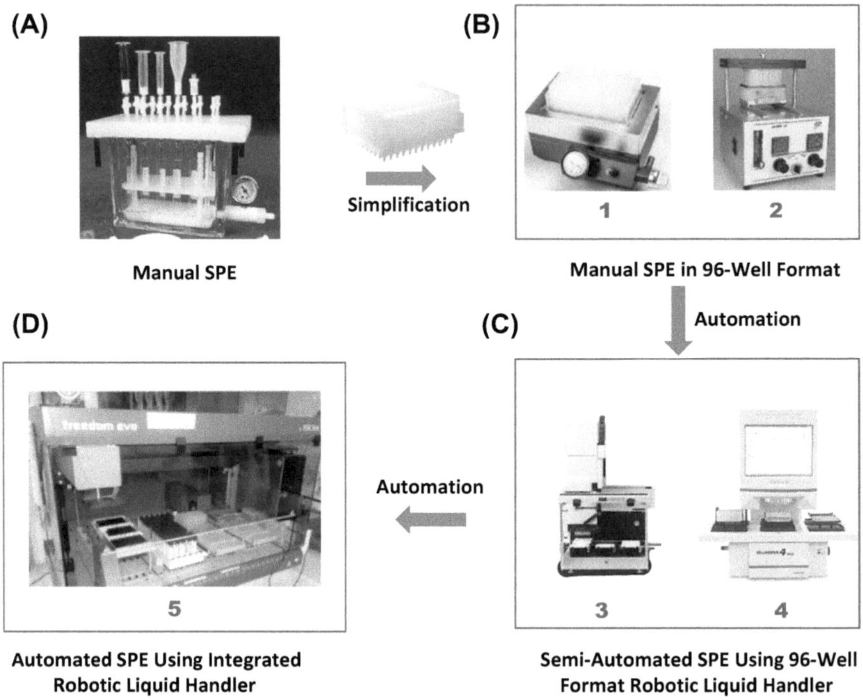


**Figure 21.2** Evolution of automated offline SPE for high throughput sample extraction. (A) Manual SPE; (B) Manual SPE using SPE accessories in 96-well plate format; (C) Semi-automated SPE using 96-well format RLH; (D) Automated SPE procedure using integrated RLH. Commonly used SPE accessories and RLHs: (**1**) 96-well plate vacuum manifold (Agela Technologies); (**2**) CEREX system 96 processor (SPEware Corporation, part of Tecan Group); (**3**) JANUS Mini RLH (PerkinElmer); (4) Tomtec Quadra 4 (Tomtec); (5) Tecan Freedom EVO (Tecan US).

**Table 21.1** Comparison of automated SPE approaches for high throughput sample extraction.

| Automated SPE approaches | SPE sorbent formats | Sample or liquid handling methods | Methods for removing the waste or eluting the samples |
|---|---|---|---|
| Offline SPE | • 96-well format SPE plate | • using an RLH | • vacuum<br>• positive pressure<br>• centrifugation |
| Online SPE | • LC column<br>• disposable SPE cartridges | • autosampler for loading the sample<br>• LC pump for delivering solvents | • LC pump for removing the waste<br>• LC pump for eluting the analytes |
| DPX | • sorbents loaded in disposable tips | • automated aspiration using an RLH | • automated dispensing using an RLH |

### 21.2.1 Automated off-line SPE approach for high-throughput extraction

Automated off-line SPE approach is the most popular high-throughput SPE method for LC-MS/MS bioanalysis in the pharmaceutical industry. It can be done using a semi-automated procedure involving the use of a standalone 96-channel pipette RLH (Fig. 21.2C) with a separate vacuum handling device, positive pressure SPE processor, or a centrifuge. However, fully off-line automated SPE is becoming more popular with the use of an integrated RLH with on deck 96-well vacuum manifold or positive pressure SPE processor [6,7,14] (Fig. 21.2D). All of the SPE extraction, including liquid handling and vacuum handling (or positive pressure control and adjust) steps can be controlled by a computer. The most popular integrated RLHs include Tecan Freedom EVO (Tecan, US) [18], JANUS Varispan (PerkinElmer) [19], and Hamilton STAR (Hamilton Company) [13] RLH platforms.

However, the extent of automation within a sample extraction method can vary from a few steps to a fully automated process. The scope of SPE automation depends on specific assay needs, availability of the RLHs, and the user's preference with using automation. As shown in Fig. 21.3, the liquid transfer from extraction plate to injection plate in 96-well format can be easily accomplished using a Tomtec Quadra 4 or PerkinElmer's JANUS Mini RLH. For other steps, manual intervention is usually required for several reasons, such as the hardware or accessories not being commercially available or not compatible with the existing automation platform. If one RLH needs to be shared in a laboratory, a method that is divided into several automated sections with fewer steps in each section may be more flexible for the scientists to share the platform. Several steps, such as centrifugation and evaporation, are typically done using standalone equipment and off the deck of an RLH (as shown in Fig. 21.3).

In general, an automated SPE workflow can be used for different assays and target compounds by modification of the sorbent chemistries of the SPE plates to achieve a satisfactory extraction recovery [11,20]. There are many commercially available SPE sorbents, such as C2, C8, C18, phenyl, polar endcapped, and polar-embedded stationary phases. To eliminate the time and efforts in modification of the sample extraction protocols for different analytes, several generic SPE plates from different manufacturers, such as Strata-X (Phenomenex, UK), Evolute ABN (Biotage, USA), and Oasis HLB (Waters, USA) can be used for multiple analytes with satisfactory recoveries using an automated sample extraction procedure without significant changes [14]. On the other hand, automated SPE method development can also be performed using an RLH, such as Hamilton MicroLab STAR to reduce the time and errors during SPE method development [21,22].

As shown in Fig. 21.3, due to the requirement of the use of 96-well plate format, although an integrated RLH is used for SPE extraction, additional manual intervention to transfer the samples from sample tubes to a 96-well plate could be time-consuming. It requires decapping and recapping of the sample tubes before sample transfer. An alternative method is using a pierceable tube for direct sampling with an 8-channel RLH [14] (Fig. 21.4) or the use of a decapper/recapper. To eliminate the need for

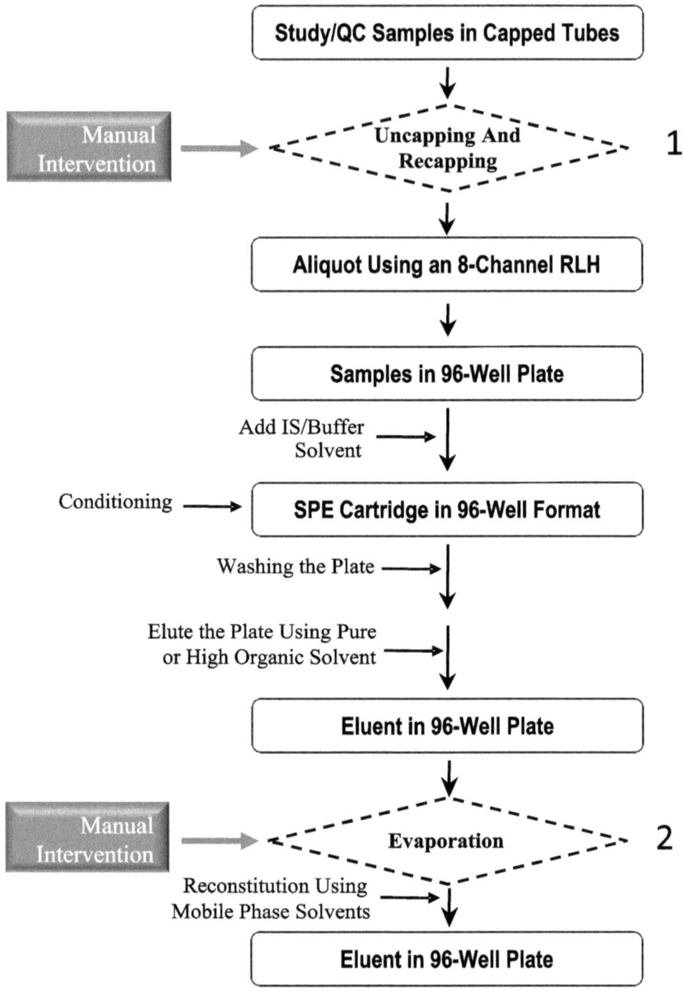


**Figure 21.3** Illustration of a typical automated SPE sample extraction process for bioanalysis using LC−MS/MS: conventional SPE sample extraction with an RLH, in which steps 1 and 2 may require manual interventions.
The figure was modified from Zheng N, Buzescu A, Pasas-Farmer S, Arnold ME, Ouyang Z, Jemal M, et al. A simplified and completely automated workflow for regulated LC-MS/MS bioanalysis using cap-piercing direct sampling and evaporation-free solid -phase extraction. J Chromatogr B. 2013;921−922:64−74. © 2013 Elsevier BV (2013).

manual intervention due to the evaporation and reconstitution steps, the evaporation-free SPE method was reported as an alternative option [14]. On the other hand, the application of micro-elution (μElution) SPE plates, such as Oasis μElution plate from Waters, only requires elution volume of as little as 25 μL, which allows for preconcentration of the samples without the need for an evaporation step.

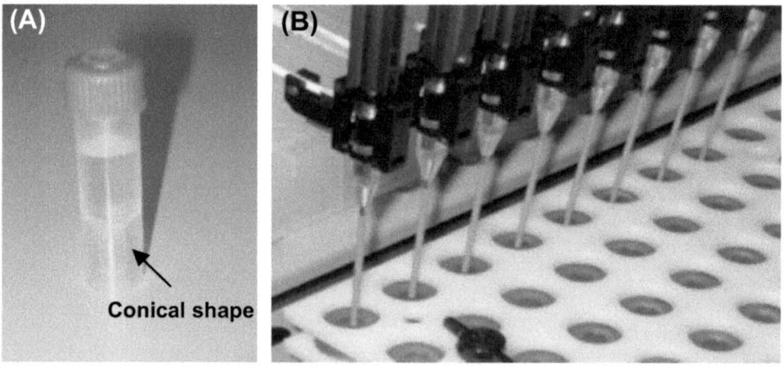


**Figure 21.4** Eliminating decapping and recapping steps using cap piercing for automated direct sample transfer. (A) A plasma sample tube with a pierceable cap; (B) Aliquots of plasma samples were transferred by Tecan with cap piercing for direct sample transfer into a 96 well plate. The figure was modified from Zheng N, Buzescu A, Pasas-Farmer S, Arnold ME, Ouyang Z, Jemal M, et al. A simplified and completely automated workflow for regulated LC-MS/MS bioanalysis using cap-piercing direct sampling and evaporation-free solid -phase extraction. J Chromatogr B. 2013;921−922:64−74. © 2013 Elsevier BV (2013).

### *21.2.1.1 Selected illustrative example*

The following is an example of a fully automated high-throughput sample preparation method for the determination of Pexacerfont (BMS-562086) and two metabolites, BMS-749241 and DPH-123554, in human plasma (adapted from Zheng et al.) [14].

Samples were pipetted and extracted using a Tecan Genesis 150 RLH equipped with an 8-channel liquid handler (LiHa) arm (fixed tip format), Te-Vac station (automated vacuum system), and a robotic manipulator arm (RoMa) from Tecan US (Durham, NC). The system was controlled by Gemini software. An Evolute ABN SPE plate (10 mg, from Biotage, Charlotte, NC) was placed onto the Te-Vac station and used for sample extraction. All SPE preconditioning, washing, and elution steps were performed automatically using the on-deck Te-Vac vacuum station with parameters that were already included in the Tecan program. The Tecan RLH deck configuration is shown in Fig. 21.5. The sample extraction steps are as follows [14]:

#### Adding IS into the mixing plate (Position 2)

- An aliquot of 30 μL of the internal standard working solution mixture containing 25/100/100 ng/mL SIL-Pexacerfont/SIL-BMS-749241/SIL-DPH-123554 in 30% MeOH in water was added by Tecan into the mixing plate (Position 2)

#### Adding plasma samples into the mixing plate (B)

- An aliquot of 100 μL of human plasma samples (blank, calibration standard, quality control, and study samples) in BMS custom tubes (Position 1) was transferred by Tecan into the mixing plate (Position 2)
- An aliquot of 300 μL of 0.01 M $Na_2CO_3$ solution was added to each well of the mixing plate (Position 2), and mixed well by offline vortexer and replaced to Tecan deck (This step can be eliminated by aspirating and dispensing by Tecan a few times).

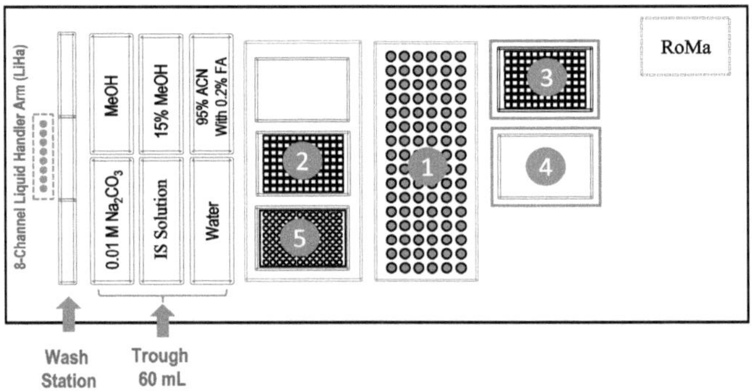


**Figure 21.5** Tecan RLH deck configuration. Labware mapping as follows. Position 1: Study samples in BMS custom tubes with pierceable caps. Position 2: Mixing plate. Position 3: SPE plate on top of Te-Vac station. Position 4: Additional Te-Vac station. Position 5: Sample collection plate.
The figure was modified from Zheng N, Buzescu A, Pasas-Farmer S, Arnold ME, Ouyang Z, Jemal M, et al. A simplified and completely automated workflow for regulated LC-MS/MS bioanalysis using cap-piercing direct sampling and evaporation-free solid -phase extraction. J Chromatogr B. 2013;921−922:64−74. © 2013 Elsevier BV (2013).

## Conditioning the SPE plate (C)

- An aliquot of 400 μL of MeOH was added into each well of SPE plate (Position 3) which is an Evolute ABN plate (10 mg), after 10 s, a 100 mbar vacuum was applied for 3 s on the Te-Vac vacuum station (Position 3)
- An aliquot of 300 μL of 0.01 M $Na_2CO_3$ solution was added into each well of the SPE plate (Position 3), after 10 s, a 150 mbar vacuum was applied for 10 s on the Te-Vac vacuum station.

## SPE extraction

- Loading Sample: The samples in the mixing plate (Position 2) were transferred onto the SPE plate (Position 3) by Tecan. After 20 s, a 100 mbar vacuum was applied for 12 s on the Te-Vac vacuum station, followed by a 550 mbar vacuum applied for 30 s on the Te-Vac vacuum station.
- Washing 1: an aliquot of 300 μL of water was added into each well of the SPE plate (Position 3), after 20 s, a 100 mbar vacuum was applied for 5 s on the Te-Vac vacuum station, followed by a 550 mbar vacuum applied for 60 s on the Te-Vac vacuum station
- Washing 2: an aliquot of 200 μL of 15% methanol in water was added into each well of the SPE plate (Position 3), after 20 s, a 100 mbar vacuum was applied for 5 s on the Te-Vac vacuum station, followed by a 550 mbar vacuum applied for 60 s on the Te-Vac vacuum station
- *The following step could be done manually if the script for the robotic manipulator arm (RoMa) was not included.* The vacuum box in Position 3 was moved by RoMa together with the SPE plate (Position 3) to the front position of the Te-Vac station (Position 4). After moving the 1 mL round 96-well collection plate (Position 5) into position 3, the vacuum box was moved together with the SPE plate by the RoMa on top of the collection plate in Position 3.
- Elution: An aliquot of 250 μL of elution solvent (0.2% formic acid in 95% of ACN) was added into each well of the SPE plate (Position 3), after 20 s, a 100 mbar vacuum was applied for 5 s on the Te-Vac vacuum station, followed by a 550 mbar vacuum applied for 60 s on the Te-Vac vacuum station.

## Reconstitution of the samples

- *The following step could be done manually if the script for RoMa was not included.* The vacuum box in Position 3 was moved by the RoMa together with the SPE plate (Position 3) to the front position of the Te-Vac station (Position 4). The 1 mL round 96-well collection plate that contained the eluate was moved by the RoMa into Position 5.
- An aliquot of 250 μL of 20 mM of $NH_4OAc$ was added into each well of the collection plate containing the eluate (Position 5).
- The final injection samples contained 10 mM $NH_4OAc$, 0.1% formic acid and 47.5% ACN, which was very close to the mobile phase at time 0 (10 mM $NH_4OAc$+0.1% formic acid + 45% ACN)
- Manually vortex-mix the samples, an aliquot of 10 μL of the sample was injected into LC-MS/MS for analysis.

For bioanalysis, the overall assay performance could be evaluated by the precision and accuracy of the calibration standards and quality control samples based on FDA guideline for bioanalytical validation [23]. However, the most important factors related to SPE extraction performance are extraction recovery and matrix effects. For the above extraction method, the extraction recoveries were 55%, 69%, and 70%, and the matrix effects were 1.17, 1.09, and 1.22 for Pexacerfont BMS-562086, BMS-749241 and DPH-123554, respectively [14]. When normalized with the matrix effects of their internal standards, there was no observable matrix effect since the normalized matrix effects were 0.98, 1.02, and 1.01, respectively. Since the evaporation step was eliminated, this evaporation-free SPE resulted in moderate extraction recoveries for the analytes with recoveries of 55%−70% for all analytes. The elimination of the sample evaporation did not impact the assay quality in terms of achieving the lower limit of quantification, as well as the assay performance as indicated by the precision and accuracy data obtained from the quality control samples [14].

For the example given above, in order to simplify the sample transfer from BMS custom tubes with pierceable caps, 8-channel fixed tips needed to be used, which require washing after each liquid transfer. In addition, due to the lack of on-deck vortexer, manual intervention for mixing the samples is required. For a Tecan platform with additional 96-channel pipette arm using disposable tips, the sample mixing can be performed by aspirating and dispensing multiple times after each liquid transfer.

In general, the procedures for automated off-line SPE methods for different analytes are similar except for the difference in the chemistry of the analytes, SPE sorbents, labware dimension, and Tecan configurations. Other than the bioanalysis of drugs and their metabolites, automated SPE methods have been used for the analysis of other analytes where high throughputs are required. For example, an automated high-throughput SPE μElution method was reported for the analysis of eight perfluorosulfonic acids, 11 perfluorocarboxylic acids, and fluorooctane sulfonamide in human serum [24]. Samples were prepared using a Tecan Freedom Evo 200 equipped with an eight-channel LiHa for conductive disposable tips, a RoMa for transport of microtiter plates and a Te-Vac station for SPE using a 96-well-μElution SPE-plate (Oasis WAX μElution Plate, 30 μm, Waters, Milford, USA). All sample preparation of 96 samples was achieved within <1 h 45 min (not including pre and postanalytical preparations) [24].

### 21.2.2 Online SPE approach for high-throughput sample extraction

Early automated online SPE systems were custom-made using a preanalytical LC column (primary column, or SPE cartridge) for sample extraction, and the eluate from the first preanalytical LC column was injected into the separation column (or secondary column) for analysis [25–27]. In this case, online automated SPE can be accomplished using a six- or ten-port switching valve for column-switching connected to an LC pumping system [26], and online automated SPE is a 2D-LC system. The sample to be extracted is first injected into the primary LC column, and the analyte(s) of interest trapped on the column, while the unwanted components are eluted to waste. After column switching, the trapped analyte fraction is eluted into the separation column for analysis by LC-MS/MS. There are two column switching modes, which includes straight-flush column switching and back-flush column switching [26]. Back-flushing technique is better than straight-flush column switching because it is able to minimize peak broadening of analytes while washing away the strongly retained matrix components. The online SPE method offers several advantages, including the ability to achieve automation in sample extraction without the additional capital investment of expensive RLH systems.

The other online SPE method is accomplished by using commercially available fully automated online SPE platforms, such as Spark Holland's Symbiosis online SPE system (Spark Holland, Netherland) [28], Prelude SPLC System using TurboFlow technology (ThermoFisher Scientific) [29] or RapidFire 365 Mass Spectrometry system (Agilent Technologies) [30]. These integrated online systems significantly simplify sample analysis by eliminating several steps, such as the collection of eluent, evaporation, or reconstitution. It is, therefore, possible to reduce potential analyte loss due to nonspecific binding as compared with off-line SPE methods. However, automated online SPE systems are highly vendor-dependent and rely heavily on vendor-provided software. In addition, online sample extraction could be more complicated in method development and pose potential carryover concerns. Recently, there have been several review articles describing the application of automated online methodologies for the direct analysis of biological samples [26,31,32].

#### 21.2.2.1 Selected illustrative example

The following is an example of automated online SPE method for high-throughput sample preparation to quantify the concentration of thiopental in plasma using automated online SPE directly coupled to LC–MS/MS (adapted from Moosavi SM et al.) [5].

A robotic online SPE system, Symbiosis Pharma System (SPARK Holland) was used to extract thiopental and its internal standard from human plasma samples [5]. The online SPE Symbiosis System consisted of a Reliance autosampler with two binary LC pumps and online SPE unit with two high-pressure solvent delivery pumps [33]. The Symbiosis Pharma System was controlled by the Symbiosis Pro for Analyst v. 2.1.0.0 and the mass spectrometer controlling software, Analyst v 1.5.1. Aliquots of human plasma (50 μL) were mixed with 50 μL of the internal standard solution (thiopental-d5 at 100 ng/mL in 0.5% v/v formic acid in water) and 100 μL of 0.5% formic acid (v/v) in water before placing in the autosampler. The purpose of acidifying the plasma samples is to convert the analyte to the unionized form for retention on the SPE cartridge (pKa = 7.5). The analyte was eluted out using a mobile phase with a high percentage of organic solvent. The online SPE sample extraction was performed automatically by Symbiosis Pharma System using C18 cartridges (HySphere, HD, 7 μm, from SPARK Holland). An X-TerraMS C18 column (2.1 mm × 150 mm, 5 μm) from Waters and a Gemini Security Guard column from Phenomenex were used for the analysis. The online SPE extraction includes the following steps [5]:

## Conditioning

- An aliquot of 1000 μL of methanol was delivered at a flow rate of 5000 μL/min by online SPE pumps

## Equilibrium

- An aliquot of 1000 μL of 50 mM $NH_4OAc$ (pH = 7) was delivered at a flow rate of 5000 μL/min by online SPE pumps

## Extraction

- An aliquot of 1000 μL of 50 mM $NH_4OAc$ (pH = 7) was delivered at a flow rate of 2000 μL/min by online SPE pumps

## Cartridge wash

- An aliquot of 1000 μL of 10% methanol in 50 mM $NH_4OAc$ (pH = 7) was delivered at a flow rate of 5000 μL/min by online SPE pumps. It was repeated twice.

## Analyte elution

- The analyte and internal standard were eluted to the analytical column using a gradient program with the mobile phases consisted of 0.1% formic acid in water as Mobile Phase A and acetonitrile as the Mobile Phase B with a flow rate of 0.55 mL/min with a total run time of 7.2 min.

## Clamp washes

- After each extraction, clamps were washed three times using water, methanol, and 0.1% formic acid in water. The total volume of each solvent was 500 μL delivered at a flow rate of 5000 μL/min by online SPE pumps.

After HPLC separation, the analyte and internal standard were detected using the multiple reaction monitoring (MRM) mode on a Sciex QTrap 5500 mass spectrometer. The assay was linear over a range of 6–600 ng/mL of thiopental in human plasma, which was fully validated and met the requirements of the 2012 European Medicines Agency (EMEA) guideline for Bioanalytical Method Validation [5]. In general, simultaneous extraction and analysis of samples using automated online SPE extraction systems minimizes analyst error and is cost-efficient. Importantly, it helps eliminate instability concerns with some analytes when using off-line SPE for sample extraction at room temperature or during sample evaporation.

### 21.2.3 Dispersive pipette extraction (DPX) approach for high-throughput sample extraction

Conventional SPE procedures usually require relatively large volumes of solvent and have a longer processing time. Dispersive Pipette Extraction (DPX) was developed as an alternative to conventional SPE procedures by combining efficient and rapid extraction with significantly less solvent consumption [16,17]. DPX is actually a dispersive SPE technique using disposable pipette tips containing loosely packed sorbent, which allows for sample mixing with high efficiency. The DPX technology has several advantages, such as rapid extraction, high analyte recoveries, minimal solvent waste, and simplification of the conventional SPE steps allowing for full automation of the extraction process. Typical sample extraction steps for conventional SPE, such as the preconditioning, sample loading, washing, and sample elution can be accomplished by the DPX approach through multiple aspirations and dispense steps. The DPX tips can be manufactured using disposable tips compatible with the commercially available RLH platforms from different vendors, such as Tecan, Hamilton, PerkinElmer, or Gerstel. This automated DPX approach was used for the extraction of catecholamine in urine by LC-MS/MS with a total sample extraction time of $<15$ min [17] and for 10 β-agonists in pork meat for LC-MS/MS analysis with total extraction time of $<20$ min [16].

#### 21.2.3.1 Selected illustrative example

The following is an example of automated DPX to quantify epinephrine, dopamine, normetanephrine, and metanephrine in human urine by LC-MS/MS (adapted from Mastrianni KR et al.) [17].

SPE was achieved by integrating DPX tips (DPX Technologies) into Hamilton's Microlab NIMBUS96 from Hamilton Robotics to increase analytical throughput and improve the assay's reproducibility [17]. The method involved the addition of diphenylboronic acid (DPBA) to form the stable and negatively charged complexes with the analytes. During DPX extraction, the complexes were aspirated into the DPX tips and retained on a styrene-divinylbenzene sorbent. After removing the interfering compounds using a buffer solution, the analyte complexes were eluted from the tips using an acidic aqueous solution [17].

The Microlab NIMBUS96 was used for all liquid transfers, and new tips were used to add each reagent and then discarded. The following is the detailed automated SPE using DPX technology for the analysis [17].

- An aliquot of 300 μL of urine sample was mixed with 10 μL of internal standard, and then manually added to each well of a 96-well assay plate.
- The assay plate containing the urine samples was loaded onto the deck of the Hamilton Microlab NIMBUS96 pipetting workstation.

### Preparation of the urine samples and reagents

- An aliquot of 600 μL of the complexing agent [0.2% (w/v) diphenyl boronic acid, 5 g/L EDTA in a 2 M $NH_4Cl/NH_4OH$ pH 8.5 buffer] was transferred by Microlab NIMBUS 96 to each well of the urine sample well plate.
- An aliquot of 500 μL wash buffer (0.2 M $NH_4Cl$, pH 8.5) was transferred by Microlab NIMBUS96 to each well of a second plate for washing.
- An aliquot of 270 μL elution solution (1M formic acid plus 30 μL methanol) was transferred by Microlab NIMBUS 96 to each well of a third plate for elution.

### Conditioning of DPX tips

- 1-mL DPX reverse phase tips were picked up by NIMBUS96, and conditioned by aspirating and dispensing 100% methanol twice and aspirating and dispensing the wash buffer once.

### Loading Samples on DPX tips

- The urine sample solution was then aspirated and dispensed four times by the DPX.

### DPX wash

- The tips were rinsed with wash buffer solution (500 μL in each well).

### DPX elution

- The analytes were eluted with the elution solution (300 μL in each well).
- The elution plate was then moved to the autosampler, where 10 μL sample volume was injected into the LC-MS/MS for analysis.

Sample extraction using the DPX approach on a Microlab NIMBUS96 was able to extract up to 96 samples in less than 15 min [17]. The extraction recoveries exceeded 96% for all analytes except dopamine, with an extraction recovery of 81%. The LC-MS/MS assay using an automated DPX approach was linear over the range from 0.5 to 1000 ng/mL, with replicate analyses resulting in CV $\leq$ 8% [17]. The other advantage of the DPX approach was the elimination of the need for possible manual intervention and time-consuming steps for automated off-line SPE that usually requires the use of a vacuum manifold, a positive pressure SPE processor, or a centrifuge. Therefore, fully automated SPE sample extraction is easily achieved using a standalone 96-channel pipetting RLH.

## 21.3 Conclusions

Automated SPE is widely used to increase the SPE throughput for qualitative or quantitative analysis in the food industry, environmental and biomedical research, as well as the pharmaceutical industry. The advantages of automated SPE include the standardization of the sample extraction procedure, increased assay throughput, improved assay reproducibility, and reduction of the overall cost. The majority of the reported SPE methods are related to off-line SPE procedures using commercially available RLHs. However, these procedures still require some manual intervention, such as the centrifugation or evaporation steps despite the use of RLHs. Automated online SPE using a custom-made system or robotic online SPE system directly connected to LC−MS/MS instrumentation can be used to eliminate the need for manual intervention. The application of automated dispersive pipette extraction (DPX) as an alternative to traditional SPE has significantly increased the speed of sample extraction and reduced solvent consumption. This new approach successfully eliminated the need for potential manual intervention and time-consuming steps related to vacuum and achieved a fully automated and high-throughput SPE sample extraction workflow using a standalone RLH.

## References

[1] Tu Y, Jeffries C, Ruan H, Nelson C, Smithson D, Shelat AA, et al. Automated high-throughput system to fractionate plant natural products for drug discovery. J Nat Prod 2010;73(4):751−4. Epub 2010/03/18.

[2] Lehotay SJ, Han L, Sapozhnikova Y. Automated mini-column solid-phase extraction cleanup for high-throughput analysis of chemical contaminants in foods by low-pressure gas chromatography-tandem mass spectrometry. Chromatographia 2016;79(17): 1113−30. Epub 2016/09/13.

[3] Sapozhnikova Y. High-throughput analytical method for 265 pesticides and environmental contaminants in meats and poultry by fast low pressure gas chromatography and ultrahigh-performance liquid chromatography tandem mass spectrometry. J Chromatogr A 2018; 1572:203−11. Epub 2018/08/27.

[4] Campone L, Piccinelli AL, Celano R, Pagano I, Russo M, Rastrelli L. Rapid and automated analysis of aflatoxin M1 in milk and dairy products by online solid phase extraction coupled to ultra-high-pressure-liquid-chromatography tandem mass spectrometry. J Chromatogr A 2016;1428:212−9. Epub 2015/11/22.

[5] Moosavi SM, Shekar K, Fraser J, Smith MT, Ghassabian S. High-throughput assay for quantification of the plasma concentrations of thiopental using automated solid phase extraction (SPE) directly coupled to LC-MS/MS instrumentation. J Chromatogr B 2016; 1038:80−7. Epub 2016/11/03.

[6] Shou WZ, Zhang J. Recent development in software and automation tools for high-throughput discovery bioanalysis. Bioanalysis 2012;4(9):1097−109. Epub 2012/05/23.

[7] Zheng N, Jiang H, Zeng J. Current advances and strategies towards fully automated sample preparation for regulated LC-MS/MS bioanalysis. Bioanalysis 2014;6(18):2441−59. Epub 2014/11/12.

[8] Janiszewski J, Schneider R, Kapinos B, West M, Funk C, Federico J, et al. Development of a high-speed, multiplexed sample-delivery instrument for LC-MS/MS bioanalysis. Bioanalysis 2012;4(9):1039−56. Epub 2012/05/23.

[9] Murphy K, Bennett PK, Duczak JrJr. High-throughput quantitation of large molecules using multiplexed chromatography and high-resolution/accurate mass LC-MS. Bioanalysis 2012;4(9):1013−24. Epub 2012/05/23.

[10] Jian W, Romm MV, Edom RW, Miller VP, LaMarr WA, Weng N. Evaluation of a high-throughput online solid phase extraction-tandem mass spectrometry system for in vivo bioanalytical studies. Anal Chem 2011;83(21):8259−66. Epub 2011/09/23.
[11] Jemal M, Ouyang Z, Xia YQ. Systematic LC-MS/MS bioanalytical method development that incorporates plasma phospholipids risk avoidance, usage of incurred sample and well thought-out chromatography. Biomed Chromatogr 2010;24(1):2−19. Epub 2009/12/18.
[12] http://www.tomtec.com/.
[13] https://www.hamiltoncompany.com/automated-liquid-handling.
[14] Zheng N, Buzescu A, Pasas-Farmer S, Arnold ME, Ouyang Z, Jemal M, et al. A simplified and completely automated workflow for regulated LC-MS/MS bioanalysis using cap-piercing direct sampling and evaporation-free solid phase extraction. J Chromatogr B 2013;921−922:64−74. Epub 2013/02/26.
[15] Wang M, Ma P, Xi X, Liu L, Wen Y, Liu K, et al. Online solid phase extraction and liquid chromatography-mass spectrometric determination of nucleoside drugs in plasma. Talanta 2016;161:278−87. Epub 2016/10/23.
[16] Mastrianni KR, Metavarayuth K, Brewer WE, Wang Q. Analysis of 10 beta-agonists in pork meat using automated dispersive pipette extraction and LC-MS/MS. J Chromatogr B 2018;1084:64−8. Epub 2018/03/24.
[17] Mastrianni KR, Kemnitzer WE, Miller KWP. A novel, automated dispersive pipette extraction technology greatly simplifies catecholamine sample preparation for downstream LC-MS/MS analysis. SLAS Technol 2019;24(1):117−23. Epub 2018/08/17.
[18] https://www.tecan.com/.
[19] http://www.perkinelmer.com/category/automation-liquid-handling-instruments.
[20] Novakova L, Vlckova H. A review of current trends and advances in modern bio-analytical methods: chromatography and sample preparation. Anal Chim Acta 2009;656(1−2):8−35. Epub 2009/11/26.
[21] Tweed JA, Gu Z, Xu H, Zhang G, Nouri P, Li M, et al. Automated sample preparation for regulated bioanalysis: an integrated multiple assay extraction platform using robotic liquid handling. Bioanalysis 2010;2(6):1023−40. Epub 2010/11/19.
[22] Li M, Chou J, King KW, Yang L. ASPECTS: an automation-assisted SPE method development system. Bioanalysis 2013;5(13):1661−76. Epub 2013/07/05.
[23] Guidance for industry, bioanalytical method validation. U.S. Department of Health and Human Services, Food and Drug Administration, Center for Drug Evaluation and Research (CDER), Center for Veterinary Medicine (CMV); 2018. Document available at: https://www.fda.gov/downloads/drugs/guidances/ucm070107.pdf.
[24] Huber S, Brox J. An automated high-throughput SPE micro-elution method for perfluoroalkyl substances in human serum. Anal Bioanal Chem 2015;407(13):3751−61. Epub 2015/03/22.
[25] Alnouti Y, Srinivasan K, Waddell D, Bi H, Kavetskaia O, Gusev AI. Development and application of a new on-line SPE system combined with LC-MS/MS detection for high throughput direct analysis of pharmaceutical compounds in plasma. J Chromatogr A 2005;1080(2):99−106. Epub 2005/07/13.
[26] Kataoka H, Saito K. Recent advances in column switching sample preparation in bioanalysis. Bioanalysis 2012;4(7):809−32. Epub 2012/04/20.
[27] Verdirame M, Veneziano M, Alfieri A, Di Marco A, Monteagudo E, Bonelli F. Turbulent flow chromatography TFC-tandem mass spectrometry supporting in vitro/vivo studies of NCEs in high throughput fashion. J Pharm Biomed Anal 2010;51(4):834−41. Epub 2009/11/10.
[28] https://www.sparkholland.com/.

[29] https://www.thermofisher.com/order/catalog/product/65000-60100.
[30] https://www.agilent.com/.
[31] Pan J, Zhang C, Zhang Z, Li G. Review of online coupling of sample preparation techniques with liquid chromatography. Anal Chim Acta 2014;815:1–15. Epub 2014/02/25.
[32] Rogeberg M, Malerod H, Roberg-Larsen H, Aass C, Wilson SR. On-line solid phase extraction-liquid chromatography, with emphasis on modern bioanalysis and miniaturized systems. J Pharm Biomed Anal 2014;87:120–9. Epub 2013/06/12.
[33] Ghassabian S, Moosavi SM, Valero YG, Shekar K, Fraser JF, Smith MT. High-throughput assay for simultaneous quantification of the plasma concentrations of morphine, fentanyl, midazolam and their major metabolites using automated SPE coupled to LC-MS/MS. J Chromatogr B 2012;903:126–33. Epub 2012/07/31.

# Design of experiments and method development

# 22

*Aline L.H. Müller, Júlia A. de Oliveira, Osmar D. Prestes, Martha B. Adaime, Renato Zanella*
Laboratory of Pesticide Residue Analysis (LARP), Chemistry Department, Federal University of Santa Maria, Santa Maria, Rio Grande do Sul, Brazil

## 22.1 Introduction

Although analytical methods and instrumentation have enjoyed great technological advances in recent decades, sample preparation remains one of the most important steps of an analytical protocol and is essential for obtaining accurate and precise results. During the development of a sample preparation method using solid phase extraction, for example, the need to evaluate the combination of the many variables has the objective of selecting the best conditions for adequate performance of the method [1].

The main and most common sample preparation methods based on solid phase extraction employ the techniques: solid phase microextraction, stir bar sorptive extraction, matrix solid phase dispersion, in-tube solid phase microextraction, and solid phase extraction in a cartridge format, among others. The variables that normally should be considered during the optimization step are summarized in Fig. 22.1. Thus, in method development, the optimization of the different variables is a critical step to find the optimum value for each factor to produce the best possible response. In an optimization, one tries to find the optimal settings or conditions for a number of factors. The factors (variables) are parameters that can be defined and redefined at certain levels [2].

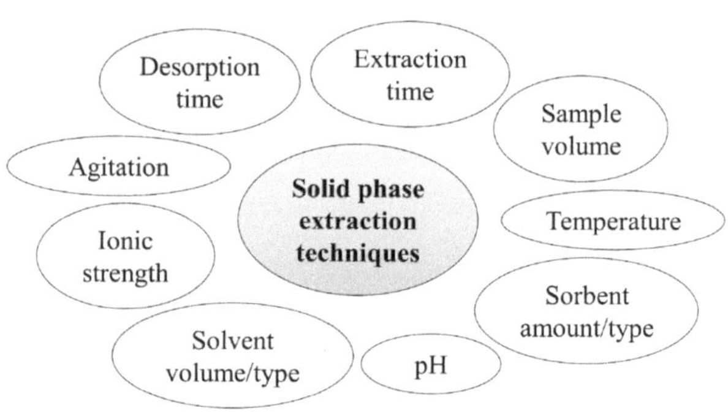


**Figure 22.1** Typical variables considered in solid phase extraction method development and optimization.

**Solid-Phase Extraction. https://doi.org/10.1016/B978-0-12-816906-3.00022-4**

In most methods, the optimization step is performed by monitoring the influence of one factor at a time on an experimental response. Consequently, there is an increase in the number of experiments necessary to conduct the study, leading to an increase in the consumption of reagents and materials, as well as time [3]. In later years, studies have employed multivariate tools with the purpose of obtaining higher performance during the development of methods [4–7]. In this context, the multivariate design of experiments (DOE) is an excellent tool, because it takes less time, effort, and resources than the univariate procedures, and allows the collection of a large amount of information while minimizing the number of experiments [8]. DOE allows simultaneously evaluation of several factors at several levels in a predefined number of experiments.

The method optimization using DOE can be divided in two steps: screening and optimization designs. The screening designs typically employ full factorial (FFD), fractional factorial (FFrD), and Plackett-Burman (PBD) designs that allow evaluation of a large number of factors in a relatively small number of experiments and are used to identify the most influencing factors, usually evaluated at two levels [9]. Response surface designs are typically used for optimization to find the optimal levels of the most important factors (previously selected from a screening design) [10].

When the aim is to find the best conditions for extracting pesticides from food samples using matrix solid phase dispersion, for example, among the factors chosen are: different types of solid dispersing support, different elution solvents, homogenization time, among others factors. In the screening step, the researcher wants to know which of these factors deserve more attention, without solving the problem completely. After selecting the factors, the next step is to select the levels (or at least the limits): amount of solid dispersing support, volume of eluting solvents, among others. Thus, after this selection, the type of DOE should be chosen, and this choice will depend on how many factors need to be studied. For example, with a few factors, it is feasible to use FFD. However, with an increasing number of factors, FFrD is better suited for screening the variables, because it provides sufficient information in fewer experiments. If there is sufficient knowledge of the process but not necessarily the particular application, it may be possible to start with the optimization design in order to obtain the response surface, and thus to obtain a more comprehensive perspective of the method [11]. In this sense, it is important to emphasize that DOE is an important step in methods development, since it allows evaluating only the variables that are important, reducing the number of experiments required.

Therefore, this chapter aims to provide an overview about the use of design of experiments as an auxiliary tool in the sample preparation process using solid phase extraction. In this sense, some concepts of DOE are reviewed and some applications are presented.

## 22.2 Methodology of design of experiments

As previously reported, one of the most common problems during method development is to determine the influence of one or more variables on an additional variable of interest. Thus, multivariate design of experiments (DOE) with suitable statistical analysis have become popular in the recent years, allowing the development of analytical methods with a reduction in the number of experiments and saving of financial resources [8].

An important role of experimental design is method optimizations, where the main objective is to identify the experimental conditions with the best possible performance. Thus, in the design of any experiment, the first step is to decide what are the *factors* and the *response* of interest. The *factors or independent variables* are experimental variables that can be changed independently of each other. The factors are the supposed or certain causes with an influence on the phenomenon studied [2]. Typical factors comprise the pH, temperature, reagents concentration (solvents), ionic strength, among others. The factors can present different values (*levels of a factor*) at which the experiments must be carried out. The factor solvent, for example, can be investigated at the volumes 5, 10, and 15 mL in the optimization of elution volume using SPE. Lastly, *responses or dependent variables* are the measured values of the results from experiments that may be affected by changes caused by the factors. Recoveries of an analyte, resolution among chromatographic peaks, among others are examples of responses [3].

During the development an analytical method using DOE, two groups are considered important, within which are contemplated the different types of design: screening design and optimization design [12] as are showed in Fig. 22.2.

## 22.2.1 Screening designs for solid phase extraction

Screening designs allow the evaluation of a relatively large number of factors with a relatively small number of experiments, since the numerous variables may affect the response of the system studied, and it is practically impossible to identify and control the small contributions from each one. The choice of the factors is essential for the efficiency of the optimization process. Thus, screening designs have been used to indicate the most important factors from those potentially influencing the considered responses [13]. Normally, the factors are evaluated at two ($2^k$, where k is factor number) or three levels ($3^k$), with two levels the most widely used for the selection

**Figure 22.2** Classification of designs of experiments.

of factors, since they are economic, efficient, and [14] allow screening of a large number of factors with a smaller number of experiments [15]. Response surface designs are used to find the optimal levels of the most important factors (which occasionally are selected from a screening design approach).

During the development of methods for solid phase extraction, a large number of factors (sorbent, solvent, temperature, pH, among others) may have a beneficial influence, while others have no significant effect during the process. Therefore, it is necessary to select those factors with major effects, in order to avoid a large numbers of experiments. For this purpose Full Factorial designs (FFD), Fractional Factorial designs (FrFD), and Plackett-Burman designs (PBD) are more frequently used as screening designs (Fig. 22.2) [16].

### 22.2.1.1 Full factorial design

When factorial design (FFD and FrFD) is used, each factor is varied between fixed settings. FFD is probably the most common and intuitive strategy of experimental design, and it is defined by all possible combinations of the factors and their settings. Assume that there are $k$ factors to investigate and that each factor can be set to $L$ different levels. The number of possible combinations of the factors and their settings will then be $N = L^k$, where N is the number of experiments [11,13]. It is noted that the number of possible combinations increases rapidly when the number of levels and the number of factors increases, such as: (A) two-factor, two-level ($N = 2^2 = 4$); (B) three-factor, two-level ($N = 2^3 = 8$); and (C) three-factor, three-level ($N = 3^3 = 27$). Fig. 22.3 shows the graphical full factorial designs with different levels. Therefore, the evaluation of experimental factors in chemical systems is usually carried out at only two levels of the factor setting. Thus, such designs permit the determination of all main effects and all interaction effects [11].

The advantage of full factorial designs is that they make a very efficient use of the data and do not mistake the effects of the parameters, so that it is possible to evaluate

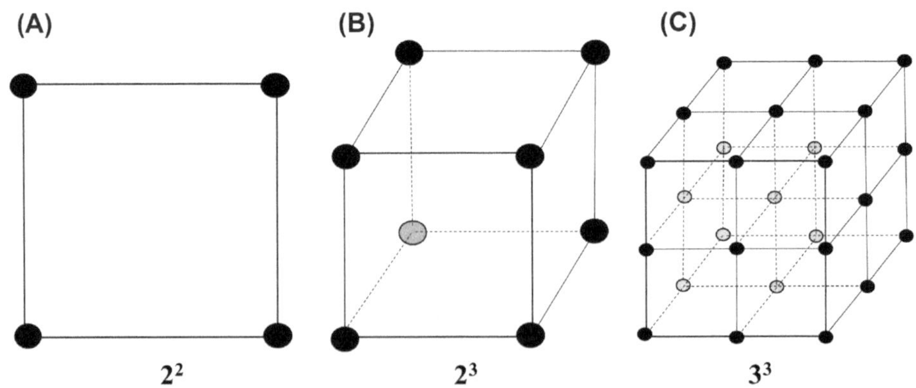


**Figure 22.3** Example of $L^k$ full factorial experimental designs: (A) Two-factor, two-level ($2^2$); (B) three-factor, two-level ($2^3$); and (C) two-factor, three-level ($3^3$) [9].

the main factors and their interactions clearly. On the other hand, it is impossible to screen more than about 16 factors due to an exponential increase in the number of assays [9,15].

Full factorial experiments of two levels are convenient when the number of factors, k, to explore is $\leq 4$. With more than four factors to explore, fractional factorial designs at two levels or other two-level designs are more convenient [11].

#### 22.2.1.2 Fractional factorial designs

The fractional factorial design (FrFD) is one of the most used screening design for solid phase extraction methods (Table 22.1). When the number of parameters increases, the FrFD becomes onerous from a practical perspective. Thus, the idea of the FrFD is to run only one subset of the FFD experiments and yet provide quite good information on the main effects and some information about their interactions. The sample size of the FrFD can be *one-half*, or *one-quarter*, and so on, of a full factorial design [9].

Considering that fractional factorial designs contain only a fraction of the experiments from a $2^k$ full factorial design, then the number of experiments for two-level fractional factorial designs can be represented as $N = 2^{k-p}$, since, $2^{k-p} = 2^k 2^{-p} = 2^k/2^p$, where p represents the number of independent design generators, chosen to fractionate the design. Thus, $1/2^p$ representing the fraction of the full factorial ($p = 1, 2, 3$, etc.) [11,17,18].

Many fractional designs can be created for a large number of factors. Thus, for a better understanding, it is possible to exemplify that when examining six factors (sorbent type, solvent volume, sample amount, sample pH, agitation time, and temperature, for example), during development of a method for solid phase extraction, a one-half fraction factorial design has $2^{6-1} = 32$ experiments, a quarter-fraction factorial design has $2^{6-2} = 16$ experiments, and an eighth-fraction factorial design requires only $2^{6-3} = 8$ experiments. Despite this, fractional designs do not enable the estimation of all major factors and their interactions separately because some of them are estimated together, in others words they are confounded [18]. Additionally, it is important to emphasize that only a fraction of a full factorial design is performed, causing loss of some information. Fig. 22.4 shows graphical examples of fractional factorial designs.

#### 22.2.1.3 Plackett-Burman design

The Plackett-Burman design (PBD) has been employed as a design of experiments for robustness evaluation during method development, once it assumes that some interactions can be completely ignored and only the main effects are calculated with a reduced number of experiments [19]. A Plackett-Burman design allows a maximum of $k = N-1$ factors to be evaluated in N experiments, where N is a multiple of four ($N = 8, 12, 16, 20$, etc.) [18].

Thus, to examine a given number of factors, different PBD can be chosen. For example, to examine six factors, a PBD with eight experiments or one with 12

**Table 22.1** Design of Experiments used for the methods development of solid phase extraction.

| Screening | Matrix | Analytes | Sample preparation | Central composite<br>Level CCD | Factors evaluated | Software used | References |
|---|---|---|---|---|---|---|---|
| – | Water | Pesticides | SPE | 5 | Sample pH and eluent composition | – | [33] |
| Plackett-Burman | Water | Benzophenone-type ultraviolet absorbers | d-μSPE | 3 | Screening: Shaking time, adsorbents drying time and temperature, derivative volume, thermal desorption time and temperature<br>Optimization: Adsorbent drying time and temperature, and thermal desorption time | Design-Expert 8.0.6 | [34] |
| – | Water | 17β-estradiol | SPE | 5 | Glutaraldehyde volume, gelatin amount, preparation time, and temperature | SPSS 18 | [35] |
| – | Water | Pharmaceuticals | SPME | 5 | Extraction time and temperature, and salt amount | Design-Expert 8.0.1 | [36] |
| Fractional factorial design | Human urine and plasma | Losartan and valsartan | SBSE | 5 | Screening: Sample pH, ionic strength, extraction time, volume of desorption solvent, desorption time and stirring speed<br>Optimization: Sample pH, extraction time, and volume of desorption solvent | Statistica 7.0 | [37] |
| Fractional factorial design | Vegetables | Pesticides | SPE (clean-up) | 5 | Screening: Sample amount, solvent volume, NaCl and $MgSO_4$ amounts<br>Optimization: Sample and NaCl amount | Minitab statistical 17 | [38] |

| | | | | | | | |
|---|---|---|---|---|---|---|---|
| – | Wine | Ochratoxin A | In-tube SPME | 5 | Solvent composition and sample load time | Statistica 10.0 | [39] |
| – | Bovine muscle, kidney, and liver | Veterinary drugs | d-SPE | 5 | Type and amount of sorbent | Statistica 7.0 | [6] |
| – | Jambolan fruit | Volatile compounds | HS-SPME | 5 | Extraction temperature and time, and salt amount | Statistica 12.0 | [40] |
| – | Water | Emerging contaminants | BAμE | 5 | Equilibrium time, stirring speed, pH, and time of ultrasonic treatment | Statistica 8.0 | [41] |
| – | Biscuits | Acrylamide | MSPE | 5 | Sorbent amount, sonication time, and sample pH | – | [25] |
| Plackett-Burman | Water, human urine, and milk | Meloxicam and piroxicam | d-SPME | 3 | Screening: Sample pH, sorbent amount, sample volume, dispersion time, extraction time, acceptor phase volume, and desorption time<br>Optimization: Sample pH, sorbent amount, and dispersion time | – | [26] |

**Box-Behnken**

| Screening | Matrix | Analytes | Sample preparation | Level BBD | Factors evaluated | Software used | References |
|---|---|---|---|---|---|---|---|
| Plackett–Burman | Water | Parathion | SPE | 3 | Screening: pH of extraction medium, stirring rate, sample volume, elution solvent volume, organic content of sample, and extraction time<br>Optimization: Elution solvent volume, organic solvent content of sample, and sample volume | | [28] |

*Continued*

**Table 22.1** Design of Experiments used for the methods development of solid phase extraction.—cont'd

| Box-Behnken | | | | | | | |
|---|---|---|---|---|---|---|---|
| **Screening** | **Matrix** | **Analytes** | **Sample preparation** | **Level BBD** | **Factors evaluated** | **Software used** | **References** |
| – | Soil | Antibiotics | DMAE-SPE | 3 | Microwave power, extraction solvent volume, and flow rate | Design-Expert 7.1.3 | [42] |
| – | Soil | Pesticides | MSPD | 3 | Solvent composition, dispersant ratio (dispersant/sample), and dispersant composition | Design-Expert 7.1.6 | [43] |
| – | Water | Polychlorinated biphenyls | MSPE | 3 | Extraction time, magnetic microspheres amount, and ionic strength | – | [44] |
| – | Water | Chlorophenoxy acid herbicides | SPE–EME | 3 | Voltage, extraction time, and pH of the acceptor phase | Stat graphics plus Package 5.1 | [45] |
| – | Human urine | Flavonoids | SPE | 3 | Sample volume and loading rate, volume of solvent elution, and elution rate | Design-Expert 8.0.6 | [7] |
| – | Water | Neonicotinoids | SPE | 3 | Type of sorbent and the type and volume of eluent | Design-Expert 8.0.6 | [46] |
| – | Berries | Monoterpenes | HS-SPME | 3 | Extraction temperature, extraction time, and equilibrium time | Design-Expert 8.0.7.1 | [29] |

| | | | | | | | |
|---|---|---|---|---|---|---|---|
| – | Water | Rhodamine B and Malachite green | PT-μSPE | 3 | pH of sample solution, surfactant concentration, ionic strength, number of extraction cycles, number of elution cycles, eluent volume, and volume of sample | – | [47] |

**Doehlert**

| Screening | Matrix | Analytes | Sample preparation | Level Doehlert | Factors evaluated | Software used | References |
|---|---|---|---|---|---|---|---|
| – | Honey and polen | Inseticides | SPE | 3 and 5 | Methanol-water ratio and dilution factor | – | [30] |
| Fractional factorial design | Wastewater | Pharmaceuticals | SPME | 3, 5 and 7 | Screening: Temperature, extraction time, sample volume, ionic strength<br>Optimization: Temperature, extraction time, sample volume | Statistica 6.0 | [48] |
| Full factorial | Water | Pesticides | SPE | 3, 5 and 7 | Screening: Elution volume, sample volume, adsorbent mass, and sorbent type<br>Optimization: Elution volume, sample volume, and adsorbent mass | Nemrod | [31] |
| – | Bovine milk | Pesticides | HS-SPME | 3 and 5 | Extraction time and temperature | – | [32] |
| Full factorial | Beer | Volatile compounds | SPDE | 3 and 5 | Screening: Extraction temperature, number of strokes, and fill/eject speed<br>Optimization: Extraction strokes and extraction temperature | Sigma Plot | [49] |
| – | Water | Organic explosives | SBSE | – | Extraction time profiles, sample volume, modifier addition, and acetic acid addition | Statistica V12 | [50] |

*Continued*

**Table 22.1** Design of Experiments used for the methods development of solid phase extraction.—cont'd

| Doehlert | | | | | | | |
|---|---|---|---|---|---|---|---|
| **Screening** | **Matrix** | **Analytes** | **Sample preparation** | **Level Doehlert** | **Factors evaluated** | **Software used** | **References** |
| Mixed-level factorial design | Oral fluid | Psychoactive substances | MEPS | 3 and 4 | Screening: Volume of washing solution, volume of eluent, sorbent type, number of aspirate-dispense cycles and sample pH<br>Optimization: Eluent volume and number of strokes (aspirate-dispense cycles) | NemrodW | [4] |
| Fractional factorial design | Human urine | Volatile organic metabolites | HS-SPME | 2 and 3 | Screening: Agitation time, incubation time, extraction time, temperature, and NaCl amount<br>Optimization: Incubation and extraction time, and temperature | Minitab 17 and SAS 9.4. | [5] |

*BAμE*, bar adsorptive microextraction; *d-μSPE*, dispersive micro-solid phase extraction; *DMAE*, dynamic microwave-assisted extraction; *d-SPME*, dispersive-solid phase microextraction; *EME*, electromembrane extraction; *HS-SPME*, headspace-solid phase microextraction; *In-tube SPME*, in-tube solid phase microextraction; *MEPS*, microextraction by packed sorbent; *MPSD*, matrix solid phase dispersion; *MSPE*, magnetic solid phase extraction; *PT-μSPE*, pipette tip micro-solid phase extraction; *SBSE*, stir bar sorptive extraction; *SPDE*, solid phase dynamic extraction; *SPE*, solid phase extraction; *SPME*, solid phase microextraction

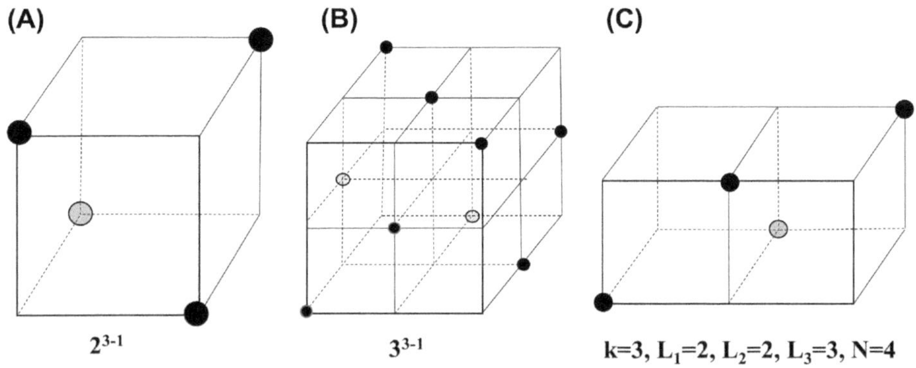


**Figure 22.4** Example of fractional factorial experimental designs [9].

experiments can be chosen. Whenever the number of factors that can, potentially, be tested in a PBD (N − 1) exceeds the number of factors to be examined (in other words, when the design is not a multiple of 4), the remaining columns are defined as dummy factor columns. The dummy factor is an imaginary variable (virtual variables) and these columns do not cause any physicochemical changes in the final design [17,18].

### 22.2.2 Optimization designs and applications for solid phase extraction

The designs employed in the screening stage only identify the statistically significant factors [20]. Based on this, an optimization step is required, where the factors will be further examined to find the optimum conditions [14]. In order to reduce the number of experiments, the optimization of procedures involving more than one factor is performed using multivariate statistic techniques [3].

The response surface methodology (RSM) is a multivariate technique based on factorial designs, which has shown great applicability for the determination of experimental conditions used in solid phase extraction, as can be seen in Table 22.1. RSM consists of a set of mathematical and statistical techniques based on the adjustment of a polynomial equation to the experimental data, which aims to describe the behavior of a group of observation to make statistical predictions [3]. The most commonly employed response surface designs in analytical chemistry are the central composite, Box-Behnken, and Doehlert design [21].

#### 22.2.2.1 Central composite design

The central composite design (CCD) is probably the most popular of all second-order designs and has been used for modeling processes with two or more factors [22]. This design consists of basically three stages: a full or fractional two-level factorial design, a star design, and a center point [3,9,20,23]. Whatever the number of factors, for a CCD to fit a second-order model, they are studied at five levels ($-\alpha$, $-1$, 0, $+1$, and $+\alpha$), except for $\alpha = 1$ [10]. The number of experiments required is defined by the expression $N = 2^k + 2k + Cp$, where k is the number of factors and Cp is the number of central points [3,21,24]. The flexibility of the CCD lies in the selection of $\alpha$ (axial distance) and Cp. The choice of these two parameters is important and can impact on the suitability of the resulting design [10]. An interesting advantage of CCD is that full two-level factorial designs can be performed not only for optimization, but also for screening of the factors [20]. Fig. 22.5 presents the full CCD for the optimization of two and three variables [3]. Table 22.1 summarizes recent applications of CCD in the solid phase extraction.

Rizzetti et al. [6] employed two CCDs with two variables at five levels to evaluate the most appropriate combination of sorbents for dispersive solid phase extraction (d-SPE) for the multiclass determination of veterinary drugs in bovine muscle, kidney, and liver by LC-MS/MS. The first CCD was performed with a combination of $C_{18}$ and Florisil and the second with a combination of diatomaceous earth and EMR-Lipid. Based on the results obtained from CCD and gravimetric tests, EMR-Lipid (140 mg $mL^{-1}$) was the selected sorbent for d-SPE. The recoveries were higher than 70% for all veterinary drugs. The use of CCD in this study facilitated the selection of the best sorbents for sample preparation.

Bagheri et al. [25] developed a dummy molecularly imprinted polymer (DMIP)-magnetic solid phase extraction (MSPE) method for the extraction of acrylamide in biscuit samples. The variables evaluated by CCD were sorbent dosage, sonication time, and sample pH, all at five different levels. Fig. 22.6 shows the response surfaces obtained for the evaluation of the recovery (ER%) of acrylamide for two independent variables. The optimum conditions were sample pH 5.0, 30 mg of DMIP, and sonication time of 20 min. Recoveries higher than 86% with RSD $\leq$4.1% were obtained for acrylamide in spiked biscuit samples.

Ghorbani et al. [26] analyzed meloxicam and piroxicam in water, human urine, and milk samples using ultrasonic assisted magnetic dispersive solid phase microextraction (SPME). A Plackett-Burman design with 16 experiments was employed to investigate seven main factors (sample pH, amount of sorbent, sample volume, dispersion time, extraction time, acceptor phase volume, and desorption time). In the next step, a

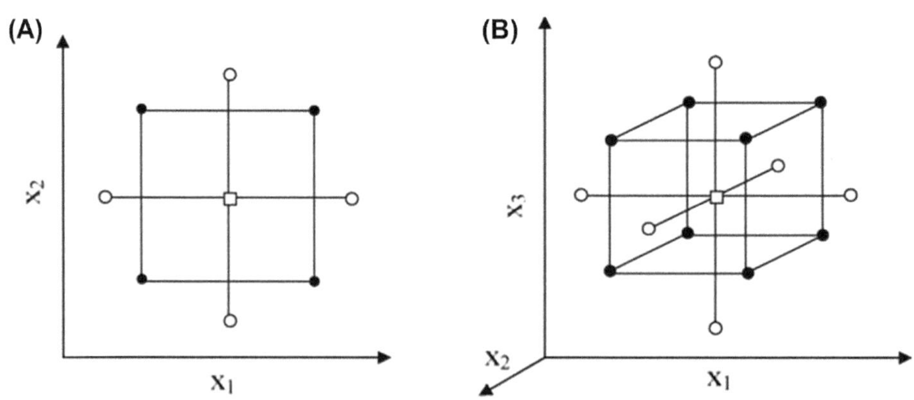


**Figure 22.5** Central composite designs for the optimization of: (A) two variables and (B) three variables. (•) Points of factorial design, (○) axial points, and (□) central point [3].

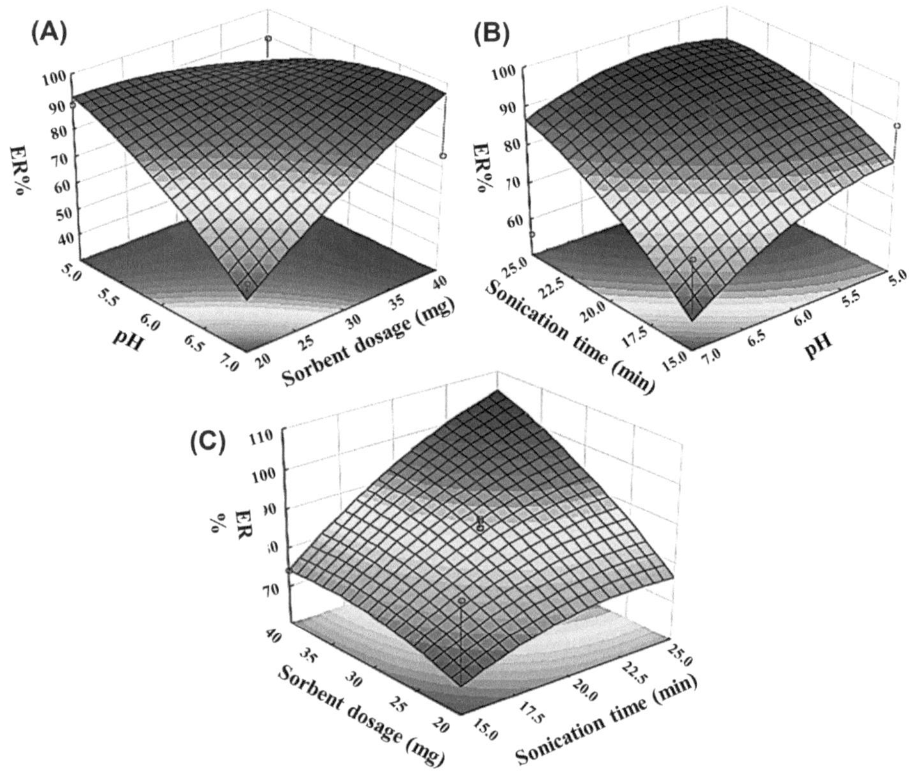


**Figure 22.6** Response surfaces obtained by CCD for (A) pH-sorbent dosage (mg), (B) sonication time (min)-pH, and (C) sorbent dosage (mg)-sonication time (min) at 25°C [25].

CCD with 20 experiments was employed to optimize the three main factors (sample pH, amount of sorbent, and dispersion) at three levels each, selected from the screening stage. The optimum values of the main factors were as follows: sample pH of 7.4, amount of sorbent of 36 mg, and dispersion time of 5.3 min.

### 22.2.2.2 *Box-Behnken design*

The Box-Behnken design (BBD) can be used for process modeling of three or more factors, all studied at three levels ($-1$, 0, $+1$) with equally spaced intervals between levels [20]. This design is based on an incomplete but balanced block design. The three variable designs consist of three blocks: in each of them, two variables are combined following a $2^2$ factorial design and the third variable is maintained at level zero. Furthermore, several center points are added. It is important to note that the BBD is a spherical design with all points in a sphere of radius $\sqrt{2}$. Moreover, this design does not contain points at the vertices of the cubic region created by the highest and lowest limits for each variable [10]. The number of experiments required to obtain the model is defined as $N = 2k\ (k - 1) + Cp$, where k is the number of factors and

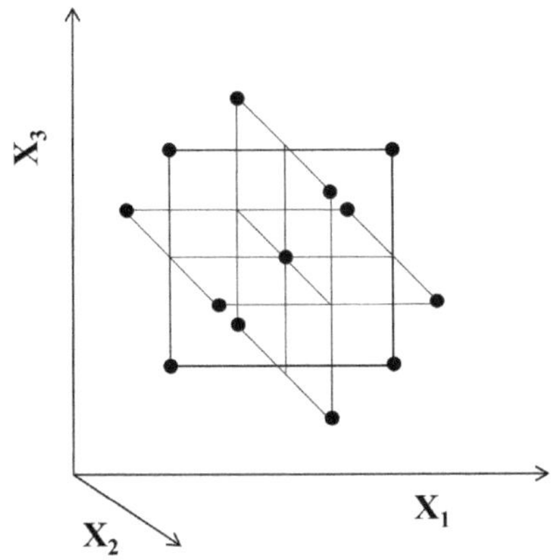


**Figure 22.7** Box-Behnken design for the optimization of three variables [3].

Cp the number of center points [27]. BBD is more efficient and economical than the corresponding $3^k$ designs, especially for a large number of variables. Fig. 22.7 shows the BBD for the optimization of three variables [3]. The BBD was applied for the optimization of several solid phase extraction variables; however, its application in analytical chemistry is sparse compared with CCD [3]. According to Table 22.1, parameters typically optimized with this design are extraction time, sample volume, and elution solvent volume.

Alizadeh et al. [28] developed a molecularly imprinted polymer (MIP)–based solid phase extraction method for the extraction of parathion from different water samples. Initially, a Plackett-Burman design was employed to screen the main factors for the extraction (pH of extraction medium, stirring rate, sample volume, elution solvent volume, organic solvent sample content, and extraction time). According to these results, elution solvent volume, organic solvent sample content, and sample volume were selected for the BBD optimization. The three variables were evaluated at three levels, for a total of 15 experiments. The optimum conditions using the desirability function method were 427 mL for sample volume, 7 mL for elution solvent volume, and 5% for organic solvent sample content with a recovery >95%.

Hou et al. [7] optimized the factors affecting the extraction efficiency of flavanoids from urine using solid phase extraction in a cartridge format with BBD. Four variables at three levels each (low, medium, and high) were evaluated: volume of sample loading (50–70 mL), sample loading rate (1–2 mL $min^{-1}$), volume of elution (1–1.2 mL) and elution rate (0.2–0.8 mL $min^{-1}$). The optimum conditions were: sample volume, 56 mL; sample loading rate, 1.25 mL $min^{-1}$; eluent solvent, methanol; elution volume, 1.1 mL; and elution rate, 0.2 mL $min^{-1}$.

Chmiel et al. [29] analyzed 12 monoterpenes in berries of different blue honeysuckle cultivars and related berry products using headspace-solid phase microextraction (HS-SPME). Three variables (extraction temperature, extraction time, and equilibrium time) were evaluated at three levels with a BBD. The optimum HS-SPME conditions were extraction temperature of 50°C, extraction time of

26 min, and equilibrium time of 29 min. The method resulted in recoveries between 70.8% and 99.2% for the monoterpenes.

#### 22.2.2.3 Doehlert design

The Doehlert design is a more practical and economical alternative compared to other second-order designs, since they require a smaller number of experiments [3,20,24]. In Doehlert designs, the number of levels is not the same for all factors. For two factors, this design is established with three and five levels, while for three factors, three, five, and seven levels are utilized [21,24]. This feature is particularly important when some factors are subject to instrumental or cost constraints or to study a factor at a higher or lower number of levels [3]. The number of experiments is defined by the relationship $N = k^2 + k + Cp$, where k is the number of factors and Cp is the number of central points [24]. Another feature of this design is that the displacement of the experimental matrix to another experimental region can be achieved while retaining the previously explored adjacent points. Fig. 22.8 represents the Doehlert design for the optimization of two variables and some possibilities for the displacement of the initial design using previous points [3].

During optimization procedures it is often necessary to employ different levels for each factor. In this case, the Doehlert design is more efficient and offers certain advantages compared with CCD and BBD. Among the sample preparation methods that use Doehlert design more are SPE and SPME, according to Table 22.1. García-Chao et al. [30] used the Doehlert design to identify the effects of two parameters on the SPE for the determination of pesticide residues in honey and pollen samples. Seven experiments were performed with all models significant at the 99% percent confidence level ($P < .05$) and the correlation coefficients were higher than 0.99. Recoveries higher than 80% were obtained for all analytes. Similarly, Doehlert design was used to represent the responses for three significant factors (Table 22.1) from a full factorial design (two levels) for the preconcentration of carbamate pesticides in water samples by SPE

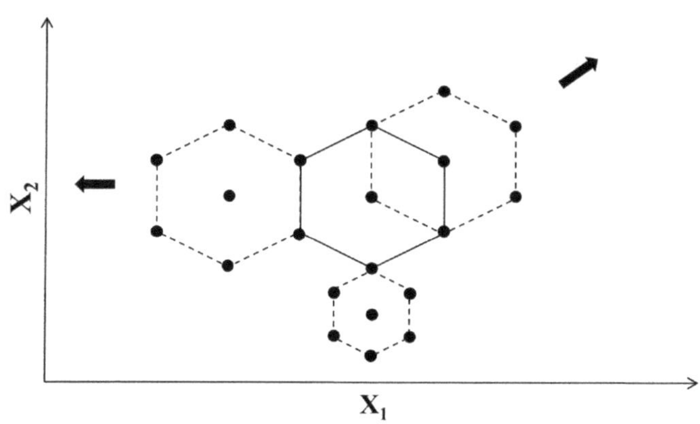


**Figure 22.8** Doehlert design for the optimization of two variables and possibilities for the displacement of the initial design using previous points [3].

[31]. Optimum working conditions were identified from 13 experiments: sample volume 143.5 mL; 200 mg of C-18 sorbent; and elution with 5.5 mL of acetonitrile. The recoveries for all carbamates exceeded 90% with relative standard deviations (RSD) in the range 3%−11%.

Merib et al. [32] used HS-SPME for the extraction of organochlorine pesticides from bovine milk samples and a Doehlert design for factor optimization. Extraction temperature (40, 60, and 80°C) and extraction time (10, 37, 65, 92, and 120 min) were evaluated. The extraction time was studied at five levels on account of its wider interval compared with the extraction temperature, which was limited to 80°C to minimize degradation of the milk samples. From the response surface design, the optimum extraction temperature was 80°C and extraction time of 90 min with recoveries >75%. Semren et al. [5] used a design of experiments approach to optimize an HS-SPME method for the determination of volatile organic metabolites in urine. A Doehlert design was used to optimize the most influential factors identified by application of a fractional factorial design. The optimum values for temperature was 60°C, extraction time 50 min, and incubation time 35 min. Volatile organic metabolites were identified in more than 70% of samples and with an RSD less than 30%.

## 22.3 Conclusions

The multivariate optimization of parameters employing DOE is increasingly used for the sample preparation step. The choice of the appropriate DOE technique depends mainly on the purpose of the experiment. If a rough estimate of the main factors is sufficient, a Plackett-Burman method would be adequate. If it is necessary to make a more precise estimate and consider some interactions among the factors, a fractional or full factorial design is more appropriate. When the goal is to optimize results with response surfaces, the CCD, BBD, or Doehlert design must be chosen. In the case of solid phase extraction, it is possible to perform a screening step to select the main factors and then optimize these using response surface techniques.

Among the main advantages of DOE are a reduction in the number of experiments with a reduced consumption of reagents and solvents and the production of statistical models which allow the best conditions for the process to be identified. Finally, the use of DOE for method development is, nowadays, largely accepted due to its advantages over the classical univariate optimization approach.

## References

[1] Moldoveanu SC, David V. Sample preparation in chromatography. Amsterdam: Elsevier; 2002.

[2] Phan-Tan-Luu R, Cela R. Experimental designs: conclusions, terminology, and symbols. In: Brown SD, Tauler R, Walczak B, editors. Comprehensive chemometrics: chemical and biochemical data analysis. Amsterdam: Elsevier; 2009.

[3] Bezerra MA, Santelli RE, Oliveira EP, Villar LS, Escaleira LA. Response surface methodology (RSM) as a tool for optimization in analytical chemistry. Talanta 2008;76: 965–77.
[4] Ares AM, Fernández P, Regenjo M, Fernández AM, Carro AM, Lorenzo RA. A fast bioanalytical method based on microextraction by packed sorbent and UPLC-MS/MS for determining new psychoactive substances in oral fluid. Talanta 2017;174:454–61.
[5] Semren TZ, Karačonji IB, Safner T, Brajenović N, Lovaković BT, Pizent A. Gas chromatographic-mass spectrometric analysis of urinary volatile organic metabolites: optimization of the HS-SPME procedure and sample storage conditions. Talanta 2018;176: 537–43.
[6] Rizzetti TM, De Souza MP, Prestes OD, Adaime MB, Zanella R. Optimization of sample preparation by central composite design for multiclass determination of veterinary drugs in bovine muscle, kidney and liver by ultra-high-performance liquid chromatographic-tandem mass spectrometry. Food Chem 2018;246:404–13.
[7] Hou X, Liu S, Zhouc P, Li J, Liu X, Wang L, Guo Y. Polymeric ionic liquid modified graphene oxide-grafted silica for solid-phase extraction to analyze the excretion dynamics of flavonoids in urine by Box-Behnken statistical design. J Chromatogr, A 2016;1456: 10–8.
[8] Brereton RG. Chemometrics: data analysis for the laboratory and chemical plant. Chichester: John Wiley & Sons; 2003.
[9] Cavazzuti M. Optimization methods: from theory to design. Berlin Heidelberg; Springer; 2013.
[10] Sarabia LA, Ortiz MC. Response surface methodology. In: Brown SD, Tauler R, Walczak B, editors. Comprehensive chemometrics: chemical and biochemical data analysis. Amsterdam: Elsevier; 2009.
[11] Carlson R, Carlson JE. The study of experimental factors. In: Brown SD, Tauler R, Walczak B, editors. Comprehensive chemometrics: chemical and biochemical data analysis. Amsterdam: Elsevier; 2009.
[12] Araujo PW, Brereton RG. Experimental design III. Quantification. Trends Anal Chem 1996;15:156–63.
[13] Miller JN, Miller JC. Statistics and chemometrics for analytical chemistry. Great Britain: Pearson Education; 2005.
[14] Candioti LV, De Zan MM, Cámara MS, Goicoechea HC. Experimental design and multiple response optimization. Using the desirability function in analytical methods development. Talanta 2014;124:123–38.
[15] Sahu PK, Ramisetti NR, Cecchi T, Swain S, Patro CS, Panda J. An overview of experimental designs in HPLC method development and validation. J Pharm Biomed Anal 2018; 147:590–611.
[16] Bruns RE, Scarminio IS, De Barros Neto B. Statistical design - chemometrics, vol. 25. Elsevier; 2006. p. 422.
[17] Dejaegher B, Heyden YV. The use of experimental design in separation science. Acta Chromatogr 2009;21:161–201.
[18] Dejaegher B, Durand A, Heyden YV. Experimental design in method optimization and robustness testing. In: Hanrahan G, Gomez FA, editors. Chemometric methods in capillary electrophoresis. New Jersey: John Wiley & Sons; 2010.
[19] Ferreira SLC, Caires AO, Borges TS, Lima AMDS, Silva LOB, Dos Santos WNL. Robustness evaluation in analytical methods optimized using experimental designs. Microchem J 2017;131:163–9.

[20] Ferreira SLC, Lemos VA, De Carvalho VS, Da Silva EGP, Queiroz AFS, Felix CSA, Da Silva DLF, Dourado GB, Oliveira RV. Multivariate optimization techniques in analytical chemistry - an overview. Microchem J 2018;140:176–82.

[21] Ferreira SLC, Silva Junior MM, Felix CSA, Da Silva DLF, Santos AS, Santos Neto JH, De Souza CT, Cruz Junior RA, Souza AS. Multivariate optimization techniques in food analysis – a review. Food Chem 2019;273:3–8.

[22] Montgomery DC. Design and analysis of experiments. New York: John Wiley & Sons; 1997.

[23] Khuri AI, Mukhopadhyay S. Response surface methodology. Wiley Interdiscip Rev: Comput Stat 2010;2:128–49.

[24] Ferreira SLC, Dos Santos WNL, Quintella CM, Neto BB, Bosque-Sendra JM. Doehlert matrix: a chemometric tool for analytical chemistry—review. Talanta 2004;63:1061–7.

[25] Bagheri AR, Arabi M, Ghaedi M, Ostovan A, Wang X, Li J, Chen L. Dummy molecularly imprinted polymers based on a green synthesis strategy for magnetic solid-phase extraction of acrylamide in food samples. Talanta 2019;195:390–400.

[26] Ghorbani M, Aghamohammadhasan M, Shams A, Tajfirooz F, Pourhassan R, Khosravi SRB, Karimi E, Jampour A. Ultrasonic assisted magnetic dispersive solid phase microextraction for preconcentration of two nonsteroidal anti-inflammatory drugs in real water, biological and milk samples employing an experimental design. Microchem J 2019; 145:1026–35.

[27] Ferreira SLC, Bruns RE, Ferreira HS, Matos GD, David JM, Brandão GC, Da Silva EGP, Portugal LA, Dos Reis PS, Souza AS, Dos Santos WNL. Box-Behnken design: an alternative for the optimization of analytical methods. Anal Chim Acta 2007;597:179–86.

[28] Alizadeh T, Ganjali MR, Nourozi P, Zare M. Multivariate optimization of molecularly imprinted polymer solid-phase extraction applied to parathion determination in different water samples. Anal Chim Acta 2009;638:154–61.

[29] Chmiel T, Kupska M, Wardencki W, Namiesnik J. Application of response surface methodology to optimize solid-phase microextraction procedure for chromatographic determination of aroma-active monoterpenes in berries. Food Chem 2017;221:1041–56.

[30] García-Chao M, Agruña MJ, Calvete GF, Sakkas V, Llompart M, Dagnac T. Validation of an off line solid phase extraction liquid chromatography-tandem mass spectrometry method for the determination of systemic insecticide residues in honey and pollen samples collected in apiaries from NW Spain. Anal Chim Acta 2010;672:107–13.

[31] El Atrache LL, Sghaier RB, Kefi BB, Haldys V, Dachraoui M, Tortajada J. Factorial design optimization of experimental variables in preconcentration of carbamates pesticides in water samples using solid phase extraction and liquid chromatography-electrospray-mass spectrometry determination. Talanta 2013;117:392–8.

[32] Merib J, Nardini G, Carasek E. Use of Doehlert design in the optimization of extraction conditions in the determination of organochlorine pesticides in bovine milk samples by HS-SPME. Anal Methods 2014;6:3254–60.

[33] Martins ML, Donato FF, Prestes OD, Adaime MB, Zanella R. Determination of pesticide residues and related compounds in water and industrial effluent by solid-phase extraction and gas chromatography coupled to triple quadrupole mass spectrometry. Anal Bioanal Chem 2013;405:7697–709.

[34] Chung WH, Tzing SH, Ding WH. Optimization of dispersive micro solid-phase extraction for the rapid determination of benzophenone-type ultraviolet absorbers in aqueous samples. J Chromatogr A 2015;1411:17–22.

[35] Hao Y, Gao R, Shi L, Liu D, Tang Y, Guo Z. Water-compatible magnetic imprinted nanoparticles served as solid-phase extraction sorbents for selective determination of trace 17beta-estradiol in environmental water samples by liquid chromatography. J Chromatogr A 2015;1396:7–16.

[36] Huang S, Zhu F, Jiang R, Zhou S, Zhu D, Liu H, Ouyang G. Determination of eight pharmaceuticals in na aqueous sample using automated derivatization solid-phase micro-extraction combined with gas chromatography–mass spectrometry. Talanta 2015;136: 198–203.

[37] Pebdani AA, Dadfarnia S, Shabani AMH, Khodadoust S, Haghgoo S. Application of modified stir bar with nickel:zinc sulphide nanoparticles loaded on activated carbon as a sorbent for preconcentration of losartan and valsartan and their determination by high performance liquid chromatography. J Chromatogr A 2016;1437:15–24.

[38] Farina Y, Abdullah MP, Bibi N, Khalik WMAWM. Determination of pesticide residues in leafy vegetables at parts per billion levels by a chemometric study using GC-ECD in Cameron Highlands, Malaysia. Food Chem 2017;224:55–61.

[39] Andrade MA, Lanças FM. Determination of ochratoxin A in wine by packed in-tube solid phase microextraction followed by high performance liquid chromatography coupled to tandem mass spectrometry. J Chromatogr A 2017;1493:41–8.

[40] Mehta PK, Galvão MS, Soares AC, Nogueira JP, Narain N. Volatile constituents of Jambolan (*Syzygium cumini* L.) fruits at three maturation stages and optimization of HS-SPME GC-MS method using a central composite design. Food Anal Methods 2018;11: 733–49.

[41] De Souza MP, Rizzetti TM, Francesquett JZ, Prestes OD, Zanella R. Bar adsorptive microextraction (BAmE) with a polymeric sorbent for the determination of emerging contaminants in water samples by ultra-high performance liquid chromatography with tandem mass spectrometry. Anal Methods 2018;10:697–705.

[42] Chen L, Zeng Q, Wang H, Su R, Xu Y, Zhang X, Yu A, Zhang H, Ding L. On-line coupling of dynamic microwave-assisted extraction to solid-phase extraction for the determination of sulfonamide antibiotics in soil. Anal Chim Acta 2009;648:200–6.

[43] Salemi A, Shafiei E, Vosough M. Optimization of matrix solid phase dispersion coupled with gas chromatography electron capture detection for determination of chlorinated pesticides in soil. Talanta 2012;101:504–9.

[44] Chen X, Ding N, Zang H, Yeung H, Zhao RS, Cheng C, Liu J, Chan TWD. $Fe_3O_4$@MOF core–shell magnetic microspheres for magnetic solid-phase extraction of polychlorinated biphenyls from environmental water samples. J Chromatogr A 2013;1304:241–5.

[45] Tabani H, Fakhari AR, Shahsavani A, Behbahani M, Salarian M, Bagheri A, Nojavan S. Combination of graphene oxide-based solid phase extraction and electro membrane extraction for the preconcentration of chlorophenoxy acid herbicides in environmental samples. J Chromatogr A 2013;1300:227–35.

[46] Zhang J, Wei Y, Li H, Zeng EY, You J. Application of Box–Behnken design to optimize multi-sorbent solid phase extraction for trace neonicotinoids in water containing high level of matrix substances. Talanta 2017;170:392–8.

[47] Hashemi SH, Kaykhaii M, Keikha AJ, Parkaz A. Application of Box–Behnken design in the optimization of a simple graphene oxide/zinc oxide nanocomposite-based pipette tip micro-solid phase extraction for the determination of rhodamine B and malachite green in seawater samples by spectrophotometry. Anal Methods 2018;10:5707–14.

[48] Gomes PCFL, Barletta JY, Nazario CED, Santos-Neto AJ, Wolff MAV, Coneglian CMR, Umbuzeiro GA, Lancas FM. Optimization of in situ derivatization SPME by experimental design for GC-MS multiresidue analysis of pharmaceutical drugs in wastewater. J Sep Sci 2011;34:436–45.
[49] Castro LF, Ross CF, Vixie KR. Optimization of a solid phase dynamic extraction (SPDE) method for beer volatile profiling. Food Anal Methods 2015;8:2115–24.
[50] Schramm S, Vailhen D, Bridoux MC. Use of experimental design in the investigation of stir bar sorptive extraction followed by ultra-high-performance liquid chromatography-tandem mass spectrometry for the analysis of explosives in water samples. J Chromatogr A 2016;1433:24–33.

# Environmental applications (water)

# 23

*Rosa Montes, José Benito Quintana, María Ramil, Rosario Rodil*
Department of Analytical Chemistry, Nutrition and Food Science, IIAA – Institute for Food Analysis and Research, Universidade de Santiago de Compostela, Santiago de Compostela, Spain

## 23.1 Introduction

Typical environmental water samples have a highly complex organic matrix and low concentration of the analytes of interest. Therefore, a sample preparation step, involving the isolation and enrichment of analytes from the sample, their transfer to the solvent/sorbent compatible with the given chromatographic technique, and the simultaneous removal of interfering substances is usually required prior to chromatographic analysis.

The scope of this chapter is to highlight solid phase–based sample preparation techniques employed in the determination of organic compounds in environmental water samples. Since the selection of the extraction procedure is dependent on the properties of the analyte, this chapter is organized according to their physicochemical properties, namely, volatile, hydrophobic, moderately polar, and highly polar compounds.

## 23.2 Volatile compounds

Volatile organic compounds (VOCs) cover a broad range of organic compounds including paraffinic, olefinic, and aromatic hydrocarbons, and various oxygen-, nitrogen-, sulfur-, and halogen-containing molecules with a vapor pressure of at least 0.01 kPa at 293.15 K [1]. The main subgroups of VOCs include halogenated organics, monocyclic aromatic hydrocarbons, organic sulfides and sulfoxides, benzene, toluene, ethylbenzene and xylene (BTEXs), trihalomethanes (THMs), acetone, and esters, whose concentrations in drinking and natural water samples are typically in the order of ng/L to μg/L. Given their volatility, the analysis of VOCs in water samples can be performed by direct extraction of the aqueous phase or by headspace extraction.

Gas chromatography (GC) is the first choice and an efficient technique for their determination and to establish their distribution in the environment. Sample-enrichment techniques can be applied to entrap a representative sample for subsequent GC analysis. For the extraction of water samples, solvent extraction has long been the method of choice [2–4], but sorbent extraction either directly from the aqueous phase

**Solid-Phase Extraction. https://doi.org/10.1016/B978-0-12-816906-3.00023-6**

or by transfer through the headspace has been applied for more than 30 years as well (see some examples in Table 23.1).

Although solid phase extraction (SPE) is the reference method for the extraction of semivolatile compounds, few applications have been reported for VOCs extraction from the aqueous phase (Table 23.1), mainly because of the risk of losses due to the high volatility of the compounds. Typical sorbents used for VOCs extraction include porous polymers, such as styrene-divinylbenzene (PS-DVB) [5] and ethylvinylbenzene-divinylbenzene copolymers [6–8]. The two main problems associated with the SPE of VOCs are losses during SPE cartridge drying and masking of the volatile compounds by the solvent peak. So, extraction techniques which avoid drying and eliminate the solvent are preferred for their analysis.

Purge and trap (PT) is still the most frequently used preconcentration method for the analysis of VOCs in water and is one of the methodologies stated in the USA EPA legislation for the determination of THMs in water [9,10]. VOCs are extracted by purging a water sample (bubbling gas) with an inert clean gas stream, which is subsequently passed through a trap where the analytes are retained by either cryofocusing or on a sorbent material. Next, the analytes are released into a GC column by heating the trap. The most common sorbents for this application are Tenax, silica gel, or carbon-based sorbents [9,11,12]. One drawback associated with this methodology is the excessive water vapor that is purged with the volatiles by the stream of inert gas. This gives rise to peak distortion, especially in the early part of the chromatogram. Moreover, complex instrumentation is required.

Microscale SPE with coated or packed needles was used to determine VOCs in water samples: in-needle capillary adsorption trap (INCAT) [13–15] which used a piece of GC capillary column inside the needle, solid phase dynamic extraction (SPDE) [16–18] which uses internally coated needles, and needle trap device (NTD) [19–21], a packed needle format. Analyte enrichment can be performed from the liquid or gas phase and samples can be introduced passively by diffusion or actively by a pump or a syringe. After extraction, the analytes are thermally desorbed from the sorbent in a GC injector. The selection of the polymeric or carbon-based material should be carefully optimized. Method detection limits (LODs) for these techniques are comparable to those obtained using PT or solid- phase microextraction (SPME) [14–18,20].

Another important group of miniaturized SPE techniques for the extraction of VOCs are those based on equilibrium between the sample matrix and the sorbent, called microextraction techniques; for example, SPME and stir-bar sorptive extraction (SBSE). Both extraction techniques are solvent-free and easy to use. Although the extraction efficiency, and consequently sensitivity, is better in SBSE [22,23] than in SPME because of the greater amount of sorbent phase, the variety of commercially available stationary phases is larger for SPME, being more difficult for SBSE to recover polar and very volatile compounds. The extraction of volatile analytes can be done by direct immersion (DI-SPME or SBSE) or by exposing the fiber/stir-bar to the headspace of the sample (HS-SPME or HS-SBSE/HSSE). Although direct immersion is the most widely used mode for semivolatile compounds, headspace can be more appropriate for volatile compounds, especially with dirty or complex

**Table 23.1** Selected examples of solid phase extraction–related applications to the determination of volatile organic compounds in water.

| Analytes | Extraction technique | Sorbent material | Determination technique | LOD (ng/L) | Reference |
|---|---|---|---|---|---|
| 27 organohalogen disinfectant byproducts | SPE | PS-DVB | GC-MS/MS | 1–1000 | [5] |
| Haloacetic acids | SPE | Lichrolut EN | Ion chromatography | 5,080–32,500 | [6] |
| Haloacetic acids | SPE | Lichrolut EN | Capillary electrophoresis | 2000–5000 | [7] |
| Haloacetic acids | SPE | Lichrolut EN | LC-ESI-MS | 200–1600 | [8] |
| VOCs | PT | 2,6-Diphenylene oxide + silica gel + coconut charcoal | GC-MS | 20–1600 | [9,10] |
| THMs | PT | Tenax + silica gel + activated carbon | GC-ECD | 20–70 | [11] |
| VOCs | PT | Carbopack B + Carbosieve III | GC-FID | 320–3290 | [12] |
| BTEXs | PT | Methylsilicone | GC-FID | 110–170 | [15] |
| VOCs | PT | VOCARB 3000 | GC-MS | 4–200 | [29] |
| VOCs | INCAT | Carbon coating | GC-FID | | [13] |
| BTEXs | INCAT | Porapack Q | GC-FID | 19–125 | [14] |
| BTEXs | INCAT | Porapack Q | GC-FID | 200–390 | [15] |
| Polar VOCs | SPDE | WAX (A) or PDMS/activated carbon (B) | GC-MS | 20–3500 (A)<br>4–4900 (B) | [16] |
| VOCs | SPDE | PDMS/activated carbon | GC-MS | 12–870 | [17] |

*Continued*

**Table 23.1** Selected examples of solid phase extraction–related applications to the determination of volatile organic compounds in water.—cont'd

| Analytes | Extraction technique | Sorbent material | Determination technique | LOD (ng/L) | Reference |
|---|---|---|---|---|---|
| Furan, benzene, toluene | SPDE | PDMS | GC-MS | 0.2–2 | [18] |
| VOCs | NTD | PDMS + Carbopack + carboxen | GC-MS | 0.15–39 | [21] |
| VOCs | HS-NTD | Tenax TA + carboxen 1000 | GC-MS | 10–60 | [19] |
| BTEXs | NTD | Carbopack X | GC-FID | 1,0,000–25,000 | [20] |
| BTEXs | SPME | PDMS/DVB | GC-FID | 27–382 | [14] |
| BTEXs | SPME | CAR/PDMS | GC-FID | 300–1000 | [20] |
| THMs | HS-SPME | CAR/PDMS | GC-ECD | 5–10 | [24] |
| VOCs | HS-SPME | CAR/PDMS | GC-ECD | 0.3–23.8 | [25] |
| VOCs | SPME | CAR/PDMS | GC-MS | 60–170 | [26] |
| VOCs | HS-SPME | CAR/PDMS | GC-MS/MS | 4–2000 | [28] |
| VOCs | SPME | DVB/CAR/PDMS | GC-MS | 8–700 | [29] |
| Malodors | HS-SPME | DVB/CAR/PDMS | GC-MS | 0.32–0.66 | [30] |
| Malodors | HS-SPME | DVB/CAR/PDMS | GC-MS | 0.1–73 | [31] |
| Malodors | SPME | PDMS | GC-MS | 0.94–9.7 | [22] |
| Malodors | HS-SPME | PA | GC-MS | – | [23] |
| Malodors | SBSE | PDMS | GC-MS | 0.09–0.18 | [22] |
| Malodors | SBSE | PDMS | GC-MS | – | [23] |

*FID*, Flame ionization detector.

matrices. After the extraction, VOCs are thermally desorbed from the stationary phase and transferred to the chromatographic column.

Several applications of SPME are reported for volatile halocarbons, monocyclic and substituted aromatic compounds, carbonyl compounds, and other VOCs (Table 23.1). The sensitivity of the method is strongly dependent upon the type of fiber selected. Many studies have been carried out in which the most suitable type of polymeric coating for the target compounds was evaluated. Among the different SPME fibers commercially available carboxen/polydimethylsiloxane (CAR/PDMS) [25–28] and divinylbenzene/carboxen/polydimethylsiloxane (DVB/CAR/PDMS) [29–31] are the most commonly used. Apart from the choice of fiber, other important variables are headspace volume, the addition of salt, sample agitation, the extraction and desorption times, and the extraction and desorption temperatures. The addition of salt and stirring during the extraction improves the transfer of volatile compounds from water to the headspace, and is widely used [24,25].

Some authors have performed a comparison of PT and SPME for analysis of VOCs in water samples [29,32–34], indicating that they are similar in precision, accuracy, and LODs, but SPME is somehow faster and simpler while PT equipment is expensive and suffers the risk of leaking.

## 23.3 Hydrophobic compounds

This section includes the most hydrophobic pollutants in this chapter, namely polyaromatic hydrocarbons (PAHs), polychlorinated biphenyls (PCBs), organochlorinated pesticides (OCPs), and brominated flame retardants (BFRs), as the most relevant classes. Most of these chemicals can be classified as persistent organic pollutants (POPs) and are known pollutants for many years, with specific legislation as regards their concentration in water and their manufacture and applications. The fact that many of them are legacy contaminants from quite some years also means that established methodologies already exist; most of them are based on liquid-liquid extraction (LLE), but also on SPE. Moreover, these chemicals constitute good targets for the development of alternative methods based on commercial devices, for example, SPME and SBSE (see below), but also to evaluate different materials (e.g., nanomaterials, metal-organic frameworks, new polymers, etc.) for SPE, SPME, or similar approaches [35]. Thus, there is an extensive literature for such hydrophobic pollutants, and we will only cover some representative examples here, focusing on commercially available techniques.

Among the different official methods, the USA EPA has published several methods for the determination of PAHs, PCBs, OCPs, and BFRs (polybrominated diphenyl ethers, PBDEs) in water (see some examples in Table 23.2) [36–39]. These methods are similar and typically consider a sample volume of 1 L which is extracted by SPE on C18 cartridges or more frequently disks. Other sorbents, for example, PS-DVB copolymers can be used, but C18 performs acceptably well [40]. The use of disks is recommended in these EPA methods, since they facilitate the analysis of large sample

**Table 23.2** Selected examples of solid phase extraction–related applications to the determination of hydrophobic pollutants in water.

| Analytes | Extraction technique | Sorbent material | Sample volume | Elution | Determination technique | LOD (ng/L) | Reference |
|---|---|---|---|---|---|---|---|
| PAHs | SPE | C18 | 1 L | Acetonitrile + dichloromethane | LC-UV/ Fluorescence | 3–2200 | [36] |
| PCBs and OCPs | SPE (or LLE) | C18 | 1 L | Acetone + dichloromethane | GC-HSD | 1–15 | [39] |
| PCBs | SPE (or LLE) | C18 | 1 L | Acetone + dichloromethane | GC-HRMS | 0.007–0.060 | [38] |
| BDEs | SPE (or LLE) | C18 | 1 L | Acetone + dichloromethane | GC-HRMS | 0.02–0.70 | [37] |
| PAHs, PCBs, and other pollutants | MEPS | C18 | 0.8 mL | Ethyl acetate | PTV-GC-MS | 0.5–32 | [42] |
| PAHs | SPME | PDMS | 4 mL | LC-interface | LC-UV | 1000–5000 | [45] |
| PAHs | SPME | PDMS & PA | 10 mL | Thermal desorption | GC-FID and GC-MS | 30–270 (PDMS; MS) | [43] |
| PCBs | SPME | PDMS | 100 mL | Thermal desorption | GC-MS | 0.3–1 | [44] |
| PCBs | SPME | PDMS | 2 mL | Thermal desorption | GC-ECD | <5 | [48] |
| OCPs | SPME | DVB/CAR/ PDMS | 4 mL | Thermal desorption | GC-ECD | <5 | [47] |
| BDEs | SPME | PDMS | 10 mL | Thermal desorption | GC-MS/MS | 0.009–0.190 | [46] |
| PAHs | SBSE | PDMS | 10 mL | Acetonitrile | LC-Fluorescence | 0.2–1.4 | [51] |

| | | | | | | | |
|---|---|---|---|---|---|---|---|
| PCBs | SBSE | PDMS | 8 mL | Thermal desorption | GC-MS | 0.05−0.15 | [50] |
| PAHs, OCPs, PCBs, and PBDEs | SBSE | PDMS | 100 mL | Thermal desorption | GC-MS | 0.01−0.2 | [52] |
| PBDEs | SBSE | PDMS | 100 mL | Thermal desorption | GC-MS | 0.006−0.200 | [53] |
| PAHs | SPME-Arrow | PDMS | 19 mL | Thermal desorption | GC-MS | 0.1−0.8 | [54] |
| PAHs | Silicon-rod | PDMS | 10 mL | Acetonitrile/Water | LC-Fluorescence | 0.1−1.2 | [55] |
| PBDEs and other BFRs | Silicon-rod | PDMS | 100 mL | Ethyl acetate | GC-MS | 0.4−10 | [56] |

*HR*, High Resolution; *HSD*, halogen-specific detector (e.g., ECD).

volumes, with less possibility of clogging. The sample is normally modified by adding 5% (v/v) methanol, to keep the C18 sorbent solvated during the whole SPE process and to improve the solubility of these hydrophobic compounds [40]. Glassware is rinsed with either acetonitrile or acetone, which is passed through the sorbent as a first elution step. Final elution is performed with dichloromethane, which is also used to rinse glassware, since the possibility of these chemicals being adsorbed to glassware is one of the major difficulties when analyzing hydrophobic chemicals in water. Some authors have evaluated different percentages of organic solvent added to the sample to improve recoveries. For instance, Kiss et al. investigated different percentages of 2-propanol for the extraction of PAHs from 50 mL water samples and found 25% to be optimal, since higher percentages led to lower recoveries of the lower molecular weight PAHs [41].

Microextraction by packed sorbent (MEPS) is a miniaturized version of SPE. Although more popular for bioanalysis, MEPS has also been applied, for instance, to the analysis of PAHs and PCBs in water. Thus, in the method proposed by Prieto et al. LODs ranging from 0.5 to 32 ng/L were obtained for only 0.8 mL of sample with 2 mg C18 packed in the syringe needle, which is finally eluted with 50 μL of ethyl acetate into a GC-mass spectrometer (MS) equipped with a programmable-temperature vaporizer (PTV) injector [42].

Besides MEPS, the most relevant microextraction techniques based on solid sorbents are SPME and SBSE. In both cases some of their early applications in water are focused on POPs, since these types of hydrophobic compounds have a great affinity for PDMS, which is the classic and most popular sorbent in both techniques.

Thus, several SPME methods have been published, with some representative examples compiled in Table 23.2 [43–48]. The conventional PDMS fiber works quite well for hydrophobic pollutants, although it has been reported that polyacrylate (PA) is preferred for lighter PAHs, though PDMS provides better results for the remaining PAHs [43], or that the triple-composition fiber DVB/CAR/PDMS provides better results for OCPs [47]. SPME is normally performed in the direct immersion mode, but some authors have shown that the headspace mode provides better sensitivity if the sample is heated [44,46]. Even though the time to reach equilibrium can be greater than 2 h, most authors adopted extraction times in the 15–90 min range, in nonequilibrium conditions, to accommodate the extraction time to the analysis time, so that one sample can be extracted while the previous sample is being analyzed. Although in some applications an SPME-liquid chromatography (LC) interface was used, GC was generally preferred, as being more straightforward. In this case, and when the detector is sensitive or sufficiently selective (e.g., an MS or electron-capture detector— ECD), LODs at the ng/L level or lower can be obtained from sample volumes lower than for SPE (ca. 2–100 mL V 1 L in SPE).

SBSE was developed in 1999 as an alternative to SPME [49], inspired by the fact that PCBs were found to bind to normal polytetrafluoroethylene (PTFE)-coated stirbars during SPME [48]. The main advantage of SBSE compared to SPME is the higher amount of PDMS, typically 100-times higher, which, on the other hand leads to longer times to reach equilibrium. Since PDMS has been the only available coating in SBSE until recently, hydrophobic pollutants in water constitute a perfect application for this technique, with some representative examples given in Table 23.2 [50–53]. SBSE can achieve LODs in the low ng/L range from 8 to 10 mL of sample in 1–2 h [50,51]. If needed, these LODs can be pushed down the sub-ng/L range by increasing the extraction time to ca. 1 day and sample volume to 100 mL. In most SBSE applications, 10%–20% (v/v) methanol is added to the sample to minimize sorption of the analytes to the sample vial walls hampering extraction to the SBSE stir-bar.

Recently the SPME patent exploited by Supelco has expired and new SPME devices have appeared in the market. As an example, the company PAL has commercialized an SPME device called "Arrow" which combines the advantages of classical SPME (e.g., direct desorption into the GC injector) with a larger amount of extraction phase. Thus, this device can be viewed as an intermediate approach between SPME and SBSE in terms of mass of PDMS that it holds and therefore, extractive capacity (and LODs) for PAHs [54].

A final possibility when working with PDMS-based devices is to use silicone rods. Such rods can be cut in the lab from longer PDMS bars to the desired size and used as in SBSE by incorporating an external agitation (e.g., a PTFE stir-bar). The main advantage is that their low cost compared with commercial SBSE and the PDMS does not need to be reused, which avoids the risk of cross-contamination. Such a possibility was used, for example, to extract PAHs, BDEs, and other BFRs [55,56], with LODs comparable to SBSE.

## 23.4 Moderately polar compounds

This section summarizes selected sorptive extraction applications for the determination of medium polar organic pollutants in the aqueous environment belonging to four main groups: pesticides (organophosphates, carbamates, pyrethroids, neonicotinoids, etc.), pharmaceuticals (including drugs and hormones), personal care products (PCPs, e.g., UV filters, parabens, disinfectants, insect repellents, etc.), and additives (perfluorinated compounds, organophosphate flame retardants, phthalates, phenols, etc.). A compilation of some recent representative applications using SPE with commercially available sorbents is presented in Table 23.3.

In recent decades, SPE has played a crucial role in sample preparation for water analysis, replacing the classic LLE of medium polarity compounds [57,58]. Some commercially available materials, such as C8, C18, hydrophilic-lipophilic balance (HLB), mixed-mode/cationic-exchange (MCX), mixed-mode/anion-exchange (MAX), and weak anion-exchange (WAX) are commonly used for SPE in water analysis [59]. More recently, a large number of new solid-phase sorbents (SPSs), such as molecularly imprinted polymers (MIPs), magnetic nanomaterials, and carbon nanoparticles have been proposed for sample preparation [60].

As regards to SPE with conventional sorbents, C18 [61], PS-DVB [62–65], or HLB [66–68] materials are used routinely in most laboratories for the extraction of organic compounds of moderate polarity [69]. Depending on the type of water, the enrichment factor will vary. Thus, for groundwater a larger sample volume can be

**Table 23.3** Recent applications of SPE using conventional sorbents for medium polarity pollutants recovered from water samples.

| Analytes | Sorbent material | Determination technique | Recovery (%) | LOD (ng/L) | Reference |
|---|---|---|---|---|---|
| 9 parabens | Strata C18-E | LC-MS/MS | 70–115 | 40–4000 | [61] |
| 10 pharmaceuticals, 1 pesticide, and 4 steroid hormones | PS-DVB | LC-ESI-MS/MS | 73–112 | 0.04–0.2 | [65] |
| 6 estrogens and 6 glucocorticoids (hormones) | PS-DVB modified with urea functional groups | LC-MS/MS | 80–110 | 0.16–2.14 | [63] |
| 9 UV filters | PS-DVB | LC-MS/MS | 76–113 | 3–10 | [62] |
| 5 insect repellents | PS-DVB | LC-(APPI)-MS/MS | – | 0.1–1.4 | [64] |
| 252 pesticides | Oasis HLB | LC-HRMS | 70-130 (86% pesticides) | 5 (LOQ) (81% pesticides) | [66] |
| 35 pharmaceuticals | Oasis HLB | LC-ESI-MS/MS | 70–132 | 0.2–67 | [68] |
| 20 compounds including preservatives, antioxidants, disinfectants, estrogens, alkyl-phenols, and bisphenol A | Supel-Select HLB | LC-MS/MS (A)<br>LC-HRMS (B) | (A) 46-136<br>(B) 57 -127 | (A) 0.3–23<br>(B) 0.04–1.4 | [67] |
| 31 herbicides | Oasis MAX<br>Bond elut Pax | LC-MS/MS | 65–126 | 0.7–30 | [75] |
| 20 drugs of abuse and pharmaceuticals | Strata-X-Drug B | LC-MS | 62–107 | 0.01–1.09 | [74] |

| | | | | | |
|---|---|---|---|---|---|
| Cathinones (drugs) | Oasis WCX | LC-HRMS | 70–100 | 0.1–0.5 | [70] |
| 5 perfluorinated compounds | OASIS WAX | LC-ICP-MS/ MS-ESI-MS | 82–145 | 0.5–0.8 | [72] |
| 13 organophosphorus compounds (flame retardants) | OASIS MAX | LC-MS/MS | 47–109 | 0.2–10 | [71] |
| 8 primary phthalate monoesters and 5 secondary phthalate monoesters | OASIS MAX | LC-MS/MS | 77–86 | 0.01–5.9 | [73] |

*HR*, High Resolution; *ICP*, Inductively coupled plasma; *LOQ*, Limit of Quantification extracted (up to 1 L) than for wastewater, which has a more complex matrix (hundreds of milliliters). The LODs also depend on the sensitivity of the detection technique. The best LODs are achieved with MS combined with a chromatographic technique and are, in general terms, at the ng/L level [69].

In addition to classical sorbents, mixed-mode polymeric sorbents [70–75] have the capability to enhance the selectivity of the extraction process [76]. Selective elution and selective washing can be performed for the removal of interfering compounds to minimize matrix effects for electrospray ionization (ESI)-MS detection. As an example, Casado et al. [66] used Oasis MCX cartridges for the determination of antimycotic drugs in environmental waters (river, raw, and treated wastewater) by LC-quadrupole-time of flight (QTOF)-MS, with recoveries in the 75%–117% range and LODs between 2 and 15 ng/L. Two consecutive washings with methanol-water (10:90) and methanol (0.1% in formic acid) were employed to remove neutral and acidic interferents. The analytes were eluted with methanol containing 2% (v/v) ammonia, so that cation-exchange interactions were disrupted.

Table 23.4 compiles selected SPE applications for medium polar water contaminants with novel sorbents. A number of examples using molecularly imprinted polymers (MIPs) were reported in the last few years [77–85]. Basically, MIPs are synthetic polymers with highly specific recognition sites for target compounds, made from monomers that have the ability to form a complex with a template, through covalent or noncovalent interactions that are then joined together by a cross-linking agent [86]. For pesticides, Binsalom et al. [87] prepared an MIP to extract chlorpyrifos. This compound was used as the template molecule, methacrylic acid as the functional monomer, and ethylene glycol dimethacrylate as the cross-linker [87]. Another example is the use of an MIP for the SPE of malathion from tap water. The MIP was prepared using malathion as a template, methacrylic acid as a functional monomer, ethylene glycol dimethacrylate as a cross-linker, azodiisobutyronitrile as an initiator, and the acetonitrile-chloroform (1:1, v/v) as a porogen, with recoveries in the range 96%–111% [88].

MIPs were also used for the determination of pharmaceuticals and PCPs. As an example, an MIP using 2,4-dihydroxybenzophenone as a template was used to extract five benzophenones from swimming pool water combined with LC-UV analysis, achieving LODs in the range of 0.12–0.68 μg/L with recoveries of 89%–113% [84]. Nonsteroidal antiinflammatory drugs (NSAIDs) are a group of compounds with several MIP applications [81,89], including multitemplate MIPs, synthesized using more than one target compound as template molecules [90]. This multitemplate approach was also employed for other groups of contaminants, such as chlorophenols [80]. Despite the fact that single and multitemplate MIPs have a great potential as selective sorbents for SPE, a limitation is the risk of template leakage into real samples during extraction in those cases where incomplete removal of the template from the cavities of the material during washing occurs. One possible solution is the use of a dummy template to synthesize the MIP. A dummy template is a compound whose physicochemical properties, especially its functional groups and chemical structure, are similar to the target analyte but does not interfere in its determination. A number of recent applications of MIP-SPE have focused on dummy templates for MIP

**Table 23.4** Selected SPE applications with novel sorbents for moderately polar water contaminants.

| Analytes | Sorbent material | Determination technique | Recovery (%) | LOD (ng/L) | Reference |
|---|---|---|---|---|---|
| 6 pyrethroids | MIP | GC-ECD | 86–96 | 17–37 | [83] |
| Triazines and triazine metabolites | MIP | LC-MS/MS | 55–123 | 0.015 | [78] |
| NSAIDs | MIP | LC-MS/MS | 72–113 | 0.5–1.1 | [81] |
| β-Blockers | MIP | LC-MS/MS | 40–110 | 0.2–6.5 | [79] |
| Estrogens | MIP | GC-HRMS | 92–114 | 0.08 (LOQ) | [85] |
| Benzophenones | MIP | LC-UV | 89–113 | 120–680 | [84] |
| 4 phthalates | MIP | LC-MS | 87–93 | 160–840 | [77] |
| Bisphenol A | MIP | LC-DAD | 82–119 | 70 | [82] |
| 6 chlorophenols | MIP | CE | 82–106 | 170–310 | [80] |
| 17-β-Estradiol | Imprinted nanoparticles | LC | 88–99 | 40 | [97] |
| Di(2-ethylhexyl)phthalate | Imprinted nanoparticles | GC-MS | 93–103 | 20 | [98] |
| Bisphenol A | Imprinted nanoparticles | LC-UV | 95–106 | 300 | [100] |
| Bisphenol A | Imprinted nanoparticles | LC-UV | 85–94 | 2.5 | [99] |
| Parabens, steroids, flavonoids, and pesticides | Nanofibers | LC-UV | 23–126 | – | [103] |
| Atrazine, deisopropylatrazine and deethylatrazine | Nanofibers | LC-UV | 81–120 | 90–120 | [101] |
| Atrazine | Nanofibers | LC-UV | 95–115 | 30 | [107] |

*Continued*

**Table 23.4** Selected SPE applications with novel sorbents for moderately polar water contaminants.—cont'd

| Analytes | Sorbent material | Determination technique | Recovery (%) | LOD (ng/L) | Reference |
|---|---|---|---|---|---|
| NSAIDs | Nanofibers | UPLC-MS/MS | 85–100 | 0.2–5 | [104] |
| Diethylstilbestrol, dienestrol, and hexestrol | Nanofibers | LC-MS/MS | >98 | 0.05–0.1 | [105] |
| Phthalate esters | Nanofibers | LC-UV | 87–102 | 2–33 | [106] |
| Bisphenol A | Nanofibers | LC-Fluorescence | 100–102 | 600 | [102] |
| Atrazine and simazine | CNTs | LC-DAD | 83–104 | 33 and 9 | [121] |
| Herbicides | CNTs/Carbon nanofibers | LC-UV | >94 | 4–30 | [108] |
| 11 organophosphorus pesticides | CNTs | LC-MS/MS | 88–125 | 0.02–1.00 | [112] |
| 13 pharmaceuticals | CNTs | LC-MS/MS | 79–119 | 0.2–103 | [111] |
| NSAIDs | CNTs | LC-MS/MS | 65–125 | 0.01–1.3 | [115] |
| β-Blockers | CNTs | LC-MS/MS | 83–96 | 0.5–1.45 | [119] |
| Benzenesulfonamides, benzotriazoles, and benzothiazoles | CNTs | LC-MS/MS | 70–116 | 30–170 | [117] |
| Carbamate pesticides | Graphene | LC-MS/MS | 81–111 | 0.5–6.9 | [116] |
| Pyrethroids | Graphene aerogel | GC-MS | 66–106 | 12–110 | [109] |
| NSAIDs | Graphene aerogel | LC-UV | 81–107 | 10–100 | [120] |
| Estrogens | Graphene | LC-UV | 91–113 | 0.25–0.52 | [114] |
| Benzotriazole UV filters | Graphene sponge | LC-UV | 89–105 | 20–80 | [118] |
| Bisphenol A/steroid hormones | Graphene aerogel | LC-UV | 76–113 | 0.01–0.11 | [110] |
| Chlorophenols | Graphene | LC-UV | 77–117 | 100–400 | [113] |

*LOQ*, Limit of Quantification.

preparation, as in the case of imidazole fungicides [91] where alpha-(2,4-dichlorophenyl)-1H-imidazole-1-ethanol was used as dummy template; climbazole [92] where miconazole replaced the target compound to prepare the MIP; bisphenol A [93] for which determination of four different MIPs prepared from bisphenol B, bisphenol F, tetrabromobisphenol A, or 4-tert-butylphenol were evaluated; or phthalate esters [94] where diethyl phthalate was used with this purpose.

In addition to SPE, MIPs have also been used in combination with supported liquid membrane techniques, such as hollow fiber—liquid phase microextraction (HF-LPME) [95], membrane-assisted solvent extraction (MASE) [96], or ultrasound-assisted emulsification combined with microsolid phase extraction, involving the packing of a small amount of MIP particles [82] for the determination of different organic pollutants, such as sulfonamides in the first case, triazines in the second case, and bisphenol A in the last example, from aqueous matrices.

Concurrently, different sorbents based on nanoparticles have been investigated for the determination of organic analytes from aqueous matrices. Some examples are imprinted nanoparticles [97—100], nylon nanofibers [101—107], or graphene-based sorbents [108—121]. MIPs prepared by nano-$TiO_2$ have excellent properties including low cost, nontoxicity, and chemical and photo stability. They have been used to determine triazine herbicides (propazine, atrazine, and simazine) in water samples [122]. MIPs have also been combined with other types of nanoparticles such as $Fe_3O_4$, for the determination of bisphenol A [100], di-(2-ethylhexyl)phthalate [98], or 17-β -estradiol [97], or even with carbon nanotubes (CNTs), for the determination of 4-nonylphenol [123].

The application of nanofibers as SPE sorbents was inspired by their large surface area on account of their small diameter which often resulted in enhanced extraction capacity and sorption kinetics [103]. A nylon 6 nanofiber sorbent was used for the extraction of phenolic compounds (parabens, steroids, flavonoids, and pesticides) from aqueous samples [103]. It was more selective than Oasis HLB for the retention of the target compounds [103].

Graphene is the basic structural unit of graphite, CNTs, and fullerenes. It is a novel two-dimensional carbonaceous material with superior chemical stability, excellent thermal stability, and ultrahigh specific surface area with a theoretical value of about 2630 $m^2/g$, suggesting a high sorption capacity and great potential for use as a sorbent for SPE [113]. Graphene and its complexes have been used for the extraction of a variety of analytes, such as phthalate esters [124] and pesticides [125]. As an SPE sorbent, single-use graphene may lead to irreversible graphene aggregation, which could then escape the SPE cartridge under high pressure and reduce the extraction efficiency [126]. The immobilization of graphene on a solid support is an effective method to solve this problem [127]. Thus, Li et al. [128] developed a graphene-derivatized silica composite as SPE sorbent combined with GC-tandem MS (MS/MS) for the determination of five polycyclic musks in aqueous samples with LODs in the range 0.3—0.8 ng/L and recoveries from tap and river water between 83% and 107% [128].

Finally, there are some other novel materials that will gain importance in the near future, such as silica monoliths. The application of silica-monolith SPE cartridges in environmental analysis is limited. The main reason is that the analysis of environmental waters generally requires large-size SPE cartridges for processing large-volume samples. However, the preparation of large silica-monolith SPE cartridges is extremely difficult because of the shrinkage and brittleness of the silica monolith [129]. Recently, Liu et al. [129] developed composite monoliths for the SPE of environmental samples. Their approach utilizes a melamine-formaldehyde (MF) sponge as a skeleton to stabilize the silica-monolith sorbents, yielding a new type of silica monolith nested in sponge (SiMNS). The high porosity of the MF sponge facilitates processing large sample volumes, while the flexible framework keeps its shape and stability even after repeated compression tests. Functionalization of SiMNS provides appropriate SPE options for selective and efficient extraction of specific contaminants.

Microextraction techniques based on solid supports are well established for the analysis of moderately polar compounds in water samples. Some representative applications are summarized in Table 23.5. Although for low-polarity compounds the most commonly used coating is PDMS, for more polar compounds there is a need for a selection of more compatible coatings. In the case of pharmaceuticals, the first applications were developed for ibuprofen [130] or mixtures of highly administered pharmaceuticals, such as paracetamol, phenazone, carbamazepine, etc. [131] and then, other applications for NSAIDs [132,133] were published. In these applications, the selected coating was mainly PA; however, other coatings such as PDMS/DVB or Carbowax (CW)/DVB also reported acceptable results for some compounds. The fiber was directly exposed to the water sample during the extraction period, since direct extraction is more efficient for these compounds than headspace sampling. For most SPME applications the compounds were analyzed using GC. After the extraction, the fiber is exposed to a derivatizing reagent, for example, a silylation reagent [131,132] and subsequently injected into the GC. However, after some years, applications based on LC analysis have grown. Here, fiber desorption was made a separate step using an LC-compatible organic solvent, such as methanol [134] or the LC mobile phase [135], this last option is also suitable when SPME is used as part of an on-line desorption system [136].

Several applications based on SPME for the analysis of PCPs, such as parabens and triclosan, in water samples [137–139] have been developed. Some authors incorporated the derivatization process in the extraction vessel (e.g., using acetic anhydride as reagent), the extracted molecules being the derivatized species (less polar than the native species) with a moderately polar coating, for example, DVB/CAR/PDMS [138]. On the other hand, when the derivatization takes place on-fiber, for example, using silylation agents, the PA coating is preferred [137,139]. With the increasing use of LC as the determination technique, the derivatization step can be avoided [140,141].

Thus, fiber selection is the key step for compounds of intermediate polarity. The fiber coatings initially commercialized can be categorized as absorbent-type (e.g., PA, PDMS) and adsorbent-type (e.g., CW/DVB, DVB/CAR/PDMS) and the extraction mechanism is based on partitioning and adsorption, respectively. Of these, PA (absorbent-type) and all other adsorbent-types, especially CW/Templated resin (TPR), are the more polar coatings for the extraction of medium polar compounds.

**Table 23.5** Selected examples of microextraction techniques in the determination of moderately polar compounds in water.

| Analytes | Extraction technique | Sorbent material | Determination technique | LOD (ng/L) | Reference |
|---|---|---|---|---|---|
| Ibuprofen | SPME | PA | GC-MS | 100 | [130] |
| Pharmaceuticals | SPME | PA | GC-MS | 200–5000 | [131] |
| NSAIDs | SPME | PA | GC-MS | 12–20 (LOQ) | [132] |
| NSAIDs | SPME | PEG-g-CNTs | GC-FID | 7–30 | [153] |
| Pharmaceuticals | SPME | DVB/CAR/PDMS | GC-MS | 20–430 | [133] |
| Pharmaceuticals | SPME | CW/TPR | LC-MS | 5–50 | [135] |
| Pharmaceuticals | SPME | PDMS/DVB | LC-UV | 500–3000 | [136] |
| Antibiotics | SPME | CW/DVB | LC-MS | 9,000–55,300 | [134] |
| Pharmaceuticals and PCPs | SPME | PA | GC-MS | 0.7–9 | [139] |
| Parabens | SPME | PA | GC-MS | 1–25 | [137] |
| PCPs | SPME | DVB/CAR/PDMS | GC-MS | 4–21 | [138] |
| Perfluorinated compounds | SPME | DVB/CAR/PDMS | GC-MS | 20–100 (LOQ) | [145] |
| OPFRs | SPME | PDMS/DVB | GC-NPD | 10–-25 (LOQ) | [146] |
| OPFRs | SPME | PDMS/DVB | GC-MS | 0.5–4 (LOQ) | [147] |
| OPFRs | SPME | GO/β-CD | GC-NPD | 1.1–60.4 | [151] |
| Phthalates | SPME | CNTs-PPy | GC-FID | 50–100 | [152] |
| Phthalates | SPME | CW/DVB | GC-MS | 5–40 | [142] |
| Phthalates | SPME | PA | GC-MS | 6–170 | [143] |
| Phthalates | SPME | MIP | GC-MS | 2.17–20.84 | [156] |

*Continued*

**Table 23.5** Selected examples of microextraction techniques in the determination of moderately polar compounds in water.—cont'd

| Analytes | Extraction technique | Sorbent material | Determination technique | LOD (ng/L) | Reference |
|---|---|---|---|---|---|
| Phthalates | SPME | SPT | GC-FID | 80–120 | [157] |
| PCPs | Online-SPME | PDMS | LC-MS | 0.38–4.67 | [141] |
| Pharmaceuticals and PCPs | Online-SPME | CW/TPR | LC-UV | 5–3112 | [140] |
| Pharmaceuticals | In-tube SPME | CAR/PLOT | LC-MS | 7–29 | [167] |
| Perfluorinated compounds | In-tube SPME | CP-PoraPLOT amine | LC-MS | 1.5–3.2 | [168] |
| UV filters | In-tube SPME | PIL-MCC/MNP | LC-UV | 40–260 | [169] |
| Pharmaceuticals | In-tube SPME | Aerogel | LC-UV | 10–50 | [170] |
| UV filters | SBSE | PDMS | GC-MS | 0.2–63 | [159] |
| Phenols | SBSE | PA | GC-MS | 0.11–25.4 | [160] |
| Pharmaceuticals | SBSE | PU | LC-UV | 400–1700 | [161] |
| Quinolones | SBSE | MADB | LC-UV | 370–560 | [162] |
| PPCPs | SBSE | Cork | LC-UV | 200–6500 | [166] |
| PPCPs | TFME | C18/SCX | LC-MS | - | [163] |

*FID*, Flame ionization detector; *LOQ*, Limit of Quantification; *MADB*, poly(methacrylic acid-3-sulfopropyl ester potassium salt-divinylbenzene); *NPD*, Nitrogen–phosphorus detector; *PIL-MCC/MNP*, polymeric ionic liquid-based monolithic capillary column doped with magnetic nanoparticles.

Different families of relatively polar additives, such as phthalates [142–144], perfluorinated compounds [145] or organophosphate flame retardants [146,147] were analyzed by SPME using these fibers.

Over the years, the development of new coatings for SPME has expanded the applicability of the technique for water contaminants of intermediate polarity. Among these new materials, the most interesting are the carbonaceous materials (graphene and CNTs), MIPs, or alternative polymers [148–150]. Due to the low polarity of graphene (hydrophobic effect), it is more frequently used for nonpolar analytes; however, it can be easily functionalized (because the carbon-carbon bonds are not saturated) to allow the extraction of more polar compounds, such as organophosphate flame retardants [151]. In this case, the SPME fiber was composed by graphene oxide and β-cyclodextrin (GO/β-CD) and exhibited good extraction capability for the studied compounds, including tris-(2-ethylhexyl)phosphate, which was not adequately extracted by conventional PDMS/DVB fibers [146]. CNTs exhibit similar characteristics, but with higher porosity. They are easily modified to tune important properties (e.g., polarity and hydrophilicity) and to introduce other specific substituents to alter their affinity toward different analytes, such as polypyrrole (PPy) for the extraction of phthalate esters [152] or polyethylene glycol (PEG) for the extraction of NSAIDs from water samples [153]. MIP-coated fibers are extremely selective materials and nowadays are less expensive and more robust than in earlier times [154,155]. As an example, He and coworkers prepared an SPME fiber using dibutyl phthalate as the template molecule for the MIP coating. Five different phthalates were successfully extracted with extraction yields much higher than those obtained with three different commercial SPME fibers [156]. Also, new polymeric materials have demonstrated their capability to extract intermediate polar pollutants, for example, nano-structured self-doped polythiophene (SPT) electrodeposited in the presence of a fluorinated organic acid was used to extract phthalate esters from water samples [157].

When SBSE was initially evaluated for the extraction of compounds of medium polarity, the PDMS sorption phase available at that time had a low affinity for these compounds. However, some applications can be found in literature adjusting some of the parameters that directly affect the partition coefficient of medium polar compounds (e.g., phthalate esters, UV filters, etc.) between water and the PDMS coating, such as ionic strength or pH [158,159]. Also, other more polar sorbent materials have been evaluated as alternatives to PDMS, such as PA [160], polyurethane (PU) foams [161], or monolithic materials [162]. In addition to SBSE, thin-film microextraction (TFME) has been applied with mixed materials, for example, C18 and a cationic exchanger for pharmaceuticals and PCPs analysis in wastewater [163]. Also, adsorptive microextraction (AμE) has emerged as a promising alternative to SBSE and capable of utilizing a wider range of sorbent materials (some of these are typically employed in SPE approaches) for the extraction of more hydrophilic compounds [164,165]. Applications of this technique are still in development. Interestingly, even natural and renewable materials, such as cork, can be used as a sorbent phase for the determination of medium polar compounds (benzophenone, triclocarban, and parabens) with satisfactory results [166].

Dynamic microextraction techniques, such as in-tube SPME, are widely used for water analysis but to a lesser extent for medium polar compounds since it typically utilizes a sorbent material with a specific affinity for more hydrophobic analytes. However, there are applications of in-tube SPME for the extraction of fluoroquinolones [167] or perfluorooctanesulfonic acid and perfluorooctanoic acid [168] even using capillary columns with hydrophobic (e.g., Carboxen) or moderately hydrophilic surfaces (e.g., PoraPLOT Amine). Magnetic nanoparticles or new polymeric sorbents are promising materials for semipolar compounds, such as UV-filters [169] or phenols [170].

## 23.5 Highly polar compounds

The extraction of highly polar compounds from water samples is a demanding task, as they possess a high affinity for the matrix. This class of contaminants includes sulfonates, sweeteners, some pesticides (as glyphosate), quaternary ammonium salts, among others, generally considered "problematic compounds" in terms of methodology development [171]. Some representative methods for different polar compounds are shown in Table 23.6.

Most of the applications for the measurement of polar contaminants in water samples use SPE. Sorbents based on reversed-phase (RP) materials, such as C18 cartridges [172,173], are not the most appropriate for highly polar compounds due to their small breakthrough volumes. Also, the use of other sorbents with a high retention capacity, such as graphitized carbon [174], leads to problems associated with the elution of compounds with organic solvents. One practical solution for some compounds is the use of ion-pairing reagents. These reagents are specific modifiers that can be added to the sample or the conditioning and elution solvents, providing ions with opposite charge to the analyte. The ions interact compensating the ionic/ionizable character of the analyte and therefore, favoring the extraction with SPE RP sorbents. For example, heptafluorobutyric acid was used as an ion-pairing reagent for the analysis of "quat" herbicides (diquat, paraquat, …) in water samples [175] or tetrabutylammonium bromide for the analysis of naphthalenesulfonates [176]. A slight modification of this approach uses admicelle-based SPE in which a surfactant, such as sodium dodecyl sulfate (SDS), is used to form surfactant-analyte ion-pairs [177]. The main drawback of ion-pairing SPE is that the recovery and repeatability is affected by the ionic strength of the sample, and can be especially problematic for some matrices, for example, wastewater or seawater.

The use of MIPs as SPE sorbents can overcome the selectivity problem that ion-pairing reagents present with complex samples. The main advantage of this type of sorbents, compared with conventional materials, is their inherent molecular selectivity, which enables MIPs to extract specified compounds from a complex mixture. In some cases, though, the MIP may show cross-reactivity to related analytes. In the case of MIP-SPE sorbents, efforts have been directed toward the synthesis of water-compatible MIPs by incorporating hydrophilic surface properties to the polymer

| Analytes | Extraction technique | Sorbent material | Determination technique | LOD (ng/L) | Reference |
|---|---|---|---|---|---|
| Sulfonates | SPE | C18 | LC-Fluorescence and GC-MS | 1-4 (LC) and 2–500 (GC) | [172] |
| Diquat and paraquat | SPE | C8 and C18 | LC-MS | 100–200 | [173] |
| Sulfonates | SPE | Graphitized carbon | LC-UV | 100–1000 | [174] |
| Quaternary ammonium compounds | Ion-pair SPE | C8 | LC-MS | 6–85 | [175] |
| Sulphonates | Ion-pair on-line SPE | Polymer labs PLRP-s | LC-Fluorescence | 10-3000 (LOQ) | [176] |
| Quaternary ammonium compounds | Admicelle-based SPE | Alumina coated with SDS admicelles | LC-MS | 10–30 | [177] |
| Sulfonates | SPE | MIP | LC-UV | – | [178] |
| Glyphosate and AMPA | SPE | MIP | LC-MS | 5 (LOQ) | [179] |
| Glyphosate and AMPA | SPE | MIP | LC-MS | 10 | [180] |
| Glyphosate and MPPA | SPE | Strata-SAX | LC-MS | – | [181] |
| Quaternary ammonium compounds | SPE | Strata-X CW | LC-MS | 0.02–0.03 | [182] |
| Short-chain perfluorinated compounds | SPE | Oasis WAX | LC-MS | 0.6–26 (LOQ) | [183] |
| Sweeteners | SPE | Oasis HLB | LC-MS | 1–500 | [184] |
| Sweeteners | SPE | Poly-Sery PWAX | LC-MS | 0.4–7.5 (LOQ) | [185] |
| Sweeteners | SPE | Oasis HLB | LC-MS | 10-500 (LOQ) | [186] |
| Acesulfame | SPME | MIP | LC-UV | 230 | [187] |
| Sulphonates | SBSE | MIP | LC-UV | 320–950 | [188] |

*LOQ*, Limit of Quantification.

in order to reduce nonspecific hydrophobic interactions. Thus, using, for example, 1-naphthalenesulfonic acid as a template molecule seven structurally related naphthalenesulfonates could be extracted [178]. The preparation of MIP-SPE sorbents for new applications is quite tedious and expensive [179]; however, there are an increasing number of commercially available MIP sorbents for specific applications, such as AFFINIMIP SPE Glyphosate-AMPA for the analysis of glyphosate and its main metabolite [180].

Ion exchangers or mixed-mode (RP/ionic exchange) sorbents remain the most applied and affordable option. Thus, there are several applications based on the use of these materials to concentrate polar analytes such as glyphosate [181], quaternary ammonium compounds [182], or short-chain perfluorinated compounds [183] from water samples. The main consideration when using these sorbents is that the conditioning and elution solvents, as well as the sample pH or ionic strength need to be exhaustively studied, since the analyte-sorbent interactions may be too weak (leading to low breakthrough volumes) or too strong (resulting in high volumes of elution solvent or even nonelution) depending on the analyte. Mixed-mode cartridges were evaluated for the extraction of sweeteners from water samples [184–186] and the recoveries compared with other sorbent types. The results are controversial, since in some cases the RP type cartridges (OASIS HLB) were selected [184,186]. However, analyzing the results for each compound, the recoveries for the most polar sweetener (acesulfame, logD −2.77) were higher with the mixed-mode cartridge while other moderately polar compounds performed better with classical RP-type sorbents.

The application of microextraction techniques based on solid sorbents for the analysis of these contaminants is limited. The main reason is that commercially available sorbents (SPME fibers, SBSE Twisters, etc.) lack sufficient affinity for highly polar compounds to extract them from water samples. However, there are some applications for materials specifically synthetized for this purpose. Molecularly imprinted sol-gel nanofibers or stir-bars were applied for SPME of acesulfame [187] and for naphthalenesulfonates [188], respectively.

## Acknowledgments

The authors acknowledge the financial support of the Spanish Agencia Estatal de Investigación (project no. CTM2017-84763-C3-2-R and CTQ2015-68660-P), the Galician Council of Culture, Education and Universities (ED431C2017/36) and FEDER/ERDF.

## References

[1] Directive 2010/75/EU of the european parliament and of the council of 25 november 2010 on industrial emissions (integrated pollution prevention and control). 2010.

[2] US-EPA. Method 551.1: Determination of chlorination disinfection byproducts, chlorinated solvents, and halogenated pesticides/herbicides in drinking water by liquid-liquid extraction and gas chromatography with electron-capture detection. U. S. Environmental Protection Agency; 1995.

[3] Golfinopoulos SK, Lekkas TD, Nikolaou AD. Comparison of methods for determination of volatile organic compounds in drinking water. Chemosphere 2001;45(3):275–84. https://doi.org/10.1016/S0045-6535(00)00553-1.

[4] Nikolaou AD, Lekkas TD, Golfinopoulos SK, Kostopoulou MN. Application of different analytical methods for determination of volatile chlorination by-products in drinking water. Talanta 2002;56(4):717–26. https://doi.org/10.1016/S0039-9140(01)00613-0.

[5] Roumiguières A, Kinani A, Bouchonnet S, Kinani S. Development and validation of a multiclass method for the determination of organohalogen disinfectant by-products in water samples using solid phase extraction and gas chromatography-tandem mass spectrometry. J Chromatogr A 2018;1579:89–98. https://doi.org/10.1016/j.chroma.2018.10.016.

[6] Barron L, Paull B. Determination of haloacetic acids in drinking water using suppressed micro-bore ion chromatography with solid phase extraction. Anal Chim Acta 2004; 522(2):153–61. https://doi.org/10.1016/j.aca.2004.06.052.

[7] Martınez D, Borrull F, Calull M. Comparative study of a solid-phase extraction system coupled to capillary electrophoresis in the determination of haloacetic compounds in tap water. J Chromatogr A 1998;827(1):105–12. https://doi.org/10.1016/S0021-9673(98)00752-3.

[8] Loos R, Barceló D. Determination of haloacetic acids in aqueous environments by solid-phase extraction followed by ion-pair liquid chromatography–electrospray ionization mass spectrometric detection. J Chromatogr A 2001;938(1):45–55. https://doi.org/10.1016/S0021-9673(01)01092-5.

[9] US-EPA. Method 5030C (SW-846): Purge-and-trap for aqueous samples. U. S. Environmental Protection Agency; 2003. Revision 3.

[10] US-EPA. Method 525.2: Measurement of purgeable organic compounds in water by capillary column gas chromatography/mass spectrometry. U.S. Environmental Protection Agency; 1992. Revision 4.1.

[11] Allonier A-S, Khalanski M, Bermond A, Camel V. Determination of trihalomethanes in chlorinated sea water samples using a purge-and-trap system coupled to gas chromatography. Talanta 2000;51(3):467–77. https://doi.org/10.1016/S0039-9140(99)00296-9.

[12] Liu H-W, Liu Y-T, Wu B-Z, Nian H-C, Chen H-J, Chiu K-H, et al. Process sampling module coupled with purge and trap–GC–FID for in situ auto-monitoring of volatile organic compounds in wastewater. Talanta 2009;80(2):903–8. https://doi.org/10.1016/j.talanta.2009.08.011.

[13] McComb ME, Oleschuk RD, Giller E, Gesser HD. Microextraction of volatile organic compounds using the inside needle capillary adsorption trap (INCAT) device. Talanta 1997;44(11):2137–43. https://doi.org/10.1016/S0039-9140(97)00093-3.

[14] Přikryl P, Kubinec R, Jurdáková H, Ševčík J, Ostrovský I, Soják L, et al. Comparison of needle concentrator with SPME for GC determination of benzene, toluene, ethylbenzene, and xylenes in aqueous samples. Chromatographia 2006;64(1):65–70. https://doi.org/10.1365/s10337-006-0800-3.

[15] Kubinec R, Berezkin VG, Górová R, Addová G, Mračnová H, Soják L. Needle concentrator for gas chromatographic determination of BTEX in aqueous samples. J Chromatogr B 2004;800(1):295–301. https://doi.org/10.1016/j.jchromb.2003.09.009.

[16] Jochmann MA, Kmiecik MP, Schmidt TC. Solid-phase dynamic extraction for the enrichment of polar volatile organic compounds from water. J Chromatogr A 2006; 1115(1):208–16. https://doi.org/10.1016/j.chroma.2006.02.061.

[17] Jochmann MA, Yuan X, Schmidt TC. Determination of volatile organic hydrocarbons in water samples by solid-phase dynamic extraction. Anal Bioanal Chem 2007;387(6): 2163–74. https://doi.org/10.1007/s00216-006-1066-1.

[18] Ridgway K, Lalljie SPD, Smith RM. Comparison of in-tube sorptive extraction techniques for non-polar volatile organic compounds by gas chromatography with mass spectrometric detection. J Chromatogr A 2006;1125(1):181–6. https://doi.org/10.1016/j.chroma.2006.06.105.
[19] Alonso M, Cerdan L, Godayol A, Anticó E, Sanchez JM. Headspace needle-trap analysis of priority volatile organic compounds from aqueous samples: application to the analysis of natural and waste waters. J Chromatogr A 2011;1218(45):8131–9. https://doi.org/10.1016/j.chroma.2011.09.042.
[20] De Crom J, Claeys S, Godayol A, Alonso M, Anticó E, Sanchez JM. Sorbent-packed needle microextraction trap for benzene, toluene, ethylbenzene, and xylenes determination in aqueous samples. J Sep Sci 2010;33(17-18):2833–40. https://doi.org/10.1002/jssc.201000213.
[21] Mesarchaki E, Yassaa N, Hein D, Lutterbeck HE, Zindler C, Williams J. A novel method for the measurement of VOCs in seawater using needle trap devices and GC–MS. Mar Chem 2014;159:1–8. https://doi.org/10.1016/j.marchem.2013.12.001.
[22] Nakamura S, Nakamura N, Ito S. Determination of 2-methylisoborneol and geosmin in water by gas chromatography-mass spectrometry using stir bar sorptive extraction. J Sep Sci 2001;25(8):674–7. https://doi.org/10.1002/1615-9314(20010801)25:8<674::AID-JSSC674>3.0.CO;2-E.
[23] Loughrin JH. Comparison of solid-phase microextraction and stir bar sorptive extraction for the quantification of malodors in wastewater. J Agric Food Chem 2006;54(9):3237–41. https://doi.org/10.1021/jf0602119.
[24] Cho D-H, Kong S-H, Oh S-G. Analysis of trihalomethanes in drinking water using headspace-SPME technique with gas chromatography. Water Res 2003;37(2):402–8. https://doi.org/10.1016/S0043-1354(02)00285-3.
[25] Antoniou CV, Koukouraki EE, Diamadopoulos E. Determination of chlorinated volatile organic compounds in water and municipal wastewater using headspace–solid phase microextraction–gas chromatography. J Chromatogr A 2006;1132(1):310–4. https://doi.org/10.1016/j.chroma.2006.08.082.
[26] Guimarães AD, Carvalho JJ, GonÇalves C, Alpendurada MF. Simultaneous analysis of 23 priority volatile compounds in water by solid-phase microextraction–gas chromatography–mass spectrometry and estimation of the method's uncertainty. Int J Environ Anal Chem 2008;88(3):151–64. https://doi.org/10.1080/03067310701642982.
[27] O'Reilly J, Wang Q, Setkova L, Hutchinson JP, Chen Y, Lord HL, et al. Automation of solid-phase microextraction. J Sep Sci 2005;28(15):2010–22. https://doi.org/10.1002/jssc.200500254.
[28] Cervera MI, Beltran J, Lopez FJ, Hernandez F. Determination of volatile organic compounds in water by headspace solid-phase microextraction gas chromatography coupled to tandem mass spectrometry with triple quadrupole analyzer. Anal Chim Acta 2011;704(1):87–97. https://doi.org/10.1016/j.aca.2011.08.012.
[29] Lara-Gonzalo A, Sánchez-Uría JE, Segovia-García E, Sanz-Medel A. Critical comparison of automated purge and trap and solid-phase microextraction for routine determination of volatile organic compounds in drinking waters by GC–MS. Talanta 2008;74(5):1455–62. https://doi.org/10.1016/j.talanta.2007.09.036.
[30] Sung Y-H, Li T-Y, Huang S-D. Analysis of earthy and musty odors in water samples by solid-phase microextraction coupled with gas chromatography/ion trap mass spectrometry. Talanta 2005;65(2):518–25. https://doi.org/10.1016/j.talanta.2004.07.014.

[31] Chen X, Luo Q, Yuan S, Wei Z, Song H, Wang D, et al. Simultaneous determination of ten taste and odor compounds in drinking water by solid-phase microextraction combined with gas chromatography-mass spectrometry. J Environ Sci 2013;25(11):2313–23. https://doi.org/10.1016/S1001-0742(12)60290-3.
[32] Wu CH, Lu JT, Lo JG. Analysis of volatile organic compounds in wastewater during various stages of treatment for high-tech industries. Chromatographia 2002;56(1):91–8. https://doi.org/10.1007/BF02590252.
[33] Fowlie P, MacGillivray B, Pawliszyn J, Sagara C. Headspace solid-phase microextraction versus purge and trap for the determination of substituted benzene compounds in water. J Chromatogr Sci 1994;32(8):317–22. https://doi.org/10.1093/chromsci/32.8.317.
[34] Zygmunt B, Jastrzecbska A, Namiesnik J. Solid phase microextraction — a convenient tool for the determination of organic pollutants in environmental matrices. Crit Rev Anal Chem 2001;31(1):1–18. https://doi.org/10.1080/20014091076668.
[35] Ncube S, Madikizela L, Cukrowska E, Chimuka L. Recent advances in the adsorbents for isolation of polycyclic aromatic hydrocarbons (PAHs) from environmental sample solutions. Trends Anal Chem 2018;99:101–16. https://doi.org/10.1016/j.trac.2017.12.007.
[36] US-EPA. EPA method 550.1. Determination of polycyclic aromatic hydrocarbons in drinking water by liquid-solid extraction and HPLC with coupled ultraviolet and fluorescence detection. Cincinnati: U. S. Environmental Protection Agency; 1990.
[37] US-EPA. EPA method 1614A. Brominated diphenyl ethers in water, soil, sediment, and tissue by HRGC/HRMS. Cincinnati: U. S. Environmental Protection Agency; 2010.
[38] US-EPA. Method 1668C. Chlorinated biphenyl congeners in water, soil, sediment, biosolids, and tissue by HRGC/HRMS. Cincinnati: U. S. Environmental Protection Agency; 2010.
[39] US-EPA. EPA method 608.3. Organochlorine pesticides and PCBs by GC/HSD. Cincinnati: U. S. Environmental Protection Agency; 2014.
[40] Marcé RM, Borrull F. Solid-phase extraction of polycyclic aromatic compounds. J Chromatogr A 2000;885(1):273–90. https://doi.org/10.1016/S0021-9673(00)00428-3.
[41] Kiss G, Varga-Puchony Z, Hlavay J. Determination of polycyclic aromatic hydrocarbons in precipitation using solid-phase extraction and column liquid chromatography. J Chromatogr A 1996;725(2):261–72. https://doi.org/10.1016/0021-9673(95)00940-X.
[42] Prieto A, Schrader S, Moeder M. Determination of organic priority pollutants and emerging compounds in wastewater and snow samples using multiresidue protocols on the basis of microextraction by packed sorbents coupled to large volume injection gas chromatography–mass spectrometry analysis. J Chromatogr A 2010;1217(38):6002–11. https://doi.org/10.1016/j.chroma.2010.07.070.
[43] Doong R-A, Chang S-M, Sun Y-C. Solid-phase microextraction for determining the distribution of sixteen US environmental protection agency polycyclic aromatic hydrocarbons in water samples. J Chromatogr A 2000;879(2):177–88. https://doi.org/10.1016/S0021-9673(00)00347-2.
[44] Llompart M, Li K, Fingas M. Solid-phase microextraction and headspace solid-phase microextraction for the determination of polychlorinated biphenyls in water samples. Anal Chem 1998;70(13):2510–5. https://doi.org/10.1021/ac971332y.
[45] Negrão MR, Alpendurada MF. Solvent-free method for the determination of polynuclear aromatic hydrocarbons in waste water by solid-phase microextraction–high-performance liquid chromatography with photodiode-array detection. J Chromatogr A 1998;823(1):211–8. https://doi.org/10.1016/S0021-9673(98)00586-X.

[46] Polo M, Gomez-Noya G, Quintana JB, Llompart M, Garcia-Jares C, Cela R. Development of a solid-phase microextraction gas chromatography/tandem mass spectrometry method for polybrominated diphenyl ethers and polybrominated biphenyls in water samples. Anal Chem 2004;76(4):1054−62. https://doi.org/10.1021/ac030292x.

[47] Raposo Júnior JL, Ré-Poppi N. Determination of organochlorine pesticides in ground water samples using solid-phase microextraction by gas chromatography-electron capture detection. Talanta 2007;72(5):1833−41. https://doi.org/10.1016/j.talanta.2007.02.025.

[48] Yang Y, Miller DJ, Hawthorne SB. Solid-phase microextraction of polychlorinated biphenyls. J Chromatogr A 1998;800(2):257−66. https://doi.org/10.1016/S0021-9673(97)01115-1.

[49] Baltussen E, Sandra P, David F, Cramers C. Stir bar sorptive extraction (SBSE), a novel extraction technique for aqueous samples: theory and principles. J Microcolumn Sep 1999;11(10):737−47. https://doi.org/10.1002/(SICI)1520-667X(1999)11:10<737::AID-MCS7>3.0.CO;2-4.

[50] Popp P, Keil P, Montero L, Rückert M. Optimized method for the determination of 25 polychlorinated biphenyls in water samples using stir bar sorptive extraction followed by thermodesorption-gas chromatography/mass spectrometry. J Chromatogr A 2005;1071(1):155−62. https://doi.org/10.1016/j.chroma.2005.01.066.

[51] Popp P, Bauer C, Wennrich L. Application of stir bar sorptive extraction in combination with column liquid chromatography for the determination of polycyclic aromatic hydrocarbons in water samples. Anal Chim Acta 2001;436(1):1−9. https://doi.org/10.1016/S0003-2670(01)00895-9.

[52] Sánchez-Avila J, Quintana J, Ventura F, Tauler R, Duarte CM, Lacorte S. Stir bar sorptive extraction-thermal desorption-gas chromatography−mass spectrometry: an effective tool for determining persistent organic pollutants and nonylphenol in coastal waters in compliance with existing directives. Mar Pollut Bull 2010;60(1):103−12. https://doi.org/10.1016/j.marpolbul.2009.08.028.

[53] Llorca-Porcel J, Martínez-Sánchez G, Álvarez B, Cobollo MA, Valor I. Analysis of nine polybrominated diphenyl ethers in water samples by means of stir bar sorptive extraction-thermal desorption-gas chromatography−mass spectrometry. Anal Chim Acta 2006;569(1):113−8. https://doi.org/10.1016/j.aca.2006.03.100.

[54] Kremser A, Jochmann MA, Schmidt TC. PAL SPME Arrow—evaluation of a novel solid-phase microextraction device for freely dissolved PAHs in water. Anal Bioanal Chem 2016;408(3):943−52. https://doi.org/10.1007/s00216-015-9187-z.

[55] Popp P, Bauer C, Paschke A, Montero L. Application of a polysiloxane-based extraction method combined with column liquid chromatography to determine polycyclic aromatic hydrocarbons in environmental samples. Anal Chim Acta 2004;504(2):307−12. https://doi.org/10.1016/j.aca.2003.10.040.

[56] Valls-Cantenys C, Villaverde-de-Sáa E, Rodil R, Quintana JB, Iglesias M, Salvadó V, et al. Application of polydimethylsiloxane rod extraction to the determination of sixteen halogenated flame retardants in water samples. Anal Chim Acta 2013;770:85−93. https://doi.org/10.1016/j.aca.2013.01.059.

[57] Huck CW, Bonn GK. Recent developments in polymer-based sorbents for solid-phase extraction. J Chromatogr A 2000;885(1):51−72. https://doi.org/10.1016/S0021-9673(00)00333-2.

[58] Raynie DE. Modern extraction techniques. Anal Chem 2010;82(12):4911−6. https://doi.org/10.1021/ac101223c.

[59] Alberti G, Amendola V, Pesavento M, Biesuz R. Beyond the synthesis of novel solid phases: review on modelling of sorption phenomena. Coord Chem Rev 2012;256(1): 28–45. https://doi.org/10.1016/j.ccr.2011.08.022.

[60] Madikizela LM, Ncube S, Chimuka L. Recent developments in selective materials for solid phase extraction. Chromatographia 2018. https://doi.org/10.1007/s10337-018-3644-8.

[61] Marta-Sanchez AV, Caldas SS, Schneider A, Cardoso SMVS, Primel EG. Trace analysis of parabens preservatives in drinking water treatment sludge, treated, and mineral water samples. Environ Sci Pollut Res 2018;25(15):14460–70. https://doi.org/10.1007/s11356-018-1583-4.

[62] Gago-Ferrero P, Mastroianni N, Díaz-Cruz MS, Barceló D. Fully automated determination of nine ultraviolet filters and transformation products in natural waters and wastewaters by on-line solid phase extraction–liquid chromatography–tandem mass spectrometry. J Chromatogr A 2013;1294:106–16. https://doi.org/10.1016/j.chroma.2013.04.037.

[63] Goh SXL, Duarah A, Zhang L, Snyder SA, Lee HK. Online solid phase extraction with liquid chromatography–tandem mass spectrometry for determination of estrogens and glucocorticoids in water. J Chromatogr A 2016;1465:9–19. https://doi.org/10.1016/j.chroma.2016.08.040.

[64] Molins-Delgado D, García-Sillero D, Díaz-Cruz MS, Barceló D. On-line solid phase extraction-liquid chromatography-tandem mass spectrometry for insect repellent residue analysis in surface waters using atmospheric pressure photoionization. J Chromatogr A 2018;1544:33–40. https://doi.org/10.1016/j.chroma.2018.02.027.

[65] Rimayi C, Odusanya D, Weiss JM, de Boer J, Chimuka L. Contaminants of emerging concern in the Hartbeespoort dam catchment and the uMngeni river estuary 2016 pollution incident, South Africa. Sci Total Environ 2018;627:1008–17. https://doi.org/10.1016/j.scitotenv.2018.01.263.

[66] Casado J, Santillo D, Johnston P. Multi-residue analysis of pesticides in surface water by liquid chromatography quadrupole-Orbitrap high resolution tandem mass spectrometry. Anal Chim Acta 2018;1025:1–17. https://doi.org/10.1016/j.aca.2018.04.026.

[67] Chen W, Huang H, Chen C-E, Qi S, Price OR, Zhang H, et al. Simultaneous determination of 20 trace organic chemicals in waters by solid-phase extraction (SPE) with triple-quadrupole mass spectrometer (QqQ-MS) and hybrid quadrupole Orbitrap high resolution MS (Q-Orbitrap-HRMS). Chemosphere 2016;163:99–107. https://doi.org/10.1016/j.chemosphere.2016.07.080.

[68] Rivera-Jaimes JA, Postigo C, Melgoza-Alemán RM, Aceña J, Barceló D, López de Alda M. Study of pharmaceuticals in surface and wastewater from Cuernavaca, Morelos, Mexico: occurrence and environmental risk assessment. Sci Total Environ 2018; 613–614:1263–74. https://doi.org/10.1016/j.scitotenv.2017.09.134.

[69] Pérez-Fernández V, Mainero Rocca L, Tomai P, Fanali S, Gentili A. Recent advancements and future trends in environmental analysis: sample preparation, liquid chromatography and mass spectrometry. Anal Chim Acta 2017;983:9–41. https://doi.org/10.1016/j.aca.2017.06.029.

[70] Fontanals N, Marcé RM, Borrull F. Solid-phase extraction followed by liquid chromatography-high resolution mass spectrometry to determine synthetic cathinones in different types of environmental water samples. J Chromatogr A 2017;1525:66–73. https://doi.org/10.1016/j.chroma.2017.10.002.

[71] García-López M, Rodríguez I, Cela R. Mixed-mode solid-phase extraction followed by liquid chromatography—tandem mass spectrometry for the determination of tri- and di-substituted organophosphorus species in water samples. J Chromatogr A 2010;1217(9): 1476—84. https://doi.org/10.1016/j.chroma.2009.12.067.

[72] Jamari NLA, Dohmann JF, Raab A, Krupp EM, Feldmann J. Novel non-targeted analysis of perfluorinated compounds using fluorine-specific detection regardless of their ionisability (HPLC-ICPMS/MS-ESI-MS). Anal Chim Acta 2019;1053:22—31. https://doi.org/10.1016/j.aca.2018.11.037.

[73] Jiang J, Mu D, Ding M, Zhang S, Zhang H, Hu J. Simultaneous determination of primary and secondary phthalate monoesters in the Taihu Lake: exploration of sources. Chemosphere 2018;202:17—25. https://doi.org/10.1016/j.chemosphere.2018.03.070.

[74] Peng Y, Gautam L, Hall SW. The detection of drugs of abuse and pharmaceuticals in drinking water using solid-phase extraction and liquid chromatography-mass spectrometry. Chemosphere 2019;223:438—47. https://doi.org/10.1016/j.chemosphere.2019.02.040.

[75] Zhang P, Bui A, Rose G, Allinson G. Mixed-mode solid-phase extraction coupled with liquid chromatography tandem mass spectrometry to determine phenoxy acid, sulfonylurea, triazine and other selected herbicides at nanogram per litre levels in environmental waters. J Chromatogr A 2014;1325:56—64. https://doi.org/10.1016/j.chroma.2013.12.021.

[76] Fontanals N, Marcé RM, Borrull F, Cormack PAG. Mixed-mode ion-exchange polymeric sorbents: dual-phase materials that improve selectivity and capacity. Trends Anal Chem 2010;29(7):765—79. https://doi.org/10.1016/j.trac.2010.03.015.

[77] Barciela-Alonso MC, Otero-Lavandeira N, Bermejo-Barrera P. Solid phase extraction using molecular imprinted polymers for phthalate determination in water and wine samples by HPLC-ESI-MS. Microchem J 2017;132:233—7. https://doi.org/10.1016/j.microc.2017.02.007.

[78] García-Galán MJ, Díaz-Cruz MS, Barceló D. Determination of triazines and their metabolites in environmental samples using molecularly imprinted polymer extraction, pressurized liquid extraction and LC—tandem mass spectrometry. J Hydrol 2010;383(1): 30—8. https://doi.org/10.1016/j.jhydrol.2009.09.025.

[79] Gros M, Pizzolato T-M, Petrović M, de Alda MJL, Barceló D. Trace level determination of β-blockers in waste waters by highly selective molecularly imprinted polymers extraction followed by liquid chromatography—quadrupole-linear ion trap mass spectrometry. J Chromatogr A 2008;1189(1):374—84. https://doi.org/10.1016/j.chroma.2007.10.052.

[80] Lu W, Wang X, Wu X, Liu D, Li J, Chen L, et al. Multi-template imprinted polymers for simultaneous selective solid-phase extraction of six phenolic compounds in water samples followed by determination using capillary electrophoresis. J Chromatogr A 2017; 1483:30—9. https://doi.org/10.1016/j.chroma.2016.12.069.

[81] Olcer YA, Demirkurt M, Demir MM, Eroglu AE. Development of molecularly imprinted polymers (MIPs) as a solid phase extraction (SPE) sorbent for the determination of ibuprofen in water. RSC Adv 2017;7(50):31441—7. https://doi.org/10.1039/C7RA05254E.

[82] Rozaini MNH, Yahaya N, Saad B, Kamaruzaman S, Hanapi NSM. Rapid ultrasound assisted emulsification micro-solid phase extraction based on molecularly imprinted polymer for HPLC-DAD determination of bisphenol A in aqueous matrices. Talanta 2017;171:252—9. https://doi.org/10.1016/j.talanta.2017.05.006.

[83] Shi X, Liu J, Sun A, Li D, Chen J. Group-selective enrichment and determination of pyrethroid insecticides in aquaculture seawater via molecularly imprinted solid phase extraction coupled with gas chromatography-electron capture detection. J Chromatogr A 2012;1227:60–6. https://doi.org/10.1016/j.chroma.2012.01.012.
[84] Sun X, Jiao X-y, Li J, Xu L. A miniaturized sorbent phase-based extraction device in the form of syringe filter holder using molecularly imprinted polymer as sorbent and its application to extract benzophenones. J Chromatogr A 2018;1543:1–13. https://doi.org/10.1016/j.chroma.2018.02.039.
[85] Zacs D, Perkons I, Bartkevics V. Determination of steroidal oestrogens in tap water samples using solid-phase extraction on a molecularly imprinted polymer sorbent and quantification with gas chromatography-mass spectrometry (GC-MS). Environ Monit Assess 2016;188(7):433. https://doi.org/10.1007/s10661-016-5435-8.
[86] He C, Long Y, Pan J, Li K, Liu F. Application of molecularly imprinted polymers to solid-phase extraction of analytes from real samples. J Biochem Biophys Methods 2007;70(2):133–50. https://doi.org/10.1016/j.jbbm.2006.07.005.
[87] Binsalom A, Chianella I, Campbell K, Zourob M. Development of solid-phase extraction using molecularly imprinted polymer for the analysis of organophosphorus pesticides-(chlorpyrifos) in aqueous solution. J Chromatogr Sep Tech 2016;7:340–5. https://doi.org/10.4172/2157-7064.1000340.
[88] Zuo HG, Zhu JX, Zhan CR, Shi L, Xing M, Guo P, et al. Preparation of malathion MIP-SPE and its application in environmental analysis. Environ Monit Assess 2015;187(7):394. https://doi.org/10.1007/s10661-015-4641-0.
[89] Guo P, Yuan X, Zhang J, Wang B, Sun X, Chen X, et al. Dummy-surface molecularly imprinted polymers as a sorbent of micro-solid-phase extraction combined with dispersive liquid–liquid microextraction for determination of five 2-phenylpropionic acid NSAIDs in aquatic environmental samples. Anal Bioanal Chem 2018;410(2):373–89. https://doi.org/10.1007/s00216-017-0727-6.
[90] Madikizela LM, Chimuka L. Determination of ibuprofen, naproxen and diclofenac in aqueous samples using a multi-template molecularly imprinted polymer as selective adsorbent for solid-phase extraction. J Pharm Biomed Anal 2016;128:210–5. https://doi.org/10.1016/j.jpba.2016.05.037.
[91] Sun X, Wang M, Yang L, Wen H, Wang L, Li T, et al. Preparation and evaluation of dummy-template molecularly imprinted polymer as a potential sorbent for solid phase extraction of imidazole fungicides from river water. J Chromatogr A 2019;1586:1–8. https://doi.org/10.1016/j.chroma.2018.11.077.
[92] Sun X, Wang M, Peng J, Yang L, Wang X, Wang F, et al. Dummy molecularly imprinted solid phase extraction of climbazole from environmental water samples. Talanta 2019;196:47–53. https://doi.org/10.1016/j.talanta.2018.12.017.
[93] Liu Y, Wang D, Du F, Zheng W, Liu Z, Xu Z, et al. Dummy-template molecularly imprinted micro-solid-phase extraction coupled with high-performance liquid chromatography for bisphenol A determination in environmental water samples. Microchem J 2019;145:337–44. https://doi.org/10.1016/j.microc.2018.10.054.
[94] Özer ET, Osman B, Yazıcı T. Dummy molecularly imprinted microbeads as solid-phase extraction material for selective determination of phthalate esters in water. J Chromatogr A 2017;1500:53–60. https://doi.org/10.1016/j.chroma.2017.04.013.
[95] Díaz-Álvarez M, Barahona F, Turiel E, Martín-Esteban A. Supported liquid membrane-protected molecularly imprinted beads for micro-solid phase extraction of sulfonamides in environmental waters. J Chromatogr A 2014;1357:158–64. https://doi.org/10.1016/j.chroma.2014.04.038.

[96] Nemulenzi O, Mhaka B, Cukrowska E, Ramström O, Tutu H, Chimuka L. Potential of combining of liquid membranes and molecularly imprinted polymers in extraction of 17β-estradiol from aqueous samples. J Sep Sci 2009;32(11):1941–8. https://doi.org/10.1002/jssc.200800659.

[97] Hao Y, Gao R, Shi L, Liu D, Tang Y, Guo Z. Water-compatible magnetic imprinted nanoparticles served as solid-phase extraction sorbents for selective determination of trace 17beta-estradiol in environmental water samples by liquid chromatography. J Chromatogr A 2015;1396:7–16. https://doi.org/10.1016/j.chroma.2015.03.083.

[98] Li C, Ma X, Zhang X, Wang R, Chen Y, Li Z. Magnetic molecularly imprinted polymer nanoparticles-based solid-phase extraction coupled with gas chromatography–mass spectrometry for selective determination of trace di-(2-ethylhexyl) phthalate in water samples. Anal Bioanal Chem 2016;408(27):7857–64. https://doi.org/10.1007/s00216-016-9889-x.

[99] Lin Z, Cheng W, Li Y, Liu Z, Chen X, Huang C. A novel superparamagnetic surface molecularly imprinted nanoparticle adopting dummy template: an efficient solid-phase extraction adsorbent for bisphenol A. Anal Chim Acta 2012;720:71–6. https://doi.org/10.1016/j.aca.2012.01.020.

[100] Wu X, Li Y, Zhu X, He C, Wang Q, Liu S. Dummy molecularly imprinted magnetic nanoparticles for dispersive solid-phase extraction and determination of bisphenol A in water samples and orange juice. Talanta 2017;162:57–64. https://doi.org/10.1016/j.talanta.2016.10.007.

[101] Cao W, Yang B, Qi F, Qian L, Li J, Lu L, et al. Simple and sensitive determination of atrazine and its toxic metabolites in environmental water by carboxyl modified polyacrylonitrile nanofibers mat-based solid-phase extraction coupled with liquid chromatography-diode array detection. J Chromatogr A 2017;1491:16–26. https://doi.org/10.1016/j.chroma.2017.02.035.

[102] Háková M, Chocholoušová Havlíková L, Chvojka J, Solich P, Šatínský D. An on-line coupling of nanofibrous extraction with column-switching high performance liquid chromatography – a case study on the determination of bisphenol A in environmental water samples. Talanta 2018;178:141–6. https://doi.org/10.1016/j.talanta.2017.08.098.

[103] Háková M, Raabová H, Havlíková LC, Chocholouš P, Chvojka J, Šatínský D. Testing of nylon 6 nanofibers with different surface densities as sorbents for solid phase extraction and their selectivity comparison with commercial sorbent. Talanta 2018;181:326–32. https://doi.org/10.1016/j.talanta.2018.01.043.

[104] Jian N, Qian L, Wang C, Li R, Xu Q, Li J. Novel nanofibers mat as an efficient, fast and reusable adsorbent for solid phase extraction of non-steroidal anti-inflammatory drugs in environmental water. J Hazard Mater 2019;363:81–9. https://doi.org/10.1016/j.jhazmat.2018.09.052.

[105] Xu Q, Wang M, Yu S, Tao Q, Tang M. Trace analysis of diethylstilbestrol, dienestrol and hexestrol in environmental water by Nylon 6 nanofibers mat-based solid-phase extraction coupled with liquid chromatography-mass spectrometry. Analyst 2011;136(23):5030–7. https://doi.org/10.1039/C1AN15494J.

[106] Xu Q, Yin X, Wu S, Wang M, Wen Z, Gu Z. Determination of phthalate esters in water samples using Nylon6 nanofibers mat-based solid-phase extraction coupled to liquid chromatography. Microchim Acta 2010;168(3):267–75. https://doi.org/10.1007/s00604-010-0290-8.

[107] Yang B-Y, Qi F-F, Li X-Q, Liu J-J, Rong F, Xu Q. Application of Nylon6/Polypyrrole core–shell nanofibres mat as solid-phase extraction adsorbent for the determination of atrazine in environmental water samples. Int J Environ Anal Chem 2015;95(12): 1112–23. https://doi.org/10.1080/03067319.2015.1085525.
[108] Boonjob W, Miró M, Segundo MA, Cerdà V. Flow-through dispersed carbon nanofiber-based microsolid-phase extraction coupled to liquid chromatography for automatic determination of trace levels of priority environmental pollutants. Anal Chem 2011; 83(13):5237–44. https://doi.org/10.1021/ac2005058.
[109] Han Q, Aydan T, Yang L, Zhang X, Liang Q, Ding M. In-syringe solid-phase extraction for on-site sampling of pyrethroids in environmental water samples. Anal Chim Acta 2018;1009:48–55. https://doi.org/10.1016/j.aca.2018.01.001.
[110] Han Q, Liang Q, Zhang X, Yang L, Ding M. Graphene aerogel based monolith for effective solid-phase extraction of trace environmental pollutants from water samples. J Chromatogr A 2016;1447:39–46. https://doi.org/10.1016/j.chroma.2016.04.032.
[111] Lalović B, Đurkić T, Vukčević M, Janković-Častvan I, Kalijadis A, Laušević Z, et al. Solid-phase extraction of multi-class pharmaceuticals from environmental water samples onto modified multi-walled carbon nanotubes followed by LC-MS/MS. Environ Sci Pollut Res 2017;25(25):20784–93. https://doi.org/10.1007/s11356-017-9748-0.
[112] Liu L, Jiang Y, Yong L, Hu B, Zhu L, Wang Y, et al. MWCNTs-solid phase extraction combined with ultra-high performance liquid chromatography-tandem mass spectrometry for the determination of eleven organophosphorus pesticides in river water. Int J Environ Anal Chem 2018;98(8):743–57. https://doi.org/10.1080/03067319.2018.1502284.
[113] Liu Q, Shi J, Zeng L, Wang T, Cai Y, Jiang G. Evaluation of graphene as an advantageous adsorbent for solid-phase extraction with chlorophenols as model analytes. J Chromatogr A 2011;1218(2):197–204. https://doi.org/10.1016/j.chroma.2010.11.022.
[114] Naing NN, Li SFY, Lee HK. Evaluation of graphene-based sorbent in the determination of polar environmental contaminants in water by micro-solid phase extraction-high performance liquid chromatography. J Chromatogr A 2016;1427:29–36. https://doi.org/10.1016/j.chroma.2015.12.012.
[115] Reinholds I, Pugajeva I, Zacs D, Lundanes E, Rusko J, Perkons I, et al. Determination of acidic non-steroidal anti-inflammatory drugs in aquatic samples by liquid chromatography-triple quadrupole mass spectrometry combined with carbon nanotubes-based solid-phase extraction. Environ Monit Assess 2017;189(11):568. https://doi.org/10.1007/s10661-017-6304-9.
[116] Shi Z, Hu J, Li Q, Zhang S, Liang Y, Zhang H. Graphene based solid phase extraction combined with ultra high performance liquid chromatography–tandem mass spectrometry for carbamate pesticides analysis in environmental water samples. J Chromatogr A 2014;1355:219–27. https://doi.org/10.1016/j.chroma.2014.05.085.
[117] Speltini A, Maraschi F, Sturini M, Contini M, Profumo A. Dispersive multi-walled carbon nanotubes extraction of benzenesulfonamides, benzotriazoles, and benzothiazoles from environmental waters followed by microwave desorption and HPLC-HESI-MS/MS. Anal Bioanal Chem 2017;409(28):6709–18. https://doi.org/10.1007/s00216-017-0627-9.
[118] Wang X, Wang J, Du T, Kou H, Du X, Lu X. Determination of six benzotriazole ultraviolet filters in water and cosmetic samples by graphene sponge-based solid-phase extraction followed by high-performance liquid chromatography. Anal Bioanal Chem 2018;410(26):6955–62. https://doi.org/10.1007/s00216-018-1301-6.

[119] Wang Z, Zhang X, Jiang S, Guo X. Magnetic solid-phase extraction based on magnetic multiwalled carbon nanotubes for the simultaneous enantiomeric analysis of five β-blockers in the environmental samples by chiral liquid chromatography coupled with tandem mass spectrometry. Talanta 2018;180:98–107. https://doi.org/10.1016/j.talanta.2017.12.034.

[120] Zhang X, Liang Q, Han Q, Wan W, Ding M. Metal–organic frameworks@graphene hybrid aerogels for solid-phase extraction of non-steroidal anti-inflammatory drugs and selective enrichment of proteins. Analyst 2016;141(13):4219–26. https://doi.org/10.1039/C6AN00353B.

[121] Zhou Q, Xiao J, Wang W, Liu G, Shi Q, Wang J. Determination of atrazine and simazine in environmental water samples using multiwalled carbon nanotubes as the adsorbents for preconcentration prior to high performance liquid chromatography with diode array detector. Talanta 2006;68(4):1309–15. https://doi.org/10.1016/j.talanta.2005.07.050.

[122] Geng HR, Miao SS, Jin SF, Yang H. A newly developed molecularly imprinted polymer on the surface of $TiO_2$ for selective extraction of triazine herbicides residues in maize, water, and soil. Anal Bioanal Chem 2015;407(29):8803–12. https://doi.org/10.1007/s00216-015-9039-x.

[123] Rao W, Cai R, Yin Y, Long F, Zhang Z. Magnetic dummy molecularly imprinted polymers based on multi-walled carbon nanotubes for rapid selective solid-phase extraction of 4-nonylphenol in aqueous samples. Talanta 2014;128:170–6. https://doi.org/10.1016/j.talanta.2014.04.087.

[124] Ye Q, Liu L, Chen Z, Hong l. Analysis of phthalate acid esters in environmental water by magnetic graphene solid phase extraction coupled with gas chromatography–mass spectrometry. J Chromatogr A 2014;1329:25–9. https://doi.org/10.1016/j.chroma.2013.12.086.

[125] Han Q, Wang Z, Xia J, Xia L, Chen S, Zhang X, et al. Graphene as an efficient sorbent for the SPE of organochlorine pesticides in water samples coupled with GC–MS. J Sep Sci 2013;36(21–22):3586–91. https://doi.org/10.1002/jssc.201300373.

[126] Luo Y-B, Zhu G-T, Li X-S, Yuan B-F, Feng Y-Q. Facile fabrication of reduced graphene oxide-encapsulated silica: a sorbent for solid-phase extraction. J Chromatogr A 2013; 1299:10–7. https://doi.org/10.1016/j.chroma.2013.05.036.

[127] Speltini A, Sturini M, Maraschi F, Consoli L, Zeffiro A, Profumo A. Graphene-derivatized silica as an efficient solid-phase extraction sorbent for pre-concentration of fluoroquinolones from water followed by liquid-chromatography fluorescence detection. J Chromatogr A 2015;1379:9–15. https://doi.org/10.1016/j.chroma.2014.12.047.

[128] Li C, Chen J, Chen Y, Wang J, Ping H, Lu A. Graphene-derivatized silica composite as solid-phase extraction sorbent combined with GC-MS/MS for the determination of polycyclic musks in aqueous samples. Molecules 2018;23(2):318. https://doi.org/10.3390/molecules23020318.

[129] Liu Z, Jiang P, Huang G, Yan X, Li X-F. Silica monolith nested in sponge (SiMNS): a composite monolith as a new solid phase extraction material for environmental analysis. Anal Chem 2019;91(5):3659–66. https://doi.org/10.1021/acs.analchem.8b05707.

[130] Huppert N, Würtele M, Hahn HH. Determination of the plasticizer N-butylbenzenesulfonamide and the pharmaceutical Ibuprofen in wastewater using solid phase microextraction (SPME). Fresenius' J Anal Chem 1998;362(6):529–36. https://doi.org/10.1007/s002160051119.

[131] Moeder M, Schrader S, Winkler M, Popp P. Solid-phase microextraction–gas chromatography–mass spectrometry of biologically active substances in water samples. J Chromatogr A 2000;873(1):95–106. https://doi.org/10.1016/S0021-9673(99)01256-X.

[132] Carpinteiro J, Quintana JB, Martınez E, Rodrıguez I, Carro AM, Lorenzo RA, et al. Application of strategic sample composition to the screening of anti-inflammatory drugs in water samples using solid-phase microextraction. Anal Chim Acta 2004;525(1): 63–71. https://doi.org/10.1016/j.aca.2004.03.026.

[133] Suchara EA, Budziak D, Martendal E, Costa LLF, Carasek E. A combination of statistical and analytical evaluation methods as a new optimization strategy for the quantification of pharmaceutical residues in sewage effluent. Anal Chim Acta 2008;613(2):169–76. https://doi.org/10.1016/j.aca.2008.02.067.

[134] Balakrishnan VK, Terry KA, Toito J. Determination of sulfonamide antibiotics in wastewater: a comparison of solid phase microextraction and solid phase extraction methods. J Chromatogr A 2006;1131(1):1–10. https://doi.org/10.1016/j.chroma.2006.07.011.

[135] Unceta N, Sampedro MC, Bakar NKA, Gómez-Caballero A, Goicolea MA, Barrio RJ. Multi-residue analysis of pharmaceutical compounds in wastewaters by dual solid-phase microextraction coupled to liquid chromatography electrospray ionization ion trap mass spectrometry. J Chromatogr A 2010;1217(20):3392–9. https://doi.org/10.1016/j.chroma.2010.03.008.

[136] Vera-Candioti L, Gil García MD, Martínez Galera M, Goicoechea HC. Chemometric assisted solid-phase microextraction for the determination of anti-inflammatory and antiepileptic drugs in river water by liquid chromatography–diode array detection. J Chromatogr A 2008;1211(1):22–32. https://doi.org/10.1016/j.chroma.2008.09.093.

[137] Canosa P, Rodríguez I, Rubí E, Bollaín MH, Cela R. Optimisation of a solid-phase microextraction method for the determination of parabens in water samples at the low ng per litre level. J Chromatogr A 2006;1125(1):3–10. https://doi.org/10.1016/j.chroma.2006.03.045.

[138] Regueiro J, Becerril E, Garcia-Jares C, Llompart M. Trace analysis of parabens, triclosan and related chlorophenols in water by headspace solid-phase microextraction with in situ derivatization and gas chromatography–tandem mass spectrometry. J Chromatogr A 2009;1216(23):4693–702. https://doi.org/10.1016/j.chroma.2009.04.025.

[139] Basaglia G, Pietrogrande MC. Optimization of a SPME/GC/MS method for the simultaneous determination of pharmaceuticals and personal care products in waters. Chromatographia 2012;75(7):361–70. https://doi.org/10.1007/s10337-012-2207-7.

[140] Kim D, Han J, Choi Y. On-line solid-phase microextraction of triclosan, bisphenol A, chlorophenols, and selected pharmaceuticals in environmental water samples by high-performance liquid chromatography–ultraviolet detection. Anal Bioanal Chem 2013; 405(1):377–87. https://doi.org/10.1007/s00216-012-6490-9.

[141] Shen JY, Chang MS, Yang S-H, Wu GJ. Simultaneous determination of triclosan, triclocarban, and transformation products of triclocarban in aqueous samples using solid-phase micro-extraction-HPLC-MS/MS. J Sep Sci 2012;35(19):2544–52. https://doi.org/10.1002/jssc.201200181.

[142] Luks-Betlej K, Popp P, Janoszka B, Paschke H. Solid-phase microextraction of phthalates from water. J Chromatogr A 2001;938(1):93–101. https://doi.org/10.1016/S0021-9673(01)01363-2.

[143] Peñalver A, Pocurull E, Borrull F, Marcé RM. Determination of phthalate esters in water samples by solid-phase microextraction and gas chromatography with mass spectrometric detection. J Chromatogr A 2000;872(1):191–201. https://doi.org/10.1016/S0021-9673(99)01284-4.

[144] Polo M, Llompart M, Garcia-Jares C, Cela R. Multivariate optimization of a solid-phase microextraction method for the analysis of phthalate esters in environmental waters. J Chromatogr A 2005;1072(1):63–72. https://doi.org/10.1016/j.chroma.2004.12.040.

[145] Bach C, Boiteux V, Hemard J, Colin A, Rosin C, Munoz J-F, et al. Simultaneous determination of perfluoroalkyl iodides, perfluoroalkane sulfonamides, fluorotelomer alcohols, fluorotelomer iodides and fluorotelomer acrylates and methacrylates in water and sediments using solid-phase microextraction-gas chromatography/mass spectrometry. J Chromatogr A 2016;1448:98–106. https://doi.org/10.1016/j.chroma.2016.04.025.

[146] Rodríguez I, Calvo F, Quintana JB, Rubí E, Rodil R, Cela R. Suitability of solid-phase microextraction for the determination of organophosphate flame retardants and plasticizers in water samples. J Chromatogr A 2006;1108(2):158–65. https://doi.org/10.1016/j.chroma.2006.01.008.

[147] Tsao Y-C, Wang Y-C, Wu S-F, Ding W-H. Microwave-assisted headspace solid-phase microextraction for the rapid determination of organophosphate esters in aqueous samples by gas chromatography-mass spectrometry. Talanta 2011;84(2):406–10. https://doi.org/10.1016/j.talanta.2011.01.055.

[148] Xu J, Zheng J, Tian J, Zhu F, Zeng F, Su C, et al. New materials in solid-phase microextraction. Trends Anal Chem 2013;47:68–83. https://doi.org/10.1016/j.trac.2013.02.012.

[149] Souza-Silva ÉA, Jiang R, Rodríguez-Lafuente A, Gionfriddo E, Pawliszyn J. A critical review of the state of the art of solid-phase microextraction of complex matrices I. Environmental analysis. Trends Anal Chem 2015;71:225–35. https://doi.org/10.1016/j.trac.2015.04.016.

[150] Portillo-Castillo Omar J, Castro-Ríos R, Chávez-Montes A, González-Horta A, Cavazos-Rocha N, de Torres Noemí HW, et al. Developments of solid-phase microextraction fiber coatings for environmental pharmaceutical and personal care products analysis. Rev Anal Chem 2018;37(2):20170018. https://doi.org/10.1515/revac-2017-0018.

[151] Deng J, Zhang P, Jin T, Zhou H, Cheng J. Graphene oxide/β-cyclodextrin composite as fiber coating for high efficiency headspace solid phase microextraction of organophosphate ester flame retardants in environmental water. RSC Adv 2017;7(86):54475–84. https://doi.org/10.1039/C7RA07903F.

[152] Asadollahzadeh H, Noroozian E, Maghsoudi S. Solid-phase microextraction of phthalate esters from aqueous media by electrochemically deposited carbon nanotube/polypyrrole composite on a stainless steel fiber. Anal Chim Acta 2010;669(1):32–8. https://doi.org/10.1016/j.aca.2010.04.029.

[153] Sarafraz-Yazdi A, Amiri A, Rounaghi G, Eshtiagh-Hosseini H. Determination of non-steroidal anti-inflammatory drugs in water samples by solid-phase microextraction based sol–gel technique using poly(ethylene glycol) grafted multi-walled carbon nanotubes coated fiber. Anal Chim Acta 2012;720:134–41. https://doi.org/10.1016/j.aca.2012.01.021.

[154] Sarafraz-Yazdi A, Razavi N. Application of molecularly-imprinted polymers in solid-phase microextraction techniques. Trends Anal Chem 2015;73:81–90. https://doi.org/10.1016/j.trac.2015.05.004.

[155] Zhang M, Zeng J, Wang Y, Chen X. Developments and trends of molecularly imprinted solid-phase microextraction. J Chromatogr Sci 2013;51(7):577–86. https://doi.org/10.1093/chromsci/bms260.

[156] He J, Lv R, Zhan H, Wang H, Cheng J, Lu K, et al. Preparation and evaluation of molecularly imprinted solid-phase micro-extraction fibers for selective extraction of phthalates in an aqueous sample. Anal Chim Acta 2010;674(1):53–8. https://doi.org/10.1016/j.aca.2010.06.018.

[157] Mehdinia A, Bashour F, Roohi F, Jabbari A, Saleh A. Preparation and evaluation of thermally stable nano-structured self-doped polythiophene coating for analysis of phthalate ester trace levels. J Sep Sci 2012;35(4):563–70. https://doi.org/10.1002/jssc.201100713.

[158] Serôdio P, Nogueira JMF. Considerations on ultra-trace analysis of phthalates in drinking water. Water Res 2006;40(13):2572–82. https://doi.org/10.1016/j.watres.2006.05.002.

[159] Rodil R, Moeder M. Development of a method for the determination of UV filters in water samples using stir bar sorptive extraction and thermal desorption–gas chromatography–mass spectrometry. J Chromatogr A 2008;1179(2):81–8. https://doi.org/10.1016/j.chroma.2007.11.090.

[160] Rodil R, von Sonntag J, Montero L, Popp P, Buchmeiser MR. Glass-fiber reinforced poly(acrylate)-based sorptive materials for the enrichment of organic micropollutants from aqueous samples. J Chromatogr A 2007;1138(1):1–9. https://doi.org/10.1016/j.chroma.2006.10.039.

[161] Silva ARM, Portugal FCM, Nogueira JMF. Advances in stir bar sorptive extraction for the determination of acidic pharmaceuticals in environmental water matrices: comparison between polyurethane and polydimethylsiloxane polymeric phases. J Chromatogr A 2008;1209(1):10–6. https://doi.org/10.1016/j.chroma.2008.08.103.

[162] Huang X, Qiu N, Yuan D, Lin Q. Preparation of a mixed stir bar for sorptive extraction based on monolithic material for the extraction of quinolones from wastewater. J Chromatogr A 2010;1217(16):2667–73. https://doi.org/10.1016/j.chroma.2009.09.072.

[163] Strittmatter N, Düring R-A, Takáts Z. Analysis of wastewater samples by direct combination of thin-film microextraction and desorption electrospray ionization mass spectrometry. Analyst 2012;137(17):4037–44. https://doi.org/10.1039/C2AN35411J.

[164] Nogueira JMF. Novel sorption-based methodologies for static microextraction analysis: a review on SBSE and related techniques. Anal Chim Acta 2012;757:1–10. https://doi.org/10.1016/j.aca.2012.10.033.

[165] Neng NR, Silva ARM, Nogueira JMF. Adsorptive micro-extraction techniques—novel analytical tools for trace levels of polar solutes in aqueous media. J Chromatogr A 2010;1217(47):7303–10. https://doi.org/10.1016/j.chroma.2010.09.048.

[166] Dias AN, da Silva AC, Simão V, Merib J, Carasek E. A novel approach to bar adsorptive microextraction: cork as extractor phase for determination of benzophenone, triclocarban and parabens in aqueous samples. Anal Chim Acta 2015;888:59–66. https://doi.org/10.1016/j.aca.2015.07.011.

[167] Mitani K, Kataoka H. Determination of fluoroquinolones in environmental waters by in-tube solid-phase microextraction coupled with liquid chromatography–tandem mass spectrometry. Anal Chim Acta 2006;562(1):16–22. https://doi.org/10.1016/j.aca.2006.01.053.

[168] Saito K, Uemura E, Ishizaki A, Kataoka H. Determination of perfluorooctanoic acid and perfluorooctane sulfonate by automated in-tube solid-phase microextraction coupled with liquid chromatography–mass spectrometry. Anal Chim Acta 2010;658(2):141–6. https://doi.org/10.1016/j.aca.2009.11.004.

[169] Mei M, Huang X. Online analysis of five organic ultraviolet filters in environmental water samples using magnetism-enhanced monolith-based in-tube solid phase microextraction coupled with high-performance liquid chromatography. J Chromatogr A 2017;1525:1–9. https://doi.org/10.1016/j.chroma.2017.09.065.

[170] Bu Y, Feng J, Tian Y, Wang X, Sun M, Luo C. An organically modified silica aerogel for online in-tube solid-phase microextraction. J Chromatogr A 2017;1517:203–8. https://doi.org/10.1016/j.chroma.2017.07.075.

[171] Raina-Fulton R. A review of methods for the analysis of orphan and difficult pesticides: glyphosate, glufosinate, quaternary ammonium and phenoxy acid herbicides, and dithiocarbamate and phthalimide fungicides. J AOAC Int 2014;97(4):965–77. https://doi.org/10.5740/jaoacint.SGERaina-Fulton.

[172] Zerbinati O, Diana I, Baiocchi C. Trace naphthalenesulphonates determination in natural water samples. Int J Environ Anal Chem 1999;74(1–4):43–54. https://doi.org/10.1080/03067319908031415.

[173] Grey L, Nguyen B, Yang P. Liquid chromatography–electrospray ionization isotope dilution mass spectrometry analysis of paraquat and diquat using conventional and multilayer solid-phase extraction cartridges. J Chromatogr A 2002;958(1):25–33. https://doi.org/10.1016/S0021-9673(02)00400-4.

[174] Altenbach B, Giger W. Determination of benzene- and naphthalenesulfonates in wastewater by solid-phase extraction with graphitized carbon black and ion-pair liquid chromatography with UV detection. Anal Chem 1995;67(14):2325–33. https://doi.org/10.1021/ac00110a002.

[175] Castro R, Moyano E, Galceran MT. On-line ion-pair solid-phase extraction–liquid chromatography–mass spectrometry for the analysis of quaternary ammonium herbicides. J Chromatogr A 2000;869(1):441–9. https://doi.org/10.1016/S0021-9673(99)01065-1.

[176] Gimeno RA, Beltrán JL, Marcé RM, Borrull F. Determination of naphthalenesulfonates in water by on-line ion-pair solid-phase extraction and ion-pair liquid chromatography with fast-scanning fluorescence detection. J Chromatogr A 2000;890(2):289–94. https://doi.org/10.1016/S0021-9673(00)00599-9.

[177] Merino F, Rubio S, Pérez-Bendito D. Evaluation and optimization of an on-line admicelle-based extraction-liquid chromatography approach for the analysis of ionic organic compounds. Anal Chem 2004;76(14):3878–86. https://doi.org/10.1021/ac049736v.

[178] Caro E, Marcé RM, Cormack PAG, Sherrington DC, Borrull F. Molecularly imprinted solid-phase extraction of naphthalene sulfonates from water. J Chromatogr A 2004;1047(2):175–80. https://doi.org/10.1016/j.chroma.2004.07.015.

[179] Puzio K, Claude B, Amalric L, Berho C, Grellet E, Bayoudh S, et al. Molecularly imprinted polymer dedicated to the extraction of glyphosate in natural waters. J Chromatogr A 2014;1361:1–8. https://doi.org/10.1016/j.chroma.2014.07.043.

[180] Claude B, Berho C, Bayoudh S, Amalric L, Coisy E, Nehmé R, et al. Preliminary recovery study of a commercial molecularly imprinted polymer for the extraction of glyphosate and AMPA in different environmental waters using MS. Environ Sci Pollut Res 2017;25(13):12293–300. https://doi.org/10.1007/s11356-017-8844-5.

[181] Wagner R, Wetzel SJ, Kern J, Kingston HMS. Improved sample preparation of glyphosate and methylphosphonic acid by EPA method 6800A and time-of-flight mass spectrometry using novel solid-phase extraction. J Mass Spectrom 2012;47(2):147–54. https://doi.org/10.1002/jms.2038.

[182] Van de Voorde A, Lorgeoux C, Gromaire M-C, Chebbo G. Analysis of quaternary ammonium compounds in urban stormwater samples. Environ Pollut 2012;164:150–7. https://doi.org/10.1016/j.envpol.2012.01.037.

[183] Janda J, Nödler K, Brauch H-J, Zwiener C, Lange FT. Robust trace analysis of polar (C2-C8) perfluorinated carboxylic acids by liquid chromatography-tandem mass spectrometry: method development and application to surface water, groundwater and drinking water. Environ Sci Pollut Res 2018. https://doi.org/10.1007/s11356-018-1731-x.

[184] Arbeláez P, Borrull F, Pocurull E, Marcé RM. Determination of high-intensity sweeteners in river water and wastewater by solid-phase extraction and liquid chromatography–tandem mass spectrometry. J Chromatogr A 2015;1393:106–14. https://doi.org/10.1016/j.chroma.2015.03.035.

[185] Gan Z, Sun H, Wang R, Feng B. A novel solid-phase extraction for the concentration of sweeteners in water and analysis by ion-pair liquid chromatography–triple quadrupole mass spectrometry. J Chromatogr A 2013;1274:87–96. https://doi.org/10.1016/j.chroma.2012.11.081.

[186] Ordóñez EY, Quintana JB, Rodil R, Cela R. Determination of artificial sweeteners in water samples by solid-phase extraction and liquid chromatography–tandem mass spectrometry. J Chromatogr A 2012;1256:197–205. https://doi.org/10.1016/j.chroma.2012.07.073.

[187] Moein MM, Javanbakht M, Karimi M, Akbari-adergani B. Molecularly imprinted sol-gel nanofibers based solid phase microextraction coupled on-line with high performance liquid chromatography for selective determination of acesulfame. Talanta 2015;134:340–7. https://doi.org/10.1016/j.talanta.2014.11.011.

[188] Hashemi SH, Kaykhaii M, Tabehzar F. Molecularly imprinted stir bar sorptive extraction coupled with high-performance liquid chromatography for trace analysis of naphthalene sulfonates in seawater. J Iran Chem Soc 2016;13(4):733–41. https://doi.org/10.1007/s13738-015-0785-7.

# Environmental applications (air) 24

*Sergio Armenta, Miguel de la Guardia, Francesc A. Esteve-Turrillas*
Department of Analytical Chemistry, Research Building, University of Valencia, Valencia, Spain

## 24.1 Solid phases for air environmental applications

Mainly, analytical instruments are designed for optimum performance for the analysis of liquid samples or solutions. Against this background, solid samples would be considered the most awkward physical state for instrumental analysis. However, it is easy to overlook the difficulties associated with managing gaseous samples, due to the need to control leaks and/or contamination, at all steps in the analysis. Additionally, the target analytes are present at low concentrations in a total volume that depends on environmental conditions [1]. Moreover, a significant characteristic of the determination of pollutants in the air is to differentiate between the concentration of the free form present in the vapor phase in the air and the concentration of the compound bound to a particulate matter. Thus, target analytes can be found in both the gas and suspended particle phases with a partition ratio that depends on their physicochemical properties and local environmental factors [2].

Particulate matter is a mixture of solid particles and liquid droplets present in the air. The particle size is typically heterogeneous and classified as $PM_{10}$, particles with an average diameter less than 10 μm; and $PM_{2.5}$, fine particles with an average particle diameter less than 2.5 μm [3]. Values for $PM_{10}$ and $PM_{2.5}$ are easily measured using filtration with 1 h short-term and 24 h long-term sampling times. Then, the amount of pollutants retained in filters is analyzed using conventional extraction methodologies, such as Soxhlet, ultrasound-assisted extraction (UAE), pressurized liquid extraction (PLE), and microwave-assisted extraction (MAE) [4].

The analysis of free dissolved compounds in air samples is a more challenging task. The simplest approach for air analysis is based on the use of Tedlar sampling bags and glass bulbs. However, sensitivity is typically inadequate for the determination of most target analytes. Thus, the current trend is to move the analytes from the dispersed or diluted form in the ambient air, to a solution or a solid phase, in which analytes are preconcentrated and stabilized. So, the advantages offered by solid-phase extraction of analytes present in the air are obvious, and the use of solid sorbents for passive or active sampling is one of the most frequently used strategies to isolate gaseous target analytes present in the air by adsorption on a solid phase.

In the above framework, sampling of the freely dissolved fraction of pollutants in indoor and outdoor air is traditionally based on the use of active or passive samplers. The main difference between these sampling strategies is the use of a controlled flow pump for active sampling, while a pumping system is not required in passive systems.

**Solid-Phase Extraction. https://doi.org/10.1016/B978-0-12-816906-3.00024-8**

In the latter systems, the target analytes flow freely through the sampling medium in a manner described by Fick's first law [5]. Both sampling methodologies are briefly discussed in the next section, highlighting the use of solid sorbents for isolation of the analytes in each case.

### 24.1.1 Active sampling

Active sampling is the method of choice for the determination of environmental contaminants in most applications. The principle of active sampling is based on the collection of the target analytes from the air by using dynamic techniques, where a specified volume of air is passed through a container, typically a tube, filled with an appropriate solid sorbent [6]. Active sampling strategies are dependent on the air flow rate, typically between 10 and 250 mL $min^{-1}$. A fast linear intake sample speed reduces interactions between the target analytes and the sorbent, while low velocities risk issues of back diffusion from the sampling tubes to the air. Thus, the linear velocity is mainly adjusted by control of the air flow rate, but also by variation of the internal diameter of the sampling tube, with smaller diameters at a constant flow rate corresponding to a higher sampling linear velocity [7]. In these cases, solid sorbents are placed inside glass or stainless steel sampling tubes. Tube material is not a significant factor, but some details must be taken into account for a proper selection. Glass tubes are fragile but inert and transparent, which facilitates locating the sorbent bed at the correct position within the sampling tube; while stainless steel tubes are mechanically robust, but are opaque and less inert; thus, it may react with sensitive analytes [8].

A large number of solid sorbents have been used for the analysis of volatile organic compounds in the air, and can be classified into three groups: (1) carbon-based adsorbents, (2) organic polymers, and (3) inorganic materials. The most commonly used solid sorbent, and their characteristic properties are summarized in Table 24.1 [9].

Carbonaceous materials were the first solid sorbents employed for active sampling and can be categorized as activated carbon, carbon molecular sieves, and graphitized carbon black. Activated carbons are heterogeneous materials that provide nonspecific and specific interactions with target analytes due to the presence of hydroxyl, carbonyl, and carboxylic functional groups on their surface, which can irreversibly bind polar analytes, such as alcohols [9]. Carbon molecular sieves, such as Carboxen, Unicarb, Carbosieve, and Carbosphere, are microporous materials of high surface area, whose adsorption is based on nonspecific interactions due to the presence of both amorphous carbon and layers of condensed aromatic rings at the surface. Carbon molecular sieves exhibit high water adsorption and appreciable reactivity with light hydrocarbons [11]. Graphitized carbon blacks are homogeneous carbon-based materials with a high degree of graphitization and a reduced amount of polar functional groups on their surface. As a consequence, these materials are hydrophobic with a specific surface area less than that of carbon molecular sieves.

Porous polymers, such as Chromosorb, Porapack, Tenax, and Amberlites (XAD) have been widely employed as solid sorbents in air monitoring. Tenax TA is one of the most commonly used sorbents for active sampling due to its high thermal stability and breakthrough volumes. It retains a wide range of hydrophobic organic compounds

**Table 24.1** Characteristic properties of commonly used solid sorbents for active sampling.

| Sorbent | Composition | Strength[a] | $T_{max}$ (°C)[b] | Area ($m^2$ $g^{-1}$)[c] |
|---|---|---|---|---|
| **Carbon- based** | | | | |
| Activated carbon | Woood or coconut charcoal | Weak | 220 | 800-1500 |
| Carbograph | Graphitized carbon black | Weak | >400 | 100-560 |
| Carbopack | Graphitized carbon black | Medium/ strong | >400 | 5-240 |
| Carbotrap | Graphitized carbon black | Medium | >400 | 5-240 |
| Carboxen 569 | Carbon molecular sieve | Strong | >450 | 485 |
| Carboxen 1003 | Carbon molecular sieve | Very strong | >450 | 1000 |
| Unicarb | Carbon molecular sieve | Strong | >400 | - |
| Carbosieve | Carbon molecular sieve | Very strong | >400 | 1160 |
| Carbosphere | Carbon molecular sieve | Strong | >400 | 1000 |
| **Organic polymers** | | | | |
| Tenax TA | Poly(2, 6-diphenylphenylene oxide) | Weak | 350 | 35 |
| Tenax GC | Poly(2, 6-diphenylphenylene oxide) | Weak | 450 | 19-30 |
| Tenax GR | 70 % Tenax TA + 30% graphite | Weak | 350 | 21 |
| Chromosorb 101 | Styrene/divinylbenzene | Medium | 275 | <50 |
| Chromosorb 102 | Styrene/divinylbenzene | Medium | 250 | 350 |
| Chromosorb 103 | Cross-linked polystyrene | Medium | 275 | 350 |
| Chromosorb 104 | Acrylonitrile/ divinylbenzene | Medium | 250 | 100-200 |
| Chromosorb 105 | Polyaromatic resin | Medium | 250 | 600-700 |
| Chromosorb 106 | Styrene/divinylbenzene | Medium | 225 | 700-800 |
| Chromosorb 107 | Polyacrylic ester resin | Medium | 225 | 400-500 |
| Chromosorb 108 | Cross-linked acrylic ester | Medium | 225 | 100-200 |

*Continued*

**Table 24.1** Characteristic properties of commonly used solid sorbents for active sampling.—cont'd

| Sorbent | Composition | Strength[a] | $T_{max}$ (°C)[b] | Area ($m^2$ $g^{-1}$)[c] |
|---|---|---|---|---|
| PoraPak N | Divinyl benzene/vinyl pyrrolidinone | Medium | 190 | 250-350 |
| PoraPak P | Styrene/divinylbenzene copolymer | Medium | 250 | 100-200 |
| PoraPak Q | Ethylvinyl benzene/ divinyl benzene | Medium | 250 | 500-600 |
| PoraPak R | Divinylbenzene/vinyl pyrrolidinone | Medium | 250 | 450-600 |
| PoraPak S | Divinylbenzene/vinyl pyridine | Medium | 250 | 300-450 |
| PoraPak T | Ethyleneglycol/ dimethacrylate | Medium | 190 | 250-350 |
| HayeSep | Divinylbenzene, ethyleneglycol/ dimethacrylate | Medium | 290 | 583 |
| XAD-2 | Styrene/divinylbenzene | Medium | 200 | 300 |
| XAD-4 | Styrene/divinylbenzene | Medium | 150 | 750 |
| XAD-7 | Polymethacrylic ester | Medium | 150 | 450 |
| XAD-8 | Polymethacrylic ester | Medium | 150 | 140 |
| Polydimethylsiloxane | Polydimethylsiloxane | Medium | 280-340 | - |
| **Inorganic- based** | | | | |
| Molecular sieve | $SiO_2$ | Very strong | >400 | 800-1000 |
| Florisil | $MgO:SiO_2$ | Medium | 650 | 100-300 |
| Quartz wool | $SiO_2$ | Very weak | >450 | - |

[a]Capacity to retain pollutants from the air.
[b]Maximum temperature for thermal desorption.
[c]Specific surface area.
Adapted from Woolfenden E. J Chromatogr A. 2010;1217:2685.

with boiling points ranging from 60 to 300°C [12]. However, it also has some disadvantages related to its high price and artifact formation when exposed to oxidizing samples [13]. Amberlite XAD sorbents are widely used as an alternative to Tenax materials for the active sampling of volatile organic compounds. Amberlite XAD is a macroreticular polymer of a high porosity that allows the efficient diffusion of target analytes through the polymer surface [12]. XAD sorbents can be classified into two general groups, XAD-2 and XAD-4, based on styrene polymers crosslinked by divinylbenzene, and XAD-7 and XAD-8, based on polymethacrylate polymers. Nevertheless, the low maximum desorption temperature of XAD sorbents discourages their use for thermal desorption approaches for analyte recovery.

Solid sorbents are usually impregnated with specific reagents for the active sampling of specific compounds, for example, the extraction of aldehydes and ketones using 2,4-dinitrophenylhydrazine (DNPH) with the formation of the corresponding chemically stable hydrazones, that are immobilized on the sorbent [14]. In a similar way, 1-naphthyl-isothiocyanate impregnated sorbents have been used for the sampling of amines with the formation of the corresponding thioureas [15] and HBr-coated Anasorb charcoal beads for sampling ethylene oxide with the formation of 2-bromoethanol [16].

One of the main problems that affects the use of solid sorbents for active sampling is related to the adsorption of water vapor by the sorbent, particularly for the sampling of atmospheres of high humidity. The presence of water may reduce the number of active sites and, in some cases, it may displace retained target compounds. Thus, hydrophilic inorganic materials, such as silica, alumina or zeolites are not often employed, while very hydrophobic adsorbents like Tenax, Carbopack, and Carbotrap are generally recommended for sampling humid atmospheres [7]. An interesting approach to increase the number of target compounds extracted is the use of multisorbent tubes (see Fig. 24.1) consisting of a single tube packed sequentially with different solid sorbents [17].

The determination of target analytes retained on the sorbent is typically based on thermal desorption, purge and trap, and headspace approaches due to the ease of use, simplicity, and high sensitivity [18]. However, the main drawbacks of thermal desorption are possible molecular rearrangements at high temperatures and the breakdown of thermally labile compounds. Furthermore, carryover effects due to incomplete desorption of less volatile compounds can be a problem [19]. Modern desorption systems allow the automatic introduction of gas-phase internal standards onto the sampling end of the sorbent bed, liquid-phase standards onto sample tubes and, more recently, the ability to quantitatively recollect any sample flow into a single conditioned sorbent to repeat the analysis [20].

The analysis of thermally labile compounds or those strongly retained by the sorbent cannot be performed by thermal desorption. In these cases, the target compounds are extracted by organic solvents. In the case of DNPH-based sorbents for the isolation

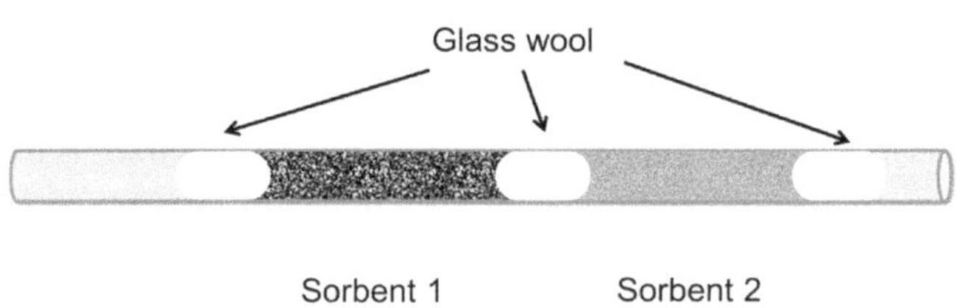


**Figure 24.1** Multisorbent approach for active sampling of air.

of aldehydes and ketones, a combination of acetonitrile as solvent and sonication is typically employed for the extraction of the colored derivatives. An additional advantage of solvent extraction methods over traditional thermal desorption is that it allows the possibility of multiple injections of the same sample for confirmation purposes. In any case, the high sensitivity achieved by thermal desorption methods is the main reason for the gradual replacement of solvent extraction techniques for the recovery of adsorbed compounds [9]. Thermal desorption is the method of choice to extract target compounds from the sorbent to be transferred to a gas chromatograph because of the 1000-fold sensitivity improvement compared to solvent extraction [9]. Moreover, the high acquisition cost of thermal desorption methods is easily recovered by the lower running costs of this technique compared to solvent extraction.

In short, the use of solid phases for the assessment of air quality facilitates the integration of sampling and analyte preconcentration steps into a single operation and favors the improvement of analytical methodologies. The main advantages of active sampling rely on (1) its capability to clearly identify particulate matter and gaseous phase contribution; and (2) precise and reliable correlation between analyte concentration in the sampler and in air samples. Although the use of active sampling to determine the concentration of organic compounds in the air is widely accepted, this technique has several drawbacks that restrict its applicability: (1) the need for bulky and expensive equipment, electrical power, and trained personnel; (2) the use of a pump to force the flow of air through the sampling medium; (3) the relatively large number of samples, which must be taken to evaluate global contamination because of the fluctuation of compound concentrations; (4) the high cost of monitoring campaigns; (5) the potential contamination of sorbed analytes due to pumps and operators; and (6) the difficulty of seasonal measurements due to the reduced sampling times. All these aspects, sometimes force the selection of sampling sites at locations different to those considered as the most appropriate. Moreover, in spite of the simplicity of active sampling strategies, several parameters need to be verified, including breakthrough volumes of sorption materials, recovery of target compounds, and sampling repeatability.

### 24.1.2 Passive sampling

The emergence of passive sampling in 1973 [21] tried to eliminate the aforementioned drawbacks of active sampling. Passive sampling was defined by Górecki and Namiesnik as a sampling technique in which the analytes flow from the sample to the collecting medium as a result of a difference in chemical potentials until equilibrium is achieved or the session is finished by the user [22]. Passive sampling approaches are nowadays widely accepted for the determination of pollutants in the air. Passive strategies do not require the use of pumps to force the flow of air through the sampling medium, and the analytes flow freely through the sampler. Moreover, passive samplers are suitable for long-term monitoring strategies and allow the determination of time-weighted average (TWA) concentrations of target compounds with particular sampling time. Inherent advantages of passive sampling strategies imply: (1) unsupervised operation without pumps, flowmeters, or power supplies; (2) easy manipulation;

(3) reduced cost of monitoring campaigns, including low acquisition costs and low maintenance requirement; (4) small size and lightweight of devices, which facilitate sampling in hard-to-access areas; (5) simultaneous sampling at different sites; (6) TWA concentration of target analytes in the air; and (7) reduced risk of nondetection of occasional emissions [23]. An indirect advantage of passive sampling is the possibilities for spatial resolution of data. While active samplers acquire data at a relatively fine temporal resolution [24], they reduce the spatial resolution due to its inherent drawbacks, namely, high cost and energy requirements. The low cost of passive sampling devices allows the deployment of a large number of samplers compared to their active counterparts and, consequently, provide substantially increased spatial resolution, albeit at a reduced temporal resolution [25]. However, the use of passive samplers requires the establishment of mathematical models to calibrate the uptake rate of each pollutant in order to establish a relationship between the analyte concentration in the air and the sampling device. Moreover, the in-situ uptake rate depends on environmental effects, such as wind speed, temperature, sunlight, and humidity. Performance reference compounds (PRCs) are typically employed to compensate for the fluctuation of environmental conditions [26]. Moreover, as in any unsupervised approach, passive samplers deployed in the field for long periods of time may disappear, suffer theft or vandalism, or be accidentally damaged or manipulated, among other incidents. In any case, active and passive sampling can be considered complementary techniques providing valuable information for the evaluation of air quality.

A brief description of the most popular passive samplers for air monitoring is provided in Table 24.2. There are a large variety of passive samplers commercially available or laboratory-made, classified as diffusive badges, diffusive tubes, diffusive radial devices, membrane devices, and others, where the design and configuration of each device are quite different. Fig. 24.2 illustrates some of the most popular sorbent-based passive samplers.

Diffusive badges, tubes, and radial devices are based on the use of a solid sorbent placed inside a holder with a diffusive barrier to limit sampling to diffusion following Fick's first law. The commonly employed solid sorbents are similar to those employed in active sampling (see Table 24.1), typically carbon-based materials and Tenax TA. Membrane-based devices are based on semipermeable membrane devices (SPMDs), where a sorptive material is placed inside a low-density polyethylene (LDPE) or polydimethylsiloxane (PDMS) membranes that act as a diffusion barrier. Sorptive materials employed in membrane devices were initially liquids, such as glyceryl trioleate. Modern alternatives, however, use solid phases as versatile, easy and rapid air monitoring (VERAM) device [27] or stir bar sorptive extraction (SBSE) adsorbents like PDMS twister bars or silicone rods [28]. Polyurethane foam (PUF) disks (typically 10–15 cm diameter and 1.2–2.2 cm thick) are also employed as passive sampling devices housed in specifically designed stainless steel chambers to homogenize wind speed during the sampling time [29].

Table 24.2 Configuration of typical devices for passive sampling of air.

| | Diffusion barrier | Adsorbent |
|---|---|---|
| **Diffusive badges** | | |
| SKC | Glass wool, glass fiber, quartz filters, PTFE and nylon rings | Tenax TA, Chromosorb 106, Anasorb, activated charcoal, Carbopack |
| Organic vapor monitor (OVM) | Charcoal wafer, PTFE | Activated charcoal, Carbopack B |
| TOPAS/TDS | Microporous PTFE membrane | Tenax TA |
| Gas absorbent badges for individual exposure (GABIE) | Stainless steel mesh | Coconut charcoal |
| Ogawa | Stainless steel mesh | Triethanolamine-coated filter |
| Sep-pak XpoSure | Polyethylene filter | DNPH-coated silica gel |
| Lanwatsu | Stainless steel mesh | Activated carbon |
| **Diffusive tubes** | | |
| Perkin–Elmer | – | Tenax TA, Tenax GR, Carbopack X, Carbopack B, Chromosorb |
| ORSA-5 | – | Activated charcoal |
| Shibata | – | Granular activated carbon |
| Steel tube badge | – | Carbopack X |
| **Diffusive radial devices** | | |
| Radiello | Porous silicone, polypropylene | Activated carbon, Carbograph 4, Tenax TA, DNPH-Florisil |
| **Membrane devices** | | |
| Semipermeable membrane devices (SPMD) | LDPE | Glyceryl trioleate |
| Polyethylene passive sampling device (PSD) | LDPE | None |
| Versatile, easy, and rapid atmospheric monitor (VERAM) | LDPE | Activated carbon, Florisil, XAD, kaolin |
| Waterloo membrane sampler (WMS) | PDMS | Anasorb 747, Carbopack B |

**Table 24.2** Configuration of typical devices for passive sampling of air.—cont'd

| | Diffusion barrier | Adsorbent |
|---|---|---|
| Membrane-enclosed sorptive coating (MESCO)<br>LDPE enclosing stir or tubing | LDPE | PDMS twister bar, silicone rod |
| **Others** | | |
| Solid-phase microextraction (SPME) | – | PDMS, Carboxen, Carbowax, divinylbenzene, polyacrylate |
| Polyurethane foam (PUF) | Stainless steel housing device | PUF, impregnated-XAD-4 PUF |
| Passive air system (PAS) | Stainless steel mesh cylinder inside a shelter | XAD-2 |
| Polymer-coated glass (POG) | Hollow glass cylinder | Ethylene vinyl acetate |

*DNPH*, 2,4-dinitrophenylhydrazine; *LDPE*, low-density polyethylene; *PDMS*, polydimethylsiloxane; *PTFE*, polytetrafluoroethylene.

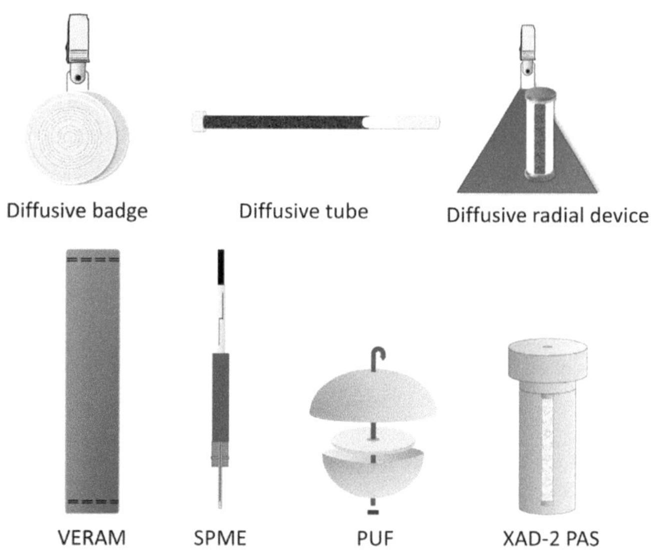


**Figure 24.2** Examples of sorbent-based passive samplers for air.

Additionally, solid-phase microextraction (SPME) devices can also be used for air sampling [30]. SPME sorbents, such as PDMS, divinylbenzene, or polyacrylate cannot be strictly considered as adsorbent, because the analytes are absorbed by partition in the polymeric material rather than by surface adsorption. The applicability of SPME for passive sampling has been widely studied but is somewhat limited in the case of air monitoring by the small amount of solid sorbent coated on the needle (in the μL range), which results in low-extraction efficiencies [31]. A recent application using SPME as an active sampling device was integrated into a handheld vacuum machine, used as a pump, for monitoring plant terpenes in the air [32].

Passive sampling strategies involve the development of calibration models to simulate the uptake process during sampling in order to move from the mass of target analytes retained in the device to their concentration in ambient air. The uptake of target analytes in a passive sampler is explained by Fick's first law and consists of three stages (1) an initial linear uptake, (2) an intermediate curvilinear stage, and (3) a final equilibrium plateau, as shown in Fig. 24.3. Thus, the sampling rate ($R_S$) can be calculated from data obtained in the linear uptake stage, and are specific for each target compound and passive sampling device. One of the simplest approaches to calibrate a passive sampler utilizes the evaluation of adsorption isotherms in a closed vessel in which a constant concentration of a volatile compound is added and the amount adsorbed by the sampling device determined for different times [33]. Other approaches are based on the use of wind tunnels [34], rooms spiked by means of a household humidifier that contained the target compounds [35], or calibration chambers [36].

The determination of target compounds retained by passive samplers is principally based on solvent stripping accelerated by the dialysis, Soxhlet, sonication MAE, and PSE approaches [23]. While the determination of volatile organic compounds is typically carried out by thermal desorption or headspace approaches. As for active sampling, the extraction of the hydrazone complexes of aldehydes and ketones is carried out by UAE with acetonitrile. In this case, the choice of elution solvent offers new possibilities for the incorporation of clean-up steps to remove matrix components that might interfere in the determination. A decision-making scheme for the most appropriate methodology for the extraction and determination of analytes in a passive sampler is shown in Fig. 24.4.

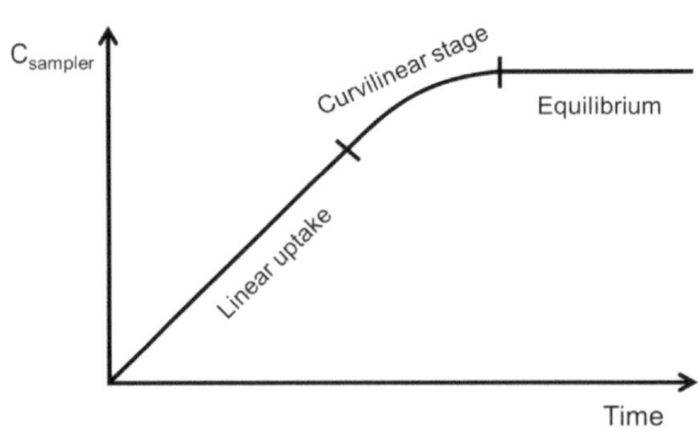


$$C_S = K_{SA} C_A \left(1 - e^{\left(-\frac{k_O A_S}{V_S K_{SA}} t\right)}\right)$$

$C_S$, concentration in sampler
$C_A$, TWA air concentration
$K_{SA}$, sampler/air partition coefficient
$k_O$, overall mass transfer coefficient
$A_S$, sampler area
$V_S$, sampler volume
$t$, sampling time

**Figure 24.3** Uptake of target compounds in a diffusive passive sampler following Fick's first law model [37].

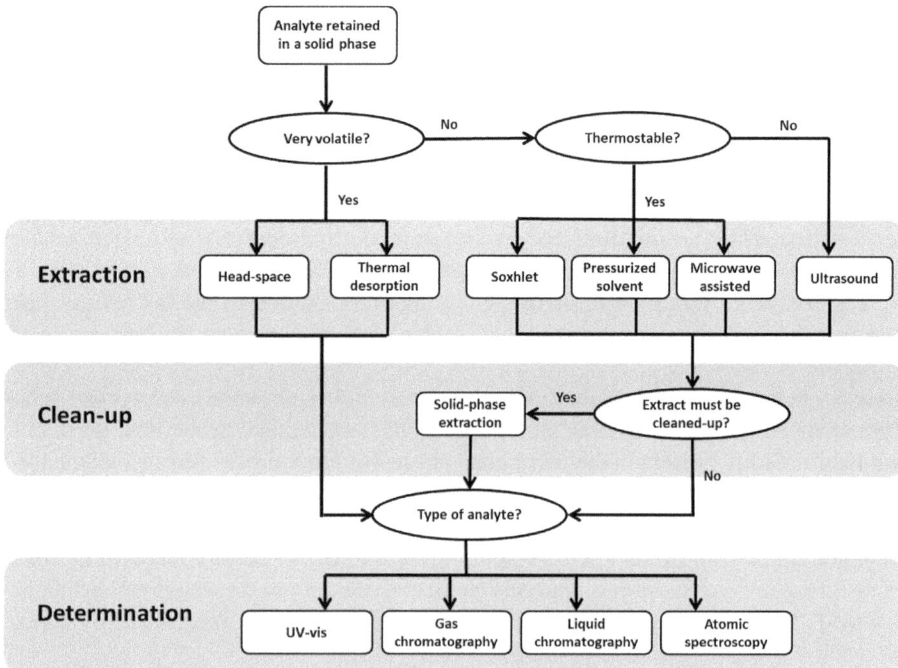


**Figure 24.4** Decision-making scheme for the selection of the most appropriate analytical method for the extraction and determination of analytes in a passive sampler.

## 24.2 Environmental applications

The main contaminants measured in environmental air applications can be classified into six groups, including: (1) particulate matter; (2) ground-level ozone; (3) carbon monoxide; (4) sulfur oxides (SOx); (5) nitrogen oxides (NOx); and (6) volatile and semivolatile compounds. The number of measurement stations to control environmental pollution has increased year on year, while the average concentrations decreased from 1995 to 2013 in developed countries [1].

Continuous analyzers located in environmental monitoring stations are the most common approach for the determination of air pollutants. Usually, NOx detection in ambient air is based on the use of chemiluminescence sensors, SOx concentration is based on fluorescence sensors, CO detection by IR spectroscopy, ozone by UV photometry, and methane and total hydrocarbons by flame ionization detection (FID) sensors, while the particulate matter is mainly determined by gravimetric methods. Additionally, the U.S. Environmental Protection Agency (U.S. EPA) has developed a list of more than 150 pollutants, often called *toxic air pollutants*, of which the most common is typically benzene. The U.S. EPA defined different benzene air concentrations depending on specified risk levels for cancer; for instance, from 13 to 45 $\mu g\ m^{-3}$ benzene increases the risk of cancer in one in 10,000; benzene concentrations of 1.3–4.5 $\mu g\ m^{-3}$ increases the risk of cancer in one in 100,000; and concentrations from 0.13 to 0.45 μg $m^{-3}$ increases the risk in one in 1,000,000 [38]. Whereas, in Europe, the maximum allowed benzene concentration in the air is 5 μg $m^{-3}$ [39]. Thus, several online, real time or near real time samplers based on Fourier transform infrared (FTIR) spectroscopy [40], automated gas chromatography [41], mass spectrometry with atmospheric pressure interfaces [42], among others, have been developed to provide temporal information for benzene concentrations.

Other than benzene, similar compounds, such as toluene, ethylbenzene and xylenes, the so-called *BTEX group*, and polycyclic aromatic hydrocarbons (PAHs) could be dangerous to the human health, and their presence in the air must be carefully monitored. PAHs are a group of semivolatile compounds emitted to the air mainly from anthropogenic combustion sources [43], which can be present in both gas and particulate phases.

Heavy metals, including cadmium, arsenic, and nickel, are another important group of toxic air pollutants. However, due to their physicochemical properties, these contaminants are only present in the particulate phase. Mercury deserves a special mention because there are three forms of atmospheric mercury: gaseous elemental, gaseous oxidized, and particle-bound mercury. Gaseous elemental mercury is typically the dominant species of atmospheric Hg [44], with a mean ambient concentration from 1.1 to 1.7 ng $m^{-3}$ as the sum of gaseous elemental mercury and gaseous oxidized mercury [45]. The residence time of gaseous elemental mercury is approximately 1 year, but much shorter for gaseous oxidized mercury [46].

The use of an automated system would obviously be advantageous for continuous data collection, which favors decision making. However, acquisition, management, and maintenance of continuous analyzers in monitoring stations are fairly costly and complex. Recent legislation, including that of the European Union, has sought to reduce the number of fixed monitoring stations in areas and conurbations where supplementary information is available [47]. Moreover, as is the case of many "hot spots," not all the sampling sites are suitable for deployment of automated monitoring stations. In these cases, passive or active sampling strategies must be adopted for monitoring of air contamination due to specific compounds or groups of compounds.

### 24.2.1 Rural air monitoring

The monitoring of rural environments is characterized by low levels of persistent organic pollutants in the air. The most sampled environmental contaminants in rural areas are mainly: PAHs and polychlorinated biphenyls (PCBs). The occurrence of polychlorinated dibenzo-p-dioxins (PCDDs) and polychlorinated dibenzofurans (PCDFs) is related to contaminations from industrialized areas around the sampling points, as well as high concentrations of pesticides due to agricultural practices. Table 24.3 illustrates some recent applications of active and passive samplers for monitoring of air pollutants in rural areas, confirmed here to pesticides, BTEX, PCDDs, PCDFs, polybrominated diphenyl ethers (PBDE), polychlorinated naphthalenes (PCN), and other common pollutants. In recent years, there is a well-established trend toward the use of PUF and XAD-2 sorbents for the sampling of semivolatile organic compounds

**Table 24.3** Applications of active and passive samplers for monitoring of air pollutants in rural areas.

| Strategy | Sampling site | Sampling device | Analytes | Analytical method | References |
|---|---|---|---|---|---|
| Active | Central Chile | PUF | Pesticides | MAE GC-MS/MS | [48] |
| | Malaysia | Multisorbent Tenax TA/ Carbograph | BTEX | TD-GC-MS | [49] |
| | Gorakhpur (India) | SKC (activated coconut shell charcoal) | BTEX | $CS_2$ GC-FID | [50] |
| | Cavaillon, (France) | PUF, XAD-2 | Pesticides | PSE GC-MS/MS | [51] |
| | Valencia (Spain) | Multisorbent PUF/XAD-2/PUF, XAD-2, XAD-4 | Pesticides | MAE LC-MS | [52] |
| | Northern Taiwan | Multisorbent PUF/XAD-2/PUF | PCNs | Soxhlet SPE GC-MS | [53] |
| | Abdali (Kuwait) | PUF | Pesticides | Soxhlet SPE GC-MS/MS | [54] |
| | North Taiwan | PUF | PCDDs, PCDFs | GC-MS | [55] |
| | France | PUF | Pesticides | PSE GC-MS LC-MS | [56] |
| Passive | Peumo (Chile) | PUF-PAS | POPs, pesticides | MAE GC-MS-MS | [48] |
| | West Canada | XAD-2 | PAHs, PCBs | Soxhlet GC-MS | [57] |
| | Alsace (France) | XAD-2-PAS | Pesticides | PSE SPME GC/MS/MS | [58] |
| | Shanghai (China) | PUF-PAS | PBDEs | Soxhlet SPE GC-MS | [59] |

*Continued*

**Table 24.3** Applications of active and passive samplers for monitoring of air pollutants in rural areas.—cont'd

| Strategy | Sampling site | Sampling device | Analytes | Analytical method | References |
|---|---|---|---|---|---|
| | Turkey | PUF-PAS | PCBs, PBDEs, OCPs | Soxhlet SPE GC-MS | [60] |
| | Düzce (Turkey) | Tenax TA | VOCs | TD-GC-FID | [61] |
| | Buenos Aires (Argentina) | PUF-PAS | PCDDs, PCDFs | Soxhlet SPE GC-MS | [62] |
| | Huntington (NY, USA) | Au-coated quartz fiber filters | Hg | - | [63] |

*BTEX*, benzene, toluene, ethylbenzene, and xylenes; *FID*, flame ionization detection; *GC*, gas chromatography; *LC*, liquid chromatography; *MAE*, microwave-assisted extraction; *MS*, mass spectrometry; *MS/MS*, tandem mass spectrometry; *PAHs*, polycyclic aromatic hydrocarbons; *PAS*, passive air samplers; *PCBs*, polychlorinated biphenyls; *PCDDs*, polychlorinated dibenzo-p-dioxins; *PBDEs*, polybrominated diphenyl ethers; *PCDFs*, polychlorinated dibenzofurans; *PCNs*, polychlorinated naphthalenes; *POP*, persistent organic pollutants; *PSE*, pressurized extraction; *PUF*, Polyurethane foam; *SPE*, solid-phase extraction; *SPME*, solid-phase microextraction; *TD*, thermal desorption; *VOCs*, volatile organic compounds.

for both active and passive samplers, with typically chromatographic and mass spectrometric systems for separation and detection.

## 24.2.2 Urban and industrialized air monitoring

The air monitoring of urban and industrialized environments is characterized by the occurrence of pollutants related to fuel combustion and vehicle traffic, such as $SO_2$, $NO_X$, and aromatic hydrocarbons (benzene, toluene, and xylenes), and those related to emissions from combustion processes, such as PAHs, PCDDs, and PCDFs, and other specific products related to industrial processes. Table 24.4 illustrates selected recent applications of active and passive samplers for the monitoring of air pollutants in urban and industrialized areas. Once again, PUF and XAD-2 based samplers are typically employed.

## 24.2.3 Indoor applications

The evaluation of contaminants in indoor areas like volatile organic compounds and other persistent pollutants, such as PAHs, PCBs, and pesticides are relevant due to their potential health effects. The evaluation of air quality in indoor compartments of homes, schools, and hospitals has also been evaluated. Additionally, aldehyde concentration levels in indoor air are often monitored to evaluate the emission from the furniture. Special attention has been paid to the evaluation of air quality inside water-pipe cafes, hookah bars, and closed areas in which e-cigarettes were employed. Smoking and alternative practices have been evaluated from measurements of the

**Table 24.4** Applications of active and passive samplers in the monitoring of air pollutants in urban and industrialized areas.

| Strategy | Sampling site | Sampling device | Analytes | Analytical method | References |
|---|---|---|---|---|---|
| Active | Beijing (China) | PUF | PCDDs, PCDFs, PCBs, PBDEs | PSE GC-MS | [64] |
| | Northern Taiwan | Multisorbent PUF/ XAD-2/ PUF | PCNs | Soxhlet SPE GC-MS | [53] |
| | Madrid (Spain) | PUF | PCDDs, PCDFs, PCBs | Soxhlet SPE GC-MS | [65] |
| | Kandur (India) | OVS (charcoal tubes) | PCBs | SPE GC-MS | [66] |
| | Tarragona (Spain) | PUF | PCDDs, PCDFs | GC-MS | [67] |
| | Kuala Lumpur (Malaysia) | Tenax TA | BTEX, biogenic VOCs | TD-GC-MS | [68] |
| Passive | Bursa (Turkey) | PUF-PAS | OPEs | Soxhlet SPE GC-MS | [69] |
| | Turkey | PUF-PAS | PCBs, PBDEs, OCPs | Soxhlet SPE GC-MS | [60] |
| | Kathmandu (Nepal) | Ogawa | $SO_2$, $NO_x$, $NH_3$, $O_3$ | LC-UV | [70] |
| | Cleveland (OH, USA) | Au-coated quartz fiber filters | Hg | - | [63] |
| | Czech Republic | Ogawa | $O_3$ | UV | [71] |
| | Lanzhou (China) | PUF-PAS | PAHs | Soxhlet GC-MS | [72] |
| | Kütahya (Turkey) | Tenax TA | VOCs | TD-GC-MS | [73] |
| | Kolkata (India) | PUF-PAS | HCHs, DDTs, PCBs, PBDEs | GC-MS | [74] |

*Continued*

**Table 24.4** Applications of active and passive samplers in the monitoring of air pollutants in urban and industrialized areas.—cont'd

| Strategy | Sampling site | Sampling device | Analytes | Analytical method | References |
|---|---|---|---|---|---|
| | Tuscany (Italy) | PUF-PAS | DDTs, PCBs | Soxhlet SPE GC-MS | [75] |
| | Strasbourg (France), Kehl (Germany) | XAD-2-PAS | PCBs, PCDDs, PCDFs | PSE SPE GC-MS | [76] |
| | Taranto (Italy) | Radiello (TEA-PE) | $NO_2$, $SO_2$ | LC-UV | [77] |
| Active/ Passive | São Paulo (Brazil) | PUF, PUF-PAS | PCDDs, PCDFs, PCBs, | Soxhlet SPE GC-MS | [78] |
| | Minnesota (USA) | XAD-4, XAD-4-PAS | PAHs | PSE GC-MS | [79] |
| | Manizales (Colombia) | PUF | PCDDs, PCDFs, PCBs | Soxhlet SPE GC-MS | [80] |
| | Tarragona (Spain) | Carbopack X | VOCs | TD-GC-MS | [81] |

*BTEX*, benzene, toluene, ethylbenzene, and xylenes; *DDTs*, dichlorodiphenyltricholoroethanes; *GC*, gas chromatography; *HCHs*, hexachlorocyclohexanes; *LC*, liquid chromatography; *MAE*, microwave-assisted extraction; *MS*, mass spectrometry; *MS/MS*, tandem mass spectrometry; *OCPs*, selected organochlorine pesticides; *OPEs*, organophosphate esters; *PAHs*, polycyclic aromatic hydrocarbons; *PAS*, passive air samplers; *PCBs*, polychlorinated biphenyls; *PCDDs*, polychlorinated dibenzo-p-dioxins; *PBDEs*, polybrominated diphenyl ethers; *PCDFs*, polychlorinated dibenzofurans; *PCNs*, polychlorinated naphthalenes; *POP*, persistent organic pollutants; *PSE*, pressurized extraction; *PUF*, Polyurethane foam; *SPE*, solid-phase extraction; *SPME*, solid-phase microextraction; *TEA-PE*, triethanolamine-coated polyethylene; *TD*, thermal desorption; *UV*, ultraviolet; *VOCs*, volatile organic compounds.

concentration of nicotine and other compounds produced in combustion and vaping practices. Table 24.5 illustrates some selected applications of air samplers used for the monitoring of indoor pollutants.

### 24.2.4 Applications in occupational exposure

The monitoring of air quality in industrial and working areas is essential to protect the health of workers and to evaluate risks from occupational exposure. In these cases, potential products and pollutant sources are typically well known. So the selection of the sampling strategy is tailored to those compounds with a particular risk for workers. Thus, strict controls of harmful compounds present in the working atmosphere must be routinely performed, generally following recommendations given in the guidelines of the U.S. National Institute for Occupational Safety and Health (NIOSH) or the European Regulation on Registration, Evaluation, Authorization, and Restriction of Chemicals (REACH). Table 24.6 illustrates selected examples of

**Table 24.5** Applications of active and passive samplers for monitoring contaminants in indoor air.

| Strategy | Sampling site | Sampling device | Analytes | Analytical method | References |
|---|---|---|---|---|---|
| Active | Waterpipe cafes | XAD-4 | Nicotine | UAE GC-FID | [82] |
| | Waterpipe cafes | SKC (PTFE) | PAHs | UAE GC-FID | [83] |
| | Swimming pool | Tenax TA | Trihalomethanes | TD-GC-MS | [84] |
| | Hookah bar | XAD-4 | Nicotine | GC-MS | [85] |
| | Room | Tenax TA | Transfluthrin | TD-GC-MS | [86] |
| | e-cigarette events | DNPH-silica gel, XAD-2, XAD-4, XAD-7 | Aldehydes, acrolein, nicotine, propylene glycol | LC-UV, GC-NPD, GC-FID | [87] |
| Passive | School buildings | PUF-PAS | PCBs, OH-PCB | Soxhlet SPE GC-MS | [88] |
| | Preschools buildings | DNPH-silica gel | Aldehydes | LC-UV | [89] |
| | Schools and homes | PUF | PCBs, SVOCs | ASE GC- MS/MS | [90] |
| | Car parks | VERAM (florisil-activated carbon), Radiello (activated carbon) | BTEX | HS-GC-MS | [27] |

*Continued*

**Table 24.5** Applications of active and passive samplers for monitoring contaminants in indoor air.—cont'd

| Strategy | Sampling site | Sampling device | Analytes | Analytical method | References |
|---|---|---|---|---|---|
| | Prenatal monitoring visit | Silicone wristbands, PUF-PAS | PAHs | GC-MS/MS | [91] |
| | Houses and apartments | PUF-PAS | PCBs, OCPs | Soxhlet SPE GC-MS/MS | [92] |
| Active/ Passive | | XAD-2 (active), PUF (passive) | PAHs | Soxhlet GC-MS/MS | [93] |

Abbreviations: *BTEX*, benzene, toluene, ethylbenzene, and xylenes; *DNPH* 2, 4-dinitrophenylhydrazine; *FID*, flame ionization detection; *GC*, gas chromatography; *HS*, head-space; *LC*, liquid chromatography; *MAE*, microwave-assisted extraction; *MS*, mass spectrometry; *MS/MS*, tandem mass spectrometry; *NPD*, nitrogen phosphorus detector; *OH-PCB*, PCB hydroxylated metabolites; *PAHs*, polycyclic aromatic hydrocarbons; *PAS*, passive air samplers; *PCBs*, polychlorinated biphenyls; *PCDDs*, polychlorinated dibenzo-p-dioxins; *PBDEs*, polybrominated diphenyl ethers; *PCDFs*, polychlorinated dibenzofurans; *PCNs*, polychlorinated naphthalenes; *POP*, persistent organic pollutants; *PSE*, pressurized solvent extraction; *PTFE*, polytetrafluoroethylene; *PUF*, polyurethane foam; *SPE*, solid-phase extraction; *SPME*, solid-phase microextraction; *TD*, thermal desorption; *UAE*; ultrasound-assisted extraction; *VOCs*, volatile organic compounds.

**Table 24.6** Applications of active and passive samplers for monitoring air pollutants in workplace environments.

| Strategy | Sampling site | Sampling device | Analytes | Analytical method | References |
|---|---|---|---|---|---|
| Active | Beauty salon | Coconut shell charcoal, (2-hydroxymethyl) piperidine-XAD-2 | BTEX, aldehydes | GC-FID | [94] |
| | Operating rooms | Anasorb 747 | Isoflurane, sevoflurane | GC-FID | [95] |
| | Residential fire | Soda lime, silica gel, XAD-7, OVS | PAHs, HCN, HBr, HF, HCl, $H_3PO_4$ | UV-vis, IC, LC-FD | [96] |
| | Construction sites | Charcoal | VOCs | GC-MS | [97] |
| | Restaurant kitchen | Tenax TA, DNPH-silica gel | VOCs | TD-GC-MS | [98] |
| Passive | Hot asphalt occupational settings | Silicone wristbands | PAHs | GC-MS | [99] |
| | Petroleum refinery | Carbopack Chi | 1,3-Butadiene, benzene | TD-GC-MS | [100] |
| | Package industries | VERAM (Florisil-activated carbon) | BTEX | HS-GC-MS | [33] |
| | Nail salon | Tenax TA | VOCs | TD-GS-MS | [101] |
| | University campus | XAD-4-coated PUF-PAS | PFASs, FRs, cVMSs | Soxhlet SPE GC-MS | [102] |
| Active/ Passive | Anatomy pathology department | DNPH-silica gel (active), SPME (on fiber PFBHA derivatization) (passive), enzyme-based badge (passive) | Formaldehyde | - | [14] |

*Continued*

**Table 24.6** Applications of active and passive samplers for monitoring air pollutants in workplace environments.—cont'd

| Strategy | Sampling site | Sampling device | Analytes | Analytical method | References |
|---|---|---|---|---|---|
| | Anatomy pathology department | DNPH tubes (active), ChemDisk badges (passive) | Formaldehyde | - | [103] |
| | Firefighters and police forensic specialists | PUF-PAS (passive), XAD-2 (active) | PAHs | GC-MS | [29] |
| | Industrial setting | Charcoal (active), OVM (passive) | Isopropyl alcohol | - | [104] |
| | Workplace | Charcoal (active and passive) | BTEX | MWTD-GC-MS | [105] |
| | Aircraft cabin and cockpit air | Tenax TA, Chromosorb 106 (active), activate charcoal clothes (passive) | OPP | GC-MS | [106] |
| | Wafer industry | Charcoal, silica gel, KOH-beaded carbon, and XAD-7 (active), OVM (passive) | VOCs, SVOCs, arsine, fluoride, phosphine | GC-FID, GFAAS, IC | [107] |
| | Alloy factory | PUF (active, passive) | PAHs | GC-MS | [108] |
| | Alloy factory and coke plant | PUF (active, passive)<br>PUF-XAD-2 (active) | PAHs | GC-MS | [109] |
| | Occupational worksetting | Activate charcoal (active), activated charcoal Radiello (passive) | BTEX | GC-MS | [110] |
| | Petrochemical industry | Tenax TA (active), Carbopack X (active), Canister (active/passive) | VOCs | GC-MS | [111] |

*BTEX*, benzene, toluene, ethylbenzene, and xylenes; *cVMSs*, cyclic volatile methylsiloxane; *DNPH*, 2,4-dinitrophenylhydrazine; *FID*, flame ionization detection; *FRs*, flame retardants; *GC*, gas chromatography; *GFAAS*, graphite furnace atomic absorption; *HS*, head-space; *IC*, ion chromatography; *LC*, liquid chromatography; *MAE*, microwave-assisted extraction; *MWTD*, microwave-assisted thermal desorption; *MS*, mass spectrometry; *MS/MS*, tandem mass spectrometry; *OPP*, organophosphate pesticides; *PAHs*, polycyclic aromatic hydrocarbons; *PAS*, passive air samplers; *PCBs*, polychlorinated biphenyls; *PCDDs*, polychlorinated dibenzo-p-dioxins; *PBDEs*, polybrominated diphenyl ethers; *PCDFs*, polychlorinated dibenzofurans; *PCNs*, polychlorinated naphthalenes; *PFASs*, polyfluoroalkyl substances; PFBHA, o-(2,3,4,5,6-pentafluorobenzyl)hydroxylamine; *POP*, persistent organic pollutants; *PSE*, pressurized extraction; *PUF*, Polyurethane foam; *SPE*, solid-phase extraction; *SPME*, solid-phase microextraction; *SVOCs*, semivolatile organic compounds; *TD*, thermal desorption; *VOCs*, volatile organic compounds.

the monitoring of air pollutants in workplace environments, including industries that manage large chemical inventories and are suspected to potentially have high concentration levels of air pollutants, for example, petroleum refineries, alloy manufacturers, and coke plants, but also other less contaminated areas, such as beauty and nail salons, restaurant kitchens, operating rooms, or university departments. In all the aforementioned sites, the main objective of air monitoring is to evaluate the occupational risk to the staff. A special case is the monitoring of compounds at fire sites, which provide interesting information from a forensic science point of view.

## 24.3 Conclusions

The use of solid sorbents for air sampling is widespread in the air quality monitoring to isolate volatile compounds and to pre-concentrate target compounds and in studies of their stability. The availability of a variety of solid sorbents with different physicochemical properties and the various designs of sampling devices provide versatility in the selection of the most appropriate sampling strategy. The use of passive sampling strategies is frequently preferred for long-term monitoring due to the lower cost associated with their deployment. For short-term monitoring, typically related to indoor or occupational exposure studies, active sampling methodologies are also widely employed. Consequently, a wide variety of devices for the monitoring of air in different environments, such as rural and urban areas, outdoor and indoor studies, or remote industrialized sites have been employed.

## References

[1] de la Guardia M, Armenta S, editors. The quality of air. Comprehensive analytical chemistry, vol. 73. Oxford, UK: Elsevier, Academic Press; 2016. p. 1−1000.

[2] Bidleman TF. Environ Sci Technol 1988;22:361.

[3] U.S. Environmental protection agency website. Particulate matter (PM) pollution. https://www.epa.gov/pm-pollution (accessed 1 April 2019).

[4] López A, Coscollà C, Yusà V, Armenta S, de la Guardia M, Esteve-Turrillas FA. J Chromatogr A 2017;1506:27.

[5] Esteve-Turrillas FA, Pastor A, de la Guardia M. In: Pawliszyn J, Bayona JM, editors. Comprehensive sampling and sample preparation. Oxford, UK: Elsevier, Academic Press; 2012. p. 201−22.

[6] Marc M, Namieśnik J, Zabiegała B. Chapter 8 − active sampling of air. In: de la Guardia M, Armenta S, editors. The quality of air. Comprehensive analytical chemistry, vol. 73. Oxford, UK: Elsevier, Academic Press; 2016.

[7] Merck website, air monitoring & air sampling, https://www.sigmaaldrich.com/analytical-chromatography/air-monitoring.html (accessed 1 April 2019).

[8] Sigma Aldrich website. Choosing the right adsorbent for your thermal desorption gas chromatography applications. https://www.sigmaaldrich.com/content/dam/sigma-aldrich/docs/Supelco/Posters/1/Adsorbent-Selection-TD-GC-Apps.pdf (accessed 1 April 2019).
[9] Dettmer K, Engewald W. Anal Bioanal Chem 2002;373:490.
[10] Woolfenden E. J Chromatogr A 2010;1217:2685.
[11] Dettmer K, Knobloch T, Engewald W. Fresenius J Anal Chem 2000;366:70.
[12] Marcillo A, Jakimovsk V, Widdig A, Birkemeyer C. J Chromatogr A 2017;1514:16.
[13] Uhde E. Application of solid sorbents for the sampling of volatile organic compounds in indoor air: organic indoor air pollutants: occurrence, measurement, evaluation. Weinheim, Germany: Wiley-VCH Verlag GmbH & Co KGaA; 2009.
[14] Dugheri S, Bonari A, Pompilio I, Colpo M, Mucci M, Arcangeli G. Safety and health at work, vol. 9; 2018. p. 479.
[15] Parmar SS, Parker K, Hueppe M. Air and waste management association - symposium on air quality measurement methods and technology, 165; 2007. p. 267.
[16] Ueta I, Saito Y, Ghani NBA, Ogawa M, Yogo K, Abe A, Shirai S, Jinno K. J Chromatogr A 2009;1216:2848.
[17] Gallego E, Teixidor P, Roca FJ, Perales JF, Gadea E. Atmos Environ 2018;182:9.
[18] Esteve-Turrillas FA, Pastor A, de la Guardia M. Environ Chem 2009;6:452.
[19] Hays L. Anal Chem 2007;79:3635.
[20] Woolfenden E. J Chromatogr A 2010;1217:2674.
[21] Palmes ED, Gunnison AF. Am Ind Hyg Assoc J 1973;34:78.
[22] Gorecki T, Namiesnik J. Trends Anal Chem 2002;21:276.
[23] Esteve-Turrillas FA, Pastor A. Chapter 9 − passive air sampling. In: de la Guardia M, Armenta S, editors. The quality of air. Comprehensive analytical chemistry, vol. 73. Oxford, UK: Elsevier, Academic Press; 2016.
[24] Gustin M, Jaffe D. Environ Sci Technol 2010;44:2222.
[25] Choi SD, Kwon HO, Lee YS, Park EJ, Oh JY. J Hazard Mater 2012;30:241.
[26] Booij K, Smedes F. Environ Sci Technol 2010;44:17.
[27] Esteve-Turrillas FA, Ly-Verdú S, Pastor A, de la Guardia M. J Chromatogr A 2009;1216:8549.
[28] Vrana B, Popp P, Paschke A, Schurmann G. Anal Chem 2001;73:1520.
[29] Strandberg B, Julander A, Sjöström M, Lewnéc M, Akdeva HK, Bigert C. Chemosphere 2018;190:35.
[30] Chen Y, Pawliszyn J. Anal Chem 2003;75:2004.
[31] Baltusen E, Sandra P, David F, Cramers C. J Microcolumn Sep 1999;11:737.
[32] Fung AG, Yamaguchi MS, McCartney MM, Aksenov AA, Pasamontes A, Davis CE. Microchem J 2019;146:407.
[33] Ly-Verdú S, Esteve-Turrillas FA, Pastor A, de la Guardia M. Anal Chim Acta 2010;677:131.
[34] Söderström HS, Bergqvist PA. Environ Sci Technol 2004;38:4828.
[35] Camino-Sánchez FJ, Bermúdez-Peinado R, Zafra-Gómez A, Ruíz-García J, Vílchez-Quero JL. J Chromatogr A 2015;1380:11.
[36] Ramos TD, de la Guardia M, Pastor A, Esteve-Turrillas FA. Sci Total Environ 2018;619:1014.
[37] Bartkow ME, Booij K, Kennedy KE, Müller JF, Hawker DW. Chemosphere 2005;60:170.
[38] U.S. Department of health and human services public health service agency for toxic substances and disease registry, toxicological profile for benzene. 2017. https://www.epa.gov/sites/production/files/2014-03/documents/benzene_toxicological_profile_tp3_3v.pdf.

[39] European commission website. http://ec.europa.eu/environment/air/quality/standards.htm (accessed 1 April 2019).
[40] Jeffers JD, Roller CB, Namjou K, Evans MA, McSpadden L, Grego J, McCann PJ. Anal Chem 2003;76:424.
[41] Tzoumaka PN, Kelessis AG, Petrakakis MJ, Samara C, Papagiannopoulos NM, Zoumakis NM, Tzourelis G. Fresenius Environ Bull 2008;17:1594.
[42] Etzkorn JM, Davey NG, Thompson AJ, Creba AS, Leblanc CW, Simpson CD, Krogh ET, Gill CG. J Chromatogr Sci 2009;47:57.
[43] Ravindra K, Sokhi R, Van Grieken R. Atmos Environ 2008;42:2895.
[44] Pandey SK, Kim KH, Brown RJC. Trends Anal Chem 2011;30:899.
[45] Driscoll CT, Mason RP, Chan HM, J Jacob D, Pirrone N. Environ Sci Technol 2013;47: 4967.
[46] Lin CJ, Pongprueksa P, Lindberg SE, Pehkonen SO, Byun D, Jang C. Atmos Environ 2006;40:2911.
[47] E.U. directive 2008/50/EC of the European parliament and of the council of 21 May 2008 on ambient air quality and cleaner air for Europe. Special edition in Croatian: [Chapter 15] Volume vol. 029 P. 169 − 212.
[48] Climent MJ, Coscollà C, López A, Barra R, Urrutia R. Sci Total Environ 2019;662:20.
[49] Hafizal H, Hamid A, Talib Latif M, Shahrul M, Nadzir M, Uning R, Firoz Khan M, Kannan N. Atmos and Health 2019;12:341.
[50] Masih A, Lall AS, Taneja A, Singhvi R. Environ Pollut 2018;242:1678.
[51] Désert M, Ravier S, Gille G, Quinapallo A, Armengaud A, Pochet G, Savelli JL, Wortham H, Quivet E. Atmos Environ 2018;192:241.
[52] López A, Coscollà C, Yusà V. Talanta 2018;189:211.
[53] Dat ND, Chang KS, Chang MB. Environ Pollut 2018;237:186.
[54] Gevao B, Porcelli M, Rajagopalan S, Krishnan D, Martinez-Guijarro K, Alshemmari H, Bahloul M, Zafar J. Sci Total Environ 2018;622:1621.
[55] Chi KH, Li YN, Hung NT. Aerosol Air Qual Res 2017;17:3151.
[56] Coscollà C, López A, Yahyaoui A, Colin P, Robin C, Poinsignon Q, Yusà V. Sci Total Environ 2017;584:856.
[57] B.A. Hussaina, J.N. Westgate, S.J. Hayward, C. Shunthirasingham, T.N. Brown, H. Hung, Y.D. Lei, F.Wania. Atmos Environ. (Available online 14 December 2018). https://doi.org/10.1016/j.aeaoa.2018.100004.
[58] Liaud C, Schwartz JJ, Millet M. J Environ Sci Health Part B 2017;52:458.
[59] Zhu YJ, Sun D, Yang NE, Ding YS, Feng WB, Hong WJ, Zhu SM, Li YF. Environ Sci Pollut Res 2017;24:10885.
[60] Kurt-Karakus PB, Ugranli-Cicek T, Sofuoglu SC, Celik H, Gungormus E, Gedik K, Sofuoglu A, Okten HE, Birgul A, Alegria H, Jones KC. Atmos Environ 2018;177:154.
[61] Bozkurt Z, Üzmez ÖÖ, Döğeroğlu T, Artun G, Gaga EO. Atmos Pollut Res 2018;9:1146.
[62] Cappelletti N, Astoviza M, Migoya MC, Colombo JC. Sci Total Environ 2016;573:1406.
[63] Huang J, Choi HD, Landis MS, Holsen TM. J Environ Monit 2012;14:2976.
[64] Hao Y, Li Y, Wang T, Hu Y, Sun H, Matsiko J, Zheng S, Wang P, Zhang Q. Environ Geochem Health 2018;40:1907.
[65] Barbas B, de la Torre A, Sanz P, Navarro I, Artíñano B, Martínez MA. Sci Total Environ 2018;624:170.
[66] Goel A, Upadhyay K, Chakraborty M. Environ Monit Assess 2016;188:278.
[67] Vilavert L, Nadal M, Schuhmacher M, Domingo JL. Arch Environ Contam Toxicol 2015; 69:241.

[68] Hamid HHA, Hazman MHM, Nadzir MSM, Uning R, Latif MT, Kannan N. IOP Conf Ser Earth Environ Sci 2019;228:012001.
[69] Kurt-Karakus P, Alegria H, Birgul A, Gungormus E, Jantunen L. Sci Total Environ 2018; 625:555.
[70] Kiros F, Shakya KM, Rupakheti M, Regmi RP, Maharjan R, Byanju RM, Naja M, Mahata K, Kathayat B, Peltier RE. Aerosol Air Qual Res 2016;16:3088.
[71] Hůnová I. Eur J Environ Sci 2017;7:125.
[72] Wang L, Zhao Y, Yi X, Wang Z, Yi Y, Huang T, Gao H, Ma J. Environ Sci Pollut Res 2017;24:12820.
[73] Artun GK, Polat N, Yay OD, Üzmez ÖÖ, Arıa A, Tuygun GT, Elbir T, Altuğ H, Dumanoğlu Y, Döğeroğlu T, Dawood A, Odabasi M, Gaga EO. Atmos Environ 2017; 150:331.
[74] Pozo K, Sarkar SK, Estellano VH, Mitra S, Audi O, Kukucka P, Přibylová P, Klánová J, Corsolini S. Chemosphere 2017;168:1430.
[75] Estellano VH, Pozo K, Přibylová P, Klánová J, Audy O, Focardi S. Environ Pollut 2017; 222:609.
[76] Guéguen F, Stille P, Millet M. Environ Sci Pollut Res 2013;20:3852.
[77] Ielpo P, Mangia C, Marra GP, Comite V, Rizza U, Uricchio VF, Fermo P. Sci Total Environ 2019;648:787.
[78] Francisco AP, Nardocci AC, Tominaga MY, da Silva CR, de Assunção JV. Atmos Pollut Res 2017;8:979.
[79] Ellickson KM, McMahon CM, Herbrandson C, Krause MJ, Schmitt CM, Lippert CJ, Pratt GC. Environ Pollut 2017;231:487.
[80] Cortés J, González CM, Morales L, Abalos M, Abad E, Aristizábal BH. Sci Total Environ 2014;491:67.
[81] Vallecillos L, Espallargas E, Allo R, Marcé RM, Borrull F. Sci Total Environ 2019;666: 235.
[82] Masjedi MR, Taghizadeh F, Hamzehali S, Ghaffari S, Fazlzadeh M, Jafari AJ, Niazi S, Mehrizi EA, Moradi M, Pasalari H, Arfaeinia H. Sci Rep 2019;9:3110.
[83] Rostami R, Zarei A, Saranjam B, Ghaffari HR, Hzrati S, Pouresh Y, Fazlzadeh M. Build Environ 2019;155:47.
[84] Nitter TB, Svendsen KVH. Sci Total Environ 2019;664:1039.
[85] Zhou S, Behrooz L, Weitzman M, Pan G, Vilcassim R, Mirowsky JE, Breysee P, Rule A, Gordon T. Tobac Contr 2017;26:40.
[86] Kwan MWC, Weisenseel JP, Giel N, Bosak A, Batich CD, Willenberg BJ. J Chromatogr A 2018;1573:156.
[87] Johnson JM, Naeher LP, Yu X, Rathbun SL, Muilenburg JL, Wang JS. Int J Hyg Environ Health 2018;221:541.
[88] Marek RF, Thorne PS, Herkert NJ, Awad AM, Hornbuckle KC. Environ Sci Technol 2017;51:7853.
[89] Persson J, Wang T, Hagberg J. Indoor Built Environ 2019;28:506.
[90] Herkert NJ, Hornbuckle KC. Environ Sci: Processes Impacts 2018;20:757.
[91] Dixon HM, Scott RP, Holmes D, Calero L, Kincl LD, Waters KM, Camann DE, Calafat AM, Herbstman JB, Anderson KA. Anal Bioanal Chem 2018;410:2059.
[92] Audy O, Melymuk L, Venier M, Vojta S, Becanova J, Romanak K, Vykoukalova M, Prokes R, Kukucka P, Diamond ML, Klanova J. Chemosphere 2018;206:622.
[93] Villanueva F, Sevilla G, Lara S, Martin P, Salgado S, Albaladejo J, Cabanas B. Microchem J 2018;142:117.

[94] Hadei M, Hopke PK, Shahsavani A, Moradi M, Yarahmadi M, Emam B, Rastkari N. J Occup Med Toxicol 2018;13:30.
[95] Jafari A, Bargeshadi R, Jafari F, Mohebbi I, Hajaghazadeh M. Int Arch Occup Environ Health 2018;91:349.
[96] Fent KW, Evans DE, Babik K, Striley C, Bertke S, Kerber S, Smith D, Horn GP. J Occup Environ Hyg 2018;15:399.
[97] Cheng WH, Huang HL, Chen KS, Chang YJ. J Environ Sci Health, Part A Environ Sci Eng Toxic Haz 2017;52:1158.
[98] Sofuoglu SC, Toprak M, Inal F, Cimrin AH. Environ Sci Pollut Res Int 2015;22:15703.
[99] O'Connell SG, Kind LD, Anderson KA. Environ Sci Technol 2014;48:3327.
[100] Strandberg B, Bergemalm-Rynell K, Sallsten G. Environ Sci: Processes Impacts 2014;16: 1008.
[101] Zhong L, Batterman S, Milando CW. Int Arch Occup Environ Health 2019;92:141.
[102] Sha B, Dahlberg AK, Wiberg K, Ahrens L. Environ Pollut 2018;241:319.
[103] Lee EG, Magrm R, Kusti M, Kashon ML, Guffey S, Costas MM, Boykin CJ, Harper M. J Occup Environ Hyg 2017;14:31.
[104] Simons AK, Handy RG, Sleeth DK, Pahler LF, Thiese MS. J Chem Health Saf 2017;24: 36.
[105] Esteve W, Langlois E. Anal Methods 2012;4:2054.
[106] Solbu K, Daae HL, Olsen R, Thorud S, Ellingsen DG, Lindgren T, Bakke B, Lundanes E, Molander P. J Environ Monit 2011;13:1393.
[107] Park H, Jang JK, Shin JA. Saf Health Work 2011;2:39.
[108] Bohlin P, Jones KC, Strandberg B. Environ Sci Technol 2010;44:749.
[109] Bohlin P, Jones KC, Levin JO, Lindahl R, Strandberg B. J Environ Monit 2010;12:1437.
[110] Campos-Candel A, Llobat-Estellés M, Mauri-Aucejo A. Talanta 2009;78:1286.
[111] Chang CP, Lin TC, Lin YW, Hua YC, Chu WM, Lin TY, Lin YW, Wu JD. Ann Occup Hyg 2016;60:348.

# Solid-phase extraction in bioanalytical applications

# 25

*Sònia Sentellas* [1], *Javier Saurina* [1], *Oscar Núñez* [1,2]
[1]Department of Chemical Engineering and Analytical Chemistry, University of Barcelona, Barcelona, Spain; [2]Serra Húnter Fellow, Generalitat de Catalunya, Barcelona, Spain

## 25.1 Introduction

Bioanalysis plays a significant role during the development of new drugs. Bioanalytical methods for the qualitative and quantitative determination of drugs and their metabolites in biological fluids are required for the assessment of the metabolite profile, bioavailability, bioequivalence and pharmacokinetic behavior of the new molecules [1,2].

Several instrumental platforms are currently used for bioanalysis mainly based on high-performance separation techniques, which are essential for acceptable resolution of analytes to avoid problems of mutual coelutions or interferences from endogenous compounds. Among the different analytical techniques available, liquid chromatography coupled to mass spectrometry (LC-MS) is the preferred option due to its separation performance, excellent sensitivity and specificity, and high sample throughput. In addition, unlike other popular detection techniques, such as UV/vis spectroscopy with limited qualitative ability, MS can generate structural information. Despite the significant advantages of LC-MS (or LC-MS/MS) over other analytical techniques, some limitations have to be overcome, such as the occurrence of important matrix effects, poor elucidation and resolution of isomers or even insufficient sensitivity (lower sensitivity than required for poorly ionized or labile compounds). These shortcomings are especially important when dealing with complex matrices, such as biological fluids (most commonly: whole blood, plasma or serum, and urine) which contain multiple endogenous components, ranging from small organic compounds to macromolecules. The presence of matrix components may affect the separation and ionization of the analytes, and thus, hinder the analysis [3–6].

The inherent complexity of these samples prevented their direct injection into the analytical system for the bioanalysis of the drug(s)/metabolite(s) of interest. For this reason, appropriate sample pretreatment is usually required before injection. The procedure for sample preparation can be considered, in fact, as the most critical and difficult step in bioanalysis aimed at enriching the analytes, eliminating interfering substances and, in some cases, also transforming the analyte into a more convenient species for separation and detection. Both analyte physicochemical properties and matrix characteristics have to be taken into account to design a proper extraction procedure. Thus, acid-base behavior, polarity, size, etc. are important analyte features

**Solid-Phase Extraction. https://doi.org/10.1016/B978-0-12-816906-3.00025-X**

while salt content, the presence of phospholipids or red blood cells are characteristics derived from the type of biological matrix to be considered. All of them determine the challenges to be overcome during sample preparation.

Different strategies can be used for sample preparation, for example, protein precipitation (PPT), liquid-liquid extraction (LLE), and solid phase extraction (SPE) being the most important. Additionally, the so-called dilute and shoot (DAS) technique can be considered as the simplest methodology consisting of the dilution of the biological fluid in a proper solvent [7]. A comparison of the main features of these strategies can be found in Table 25.1. In DAS, the analyte is not extracted from the matrix or enriched. Matrix effects may be reduced notably due to the lower content of interfering compounds on the sample while the highly sensitive instruments available today can compensate for the dilution of the analyte. It is worthy of mention that, although some applications can be found applying DAS to plasma samples [8,9], this methodology is mainly used when analyzing urine [10–12].

PPT is mainly used for the determination of drugs and their metabolites in plasma or blood samples. The procedure is quite simple, based on the interaction of a precipitating agent with the proteins present in the sample. Such interaction causes a decrease in protein solubility. Then, the solid residue is separated by centrifugation, and the supernatant phase containing analytes (drug and metabolites) is recovered. Sample solutions obtained in this way can be analyzed directly or can be subjected to furthermore exhaustive cleanup procedures [13]. Although organic solvents, especially acetonitrile, are the most widely used precipitating agents, other reagents, such as

**Table 25.1** Comparison of features of the principal sample treatment for bioanalysis, including dilute and shoot (DAS), protein precipitation (PPT), liquid-liquid extraction (LLE), and solid phase extraction (SPE).

| | DAS | PPT | LLE | SPE |
|---|---|---|---|---|
| Matrix | Urine | Plasma/blood | Urine, plasma/blood | Urine, plasma/blood |
| Optimization | Simple | Simple | Elaborate | Elaborate |
| Manipulation | Low | Low | High | High |
| Handling time | <5 min | 10–15 min | 20–30 min | 20–30 min |
| Automatable | Yes | Filtration microplates | 96-well format (SLE) | 96-well format |
| Clean-up | No | Limited (protein removal) | Yes | Yes |
| Analyte enrichment | No (sample dilution) | No (sample dilution) | Yes | Yes |
| Cost | Low | Low | Medium | Medium-high |

strong acids (e.g., trichloroacetic or m-phosphoric acids), metal ions (e.g., zinc sulfate) or salts at high concentrations (e.g., aluminum chloride, ammonium sulfate) are also effective for the removal of proteins [14]. In addition, the precipitation efficiency depends, not only on the selected reagent but also on the plasma/blood:reagent ratio. In this regard, Polson and coworkers calculated the protein removal efficiency using various precipitating agents at different proportions. For instance, a ratio of 0.5:1 (acetonitrile:human plasma) resulted only in 4% of protein removal while a ratio of 1:1 increased the efficiency up to 89% and reached 95% at a ratio of 4:1 [14]. In summary, PPT is a simple and rapid procedure widely used, although the concomitant sample dilution and the limited sample cleanup are the main drawbacks.

Generally speaking, LLE has been widely used for sample preparation when dealing with aqueous matrices. In a similar way, drugs and metabolites can be extracted from biological fluids using an immiscible organic solvent as the extractant [7,15]. The efficiency of analyte extraction depends on analyte-solvent interactions. Thus, the selection of the solvent is based on the physicochemical properties of the analytes. For instance, for basic compounds solvents with high hydrogen bonding donor capacity, such as chloroform are preferred. Reversely, hydrogen bonding acceptor solvents (ex. diethyl ether, methyl-tertbutyl ether) show better efficiency for acidic compounds. Dipolar solvents (e.g., dichloromethane, ethyl acetate) or aromatic solvents (e.g., toluene, p-xylene) are selected for polar or aromatic molecules, respectively. In addition, as compounds are only extracted in neutral (uncharged) form, their pKa values are important features to consider, and the adjustment of the sample pH is of vital importance. Nevertheless, the formation of an ion-pair using an appropriate reagent can provide a suitable alternative. After extraction, if required, the solvent can be evaporated and the residue re-dissolved in a smaller volume of an appropriate solvent, compatible with the analytical platform, thus resulting in the concentration of the analyte(s). Although LLE is, in general, efficient for sample cleanup removing salts and macromolecules, other compounds co-extracted with the analyte of interest can cause matrix effect in LC-MS. However, from our point of view, the main drawback of the technique is the large amount of sample usually required, and the tedious handling. Later, these have been overcome by working on a 96-well format using supported liquid extraction (SLE).

Biological fluids or pretreated samples resulting from preliminary DAS, PTT, and LLE processes are often subjected to SPE for additional cleanup. The wide variety of stationary phases offers almost unlimited possibilities for the treatment of bioanalytical samples. In general, the selection of the most appropriate stationary phase is mainly based on size (molecular mass), acid-base behavior, and polarity of the analytes. Reversed-phase (RP) is currently the most popular mode for a wide range of target compounds of low and intermediate polarity. Reasonably, charged compounds can be extracted by ion exchange. Pure ionic interactions can be accomplished with strong cation (SCX) and anion (SAX) exchangers based on sulfonic and quaternary ammonium groups, respectively. Strong exchangers are active at any pH and are generally recommended for physiological media. Conversely, weak exchangers consisting of carboxylic acid and amino groups offer a more limited performance at pH values around 7. Recently, mixed phases have been introduced to deal with components exhibiting high polarity and diversity of functional groups, in which several interaction mechanisms (hydrophilicity, hydrophobicity and/or ion exchange) may occur simultaneously.

As mentioned previously, the selected strategy depends mainly on the characteristics of the analyte(s) and matrix, the sensitivity required, and the purpose of the study. Fig. 25.1 shows a scheme for the analytical work-flow from sampling to analysis. In this regard, if the methodology is focused on the quantitative determination of a given (known) compound, more specific and optimized methodologies can be used. However, for qualitative purposes, such as for the identification of metabolites, a less restrictive procedure is generally chosen to avoid the loss of unknown metabolites which may have quite different physicochemical properties compared to the parent drug.

## 25.2 Conventional SPE approches in bioanalytical application

Solid-phase extraction is gaining popularity in sample pretreatment, replacing classic LLE approaches in many application [16], due to its capability to overcome several of the drawbacks of LLE, reducing solvent consumption, intrinsic costs and processing time, especially when online and totally automated strategies are employed. Although several high throughput formats and procedures are nowadays available in SPE, and more specifically in the bioanalytical field, conventional SPE approaches, both off-line and online, are still used in bioanalysis. Table 25.2 illustrates some selected bioanalytical applications employing SPE with conventional off-line or online SPE approaches [17–30].

Reversed-phase SPE with C18 or polymeric-based sorbents remains the most employed stationary phase in the bioanalytical field. For instance, Aboul-Enein et al. [17] proposed a rapid and sensitive LC-UV method using a monolithic silica column for the simultaneous determination of ketamine (KE) and its two main metabolites, norketamine (NK) and dehydronorketamine (DHNK) in human plasma. The authors compared the performance of several SPE sorbents (Oasis HLB and Sep-Pak C18, C8, and CN cartridges) for the extraction of the targeted drugs from the human plasma matrix. Both Sep-Pak C8 and CN showed absolute recoveries of 60%–75%, while the Oasis HLB cartridge showed recoveries above 80% and Sep-Pak C18 cartridges above 90%. Recently, Papoutsis et al. [29] described a sensitive GC-MS method for the determination of amisulpride, a substituted benzamide derivative employed as an antipsychotic drug, in whole blood. The authors compared LLE with different mixtures of solvents and SPE using different sorbents (Bond Elut LRC Certify and Certify II, HF Bond Elut LRC C18, and ABS Elut Nexus LRC). The first two provided poor recoveries (lower than 60%). High extraction efficiency was achieved with ABS Elut Nexus LRC sorbent although extracts were not clean enough for GC-MS analysis witnessed by some baseline interferences. In contrast, HF Bond Elut LRC C18 sorbent provided clear advantages over the other SPE columns. It was shown

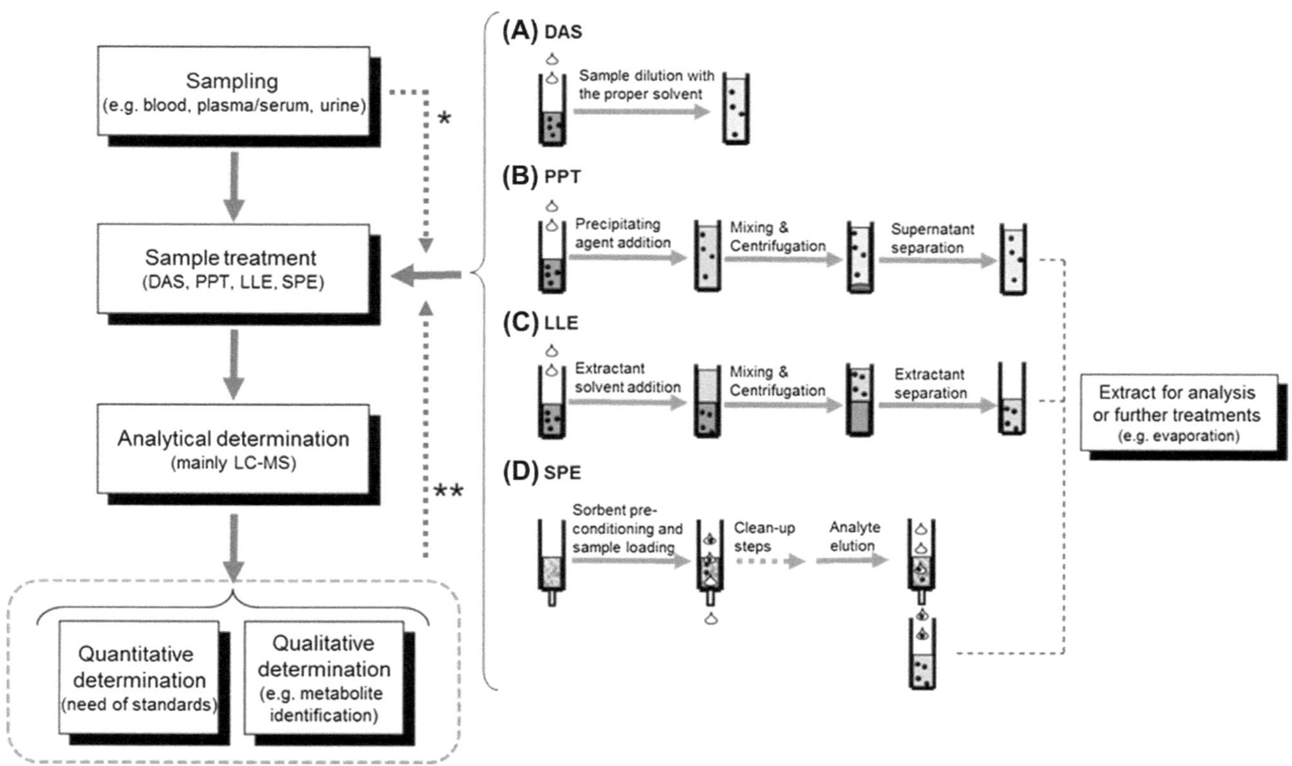


**Figure 25.1** Schematic flow-chart focusing on the sample treatment in bioanalysis. The final choice will depend on the sample complexity and the purpose of the study. (*) Simpler samples may be treated with simple procedures, such as DAS or PPT; more complex samples may require more sophisticated procedures and even combinations of techniques. (**) Quantitative analysis will be focused on specific treatments depending on the target analytes and the required sensitivity; qualitative analysis for screening new compounds will be addressed under less restrictive conditions to avoid losses of unknown species.

**Table 25.2** Selected applications based on conventional SPE procedures.

| Analytes | Sample | SPE cartridges | Analytical method | Reference |
|---|---|---|---|---|
| Ketamine and its metabolites | Human plasma | Sep-Pak C18<br>Elution: 2 × 500 μL methanol with 1% TEA | LC-UV<br>- Chromolith reversed phase monolithic silica column (100 × 4.6 mm I.D.); isocratic elution: 30 mM monobasic sodium phosphate:acetonitrile 75:25 (v/v); 3.0 mL/min; λ 220 nm. | [17] |
| Ciprofloxacin | Plasma samples | Supelclean C18 SPE cartridges<br>Elution: 1 mL potassium dihydrogenphosphate (0.05 M; pH 3):acetonitrile (1:1 v/v) | LC-FL<br>- Discovery HS F5 column (205 × 4.6 mm I.D., 5 μm); isocratic elution: Potassium dihydrogenphosphate (0.05 M, pH 3): acetonitrile (9:1 v/v); 1 mL/min; $\lambda_{exc}$ 280 nm, $\lambda_{em}$ 446 nm | [18] |
| 43 Benzodiazepins and their metabolites, zolpidem and zopiclone | Human plasma | SPE with Focus™ column (Varian, Inc., CA, USA)<br>Elution: 0.75 mL 0.1% TFA in ACN and 0.75 mL ammonia in ACN | LC-MS<br>- UPLC BEH C18 column (100 × 2.1 mm, 1.7 μm); gradient elution: (A) 0.05% formic acid in water and (B) 0.05% formic acid in acetonitrile; 0.4 mL/min.<br>- Single quadrupole MS, ESI (+), SIM | [23] |
| Tiapride and its phase I metabolite | Rat blood plasma | DSC-PH (100 mg/1 mL) SPE cartridges<br>Elution: 1 mL methanol | HPLC-UV-FL<br>- Discovery HS-F5 (250 × 4 mm I.D., 5 μm) column; gradient elution: (A) acetonitrile: 0.01 M phosphate buffer pH 3 (6:4 v/v) and (B) aceotonitrile:0.01 M phosphate buffer pH 3 (3:7 v/v); 1 mL/min; UV: λ 212 nm; FL $\lambda_{exc}$ 232 nm, $\lambda_{em}$ 334 nm | [24] |

| | | | | |
|---|---|---|---|---|
| Atazanavir | Human plasma | LiChrosep Sequence (1 $cm^3$, 30 mg) SPE cartridges<br>Elution: 1 mL mobile phase | LC-MS/MS<br>- Hypersil Gold C18 column (50 × 4.6 mm I.D., 5 μm); isocratic elution: 5 mM ammonium formate in water:methanol (10:90 v/v); 0.7 mL/min<br>- QqQ MS, ESI(+), MRM | [25] |
| TAK-448 (nonapeptide analogue and metastin receptor agonist) | Human plasma | OASIS WCX (30 ng/1 mL, 20 μm)<br>Elution: 1 mL methanol:formic acid (98:2 v/v) | LC-MS/MS<br>- Acquity UPLC BEH Phenyl (50 × 2.1 mm I.D., 1.7 μm); gradient elution: (A) water/methanol/formic acid (900:100:1 v/v/v) and (B) water/methanol/formic acid (100:900:1 v/v/v); 0.4 mL/min<br>- QqQ MS, ESI(+), MRM | [26] |
| Indapamide | Human whole blood | Automated SPE using Xtrata-X-Drug N Polymer RP (60 mg/3 mL) cartridges<br>Elution: 1 mL methanol | LC-MS/MS<br>- Kinetex C18 (100 × 2.1 mm I.D., 1.7 μm) column; isocratic elution: 2 mM ammonium acetate (with added 0.5 mL formic acid in 1 L buffer) and acetonitrile in a 10:90 (v/v) ratio; 0.2 mL/min<br>- QqQ MS, ESI (+), MRM | [27] |
| Raltegravir | Human plasma | Oasis HLB SPE cartridges (30 mg/1 mL)<br>Elution: 1 mL mobile phase | LC-MS/MS<br>- Supelco C18 (50 × 4.6 mm, 3.5 μm) column; isocratic elution: acetonitrile:0.1% formic acid (90:10 v/v); 0.8 mL/min<br>- QqQ MS, ESI (+), MRM | [28] |
| Amisulpride | Whole blood | HF bond elut C18 SPE cartridges<br>Elution: 1.5 mL dichloromethane: isopropanol:ammonium hydroxide (85:15:2 v/v/v)<br>Derivatization with hetafluorobutyric anhydride (HFBA) | GC-MS<br>- HP-5MS (5% phenyl-methylsilicone, 30 m × 0.25 mm I.D., 0.25 μm film thickness); Helium flow rate: 1.0 mL/min<br>- Quadropole MS, EI, SIM | [29] |

*Continued*

**Table 25.2** Selected applications based on conventional SPE procedures.—cont'd

| Analytes | Sample | SPE cartridges | Analytical method | Reference |
|---|---|---|---|---|
| Zolpidem and its metabolite | Whole blood and oral fluid | Strata X–C SPE columns (strong cation exchanger)<br>Elution: 2 mL of 10% ammonia in 90% acetonitrile/methanol (1:1) | LC-MS<br>- Phenomenex Synerty Fusion-RP column (100 × 4.6 mm I.d., 4 μm); isocratic elution: 100 mM ammonium acetate (pH 5.8) and methanol (30:70 v/v); 0.4 mL/min<br>- Single quadrupole MS, ESI (+), SIM | [30] |
| Monomethyl fumarate | Human plasma | Strata-X 33 μm polymeric sorbent SPE cartridges (30 mg/mL)<br>Elution: 0.5 mL LC mobile phase | LC-MS/MS<br>- Zodia C18 column (50 × 46 mm I.D., 3 μm); isocratic elution: 0.1% formic acid:acetonitrile 25:75 v/v; 0.6 mL/min<br>- QqQ MS, ESI (+), MRM | [19] |
| Monomethyl fumarate | Human plasma | Strata X SPE cartridge (33 mg/mL)<br>Elution: 0.5 mL LC mobile phase | LC-MS/MS<br>- Zodia C18 column (50 × 46 mm I.D., 3 μm); isocratic elution: 0.1% formic acid:acetonitrile 30:70 v/v; 0.5 mL/min<br>- QqQ MS, ESI (+), MRM | [20] |
| Amphotericin B, fluconazole and fluorocytosine | Human plasma and cerebrospinal fluid | Online-dual-SPE method with Oasis HLB cartridge (2.1 × 20 mm, 15 μm) and HyperSep Hypercarb cartridge (2.1 × 20 mm, 30 μm)<br>Elution: Mobile phase | LC-HRMS<br>- Protect column (20 × 4.6 mm, 5 μm); gradient elution: (A) acetonitrile and (B) 10 mM ammonium formate pH 4; 1 mL/min<br>- Q-Exactive hybrid quadrupole-Orbitrap instrument, ESI (+), t-SIM | [21] |
| Amisulpride | Human plasma | Online SPE with restricted-access media (RAM) column (Capcell Pak MF Ph-1, 10 × 4.0 mm I.D.) | HPLC-UV<br>- Luna C18 (250 × 4.6 mm I.D., 5 μm) column; gradient elution: (A) acetonitrile and (B) 100 mM ammonium acetate; 1.0 mL/min; λ 280 nm | [22] |

that an alkaline eluting mixture of dichloromethane:isopropanol:ammonium hydroxide (85:15:2 v/v/v) provided high and reproducible extraction values (higher than 93%) and GC-MS chromatograms free of interferences.

Ionic-exchange SPE cartridges are also quite popular for bioanalytical applications. Piotrowski et al. [30] proposed the simultaneous determination of zolpidem, a popular sleep agent used in the short-term treatment of insomnia, and its metabolite in whole blood and oral fluid samples by SPE and LC-MS for clinical and forensic purposes. Off-line SPE was carried out using a polymeric strong cation-exchange (SCX) sorbent (Strate X-CE SPE). The active site for ion exchange is a sulfonic group bonded to an aromatic ring. After conditioning the sorbent with methanol and HCl, whole blood (after protein precipitation with acetonitrile) was acidified with 0.1 M HCl and passed through the cartridge. After washing, the target analytes were eluted with 10% $NH_3$ in 90% acetonitrile:methanol (1:1). The eluate was then evaporated to dryness and reconstituted in the mobile phase before LC-MS analysis. SPE recoveries in the range 79.9%−104.1% for blood samples and 80.2%−103.8% for oral fluid samples were achieved.

Solid-phase extraction sorbents combining different interaction mechanisms are increasingly exploited to improve the retention of a wide variety of analytes for bioanalytical applications. For example, Ishida et al. [23] described the rapid and quantitative screening of 43 benzodiazepines and their metabolites, zolpidem and zopiclone, in human plasma with SPE preconcentration on a Focus column from Varian Inc. In this column, four different interactions are possible to preconcentrate the analytes:hydrogen bond interaction (donor and acceptor), dipole-dipole interactions, and hydrophobic interactions. The usefulness of this Focus column for the extraction of abused drugs from biological fluids was previously demonstrated [31,32]. In this application, all the targeted drugs, with a wide variety of pKa values, were retained by the column in slight acid conditions (using 1 M acetic acid buffer at pH 5.0), and high average recoveries (70.1% at 10 ng/mL and 87.1% at 100 ng/mL) were achieved using elution with 0.1% trifluoroacetic acid (TFA) and 0.1% ammonia in acetonitrile. Sample analysis was then performed by reversed-phase LC-MS with gradient elution and single MS quadrupole detector in the positive ESI mode.

The proposed method was not only applied to plasma samples but also to whole blood and urine samples in forensic cases. As an example, Fig. 25.2 illustrates the MS chromatograms of the extracts obtained from whole blood for two autopsy cases where the ingestion of flunitrazepam (given before strangulation in case 1 and ingested for suicide attempt in case 2) was confirmed. In case 1, only the metabolite 7-aminoflunitrazepam was detected, while both flunitrazepam and the corresponding 7-aminoflunitrazepam were detected in case 2, showing the usefulness of the proposed methodology for forensic practices.

Online SPE preconcentration procedures are gaining acceptance in the bioanalytical field due to the reduction in both analysis time and sample handling. For instance, Qu et al. [21] described the utilization of an online-dual-SPE-LC procedure combined with high resolution mass spectrometry (HRMS) in a hybrid Q-Orbitrap instrument for the simultaneous quantification of amphotericin B (AMB), fluconazole (FZ) and fluorocytosine (FC), recommended as preferred antibiotics for HIV-associated

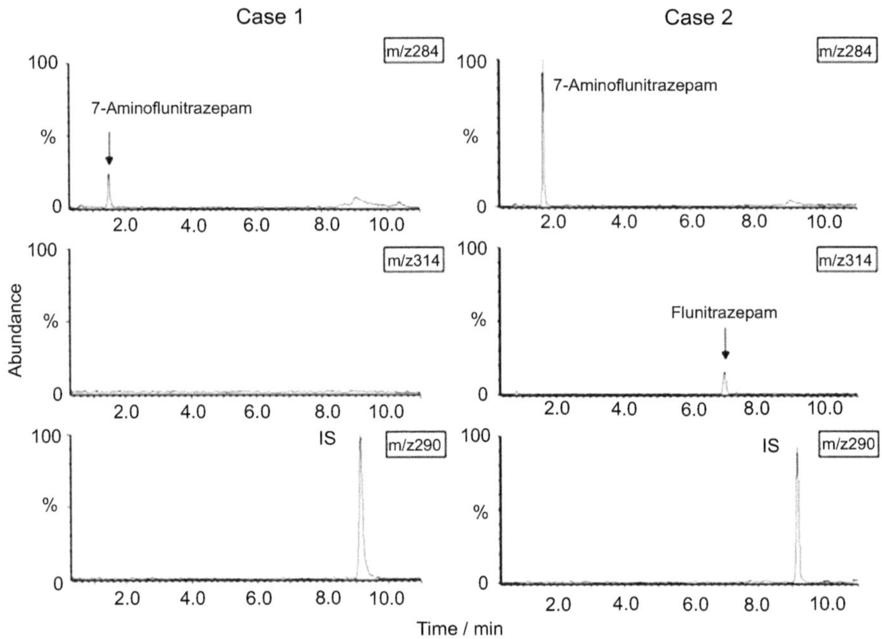


**Figure 25.2** LC-MS chromatograms of the extracts from the whole blood of two autopsy cases. In Case 1, flunitrazepam was given before strangulation. In case 2, flunitrazepam was ingested for a suicide attempt.
Reproduced with permission from Ishida T, Kudo K, Hayashida M, Ikeda N. Rapid and quantitative screening method for 43 benzodiazepines and their metabolites, zolpidem and zopiclone in human plasma by liquid chromatography/mass spectrometry with a small particle column. J Chromatogr B 2009;877:2652–2657. https://doi.org/10.1016/j.jchromb.2009.05.008. Copyright (2009) Elsevier.

cryptococcal meningitis, in human plasma and cerebrospinal fluid (CSF). Two SPE sorbent materials were proposed to ensure suitable retention of the targeted antibiotics. Therefore, an Oasis HLB SPE cartridge was employed for the preconcentration of AMB, FZ and carbamazepine (CBZ, used as internal standard) while a HyperSep Hypercarb cartridge provided suitable retention of FC and 5-methylcytosine hydrochloride (MC, used as internal standard). Fig. 25.3 illustrates the work-flow of the proposed online-dual-SPE-LC-HRMS system.

The full SPE cleanup and preconcentration procedure consisted of four steps to isolate the antibiotics from endogenous interfering components. In Step 1 (cleanup step) the plasma sample is loaded, and the matrix interferences are washed away while AMB and FZ are retained on the Oasis HLB cartridges and FC in the HyperSep Hypercarb cartridge. At the same time, the protection column (where chromatographic separation will take place) is being conditioned with the initial mobile phase for the separation. Next, the elution steps take place. First, AMB and FZ are eluted from the Oasis HLB cartridge to the protection column (Step 2), followed by the elution of FC from the HyperSep Hypercarb cartridge to the protection column (Step 3).

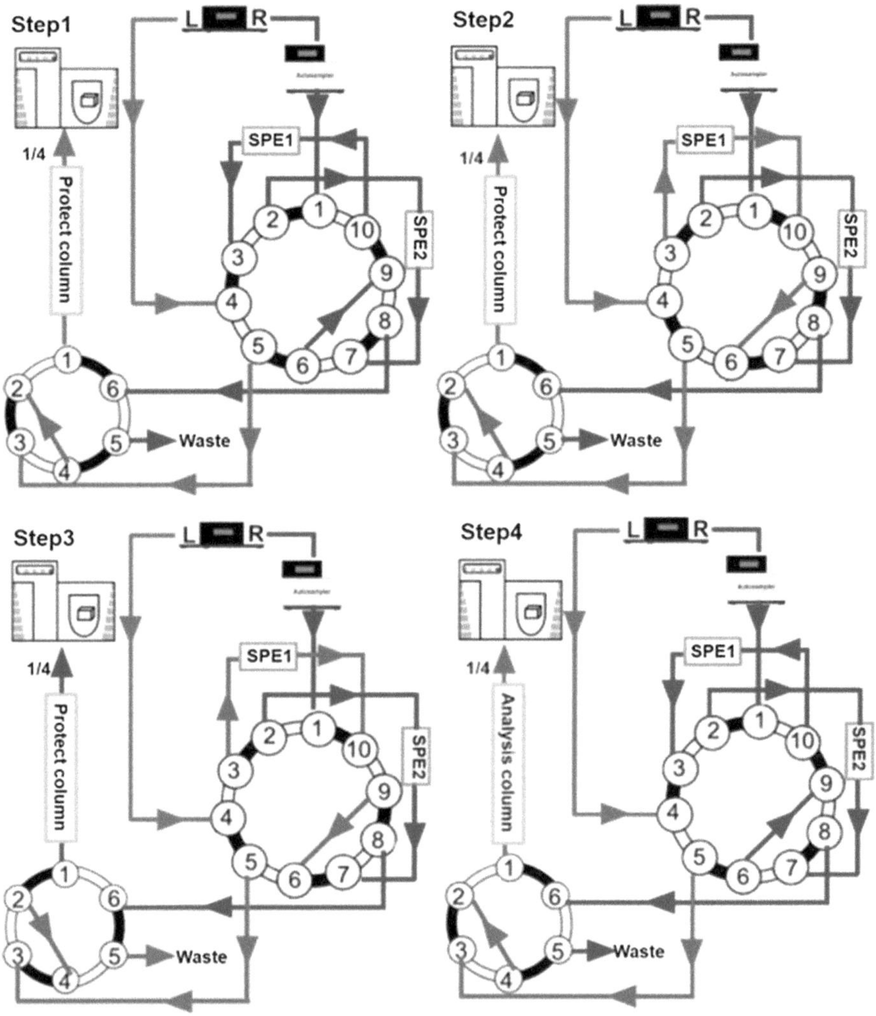


**Figure 25.3** Schemat of the online-dual-SPE-LC-HRMS methodology for the analysis of AMB, FZ, and FC in human plasma and CSF samples. *SPE1*, Oasis HLB cartridge; *SPE2*, HyperSep Hypercarb cartridge.
Reproduced with permission from Qu L, Qian J, Ma P, Yin Z. Utilizing online-dual-SPE-LC with HRMS for the simultaneous quantification of amphotericin B, fluconazole, and fluorocytosine in human plasma and cerebrospinal fluid. Talanta 2017;165:449–457. https://doi.org/10.1016/j.talanta.2016.12.052. Copyright (2017) Elsevier.

Finally, in the rebalance step (Step 4), analytes are separated in the protection column and detected by HRMS. Quantification by Q-Exactive Hybrid Quadrupole-Orbitrap with targeted-selected ion monitoring (t-SIM) mode was employed to simultaneously determine the concentrations of the three antibiotics. The total analysis time was less than 7 min. Fig. 25.4 shows the chromatograms of a blank and spiked plasma and CSF

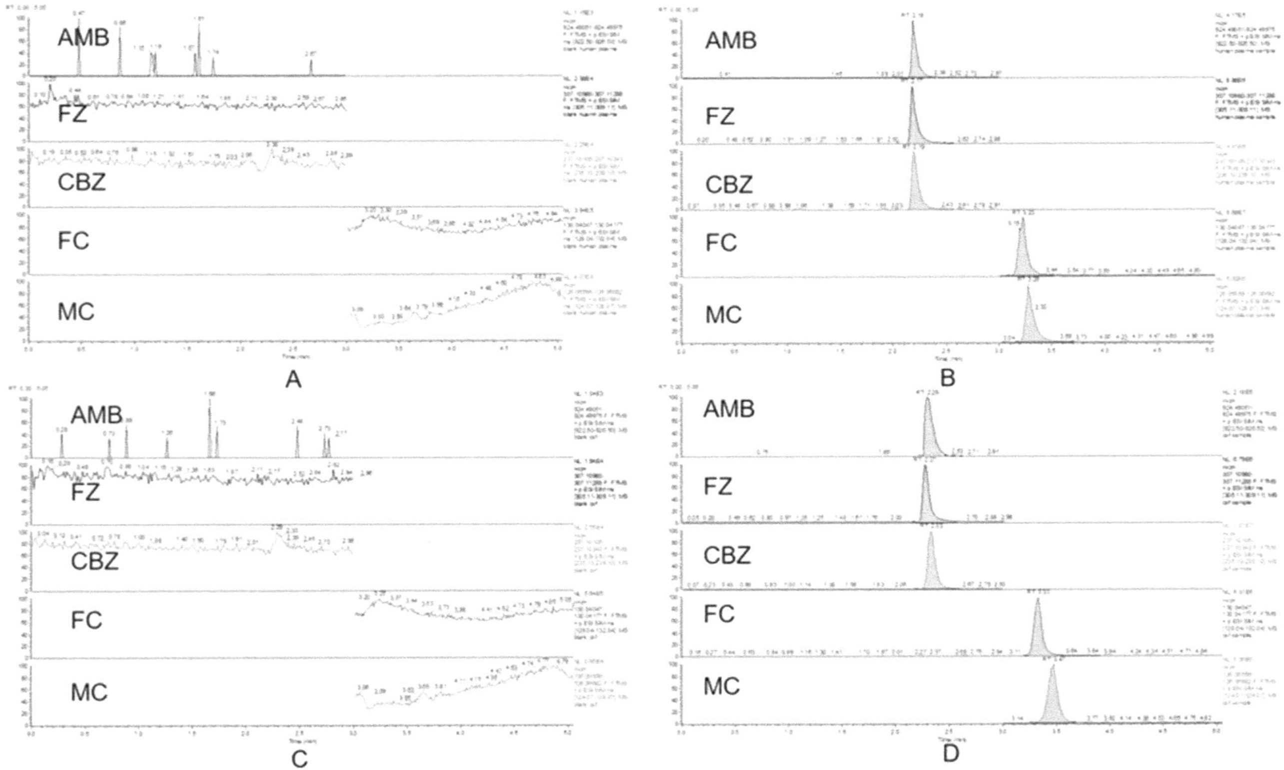


**Figure 25.4** Chromatograms of blank plasma sample (A), plasma spiked with AMB, FZ, CVZ, FC, and MC (B), blank CSF sample (C), and CSF spiked with AMB, FZ, CBZ, FC, and MC (D).
Reproduced with permission from Qu L, Qian J, Ma P, Yin Z. Utilizing online-dual-SPE-LC with HRMS for the simultaneous quantification of amphotericin B, fluconazole, and fluorocytosine in human plasma and cerebrospinal fluid. Talanta 2017;165:449–457. https://doi.org/10.1016/j.talanta.2016.12.052. Copyright (2017) Elsevier.

samples. The method was fully validated showing good performance, with the lowest limits of quantification (LLOQ) in the range 0.04–0.4 μg/mL, and good linearity ($r^2 > 0.99$), intra-day and inter-day precisions (RSD < 4.32% and <4.06%, respectively), and recoveries (89.93%–93.28% and 90.09%–93.58% for plasma and CSF samples, respectively).

In a recent application, a restricted-access media (RAM) column was proposed for the automated online SPE determination of amisulpride, an antipsychotic drug, in human plasma [22]. RAM materials are attractive for the analysis of biological fluids, especially when dealing with online SPE procedures [33–36]. The presence of macromolecules, such as proteins, in biological fluids, can easily cause clogging of SPE sorbents. This undesired phenomenon may affect the robustness, especially of online SPE methodologies, when applied to biological fluids, and mainly for complex plasma samples. RAM materials are then a good alternative to traditional sample preparation techniques that combine a restrictive outer surface to exclude the retention of large biomolecules, thus removing these interferences, with retentive inner pores or phases to capture the target compounds. He at al [22] employed, for the first time, a RAM column for the analysis of amisulpride. Fig. 25.5 illustrates the scheme of the automated online SPE-HPLC system, the extraction and cleanup principle of a RAM column, and the proposed SPE-HPLC procedure. Briefly, the sample is loaded on the autosampler at the beginning of each SPE-HPLC cycle. Within the first 2 min, both conditioning of the analytical column and the sample extraction and washing steps are carried out. Proteins and other macromolecules are excluded from the RAM column to waste while amisulpride is retained by hydrophobic interactions (Fig. 25.5B). Then the analyte is eluted from the RAM column to the separation column. The total time for the full online SPE-HPLC procedure is 10 min. Representative chromatograms for plasma samples (blank and spiked) are shown in Fig. 25.6. A low limit of detection of 0.0035 μg/mL, and satisfactory precision and accuracy (recoveries in the range 98.1%–104.6% with RSD values below 4.6%) were achieved, illustrating the potential of the proposed method for therapeutic drug monitoring as well as large-scale bioanalytical and clinical studies.

## 25.3 High throughput sample analysis using 96-well SPE microplates in bioanalytical applications

Sample throughput in LC-MS/MS bioanalytical methods is often limited by the complexity and time required for sample preparation procedures. The labor-intensive and time-consuming steps typically associated with manual sample preparation, often limit the number of samples that can be analyzed in drug discovery and development applications. Consequently, in these fields, the use of parallel sample processing in a 96-well SPE format by means of robotic liquid handler systems has been widely adopted for routine high throughput analysis. Table 25.3 illustrates some selected bioanalytical applications where 96-well SPE microplates are employed [37–43]. Different SPE sorbents are used depending on the physicochemical

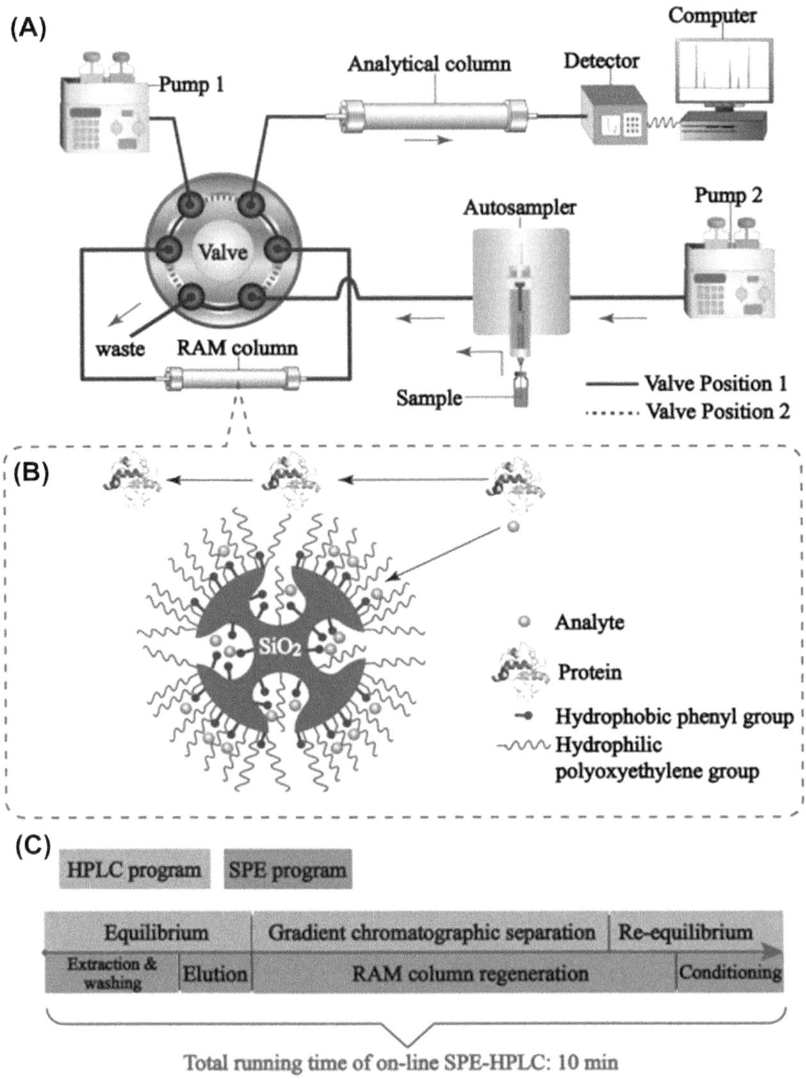


**Figure 25.5** Schema of the automated online SPE-HPLC system (A), the extraction and cleanup principle of a RAM column (B), and the SPE-HPLC procedure (C).
Reproduced with permission from He J, Yuan J, Du J, Chen X, Zhang X, Ma A, et al. Automated on-line SPE determination of amisulpride in human plasma using LC coupled with restricted-access media column. Microchem J 2019;145:154–161. https://doi.org/10.1016/j.microc.2018.10.029. Copyright (2019) Elsevier.

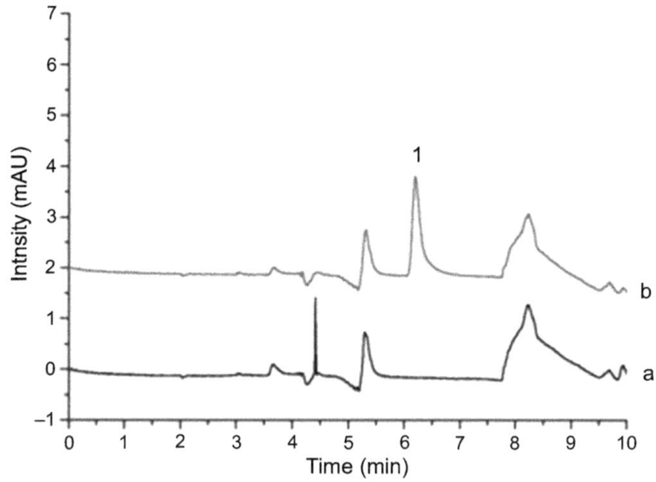


**Figure 25.6** Chromatograms of plasma samples analyzed by the proposed automated online SPE-HPLC methods, including a blank sample (A) and a sample spiked with amisulpride (peak 1) at 0.7140 μg/L (B).
Reproduced with permission from He J, Yuan J, Du J, Chen X, Zhang X, Ma A, et al. Automated on-line SPE determination of amisulpride in human plasma using LC coupled with restricted-access media column. Microchem J 2019;145:154–161. https://doi.org/10.1016/j.microc.2018.10.029. Copyright (2019) Elsevier.

properties of the target analytes. Moreover, these 96-well plates can be used both for cleanup and preconcentration purposes.

As an example, Nilsson et al. [41] proposed the use of Ostro 96-well filter plates for protein and phospholipid removal, in combination with a semi-automated 96-well SPE application (using HLB sorbent) for the determination of budesonide (a synthetic corticosteroid used for treatment of asthma, allergic rhinitis, autoimmune hepatitis, and inflammatory bowel diseases), in human plasma. The removal of phospholipids from the sample matrix was found to be necessary in order to reduce ion suppression effects in the electrospray ionization of the analyte and to increase method sensitivity. The absolute recovery obtained over the sample preparation procedure was around 67%.

Zheng et al. [40] outlined some of the advantages of 96-well SPE microplates in comparison to conventional SPE approaches in clinical drug studies. In this case, a number of samples obtained from different sites is normally processed and, as a consequence, it is not always very practical to collect samples in 96-well plates (operational errors, sample contamination or concerns regarding sample traceability with respect to each sample handling step may arise). In these cases, sample collection in capped polypropylene tubes is preferred and, regardless of the sample extraction procedure, biofluid subsampling to a 96-well plate is typically accomplished by manually uncapping and recapping of the sample tubes, and is the bottleneck for automated sample extraction strategies (see, as an example, the schematic of a typical SPE sample extraction strategy in Fig. 25.7A).

**Table 25.3** Selected applications based on 96 well-SPE microplates.

| Analytes | Sample | SPE procedure | Analytical method | Reference |
|---|---|---|---|---|
| Bisphosphonates | Human serum and urine | Tomtec Quadra 96-well plate (silica-based anion exchange sorbent)<br>Derivatization with diazomethane<br>Elution: 0.5 mL methanol | LC-MS/MS<br>- Zorbax 300-SCX column (50 × 3.1 mm I.D., 5 μm); isocratic elution: Acetonitrile:ammonium formate buffer pH 2.5 75:25 v/v; 1 mL/min<br>- QqQ MS, ESI (+), MRM | [37] |
| Lopinavir and ritonavir | Plasma samples | 384-Well formatted SPE plate (C18, 8 mg packing)<br>Elution: 2-Propanol with 2% ammonia or 0.005 N sodium hydroxyde | HPLC-MS/MS<br>- C18 (150 × 3.9 mm I.D., 3.9 μm) column; isocratic elution: Acetonitrile/aqueous 2 mM ammonium acetate with 0.01% formic acid in a 70:30 (v/v) ratio; 1.0 mL/min<br>- QqQ MS, APCI (+), MRM | [38] |
| 7 psychotropic drugs and 4 metabolites | Human plasma | 96-well SPE plate (Oasis MCX support, 10 mg)<br>Loading: 1 mL<br>Elution: 500 μL methanol-ammonium hydroxide 25% (94:6 v/v) | HPLC-MS<br>- C18 (10 × 201 mm I.D., 3.5 μm) column; gradient elution: (A) 25% ammonium hydroxide and (B) acetonitrile; 0.3 mL/min<br>- Single quadrupole MS, ESI (+), SIM | [39] |
| CNS drugs and metabolites | Dog, monkey and human plasma | 96-well SPE plates: Strata-X (10 mg), Isolute-96 C2 (EC, 25 mg) Evolute ABN (10 mg) | LC-MS/MS<br>- Atlantis dC19 (50 × 2 mm I.D., 5 μm) column; gradient elution: (A) 0.1% formic acid in water and (B) 0.1% formic acid in acetonitrile:water (95:5 v/v); 0.3 mL/min<br>- QqQ MS, ESI (+), MRM | [40] |

| | | | | |
|---|---|---|---|---|
| Budesonide | Human plasma | Ostro 96-well filter plate (for protein and phospholipid removal)<br>96-well SPE plate (HLB 30 mg/well)<br>Elution: 500 μL acetonitrile | LC-MS/MS<br>- Acquity UPLC BEH (100 × 2.1 mm I.D., 1.7 μm) column; gradient elution: (A) 0.1% formic acid in water and (B) acetonitrile; 0.4 mL/min<br>- QqQ MS, ESI (+), MRM | [41] |
| Gemcitabine | Human plasma | 96-well plate (hybrid SPE-precipitation 50 mg) | LC-MS/MS<br>- Zorbax RX-SIL rapid resolution HT HILIC (100 × 2.1 mm I.D., 1.8 μm) column; isocratic elution: 0.1% formic acid in acetonitrile/0.1% formic acid in water (80:20 v/v); 0.25 mL/min<br>- QqQ MS, ESI (+), MRM | [42] |
| Ketamine and its main metabolites | Pig plasma | 96-well microplates OASE™ | LC-MS/MS<br>- Atlantis™ T3C18 (100 × 2.1 mm I.D., 3 μm) column; gradient elution: (A) 0.1% formic acid in water and (B) 0.1% formic acid in methanol; 0.5 mL/min<br>- QqQ MS, ESI (+), MRM | [43] |

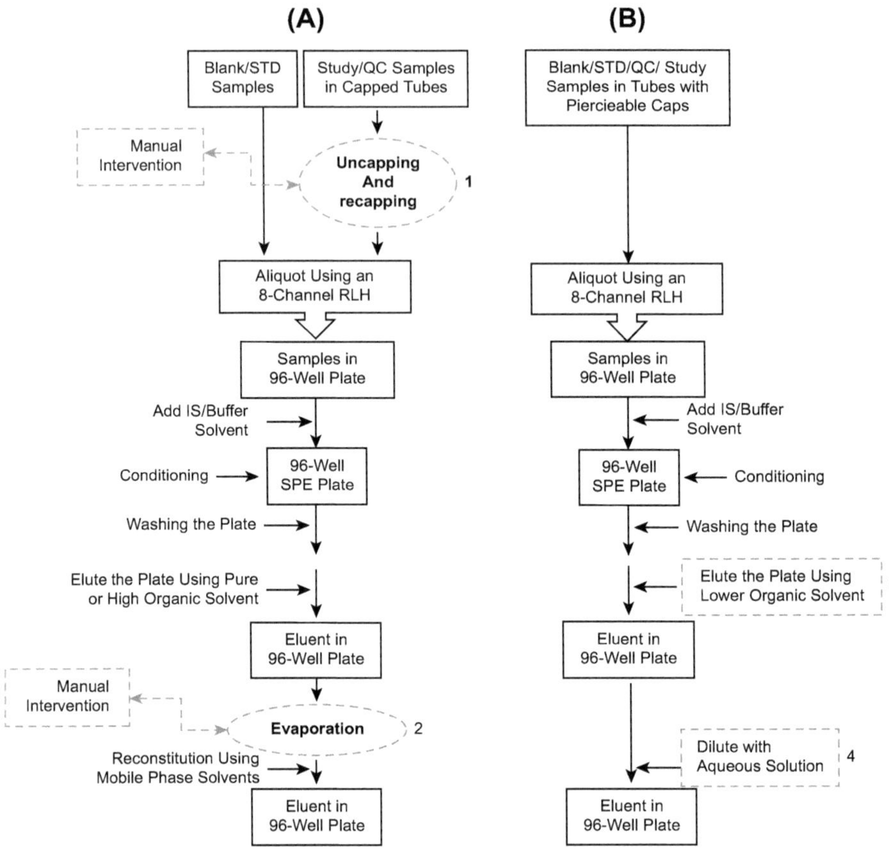


**Figure 25.7** Illustration of two SPE sample extraction processes for bioanalysis using LC-MS/MS: (A) conventional SPE sample extraction with a robotic liquid handler (RLH), in which steps 1 and 2 required manual interventions (B) direct biofluid transfer with evaporation-free SPE sample extraction, in which no manual interventions are required for steps 1 and 2. The evaporation step was eliminated by a modification of the elution solvent followed by dilution (steps 3 and 4).
Reproduced with permission from Zheng N, Buzescu A, Pasas-Farmer S, Arnold ME, Ouyang Z, Jemal M, et al. A simplified and completely automated workflow for regulated LC-MS/MS bioanalysis using cap-piercing direct sampling and evaporation-free solid phase extraction. J Chromatogr B 2013;921–922:64–74. https://doi.org/10.1016/j.jchromb.2013.01.028. Copyright (2013) Elsevier.

In contrast, Zhen et al. [40] proposed a simplified and completely automated workflow (see schematic in Fig. 25.7B) for LC-MS/MS bioanalytical applications using cap-piercing direct sampling and evaporation-free SPE. First, the use of pierceable caps for sample collection tubes was employed to eliminate the need for uncapping and recapping during the sample analysis (as indicated in Fig. 25.7B) [44]. Moreover, the authors designed a sample tube with a conical shape at the bottom to enable the

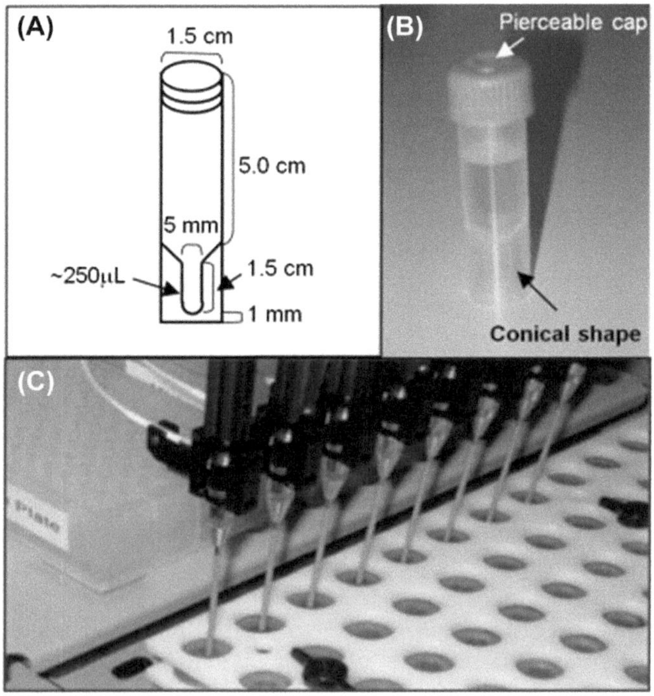


**Figure 25.8** (A) Size and dimension of a plasma sample tube; (B) a plasma sample tube with a pierceable cap; (C) aliquoting of plasma samples by TECAN Genesis fixed tips via cap piercing for subsequent transfer to a 96-well plate.
Reproduced with permission from Zheng N, Buzescu A, Pasas-Farmer S, Arnold ME, Ouyang Z, Jemal M, et al. A simplified and completely automated workflow for regulated LC-MS/MS bioanalysis using cap-piercing direct sampling and evaporation-free solid phase extraction. J Chromatogr B 2013;921–922:64–74. https://doi.org/10.1016/j.jchromb.2013.01.028. Copyright (2013) Elsevier.

robotic liquid handler to accurately take small sample aliquots typically employed in preclinical (animal) and clinical (human) studies (Fig. 25.8).

In addition, to achieve fully automated sample extraction, the authors proposed elution with an LC-MS/MS compatible solvent to overcome solvent evaporation and reconstitution steps, thus achieving an evaporation-free SPE methodology with automated sample treatment and analysis for regulated bioanalysis (Fig. 25.7A). The applicability of the method was demonstrated by three LC-MS/MS assays in CNS drug development programs for monkey, dog, and human plasma samples [40], using Strata-X, Isolute-96 C2 and Evolute ABN 96-well SPE plates. This methodology eliminated all manual interventions achieving a fully automated sample preparation without compromising assay performance. In addition, exposure of the bioanalyst to possibly infectious biofluids during sample preparation was avoided.

## 25.4 Microextraction by packed sorbents (MEPS) in bioanalytical applications

Microextraction by packed sorbent (MEPS) is a miniaturized form of SPE considered within the green sample pretreatment methodologies [45]. MEPS is suitable for small sample volumes and can easily be interfaced with different chromatographic separation techniques without modification. The sorbent bed in MEPS is integrated into a liquid handling syringe that allows for low void volume sample manipulations either manually or in combination with automated laboratory robotics. Nowadays, MEPS is gaining acceptance in many fields, including bioanalytical applications. Some selected examples are summarized in Table 25.4 [46–50].

Different MEPS sorbent materials can be employed depending on the analytes to be extracted and/or the matrix interferences to be removed [45]. Moreover, MEPS technology can be implemented in a semi or fully automated way where the sample-processing, extraction, and injection into the analytical instrument can be addressed online within the same syringe. The main difference of MEPS in comparison with SPE, where sample enrichment is addressed in one direction (normally up to down), is that in MEPS enrichment this is achieved twice (up, when loading the syringe, and down, when emptying the syringe).

As an example, Elmongy et al. [48] described the determination of metoprolol enantiomers in human plasma and saliva by MEPS using a C18 sorbent and LC-ESI-MS/MS. Fig. 25.9 shows the MEPS syringe employed.

In this application, plasma and saliva samples were drawn through the MEPS sorbent (draw-eject in the same vial). Moreover, the samples were drawn four times for preconcentration of the analytes. After washing the sorbent with water:methanol (95:5 v/v) to remove proteins, lipids and other interfering materials, analytes were eluted with 200 μL isopropanol.

It was also shown that the MEPS sorbent could be reutilized after washing four times with isopropanol and four times with water to eliminate possible carry-over. Analytes were then determined by LC-ESI-MS/MS using a cellulose-SB column with isocratic elution using 0.1% ammonium hydroxide in hexane-isopropanol (80: 20 v/v). Postcolumn addition of 0.5% formic acid in isopropanol was applied to improve the MS ionization of the metoprolol enantiomers. Mean recoveries in both plasma and saliva samples were in the range 93%–97%.

Moreno et al. [49] proposed the use of a mixed mode (C8 and strong cation exchange) MEPS for the determination of ketamine, a hallucinogenic abused drug, and its major metabolite, norketamine, in urine and plasma. After enrichment, determination by GC-MS/MS was proposed without any derivatization step. Again, several sample aspirations were employed during sample enrichment (8 times for urine and 26 times for plasma samples). The analytes were then eluted with ammonia solutions in methanol. Recoveries ranged from 63% to 101%, while precision and accuracy were below 15% for ketamine and norketamine. The proposed method was fast since no derivatization step was required, and cost-effective for the quantification of ketamine and norketamine in biological specimens.

**Table 25.4** Selected applications based on microextraction by packed sorbents (MEPS).

| Analytes | Sample | Extraction procedure | Analytical method | Reference |
|---|---|---|---|---|
| Local anesthetics | Human plasma | MEPS (1 mg silica based benzenesulphonic acid cation exchanger) | LC-MS/MS<br>- C8 (10 × 1 mm I.D.) column; Isocractic elution: Methanol/water (1:1, v/v), with 0.1% formic acid; 0.20 mL/min<br>- QqQ MS, ESI (+), MRM | [46] |
| Remifentanil | Human plasma | MEPS (M1 sorbent: C8 and SCX) | LC-MS/MS<br>- Kinetex C18 (50 × 2.1 mm I.D., 2.6 μm) column; gradient elution: (A) 0.1% formic acid aqueous solution and (B) methanol; 0.3 mL/min<br>- QqQ MS, ESI (+), MRM | [47] |
| Metoprolol enantiomers | Human plasma and urine samples | MEPS (C18 sorbent) | LC-MS/MS<br>- Cellulose-SB (150 × 4.6 mm, 5 μm) column; isocratic elution: 0.1% ammonium hydroxide in hexane-isopropanol (80:20 v/v); 0.8 mL/min<br>- QqQ MS, ESI (+), MRM<br>Postcolumn solvent-assisted ionization with 0.5% formic acid in isopropanol at 0.2 mL/min | [48] |
| Ketamine and its major metabolite, norketamine | Human urine and plasma samples | MEPS (C8, 80%, and SCX, 20%, mixed mode, 4 mg) | GC-MS/MS<br>- HP-5MSfused-silica capillary column (30 m × 0.25 mm I.D., 0.25 μg film thickness). Helium flow rate; 0.8 mL/min<br>- QqQ MS, EI, MRM | [49] |
| 5 pyrethroid metabolites | Human urine samples | MEPS (C18 sorbent) coupled to large volume injection (LVI) | LVI-GC-MS<br>- VF-5 ms low bleed capillary column (30 m × 0.25 mm I.D., 0.25 μm film thickness). Helium flow rate: 1.0 mL/min<br>- Ion trap mass analyzer, EI, SIM | [50] |

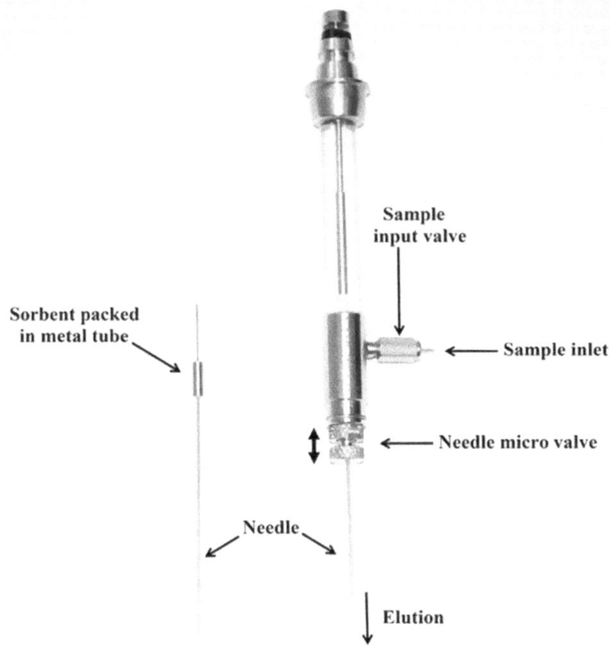


**Figure 25.9** Example of a MEPS syringe.
Reproduced with permission from Elmongy H, Ahmed H, Wahbi AA, Amini A, Colmsjö A, Abdel-Rehim M. Determination of metoprolol enantiomers in human plasma and saliva samples utilizing microextraction by packed sorbent and liquid chromatography−tandem mass spectrometry. Biomed Chromatogr 2016;30:1309−1317. https://doi.org/10.1002/bmc.3685. Copyright (2018) John Wiley & Sons, Ltd.

## 25.5 General conclusions and future perspectives

Solid-phase extraction offers great possibilities for the preparation of bioanalytical samples. SPE is almost indispensable for dealing with complex matrices, such as blood and plasma that contain a wide variety of endogenous compounds, from small organic molecules to biopolymers. This set of components needs to be separated from the analytes to avoid interferences. The vast majority of drugs and their metabolites are compatible with reversed phase sorbents, with octadecylsiloxane-bonded silica (C18) the most widely used for the treatment of biological fluids and related samples. Ion exchange modes are also well established in the context of bioanalysis. Phosphorylated drugs, such as nucleotide and oligonucleotide analogs, with negative charges, are retained and purified by SAX while aminated drugs, positively charged in most physiological media, can be preconcentrated by SCX. The range of sorbents currently available (sometimes not entirely described to protect them for commercial reasons) is exceptional, among which hybrid materials stand out because of their integration of several interaction mechanisms. In this regard, the advances in stationary phases for liquid chromatography may be extended to their SPE counterparts. In the near future, the generalization of novel SPE trends, such as those based on stereo-affinity, antigen-antibody recognition, molecularly imprinted polymers to deal with highly specific bioanalytical problems are expected to emerge.

## Acknowledgments

This work was supported by the Agency for the Administration of University and Research Grants (Generalitat de Catalunya, Spain) under the projects 2017 SGR-171 and 2017 SGR-310.

## References

[1] Saurina J, Sentellas S. Liquid chromatography coupled to mass spectrometry for metabolite profiling in the field of drug discovery. Expert Opin Drug Discov 2019;14:469–83. https://doi.org/10.1080/17460441.2019.1582638.

[2] Pandey S, Pandey P, Tiwari G, Tiwari R. Bioanalysis in drug discovery and development. Pharm Methods 2010;1:14–24.

[3] Álvarez-Sánchez B, Priego-Capote F, Castro MDL de. Metabolomics analysis II. Preparation of biological samples prior to detection. TrAC Trends Anal Chem 2010;29:120–7. https://doi.org/10.1016/j.trac.2009.12.004.

[4] Darvesh AS, Carroll RT, Geldenhuys WJ, Gudelsky GA, Klein J, Meshul CK, et al. In vivo brain microdialysis: advances in neuropsychopharmacology and drug discovery. Expert Opin Drug Discov 2011;6:109–27. https://doi.org/10.1517/17460441.2011.547189.

[5] David A, Abdul-Sada A, Lange A, Tyler CR, Hill EM. A new approach for plasma (xeno) metabolomics based on solid-phase extraction and nanoflow liquid chromatography-nanoelectrospray ionisation mass spectrometry. J Chromatogr A 2014;1365:72–85. https://doi.org/10.1016/j.chroma.2014.09.001.

[6] Saunders KC, Ghanem A, Boon Hon W, Hilder EF, Haddad PR. Separation and sample pre-treatment in bioanalysis using monolithic phases: a review. Anal Chim Acta 2009;652:22–31. https://doi.org/10.1016/j.aca.2009.05.043.

[7] Pedersen-Bjergaard S, Gjelstad A, Halvorsen TG. Sample preparation in bioanalysis of pharmaceuticals: sample preparation, separation techniques and mass spectrometry. 1st ed. John Wiley & Sons; 2015. Steen Honoré Hansen and Stig Pedersen-Bjergaard.

[8] McCauley-Myers D, Eichhold T, Bailey R, Dobrozsi D, Best K, Hayess IIII, et al. Rapid bioanalytical determination of dextromethorphan in canine plasma by dilute-and-shoot preparation combined with one minute per sample LC-MS/MS analysis to optimize formulations for drug delivery. J Pharm Biomed Anal 2000;23:825–35. https://doi.org/10.1016/S0731-7085%2800%2900346-0.

[9] Esposito S, Bracacel E, Nibbio M, Speziale R, Orsatti L, Veneziano M, et al. Use of “dilute-and-shoot” liquid chromatography-high resolution mass spectrometry in preclinical research: application to a DMPK study of perhexiline in mouse plasma. J Pharm Biomed Anal 2016;118:70–80. https://doi.org/10.1016/j.jpba.2015.10.004.

[10] Deventer K, Pozo OJ, Verstraete AG, Van Eenoo P. Dilute-and-shoot-liquid chromatography-mass spectrometry for urine analysis in doping control and analytical toxicology. TrAC Trends Anal Chem 2014;55:1–13. https://doi.org/10.1016/j.trac.2013.10.012.

[11] Agostini M, Renzoni C, Pierini E, Piergiovanni M, Termopoli V, Famiglini G, et al. Rapid, hydrolysis-free, dilute-and-shoot method for the determination of buprenorphine, norbuprenorphine and their glucuronides in urine samples using UHPLC-MS/MS. J Pharm Biomed Anal 2019;166:236–43. https://doi.org/10.1016/j.jpba.2019.01.014.

[12] Yan J, Kuzhiumparambil U, Bandodkar S, Solowij N, Fu S. Development and validation of a simple, rapid and sensitive LC-MS/MS method for the measurement of urinary neurotransmitters and their metabolites. Anal Bioanal Chem 2017;409:7191–9. https://doi.org/10.1007/s00216-017-0681-3.

[13] Taevernier L, Wynendaele E, Dhondt M, De Spiegeleer B. Analytical quality-by-design approach for sample treatment of BSA-containing solutions. J Pharm Anal 2015;5:27–32. https://doi.org/10.1016/j.jpha.2014.06.001.

[14] Polson C, Sarkar P, Incledon B, Raguvaran V, Grant R. Optimization of protein precipitation based upon effectiveness of protein removal and ionization effect in liquid chromatography-tandem mass spectrometry. J Chromatogr B 2003;785:263–75. https://doi.org/10.1016/S1570-0232(02)00914-5.

[15] Ashri N, Abdel-Rehim M. Sample treatment based on extraction techniques in biological matrices. Bioanalysis 2011;3:2003–18.

[16] Płotka-Wasylka J, Szczepańska N, de la Guardia M, Namieśnik J. Modern trends in solid phase extraction: new sorbent media. TrAC Trends Anal Chem 2016;77:23–43. https://doi.org/10.1016/j.trac.2015.10.010.

[17] Aboul-Enein HY, Hefnawy MM. Liquid chromatographic high-throughput analysis of ketamine and its metabolites in human plasma using a monolithic silica column and solid phase extraction. Talanta 2005;65:67–73. https://doi.org/10.1016/j.talanta.2004.05.021.

[18] Vybíralová Z, Nobilis M, Zoulova J, Květina J, Petr P. High-performance liquid chromatographic determination of ciprofloxacin in plasma samples. J Pharm Biomed Anal 2005;37:851–8. https://doi.org/10.1016/j.jpba.2004.09.034.

[19] Pasha S, ImamVaranasi MB, Mohammed I. Bioanalysis of monomethyl fumarate in human plasma by a sensitive and rapid LC-MS/MS method and its pharmacokinetic application. J Pharm Biomed Anal 2017;146:109–16. https://doi.org/10.1016/j.jpba.2017.08.015.

[20] Satish Ramanatham V, Venkateswarlu P, Silva Sankar Rao Y, Ravi Kumar K. Bioanalytical method development and validation of monomethyl fumarate by liquid chromatography tandem mass spectrometry and its application to a pharmacokinetic study. Eur J Biomed Pharm Sci 2017;4:515–23.

[21] Qu L, Qian J, Ma P, Yin Z. Utilizing online-dual-SPE-LC with HRMS for the simultaneous quantification of amphotericin B, fluconazole, and fluorocytosine in human plasma and cerebrospinal fluid. Talanta 2017;165:449–57. https://doi.org/10.1016/j.talanta.2016.12.052.

[22] He J, Yuan J, Du J, Chen X, Zhang X, Ma A, et al. Automated on-line SPE determination of amisulpride in human plasma using LC coupled with restricted-access media column. Microchem J 2019;145:154–61. https://doi.org/10.1016/j.microc.2018.10.029.

[23] Ishida T, Kudo K, Hayashida M, Ikeda N. Rapid and quantitative screening method for 43 benzodiazepines and their metabolites, zolpidem and zopiclone in human plasma by liquid chromatography/mass spectrometry with a small particle column. J Chromatogr B 2009;877:2652–7. https://doi.org/10.1016/j.jchromb.2009.05.008.

[24] Nobilis M, Vybíralová Z, Szotáková B, Sládková K, Kuneš M, Svoboda Z. High-performance liquid chromatographic determination of tiapride and its phase I metabolite in blood plasma using tandem UV photodiode-array and fluorescence detection. J Chromatogr B 2011;879:3845–52. https://doi.org/10.1016/j.jchromb.2011.10.032.
[25] Yadav M, Trivedi V, Upadhyay V, Shah G, Baxi GA, Goswami S, et al. Comparison of extraction procedures for assessment of matrix effect for selective and reliable determination of atazanavir in human plasma by LC-ESI-MS/MS. J Chromatogr B 2012; 885–886:138–49. https://doi.org/10.1016/j.jchromb.2011.12.031.
[26] Kuze Y, Jinno F, Kondo T, Asahi S. Highly sensitive liquid chromatography-tandem mass spectrometry method for quantification of TAK-448 in human plasma. J Pharm Biomed Anal 2013;83:75–81. https://doi.org/10.1016/j.jpba.2013.04.023.
[27] Nakov N, Mladenovska K, Labacevski N, Dimovski A, Petkovska R, Dimitrovska A, et al. Development and validation of automated SPE-LC-MS/MS method for determination of indapamide in human whole blood and its application to real study samples. Biomed Chromatogr 2013;27:1540–6. https://doi.org/10.1002/bmc.2957.
[28] Matta MK, Rao Pilli N, Seshagiri Rao JVLN. Bioanalysis of raltegravir, an integrase inhibitor in human plasma by novel SPE-ESI-LC-MS/MS method and its pharmacokinetic application. Am J PharmTech Res 2013;3:575–86.
[29] Papoutsis I, Rizopoulou A, Nikolaou P, Pistos C, Spiliopoulou C, Athanaselis S. A validated GC/MS method for the determination of amisulpride in whole blood. J Chromatogr B 2014;947–948:111–6. https://doi.org/10.1016/j.jchromb.2013.12.024.
[30] Piotrowski P, Bocian S, liwka K, Buszewski B. Simultaneous analysis of zolpidem and its metabolite in whole blood and oral fluid samples by SPE-LC/MS for clinical and forensic purposes. Adv Med Sci 2015;60:167–72. https://doi.org/10.1016/j.advms.2015.01.010.
[31] Ishida T, Kudo K, Naka S, Toubou K, Noguchi T, Idea N. Rapid diagnosis of drug intoxication using novel NAGINATA gas chromatography/mass spectrometry software. Rapid Commun Mass Spectrom 2007;21:3129–38. https://doi.org/10.1002/rcm.3194.
[32] Kudo K, Ishida T, Hara K, Kashimura S, Tsuji A, Ikeda N. Simultaneous determination of 13 amphetamine related drugs in human whole blood using an enhanced polymer column and gas chromatography-mass spectrometry. J Chromatogr B 2007;855:115–20. https://doi.org/10.1016/j.jchromb.2007.03.002.
[33] Souverain S, Rudaz S, Veuthey JL. Restricted access materials and large particle supports for on-line sample preparation: an attractive approach for biological fluids analysis. J Chromatogr B 2004;801:141–56. https://doi.org/10.1016/j.jchromb.2003.11.043.
[34] Yang SH, Fan H, Classon RJ, Schug KA. Restricted access media as a streamlined approach toward on-line sample preparation: recent advancements and applications. J Sep Sci 2013;36:2922–38. https://doi.org/10.1002/jssc.201300595.
[35] Rogeberg M, Malerod H, Roberg-Larsen H, Aass C, Wilson SR. On-line solid phase extraction-liquid chromatography, with emphasis on modern bioanalysis and miniaturized systems. J Pharm Biomed Anal 2014;87:120–9. https://doi.org/10.1016/j.jpba.2013.05.006.
[36] Nazario CED, Fumes BH, da Silva MR, Lanças FM. New materials for sample preparation techniques in bioanalysis. J Chromatogr B 2017;1043:81–95. https://doi.org/10.1016/j.jchromb.2016.10.041.
[37] Zhu LS, Lapko VN, Lee JW, Basir YJ, Kafonek C, Olsen R, et al. A general approach for the quantitative analysis of bisphosphonates in human serum and urine by high-performance liquid chromatography/tandem mass spectrometry. Rapid Commun Mass Spectrom 2006;20:3421–6. https://doi.org/10.1002/rcm.2755.

[38] Chang MS, Kim EJ, El-Shourbagy TA. Evaluation of 384-well formatted sample preparation technologies for regulated bioanalysis. Rapid Commun Mass Spectrom 2007;21: 64–72. https://doi.org/10.1002/rcm.
[39] Choong E, Rudaz S, Kottelat A, Guillarme D, Veuthey JL, Eap CB. Therapeutic drug monitoring of seven psychotropic drugs and four metabolites in human plasma by HPLC-MS. J Pharm Biomed Anal 2009;50:1000–8. https://doi.org/10.1016/j.jpba.2009.07.007.
[40] Zheng N, Buzescu A, Pasas-Farmer S, Arnold ME, Ouyang Z, Jemal M, et al. A simplified and completely automated workflow for regulated LC-MS/MS bioanalysis using cap-piercing direct sampling and evaporation-free solid phase extraction. J Chromatogr B 2013;921–922:64–74. https://doi.org/10.1016/j.jchromb.2013.01.028.
[41] Nilsson K, Andersson M, Beck O. Phospholipid removal combined with a semi-automated 96-well SPE application for determination of budesonide in human plasma with LC-MS/MS. J Chromatogr B 2014;970:31–5. https://doi.org/10.1016/j.jchromb.2014.08.035.
[42] Mano Y, Sakamaki K, Ueno T, Kita K, Ishii T, Hotta K, et al. Validation of a hydrophilic interaction ultra-performance liquid chromatography-tandem mass spectrometry method for the determination of gemcitabine in human plasma with tetrahydrouridine. Biomed Chromatogr 2015;29:1343–9. https://doi.org/10.1002/bmc.3429.
[43] Ramiole C, D'Hayer B, Boudy V, Legagneux J, Fonsart J, Houzé P. Determination of ketamine and its main metabolites by liquid chromatography coupled to tandem mass spectrometry in pig plasma: comparison of extraction methods. J Pharm Biomed Anal 2017;146:369–77. https://doi.org/10.1016/j.jpba.2017.09.001.
[44] Teitz DS, Khan S, Powell ML, Jemal M. An automated method of sample preparation of biofluids using pierceable caps to eliminate the uncapping of the sample tubes during sample transfer. J Biochem Biophys Methods 2000;45:193–204. https://doi.org/10.1016/S0165-022X(00)00117-2.
[45] Yang L, Said R, Abdel-Rehim M. Sorbent, device, matrix and application in microextraction by packed sorbent (MEPS): a review. J Chromatogr B 2017;1043:33–43. https://doi.org/10.1016/j.jchromb.2016.10.044.
[46] Altun Z, Abdel-Rehim M, Blomberg LG. New trends in sample preparation: on-line microextraction in packed syringe (MEPS) for LC and GC applications: part III: determination and validation of local anaesthetics in human plasma samples using a cation-exchange sorbent, and MEPS-LC-MS-MS. J Chromatogr B 2004;813:129–35. https://doi.org/10.1016/j.jchromb.2004.09.020.
[47] Said R, Pohanka A, Andersson M, Beck O, Abdel-Rehim M. Determination of remifentanil in human plasma by liquid chromatography-tandem mass spectrometry utilizing micro extraction in packed syringe (MEPS) as sample preparation. J Chromatogr B 2011; 879:815–8. https://doi.org/10.1016/j.jchromb.2011.02.002.
[48] Elmongy H, Ahmed H, Wahbi AA, Amini A, Colmsjö A, Abdel-Rehim M. Determination of metoprolol enantiomers in human plasma and saliva samples utilizing microextraction by packed sorbent and liquid chromatography–tandem mass spectrometry. Biomed Chromatogr 2016;30:1309–17. https://doi.org/10.1002/bmc.3685.
[49] Moreno I, Barroso M, Martinho A, Cruz A, Gallardo E. Determination of ketamine and its major metabolite, norketamine, in urine and plasma samples using microextraction by packed sorbent and gas chromatography-tandem mass spectrometry. J Chromatogr B 2015; 1004:67–78. https://doi.org/10.1016/j.jchromb.2015.09.032.
[50] Klimowska A, Wielgomas B. Off-line microextraction by packed sorbent combined with on solid support derivatization and GC-MS: application for the analysis of five pyrethroid metabolites in urine samples. Talanta 2018;176:165–71. https://doi.org/10.1016/j.talanta.2017.08.011.

# Index

*Note:* 'Page numbers followed by "f" indicate figures and "t" indicate tables'.

**D**

**E**

## I

**P**

## Z